3rd International Conference on Mechanical, Aerospace Technology and Materials Application (MATMA 2024)

Journal of Physics: Conference Series Volume 2879

Hohhot, China
30 August – 1 September 2024

ISBN: 979-8-3313-0808-7
ISSN: 1742-6588

Printed from e-media with permission by:

Curran Associates, Inc.
57 Morehouse Lane
Red Hook, NY 12571

Some format issues inherent in the e-media version may also appear in this print version.

Institute of Physics
Dirac House, Temple Back
Bristol BS1 6BE UK

Phone: 44 1 17 929 7481
Fax: 44 1 17 920 0979

techtracking@iop.org

Additional copies of this publication are available from:

Curran Associates, Inc.
57 Morehouse Lane
Red Hook, NY 12571 USA
Phone: 845-758-0400
Fax: 845-758-2633
Email: curran@proceedings.com
Web: www.proceedings.com

TABLE OF CONTENTS

Journal of Physics: Conference Series **2879** (2024) 011001 doi:10.1088/1742-6596/2879/1/011001

Preface

It is our honor to present the Proceedings of the 2024 3rd International Conference on Mechanical, Aerospace Technology and Materials Application (MATMA 2024). MATMA 2024 is an event that brings together the world's leading experts and scholars in the field of mechanical and aerospace technology, and is also an excellent platform for sharing the latest research results and discussing the application of cutting-edge technologies in related domains.

The conference aims to discuss the latest progress and challenges in mechanical engineering, aerospace technology and materials applications, and to promote the innovation of related theories and technological breakthroughs. By building this international exchange platform, we expect to promote extensive cooperation among academics, strengthen the integration of industry, academia and research between universities and enterprises, jointly address the key issues of industry development, and promote the sustained and healthy development of relevant fields.

MATMA 2024 brought together top scholars and researchers from all over the world, who shared their latest research results and practical experiences in mechanical engineering, aerospace technology and materials applications during this conference. It is worth mentioning that the conference specially invited Duc Truong Pham (Professor from University of Birmingham, UK and Member of the Royal Academy of Engineering), Eckart Meiburg (Professor from University of California Santa Barbara, US), Alam Md. Mahbub (Professor from Harbin Institute of Technology (Shenzhen), China), Yufei Gao (Professor from Shandong University, China), Caixu Yue (Professor from Harbin University of Science and Technology, China), Taimoor Asim (Robert Gordon University, Aberdeen, UK), etc., to give the keynote speeches. In-depth exchanges and discussions around aerodynamics and aeroelasticity, aerospace design, testing and performance, numerical control technology, noise vibration and harshness, combustion and emission control, advanced manufacturing technology, as well as deformation and fracture of materials, materials processing, high-performance materials, etc. were carried out during the conference. Through these communications, we have jointly explored the current and future development direction of mechanical engineering, aerospace technology and materials applications, stimulated innovative thinking, and promoted academic exchanges and cooperation.

This Proceedings of MATMA 2024 contains excellent academic papers submitted from all around the world, which not only cover the basic theories, structure design and analysis, and operation and evaluation systems of aerospace, but also discuss in depth the applications and innovations of mechanical technologies and high-performance materials. Each paper has been strictly reviewed and screened to ensure its academic value and practical application value.

We would like to express our sincere thanks to all the participants for their enthusiastic participation and active contributions, to the organizers and committee members for their careful organization and thoughtful arrangements, and to all the experts and scholars who took the time out of their busy schedules to attend the conference and bring us wonderful speeches and profound insights. We believe that with the joint efforts of all of you, MATMA 2024 will become an important milestone in promoting the development of mechanical and aerospace technology and materials applications.

The Committee of MATMA 2024

Journal of Physics: Conference Series **2879** (2024) 011001 doi:10.1088/1742-6596/2879/1/011001

Committee Member

General Conference Chair

Prof. Junxi Bi, Inner Mongolia University of Technology, China
Prof. Eckart Meiburg, University of California Santa Barbaray, USA
Prof. Alam Md. Mahbub, Harbin Institute of Technology (Shenzhen), China

Technical Program Committee Chair

Prof. Yufei Gao, Shandong University, China

Publication Chair

Prof. Caixu Yue, Harbin University of Science and Technology, China

Organizing Committee

Prof. Murat Tolga OZKAN, Gazi University, Turkey
Prof. Prashanth Konda Gokuldoss, Tallinn University of Technology, India
Prof. Gang Zhang, Harbin Institute of Technology, China
Prof. Sirong Yu, China University of Petroleum, China
Prof. Tao Yu, Tongji University, China
Prof. Weibing Zhu, Xihua University, China
Assoc. Prof. Yun Bai, Inner Mongol University of Technology, China
Assoc. Prof. Jiandong He, Inner Mongol University of Technology, China
Assoc. Prof. Xin Liu, Inner Mongol University of Technology, China
Assoc. Prof. Biao Li, Inner Mongol University of Technology, China
Dr. Jin Chai, Inner Mongol University of Technology, China
Dr. Liang Zhang, Jiangsu Normal University, China
Dr. Hock Jin Quah, Universiti Sains Malaysia, Malaysia

Technical Program Committee

Assoc. Prof. SANDIP KUNAR, Aditya Engineering College, India
Assoc. Prof. Engin Derya Gezer, Karadeniz Technical University, Turkey
Dr. AHMED LEGROURI, International University of Grand-Bassam, Morocco
Prof. Junxi Bi, Inner Mongolia University of Technology, China
Prof. Ruixiang Bai, Dalian University of Technology, China
Prof. Dr. Murat Tolga OZKAN, Gazi University, Turkey
Prof. Prashanth Konda Gokuldoss, Tallinn University of Technology, India
Dr. Hock Jin Quah, Universiti Sains Malaysia, Malaysia
Prof. Gang Zhang, Harbin Institute of Technology, China
Assoc. Prof. Jie Jia, Chang'an University, China
Dr. Yongmao Xiao, Xinyang Normal University, China
A. Prof. Rongyun Zhang, Anhui Polytechnic University, China
Dr. Shichang Xiao, Shanghai Maritime University, China
Prof. Tao Yu, Tongji University, China
Prof. Sirong Yu, China University of Petroleum, China
Dr. Liang Zhang, Jiangsu Normal University, China
Prof. Weibing Zhu, Xihua University, China

Journal of Physics: Conference Series **2879** (2024) 011002 doi:10.1088/1742-6596/2879/1/011002

Peer Review Statement

All papers published in this volume have been reviewed through processes administered by the Editors. Reviews were conducted by expert referees to the professional and scientific standards expected of a proceedings journal published by IOP Publishing.

- **Type of peer review:** Single Anonymous
- **Conference submission management system:** Morressier
- **Number of submissions received:** 146
- **Number of submissions sent for review:** 111
- **Number of submissions accepted:** 78
- **Acceptance Rate (Submissions Accepted / Submissions Received × 100):** 53.4
- **Average number of reviews per paper:** 3
- **Total number of reviewers involved:** 20
- **Contact person for queries:**
  Name: Xuexia
  Email: xx.ye@keoaeic.org
  Affiliation: AEIC Academic Exchange Information Centre

Journal of Physics: Conference Series 2879 (2024) 012001 doi:10.1088/1742-6596/2879/1/012001

Porous dielectric design for high-performance flexible shear sensors

Yingxuan Bu*

Harbin Institute of Technology, Weihai 264209, China.

*Corresponding author's e-mail: buyingxuan@126.com

Abstract: Flexible sensors, renowned for their exceptional adaptability, are widely utilized in areas such as robotics and healthcare monitoring. Currently, understanding the relationship between the structure and performance of flexible shear sensors has become increasingly important. In this study, the structural characteristics of porous flexible shear sensors and their impact on performance were analyzed using Finite Element Method (FEM). The investigation focused on three factors: porosity, pore size, and the shape of elliptical pores. The findings suggest that augmenting the porosity substantially boosts the shear sensitivity of the sensor, whereas the pore size exerts a relatively minor influence on sensitivity. Furthermore, compared to circular ones, elliptical pores can further increase the shear sensitivity of the sensor, particularly when the ellipse's major axis is oriented horizontally. These findings are of significant reference value for designing and optimizing high-performance flexible shear sensors.

1. Introduction

Flexible sensors, distinguished by their flexibility and extensibility, can closely conform to the surfaces of objects and are widely implemented in various fields, such as robotics [1] and medical health monitoring [2, 3]. Capacitive flexible sensors are widely used, especially because they offer advantages such as affordability, minimal power usage, excellent temperature stability, heightened responsiveness, and reduced hysteresis [4]. These sensors typically feature a sandwich structure comprising top and bottom electrodes with a dielectric layer sandwiched in between. When an external force is applied, the dielectric material deforms, which changes the gap or the facing area between the electrodes, resulting in a variation in capacitance. By monitoring the alteration in capacitance, the intensity and orientation of the applied force can be precisely deduced. However, the deformation capability of solid dielectric layers is limited and insufficient to meet the high sensitivity requirements of the sensors. To address this, researchers have significantly enhanced the dielectric layer's deformability and sensor performance by designing structures such as surface microstructures [5], spacers [6], nanowires structures [7], porous structures [8], and fabric structures [9]. Specifically, flexible capacitive sensors with a porous structure, which utilize porous materials for the dielectric layer, have attracted significant interest due to attributes such as a large specific surface area, superior elasticity and reversibility, reduced weight, and their lightweight nature.

However, current research primarily focuses on optimizing porous pressure sensors, with relatively less exploration into porous shear flexible sensors, especially the relationship between porous structures and shear response. Bilent [10] prepared porous PDMS as the dielectric layer using a sugar templating method and conducted shear tests on the sensor, but the specific impact of porosity was not sufficiently explored. Mu [11] developed a flexible sensor utilizing porous polydimethylsiloxane (PDMS), carbon nanotubes, and graphene oxide nanoplatelets, designed to emulate the human skin to detect shear forces. However, this study also did not deeply investigate the specific effects of the pore structure. This paper seeks to explore how the structural features of the porous dielectric layer—like porosity, the size of the pores, and the shape ratio of elliptical pores—affect the performance of shear sensors. The goal is to offer theoretical insights that can aid in the development of superior shear sensors in the future.

2. Materials and methods

2.1. Finite element method

The FEM model is depicted in Figure 1. It features a porous flexible shear sensor comprising several components. The components encompass an upper electrode, a polyimide shield layer, a porous dielectric layer, and a lower electrode. Blue regions represent the elastic polymer in the porous dielectric layer, while circular areas indicate the pores. Copper foil is selected as the material for the electrodes. The characteristics of the porous elastic polymer are specified as a Young's modulus of 3 MPa, a Poisson's ratio of 0.49, and a relative permittivity of 3. The model measures 2 mm in height and 4 mm in length. Regarding the boundary conditions of the FEM, a horizontal load and a 1V voltage are applied to the top electrode. The bottom electrode is subjected to fixed and grounded conditions. Additionally, a circular area with a radius of 50 mm is designed to simulate the air domain. Furthermore, to ensure simulation accuracy, the mesh is carefully refined. The maximum mesh size is kept within 0.01 mm.

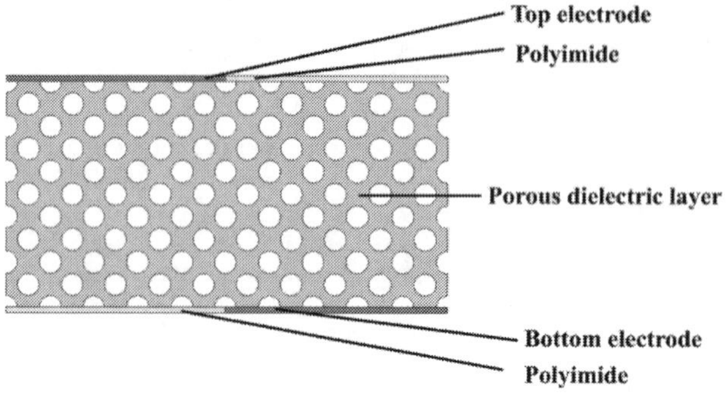

Figure 1. Schematic diagram of sensor structure.

3. Results and discussion

3.1. Sensor principle

The operational mechanism of the sensor is depicted in Figure 2, utilizing the equation for a capacitor with parallel plates:

$$C = \frac{\varepsilon_0 \varepsilon_r A}{d} \tag{1}$$

In this equation, the term C denotes the sensor's capacitance. ε_0 stands for the permittivity of a vacuum. The relative permittivity of the porous dielectric material is indicated by ε_r. A refers to the area facing each other across the two electrodes. Lastly, d signifies the gap between these electrodes.

When the sensor is subjected to shear force, the porous dielectric layer undergoes shear deformation. This deformation leads to changes in the opposite area between the electrodes, thereby causing variations in capacitance. We can accurately calculate the magnitude of the shear force exerted on the sensor by measuring these capacitance changes.

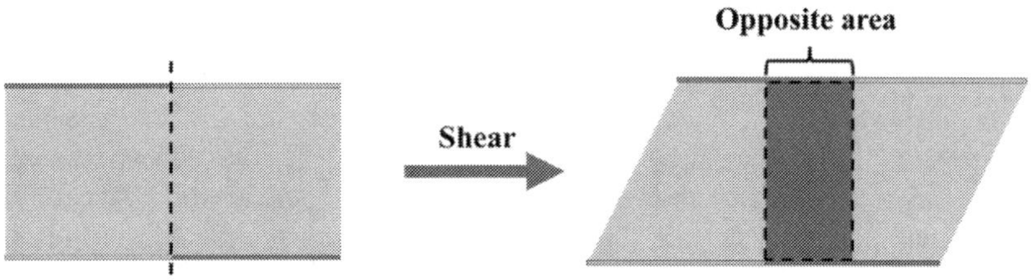

Figure 2. Flexible shear sensor operation principle schematic.

3.2. Effect of porosity on the performance of shear flexure sensor

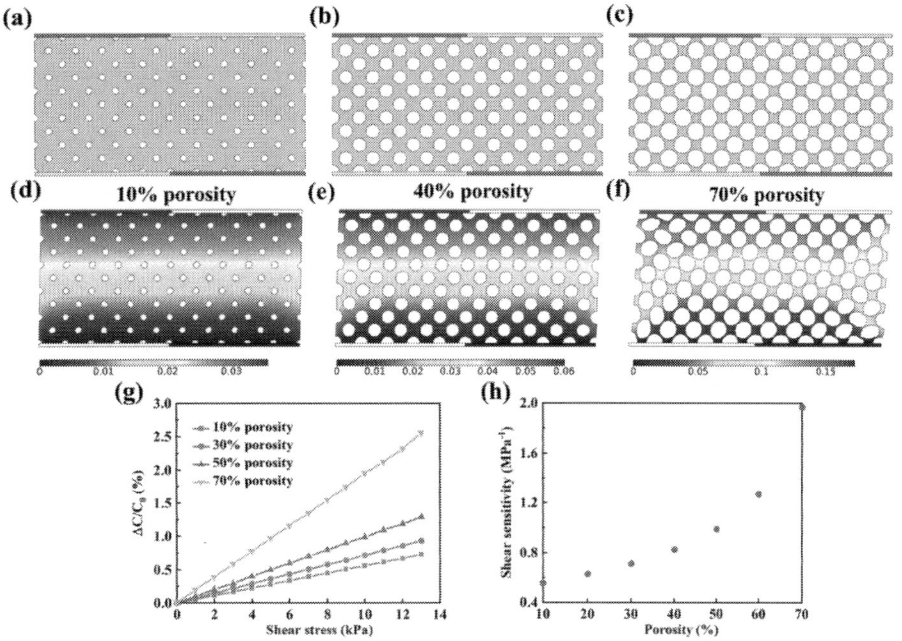

(a) Dielectric layer structure at 10% porosity. (b) Dielectric layer structure at 40% porosity. (c) Dielectric layer structure at 70% porosity. (d) Displacement of the sensor with 10% porosity under shear force. (e) Displacement of the sensor with 40% porosity under shear force. (f) Displacement of the sensor with 70% porosity under shear force (g) Trend of relative capacitance change versus shear stress. (h) Sensor shear sensitivity in response to changes in porosity levels.

Figure 3. The effect of porosity on the performance of shear sensors.

The sensor's performance is predominantly determined by its porosity. We modified the porosity of the dielectric layer by altering the pore size, maintaining a consistent number of pores. Porosity is determined by the proportion of the area covered by the white circular regions relative to the entire area of the rectangular dielectric layer. The figures in 3(a)-(c) depict structures with porosities of 10%, 40%, and 70%, respectively. After applying a certain shear force to the top electrode, the deformation of the sensor at different porosities was illustrated in Figures 3(d)-(f). It could be observed that with increased porosity, the deformation under the same shear force became more pronounced. Additionally, we plotted the curve of the sensor's relative capacitance change versus shear stress (Figure 3(g)). From this, it was evident that the magnitude of relative capacitance change increased linearly with increasing shear stress, and the slope of relative capacitance change was greater at higher porosities. According to the formula for the shear sensitivity of the sensor:

$$S = \frac{\Delta C}{\Delta P C_0} \tag{2}$$

In the formula, ΔC denotes the variation in the sensor's capacitance, C_0 signifies the sensor's initial capacitance, and ΔP indicates the shear stress exerted on the sensor. We found that shear sensitivity increased with increasing porosity. This was because a higher porosity lowered the shear modulus of the dielectric layer, allowing the sensor to undergo greater deformation under the same shear force, thereby resulting in larger capacitance changes and enhanced shear sensitivity. Ultimately, we examined the correlation between the sensor's shear sensitivity and porosity, as depicted in Figure 3(h). The data in the figure indicates that the sensor's sensitivity rose significantly, following an approximately exponential trend in response to the increase in porosity. The higher the porosity, the greater the effect on the sensor sensitivity.

3.3. *Effect of pore size on the performance of shear flexure sensor*

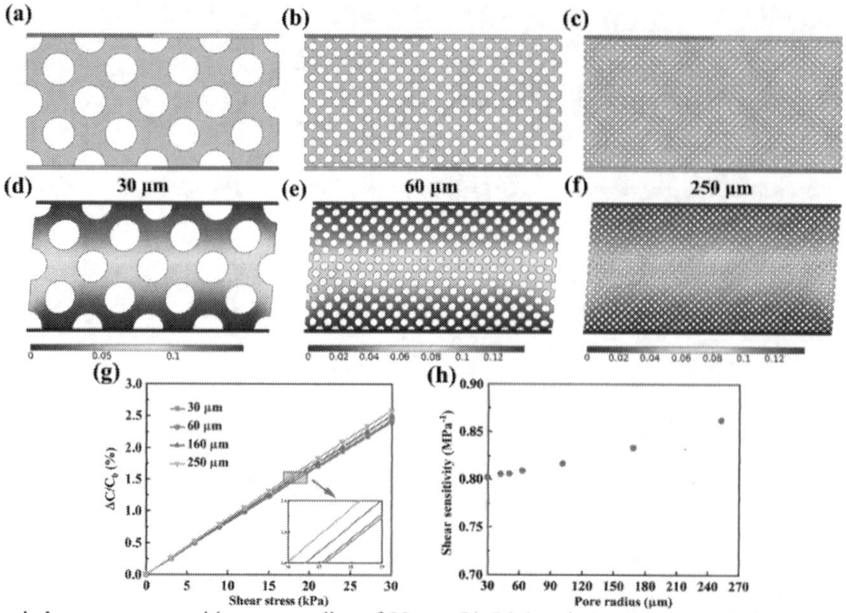

(a) Dielectric layer structure with a pore radius of 30μm, (b) Dielectric layer structure with a pore radius of 60μm, (c) Dielectric layer structure with a pore radius of 250μm, (d) The sensor's displacement variation in response to a shear force, featuring a 30μm pore radius. (e) The sensor's displacement variation in response to a shear force, featuring a 60μm pore radius. (f) The sensor's displacement variation in response to a shear force, featuring a 250μm pore radius. (g) The trend of the relative capacitance change as a function of shear stress for sensors with different pore radii (h) The trend of the sensor's shear sensitivity with changes in pore radius.

Figure 4. The influence of pore size on the performance of flexible shear sensors.

Subsequently, we conducted a study on the impact of pore size on the shear performance of the sensor. While keeping the porosity constant, we adjusted the number of pores to obtain dielectric layer structures with different pore sizes, as shown in Figures 4(a)-(c). All three dielectric layer structures maintained a porosity of 40% but with varying pore sizes of 30μm, 60μm, and 250μm. After exerting a defined shear force on the electrode's surface, the resulting sensor deformation is shown in Figures 4(d)-(f). The data indicated that the deformation patterns of the dielectric layers, despite varying pore dimensions, were largely consistent when subjected to the same shear force. In Figure 4(g), we have presented the variations in the relative capacitance of sensors with different pore sizes. The figure's data suggests a direct proportionality between the increase in relative capacitance and the escalation of shear stress. However, the difference in the slope of relative capacitance change among sensors with different pore sizes is small. As shown in Figure 4(h), although an increase in pore size slightly enhances the shear sensitivity of the sensor, this enhancement is very limited. This indicates that the impact of pore size on the shear sensitivity of the sensor is relatively minor.

3.4. Effect of elliptical pores on sensor performance

We conducted a study on the impact of elliptical pores, maintaining a porosity of 40%. We investigated how the aspect ratio of the elliptical axes affects the sensor's performance, defining "Ratio" as the relationship between the dimensions of the major and minor axes. We selected ratios 1.5 and 2, as shown in Figures 5(a)-(b) for sensors with horizontal major axes and in Figures 5(c)-(d) for sensors with vertical major axes. After applying the shear force, the deformed image of the sensor is obtained, as shown in Figures 5(e)-(h). Specifically, with a 2:1 aspect ratio, we assessed the performance of elliptical pores oriented with either the major axis horizontal or vertical, in contrast to circular pores, all under the same shear stress, as depicted in Figure 5(e). The findings demonstrated that the proportional change in capacitance for sensors with elliptical pores rose in direct proportion to the increase in shear stress. Moreover, sensors with elliptical pores showed higher shear sensitivity than those with circular pores, with horizontal major axes exhibiting higher sensitivity than vertical ones. Further, we explored how changing the ratio impacts sensor sensitivity when the major axis is horizontal versus vertical. As demonstrated in Figure 5(i), increasing the ratio in sensors with a horizontal major axis led to a steeper slope in relative capacitance change, thus enhancing sensitivity. Similarly, increasing the ratio also improved sensitivity for the vertical major axis structure. Subsequently, we analyzed the sensor's shear sensitivity trend as the aspect ratio (Ratio) changed, as shown in Figure 5(l). The research showed that the sensor's sensitivity gradually improved as the ratio increased. Notably, the increase in sensitivity was more significant when the major axis was aligned horizontally compared to when it was aligned vertically.

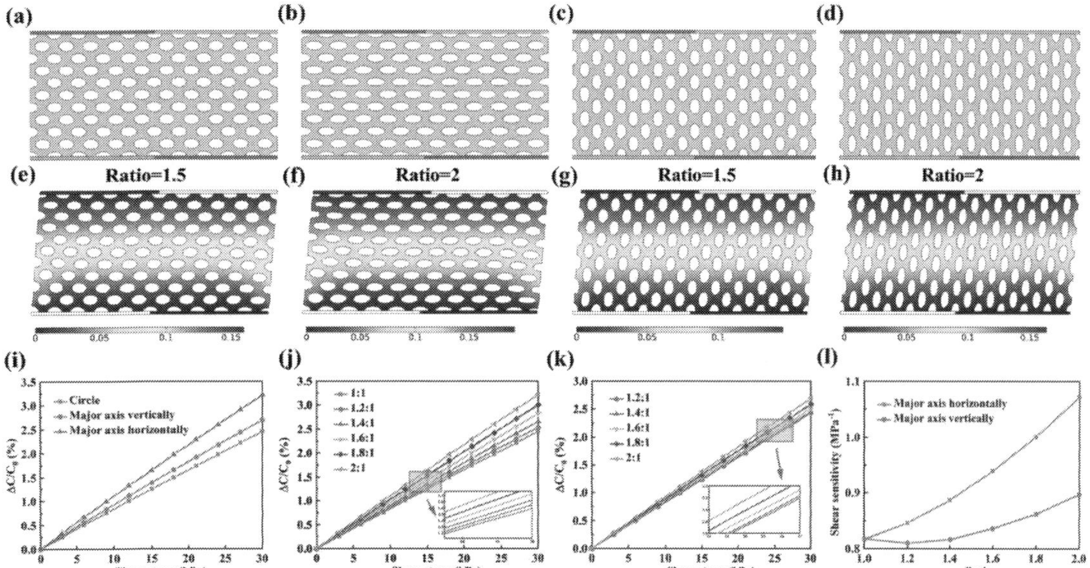

(a) A dielectric layer with a 1.5 aspect ratio, featuring a horizontally aligned major axis. (b) A dielectric layer with a 2:1 aspect ratio, featuring a horizontally aligned major axis. (c) A dielectric layer with a 1.5 aspect ratio, featuring a vertically aligned major axis. (d) A dielectric layer with a 2:1 aspect ratio, featuring a vertically aligned major axis. (e) Displacement change of the sensor with an elliptical pore of aspect ratio 1.5 and major axis horizontal under shear force. (f) Displacement change of the sensor with an elliptical pore of aspect ratio 2 and major axis horizontal under shear force. (g) Displacement change of the sensor with an elliptical pore of aspect ratio 1.5 and major axis vertical under shear force. (h) Displacement change of the sensor with an elliptical pore of aspect ratio 2 and major axis vertical under shear force. (i) Relative capacitance change of sensors with circular pores, elliptical pores with horizontal major axes, and elliptical pores with vertical major axes under shear force. (j) Relative capacitance change of sensors with elliptical pores and a horizontal major axis at different aspect ratios. (k) Relative capacitance change of sensors with elliptical pores and a vertical major axis at different aspect ratios. (l) The trend of the sensor's shear sensitivity with changes in ratio.

Figure 5. The impact of elliptical pores on the performance of flexible shear sensors.

4. Conclusions

This study delved into the relationship between the structural features of porous shear sensors and their performance using the Finite Element Method (FEM). Three principal factors are focused on: the level of porosity, the sizes of the pores, and the aspect ratio of elliptical pores. The research reveals that elevated porosity markedly enhances the sensor's capacity to sense shear forces while the pore size's impact on sensitivity is less pronounced. Moreover, compared to circular pores, elliptical pores further increase the shear sensitivity of the sensor, especially when the ellipse's major axis is horizontal. In conclusion, this study offers vital theoretical insights and practical direction for the prospective development and refinement of advanced flexible shear sensors. By finely adjusting the parameters of the porous structure, such as porosity and pore shape, the sensor's performance can be significantly improved to meet the sensitivity and reliability requirements of various application fields. Implementing these designed porous sensors in robots' end effectors can provide precise tactile feedback, thereby enhancing operational flexibility and adaptability. Additionally, their excellent elasticity and durability help maintain high reliability and extend the lifespan of robots performing repetitive tasks in complex environments. Future research will further explore the impact of different materials and composite structures on sensor performance, aiming to broaden its adaptability in a broader range of applications.

References

[1] Guo, Y.; Zhong, M.; Fang, Z.; Wan, P.; Yu, G. A Wearable Transient Pressure Sensor Made with MXene Nanosheets for Sensitive Broad-Range Human–Machine Interfacing. Nano Letters 2019, 19, 1143-1150.

[2] Kim, K.; Kim, H. J.; Zhang, H.; Park, W.; Meyer, D.; Kim, M. K.; Kim, B.; Park, H.; Xu, B.; Kollbaum, P.; et al. All-printed stretchable corneal sensor on soft contact lenses for noninvasive and painless ocular electrodiagnosis. Nature Communications 2021, 12, 1544.

[3] Li, L.; Zheng, J.; Chen, J.; Luo, Z.; Su, Y.; Tang, W.; Gao, X.; Li, Y.; Cao, C.; Liu, Q.; et al. Flexible Pressure Sensors for Biomedical Applications: From Ex Vivo to In Vivo. Advanced Materials Interfaces 2020, 7, 2000743.

[4] Kurup, L. A.; Cole, C. M.; Arthur, J. N.; Yambem, S. D. Graphene Porous Foams for Capacitive Pressure Sensing. ACS Applied Nano Materials 2022, 5, 2973-2983.

[5] Duan, Y.; Wu, J.; He, S.; Su, B.; Li, Z.; Wang, Y. Bioinspired Spinosum Capacitive Pressure Sensor Based on CNT/PDMS Nanocomposites for Broad Range and High Sensitivity. Nanomaterials 2022, 12.

[6] Huang, B.; Feng, J.; He, J.; Huang, W.; Huang, J.; Yang, S.; Duan, W.; Zhou, Z.; Zeng, Z.; Gui, X. J. A. A. M.; et al. High Sensitivity and Wide Linear Range Flexible Piezoresistive Pressure Sensor with Microspheres as Spacers for Pronunciation Recognition. ACS Applied Materials & Interfaces 2024, 16, 19298-19308.

[7] Zhang, J.; Sun, T.; Chen, Y.; Liu, L.; Zhao, H.; Zhang, C.; Meng, X.; Wang, D.; Hu, Z.; Zhang, H. J. A. M. T. Nanowires in flexible sensors: structure is becoming a key in controlling the sensing performance. Advanced Materials Technologies 2022, 7, 2200163.

[8] Bu, Y.; Wu, J.; Zhang, Z.; Wei, Q.; Su, B.; Wang, Y. Design and Analysis of Porous Elastomeric Polymer Based on Electro-Mechanical Coupling Characteristics for Flexible Pressure Sensor. Polymers 2024, 16, 701.

[9] Li, Z.; Lu, L.; Xie, Y.; Wang, W.; Lin, Z.; Tang, B.; Lin, N. J. A. E. M. Preparation of Laser-Induced Graphene Fabric from Silk and Its Application Examples for Flexible Sensor. Advanced Engineering Materials 2021, 23, 2100195.

[10] Bilent, S.; Dinh, T. H. N.; Martincic, E.; Joubert, P. Y. Influence of the porosity of PDMS foams on the performances of flexible capacitive stress sensors. In Proceedings of the 2019 Symposium on Design, Test, Integration & Packaging of MEMS and MOEMS (DTIP), 12-15 May 2019, 2019; pp. 1-4.

[11] Mu, C.; Song, Y.; Huang, W.; Ran, A.; Sun, R.; Xie, W.; Zhang, H. Flexible Normal-Tangential Force Sensor with Opposite Resistance Responding for Highly Sensitive Artificial Skin. Advanced Functional Materials 2018, 28, 1707503.

Journal of Physics: Conference Series **2879** (2024) 012002 doi:10.1088/1742-6596/2879/1/012002

Effect of alloying elements on coexisting phases and microstructure of M174 heat-resistant aluminum alloy

Jinjun Tang[1]*, Cui Liang[2] and Jiqiang Li[1]

[1]NingboTech University, Ningbo 315100, Zhejiang, China
[2]TUV Rheinland /CCIC (Ningbo) Co., Ltd., Ningbo 315103, Zhejiang, China

*Corresponding author's e-mail: jinjun.tang@163.com

Abstract: This article mainly focuses on high-power density diesel engine piston alloys, with the main component system being M174 heat-resistant aluminum alloy. On this basis, the main heat-resistant alloying elements Cu, Ni, and Fe were optimized and designed. This article is based on Thermo-Calc thermodynamic software, and comprehensively calculates and analyzes the multi-component alloy equilibrium phase diagram and Scheil non-equilibrium solidification of Al-(4-6)Cu-(1.5-3)Ni-(0.7-1.1Fe)-12.5Si-0.85Mg (mt.%) heat-resistant aluminum alloy system. Through the orthogonal experimental method of alloy composition, the phase distribution and microstructure composition of different alloy compositions are obtained. Research has found that Fe-rich phases such as Al_5FeSi and Al_9FeNi, as well as Ni-rich phases such as Al_3CuNi, Al_7Cu_4Ni, and Al_3Ni, are the main heat-resistant phases of the alloy in the temperature range of 350°C-420°C, providing important support for the high-temperature mechanical properties and creep damage resistance of piston aluminum alloy. At the same time, this article further studies the solidification history and microstructure precipitation of alloys. Fe-rich heat-resistant phases such as Al_5FeSi and Al_9FeNi have a higher solidification sequence. As primary and secondary phases, it is easy to generate coarse solidified structures. The organizational content is sensitive to the alloy composition. Al_3CuNi and Al_7Cu_4Ni mainly precipitate in a multi-component eutectic structure, with layered and petal-shaped morphologies. The research results of this article have important theoretical significance and application value for the optimization design of high-power density piston aluminum alloy composition and the coordinated control of multi-phase structure.

1. Introduction

With the advancement of modern science and technology, the emergence of high-pressure conjugate injection and efficient turbocharging technology in diesel engines, and the continuous increase in power, the key components of the diesel engine combustion chamber are subjected to high temperature, high pressure, and high operating frequency, causing the maximum working temperature of the piston alloy to rise to the range of 350°C-420°C [1]. To meet the requirements of environmental protection, energy conservation, emission reduction, and improving fuel utilization have become goals pursued by countries around the world, especially in the manufacturing industry represented by the automotive industry. Among the many measures proposed with the goal of reducing body weight and improving fuel utilization, material light-weighting is one of the most effective measures [2-3]. Further improving the output power of the engine is an important way to solve environmental and energy problems [4]. Improving the high-temperature mechanical properties of piston aluminum alloy is a key technical approach to enhancing engine output power [5].

Therefore, it is urgent to develop new high-strength and heat-resistant aluminum alloy piston materials to meet the high-temperature performance requirements of 300°C-400°C. M174 piston alloy belongs to the Al-Si-Cu-Ni-Mg-Fe alloy composition system [6]. The content and microstructure distribution of Fe-rich and Ni-rich heat-resistant phases in alloys have a significant impact on the material's performance [7]. Especially the control of Fe, Ni, and Cu elements has a significant impact on the precipitation types and contents of the main heat-resistant phases in the alloy, the solidification history and morphological characteristics of the microstructure, as well as the content of characteristic heat-resistant structures [8]. Previous studies have shown that the heat resistance, fatigue performance, and impact resistance of piston aluminum alloys in their high-temperature and high-pressure working environment are closely related to the types and contents of heat-resistant phases, especially Fe-rich and Ni-rich heat-resistant phases, as well as the characteristic microstructure [9]. Further research is urgently needed on the thermodynamic calculations, phase composition, and microstructure of M174 piston aluminum alloy material [10].

2. The alloy materials and research methods

This article mainly focuses on the composition system of Al-(4-6)Cu-(1.5-3.0)Ni-(0.7-1.1Fe)-12.5Si-0.85 Mg (mt.%) heat-resistant piston aluminum alloy, and uses Thermo Calc software to perform thermodynamic and kinetic calculations and analysis of multi-component alloy materials. We obtain the multicomponent equilibrium phase diagram of piston aluminum alloy and Scheil non-equilibrium solidification path data. The changes in the composition of alloys such as M174 and M142 not only alter the types and contents of precipitated strengthening phases but also contribute to the types and contents of solidified structures. This article uses thermodynamics and phase diagram calculations to quantitatively label and analyze the types, contents, and phase transition conditions of alloy solidification structures, and explore differences and commonalities.

To better study the thermodynamic state of the alloy system, this article sets three composition points for Cu element based on the sub-eutectic composition range of M174 aluminum alloy, namely 4.0 mt.%, 5.0 mt.%, and 6.0 mt.%. We set the Ni element to three composition points: 1.5 mt.%, 2.2 mt.%, and 3.0 mt.%. We set the Fe element to three composition points: 0.7 mt.%, 0.9 mt.%, and 1.1 mt.%. The Si element is fixed at 12.5 wt.% and the Mg element is fixed at 0.85 mt.%. The remaining Ti is added as a refining agent, while Mn and Zn are impurity elements, and their content should be strictly controlled within 0.3 mt.%. Table 1 shows the M174 heat-resistant aluminum alloy three-factor and three-level orthogonal experimental composition design. Si and Mg are invariant elements. Mn, Ti, and Zn are impurity elements. Table 2 shows M174 alloy residual microstructure content changes with alloy elements of the Al-Cu-Ni-Fe-Si-Mg alloy system. In M174 alloy, the change of Cu element mainly affects the content of solidification structures such as Si, E1, E2, and P1. At the same time, when Cu increases to 6%, E3, E3, XE3, E6, P3, and E7 structures will disappear. The change of Ni element mainly affects the solidification structure of M174 alloy E1, E2, E3, P1, etc.

Table 1. M174 heat-resistant aluminum alloy three-factor and three-level orthogonal experimental composition design (mt.%).

	Cu	Ni	Fe	Si	Mg	Mn	Ti	Zn	Al
M174-1	4.0	1.5	0.7	12.5	0.85	<0.3	<0.2	<0.3	The rest
M174-2	4.0	2.2	0.9	12.5	0.85	<0.3	<0.2	<0.3	The rest
M174-3	4.0	3.0	1.1	12.5	0.85	<0.3	<0.2	<0.3	The rest
M174-4	5.0	1.5	0.9	12.5	0.85	<0.3	<0.2	<0.3	The rest
M174-5	5.0	2.2	1.1	12.5	0.85	<0.3	<0.2	<0.3	The rest
M174-6	5.0	3.0	0.7	12.5	0.85	<0.3	<0.2	<0.3	The rest
M174-7	6.0	1.5	1.1	12.5	0.85	<0.3	<0.2	<0.3	The rest
M174-8	6.0	2.2	0.7	12.5	0.85	<0.3	<0.2	<0.3	The rest
M174-9	6.0	3.0	0.9	12.5	0.85	<0.3	<0.2	<0.3	The rest

Table 2. M174 alloy residual microstructure content changes with alloy elements.

Composition	M174	Cu4	Cu6	Ni1.5	Ni3	Fe0.7	Fe1.1
(si)	0.018	0.013	0.029	0.019	0.016	0.007	-
E1	0.329	0.36	0.296	0.352	0.251	-	-
E2	0.031	0.027	0.017	0.122	0.185	0.312	0.322
P1	0.225	0.246	0.218	0.156	0.263	0.231	0.239
P2 (E)	0.209	0.169	0.251	0.158	0.27	0.209	0.21
E3	0.044	0.092	-	0.022	0.071	0.044	0.045
E4	0.013	0.011	-	0.015	0.008	0.013	0.011
XE5	0.044	0.026	-	0.045	0.037	0.041	0.045
E6	0.006	0.004	-	0.008	0.008	0.006	0.008
P3	0.022	0.015	-	0.03	0.017	0.025	0.021
E7	0.007	0.004	-	0.009	0.005	0.007	0.003
P4	0.008	0.004	0.007	0.006	0.008	0.008	0.015
E8	0.007	0.005	0.013	0.009	0.009	0.006	0.005
XP4	0.008	0.008	0.022	0.015	0.009	0.009	0.011
E9	0.033	0.026	0.058	0.046	0.036	0.039	0.033
Fe0.7+1	-	-	-	-	-	0.015	0.037
Fe1.1+1	-	-	-	-	-	-	0.006

The solidification phase transformation process of M174 heat-resistant aluminum alloy is as follows:

$$L1 \rightarrow (Si) \qquad 581°C\text{-}557°C \qquad (1)$$
$$L2 \rightarrow (Al)+(Si) \qquad 557°C\text{-}547.5°C \qquad (2)$$
$$L3 \rightarrow (Al)+(Si)+\beta \qquad 547.5°C\text{-}547°C \qquad (3)$$
$$L4 \rightarrow (Al)+(Si)+\beta+T \qquad 547°C\text{-}546.8°C \qquad (4)$$
$$L5+\beta \rightarrow (Al)+(Si)+T \qquad 546.8°C\text{-}535.5°C \qquad (5)$$
$$L6 \rightarrow (Al)+(Si)+T+\delta \qquad 535.5°C\text{-}527°C \qquad (6)$$
$$L7 \rightarrow (Al)+(Si)+T+\delta+Q \qquad 527°C\text{-}525.1°C \qquad (7)$$
$$Q+L8 \rightarrow (Al)+(Si)+T+\delta+\pi \qquad 525.1°C\text{-}524.9°C \qquad (8)$$
$$T+L9 \rightarrow (Al)+(Si)+\delta+\pi \qquad 524.9°C\text{-}524.8°C \qquad (9)$$
$$L10 \rightarrow (Al)+(Si)+\delta+\pi+Q \qquad 524.8°C\text{-}523°C \qquad (10)$$

From thermodynamic calculations and phase diagram analysis of M174 alloy, the initial precipitation temperature of the Si phase is 581°C, which is 16°C higher than that of M142. The temperature range for Si phase precipitation is 24°C, which is 20°C higher than that of M142, indicating that M142 is more inclined towards eutectic composition. M174 tends to have a hypereutectic composition, resulting in an increase in Si content during crystallization. M174 disappears from the liquid phase at 523°C and ends solidification, 5°C lower than M142. The solidification of M174 alloy is more complex than other aluminum alloys. M174 aluminum alloy has multiple eutectic and peritectic solidification reaction platforms, which have a significant impact on the solidification microstructure morphology.

3. Results and discussion

The phase diagram of M174 alloy with changes in Fe element is similar to that of M142 alloy, and the corresponding reaction platform temperature decreases. The phase diagram of M174 alloy with changes in Fe element is similar to that of M142 alloy, and the corresponding reaction platform temperature decreases. As shown in Figure 1, according to the multi-component alloy phase diagram and the calculation of the far equilibrium solidification structure, at Fe=0.1 wt%, there is no β phase, and the Fe-rich structure is mainly T phase. When Fe=0.3 mt%, the alloy composition is at the Fe-rich angle, and the T phase content reaches its maximum value. It is also possible that the fluctuation of the composition may cause trace amounts of β phase. When Fe=0.5 mt%, the β phase precipitates, but the β phase and Al phase belong to symbiotic precipitates, and the Fe-rich structure is symbiotic β phase and T phase. We mainly study the microstructure of the primary β Fe-rich phase under symbiotic precipitation conditions; When Fe=0.9 mt%, the β phase precipitates before the Al phase, and the Fe-rich structure is mainly composed of the primary β phase.

The calculation phase diagram of the M174-8 multi-component alloy with Mg variation is shown in Figure 2, and its phase region is more complex than that of the M142 alloy. When Mg content is less than 0.3%, the Q ($Al_5Cu_2Mg_8Si_6$) phase disappears. When the Mg element is greater than 1.6%, the precipitate phase will contain the Mg_2Si phase.

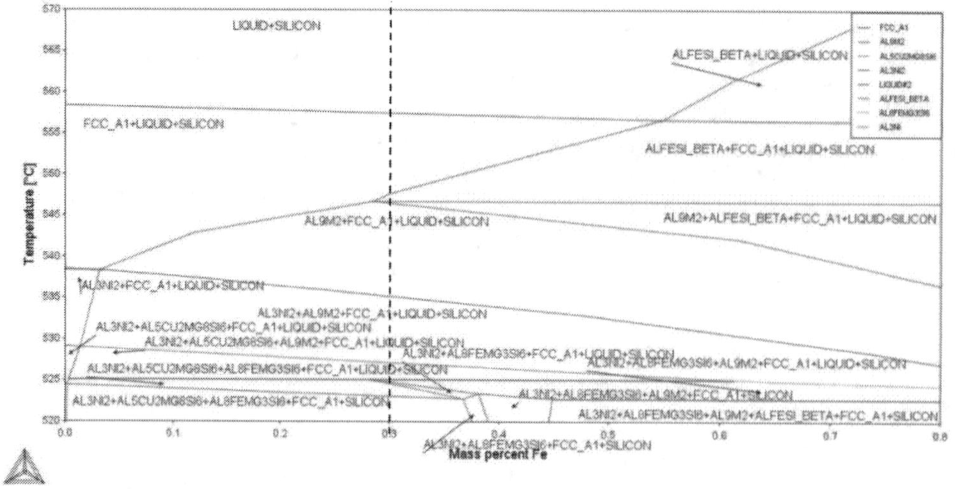

Figure 1. Calculation phase diagram of M174-5 multi-component alloy with Fe variation.

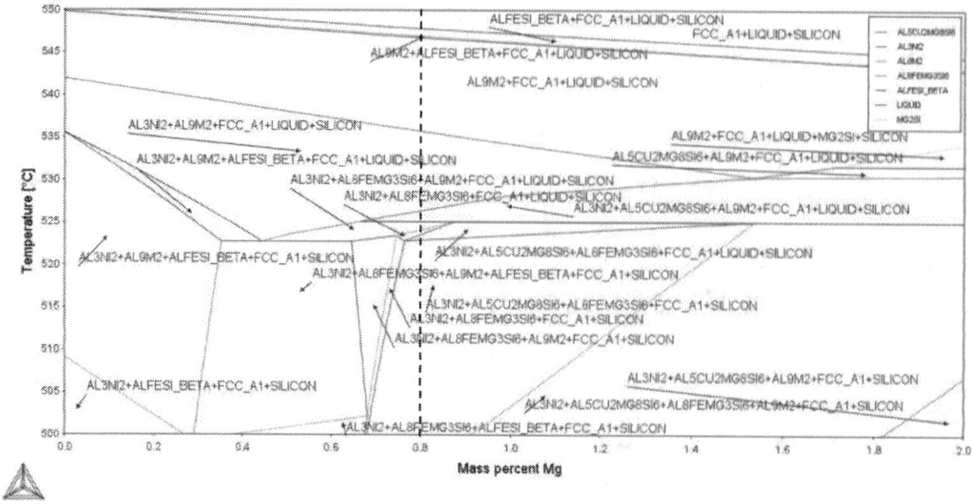

Figure 2. Calculation phase diagram of M174-8 multi-component alloy with Mg variation.

The effect of Ni element content on the solidification structure and phase transformation of M174 alloy is shown in Figure 3. Compared to M142 alloy, the temperature range for Si phase precipitation increases, and the critical composition for Al_3Ni precipitation increases (from 2.6% to 3.8%). When Ni element<1.8%, except for β, in addition to increasing the temperature range for phase precipitation, there will also be β, T-rich iron coexisting phase. When Ni element>2.5%, the T phase will replace β as a primary phase rich in iron, and eutectic precipitation occurs. When Ni element>3.8%, Al_3Ni and Al_3Ni_2 coexist in the matrix structure.

The Al_3Ni phase in M174 piston aluminum alloy tends to exhibit a block-like morphology, as shown in Figure 4. The Al_3CuNi phase tends to exhibit a strip-shaped morphology, while the supplementary Al_7Cu_4Ni phase shows a tendency towards skeletonization. Affected by the solidification sequence and eutectic platform of M174 piston aluminum alloy, the primary Al_3Ni phase is mainly distributed in point or block form in the Ni high composition zone. Due to the high content of the Al_3CuNi phase and its eutectic platform, it is mainly distributed in the aluminum matrix in layered and petal-shaped structures. The content of the Al_7Cu_4Ni phase is relatively low, mainly distributed in elongated and rod-like structures.

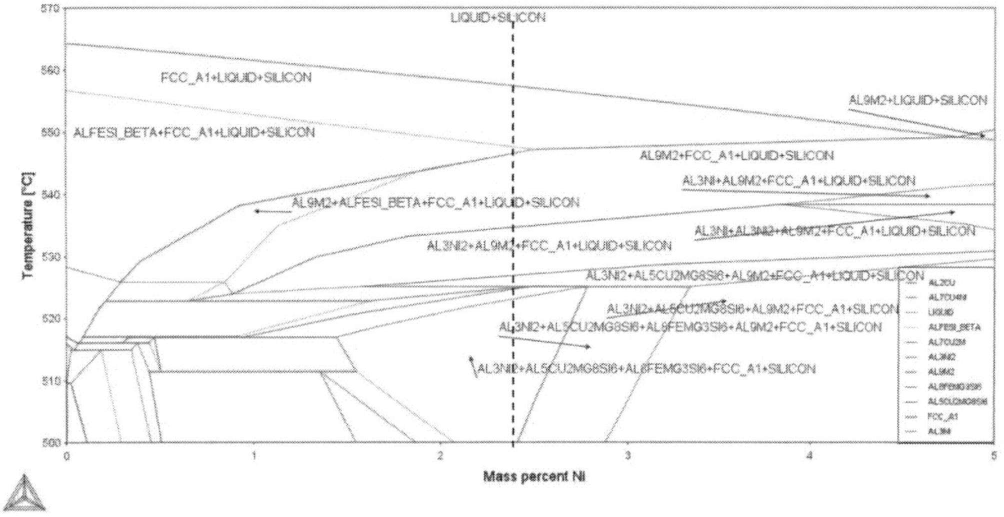

Figure 3. Calculation phase diagram of M174-6 multi-component alloy with Ni variation.

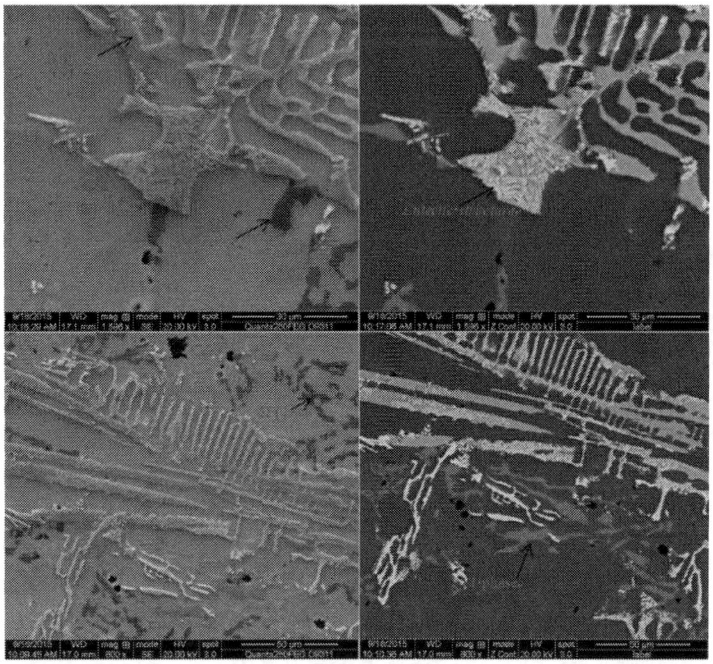

Figure 4. Coupled microstructure of heat-resistant and eutectic phases in M174 heat-resistant aluminum alloy.

As shown in Figure 4, the microstructure image of M174 piston aluminum alloy is presented. After extrusion casting of the alloy sample, it undergoes a solution treatment at 510°C for 6 hours and is quenched in water at 60-85°C to room temperature. The aging temperature is 260°C for 3 hours, and it is air-cooled to room temperature. The black block-like phase in the picture is the eutectic silicon phase. The petal-shaped phase is the Al_3CuNi phase, which plays a positive role in the high-temperature mechanical properties of the alloy. The $Al_3CuNi+Al_5Cu_2Mg_8Si_6$ eutectic structure appeared in the alloy matrix, mainly formed during solid solution and failure processes. The skeletal phase in the picture is rich in Ni, while the gray blocky phase is rich in the T phase and β-Fe phase. The rich Fe phase plays an important role in the high-temperature mechanical properties and creep resistance of alloys.

4. Summary

In the M174 piston aluminum alloy composition system, there are multiple eutectic reaction platforms such as E1-E9, as well as multiple eutectic reaction platforms such as P1-P4. The content of heat-resistant phases precipitated during alloy solidification and their microstructure morphology are closely related to their solidification history and precipitation reaction type. The Fe-rich heat-resistant phase appears at approximately Fe=0.3 mt% position, with a Fe-rich angle. Within the range of Fe<0.3 mt% in the low-composition region, there is no precipitation of the β-rich Fe phase. The content of Fe rich phase is relatively low, mainly precipitated in the form of the T phase. Within the range of Fe>0.3 mt% in the high composition region, β- rich Fe phase appears in the alloy structure.

In terms of the solidification structure of piston aluminum alloy, the main structures are E1, P1, and P2. With the Cu element increasing, the E1, E2, and P1 structures decrease, while the P2 structure increases by 25.1%. With the increase of Ni element, the eutectic structure of E1 and E2 decreases. The crystal structures of P1 and P2 increased to 26.2% and 27%, respectively. Meanwhile, as the Fe element increases, the E2 structure also increases, which has a positive effect on the heat resistance of the alloy.

Acknowledgments

This research is supported by the Key Research and Development Project of Ningbo (2020Z060, 2022Z046, and 2023Z094).

References

[1] J. Wang and X. Xiao. Thermodynamic, kinetic, and strength calculation of high zinc containing Al-Zn-Mg-Cu alloys, Crystals 12 (2022): 181-192.
[2] S. S. Liang, J. R. Ma, and K. H. Guo. Effect of Er and Si co-microalloying on mechanical properties and microstructures of AlCuMg alloys, J. Mater. Res. Technol. 24 (2023:) 430-439.
[3] S. Bai, X. Yi, and Z. Liu. On the role of Ag additions on the initial solute hardening and competitive precipitation of Al-Cu-Mg alloys, J. Alloys Compd. 945 (2023): 169-179.
[4] S. L. Yang, N. Wilson, and J. F. Nie. Revisit of the structure of Ω precipitate in Al-Cu-Mg-Ag alloys, Scripta Mater. 205 (2021), 116-127.
[5] M. M. Haque A. Sharif Study on wear properties of aluminium-silicon piston alloy. J. Mater. Process. Technol. 118 (2001): 69-73.
[6] N. A. Belov, D. G. Eskin, and N. N. Avxentieva. Constituent phase diagrams of the A1-Cu-Fe-Mg-Ni-Si system and their application to the analysis of aluminum piston alloys. Acta Mater. 53 (2005): 4, 709-4, 722.
[7] S. Jain, R. Jain, M. R. Rahul, and S. Samal. Phase equilibria and mechanical properties in multicomponent Al-Ni-X (X = Fe, Cr) alloys, Trans. Indian Inst. Met. 71 (2018): 2, 819-2, 825.

[8] P. Pandey, S. Kashyap, and C. S. Tiwary. Development of high-strength high-temperature cast Al-Ni-Cr alloys through the evolution of a novel composite eutectic structure, Metall. Mater. Trans. A Phys. Metall. Mater. Sci. 48 (2017): 5, 940-5, 950.
[9] J. M. Park, N. Mattern, U. Kühn, and J. Eckert. High-strength bulk Al-based bimodal ultrafine eutectic composite with enhanced plasticity, J. Mater. Res. 24 (2009): 2, 605-2, 609.
[10] C. Mondal, A. K. Mukhopadhyay, and T. Raghu. Tensile properties of peak aged 7055 aluminum alloy extrusions, Mater. Sci. Eng. 454 (2007): 673-678.

Detection and characterization of hole processing defects in carbon fiber-reinforced polymer

Shanling Han[1], Xiaodong Li[1], Zhen Cui[1] and Yong Li[1,*]

[1]College of Mechanical and Electronic Engineering, Shandong University of Science and Technology, Qingdao 266590, Shandong, China

*Corresponding author's e-mail: hexidian@sdust.edu.cn

Abstract: The defects of carbon fiber-reinforced polymer (CFRP) during the drilling process will affect the mechanical properties and service life of the components. We investigate the detection and characterization of delamination and burr defects after drilling CFRP. The drilling test was conducted on CFRP specimens, followed by image processing software to detect and extract the resulting defects. Subsequently, a more accurate two-dimensional quantitative characterization of the extracted delamination and burr defects was performed. Finally, by comprehensively considering the quality indexes of delamination and burr defects, a comprehensive evaluation factor that can fully reflect the hole-making defects is established to provide a theoretical basis and technical support for enhancing CFRP's hole-making quality.

1. Introduction

CFRP are widely used in the field of aerospace due to their high specific strength and specific stiffness. The drilling quality of CFRP seriously affects the mechanical properties and reliability of connectors [1-2]. For carbon fiber composite materials, due to its special material properties, the evaluation criteria need to be more detailed. In addition to dimensional accuracy and surface roughness, it is also necessary to pay attention to specific defects such as burrs, tears, and delamination [3-4]. These defects have a significant impact on the mechanical properties and service life of the material.

At present, there is no unified standard for the detection and characterization of hole defects. Many scholars have proposed evaluation criteria for the processing quality of CFRP. Tsco and Hocheng [5] proposed to use the ratio of the maximum damage diameter of the delamination defect to the theoretical diameter of the hole as a delamination factor. Sugita et al. [6] conducted a hole-making test on CFRP and utilized a coordinate machine to measure the height of the burr, which served as an evaluation criterion for defects. Kim et al. [7] proposed a new process for deburring drilled holes in CFRP composites based on large pulsed electron beam (LPEB) irradiation, and the length of the burr is used as the evaluation index. These evaluation methods mainly focus on one-dimensional and unilateral defects [8], and these evaluation methods cannot fully reflect the overall quality of the material. These traditional methods usually evaluate the quality of hole processing by analyzing a single factor and mainly focus on one-dimensional evaluation criteria, and the evaluation results are often not accurate enough. Due to the multi-layer structure and heterogeneity of CFRP materials, multiple defects often occur at the same time during hole processing, which makes it difficult for traditional evaluation methods to fully reflect the processing quality. To solve this problem, this study uses a two-dimensional evaluation factor to characterize the defects and proposes an innovative comprehensive evaluation factor. By comprehensively considering the quality indicators of delamination and burr defects, a more comprehensive and accurate evaluation method is provided, which plays a positive role in promoting the further application and development of CFRP.

2. Defect image extraction and processing

The carbon fiber workpiece material used in this drilling test is T300 carbon fiber reinforced epoxy resin composite laminate. The layering method is [0°/45°/90°/-45°], and the thickness of the laminate is 2 mm. The tool used in the test is a cemented carbide twist drill with a diameter of 6 mm. The defect morphology of CFRP after drilling was observed and measured by electron microscope. The shape of the hole defect after drilling is shown in Figure 1.

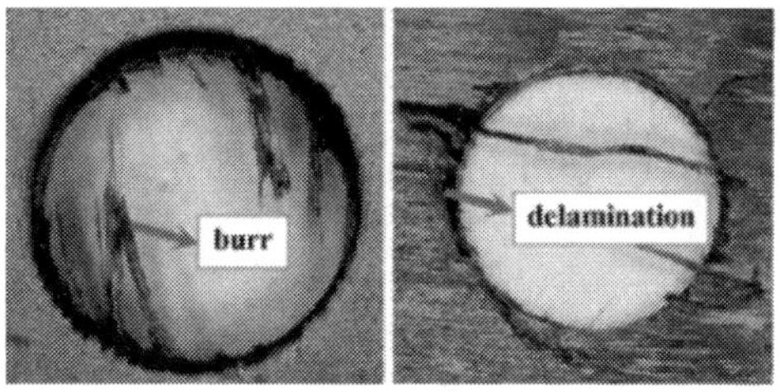

Figure 1. CFRP hole defect imaging diagram.

Digimizer and ImageJ image processing software were used to measure the morphology of drilling holes. Both software are powerful image measurement and analysis software [9]. By utilizing Digitizer software, both the diameter of the drilling hole and the maximum layered area can be accurately measured, along with determining the actual area of layering. The software offers two modes for measurement: automatic and manual. To enhance measurement precision in this study, manual measurement mode was chosen, and the red part shown in Figure 2 (a) is the recognition area of the delamination defect. Subsequent characterization of burrs defects requires representation through area ratio; hence ImageJ image processing software's area pixel ratio function can be employed. The acquired defect image is imported into ImageJ software, where it undergoes grayscale conversion followed by threshold binarization to depict defect distribution. In this process, a threshold is set by the Set Measurements function of ImageJ, and each pixel in the defect image is converted to black or white according to its gray value (brightness). The color image of the defect is converted into a black-and-white binary image, that is, only two colors of black and white, as shown in Figure 2 (b). The burr distribution in the hole is black. Dividing measured area value by total image area (total number of pixels) enables direct calculation of the defect's area ratio. Utilizing this two-image processing software for characterizing defects reveals that results obtained from automatic identification measurement, manual identification measurement, and pixel proportion calculation are nearly identical. Therefore, we should ensure the authenticity of output measurements while reducing errors significantly and guaranteeing more accurate data.

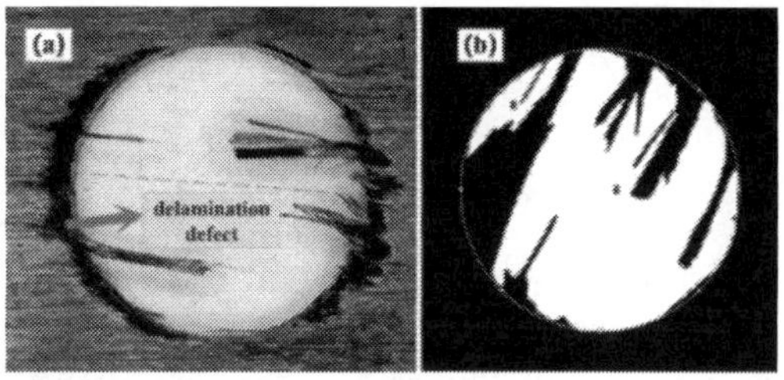

Figure 2. Delamination and burr detection image.

3. Characterization of delamination and burr defects

Many scholars have proposed the evaluation criteria for the processing quality of carbon fiber composite materials. However, due to the limitations of early measurement equipment and the convenience of low cost, these evaluation methods mainly focus on one-dimensional and unilateral defects, but this evaluation method cannot fully reflect the overall quality of the material. The two-dimensional evaluation criterion optimizes the one-dimensional evaluation criterion and solves the shortcoming that the one-dimensional evaluation criterion is not accurate and comprehensive. Through the two-dimensional evaluation criterion, the shape and distribution of defects can be described more accurately, to improve the accuracy and reliability of the evaluation, which can better reflect the actual processing quality of materials. Therefore, this study proposes a more comprehensive evaluation standard for pore quality defects.

3.1. Quantitative characterization of delamination defects

For delamination defects, the one-dimensional delamination factor is the most used metric. This factor is defined as the ratio of the maximum diameter of the delamination area to the nominal diameter of the hole, serving as a criterion for evaluating delamination defects. The formula for this factor is as follows:

$$F_d = \frac{D_d}{D_0} \tag{0}$$

where D_d denotes the maximum diameter of the stratified region; D_0 represents the nominal diameter of the hole. The diameter stratification factor F_d is a one-dimensional simple evaluation index. However, in certain specific cases of delamination, such as instances where only a small portion of the fibers undergo extensive peeling while other regions surrounding the pore size do not exhibit significant delamination, localized delamination occurs exclusively and thus renders the application of F_d invalid.

As shown in Figure 3, using the one-dimensional stratification criterion to evaluate delamination defects, the results of the two cases are the same, but the actual delamination defects in the two graphs are very different. In this case, it is unreasonable to use F_d to evaluate the degree of stratification [10]. In contrast, the area stratification factor F_a considers the actual area and inhomogeneity of the stratification, and it evaluates the degree of stratification by the ratio of the actual area of the stratification to the nominal area of the hole. Therefore, the area stratification factor F_a is more accurate than the diameter stratification factor F_d to evaluate the actual situation of delamination defects, especially in the case of uneven stratification. The expression of the area stratification factor F_a is as follows:

$$F_a = \frac{A_d}{A_0} \tag{1}$$

The area stratification factor is recorded as φ_1, where A_d represents the actual stratification area and A_0 represents the nominal area of the hole.

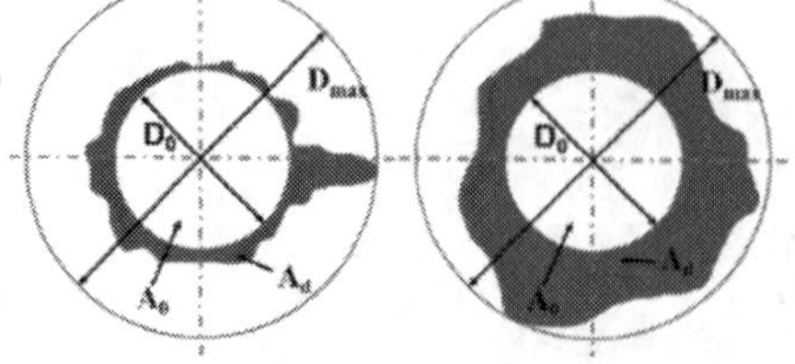

Figure 3. Diameter stratification factor and area stratification factor.

As shown in Figure 4, the delamination defect map of the CFRP drilling outlet under the processing parameters of spindle speed of 4, 500 r/ min and feed rate of 0.04 mm/ r is shown. Using Digimizer's automatic identification tool to identify the area of the export delamination defect, as shown in Figure 4 (a), the maximum recognition area is 3.51 mm^2, the minimum recognition area is 0.24 mm^2, the standard deviation is 1.489, the average number is 1.3393, and the recognition number is 5. Therefore, the total area of the delamination defect is $6.70\,mm^2$. The nominal area of the hole is $28.27\,mm^2$, and the stratification factor obtained by using two-dimensional stratification is 0.237. The stratification factor obtained by using one-dimensional stratification in Figure 4 (b) is 1.43, and the results obtained by the two are very different. The results obtained by the diameter stratification factor do not take into account the inhomogeneity of the stratification and cannot accurately reflect the real defect condition.

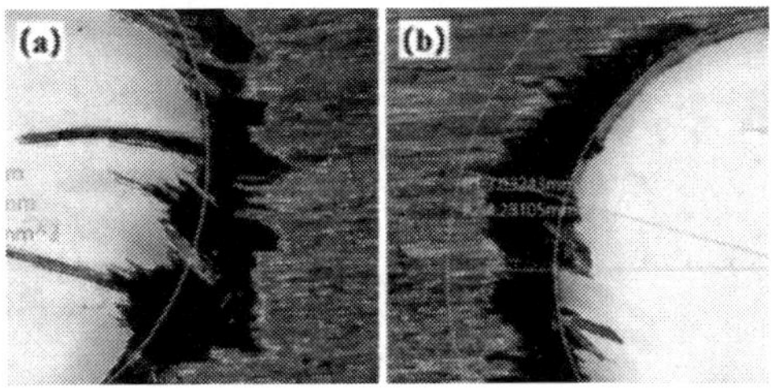

Figure 4. Delamination characterization image.

3.2. Quantitative characterization of burr defects

The burrs usually appear on the inner side of the hole exit circle, showing irregular long strips, and sometimes accompanied by small burrs produced by tearing. In the past, the criteria for evaluating burrs were mostly based on their width [11]. Specifically, burrs with a width greater than 0.1 mm were calculated according to the area, while burrs with a width less than 0.1 mm were calculated according to the length, or only based on the ratio of the total length of the burr to the aperture or the ratio of the burr area to the nominal area of the hole. Although the small burrs associated with tearing are usually ignored, these small burrs also affect the roundness of the workpiece, so they cannot be ignored. In this paper, the burr factor B_u is proposed, and all burrs are included in the evaluation range by area evaluation method, to evaluate burr defects more comprehensively. The evaluation factor of burr is recorded as φ_2, and the formula is as follows:

$$\varphi_2 = B_u = \frac{A_{nor} - A_1}{A_{nor}} = \frac{A_{bi}}{A_{nor}} \tag{2}$$

where A_1 represents the actual measured area in the hole circle; A_{bi} denotes the area of the burr; A_{nor} denotes the theoretical area of the hole. Considering the needs of actual production, convenient observation and measurement is an important factor. Therefore, this study only extracts and evaluates the stratification at the exit when conducting stratification analysis.

4. The establishment of a comprehensive defect evaluation factor

There are some contingencies and irregularities in the damage evaluation of orifice defects. To evaluate the hole defect damage more accurately, it is not only necessary to rely on the evaluation method of a single type of defect but also to consider a variety of hole defects, such as delamination and burr. By summing up the unilateral evaluation coefficients of these defects, a comprehensive defect evaluation damage coefficient φ is formed. This method can more fully reflect the quality of the composite material during the drilling process, make the evaluation of the hole defect damage more systematic and comprehensive, and significantly improve the accuracy and reliability of the evaluation.

$$\varphi = a\varphi_1 + b\varphi_2 \tag{3}$$

where φ represents the comprehensive damage coefficient; a and b represent the weighting factors of the damage coefficients of delamination and burr, respectively. Since the two types of defects of delamination and burr are considered, $a + b = 1$. Because the influence of the two defects on the quality of the hole is not the same, it is necessary to determine the value of the weighting factor of the damage coefficient of the delamination and burr according to the specific actual situation. Delamination defects and burr defects have different effects on the strength and assembly accuracy of the hole. The effects of the two defects on the machining quality of the hole are shown in Table 1. Delamination defects can affect the strength and assembly accuracy of the hole at the same time. This is because delamination will cause the internal structure of the material to become unstable, thus weakening the overall bearing capacity of the hole, and may cause deviation in the assembly process, which is an important reason for the scrapping of CFRP-processed parts. The burr defect mainly affects the assembly accuracy. Because the burr forms irregular protrusions at the edge of the hole, these protrusions will interfere with the precise assembly process, resulting in assembly position offset or poor engagement. However, the burr has little effect on the strength of the hole, because it does not significantly change the structural integrity of the hole. In summary, the delamination defect not only affects the strength of the hole but also affects the accuracy, while the burr defect only affects the accuracy of the hole, so the number of factors affecting the quality of the hole is twice that of the burr defect. Because the two defects interact with each other and affect the processing quality together, the sum of the two weights is equal to 1. Based on the above principles, the weighting factor $a = 2/3$ for the delamination defect and $b = 1/3$ for the burr defect is taken. Therefore, it is concluded that the comprehensive defect evaluation coefficient of hole processing is as follows:

$$\varphi = \frac{2}{3}\varphi_1 + \frac{1}{3}\varphi_2 \tag{4}$$

Table 1. Effect of delamination and burr defects on hole machining quality.

Defect type	Whether it affects the strength of the hole	Whether it affects the accuracy of the hole
Delamination	Yes	Yes
Burr	No	Yes

The drilling test of CFRP was carried out under the machining parameters of spindle speed of 3, 000 r/ min, 3, 750 r/ min, 4, 500 r/ min, 5, 250 r/ min, and 6, 000 r/ min and feed rate of 0.04 mm/ r. The curve of the calculated comprehensive defect factor with the spindle speed is shown in Figure 5. The change in spindle speed influences the hole defects, and the comprehensive defect factor decreases with the increase in spindle speed. The reason is that with the increase of spindle speed, the drilling axial force becomes smaller, and the delamination and burr defects show a downward trend, so the comprehensive defect factor also decreases. This phenomenon is in line with the actual processing situation, which verifies the universality of the comprehensive defect factor proposed above.

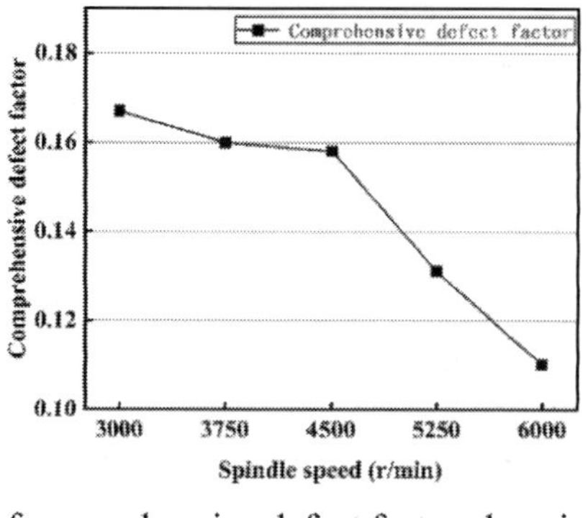

Figure 5. The curve of comprehensive defect factor changing with spindle speed.

5. Conclusion

This study investigates the detection and characterization of delamination and burr defects in CFRP following drilling. Drilling tests were conducted on these composites, and image-processing techniques were employed to identify and highlight the defects. Three different methods were used to calculate the delamination and burr pixels after threshold binarization. The results indicate that the defect measurements obtained through these methods are nearly identical, confirming the reliability of the measurement outputs. A two-dimensional evaluation criterion was established to quantitatively characterize the extracted delamination and burr defects. Finally, considering the impact of drilling defects on the strength and precision of machined parts, a comprehensive evaluation factor was developed, which more accurately reflects the severity of these defects. This provides a systematic theoretical foundation for enhancing the quality of hole-making in CFRP. In the future, this method can be combined with the online recognition system to identify and quantify the specific impact of different defects on the hole quality and help to monitor and adjust the process parameters in real time during the CFRP processing process, to more effectively control the processing quality.

Acknowledgments
This work was supported by the Qingdao Natural Science Foundation (23-2-1-218-zyyd-jch).

References

[1] Mishra R., Malik J., Singh I., and Davim J. P. (2010). Neural network approach for estimating the residual tensile strength after drilling in uni-directional glass fiber reinforced plastic laminates. J. Materials Design 31, 2, 790-2, 795.
[2] Hongguang L., Mehdi C., Madalina C., Hélène B., Frédéric R., Gérard P., and Yessine A. (2023). Progressive damage-induced degradation of mechanical properties in the hole surfaces during drilling processes of CFRP. J. CIRP Annals - Manufacturing Technology 72, 65-68.
[3] Mohammadi R., Assaad M., Imran A., and Fotouhi M. (2024). Fractographic analysis of damage mechanisms dominated by delamination in composite laminates: A comprehensive review. J. Polymer Testing 134, 108, 441.
[4] Ze G. K., Pramanik A., Basak A. K., Prakash C., Shankar S., and Radhika N. (2023). Challenges associated with the drilling of carbon fiber reinforced polymer (CFRP) composites-A review. J. Composites Part C: Open Access 11, 100, 356.
[5] Tsao C. C. and Hocheng H. (2004). Taguchi analysis of delamination associated with various drill bits in the drilling of composite material. J. International Journal of Machine Tools Manufacture 44, 1, 085-1, 090.
[6] Sugita N., Shu L., Kimura K., Arai G., and Arai K. (2019). Dedicated drill design for a reduction in burr and delamination during the drilling of composite materials. J. CIRP Annals 68, 89-92.
[7] Kim J., Seo J. W., and Park H. W. (2019). Deburring drilled holes in CFRP composites with large pulsed electron beam (LPEB) irradiation. J. Journal of Manufacturing Processes 40, 68-75.
[8] Karnik S. R., Gaitonde V. N., Rubio J. C., Correia A. E., Abrão A. M., and Davim J. P. (2008). Delamination analysis in high-speed drilling of carbon fiber reinforced plastics (CFRP) using artificial neural network model. J. Materials & Design 29, 1, 768-1, 776.
[9] Salvarzi E., Choobineh A., Jahangiri M., and Keshavarzi S. (2020). Application of Digimizer Image Analysis Software In Facial Anthropometry. J. Iranian Journal of Ergonomics, 8 (2), 61-67.
[10] Neslihan O., Secil E., and Murat O. (2023). Cutting parameters optimization of hybrid fiber composite during. J. Materials Testing 65, 291-302.
[11] Shahabaz S. M., Shetty N., Sharma S., Shetty S. D., and Naik N. (2024). Optimization of drilling parameters on delamination and burr formation in drilling of neat CFRP and hybrid CFRP nano-composites. J. Materials Research Express 11 (3), 035, 006.

Design and aerodynamic performance study of subsonic Y-shaped inlet

Ziguang Yan*, Zhifang Zhao, Linjie Li and Mengwei Wu

Aviation Technical Service Professional Group, Beijing Polytechnic, Beijing, 100176, China

*Corresponding author's e-mail: yzg051565@163.com

Abstract. A Y-shaped inlet with a quadrate import was designed by the super-ellipse method in this paper, and according to the variation pattern of the cross-section of the inlet, the variation pattern of the short and long axis of the super-ellipse was deduced. For many different flight situations, the numerical simulations of the Y-shaped inlet were made. The result shows (1) If $0.2 \leq Ma \leq 0.8$, the total pressure recovery coefficient exceeds 0.96, and the total pressure distortion coefficient is close to 0.1, meeting the criterion of the design; (2) When Ma=0.8, for many different attack angle α=-4°~22°, the total pressure recovery coefficient is close to 0.97, reaching the level of practical application, and it has little impact on total pressure recovery coefficient and the distortion coefficient. (3) These results demonstrate high total pressure recovery and low distortion coefficients across various flight conditions, indicating its potential for practical applications and future inlet designs.

1. Introduction

The air intake is an important part of the aircraft propulsion system and also one of the three major radar wave-strong scattering sources of the aircraft. For aircraft with stealth requirements, to reduce the radar scattering area of the air intake, the S-shaped air intake design is usually adopted to effectively shield the inlet end face of the engine with its curved structure. In 2011, to improve the efficiency of S-shaped inlets, Li et al. [1] combined the area distribution ratio and shape index of S-shaped inlets and proposed a more perfect design method for super-elliptic S-shaped inlets. Through calculation and analysis, it is concluded that the S-shaped inlet designed by this method has good aerodynamic performance. In 2020, Wang et al. [2] analyzed the internal flow field structure of an S-shaped inlet of the unmanned aerial vehicle (UAV) and studied the flow field characteristics and outlet parameters of the outlet section. The influence of the offset distance of the S-bend and the length of the inlet on the aerodynamic characteristics was calculated. The results show that increasing the length of the S-bend can significantly reduce the vortex structure area and separation area, improve the flow field structure, and simultaneously increase the total pressure recovery coefficient at the outlet and control the outlet distortion. In 2022, taking the UAV with an S-shaped inlet as the research object, Yao et al. [3] analyzed the aerodynamic drag on the whole aircraft. A certain flight with in-air shutdown status of the UAV was selected as the research object, and two main factors, flow coefficient and Mach number, were identified as having a significant impact on the spillage drag. When the flow coefficient decreases, both the additional drag and the spillage drag increase; when the flight Mach number increases, the spillage drag decreases.

The engine inlet mainly includes four types: annular inlet, single scoop inlet (or jaw inlet), double scoop inlet (or bifurcated inlet, kidney-shaped inlet), and triple scoop inlet (triple-fork inlet). All the above research papers focus on the numerical simulation of the single scoop inlet, while the offset distance of the S-bend is limited to the XY plane, and the length of the latter half of the inlet is relatively short, which is not conducive to the total pressure recovery. Based on this foundation, this paper designs a Y-shaped inlet (or bifurcated inlet) by introducing an additional eccentricity in the S-bend along the YZ direction and modifying the section preceding the inlet. This design prevents the compressor from being directly exposed and enhances stealth capabilities. The superellipse method is employed to simplify the design process and ensure the smoothness of the internal flow path of the inlet, meeting the requirements of a high total pressure recovery coefficient and a low distortion index. VC language programming is used to shape the middle section of the Y-shaped inlet, and a geometric model of the inlet is established by using CAD software. Subsequently, numerical simulations are conducted by using Fluent to analyze the flow field and aerodynamic characteristics of the inlet.

2. Design methodology

2.1. Calculation of the inlet area

The formula for calculating the throat area is defined as

$$A_{th} = \frac{Kq_{mc}\sigma}{242q(\lambda_{th})} \tag{1}$$

where $q(\lambda_{th})$ is the flow function of the throat, λ_{th} is the speed coefficient of the throat, σ is the total pressure recovery coefficient of the inlet, K is the correction coefficient, and q_{mc} is the converted flow rate of the engine.

2.2. Design of the inlet lip

A well-designed lip profile can significantly enhance the aircraft's maneuverability. The flow separation and secondary flow in the S-shaped inlet cause serious pressure loss and flow distortion at the inlet-outlet, which seriously affects the working performance of the engine [4]. It is necessary to meet the requirements of low drag and minimal lip losses. Typically, in subsonic inlets, to shorten the inlet length, the lip also serves the function of diffusing, resulting in a lip profile that first contracts and then expands. During the aircraft's maneuvering flight, the flow turning angle along the inner wall of the outlet is relatively large, which can easily lead to flow separation [5]. Therefore, the lip design adopted in this paper is a purely contracting profile, with the lip shape being the standard airfoil NACA-1 [6], as shown in Figure 1.

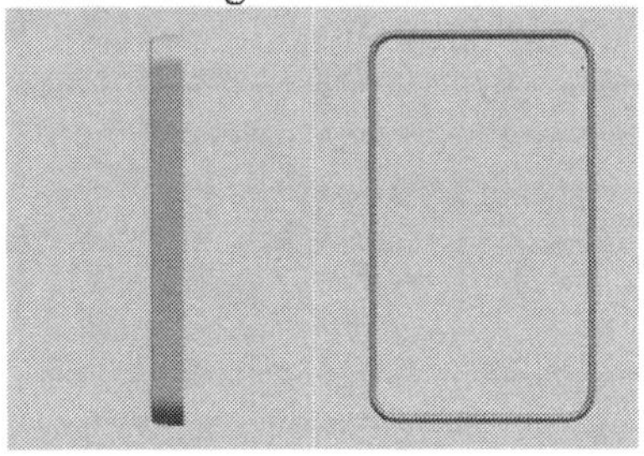

Figure 1. Inlet lip.

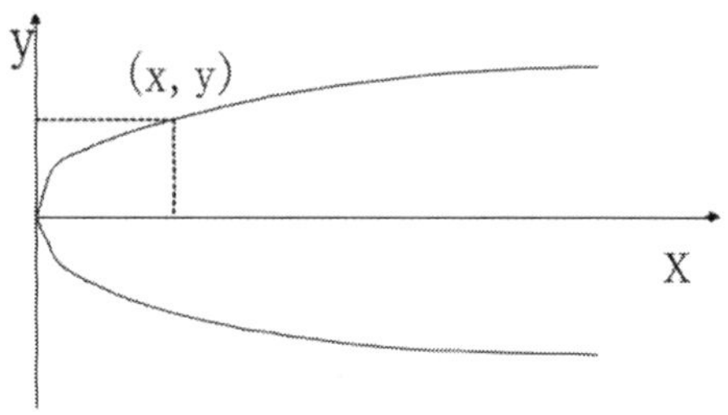

Figure 2. Inlet lip profile.

The inlet profile is rectangular, with the opposite sides being rounded to avoid sharp corners, which can cause discontinuities in the first derivative and singularities in the flow. Figure 2 shows the coordinate diagram of the inlet lip profile.

2.3. Design of the diffuser section

The design of the diffuser section includes the design of the centerline variation and the area variation. Among them, the centerline variation determines the lateral pressure gradient of the airflow within the S-shaped duct, which is characteristic of the secondary flow; the area variation determines the change in the flow direction expansion ratio, and the change in the expansion ratio should not be too large, as it can easily lead to airflow separation. These two parameters need to be determined first when designing an S-shaped inlet. There are three types of centerline variation and area variation curves: fast-to-slow, slow-to-fast, and comparable rates of change [7]. In this paper, a combination of a fast-to-slow centerline variation and a comparable rate of area variation is selected, as shown in Figure 3 and Figure 4.

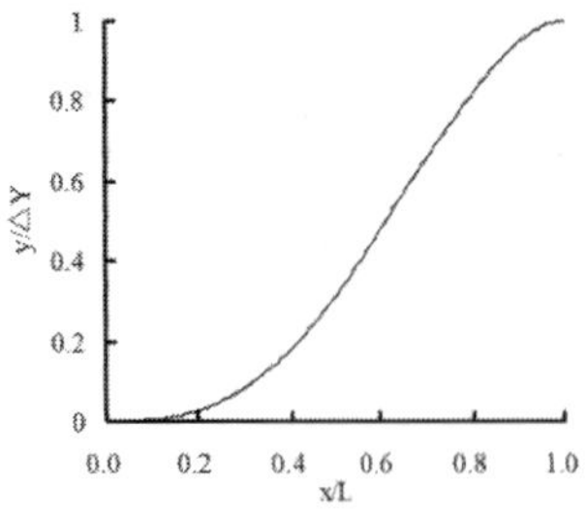

Figure 3. The variation of the centerline.

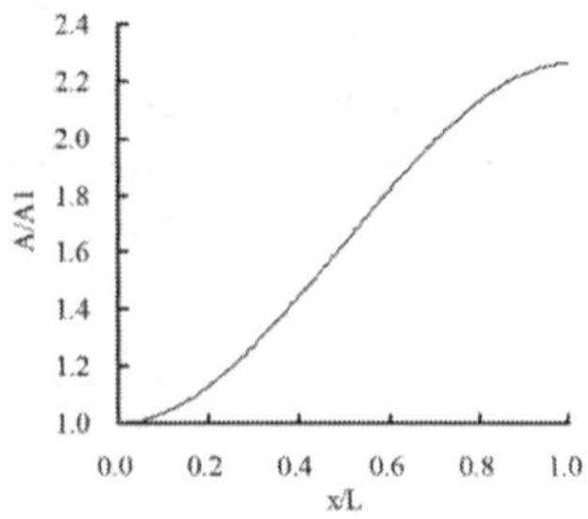

Figure 4. The variation of the area.

In the design of the inlet's inner duct, the superellipse design method is applied. The superellipse formula boasts advantages such as simplicity in expression, strong symmetry of the curve, and continuous curvature changes between curves without abrupt changes. The formula of a standard superellipse [8] is defined as

$$(\frac{x^2}{a^2})^s + (\frac{y^2}{b^2})^s = 1 \tag{2}$$

where a, b, and s ($s = 0.5n$) are real numbers greater than zero. The formula for the variation rule of the major and minor axes is defined as

$$\begin{bmatrix} a^2(x) \\ b^2(x) \end{bmatrix} = B \bullet \begin{bmatrix} r_2^2 - a_1^2 \\ r_2^2 - b_1^2 \end{bmatrix} + \begin{bmatrix} a_1^2 \\ b_1^2 \end{bmatrix} \tag{3}$$

where $B = 3(\frac{x}{L})^4 - 2(\frac{x}{L})^3$, a_1 and b_1 are the length and width of the rectangular inlet, respectively.

The inlet designed in this paper has a rectangular inlet and a semicircular outlet, with parameters implemented through computer programming. The expansion section of the inlet can simultaneously have eccentricities in both the XY and YZ planes, reducing the scattering area of the engine inlet section on external radars and enhancing stealth capabilities.

3. Model and mesh

3.1. Model

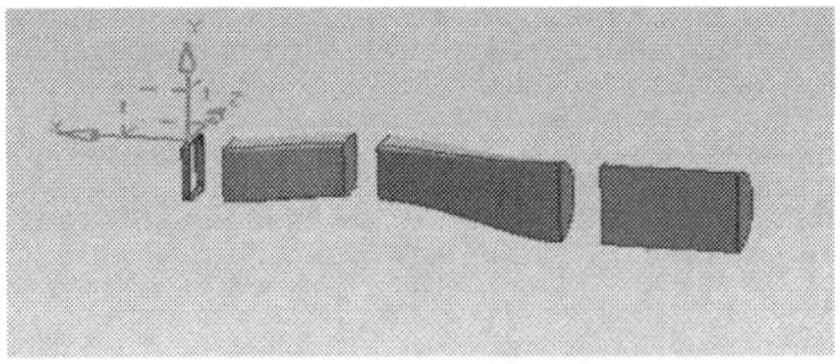

Figure 5. Inlet model.

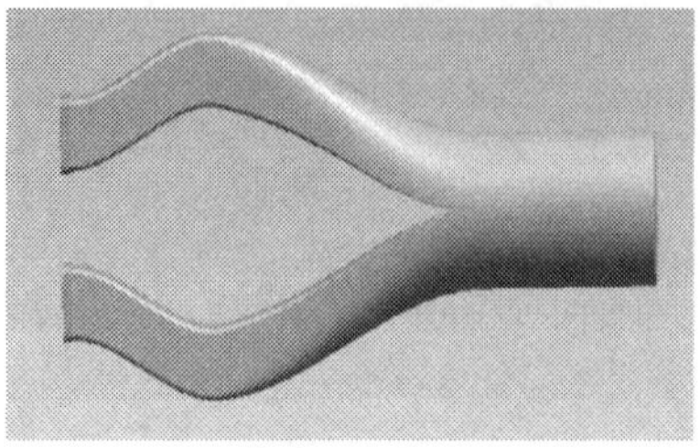

Figure 6. The whole model.

As shown in Figure 5, the geometric model consists of four parts, including the lip, inlet section, diffuser section, and extension section. The lip is described previously; the inlet section is rectangular with a constant geometric area and a length of 800 mm. The model generates an eccentricity in the z-direction with a distance of 300 mm; the diffuser section transitions from a rectangular shape to a semicircular shape with a length of 1500 mm and an eccentricity of 700 mm. The extension section is directly extended in the -x direction with a length of 1000 mm. The final model is shown in Figure 6.

3.2. Mesh

As shown in Figure 7, the numerical calculation adopts a partitioned hexahedral structured grid. The inlet is divided into two regions. The front section of the inlet is divided into 10 regions. The outer flow field is the entire cuboid, and each part is further divided into several sub-regions. The outer flow field varies from dense to sparse from the inside to the outside, with denser grids closer to the inlet lip. The total number of grids is approximately 530, 000 (with 1.06 million for the whole model).

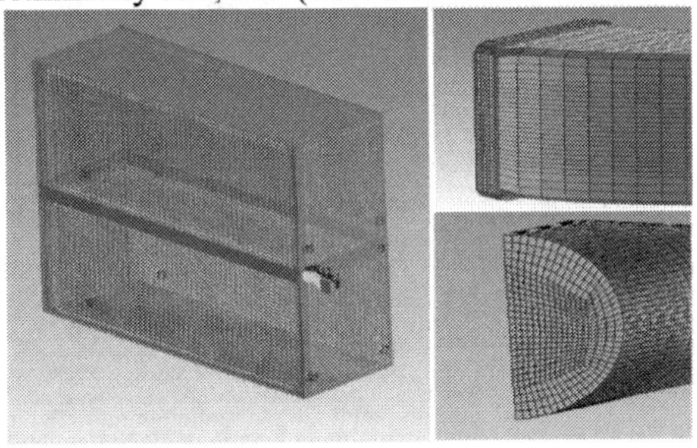

Figure 7. Mesh of the model.

3.3. Computational domain and boundary conditions

For the numerical simulation of subsonic aircraft, pressure far-field boundary conditions are typically used. To satisfy these conditions, the computational domain in this paper is a cuboid with a length of 10 L, a height of 7 L, and a width of 5 L (where L is the length of the inlet). The outlet of the external fluid domain is set as a pressure far-field, the cross-section of the extension segment and the rectangular side are set as symmetry boundaries, the outlet cross-section is the semicircular outlet of the inlet, and the remaining four faces of the cuboid are set as pressure far-field boundaries.

The Navier-Stokes relations transport formulas are used to simulate flow on a multiblock structural grid. The flow field calculations are performed in commercial computational fluid dynamics (CFD) software, utilizing the second-order upwind scheme for discretizing the temporal and spatial terms. The realizable k-ε turbulence model is selected for the flow computations, which relates the turbulent viscosity to the strain rate, making the flow more physically realistic and suitable for simulating jet flows, boundary layer flows, flows with separation, etc. The primary computational boundary conditions include far-field pressure conditions, pressure outlet conditions, symmetric boundaries, and no-slip adiabatic solid wall conditions [9].

3.4. Operating conditions for calculation

Table 1. Different conditions.

Mach number	Altitude (km)	AOA (°)	Yaw angle (°)
0.2	0	0	0
0.6	11	0	0
0.8	11	0	0
0.8	11	-4°~22°	-10°~10°

Different conditions are given in Table 1.

4. Calculation results and analysis

4.1. Speed characteristics

Figure 8 shows the variation of the total pressure recovery coefficient σ and distortion index D of the inlet with the incoming Mach number ($D = (P_{P\max} - P_{P\min}) / P_{FAV}$).

As can be seen from Figure 8, the total pressure recovery coefficient σ decreases with the increase of the inlet Mach number. When Ma ≤ 0.6, the change is small, and the value is always greater than 0.98. The total pressure recovery coefficient σ changes slightly more between Ma = 0.6 and 0.8. At Ma = 0. 8, its value is the smallest, about 0.965. The distortion index D gradually increases with the increase of the inlet Mach number, and the change is relatively rapid. When Ma = 0.2, the value of the distortion index D is the smallest in the calculation domain (D = 0.01); when Ma = 0.8, the distortion index D reaches the maximum value in the calculation domain (D = 0.099). The increase of Ma has a great influence on the inlet-outlet pressure, especially when the inlet flow velocity is close to the speed of sound.

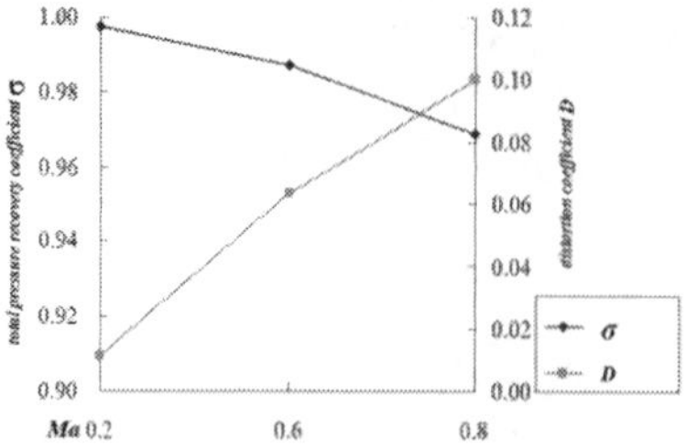

Figure 8. Variation of σ and D with Ma.

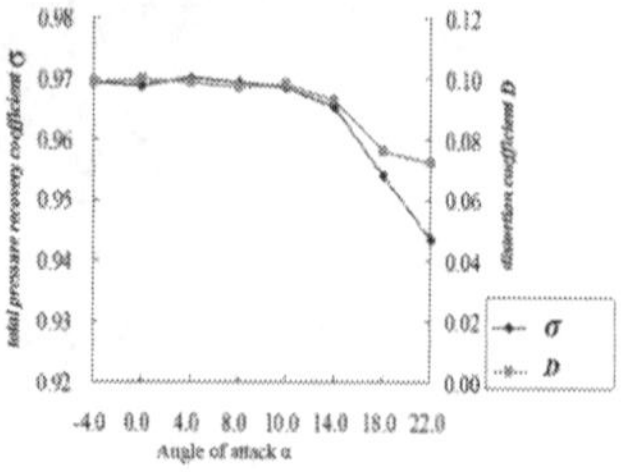

Figure 9. Ma=0.8, β=0°.

4.2. Angle of attack (AOA) characteristics

As can be seen from Figure 9, when Ma=0.8 and β=0°, the variation patterns of the total pressure recovery coefficient σ and distortion index D with the angle of attack α are presented. Between α=-4° and 0°, σ follows a slightly descending straight line as α increases. Between α=0° and 4°, the total pressure recovery coefficient σ slightly increases with the increase of α and reaches its maximum value (σ =0.9701) within the computational domain at α=4°. Between α=4° and 10°, σ slightly decreases with the increase of α, and between α=10° and 14°, σ decreases slowly with the increase of α. Between α=14° and 22°, σ decreases sharply with the increase of α and reaches its minimum value (σ =0.9434) within the computational domain at α=22°. Therefore, flying with a small positive angle of attack is beneficial to the pressure at the inlet-outlet. When α=14°, the airflow inside the inlet duct separates near the throat, and the separation reaches its maximum when α=22°, which explains the sharp decrease in the total pressure recovery coefficient when α≥14°. Figure 10 shows the internal streamline diagram of the inlet at α=22°.

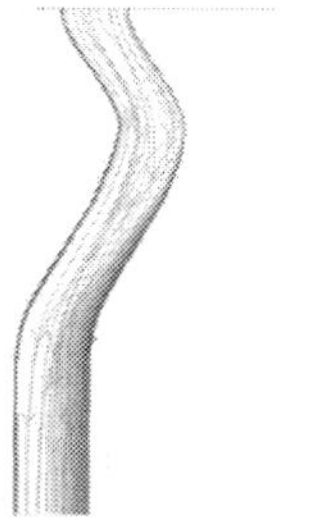

Figure 10. Internal flow line.

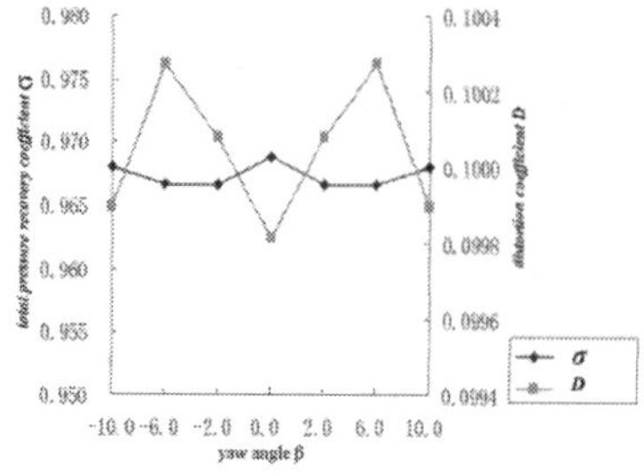

Figure 11. Ma=0.8 and α=0°.

The distortion index D changes relatively smoothly with the increase of α. Between α=-4° and 10°, the variation curve is almost a straight line. Between α=10° and 14°, the distortion index D begins to decrease slowly. Between α=14° and 18°, the distortion index D starts to decrease sharply. Between

α=18° and 22°, the distortion index D begins to decrease slowly again with the increase of α. At α=0°, its value is the maximum within the computational domain; at α=22°, its value is the minimum. From the above analysis, it can be seen that the increase in the angle of attack α has a relatively small impact on the distortion index. Furthermore, the analysis indicates that the selected centerline with a steep front and gradual rear section meets the requirements, and the extended section increases the length of the inlet-outlet, enhancing the quality of the airflow entering the engine.

4.3. Yaw angle characteristics

Figure 11 shows the variation curves of the total pressure recovery coefficient σ and the distortion index D with the yaw angle β when Ma=0.8 and α=0°. The total pressure recovery coefficient σ first decreases with the increase of the yaw angle β and then begins to increase at β=-2° with the entire curve being symmetric about β= 0°. Between β= -10° and -6°, the variation curve decreases slowly. Between β= -6° and -2°, the total pressure recovery coefficient σ starts to decrease slightly with the increase of the angle, reaching its minimum value at β= ±2°. Between β= -2° and 0°, the total pressure recovery coefficient σ increases sharply and reaches its maximum value.

The distortion index D first increases with the increase of the yaw angle β, and then begins to decrease at β= -6°, with the entire curve also being symmetric about β=0°. Between β= -10° and -6°, the curve increases relatively sharply. Between β= -6° and -2°, the distortion index D starts to decrease slowly. Between β= -2° and 0°, the distortion index D decreases sharply, reaching its minimum value within the computational domain at β=0°. At β= ±6°, its value is the maximum within the computational domain.

As the yaw angle β increases, the distortion index D changes smoothly, with its value remaining around 0.1. This is primarily due to the second gentle transition section, which enhances the uniformity of the airflow at the inlet-outlet.

5. Conclusions

Through the design and flow field analysis of the Y-shaped inlet, the following conclusions are drawn in this paper:

(1) As the incoming flow velocity increases, the total pressure recovery of the inlet shows an increasing trend, and the increase rate decreases as the Mach number increases; the total pressure distortion shows a decreasing trend, and the increase rate increases as the Mach number increases.

(2) Flying with a small positive angle of attack is beneficial to the outlet pressure of the inlet; when the angle of attack A=14°, the airflow separates, and the total pressure recovery coefficient drops sharply; the increase in the angle of attack has little effect on the total pressure distortion index at the inlet-outlet.

(3) The yaw angle has little effect on the outlet pressure of the inlet. Although the total pressure recovery coefficient shows a downward trend, the change is relatively small, remaining between 0.965 and 0.97; the distortion index is insensitive to the increase in the yaw angle, remaining around 0.1.

(4) The inlet designed with a sharp front and gentle rear centerline can reduce the pressure distortion of the inlet-outlet section.

This paper presents the design and aerodynamic performance analysis of a subsonic Y-shaped inlet by using the super-ellipse method, demonstrating high total pressure recovery and low distortion coefficients across various flight conditions, indicating its potential for practical applications and future inlet designs.

Acknowledgment

This research was financially supported by the Key Scientific Research Projects of Beijing Polytechnic in 2024 (CJGX2024-KY-056).

Journal of Physics: Conference Series **2879** (2024) 012004 doi:10.1088/1742-6596/2879/1/012004

References

[1] Li Yuefeng, Yang Qingzhen, Sun Zhiqiang. Design and Aerodynamic Performance Research of Superellipse S-shaped Inlet [J]. Computer Simulation, 2011, 03.
[2] Wang Yunfei, Shen Tianrong, et al. Research on the Influence of Geometric Parameters on the Aerodynamic Characteristics of S-shaped Inlet [J]. Machinery Manufacturing & Automation, 2020, 49(4): 128-132.
[3] Yao Jieke, Chen Bin, Shang Yinhui, et al. Numerical Simulation of the Whole Aircraft Overflow Drag Based on S-shaped Inlet [J]. Aeroengine, 2022, 48(6): 18-27.
[4] Liu Rubing, ZengJianhong, et al. Experimental Study on Active Control of Flow Separation in S-shaped Inlet Using AC Dielectric Barrier Discharge Plasma [J]. Journal of Propulsion Technology, 2023, 44(10), 38-50.
[5] Peng Wenqiang, Luo Zhenbing, Zhou Xinrun, et al. Analysis of flow field control characteristics of synthetic dual-jet in S-shaped inlet [J]. Journal of National University of Defense Technology, 2023, 45(6): 215-224.
[6] Richard J R. An investigation of several NACA 1-series inlets at Mach number from 0.4 to 1.29 for mass flow ratio near 1.0 [R]. NASA TM-X-3324, 1975.
[7] Lee C C, Boekicker C. Subsonic diffuser design and performance for advanced fighter aircraft [R]. AIAA-85-3073, 1985.
[8] Wang K L, Zhou Z, et al. Propulsive/aerodynamic coupled characteristics of distributed-propulsion-wing during forward flight [J]. Acta Aeronautica et Astronautica Sinica, 2024, 45(2): 128643 (in Chinese).
[9] Xiao Yi, Ma Jingzhong, Shen Liang, et al. Numerical Simulation Study on Aerodynamic Characteristics of Embedded Inlet for Unmanned Aerial Vehicle [J]. Aerodynamic Research and Experiment, 2024, 2(3): 81-86.

Journal of Physics: Conference Series 2879 (2024) 012005 doi:10.1088/1742-6596/2879/1/012005

Design and optimization of the exhaust system for an aviation piston engine

Jinqian Chen[1], Chen Zhang[2,*], Wei Chen[1,3] and Zhecheng Zhang[1]

[1]School of Aeronautics, Chongqing Jiaotong University, Chongqing, 400074, China
[2]Jiangsu Xinyuan Aviation Technology Co., Ltd., Wuxi, Jiangsu, 214187, China
[3]Chongqing Key Laboratory of Green Aviation Energy and Power, Chongqing Jiaotong University, Chongqing, 400074, China

*Corresponding author's e-mail: nhzhangchen@126.com

Abstract. To further enhance the performance of an aviation piston engine, an engine performance simulation analysis model was constructed by using GT-Power software to design improvements to its exhaust system. The impacts of different exhaust system schemes on the exhaust process, the in-cylinder combustion, and the heat release process at a cruising speed of 5500 RPM were analyzed. Based on the results, under cruising speed conditions, a new exhaust system was designed for the engine by referring to the structural form of a resonant exhaust system. The D-optimal Latin hypercube sampling method was used to obtain sample data for the experimental design of the structural parameters of the designed exhaust system. The response surface approximation models characterizing power and fuel consumption rate were established by using the least squares method. The significance of the impact of each structural parameter on engine performance was analyzed through a multifactor analysis of variance. The key structural parameters of the designed exhaust system were then subjected to multi-objective optimization. The results indicate that at 5500 RPM, the designed exhaust system can increase engine power by 2.88%; after optimization, engine power can be further increased by 3.15%, and the fuel consumption rate is reduced by 4.07%.

1. Introduction

Two-stroke piston engines, due to their high power-to-weight ratio, simple structure, low cost, reliable operation, and compact design, hold significant practical application value in the aviation field [1]. The structural parameters of the exhaust system in two-stroke engines critically impact their performance [2, 3, 4, 5]. By reasonably designing the structure and parameters of the exhaust system, it is possible to reduce scavenging resistance, improve scavenging efficiency, and consequently enhance engine performance [6, 7, 8, 9].

Developing engines with superior performance within a short period and at a lower economic cost has become a common goal for researchers and designers both domestically and internationally. Many researchers have utilized software such as GT-Power and AVL Boost to study the structure and parameters of engine exhaust systems to improve engine performance [10, 11, 12]. Additionally, researchers have employed the Design of Experiments (DOE) method to optimize the structural parameters of engine exhaust systems, aiming to achieve exhaust systems with higher power output and lower fuel consumption rates [13, 14, 15, 16, 17].

This paper aims to enhance the power output of a certain aviation piston engine and reduce its fuel consumption rate. To achieve this goal, this paper refers to the basic structure of a resonant exhaust system and combines it with the engine's dimensional characteristics to redesign and optimize the original exhaust system. By using GT-Power software, an engine performance simulation model was built to calculate and analyze the designed exhaust system, revealing its specific performance improvements in enhancing engine performance. Furthermore, the DOE analysis method was employed to evaluate and compare possible structural parameter combinations, identifying the optimal set of structural parameters that achieve dual optimization of power performance and fuel economy.

2. Engine model construction and validation

2.1. Simulation model establishment

This paper investigates a two-stroke twin-cylinder horizontally opposed piston engine. The main parameters of the engine are shown in Table 1.

Table 1. Main parameters of the two-stroke aircraft piston engine.

Parameters	Value
Effective compression ratio	9.6
Crankcase compression Ratio	1.4
Cylinder bore (mm)	66
Stroke (mm)	50
Scavenging method	Loop scavenging
Exhaust port opening crank angle (°CA)	102
Exhaust port closing crank angle (°CA)	258
Scavenging port opening crank angle (°CA)	118
Scavenging port closing crank angle (°CA)	242

The prototype exhaust system equipped on this engine is shown in Figure 1. The upper end of the exhaust system is connected to the exhaust port outlet, while the lower end is connected to the atmosphere. Based on the exhaust system, the engine performance simulation model established through GT-Power is shown in Figure 2. This model mainly consists of modules such as the intake system, scavenging system, crankcase, cylinder, engine body, and exhaust system. When calculating the heat transfer in the crankcase and cylinder, the WoschniGT heat transfer model was chosen; the combustion process is represented by the quasi-dimensional predictive combustion model SITurb.

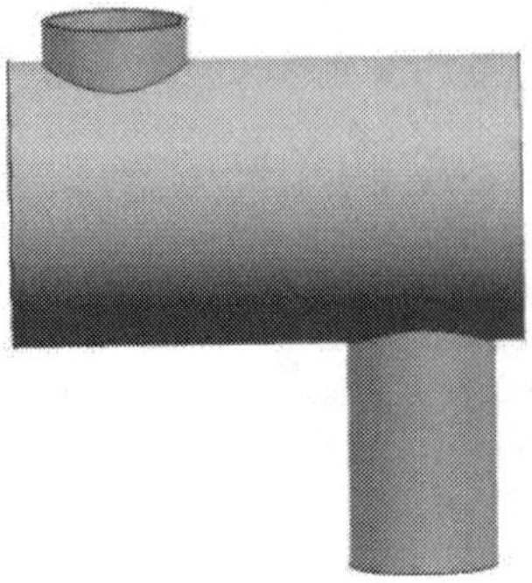

Figure 1. Prototype exhaust system.

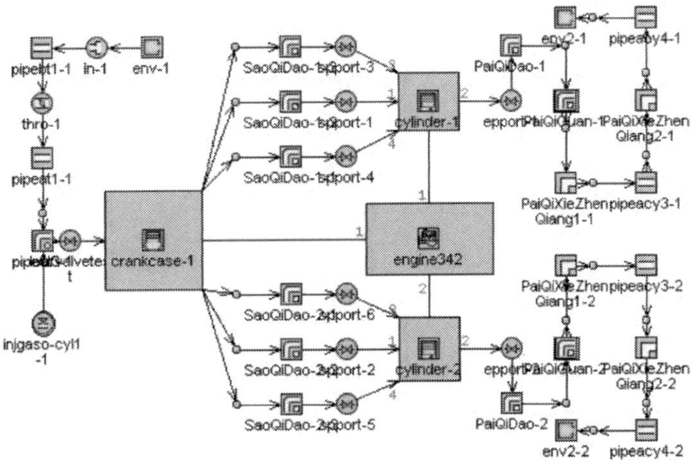

Figure 2. Engine performance simulation model.

2.2. Model validation

The established model is validated by experimentally measuring the power and fuel consumption rate at different speeds under a specific load condition. The external characteristic data of power and fuel consumption rate measured from the engine bench test are compared with the numerical simulation results, as shown in Figures 3 and 4. From the comparison figures, it can be seen that the simulation curves for engine power and fuel consumption rate at different speeds are quite close to the experimental values, with a maximum error of 5%, which is within the normal error range, indicating that the established simulation model meets the consistency requirements of the engine.

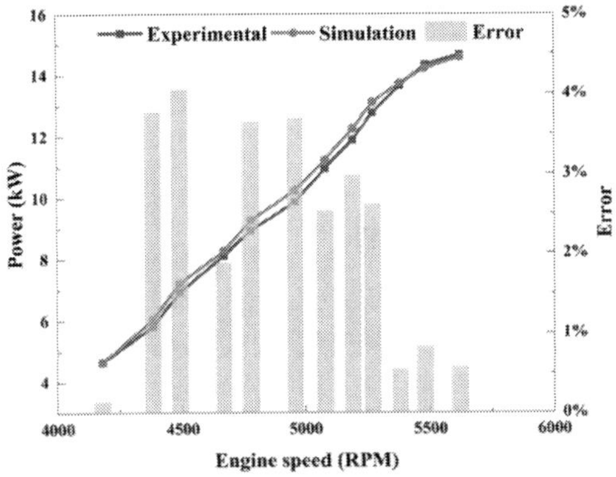

Figure 3. Engine power comparison.

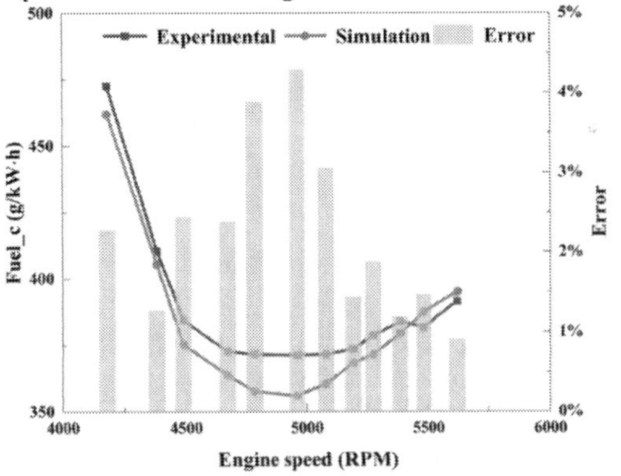

Figure 4. Engine fuel consumption rate comparison.

2.3. Multifactor analysis of variance

When two or more variable factors influence the response, a multifactor analysis of variance (ANOVA) can be used to determine whether multiple variable factors have a significant impact on the response. In multifactor ANOVA, the response is affected by three aspects:

1. The effect of individual variable factors: it refers to the impact of a single variable factor on the response.

2. The effect of interactions between variable factors: it refers to the impact on the response when multiple variable factors interact with each other.

3. The effect of random factors: it refers to the impact of sampling errors.

Therefore, multifactor ANOVA decomposes the total variation in the response into (for example, with two variable factors):

$$SST=SSA+SSB+SSAB+SSE \tag{1}$$

where SST is the total variation of the response; SSA and SSB are the variations caused by the individual effects of variable factors A and B, respectively; SSAB is the variation caused by the interaction between variable factors A and B; SSE is the variation caused by random factors. SSA and SSB are usually referred to as main effects, SSAB as N-way interaction effects, and SSE as residual effects. The P-value in ANOVA is a key statistical metric used to determine whether the experimental results are statistically significant. Specifically, the P-value represents the probability of observing the current data or more extreme results under the assumption that the null hypothesis is true. In ANOVA, the P-value is derived from the calculated F-value and the corresponding degrees of freedom.

3. Exhaust system design and simulation analysis

3.1. Exhaust system design

To further improve engine performance, a design improvement was carried out on the engine's exhaust system. Three exhaust systems were designed for the study. For ease of description, the original engine exhaust system is referred to as Type-A, while the three improved exhaust systems are referred to as Type-B, Type-C, and Type-D, respectively. Figure 5 shows the Type-B exhaust system, whose design is simple and lightweight, and can be easily mounted onto the engine. The exhaust system directly expels gases from the cylinder.

Figure 5. Type-B exhaust system.

The Type-C and Type-D exhaust systems are designed in the form of resonant exhaust pipes, as shown in Figure 6. It consists of an inlet section (a), a gradually expanding section (b), an expansion section (c), a gradually contracting section (d), and a tailpipe section (e). After the exhaust begins, the ideal exhaust waveform is to quickly form a long-lasting negative pressure zone at the exhaust port, which helps a large amount of exhaust gas to be drawn out; just before the exhaust port closes, a positive pressure zone forms at the exhaust port to prevent the fresh charge that has entered the cylinder from being expelled. During the exhaust process, the pressure waves will form reverse negative pressure waves and positive pressure waves when passing through the gradually expanding and contracting sections, respectively. This structure can fully utilize the pressure fluctuations during the exhaust process to improve engine performance.

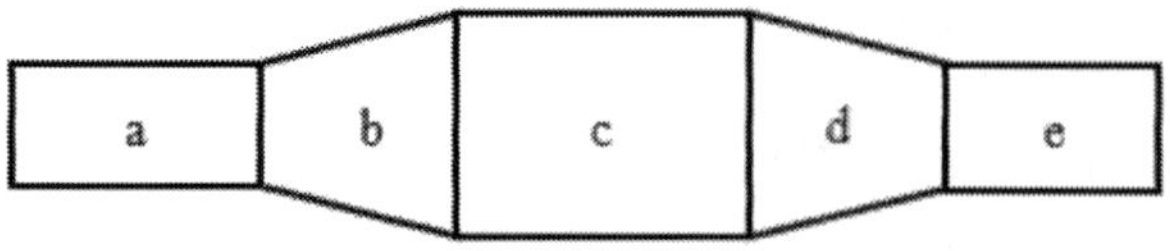

Figure 6. Resonant exhaust system structure.

The engine studied in this paper primarily operates at a cruising speed of 5500 RPM. According to fluctuation effects and resonance theory [18], the structural parameter values of each part of the Type-C exhaust system can be obtained through calculations. However, the exhaust system designed based on the wave effect and resonance theory is too large and incompatible with this engine. Referring to the smaller size and lighter weight of the Type-A and Type-B exhaust systems, and according to the third similarity law: when single-value conditions (such as system geometry, physical properties of the medium, initial conditions, and boundary conditions) are similar, and the similarity criteria composed of the physical quantities of the single-value conditions are numerically equal, phenomena with the same characteristics will occur. The Type-C exhaust system was scaled down proportionally to obtain the structural parameters of each part of the Type-D exhaust system. The structural parameter values of the Type-C and Type-D exhaust systems are shown in Table 2.

Table 2. Structural design parameters of Type-C and Type-D exhaust systems.

Structural parameters	Type-C design value	Type-D design value
Inlet section length La (mm)	330	75
Inlet section diameter Da (mm)	33	33
Gradually expanding section length Lb (mm)	165	37
Expansion section length Lc (mm)	395	89
Expansion section diameter Dc (mm)	60	39
Gradually contracting section length Ld (mm)	110	25
Tailpipe section length Le (mm)	150	34
Tailpipe section diameter De (mm)	20	36

The power and fuel consumption rates of the four exhaust systems at different speeds are shown in Figures 7 and 8. At 5500 RPM, the power of the Type-C exhaust system is the biggest, reaching 22.44 kW, which is significantly greater than the other three exhaust systems. However, its performance at other RPMs is relatively poor. The power of the Type-D exhaust system is bigger than that of the Type-A and Type-B exhaust systems at four different RPMs and is only lower than the Type-C exhaust system at 5500 RPM. The Type-D exhaust system improves engine power based on the existing exhaust systems. Figure 8 shows that at 5500 RPM, the fuel consumption rate of the Type-C exhaust system is the lowest, at 347.60 g/kW·h. The fuel consumption rate of the Type-D exhaust system at various RPMs is between the Type-A and Type-B exhaust systems. Combined with the previous power analysis, the Type-C exhaust system has the best engine performance at 5500 RPM, and the Type-D exhaust system has better engine performance than the Type-A and Type-B exhaust systems at all RPMs, indicating that the designed Type-D exhaust system can significantly improve engine performance.

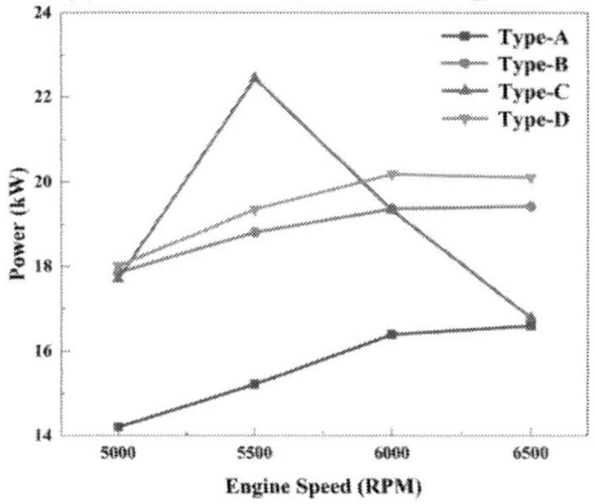

Figure 7. Power comparison.

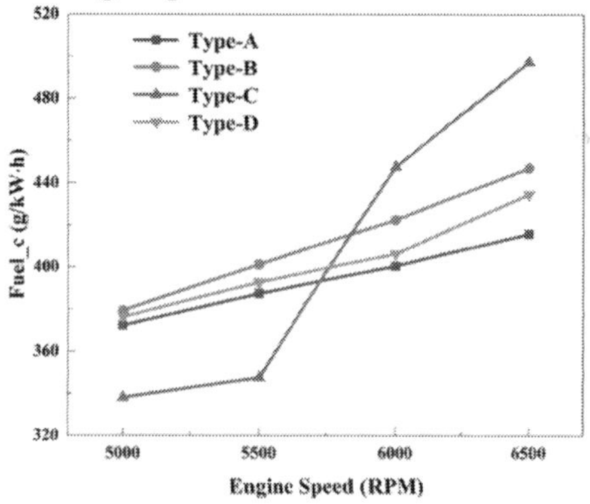

Figure 8. Comparison of fuel consumption rates.

3.2. The impact of different exhaust systems on the exhaust process

When the engine primarily operates at the cruising speed of 5500 RPM, the effects of the four exhaust systems on the engine's exhaust process were studied under the conditions of 5500 RPM and a throttle opening of 55°. Figure 9 shows the pressure fluctuations in the exhaust ducts of the four exhaust systems under these conditions. The four fine lines in the figure represent the scavenging phases, with EVO and EVC indicating the opening and closing moments of the exhaust valve, and SVO and SVC indicating the opening and closing moments of the scavenging valve. When the exhaust valve opens, the cylinder pressure is very high, and the exhaust process occurs in a supercritical state, with the exhaust gases being expelled at the speed of sound, causing the exhaust duct pressure to reach its peak quickly.

During the exhaust process, gases in the exhaust system form reverse expansion waves and compression waves. Reverse-expansion waves can promote the exhaust process, while reverse-compression waves can retain fresh charge in the cylinder during the late exhaust stage, reducing short-circuit losses. From Figure 9, it can be seen that near the bottom dead center (BDC), the exhaust passage pressures from low to high are as follows: Type-C, Type-D, Type-B, and Type-A. At this point, the exhaust port opening area is at its maximum, and lower pressure is more conducive to the exhaust process. During the late exhaust phase (from SVC to EVC stages), the exhaust passage pressures from high to low are as follows: Type-C, Type-D, Type-B, and Type-A. At this point, the scavenging port is closed, and the substance flowing out of the exhaust port is mainly a fresh charge. Higher pressure at this stage is more beneficial for sealing the fresh charge and reducing short-circuit losses.

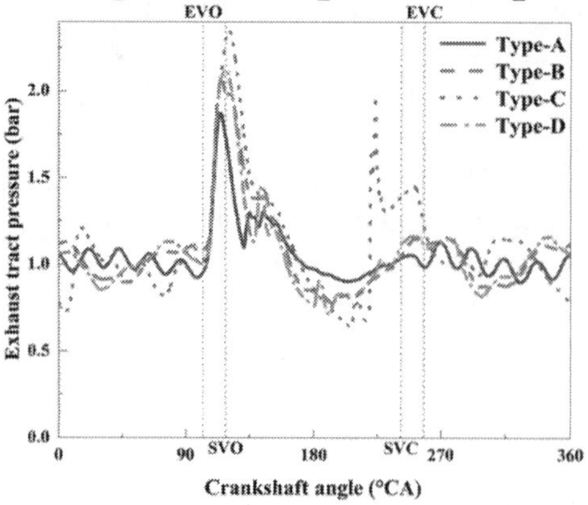

Figure 9. Pressure fluctuations in exhaust ducts of different exhaust systems.

Figures 10, 11, and 12 show the mass flow rate at the exhaust port for four types of exhaust systems under the condition of 5500 RPM and a throttle opening of 55°. From Figure 10, it can be seen that all four exhaust systems exhibit backflow phenomena in the mid to late stages of the exhaust process. Among them, the Type-C exhaust system has the largest backflow, the Type-A exhaust system has a slight backflow, and the backflow rates of the Type-B and Type-C exhaust systems are quite similar.

Figures 11 and 12 indicate that the flow rate of burnt exhaust gases at the BDC significantly decreases, and a considerable amount of unburned air flows out of the cylinder at this point. The backflow material before the exhaust port closes mainly consists of fresh charge lost through short-circuiting. The exhaust systems, ranked by the amount of unburned air flowing back into the cylinder during the mid to late exhaust stages from most to least, are as follows: Type-C, Type-D, Type-B, and Type-A. From the above analysis, it is evident that at 5500 RPM, the Type-C exhaust system exhibits the best exhaust performance, while the Type-D exhaust system also performs well. Both systems can push back the fresh charge lost to short-circuiting into the cylinder during the late exhaust phase, thereby reducing fuel consumption and enhancing engine performance.

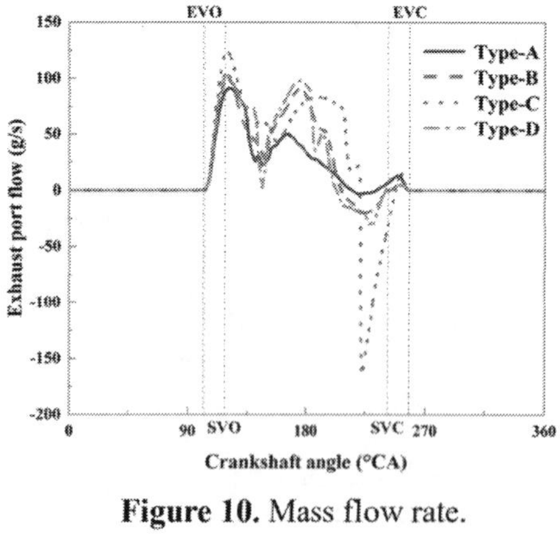

Figure 10. Mass flow rate.

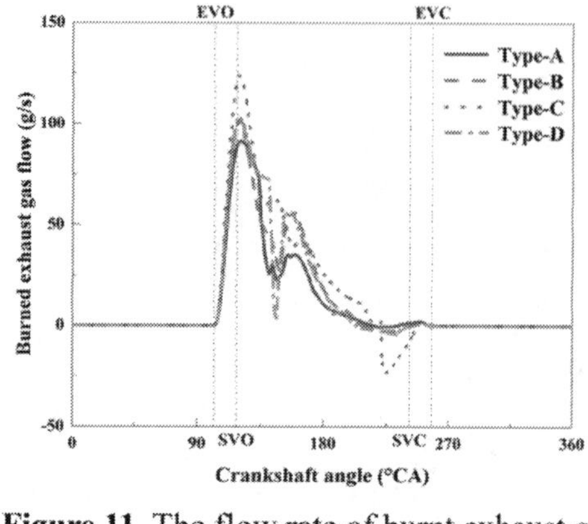

Figure 11. The flow rate of burnt exhaust gases.

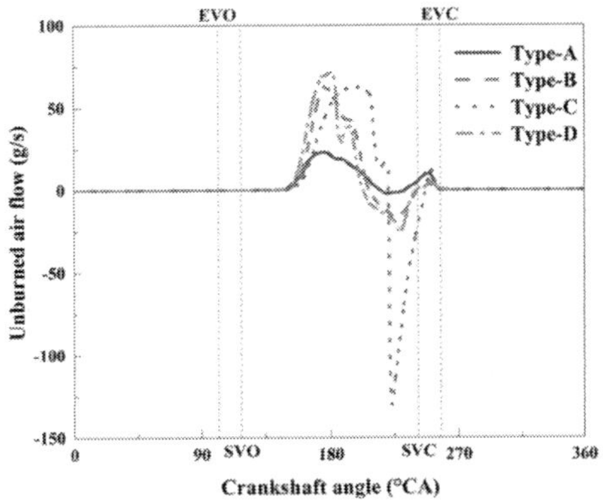

Figure 12. The flow rate of unburned air.

3.3. The impact of different exhaust systems on the in-cylinder combustion process

Figures 13 and 14 show the in-cylinder mass and cumulative heat release of four exhaust systems at 5500 RPM and a throttle opening of 55°. At the end of the scavenging process, the in-cylinder mass retained by the type A, B, C, and D exhaust systems is 0.1264 g, 0.1420 g, 0.1710 g, and 0.1458 g, respectively. The in-cylinder mass retained by the Type-D exhaust system is higher than that of the Type-A and Type-B exhaust systems. Combining the analysis of the exhaust process, the reason for the difference in retained mass is the significant variation in the mass pushed back into the cylinder by the reverse compression waves in the mid to late stages of the exhaust process for the four exhaust systems.

From Figure 14, it can be seen that the cumulative heat release of the four exhaust systems, from highest to lowest, is Type-C, Type-D, Type-B, and Type-A. From the previous analysis, the difference in the in-cylinder retained mass leads to different cumulative heat releases. Therefore, at 5500 RPM, the Type-C exhaust system has the highest power, followed by the Type-D exhaust system, and the Type-B and Type-A exhaust systems have the lowest power.

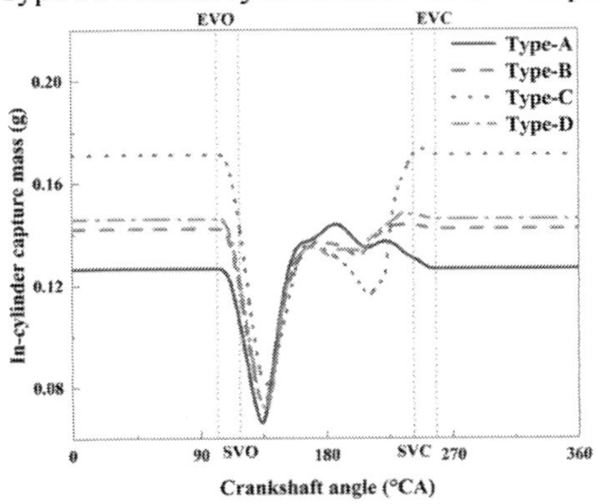

Figure 13. In-cylinder capture amounts.

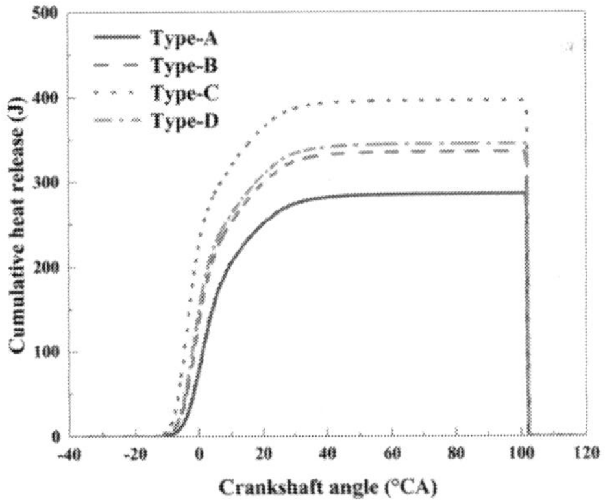

Figure 14. In-cylinder cumulative heat release amounts.

Based on the above analysis, it is evident that the Type-C exhaust system performs best at cruising speed but has poorer performance at other speeds. In contrast, the Type-D exhaust system exhibits excellent engine performance across all speeds. Therefore, we will optimize the Type-D exhaust system at cruising speed to enhance the engine performance to be closer to that of the Type-C exhaust system.

4. Optimization of structural parameters of exhaust systems

To maximize the performance of the designed Type-D exhaust system, optimization was performed based on the structural parameters calculated through the third law of similarity. Given the numerous structural parameters of the Type-D exhaust system, the DOE method was employed to obtain a sample matrix of exhaust system structural parameters, aiming to find the optimal results in the design space established by multiple parameters[19]. An approximate model was built based on the sample points to carry out multi-objective optimization of the structural parameters of the Type-D exhaust system. This optimization ensures that the results meet the optimal objectives while considering the interrelationships among various parameters. The La, Lc, Dc, Le, and De of the Type-D exhaust system were taken as variable factors. The range of values for each variable factor, constrained by the structural parameters of the Type-B exhaust system, is shown in Table 3.

Table 3. Range of values for each variable factor.

Variable factors	Initial design values	Optimization range
La (mm)	75	60~90
Lc (mm)	89	70~110
Dc (mm)	39	38~40
Le (mm)	34	20~50
De (mm)	36	21~38

The D-Optimal Latin Hypercube (D-OLH) sampling method maximizes the determinant of the design matrix for a given sample size and dimension, thereby improving the efficiency and representativeness of the sampling design, which leads to more accurate results in complex problems. In this paper, the D-OLH sampling method was employed, with a total of 150 test iterations set for simulation optimization at a condition of 5500 RPM and a throttle opening of 55°. The primary aim is to enhance engine performance, with power output and fuel consumption rate selected as the response variables for multi-objective optimization.

4.1. Response surface fitting and quality evaluation

With the sample points obtained from DOE, quadratic polynomial fitting of power and fuel consumption rate was performed by using the Least Squares method. Figures 15 and 16 show the response surfaces of the Dc and De to power and fuel consumption rate under the conditions of 5500 RPM and a throttle opening of 55°. From the figure, it can be seen that as Dc increases, the engine power remains essentially unchanged, while the fuel consumption rate decreases. When De increases, the engine power increases, the fuel consumption rate first decreases, then increases, and finally decreases again.

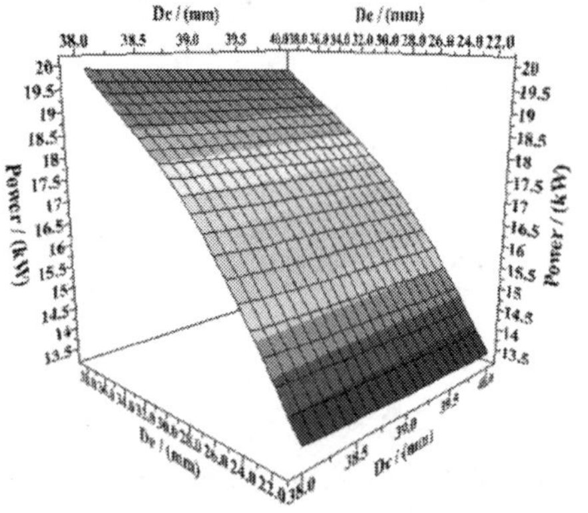

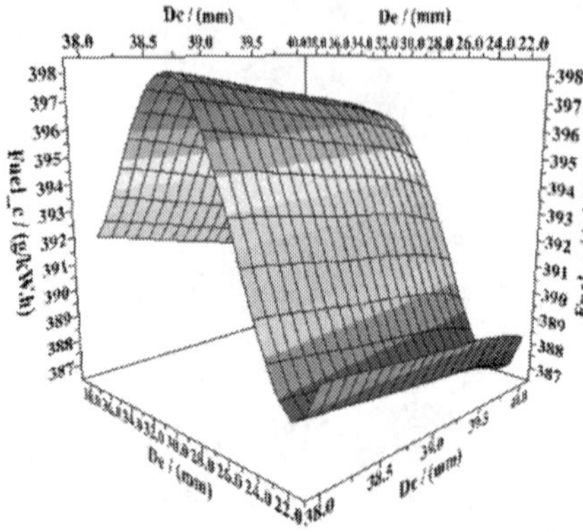

Figure 15. Response surface to power. **Figure 16.** Response surface to fuel consumption rate.

Figures 17 and 18 show the residual plots for power and fuel consumption rates. The dashed line in the middle represents zero error; the closer the points are to the dashed line, the closer the predicted values are to the observed values, indicating that the smaller the error is, the higher the accuracy of the response fit. The figures show that the fitting accuracy for power and fuel consumption rates is very high. It can be considered that the fitted response surface meets the required accuracy, making it applicable for subsequent multi-objective optimization design.

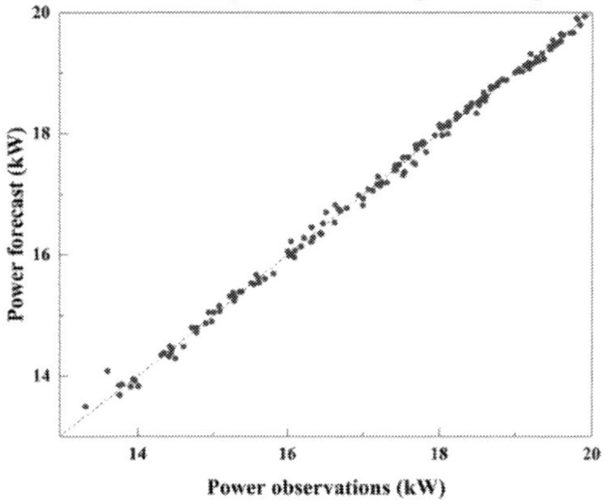

Figure 17. Residual plot for power response.

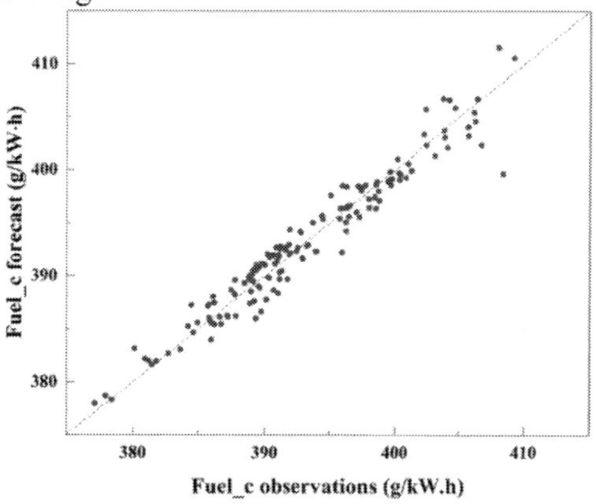

Figure 18. Residual plot for fuel consumption rate response.

4.2. Analysis of the influence of variable factors

By analyzing the DOE sample data using multifactor ANOVA, the assumption is that there is no significant difference in the mean responses at different levels of each variable factor and the effects and interaction effects of variable factors are simultaneously zero, i.e., La, Lc, Dc, Le, De and their interactions have no significant impact on power and fuel consumption rate. The P-values for power and fuel consumption rate calculated from DOE sample data are 2.4761×10^{-139} and 4.5249×10^{-61}, respectively, both of which are much smaller than the given significance level (0.01). Therefore, the hypothesis does not hold, indicating that the variable factors have a significant impact on the response. The main effect and interaction effect plots for power and fuel consumption rate under the conditions of 5500 RPM and a throttle opening of 55° are shown in Figures 19 and 20. The positive and negative values indicate the correlation of the variable factor with the response, and the magnitude of the values indicates the extent of the impact.

The structural parameters of the Type-D exhaust system have different effects on power and fuel consumption rates. Among them, De has the most significant impact on power; the significance order for fuel consumption rate is La> Le > De > Lc > Dc. Moreover, there are parameters (such as Dc and De) for which optimizing one objective comes at the expense of others. Figure 20 shows that the effects of the various structural parameters of the exhaust system on the objective functions are mutually coupled. The complex relationship of multi-objective mutual constraints and coupling requires the use of multi-objective and multi-variable optimization algorithms to determine the optimal structural parameters of the exhaust system, thereby improving engine performance.

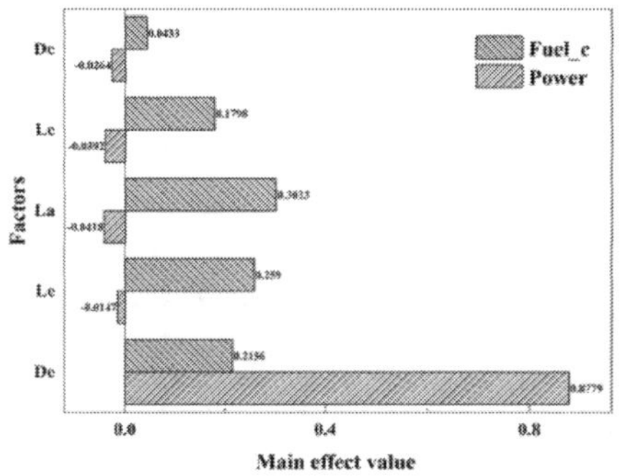

Figure 19. Main effect plot.

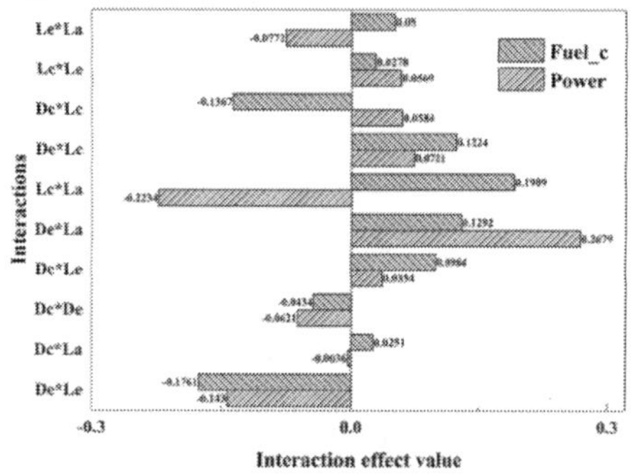

Figure 20. Interaction effect plot.

4.3. Multi-objective optimization of exhaust system structural parameters

Under the condition of 5500 RPM and 55° throttle opening, by selecting La, Lc, Dc, Le, and De as variable factors and using the objectives of maximizing power and minimizing fuel consumption rate, the multi-objective Pareto optimization algorithm is used for solving. The obtained Pareto front is shown in Figure 21. This study primarily focuses on improving engine power, and the point with the maximum power on the Pareto front is selected as the optimal solution.

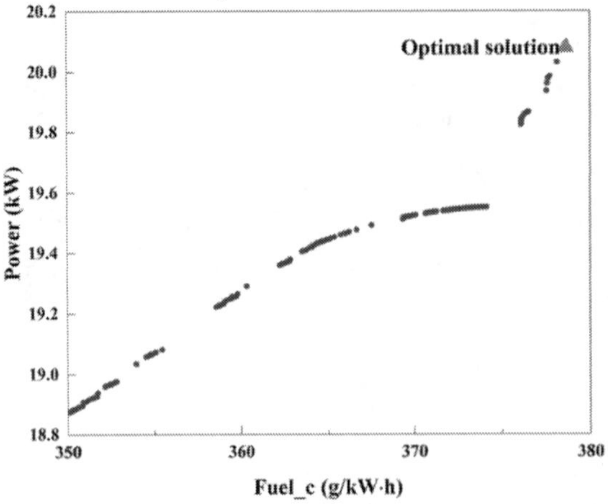

Figure 21. Pareto front of power and fuel consumption rate.

The structural parameters of the exhaust system before and after optimization are shown in Table 4. The optimized structural parameters of the exhaust system were input into the established simulation model for calculation. The engine performance before and after optimization of the Type-D exhaust system is shown in Table 5. The optimized Type-D exhaust system increases engine power by 3.15% and reduces the fuel consumption rate by 4.07%, further enhancing engine performance.

Table 4. Structural parameters of the exhaust system before and after optimization.

Structural parameters	Before optimization	After optimization	Percentage of change
Dc (mm)	39	38.1	-2.31%
De (mm)	36	37.9	5.28%
Lc (mm)	89	70.8	-20.45%
Le (mm)	34	23.3	-31.47%
La (mm)	75	60.6	-19.2%

Table 5. Engine performance comparison before and after optimization.

	Before optimization	After optimization	Percentage of change
Power (kW)	19.35	19.96	3.15%
Fuel consumption rate (g/kW·h)	393	377	-4.07%

5. Conclusion

Based on the design of the exhaust system for a certain aviation piston engine, this paper conducted a multi-objective optimization of its structural parameters, leading to the following conclusions.

1. To fully utilize the pressure wave effect of the exhaust system, a resonant exhaust system was designed. Based on the theoretically sized resonant exhaust system and using the third similarity law, a resonant exhaust system suitable for the current dimensions of the engine's exhaust system was designed, which can increase the engine power by 2.88% at cruising speed.

2. The exhaust system essentially influences the pressure fluctuations during the engine exhaust process, which in turn affects the exhaust process and the quality of the working medium retained in the cylinder. This ultimately impacts the combustion heat release within the engine cylinder, thereby affecting the engine's performance.

3. For this aviation piston engine at 5500 RPM cruising speed, sample data was obtained through DOE, and a similar model was established. Then, the structural parameters of the resonant exhaust system, suitable for the current exhaust system size of the engine, were optimized by using the multi-objective Pareto algorithm. The optimized resonant exhaust system, suitable for the current exhaust system size of the engine, resulted in a 3.15% increase in engine power and a 4.07% reduction in fuel consumption rate.

Acknowledgments

This research was supported by the Natural Science Foundation of Chongqing, China (No. CSTB2022NSCQ-LZX0068) and the project (KF1905) of Jiangsu Xinyuan Aviation Technology Co., Ltd. The authors acknowledge that, for such a significant research topic, the achievements made are limited due to constraints of time and space. The authors express their gratitude to all those who have lent him a helping hand throughout the process.

References

[1] Guo D, Qian J C, Zhao Z F, et al. The scavenging process of variable exhaust valve in aviation piston two-stroke engines[J]. Journal of Aerospace Power, 2020, 35(03): 471-481.

[2] Jang J, Woo Y, Jung Y, et al. Research for intake and exhaust system parameterization of 2-cylinder gasoline engine for RE-EV[J]. International Journal of Energy Research, 2018, 42(13): 4256-4256.

[3] Wang H L, Wang F, Li Y, et al. Study on the influence of intake and exhaust boundary conditions on engine performance parameters[J]. Internal Combustion Engine, 2022, 38(06): 13-18.

[4] Hu P. Analysis of the influence of structural parameters of intake and exhaust systems on combustion parameters in turbocharged gasoline engines[J]. Internal Combustion Engine, 2023, 39(03): 31-35.

[5] Xu Y Y. Research on the influence of intake and exhaust port throats on engine performance in high boost ratio engines[D]. Jiangsu University of Science and Technology, 2022.

[6] Binqi L, Ya-Ping H, Zheng-Yan Q, et al. Research on performance matching of intake and exhaust ports of marine medium speed dual fuel engine[J]. Energy Reports, 2021, 7(S1): 72-83.

[7] Wang L, Zhao Z F, Yu C C, et al. Optimization study on the height of the exhaust port in aviation piston two-stroke gasoline engines[J]. Journal of Aerospace Power, 2019, 34(02): 442-450.

[8] Bai J, Sun S N, Wang Q. Numerical simulation study on the influence of intake and exhaust structural parameters on the scavenging process of a micro free-piston engine[J]. Journal of Internal Combustion Engine Engineering, 2021, 42(05): 42-51.

[9] Sipeng Z, Yujun T, Dawei W, et al. Numerical studies on the flow characteristic of the marine two-stroke engine integrated with the high-pressure exhaust gas recirculation system[J]. Case Studies in Thermal Engineering, 2024, 53103958-.

[10] Zhang K, Chen H T, Deng K Y, et al. Calculation method for unsteady flow with variable grid in intake and exhaust of internal combustion engines[J]. Journal of Harbin Engineering University, 2022, 43(06): 809-816, 920.

[11] Lee D, Yoon K, Ryu S, et al. Study on the Simulation of the Intake and Exhaust Systems of a Gasoline Engine Using BOOST[J]. Transaction of the Korean Society of Automotive Engineers, 2013, 21(4): 23-32.

[12] Kim J, Yoon K, Woo S. Study on the Modeling of the Intake and Exhaust Systems of an SI Engine Using GT-POWER[J]. Journal of the Korean Society of Marine Engineering, 2011, 35(6): 779-785.

[13] Hu C M and Hao M M. Optimization of the intake and exhaust systems of UAV piston engines[J]. Journal of Aerospace Power, 2018, 33(04): 1009-1016.
[14] Tian M Y. Simulation study on the scavenging process of two-stroke engines based on multi-objective optimization[D]. Tianjin University, 2021.
[15] Li B. Numerical simulation and optimization research on the intake and exhaust systems of high-speed gasoline engines[D]. Chongqing University of Technology, 2020.
[16] Luo Z. Study on the influence characteristics of intake and exhaust system parameters and multi-objective optimization of turbocharged direct injection engines[D]. Chongqing University of Technology, 2021.
[17] Wang J T. Optimum design for intake and exhaust system of a heavy-duty diesel engine by using DFSS methodology[J]. Journal of Mechanical Science and Technology, 2018, 32(7): 3465-3472.
[18] Yang G X, Ye S Y, Cheng S B. Principles and Design of Motorcycle Engines[M]. Hubei: Wuhan University of Surveying and Mapping Science and Technology Press, 1993.
[19] Fu S L, Cheng H S, Yi K L, et al. Performance Analysis and Optimization Design of Exhaust System for Turbocharging Diesel Engines[J]. International Journal of Automotive Technology, 2021, 22(3): 735-745.

Effect of tempering process on the structure and properties of M51/X32 bimetal band saw blade

Gusiyang Zhao[1], Yan Liu[1,*], Zhuo Song[1], Guimao Li[1], Bo Fu[2], Changsheng Xia[2] and Qu Ning[3]

[1]School of Mechanical Engineering, Liaoning University of Science and Technology, Benxi Liaoning 117004, China

[2]Benxi Tools Co., Ltd., Benxi Liaoning 117004, China

[3]School of Materials Science and Engineering, Shenyang University of Technology, Shenyang Liaoning 110000, China

[*]Corresponding author's e-mai: 87191441@qq.com

Abstract: The mechanical properties of band saw have a great influence on the cutting life. In this paper, through the systematic study of M51/X32 bimetallic band saw blade with different tempering temperatures and times, the metallography of the bimetallic band saw blade under different heat treatment conditions was analyzed. The mechanical properties of the bimetallic band saw blade under different conditions were tested. The results showed that the carbide in the metallography of the tooth tip was uniform and tended to be spherical when tempering at 600°C for five times. The mechanical properties such as hardness, tensile strength, and fatigue properties are well matched, which can extend the service life of bimetallic band saw blades and ensure the cutting quality.

1. Introduction

The bimetal band saw blade is a commonly used cutting tool; its performance requirements are increasingly high; it not only needs high cutting efficiency but also a long service life, so it must continue to improve the bimetal band saw blade in order not to be eliminated by the market. Improvement of bimetallic band saw blades, on the one hand, can be improved through the improvement of tooth shape to improve cutting efficiency [1-2]. On the one hand, the heat treatment process improves the mechanical properties of bimetallic band saw blades to improve service life [3-5].

The bimetallic band saw blade tooth tip and tooth back have two different materials: Tooth tip material is generally used high-speed steel, and tooth back material is generally used spring steel. In this paper, the M51-X32 bimetal band saw blade is taken as the main research object. The original experimental parameters of the enterprise are compared and improved. The influence of tempering process parameters on the microstructure and properties of the tooth tip and back material is explored. The tempering process parameters are optimized. The metallographic structure of the tooth tip and back of the bimetal band saw blade is observed. The morphology of the martensite structure at different tempering temperatures is observed. The diameter and area percentage of the carbide at the tooth tip are compared. The hardness at different tempering temperatures was analyzed to obtain the best heat treatment process parameters. The best heat treatment process is used to obtain the ideal structure and excellent performance of the band saw blade, which provides a reference for the factory to optimize the process, improve the quality and service life of the bimetallic saw blade, provide the core competitiveness for the enterprise's products, and create greater profits.

2. Experimental methods

2.1 Hardening process

The bimetallic band saw blade is composed of two steel grades with very different chemical compositions, so its heat treatment process should ensure the hardness and alloying degree of the tooth tip material while effectively controlling the overgrowth of the tooth surface tissue. In this experiment, high-pressure nitrogen was used for quenching. The process involved heating the material to 990°C and maintaining this temperature for three minutes. Then, the temperature was increased to 1145°C and maintained for three minutes, followed by further heating to 1195°C with a three-minute preservation period. Subsequently, the material was heated to 1209°C and preserved for one minute, and then to 1214°C with a one-minute preservation.

2.2 Tempering process

In the study of the quenched tooth tip M51 high-speed steel tempering process, we based our final selection of tempering temperatures on the material's chemical composition, actual production experience, and previous experimental data. We chose to temper at 540°C, 570°C, 600°C, and 630°C for one to seven tempering cycles, followed by air cooling.

2.3 Organization analysis and performance testing

After mounting the samples, the bimetallic band saw blade tooth tip and the tooth back part were ground and polished. Then, they were subjected to corrosion using 4% nitric acid alcohol for 15 seconds. The resulting corrosion was observed under the Zeiss high-magnification metallurgical microscope.

The subject was tested for hardness, fatigue performance, and wear resistance of the tempered samples, respectively. The digital micro Vickers hardness tester is used to determine the microhardness with a loading force of 1kg and a loading time of 10s. For each sample, the tooth tip and tooth back were measured with three hardness data and the average value was used to characterize the state of the hardness of the sample.

The fatigue performance test is completed in the fatigue failure times on the tester. The length of the band saw blade tested is 1650mm, and the number of fatigue failures is determined by subjecting the bimetallic band saw blade to repeated tensile stress loads, which ultimately leads to its fracture. The results are determined by the number of alternating tensile actions until the blade breaks.

Abrasion resistance is the use of a pin disk friction wear testing machine on the tooth tip part of the wear test. The wear resistance was determined by measuring the weight loss of the specimen after wear. The loading load of the experimental process was 20N, rotational speed was 240r/min, friction radius was 25mm, and time was 10min.

3. Experimental results and discussion

3.1 Metallographic organization of the tip and back of band saw blade teeth under different tempering temperatures

In Figure 1(a), the metallographic analysis of M51 high-speed steel after various tempering temperature treatments shows the presence of tempered martensite, residual austenite, and spherical carbides [3]. When M51 high-speed steel is tempered at different temperatures, there are significant differences in its microstructure. With the increase in tempering temperature, the organization's precipitated carbide area increases, and the volume gradually accumulates and grows and is not uniformly distributed. At 540°C and 570°C, high-speed steel undergoes alloy carbide secondary precipitation with less dispersion. At 600°C, residual austenite decreases and leads to more uniform carbide size and growth. Furthermore, at 630°C, the carbide diffusion capacity increases, resulting in carbide aggregation and coarsening.

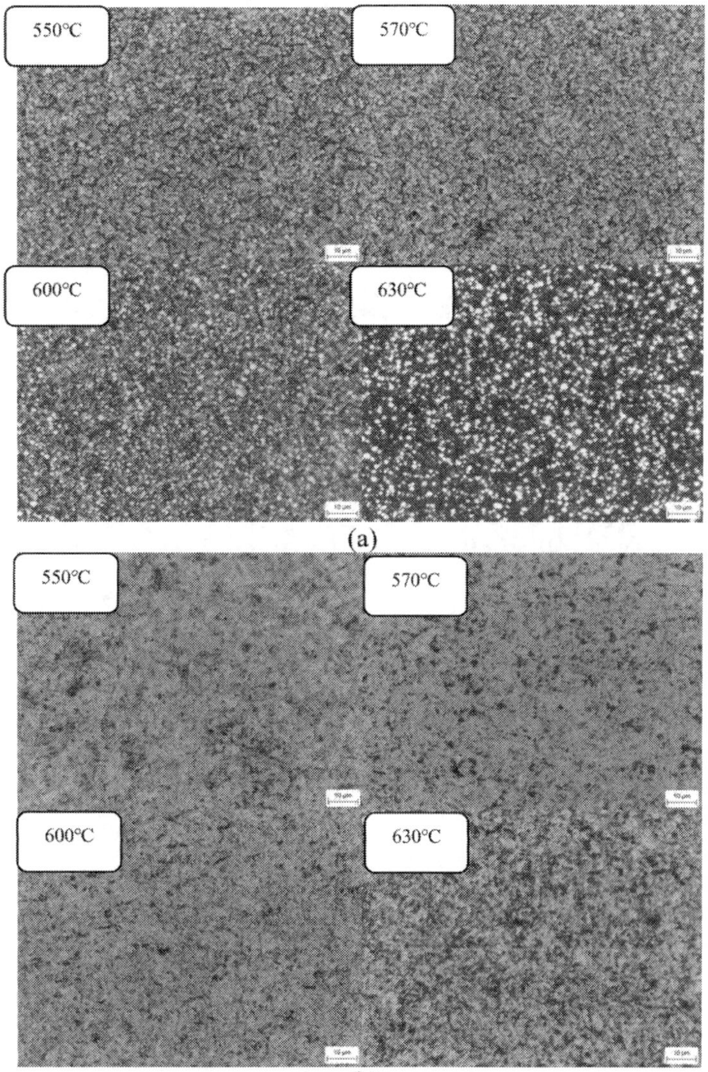

(b)

Figure 1. Microstructure of band saw blade in different tempering temperatures. (a) Tooth tip, (b) Tooth back.

Figure 1(b) illustrates the metallographic structure of the tooth back of material X32 after different tempering temperature treatments. The structure mainly consists of tempered martensite with a small amount of residual austenite. In the figure, gray represents tempered martensite, black represents carbide, and ferrite is also present. As the tempering temperature increases, the carbon in the tempered martensite also dissolves and precipitates and gathers and grows, the organization turns into ferrite and carbide organization. The flexural increases. When it reaches 630°C, the spacing between the flexural lamellae becomes larger, and the emergence of Soxhlet leads to the reduction of the hardness of the material [3].

Journal of Physics: Conference Series **2879** (2024) 012006 doi:10.1088/1742-6596/2879/1/012006

3.2 Metallographic organization of the tip and back of the teeth of band saw blades under different tempering times

In Figure 2(a), the metallographic structure of the M51/X32 bimetallic band saw blade, which has been tempered once to seven times at the tooth tip, is depicted. It is evident that with a lower number of tempering times, the carbide distribution is discrete and the particle size is large. As the number of tempering times increases, secondary carbide precipitation occurs, resulting in finer and more diffuse particles. However, excessive tempering times lead to the annexation of fine carbide particles by larger ones, causing the carbides to grow, which is detrimental to the improvement of the mechanical properties [4].

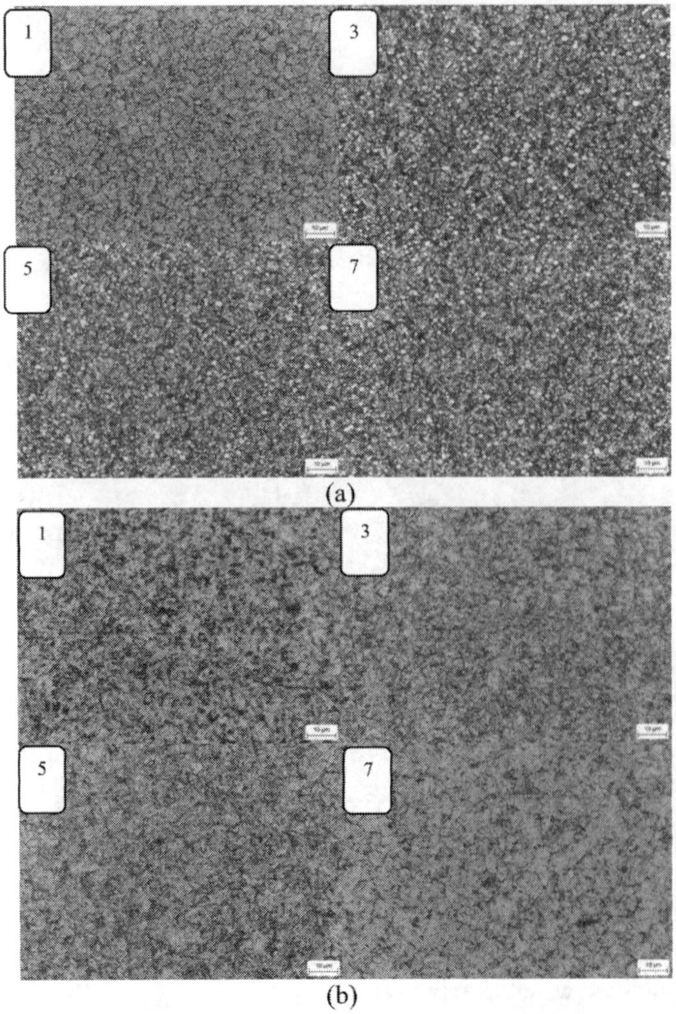

Figure 2. Microstructure of band saw blade in different tempering times. (a) Tooth tip, (b) Tooth back.

Figure 2(b) shows the metallographic organization of the tooth back of the M51/X32 bimetallic band saw blade tempered once to seven times at 630°C. It can be seen that most of the metallographic organization at 630°C is flexural. With an increase in the number of tempering times, the flexural organization continues to decompose into ferrite and carbides, the aggregation grows, and the spacing of the lamellae also increases [5].

In Figure 3, we can observe the trend of carbide diameter at the tip of the M51 high-speed steel tempered at different temperatures. It is evident from the figure that as the number of tempering increases, the carbide diameters initially decrease and then increase. This phenomenon occurs because as the tempering time increases, it favors the precipitation and growth of secondary carbides. Initially, the overall carbide particles become smaller due to the precipitation of secondary carbides. However, with a further increase in tempering time, the carbides grow and aggregate again. When the number tempering, the carbide will gather and grow up; when the tempering temperature is increased from 540°C to 630°C, the carbide diameter becomes thin. This is mainly because the high-temperature tempering of the initial carbide particles is small, so the overall follow-up with the increase in the amount of tempering is relatively small.

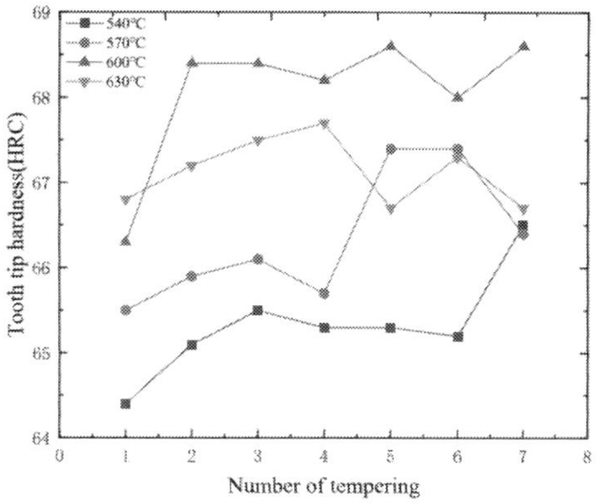

Figure 3. Diameter of carbide after different tempering temperatures and times.

3.3 Tempering treatment on the M51/X32 bimetallic band saw blade hardness effect

Table 1 shows that for the M51/X32 bimetallic band saw blade tooth tip of M51 high-speed steel, after tempering the average hardness value, the hardness of the material after tempering is significantly higher than that of the quenched state. With the increase of tempering temperature, the hardness of the material increases first and then decreases. During tempering at 540°C ~ 600°C, the carbide of vanadium-tungsten alloy is precipitated, and the content of W, V, and Cr in solid solution is reduced. The hardness of high-speed steel increases due to the very fine-grained distribution of V and W carbides in the martensite matrix. When the tempering temperature is above 600°C, a large number of carbon atoms in martensite are precipitated, and carbides are formed at the boundary of the alloy. The precipitation process accelerates with the increase of tempering temperature. At the same time, the alloy carbides begin to grow and aggregate so that the hardness gradually decreases. Because 600°C is the ideal tempering temperature, the hardness of the metal strip serrated tip is the best.

Table 1. Hardness of tooth tip after tempering.

	Tooth tip hardness/HRC			
Number of tempering cycles	540°C	570°C	600°C	630°C
1	64.4	65.5	66.3	66.8
2	65.1	65.9	68.4	67.2
3	65.5	66.1	68.4	67.5
4	65.3	65.7	68.2	67.7
5	65.3	67.4	68.6	66.7
6	65.2	67.4	68	67.3
7	66.5	66.4	68.6	66.7

3.4 Effect of tempering treatment on the tensile strength of saw bands

The influence curve of process parameters on tensile strength is shown in Figure 4. It can be seen from the chart that the tensile strength increases with the increase of tempering times at 540°C and 570°C. At 600°C and 630°C, the tensile strength increases first and then decreases, while it increases slightly at 5 ~ 7 times of tempering and reaches the maximum at twice of tempering. At 600°C, the tensile strength at 630°C and 600°C is basically the same, but it is still lower than that at 600°C. After tempering at 630°C, the grain size of the steel plate becomes coarse, and the tensile strength and elongation are significantly reduced. The results show that the more times of tempering at 540°C and 570 °C are, the more times of tempering are beneficial to improve the tensile strength of the strip. However, at 600°C and 630°C, the tensile strength decreases after multiple tempering, which is unfavorable to improving the performance of the strip saw blade.

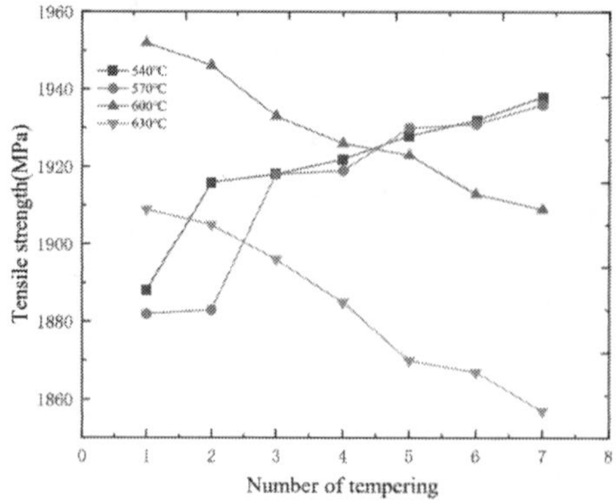

Figure 4. Tensile strength of the band, saw blade under different tempering times.

3.5 Fatigue performance testing of M51/X32 bimetal band saw blades by tempering treatment

As shown in Figure 5, the sawing process of the bimetallic band saw blade is repeated. Most of the stress is on the back of the tooth, so it is very important to determine the fatigue life of the back material of the bimetal band saw blade. From the experimental results of the fatigue time curve of the bimetallic band saw blade at different tempering temperatures and tempering times, it can be seen that the fatigue time at 600 °C and 630 °C is generally higher than that at 540 °C and 570 °C. In addition, the fatigue time fluctuates with the number of tempering cycles. The fifth tempering cycle marks the point at which the fatigue performance reaches an inflection point, after which the fatigue time begins to decrease. This indicates that the fatigue resistance of the bimetal band saw blade is optimal after five tempering cycles. However, the fatigue performance of tempering at 630°C has been decreasing, indicating that the tempering temperature is too high.

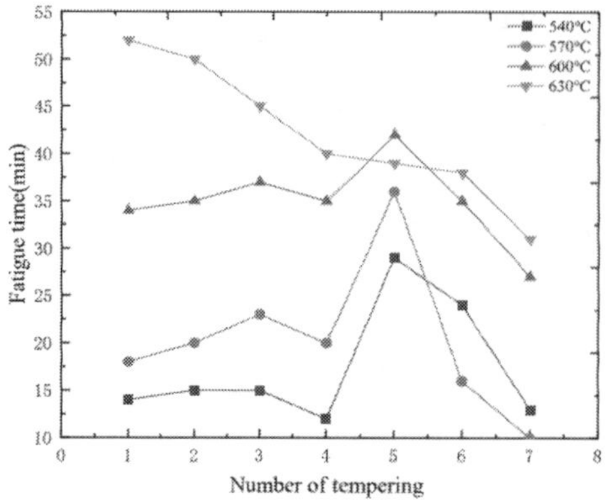

Figure 5. Fatigue strength of band saw blade under different tempering times.

4. Conclusion

(1) The study revealed that at 540°C and 570°C, carbide precipitation in high-speed steel is discrete, while at 600°C, the carbides become more uniform, fine, and tend to be spherical in shape. However,

at 630°C, the carbides begin to aggregate and coarsen significantly. With the increase of tempering times, secondary carbides precipitate and the particles are dispersed finely. However, if the tempering times are too many, the fine particles of carbides are swallowed up by large particles. The carbides grow up, which is not conducive to the improvement of mechanical properties.

(2) With the effect of the tempering process on the M51-X32 bimetallic band saw blade, it is found that the hardness, tensile strength, and fatigue performance of the band saw blade tempered at 600°C for five times are better. Determining the optimal parameters of the heat treatment process has an indispensable reference value for the manufacturing of production equipment, the selection of thermal cycle components, and the assembly line tooling settings.

(3) With the expansion of its application in the fields of cutting and precision machining of different materials, the optimal process parameters obtained make the microstructure of steel more stable, reduce the brittleness of saw blade, enhance toughness, reduce thermal stress, and greatly improve the quality and service life of the strip. It is expected to break through the existing limitations and meet the broader market demand.

Acknowledgments

The work of this paper has obtained the innovation and entrepreneurship project of college students in Liaoning University of Science and Technology in 2024 (project number: 202411430056); 2022 Department of Education Service Local Project (Project No. : LJKFZ20220279); general Project of Education Department in 2023 (2023J0904); 2023 Pioneer Research Team Project (XKT202303); 2024 Liaoning Provincial Department of Education Research Platform Construction Project (2024JYTKYPT04); 2024 A series of funding for the Science and Technology Innovation Team Project of Liaoning Provincial Department of Education (2024JYTKYTD19 & 2024JYTKYTD20).

References

[1] Jinbin Chen, Yan Liu, Guimao Li, et al. Optimized design of tooth parameters of 1.4/2.0T bimetallic band saw blade [J]. Tool Technology, 2022, 56(8): 77-80.

[2] Linlin Chen, Guimao Li, Hanwen Liu, et al. Optimization of tooth shape of combined toothed bimetallic band saw blades [J]. Tool Technology, 2022, 56(12): 122-124.

[3] Dongdong Wang. Influence of heat treatment process on the organizational properties of M42_D6A bimetallic band saw blade [D]. Hunan University: School of Materials Science and Engineering, 2013.

[4] Liu Liu, Guimao Li, Rui Wang, Zhuang Ma, Mingzhi Tan. Effect of tempering process on the organization and properties of M42-X32 bimetallic band saw blade [J]. Thermal Processing Technology, 2019, 48(10): 217-219+223.

[5] Liangjun Li, Jiamin Zhang, HongXiao Chi, Jian Zhou. Effect of heat treatment process on the organization and properties of M2 high-speed steel [J]. Metal Heat Treatment, 2021, 46(09): 72-79.

Journal of Physics: Conference Series **2879** (2024) 012007 doi:10.1088/1742-6596/2879/1/012007

Reaction chamber design for an organic molecular layer deposition instrument

Shuai Zhu[1], Pengfei Li[1,2,*], Dada Yang[1], Yang Hou[1] and Qiaorui Sheng[1]

[1]School of Mechanical Engineering, Liaoning Technical University, Fuxin 123000, China

[2]Mining Hydraulic Technology and Equipment Engineering Research Center, Liaoning Technical University, Fuxin 123000, China

[*]Corresponding author's e-mail: 13324237703@163.com

Abstract: In the molecular layer deposition (MLD) process, the reaction chamber plays a crucial role as the deposition carrier. The geometry of the reaction chamber directly affects the deposition processes of delivery, heating of reactive gases, and purging of excessive molecules. In this study, we designed an MLD instrument for the preparation of pure organic polymer layered coatings and investigated key structural parameters of its reaction chamber. By the Computational Fluid Dynamics (CFD) method, precursor delivery, nitrogen purge, vacuuming, and oil bath heating in the deposition process were modeled. The flow field distribution under different inlet and outlet heights and the temperature field distribution under different axial heights of the chamber were analyzed, and the most suitable parameters were thus determined. The results showed that when the heights of precursor inlet, nitrogen inlet, and vacuum pumping outlet were selected as 50 mm, 45 mm, and 50 mm, respectively, the chamber could serve better deposition and purge efficiency. When the axial height of the chamber is 185 mm, both the chamber and the fluoro rubber sealing ring meet the temperature requirements.

1. Introduction

Molecular Layer Deposition (MLD) is a thin film deposition technique developed as a subcategory of atomic layer deposition (ALD), based on the principle of self-limiting surface chemical reactions through the gradual growth of organic single molecules in a specified molecular arrangement [1]. Typical MLD processes include at least two types of precursor molecules. After one type of molecule is deposited, a purge step by inert gas is immediately conducted, followed by the deposition of the other type of molecules and another purge step [2]. By means of suitable processes, it is possible to realize organic polymer films of a certain thickness with a near-conformal coverage on the surface of the substrate [3-4].

In recent years, scholars have designed a variety of MLD devices for different MLD process schemes. Higgs et al. [5] developed a rotating cylinder reactor, mounted in a customized oven enclosure, and successfully realized the application on flexible polyethylene terephthalate substrate. Selvaraj et al. [6] developed a reactor that has been successfully tested for molecular layer deposition of titanium oxide (TiO2) using specific precursors and water vapor and the results showed good growth rate and film purity. These studies demonstrated the superior performance of different MLD reactor designs for specific MLD. As the design of new MLD reactors, more substrates and precursor materials can be applied.

Currently, available MLD devices are expensive to produce and are mostly used for inorganic molecule deposition. During the deposition of organic polymer films, owing to the SHB [7] effect, organic precursor molecules are apt to adhere to the tube wall, which may be revolatilized and participate in the follow-up deposition steps, leading to contamination of the prepared film.

In this paper, the reaction chamber of a self-developed molecular layer deposition device has been designed and the structural parameters of that chamber have been investigated and optimized based on the process principles of MLD and the process requirements needed for the deposition reaction chamber. The optimized chamber has a simple structure, easy assembly, and high hardware compatibility. COMSOL [8] has been used to perform parametric simulations for the axial height and gas inlet and outlet heights of the chamber. By analyzing the velocity distribution of the airflow through the surface of the substrate, as well as the temperature distribution of the chamber undergoing oil bath heating, those height parameters have been optimized and selected. The results may provide references for the structure design of MLD reaction chambers in practical engineering applications.

2. Chamber design and process

2.1. Chamber design scheme

The structure of the reaction chamber is presented in Figure 1. The main body of the chamber is constructed of a cylindrical tube, a steel plate is welded at the bottom of the tube for sealing, and the top end of the tube is welded with an ISO-K flange and covered with a blind flange. Fluoro rubber rings are provided between those flanges for sealing. A viewing window and a temperature measurement window are set up on the blind flange by welding. A cylindrical underlay is welded at the inside bottom of the chamber, which is smaller than the substrate for film deposition.

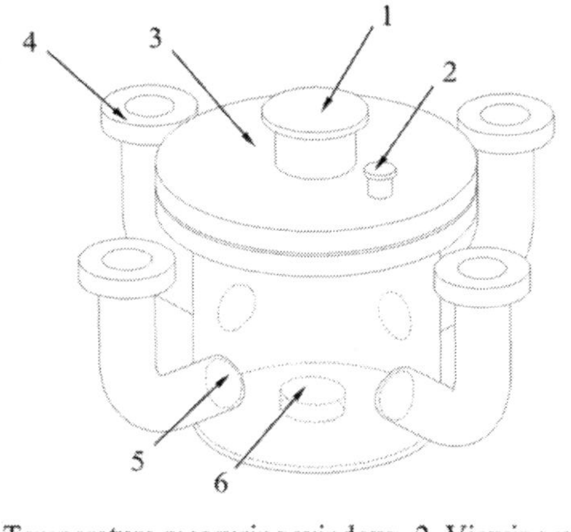

1_Temperature measuring window; 2_Viewing window;
3_ISO-K Flange; 4_CF Flange; 5_Airflow inlet and outlet;
6_Underlay;

Figure 1. Schematic diagram of chamber structure.

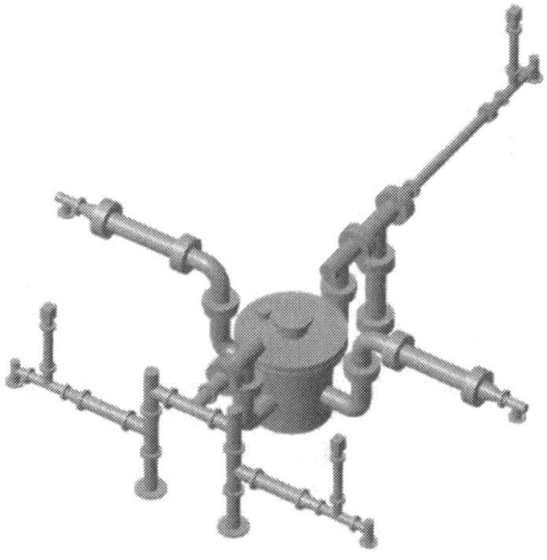

Figure 2. Overview of the self-developed MLD device.

Four right-angle bending tubes are welded at the cylindrical main body of the chamber [9]. The other ends of those four bending tubes are welded with a CF flange. The CF flange can be used to connect various modules, allowing the construction of a complete deposition device. As shown in Figure 2, a 3D model of the MLD device is constructed by the chamber and various modules. Those modules are composed of a large number of standardized parts, which are easy to assemble, disassemble, and replace.

The four right-angle bending tubes are used as inlets or outlets of the chamber. Among them, two are precursor delivery inlets, one is a nitrogen inlet, and the other is a vacuum pumping outlet. Those four bending tubes are DN35 tubes, and the flanges welded on them are CF35 flanges.

2.2. Deposition process program

First, the gas with a vapor pressure of 0.5 KPa [10-11] can be released and delivered to the chamber by heating the organic precursor. Thus, 0.5 KPa is used as the nominal pressure for gas delivery. Second, vacuum pumping and nitrogen delivery work together during the nitrogen purge process to remove excessive molecules on the surface of the substrate. Then, the vacuum pump evacuates the chamber separately to meet the vacuum degree needed for molecular deposition of the next round.

The chamber is submerged in oil, which is used for external oil bath heating. As shown in Figure 3, a safety gap is kept between the oil surface and the flange welded on the right-angle bending tube, to avoid oil seeping into the chamber as oil fluctuates or boils.

Figure 3. On-site graph of the safety gap between the oil-bath surface and flange of the right-angle bending tube.

As the core unit of the MLD device, the chamber needs to meet the following specifications:

(1) The precursor gas flows through the substrate on the underlay with a suitable velocity during precursor deposition and nitrogen purge.

(2) The temperature around the substrate should be high enough to meet the reaction temperature between the precursor and the substrate surface. Meanwhile, the temperature around the fluoro rubber ring should be lower than its extreme temperature values.

The requirements for the actual machining of the chamber include:

(1) Chamber feature height range of 170~190 mm ensures structural support in vacuum environments, enhancing stability and durability.

(2) The inlet and outlet feature height is not less than 35 mm to ensure that the right-angle bend has enough welding space.

3. Simulation analysis setup

3.1. Modeling and meshing

To improve the calculation efficiency and reduce the calculation time, according to the characteristics of the above chamber model, under the premise of ensuring the simulation accuracy, the chamber is simplified as:

(1) We remove the flanges welded to the four right-angle bending tubes and the blind flange on the top of the chamber.

(2) The fluoro rubber ring is replaced by a circular L-plane at the same position.

(3) A substrate for the deposition with a radius of 50 mm and a thickness of 1 mm is added to the underlay.

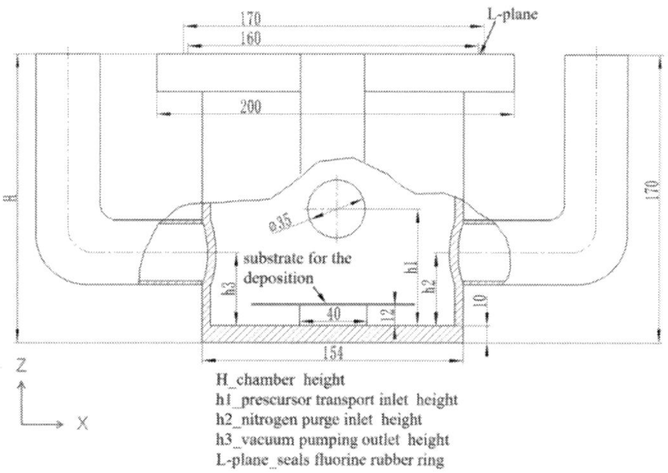

Figure 4. 2D Dimensional diagram of the Z-X view of the chamber.

The simplified chamber is shown in Figure 4, where H, h1, h2, and h3 are used to denote chamber axial height, precursor delivery inlet height, nitrogen purge inlet height, and vacuum pumping outlet height, respectively, and their tentative parameters are 170 mm, 60 mm, 50 mm, and 50 mm, respectively.

Creo Parametric software is used to construct the 3D model of the chamber, as shown in Figure 5. That 3D model was imported into COMSOL software to construct the fluid domain inside the chamber, as shown in Figure 6.

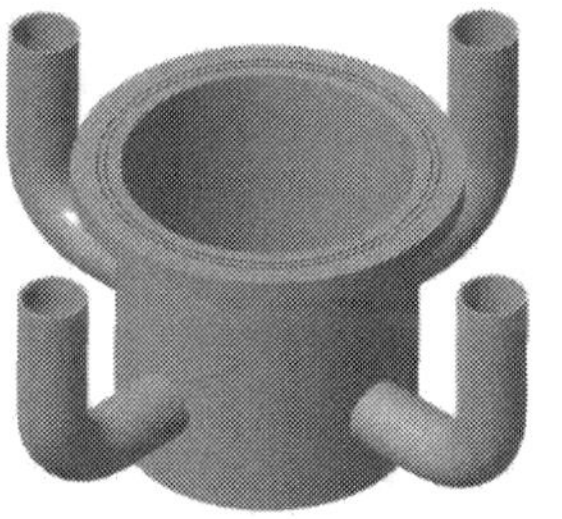

Figure 5. 3D model of the chamber.

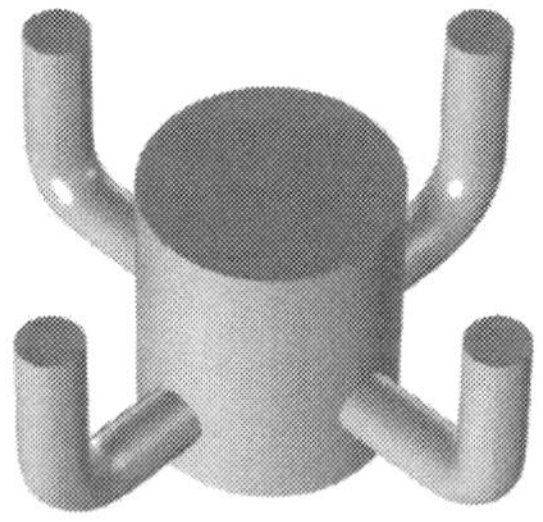

Figure 6. The fluid domain inside the chamber figure.

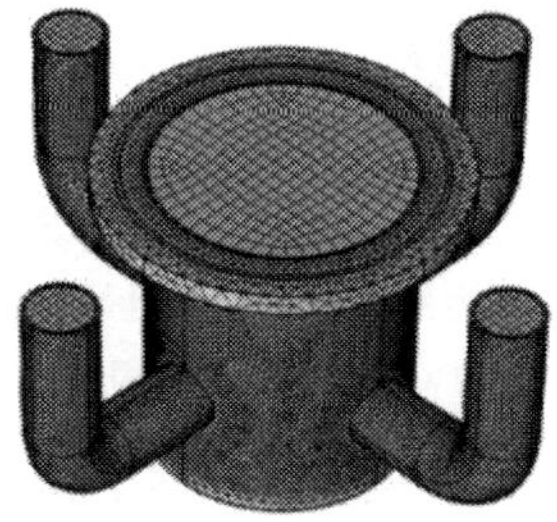

Figure 7. The meshing of a 3D model and fluid domain.

The fluid domain inside the chamber is meshed by hexahedral meshes and the 3D model of the chamber is meshed by using unstructured tetrahedral meshes, as shown in Figure 7. The COMSOL software can automatically update the mesh for changes in different structural parameters. The number of meshes is controlled between 200, 000 and 300, 000 meshes. The proportion of hexahedral meshes is kept at 50% to 70%, to improve the calculation efficiency.

The material of the chamber is set as [Steel AISI 4340], nitrogen gas as [Nitrogen gas], precursor as [C2H5OH gas], and the substrate for the deposition as [Silicon solid bulk].

Journal of Physics: Conference Series **2879** (2024) 012007 doi:10.1088/1742-6596/2879/1/012007

3.2. Boundary condition setup

The simulation is modeled by using COMSOL's built-in control functions: precursor delivery using the "turbulence, k-ε" model, nitrogen purge and vacuum pumping using the "laminar flow" model, and oil bath heating using a coupled "solid and fluid heat transfer" and "turbulent flow" model. The simulation model boundary conditions are listed in Table 1.

Table 1. Model boundary conditions.

Model	Boundary conditions
Suitable for all models	Gases are compressible; No slippage of chamber walls; The gravitational force G is 9.8 N/kg; Ambient temperature of 20°C.
Precursor delivery	Pressure inlet 0.5 Kpa; Velocity exit 0 m/s.
Nitrogen purge and vacuum pumping	Mass inlet 300 mL/min; Pressure outlet -1 atm.
Chamber oil bath heating	Oil bath immersion depth of 140 mm; Oil bath temperature of 250°C; Chamber surface heat transfer coefficient 7 W/(m^2·K); Total heating time in oil bath 70 min.

4. Results and discussion

4.1. The effects of inlet and outlet height

The two precursors provided in this experiment have the same vapor pressure, so the simulation analysis of only one precursor delivery for the deposition is used to explore and confirm the height parameters for the two precursor delivery inlets. The simulation of vacuum pumping together with a nitrogen purge is used to explore and confirm the height parameters of the nitrogen purge inlet and vacuum pumping outlet.

4.1.1. Analysis of precursor delivery. A low flow velocity of the precursor gas across the substrate surface can result in insufficient precursor supply on the substrate, which should reduce the deposition efficiency; On the other hand, if the flow velocity is too high, the substrate on the underlay may be shaken or blown over. Therefore, during the precursor deposition process, the gas flow rate passing over the substrate is limited to the index range of 0.8~1.6 m/s.

Figure 8 shows the Z-Y cross-section of the chamber and the velocity boundary setup during precursor deposition simulation. In the figure, Point A is located 2 mm over the substrate. Figure 9 shows the change in velocity of the precursor flow passing through Point A for various h1 values. The flow velocity is within the index range while h1 is 55 mm, 50 mm, 45 mm, and 40 mm.

Figure 10 then gives the streamlined distribution of precursor delivery. As h1 decreases, the number of streamlines above the substrate surface gradually increases and the spacing of streamlines gradually decreases. It can be known by comparing those four graphs in Figure 10:

(1) Figure 10 (h1=55 mm) shows a low number of flow lines above the surface of the substrate with a large spacing, which may reduce the deposition efficiency.

(2) Figure 10 (h1=50 mm) shows a higher number of flow lines above the substrate with a smaller spacing, which may improve the deposition efficiency.

(3) Figure 10 (h1=45 mm) shows the presence of a vortex on the anterior side of the substrate, which may lead to shakes of the substrate.

(4) Figure 10 (h1=40 mm) shows a portion of the gas flow impinging directly on the underlay, resulting in the gas flow being reflected and impinging upward on the substrate, potentially blowing over the substrate.

In summary, when h1 is finally selected to be 50 mm, it not only meets the requirement of velocity index but also effectively avoids vortex and impacting on the substrate.

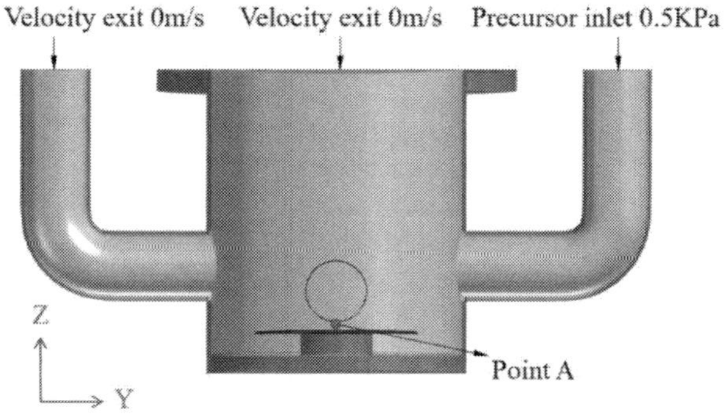

Figure 8. Z-Y cross-section of the chamber.

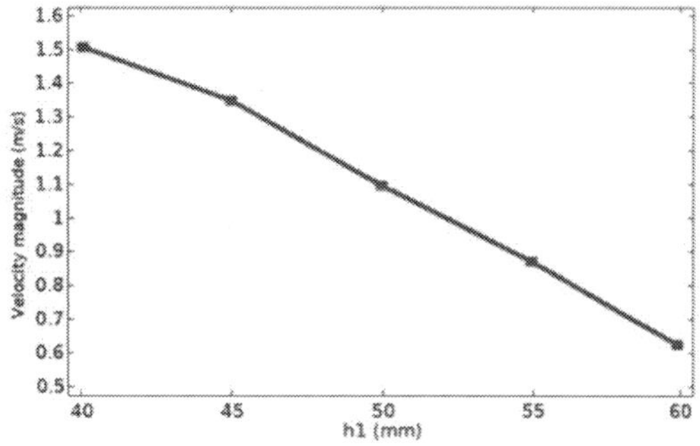

Figure 9. Velocity change of precursor gas flow through point.

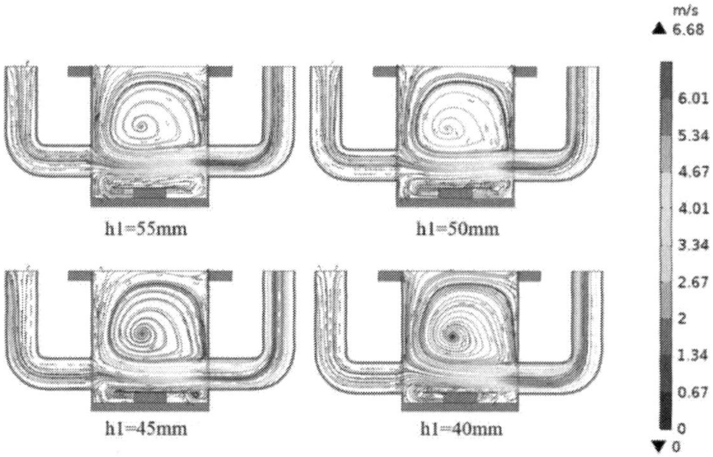

Figure 10. Velocity streamlines distribution graphs in Z-Y cross sections for precursor deposition at different h1s.

4.1.2. Analysis of nitrogen purge and vacuum pumping results. Nitrogen purge with slow nitrogen flow velocity over the substrate surface can result in an insufficient cleanup for the excessive molecules which in turn affects the quality of the film. On the other hand, if the flow velocity is too high, the substrate on the underlay may be shaken or blown over, it even causes an energy shock to the film, destroying the film structure. Therefore, the flow velocity should be limited to the range of 6-8 m/s during nitrogen purge together with vacuum pumping.

Figure 11 shows the Z-Y cross-section of the chamber and the velocity boundary setup during the nitrogen purge together with vacuum pumping. In the figure, Point B is located 2 mm over the substrate. The initial outlet pressure for vacuum pumping is set as -1 atm, since the chamber should have been filled with precursor molecules before this process. The flow velocity of the nitrogen inlet is set at 300 mL/min, according to the rated airflow of the nitrogen generator. Table 2 shows the simulated result of the velocity of the nitrogen passing through Point B for various h2 and h3 values. As shown in Table 2, when h3 is 50 mm and h2 is 40 mm, 45 mm, and 50 mm, respectively, the nitrogen flow velocities are all within the index range of 6-8 m/s.

Figure 12 gives the distribution of nitrogen flow streamlines for the nitrogen purge process. It can be known by comparing those four graphs in Figure 12:

(1) Figure 12 (h2=50 mm) shows the presence of a significant vortex on the anterior side of the substrate, which may lead to tremors of the substrate.

(2) Figure 12 (h2=45 mm) and (h2=40 mm) shows streamlines with the ideal number and density of streamlines above the substrate, as well as the absence of a vortex. It may enhance the efficiency of nitrogen purge.

In summary, when h3 is 50 mm and the corresponding h2 is 45 mm and 50 mm, the nitrogen purge process would be efficient and reliable. The final selection of h3 is 50 mm.

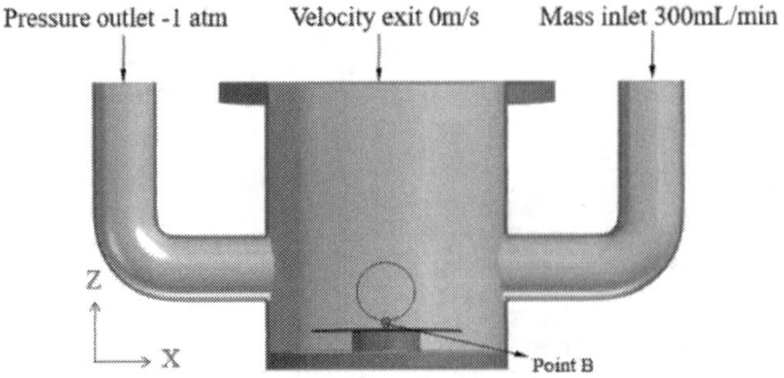

Figure 11. Schematic diagram in Z-X cross-section of nitrogen purge together with vacuum pumping.

Table 2. Flow velocity results under different h2 and h3 values (m/s).

	h3=50 mm	h3=45 mm	h3=40 mm
h2=50 mm	6.7	8.2	9.2
h2=45 mm	7.0	8.1	8.5
h2=40 mm	7.36	9.3	9.5

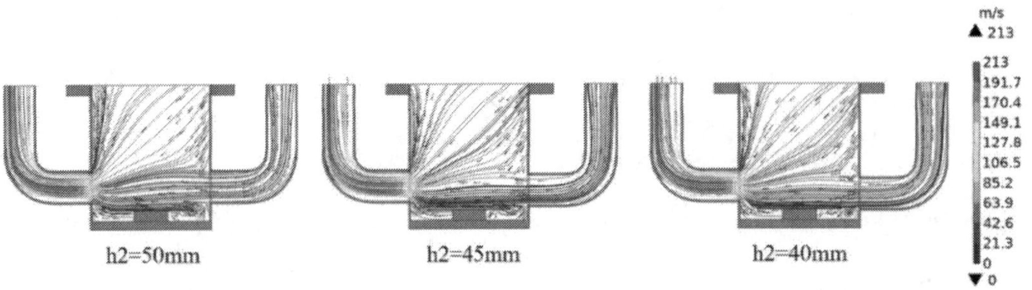

Figure 12. Velocity streamlines distribution graphs in Z-X cross sections during nitrogen purge together with vacuum pumping while h3 = 50 mm with valorous h2 values.

4.2. Chamber height analysis

Low temperature can lead to condensation of the precursor on the walls of the chamber, while the excessive temperature can cause thermal damage to the fluoro rubber ring. The bearing temperature of the fluoro rubber ring should be lower than 190°C, while the wall temperature is required to be higher than 180°C.

Figure 13 shows a schematic model of the chamber submerged in an oil bath, wherein the part of the chamber is exposed to the air above the soil surface, and the part is submerged in the oil bath below the soil surface. Since the oil bath tank of the oil bath device is fixed, the height of the oil bath surface is fixed. When the H of the chamber increases, the height of the exposed part of the chamber to the air will be increased. The oil bath immersion depth was maintained at 140 mm, and the oil bath heating temperature reached 250°C after 40 min and was kept at 250°C for 30 min.

Figure 14 gives the temperature variations of M being 30, 35, 40, 45, and 50 mm, respectively, corresponding to H = 170, 175, 180, 185, and 190 mm, respectively. The legend temperature range is 174 to 250°C for all five figures. When H is 190 mm and M is 50 mm, a region with a minimum temperature of 174°C appears, failing to meet temperature requirements, resulting in condensation of precursors on the chamber wall.

Figure 15 shows a circular L-plane representing the position of the fluoro rubber ring. Figure 16 gives the variation of the L-plane temperature as the H-parameter is raised. As seen in Figure 16, the temperature at the L-plane decreases gradually as H increases. When the H value is 185 mm or 190 mm, the temperature of the fluoro rubber ring meets the requirement (under 190°C).

In summary, to ensure that the precursor entering the chamber maintains a good temperature, the H parameter with a higher wall temperature should be selected, combined with the temperature requirements of the fluoro rubber ring, and the final H selected 185 mm.

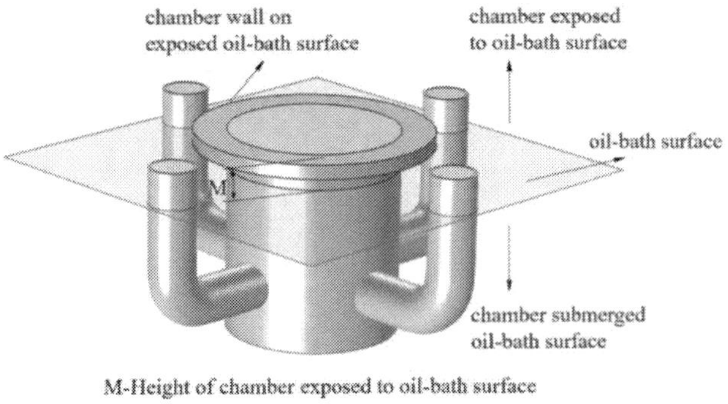

Figure 13. Schematic model of oil bath of the reaction chamber.

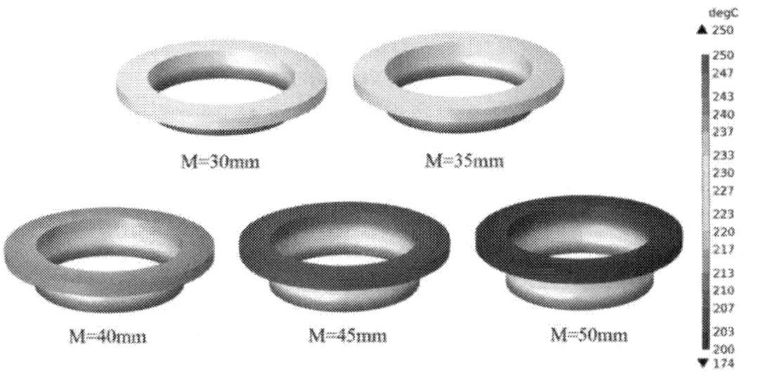

Figure 14. Wall temperature variation with different M values for the part exposed upon the soil surface.

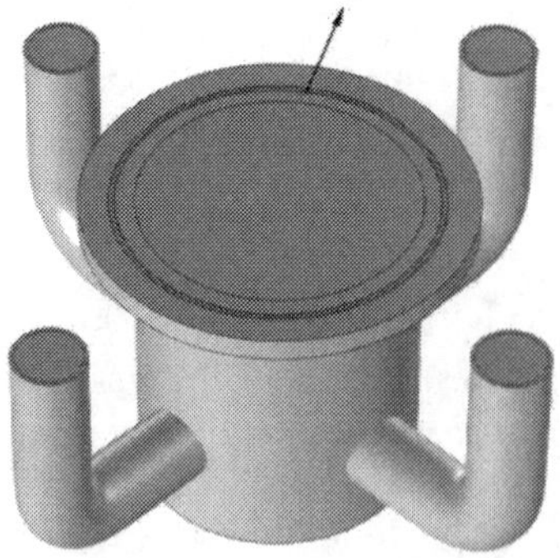

Figure 15. The location of the L-plane on the chamber representing the fluoro rubber ring.

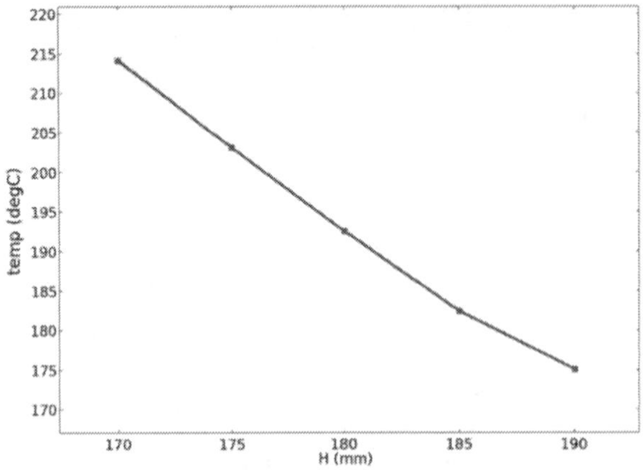

Figure 16. Variation of the L-plane temperature as H differs.

5. Conclusion

A molecular layer deposition chamber for the preparation of organic polymer films was designed according to the working principle of the MLD device and the requirements of the deposition process. Through modeling and simulation analysis of structural parameters, relevant studies have shown that:

(1) The lower the precursor inlet position is, the more efficient the deposition reaction is, but at the same time, it also increases the possibility of vortex or substrate shake.

(2) Vacuum pumping-assisted nitrogen purge significantly improves purge efficiency, but at the same time, it also increases the possibility of vortex or substrate shake.

(3) Elevating the axial height of the chamber can effectively reduce the temperature of the fluoro rubber ring, but at the same time, this causes the temperature decrease of the chamber wall, thereby increasing the risk of precursor condensation.

Four structural parameters were finally selected: the height of the two precursor delivery inlets was selected to be 50 mm; the height of the nitrogen purge inlet was 45 mm; the height of the vacuum pumping outlet was 50 mm; the axial height of the chamber was selected to be 185 mm. These parameters are chosen to optimize the deposition performance and to ensure the effectiveness and stability of the deposition process.

References

[1] Lee S. H., Jeong H. J., Han K. L., Baek G., and Park J. S. (2021). An organic-inorganic hybrid semiconductor for flexible thin film transistors using molecular layer deposition. Journal of Materials Chemistry C, 9 (12), 4, 322-4, 329. [CrossRef]

[2] Bo Z. and Jin L. (2023). Atomic/molecular layer deposition technology and its application in advanced semiconductor processes [J/OL]. Materials Herald, 1-27 [2024-07-22]. [CrossRef]
[3] Koponen S. E., Gordon P. G., and Barry S. T. (2016). Principles of precursor design for vapor deposition methods. POLYHEDRON, 108. 59-66. [CrossRef]
[4] Zhang B., Wang Z., Wang J., and Chen X. (2024). Recent Achievements for Flexible Encapsulation Films Based on Atomic/Molecular Layer Deposition. Micromachines, 15 (4), 478. [CrossRef]
[5] Higgs D. J., DuMont J. W., Sharma K., and George S. M. (2018). Spatial molecular layer deposition of polyamide thin films on flexible polymer substrates using a rotating cylinder reactor. JOURNAL OF VACUUM SCIENCE & TECHNOLOGY A, 36 (1). [CrossRef]
[6] Selvaraj S. K., Jursich G., and Takoudis C. G. (2013). Design and implementation of a novel portable atomic layer deposition/chemical vapor deposition hybrid reactor. Review of Scientific Instruments, 84 (9). [CrossRef]
[7] Xiaojuan W., Yihong D., and Xiao T. (2024). Advances in the study of strong hydrogen bonding %J Liaoning Chem. 53 (02), 250-253. [CrossRef]
[8] Bhat G. S., Qayoum A., and Saleem S. S. (2024). Multi-parameter assessment of fluid flow and heat transfer in a spinning disk: <i>COMSOL Multiphysics</i>® simulation and experimental insights. MULTISCALE AND MULTIDISCIPLINARY MODELING EXPERIMENTS AND DESIGN. [CrossRef]
[9] Junliang F., Li X., Yueting L., Gang C., Xiaolin Z., Yi T., and Hengxiang G. (2023). Optimized design and implementation of an atomization-assisted CVD chamber % J. Journal of Engineering Design, 30 (02), 182-188. [CrossRef]
[10] Yuan Y., Ping H., Lee D. T., Corkery P., Zhang Z., Liu C., Xue S. M., Zhuang L., and Tsapatsis M. (2023). CFD Simulation of the Dosing Behavior within the Atomic Layer Deposition Feeding System. INDUSTRIAL & ENGINEERING CHEMISTRY RESEARCH 62 (23), 9, 335-9, 347. [CrossRef]
[11] Yun S., Tom M., Ou F., Orkoulas G., and Christofides P. D. (2022). Multiscale computational fluid dynamics modeling of thermal atomic layer etching: Application to chamber configuration design. Computers & Chemical Engineering, 161, 107, 757. [CrossRef]

Journal of Physics: Conference Series **2879** (2024) 012008 doi:10.1088/1742-6596/2879/1/012008

Injection over-molding process of PEEK composites

Yufan Yang[1,2], Nan Hu[1], Zijie Wu*[1], Jie Hao*[1] and Lili Qiu[2]

[1]Aerospace Institute of Advanced Materials Processing Technology, Beijing, 100074, China

[2]Beijing Institute of Technology, Beijing, 100081, China

*Corresponding author's e-mail: zijie.wu@outlook.com(Zijie Wu); haoj15@tsinghua.org.cn(Jie Hao)

Abstract: This study utilizes experimental and microscopic characterization methods to investigate the prospects and feasibility of using the Injection Over-Molding (IOM) process to replace traditional adhesives in the manufacturing of complex ribbed panel structures with Polyether Ether Ketone (PEEK) composites. The results indicate that the IOM process can achieve the optimal bonding strength of adhesives while significantly reducing the processing cycle, demonstrating substantial potential. Additionally, enhancing the melt temperature and mold temperature can further improve interface bonding quality and increase bonding strength. Based on the experimental results, it is inferred that increasing the melt temperature can elevate the interface temperature while raising the mold temperature can slow down the interface cooling rate, suggesting mechanisms by which these parameters influence interface bonding quality.

1. Introduction

In recent years, carbon fiber-reinforced thermoplastic composites (CFRTP), particularly Polyether Ether Ketone (PEEK), have shown promise in aerospace applications due to their strong mechanical properties and stability [1, 2]. PEEK has already been used in aerospace products such as the Airbus A380 [3] and Westland helicopters.

Stiffened panels are often used in aerospace vehicles as structural components. The traditional manufacturing method involves forming the panel and the stiffener structure separately, followed by assembly and connection [4]. Currently, adhesive bonding is the prevalent connection method. Compared to mechanical connections such as riveting, adhesive bonding does not damage the internal fiber structure of the material [5], reduces the likelihood of stress concentration [6], and adds minimal weight, which is beneficial for reducing the aircraft's weight. In recent years, for complex structures like stiffened panels, researchers have proposed the Injection Over-Molding (IOM) technique for thermoplastic composites. IOM is a relatively novel forming and bonding process that combines hot press forming and injection molding, leveraging the advantages of thermoplastic materials, such as rapid forming and suitability for complex structures. In IOM production, continuous fiber-reinforced composite hot press forming is typically used as the baseplate to ensure the overall mechanical performance of the specimen. The baseplate is then embedded into an injection mold, where short fiber-reinforced composites are rapidly molded onto its surface through injection molding, forming complex structures. The connection between the complex structure and the baseplate is completed in a single injection step, significantly shortening the manufacturing process cycle. For the process itself, it is evident that a critical issue in IOM is the interface bonding quality between the insert and the injection-molded component.

Currently, a significant amount of research has been conducted by scholars on the IOM (Insert Over-Molding) forming process, primarily characterizing the interface bonding quality of IOM processes through methods such as single-lap joint tests or three-point bending tests and exploring the effects of process parameters on bonding quality. Lafranche et al. [7] investigated the impact of melt temperature and mold temperature on the bonding quality of IOM specimens based on a polypropylene (PP) material system. Lu et al. [8] also worked with a PP material system and studied the effects of melt temperature, mold temperature, and holding pressure on bonding strength. Fu et al. [9] used continuous fiber-reinforced polypropylene as the thermoformed preform and injection-molded short fiber-reinforced nylon 66 (PA66) onto it, investigating the effects of melt temperature, preform heating temperature, and mold temperature on the interface bonding quality of IOM specimens. Neveu et al. [10] examined the tensile shear properties of lap joint specimens under different resin systems, including polyetherimide (PEI)/polyphenylene sulfide (PPS), PPS/PPS, and PEI/PEEK, discovering that heterogeneous interfaces exhibited weaker bonding than those of the same material. Sun et al. [11], based on polymethyl methacrylate (PMMA) and polycarbonate (PC) material systems, found that appropriately increasing the holding pressure could enhance the IOM interface bonding. Research by Islam et al. [12] indicated that increasing the surface roughness of specimens effectively improved the bonding between PC and ABS. Boros et al. [13] enhanced the IOM bonding strength between an ABS insert and PA6 by performing plasma activation treatment on the surface of the ABS insert.

Most current research focuses on general engineering plastics such as PP, PA6, and PA66, with relatively few studies on the specialized engineering plastic PEEK, which is commonly used in the aerospace field. This paper aims to explore the application prospects and feasibility of replacing traditional adhesive bonding with the IOM forming process for PEEK material systems in the molding of stiffened panel structures. In this study, the connection between the stiffener and the panel in engineering is simplified to a single lap joint specimen for characterization. The impact of process parameters on the bonding quality of specimens is studied using tensile shear tests combined with microscopic characterization., thereby confirming the feasibility of this process for engineering applications

2. Experimental study on injection over-molding of PEEK

2.1. Experimental material
Continuous fiber reinforced polyether ether ketone prepreg (TZ700S/PAEK): Heilongjiang Yingchuang Materials Co., LTD. Short fiber reinforced polyether-ether ketone injection plastic (330G): Jilinzhongyan Polymer Materials Co., LTD. Adhesive (J133, J135, J47): Heilongjiang Petrochemical Research Institute.

2.2. Experimental equipment
Injection molding system (MA5300II/4000): Ningbo Haitian Plastic Machinery Group Co., Ltd.; Hot press: Xuzhou Pressure Machinery Co., Ltd.; Material mechanics test system (TSE 105D): Wantse Company; Field emission scanning electron microscope (Sigma300): Zeiss Group, Germany; Metallographic microscope (DM4M): Leica, Germany.

2.3. Preparation of IOM sample
For complex stiffened panel structures, the bonding between the stiffener and the panel is characterized using tensile shear lap joint specimens. The IOM specimens are designed according to the GB/T7124-2008 standard for determining the tensile shear strength of adhesives, as shown in Figure 1(a). The dark-colored section represents the insert part, which is made using continuous fiber-reinforced polyether ether ketone (PEEK) prepreg through thermoforming, with a layup sequence of $[0°]_{14}$. After forming, the insert is embedded into an injection mold for injection molding, as shown in Figure 1(b), resulting

in the IOM specimen depicted in Figure 1(a). Both parts of the specimen have dimensions of 100×24×2 mm, with an overlap area of 25×12 mm. The injection molding experiments are conducted using varying process parameters, including melt temperature and mold temperature, to investigate the effects of these parameters on the bonding quality of the IOM.

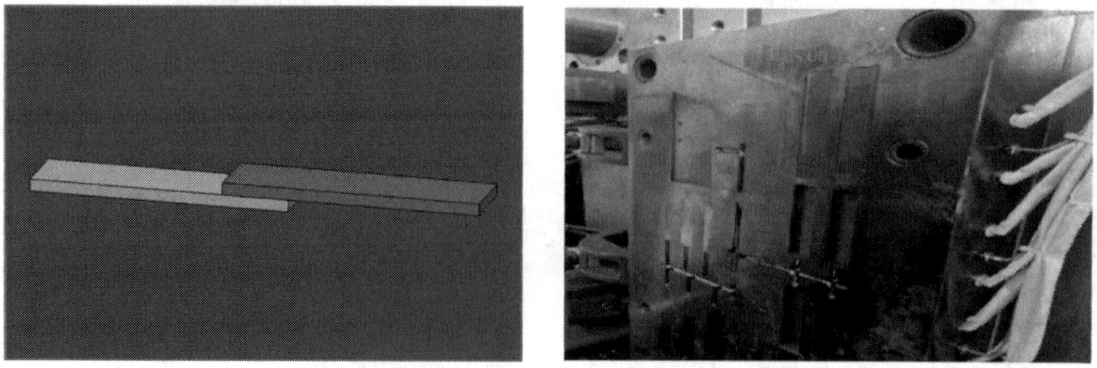

Figure 1. Injection over-molding. (a) Experimental sample; (b) Experimental process.

2.4. Preparation of adhesive bonding sample

Adhesive bonding specimens are prepared for comparison with the bonding strength of the IOM specimens to explore the feasibility of replacing adhesive bonding with the IOM process. Preformed injection-molded and thermoformed specimens are cleaned with acetone, with dimensions identical to the IOM specimens. Three types of epoxy adhesives commonly used in the aerospace field are selected. After mixing the adhesives in the specified ratios, they are evenly applied to the bonding interface, with the interface area being the same as in Section 2.3. Finally, the bonded area is clamped with a dovetail clip and wrapped with high-temperature tape for fixation. The specimens are cured at room temperature or elevated temperatures according to the specifications of the different adhesives.

2.5. Testing and characterization

2.5.1. SEM microstructural characterization. We select the bonding interface surface of the specimens after tensile shear failure. After gold sputtering treatment, the specimens are adhered to the sample stage with conductive adhesive and placed into the scanning electron microscope (SEM). Microstructural observations are conducted under vacuum conditions with an accelerating voltage of 10 kV.

2.5.2. Metallographic microscope interface morphology characterization. We cut the bonding area of the IOM specimens for use as samples. The interface bonding between the insert and the injection-molded part is observed using a metallographic microscope (10X). Mix epoxy and curing agent in a specified ratio, place the sample in a mold, and pour the prepared resin. The sample is then cured at elevated temperatures in an oven. After curing, the sample is ground and polished sequentially with sandpaper from coarse to fine grit, followed by polishing with a polishing cloth. Finally, the prepared sample can be used for metallographic observation.

2.5.3. Tensile shear strength testing. The TSE105D material mechanics testing system is used to apply a tensile force parallel to the bonding surface and along the main axis of the tensile shear specimens, testing the lap joint interface strength of the material. The lap joint interface dimensions of the specimen are 12.5×25 mm. The formula for calculating tensile shear strength is as follows:

$$\sigma_S = \frac{F}{S} \tag{1}$$

where σ_s is the tensile shear strength; F is the failure load; and S is the lap area.

3. Results and discussion

3.1. Adhesive bond strength

Table 1 lists the tensile shear strength test results and failure modes of adhesive bonding specimens with different grades of adhesives.

From the test results, it can be seen that the bonding strength of adhesives to PEEK specimens ranges from approximately 6 to 11 MPa, with J47 and J135 adhesives performing better, achieving tensile shear strengths around 10 MPa. All specimens exhibited adhesive failure modes, suggesting that due to the inert nature and low surface energy of the PEEK material, the adhesives could not achieve good bonding with the PEEK surface. The failure occurred at the interface between the adhesive and the PEEK material.

Table 1. Adhesive tensile shear strength.

Adhesive	Bonding strength/MPa	Failure mode
J47	9.79±0.4	Adhesive failure
J135	11.12±1	Adhesive failure
J133	6.13±1.4	Adhesive failure

3.2. Effect of melting temperature on interface joining quality of IOM

Based on the adhesive process data in Section 3.1, IOM experiments are conducted using this as a standard. The study explores the impact of process parameters on the bonding quality of IOM and the feasibility of replacing the adhesive process with the IOM process.

By varying the melt temperature during injection molding, a single-factor experiment is conducted to verify the effect of melt temperature on the interface bonding quality in the IOM process. PEEK is a type of high-temperature-resistant specialty engineering plastic with a melting point of 343°C. The typical melt temperature range for injection molding is around 370-420°C. In this section, the melt temperatures of 380°C, 390°C, 400°C, and 410°C are selected for secondary molding experiments, with other process parameters kept constant, as shown in Table 2.

Table 2. Injection over-molding process parameters.

Parameter	Value
Mold tempreture/°C	180
Injection pressure/MPa	10
Holding pressure/MPa	4.5
Holding time/s	30

Five sets of data were collected for each melt temperature, and the average value was taken as the final result, as shown in Figure 2. The standard error and standard deviation of the data are indicated in Appendix A, to demonstrate the reliability of the data. Figure 3 displays the state of the tensile shear specimens before and after testing, as well as the different failure modes.

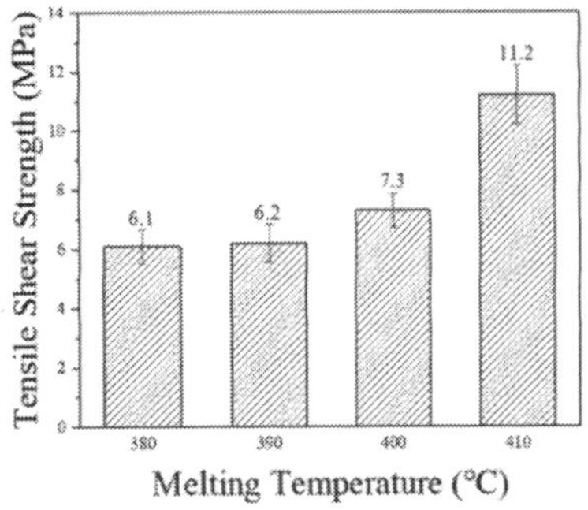

Figure 2. Tensile shear strength-melting temperature.

From Figure 2, it can be seen that with the increase in injection temperature, the tensile shear strength of the IOM specimens shows an upward trend, reaching the highest bonding strength of 11.2 MPa at 410°C. When the melt temperature increased from 380°C to 410°C, the interface bonding quality improved by 78%. Increasing the melt temperature of the injection resin in the IOM process can effectively enhance the interface bonding quality of IOM specimens. The experimental results for the IOM process indicate that increasing the melt temperature is beneficial for interfacial bonding. In the current system, the PEEK melt temperature should be set to 410°C, as it can provide an appropriate interface temperature for molecular chain entanglement, resulting in a robust bonding interface.

At lower melt temperatures, the interface bonding quality is poor. The tensile shear specimens exhibit cohesive failure modes, breaking into two parts at the bonding interface, as shown in Figures 3(c) and 3(d). However, when the temperature rises to 410°C, some tensile shear specimens exhibit a transition from cohesive failure to substrate failure, as shown in Figure 3(b). This indicates that the bonding strength at the interface exceeds the tensile strength of the substrate, further proving that increasing the injection temperature can effectively promote the interface bonding quality in the IOM process.

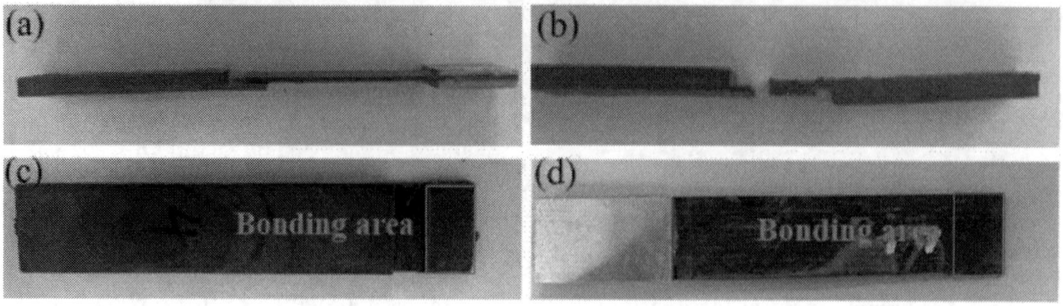

Figure 3. Tensile shear sample. (a) Before test; (b) Matrix failure after test; (c) Cohesive failure of injection part; (d) Cohesive failure of insert part.

The bonded regions of the tensile shear specimens were cut after testing. The surface morphology of the lap joints after failure was observed using SEM. The results are shown in Figure 4. From the SEM images, it can be seen that at lower temperatures, the fracture surfaces on both the injection molding side and the insert side of the IOM specimens are relatively flat and smooth, displaying only a few pits formed due to resin bond failure. The interface bonding of the short-cut fibers on the surface is also not evident. As the temperature increases, as shown in Figure 4(b), more short-cut fibers can be observed at the bonding interface on the injection molding side. This is because increasing the melt temperature reduces the viscosity of the melt, making it easier for the short-cut fibers within the injection material to move to the bonding interface. As shown in Figures 4(c) and (d), when the melt temperature rises from 380°C to 410°C, the surface roughness of the bonding area on the insert significantly increases. This is because the higher temperature of the injection material leads to a tighter melt bond with the resin on the insert surface, resulting in more pronounced defects and undulations upon failure. Additionally, more interspersed short-cut fibers can be observed on the insert surface, including fibers remaining on the surface and pits left by fibers being pulled out. This confirms that the two-phase resins have bonded more fully, with more short-cut fibers from the injection material moving to the IOM specimen's bonding interface, corroborating the results shown in Figure 4(b).

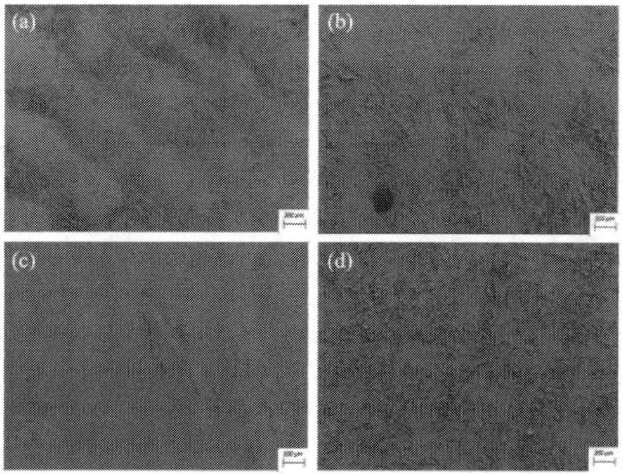

Figure 4. Surface morphology of sample failure. (a) 380°C injection part; (b) 410°C injection part; (c) 380°C insert part; (d) 410°C insert part.

Figure 5 shows the lap joint interfaces of IOM lap joint specimens observed through a metallographic microscope at low and high melt temperatures, respectively.

In Figure 5(a), a distinct black region can be observed at the interface between the injection-molded part (upper) and the hot-pressed insert part (lower). This is due to insufficient resin flow at the insert interface at lower melt temperatures, resulting in poor fluidity. As a result, it is difficult to adequately cover the lap joint area and melt enough of the insert surface resin to form a bonding layer. This leads to gaps and bonding defects, which appear as black regions in the metallographic image. Correspondingly, the interface at low temperatures, as shown in Figure 5, is relatively smooth, without obvious signs of resin melting and short fiber interspersion. As the temperature increases, the resin bonding at the interface gradually improves. At a temperature of 410°C, as shown in Figure 5(b), the defects at the interface have essentially disappeared. A good interface bond has formed between the injection-molded part and the insert surface. The resin is fully bonded, with no obvious gaps or voids.

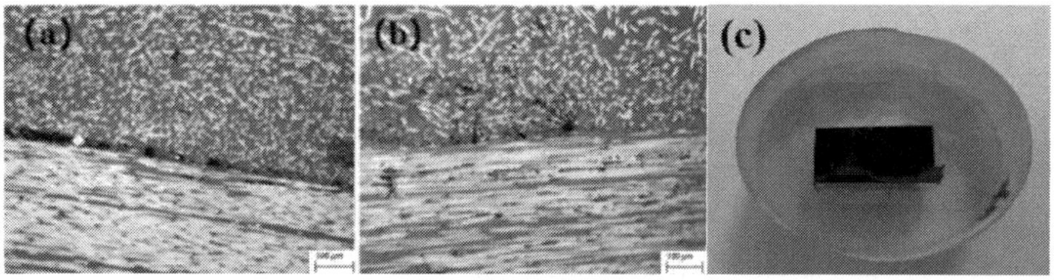

Figure 5. Overlap interface morphology of injection over-molding sample. (a) 380°C; (b) 410°C; (c) Metallographic sample.

Based on the microscopic characterization and experimental results, it is hypothesized that the mechanism by which melt temperature affects the quality of interfacial bonding operates through two main aspects. Firstly, increasing the melt temperature can reduce the viscosity of the injection material, making it easier for short fibers within the melt to move to the interface surface and create an interlocking effect. Additionally, the reduction in viscosity allows the melt to more effectively wet the surface resin of the insert, increasing the contact area and decreasing the likelihood of forming interfacial voids. Secondly, raising the melt temperature can increase the temperature at the secondary molding interface. The elevated interface temperature can effectively promote the thermal motion of molecular chains at the interface, accelerating the entanglement of molecular chains between the two-phase resins, ultimately forming a strong bonding interface.

3.3. Effect of mold temperature on interface quality of IOM

By varying the mold temperature in the IOM process, a single-factor experiment was conducted to verify the effect of mold temperature on the interface bonding quality in the IOM process. The mold temperatures were set at 170°C, 180°C, 190°C, and 200°C. The melt temperature was set at 390°C, with other parameters kept constant.

Five sets of data were collected for each mold temperature. The average value was taken as the final result, as shown in Figure 6(a). The standard error and standard deviation of the data are indicated in Appendix A, demonstrating the reliability of the data. As the mold temperature increases, the tensile shear strength of the IOM specimens shows an upward trend, demonstrating that increasing the mold temperature can effectively improve the interface bonding between the insert and the injection material in the IOM process. When the mold temperature is 170°C, the tensile strength is only 3.5 MPa, indicating poor interface bonding quality. When the temperature rises to 200°C, the tensile shear strength of the specimens reaches 11.2 MPa, an increase of 220% compared to 170°C. This indicates that the interface bonding quality in the IOM process is quite sensitive to mold temperature. Therefore, in the injection molding process, it is essential to increase the mold temperature as much as possible to improve interfacial bonding. The optimum mold temperature identified in this research is 210°C.

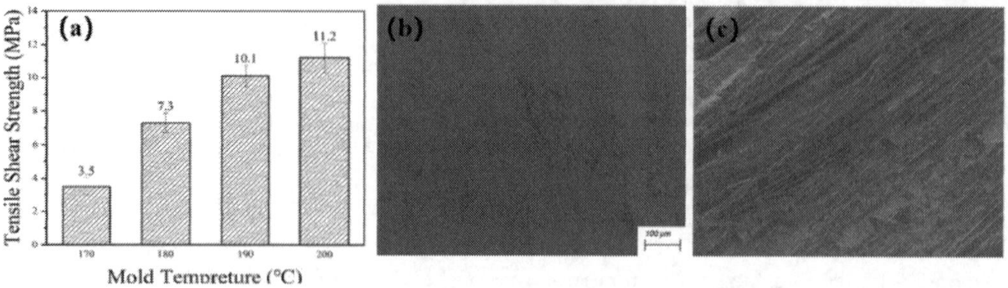

Figure 6. Experimental results. (a) Tensile shear strength-mold temperature; (b) Surface morphology of 170°C sample failure; (c) Surface morphology of 200°C sample failure.

The fracture surfaces of the tensile shear specimens were cut and observed using SEM to examine the surface morphology after failure, as shown in Figure 6(b) and Figure 6(c). When the mold temperature was 170°C, the fracture interface on the surface of the IOM specimen's insert was smooth and flat, with no obvious pits or defects left by resin peeling, indicating that the resin bonding at the interface was insufficient at this temperature, resulting in poor bonding quality. When the mold temperature was 200°C, pits caused by resin peeling and exposed continuous fibers within the resin could be observed at the interface. This demonstrates that at this mold temperature, the resin on the surface of the insert and the injection resin had melted well. During the tensile shear test, the resin bonded to the surface of the insert, and the injection part was pulled away, exposing the fibers beneath.

The results of metallographic observations are shown in Figure 7 and follow the same trend as the melt temperature results. When the mold temperature increased from 170°C to 200°C, the gap defects at the interface between the insert and the injection-molded specimens significantly decreased. The resins at the interface fully bonded and melted, forming a strong bonding interface, which is consistent with the trends observed in tensile shear strength and SEM results.

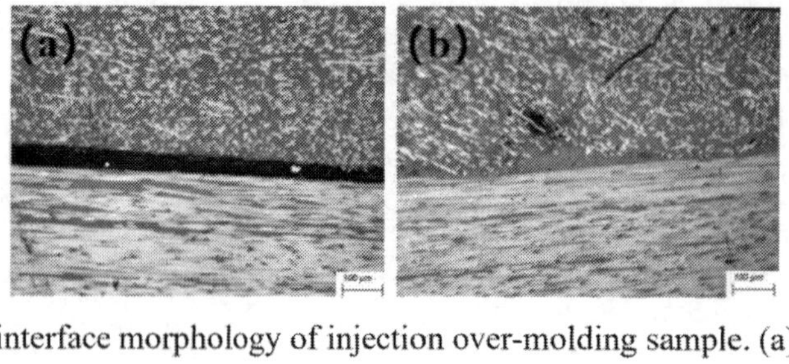

Figure 7. Overlap interface morphology of injection over-molding sample. (a) 170°C; (b) 200°C.

Both microscopic characterization and experimental results indicate that increasing the mold temperature is beneficial for IOM interfacial bonding. It is hypothesized that the reason is that in the current process system, the thermal conductivity between the mold metal and the material is greater than that between the materials. Increasing the mold temperature is more conducive to maintaining the interface temperature in the IOM process, extending the window period for resin molecular entanglement, and forming a strong interface bond.

3.4. Summary and prospect of the IOM process

Based on the above experimental validation, the IOM process has demonstrated feasibility for engineering applications. Table 3 compares the IOM process with the adhesive bonding process, highlighting its excellent application potential.

Table 3. Comparison of IOM and adhesive processes.

	Bonding time	Bonding strength	Additional weight	Flexibility
IOM	30s	11MPa	None	High
Adhesive	6h	11Mpa	Light	Low

The bonding in the IOM process is completed in one step through injection molding, with the bonding interface forming immediately after the injection and holding pressure phase, typically taking only about 30 seconds. In contrast, the adhesive bonding process requires an additional 6 hours for the adhesive to cure and form an effective bond after application. In the material system studied, both processes achieve an optimal bonding strength of around 11 MPa. Still, there is a significant difference in production efficiency, with the advantages of the IOM process being evident. Moreover, the traditional advantage of lightweight additional mass in the adhesive bonding process no longer holds compared to the IOM process. The bonding in the IOM process is directly formed by the melting and cooling of the material, with no added weight, which is more beneficial for reducing the weight of aircraft. Furthermore, the IOM process offers greater flexibility, is applicable to any injection molding material system, and allows for the enhancement of interface bonding strength by adjusting process parameters.

However, the IOM process also faces several challenges in practical applications. Firstly, the IOM process requires different specialized injection molds for manufacturing products of various shapes, which may pose cost issues when demand is low. Secondly, the IOM process has high requirements for molds; to ensure that the injection melt can fully wet the bonding interface, the design of mold runners, gates, and cooling channels must be optimized. Currently, the application of the IOM process is limited, and related mold design experience is lacking, necessitating further research and application trials. Finally, during the ejection and demolding of IOM products, there is a risk of the product sticking to the mold, leading to damage. To prevent this, it is necessary to design the draft angles of the mold appropriately and apply high-temperature release agents between the insert and the mold.

4. Conclusion

This study experimentally and numerically verified the application prospects of the IOM process in manufacturing complex stiffened panel structures in the aerospace field. The impact of process parameters on the bonding quality of IOM was investigated through single-factor experiments. The mechanisms were analyzed using characterization techniques. The following conclusions were drawn from this study:

1. For the PEEK material system, the IOM process can achieve the optimal bonding strength of 11 MPa, comparable to traditional adhesives. Moreover, compared to traditional adhesives, the IOM process eliminates the need to wait for adhesive curing, achieving molding and bonding in a single step, which saves 5-6 hours of manufacturing time and demonstrates excellent application potential.

2. Increasing the melt temperature of the injection material and the mold temperature can both effectively improve the bonding quality of the IOM process. This results in more resin molecules intertwining and bonding at the interface, leading to a tighter combination of the two phases.

3. Combining microscopic characterization with experimental results, it is evident that increasing the melt temperature primarily enhances molecular entanglement by raising the interface temperature, thereby improving bonding strength. Increasing the mold temperature, on the other hand, reduces the cooling rate at the interface, providing a longer window for molecular chain entanglement and forming a better interface bond.

Acknowledgment
I would like to express my deepest gratitude to the National Excellence Engineer Program for providing me with the invaluable opportunity to learn and gain practical experience in an enterprise setting. This program has significantly enriched my knowledge and skills, bridging the gap between academic theories and real-world applications. The support and resources offered by the program have been instrumental in my professional development. I am profoundly thankful for this extraordinary educational experience.

Appendix A

Table 4. Reliability analysis of melt temperature-bonding strength.

Melt temperature/°C	Bonding strength/MPa	SE	SD
170	6.1	0.55	1.23
180	6.2	0.63	0.58
190	7.3	0.57	1.28
200	11.2	1	2.24

Table 5. Reliability analysis of mold temperature-bonding strength.

Mold tempreture/°C	Bonding strength/MPa	SE	SD
170	3.5	0.53	0.71
180	7.3	0.58	1.3
190	10.1	0.65	1.45
200	11.2	0.91	2.03

Tables 4 and 5 present the reliability analysis of the data shown in Figures 2 and 6(a). As seen in the tables, the eight sets of data exhibit low dispersion, indicating high repeatability and reliability.

Reference

[1] Aamir Muhammad, Tolouei-Rad Majid, Giasin Khaled, Nosrati Ataollah. Recent advances in drilling of carbon fiber–reinforced polymers for aerospace applications: a review. *The International Journal of Advanced Manufacturing Technology*. 2019; **105**: 2289-2308. [CrossRef]

[2] Xu Jinyang, Huang Xianghui, Davim J. Paulo, Ji Min, Chen Ming. On the machining behavior of carbon fiber reinforced polyimide and PEEK thermoplastic composites. *Polymer Composites*. 2020; **41**: 3649-3663. [CrossRef]

[3] Chen Yali. Application of thermoplastic composites on aircraft. *Aviation Maintenance & Engineering*. 2003(03): 28-30. [CrossRef]

[4] Deng yunhui, Gu Chunhui, Yin Jin, et al. Design technology research of Heat-resistant composites for reusable aerospace vehicles structure. *Aerospace Materials & Technology*. 2024; **54**(02): 11-17. [CrossRef]

[5] Wei Liu, Ji Yang, Chunhui Yuan, et al. Study on interface bonding properties of CFRP-steel bonding at different temperatures *Composites Science and Engineering*. 2024: 1-8. [CrossRef]

[6] Gamdani Farid, Boukhili Rachid, Vadean Aurelian. Tensile behavior of hybrid multi-bolted/bonded joints in composite laminates. *International Journal of Adhesion and Adhesives*. 2019. [CrossRef]

[7] Lafranche Eric, Renault Thierry, Krawczak Patricia. Effect of the interdiffusion at the polymer/polymer interface on the flexural properties of over-moulded short glass fibre/glass fabric reinforced PA6 composites. *Key Engineering Materials*. 2014; **611**: 821-828. [CrossRef]
[8] Lu Yi, JIANG Bingyan, FU Liang, et al. Interfacial bonding strength between polypropylene and perform on in-mold hybrid injection molding technology. *Journal of Central South University (Science and Technology)*. 2018; **49**(07): 1626-1633. [CrossRef]
[9] FU Liang, JIANG Bingyan, WU Wangqing, et al. Experimental research on in-mold hybrid molding of glass fiber reinforced polypropylene and nylon66. *Journal of Central South University (Science and Technology)*. 2019; **50**(05): 1075-1081. [CrossRef]
[10] Neveu F, Cornu C, Olivier P, Castanié Bruno. Manufacturing and impact behaviour of aeronautic overmolded grid-stiffened thermoplastic carbon plates. *Composite Structures*. 2022; **284**: 115228. [CrossRef]
[11] Sun Qiwei, Chen Yuhong, Yan Yue, et al. Effect of overmolding injection parameters on warpage of PMMA-PC composite board. *Journal of Materials Engineering*. 2018; **46**(04): 134-139. [CrossRef]
[12] Islam Aminul, Hansen Hans Nørgaard, Bondo Martin. Experimental investigation of the factors influencing the polymer–polymer bond strength during two-component injection moulding. *The International Journal of Advanced Manufacturing Technology*. 2010; **50**: 101-111. [CrossRef]
[13] Boros Róbert, Tatyana Ageyeva, Golcs Ádám, Krafcsik Olga H, Kovács József Gábor. Plasma treatment to improve the adhesion between ABS and PA6 in hybrid structures produced by injection overmolding. *Polymer Testing*. 2022; **106**: 107446. [CrossRef]

Welding residual stress analysis based on ABAQUS direct coupling

Yunping Pan*, **Cheng Zhang and Zhongyin Feng**

School of Mechanical and Electrical Engineering, Wuhan University of Technology, Wuhan, 430000, China

*Corresponding author's e-mail: panyunping@whut.edu.cn

Abstract. Taking the welded joints of the drive axle shell and half shaft casing as the research object, based on the ABAQUS engineering software and its DFLUX subroutine, numerical simulation of the residual stresses of the welded joints before and after the heat treatment is carried out by adopting the direct coupling of temperature-displacement and the method of birth-death unit, to analyze the improvement effect of post-weld heat treatment on the residual stress of welded joints. The results show that the heat treatment process significantly reduces the equivalent peak stress of welded joints, from the original 195.25 MPa to 145.21 MPa, with a decrease of 25.6%. The axial peak stress of the welded joints decreased from 170.94 MPa to 130.84 MPa, with a decrease of 23.5%.

1. Introduction

The drive axle housing is one of the important parts of the drive axle. Drive axle housing is both the differential, main reducer, and wheel drive housing, but also the force transmitting parts, such as Figure 1 [1]. In this paper, we will study the Hubei axle—a certain light truck drive axle housing, which is a drive axle housing for the punch-welded monolithic axle housing. In this type of axle housing body for the 510 L steel plate stamping molding, the half shafts on both sides are connected to the axle housing body by gas-shielded welding. Punch-welded integral axle housing not only has a low production cost, but this type of axle housing has great bending, torsion, tensile strength, and stiffness. Its main advantage is that it is light in mass but has a higher reliability than the casting of the bridge [2].

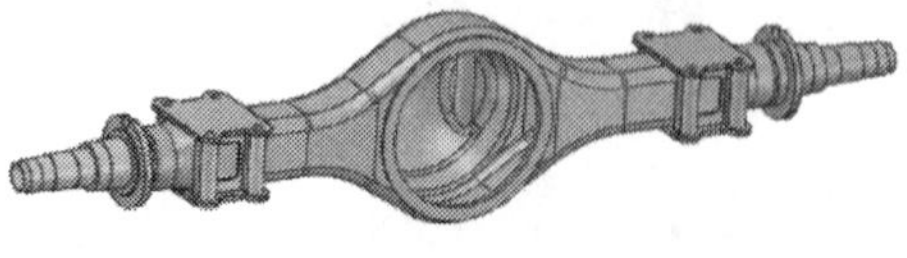

Figure 1. Three-dimensional modeling of the drive axle housing.

The welded parts of the drive axle shell are prone to welding residual stresses, stress concentrations, and various types of welding defects, which have an impact on the strength, stiffness, and life of the axle shell, and are therefore more susceptible to damage during use [3]. In this paper, with the ABAQUS software combined with the actual engineering on the half shaft casing and the bridge shell body welding process for numerical simulation, the study of welded joints welding residual stresses. In the process of actual production of punch-welded whole axle casing, the residual stress after welding is not eliminated. On this basis, the post-weld heat treatment of welded joints was carried out by ABAQUS software to study the effect of heat treatment on residual stresses. For the follow-up of the actual production process to join the heat treatment process to provide a reference, ABAQUS's unique powerful nonlinear analysis function for the study of complex welding process provides a convenient [4].

2. Numerical simulation of welding

2.1 Geometric modeling and meshing

The drive axle shell ring weld butt joint studied in this paper is welded by the axle shell body and half shaft casing Y-type butt bevel, as shown in Figure 2. Welded joints are symmetrically modeled and can be simplified to improve computational efficiency. The 1/4 geometric model of the welded joint is established and meshed using ABAQUS analysis software, refinement of the mesh near the center of the weld and the heat-affected zone, and thinning of the mesh away from the weld [5]. The whole welding process is divided into three main parts: bottom welding, intermediate welding, cover welding, and welding parameters in Table 1. The division of the mesh is processed accordingly, as shown in Figure 3.

Table 1. Welding process parameters.

Weld channel	Welding current (A)	Welding voltage (V)	Welding speed (mm/s)	Cooling time (s)	Protective gas
Bottom Welding	280~300	28~30	12	30	
Intermediate Welding	280~300	28~30	10	30	90%Ar+10%CO2 gas mixture
Cover Welding	280~300	28~30	8	To room temperature	

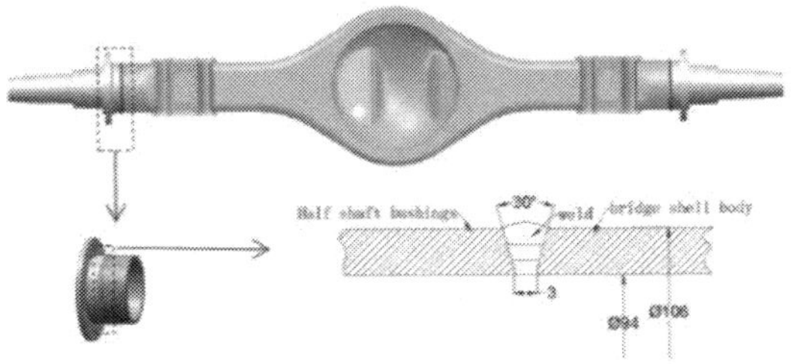

Figure 2. Welded joints for drive axle shells.

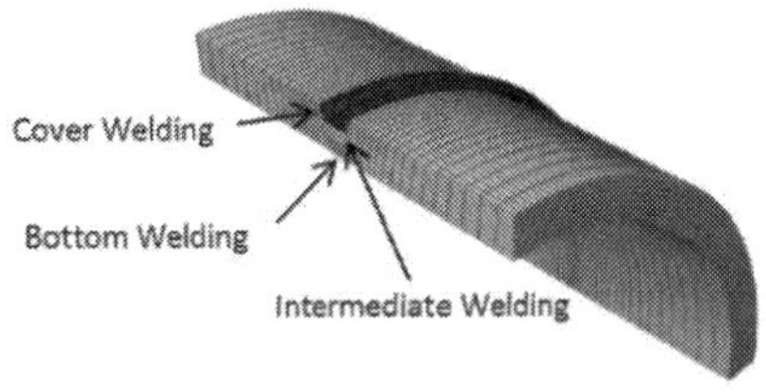

Figure 3. Meshing.

2.2 Methods of Analysis

Temperature-displacement direct coupling is used for simulation and analysis. The welding process considers the interaction of temperature and stress fields, and the grid cell used in the analysis process has both temperature and displacement degrees of freedom and uses the C38RT cell type. This method is more accurate than the sequential temperature field-displacement coupling method, which calculates the temperature field and then the stress field [6].

To make the simulated welding process more realistic, this simulation utilizes the life-death cell method to simulate the filling of the welding wire. While the welding heat source is moving, the corresponding analysis step is set up according to the welding speed, and when the welding heat source is loaded into the weld position in front of the moment, the script written in Python language will activate the cell set of the weld area step by step in sequence and activate the corresponding set of the weld metal to fill in [7-8].

2.3 Welding heat source

Welding heat input is highly concentrated, and in numerical simulation of the welding process, the heat input is usually equated to a heat source model that can be expressed as a mathematical expression. For different welding processes and welding structures, the use of appropriate heat source models is the key to accurate calculation of numerical simulation. At present, the commonly used heat sources are Gaussian plane heat sources for thin plate welding, cone heat sources for high-energy beam welding such as laser welding, particle beam, and electron beam welding, as well as double ellipsoid heat source models for large melting depth of melting electrode gas-shielded welding, hand electric arc welding, tungsten-arc welding and so on. The welding of the axle shell and axle tube uses the melting electrode gas shielded welding (MAG) welding method, and the heat flow density before and after the arc is inconsistent during the welding process, and its steep heat flow density curve in the first half and gentler in the second half. The double ellipsoid heat source model conforms to the actual melt pool shape of the welded joint and is therefore used [9]. The heat source model is shown in Figure 4.

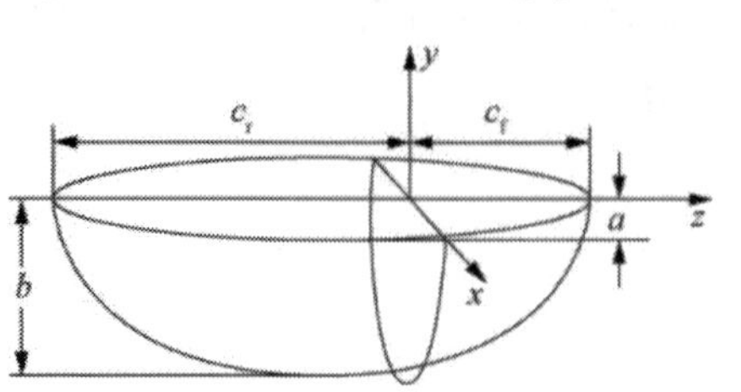

Figure 4. Dual ellipsoid heat source model.

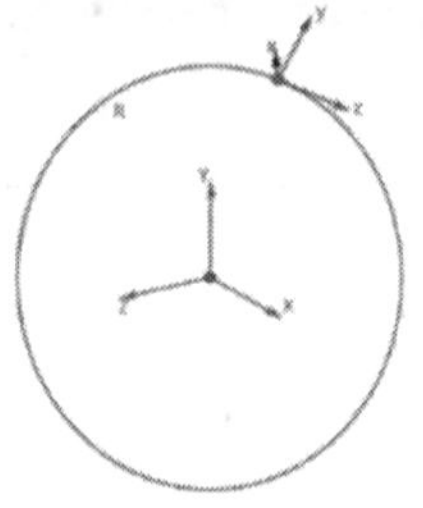

Figure 5. Coordinate system conversion.

The expression for the density distribution of heat flow in the front and rear semi-ellipsoids is given by:

$$\begin{cases} q_f(x,y,z) = \dfrac{6\sqrt{3} f_1 Q_W}{abc_f \pi^{3/2}} e^{\frac{-3x^2}{a^2}} \cdot e^{\frac{-3y^2}{b^2}} \cdot e^{\frac{-3z^2}{c_f^2}} \\ q_r(x,y,z) = \dfrac{6\sqrt{3} f_2 Q_W}{abc_r \pi^{3/2}} e^{\frac{-3x^2}{a^2}} \cdot e^{\frac{-3y^2}{b^2}} \cdot e^{\frac{-3z^2}{c_r^2}} \end{cases} \tag{1}$$

In the formula, f_1, and f_2 are the heat flow density distribution coefficients of the front and rear semi-ellipsoids, and $f_1+f_2=2$; Q_W is effective heat input; a_f, a_r are the lengths of the anterior and posterior semiaxes of the double ellipsoid, b is half the width of the melt pool, and c is the depth of the melt.

Currently, ABAQUS does not support direct application of external heat loads, requiring secondary development using a Fortran DFLUX subroutine to implement the loading of a moving heat source. Before writing the subroutine, it is essential to first integrate Visual Studio with ONEAPI and ABAQUS, then determine the actual welding parameters such as voltage, current, thermal efficiency, welding paths, and heat source characteristics. The movement of the heat source is controlled in real time based on the position of the weld point. As welding progresses, the path of the welding movement changes over time, ultimately completing the entire welding process.

The key to the DFLUX subroutine is the movement path of the heat source. In the realization of welding heat source movement, it is necessary to move the double ellipsoid along the welding path. The position of the double ellipsoid changes with time along the path, and the welding path is circular. In this context, the model's center coordinate system serves as the global coordinate system, while the heat source's coordinate system functions as the local coordinate system. The critical aspect is the conversion of the model's spatial coordinates (integration points) from the global coordinate system to the heat source's local coordinate system. The z-axis in the local coordinate system indicates the direction of instantaneous welding movement, the y-axis indicates the melt depth, and the x-axis indicates the direction of the melt pool width. In the global coordinate system, the point (X, Y, Z) is the coordinate value of the integration point, and the coordinate value of this integration point in the local coordinate system is (x, y, z). According to Figure 5, we establish the relationship equation for the coordinate conversion to establish the moving path of the heat source, which is subsequently added to the subroutine of the heat source. Welding is a three-layer welding. Each layer of the weld needs to be cooled after the end of the weld, and the heat source subroutine needs to be based on the total welding time for conditional judgment, indicating that when entering the cooling analysis step, the heat source heat input needs to be set to 0.

$$\begin{cases} x = -X \\ y = -Z\sin\theta + Y\cos\theta - R \\ z = -Z\cos\theta - Y\sin\theta \end{cases} \tag{2}$$

In the formula, θ is the angle of rotation of the heat source center from the initial position, and R is the radius of the heat source moving path circle.

3. Heat treatment process parameters

During the welding process, the temperature distribution of the weld and heat-affected zone is extremely uneven. Due to thermal expansion and contraction and other reasons, the weld zone will produce large residual stresses. Engineering often uses post-weld heat treatment operation to reduce the welding residual stresses and improve the mechanical properties of welded joints. The weld is heat-treated by localized heating, and the heating band should contain the weld area and heat-affected zone, as well as the adjacent part of the base metal, and the heating range of each side of the weld should be not less than three times the width of the weld, and the whole welded joint is heat-treated in this study. The post-weld heat treatment process parameters include inlet temperature, heating rate, heat treatment temperature, holding time, cooling rate before discharge, and final discharge temperature. To ensure the improvement, we set 600°C as the heat treatment holding temperature, the holding time was set to 4 h, and a uniform rate was used in the warming and cooling process. When performing heat treatment analysis, the simulation results of the welding process are used as the initial state for post-weld heat treatment simulation [10].

4. Analyzing the results

4.1 Temperature field

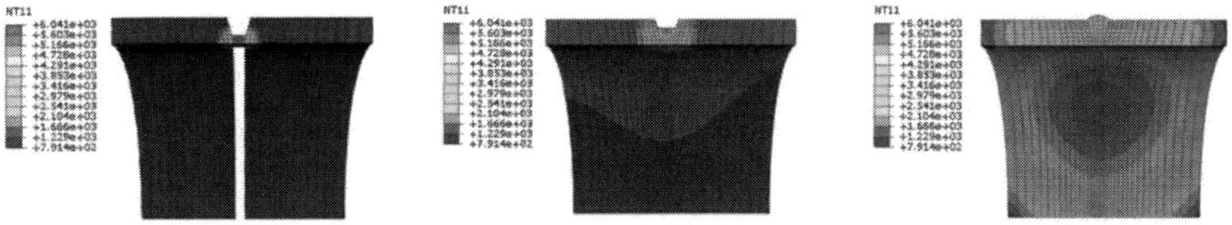

(a) Underwelding started. (b) Intermediate welding. (c) Surface welding.

Figure 6. Temperature field at different welding moments.

As shown in Figure 6, with the heat source moving forward, the metal material in front of the heat source is continuously heated by the heat source, and the temperature of the metal material gradually rises until it exceeds the melting point and forms a stable molten pool. The temperature of the metal material behind the heat source gradually decreases due to heat transfer to the base material, and the difference in temperature distribution between the weld zone and the base material gradually increases. Eventually, the heat in the weld zone is transferred to the base metal by heat conduction and to the air by heat convection heat radiation. The temperature is highest at the center of the heat source's area of action, about 6000°C, and decreases the farther away from the center of the heat source.

4.2 Distribution of residual stresses in the welded joints before and after the heat treatment process

The equivalent residual stresses (Mises) cloud plots of the outer and inner walls of the weldment before and after heat treatment are shown in Figure 7. From the view of the outer wall of the weldment, higher stress in the center of the weld, and as the distance from the center of the weld increases, the values of stress decrease and then increase, and then decrease and then tend to stabilize. From the inner wall of the joint, the equivalent force distribution shows a different trend from that of the outer wall. Less stress in the center of the weld, and the equivalent force on both sides of the weld reaches the peak, and then gradually decreases and then levels off. As shown in Figure 7, the equivalent stress distribution in the center of the weld presents a high-stress state, with a peak stress of about 195.25 MPa. However, after heat treatment, the equivalent peak stress decreases significantly to about 145.21 MPa, with a decrease of 25.6%. After the post-weld heat treatment process, although the location of the maximum stress value did not change, the stress distribution of the joint was changed to a certain extent, and the stress distribution became more uniform in small magnitude, and the overall residual stress number in the weldment was reduced. A comparison of the stress distribution on both sides of the joint reveals that the stress peak on the outside is situated at the toe of the weld, while the stress peak on the inside is located on both sides of the weld.

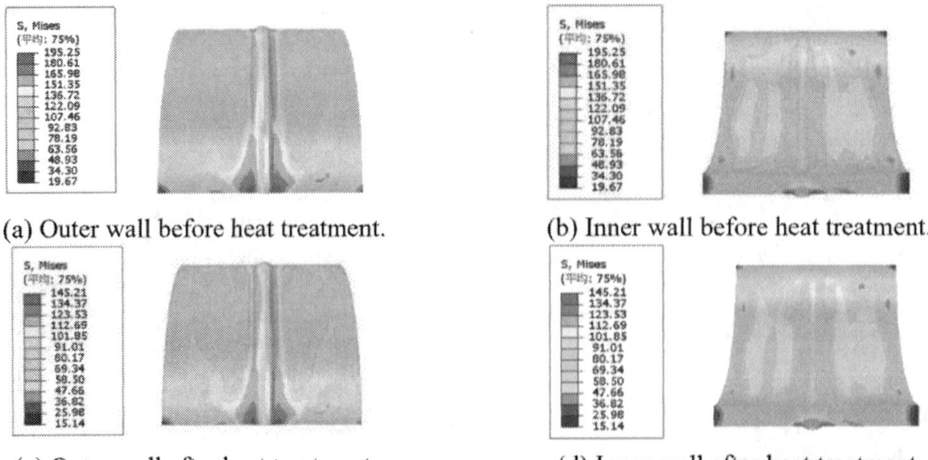

(a) Outer wall before heat treatment. (b) Inner wall before heat treatment.

(c) Outer wall after heat treatment. (d) Inner wall after heat treatment.

Figure 7. Equivalent residual stress clouds.

A cloud of axial residual stresses (S11) in the outer and inner walls of the welded joints before and after heat treatment is shown in Figure 8. From the view of the outer wall of the weldment, higher stress at the weld toe, and the stress is negative for compressive stress. As the distance from the center of the weld increases, the axial residual stress decreases gradually, the transition to tensile stress, and the value of tensile stress increases, then decreases, and finally smooths out. From the view of the inside of the joint, the axial residual stress distribution has a different trend than the external wall. The main residual axial residual stresses on the outer side are compressive stresses, while the main axial residual stresses on the inner side are tensile stresses. The stress numbers at the center and toe of the weldment on the inner side are small, and the axial residual stresses on both sides of the weld reach the peak, which then gradually decreases and levels off. As can be seen in Figure 8, higher axial residual stresses are observed at the weld toe, with a peak stress of about 170.94 MPa. However, after heat treatment, the equivalent peak stress is significantly reduced to about 130.84 MPa, with a decrease of 23.5%. After the post-welding heat treatment process, although the location of the maximum value of the stress has not been changed, the stress distribution of the joint was changed to a certain extent, the stress distribution of small magnitude became more uniform, and the total amount of residual stress in the weldment has been reduced. Comparison of the stress distribution between the inner side of the weldment and the outer side of the weldment reveals that the stress peak on the outer side is located at the toe of the weldment, while the stress peak on the inner side is located on both sides of the weldment.

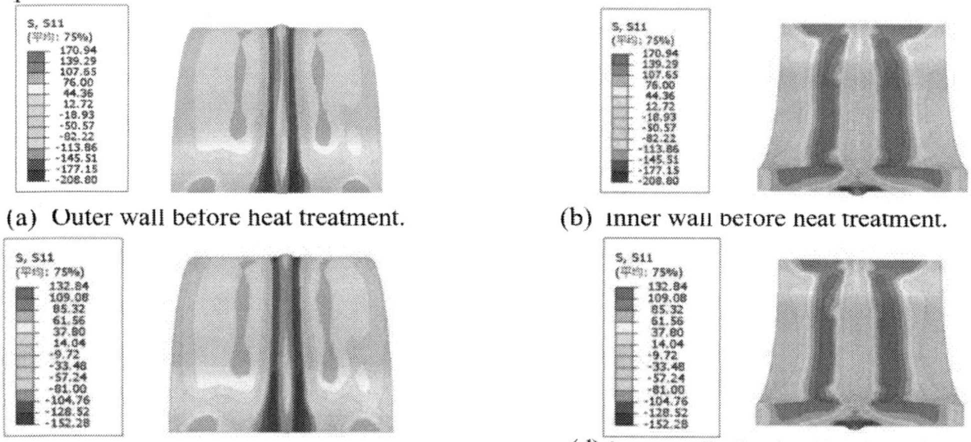

(a) Outer wall before heat treatment. (b) Inner wall before heat treatment.

(c) Outer wall after heat treatment. (d) Inner wall after heat treatment.

Figure 8. Clouds of axial residual stresses.

We establish the path shown in Figure 9 to study the residual stress change curves and change rules on both sides of the joint before and after heat treatment. Figures 10 and 11 show the variation curves of equivalent stresses and axial residual stresses on the inside and outside of the weld before and after post-weld heat treatment, respectively, and it is visible that there is a great difference in the stress distribution between the two sides. At the weld, the axial residual stress of the outer wall of the weld is obviously larger than that of the inner side of the weld, and both sides of the axial residual stress, with the increase of the distance from the middle of the weld, the stress gradually decreases and eventually reduced to 0 Mpa. The curves show that, after the heat treatment, the equivalent stress and axial residual stress on both sides of the weld are reduced to a certain extent as a whole.

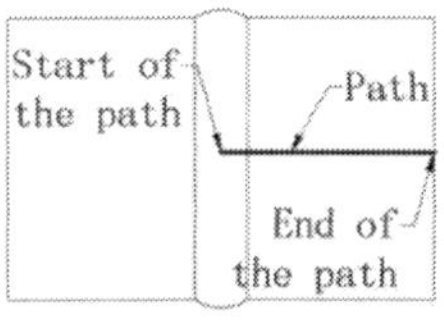

Figure 9. Path.

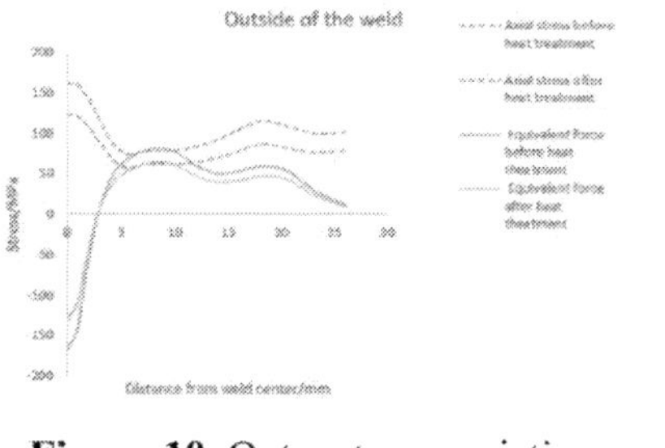

Figure 10. Outer stress variation curve.

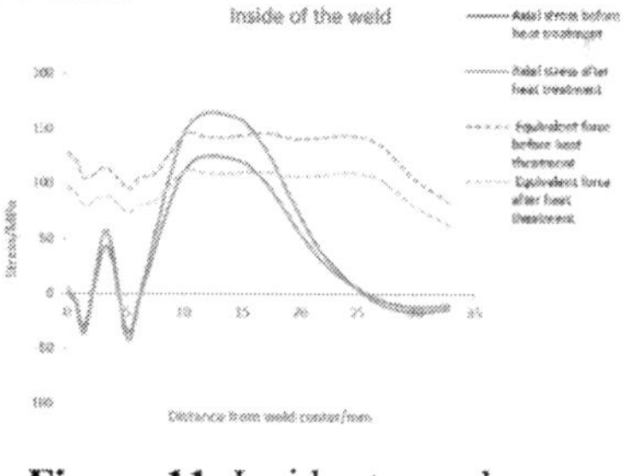

Figure 11. Inside stress change curve.

5. Model validation

The blind hole method was used to measure the residual stress and deformation of the axle housing and the half-shaft casing after welding to verify the accuracy of simulation results. The inner wall of the weld joint is inconvenient to measure by punching, and only the equivalent stress measurement is carried out on the outer wall of the joint after heat treatment, and the triaxial strain gauge and LM-12 stress detector are used for measurement, as shown in Figure 12. According to the path established by the above model, the hole is measured at 0 mm, 3 mm, 6 mm, 10 mm, 15 mm, and 25 mm from the starting point of the welding path, and the measurement data is shown in Figure 13. Although there are errors between the measurement results and the simulation results, the changing trend of the data is roughly consistent, which verifies the correctness of the simulation results to a certain extent.

Figure 12. Measure stress.

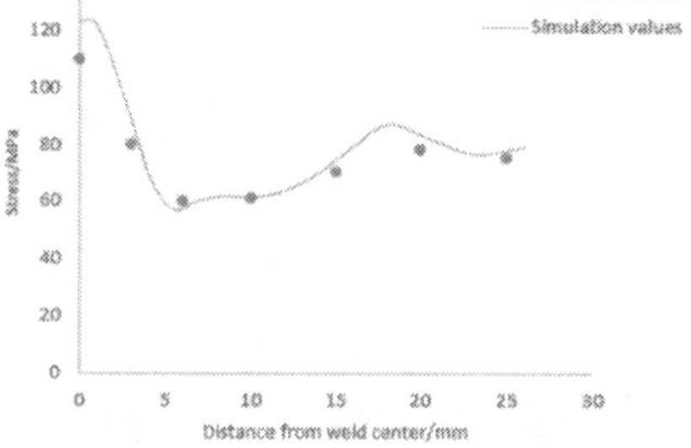

Figure 13. Stress comparison.

6. Conclusion

Based on ABAQUS software and subroutines, the calculation method of temperature-displacement direct coupling and the birth-death unit technology are used to simulate the welding process and post-weld heat treatment of the welded joints of the drive axle shell and half-shaft casing and to obtain the changes of its stress and temperature state. The equivalent peak stress and axial peak stress of the welded joint were significantly reduced by heat treatment. In the follow-up study, the specific effects of welding voltage, current, welding speed, and different heat treatment parameters on residual stress will be explored.

References

[1] Yu B. Numerical simulation study on the molding process of stamped and welded drive axle shell. [D] Shandong University, 2007.
[2] Tang P. Safety analysis and optimization design of punch-welded drive axle housing for medium trucks. [D] Anhui University of Technology, 2021.
[3] Li S S. Numerical simulation of drive axle shell welding process and its fatigue life [D]. Wuhan University of Technology, 2022.
[4] Xiao Y H. Optimization and validation of plasma cladding residual stress and its elimination process by heat treatment for marine gas valves [D]. Jiangsu University of Science and Technology, 2020.
[5] Chen Y, Xu Y L, Li Q T, Wang Y F, Zhao X R and Zhang M G. Finite element simulation of post-welding heat treatment for eliminating residual stresses in welded pipe rings[J]. Welded Pipe,2022,45(05):41-47.
[6] Li Z H and Yu L. Simulation analysis of welding and post-welding heat treatment process of pipeline circumferential weld[J]. Chemical Machinery, 2022, 49(03): 445-449.
[7] Zhang L, Guo Y L and Qi L. Influence of weld channel filling method on numerical simulation of residual stress in ring weld of X65 pipeline steel [J]. Welding Technology, 2020, 49(6):5-8.
[8] Li Z H and Yu L. Simulation analysis of pipeline circumferential weld welding and post-welding heat treatment process [J]. Chemical Machinery, 2022, 49(3): 445-449.
[9] Bai C X. Numerical simulation of welding residual stress and hydrogen diffusion in X80 pipeline[D]. Northeast Petroleum University, 2023.
[10] Zhang H, Li X Z, Wei Z X and Xu L. Numerical simulation study on improvement of residual str esses in straight welds of pipelines by post-weld heat treatment[J]. Welded Pipe, 2023, 46(0 9): 14-20.

Study on the influence of thermal properties of paraffin on flexible PTC materials

Jianmin Hao, Yinfa Yang, Hua Chen and Wenlong Cheng*

University of Science and Technology of China, Hefei, Anhui, 230026, China

*Corresponding author's email: wlcheng@ustc.edu.cn

Abstract: This paper experimentally studied paraffin's influence on flexible PTC materials' temperature resistance and flexibility. Their resistance-temperature characteristics are notably influenced by PA's phase transition range thermal properties. Specifically, the Curie temperature is intricately linked and determined by the PA's phase transition temperature. PA's phase state and proportion in the composite directly impact PTC material flexibility significantly. Specifically, a lower PA phase transition temperature results in better flexibility of PTC materials at room temperature. Additionally, a lower PA content minimizes interference with the overall flexible mesh structure, further enhancing material flexibility. The PTC of this series of formulations showed good adaptive temperature control ability. The thermal control performance can be customized by adjusting the PA.

1. Introduction

With the rapid advancements in electronic technology and precision measurement techniques, the demands for temperature control in electronic devices have increased significantly [1]. In modern microelectronics, sensors, and aerospace industries, the development of a miniaturized, adaptive temperature control system has become paramount. PTC materials, renowned for rapid resistivity increase with temperature, have extensive applications in active temperature control [2].

In recent years, polymer-based PTC materials have garnered considerable attention due to their exceptional processability, and flexibility, coupled with pronounced PTC behavior. Flexibility enables seamless integration into complex electronic device geometries, enhancing performance and reliability. Wang and Cheng [3] pioneered flexible PTCs with room-temperature Curie points, leveraging OBC as a scaffold. Hao et al. [4] expanded this frontier by crafting highly flexible and thermally stable PTC heaters using silicone rubber and paraffin, resolving leakage issues and showcasing excellent reproducibility. Polymer-based PTC materials frequently leverage low-melting-point phase change materials to adjust their Curie points [5]. Therefore, a thorough investigation into the influence of the thermal properties of different raw materials on their performance holds significant scientific merit and practical value for advancing related technologies. However, there is no in-depth study on the properties of PTC materials according to the thermal physical properties of raw materials.

This study experimentally develops novel flexible paraffin-based PTC material and examines how their thermal properties affect resistance-temperature characteristics, flexibility. The adaptive temperature control ability of PTC material is substantiated through experimental verification. It provides theoretical guidance for designing PTC material with room-temperature Curie points, fostering their use in intelligent thermal management and other fields.

Journal of Physics: Conference Series **2879** (2024) 012010 doi:10.1088/1742-6596/2879/1/012010

2. PTC material preparation and characteristic experimental

2.1 Synthesis process

The new PTC material is first stirred by mixing support material (SM) and paraffin (PA) in a vacuum stirring tank. Subsequently, conductive filler is added to the mixture and thoroughly mixed. The mixture is compressed into the 1 mm-thick sheet. In order to apply the PTC material in practice, the material is further encapsulated into a heating sheet. Both sides of the sample are evenly coated with silver paste as electrodes. Copper foil is utilized as electrode lead. Following this, an evenly distributed outer layer of silicone rubber serves as an insulating lead. Figure 1 presents the actual sample and its corresponding packaging. In total, five distinct PTC samples were fabricated, with their respective serial numbers and compositions detailed in Table 1.

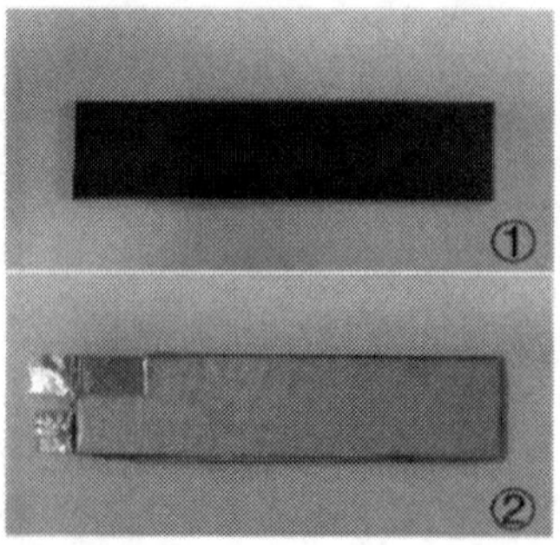

Figure 1. Sample diagram and package scheme.

Table 1. Components of the novel PTC samples.

No.	PA	conductive filler	SM/PA
1	40#	36%	70:30
2	45#	36%	70:30
3	48#	36%	70:30
4	40#	36%	65:35
5	40#	36%	75:25

2.2 Electrical characterization

PTC material is fundamentally a temperature-sensitive resistor. Its electrical properties are analyzed through the study of resistivity-temperature variation trends. The resistivity of the sample, measured along its thickness direction, was determined using the ohm method and varying temperature. After packaging the sample, connect the controlled power supply, and paste platinum resistance (PT100, 0.1°C) in the center of the sample surface. Record the surface temperature changes, the voltage, and the current at both ends of the sample. The sample is placed in the ambient temperature chamber, and the temperature of the sample is changed by adjusting the internal temperature of the ambient temperature chamber. The material's volume resistivity is calculated utilizing the following formula.

$$\rho = \frac{U}{A} \times \frac{S}{L} \tag{1}$$

where U represents the voltage present at both ends of the material, A denotes the current flowing through the material, S signifies the cross-sectional area associated with the direction of current flow, and L refers to the length of the flow direction.

The PTC intensity serves as a critical reference index for PTC composites. It is defined as:

$$I_{PTC} = \log \frac{\rho_{max}}{\rho_{max}} \tag{2}$$

where ρ_{max} represents the peak resistivity in the experimental range, and ρ_{min} represents the resistivity in the low-temperature region.

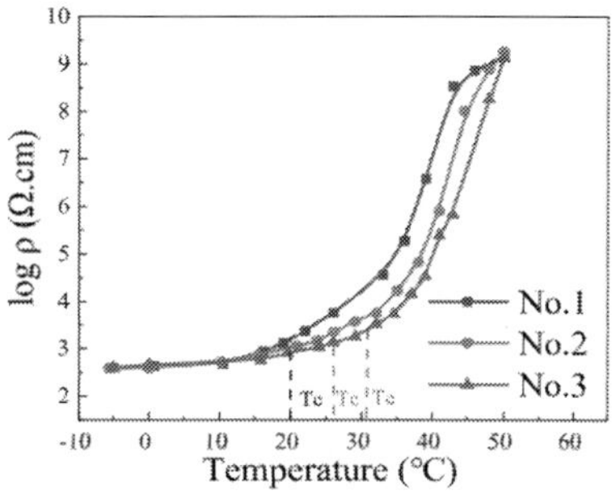

Figure 2. Resistivity - temperature curve of sample No. 1-No. 3.

Table 2. Solid-solid phase transition temperature and Curie temperature.

No.	PA	solid-solid phase transition temperature	Curie temperature
1	40#	10°C	20°C
2	45#	22°C	25°C
3	48#	26°C	31°C

The resistivity-temperature curve of samples No. 1-No. 3 is depicted in Figure 2. All curves exhibit a similar trend: initially, the resistance of the sample remains low, and upon surpassing the Curie point, the resistivity experiences a sudden increase. We believe that all the samples have an obvious PTC effect. Table 2 shows the solid-solid phase transition temperature and Curie temperature of different types of PA. The Curie temperatures of the three PTC samples are all in the range of room temperature and slightly higher than the solid-solid phase transition temperature. Figure 3 (a) is DSC thermograms of pure PAs for preparation of PTC materials. The mechanism of the PTC effect was analyzed using Sample No.1 as a case study. Sample No.1 employs 40# PA with a solid-solid phase transition at 10°C and solid-liquid transition completion at 40°C. The resistivity rise in the PTC sample aligns with PA's solid-solid transition. Limited PA content minimally disrupts the conductive network, causing a gradual resistivity shift. The Curie temperature exceeds the solid-solid transition temperature, which stands at 20°C.

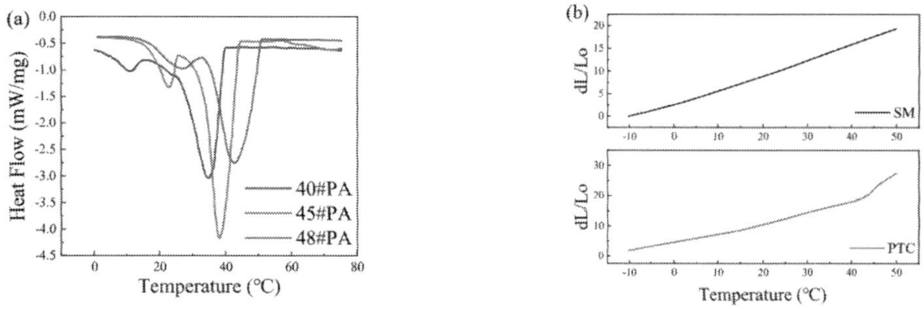

Figure 3. (a) DSC thermograms of pure PAs, (b) Thermal expansion coefficient of PTC sample (No. 1) and pure SM.

The resistivity of PTC material experiences a significant increase beyond 40°C, suggesting that the phase transition of PA is not the exclusive cause for the observed resistivity change. To further elucidate this point, thermal expansion coefficients (TECs) were measured for PTC material and pure SM. The results are depicted in Figure 3 (b). Given the difficulties associated with measuring the TEC of PA solids, we present only the test results for pure SM and PTC. Comparing the thermal expansion coefficient of pure SM and PTC, it can be found that the thermal expansion coefficient of PTC is influenced by additional factors. It can be considered that PA will also produce thermal expansion during the heating process. The thermal expansion coefficient notably increases when temperatures surpass 40°C. The thermal expansion of a material disrupts the conductive network within it, leading to an increase in resistivity. This effect is particularly pronounced when the temperature exceeds 40°C, significantly contributing to the PTC effect. Comprehensive analyses using DSC and TECs measurements confirm that the change in sample resistivity is related to PA phase transition and thermal expansion.

Upon examining the comparative resistivity and temperature curves of distinct formulations, it is evident that the low-temperature resistivity and PTC strength of these materials remain largely unaltered. However, a significant shift is observed in the Curie temperature of the PTC material. Due to the minimal coefficient of thermal expansion and absence of phase transitions within the matrix at low temperatures, the resistivity remains stable, governed by uniformly distributed conductive fillers. Upon reaching 50°C, the phase change material liquefies completely, significantly increasing the matrix's coefficient of thermal expansion. This severe disruption of the conductive network renders the sample non-conductive, as thermal expansion's impact on resistivity becomes substantial. Between 10°C-50°C, PA undergoes solid-solid and solid-liquid transitions, significantly altering the matrix and disrupting the internal conductive network. The sample's PTC strength can attain a value of 6.3 due to the discernible change in resistivity. The influence of PA's thermal properties on the PTC material's resistivity-temperature curve is primarily manifested in the Curie point temperature.

2.3 Flexibility of material

In order to adapt to the more complex shape requirements of thermal control components, the new PTC material has good flexibility. The minimum bending diameter is used to characterize the flexibility of the material. As shown in Figure 4, the prepared samples (thickness 1 mm) have good flexibility at room temperature. The bending diameters of samples No. 1 and No. 2 are only 2 mm. The bending diameter of sample No. 3 is 2.5 mm.

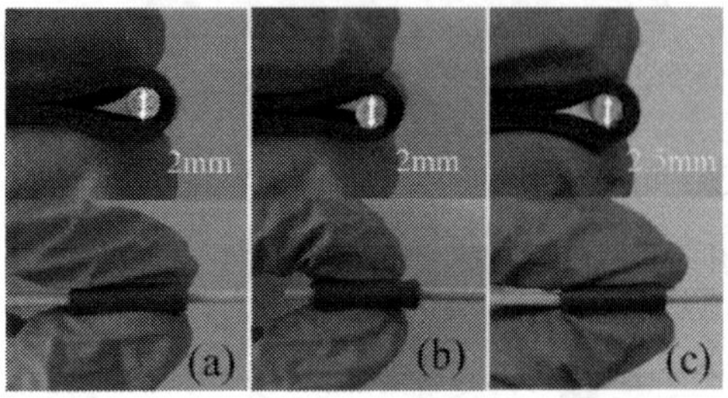

Figure 4. (a) Sample No. 1-40#PA, (b) Sample No. 2-45#PA, (c) Sample No. 3-48#PA.

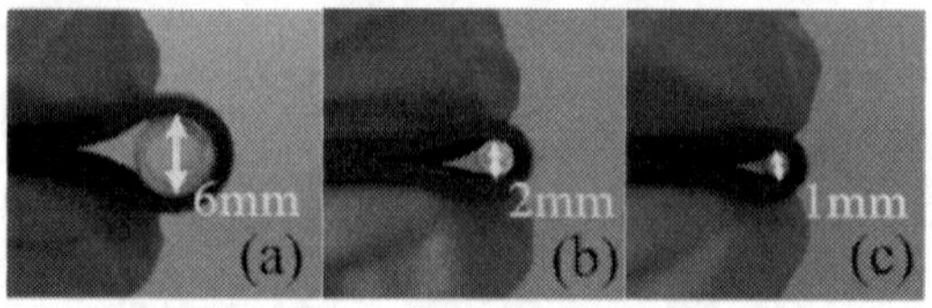

Figure 5. (a) Sample No. 4-35%PA, (b) Sample No. 1-30%PA, (c) Sample No. 5-25%PA.

Based on the properties of the raw materials, SM constructs a three-dimensional network structure, providing a resilient support framework for the entire polymer. PA is encapsulated within this framework. The stable three-dimensional mesh structure formed by SM as the support matrix enhances the flexibility of the entire PTC material. Upon the addition of PA and conductive filler to the flexible SR network structure, the resulting PTC material retains exceptional flexibility.

PTC material flexibility varies with PA formulations, linked to PA's solid-solid phase shift. This arises from n-alkanes' two solid phases: low-temp monoclinic (rigid, brittle) and high-temp disordered rotating (soft plastic near melting). When the temperature approaches the melting point, the hexagonal phase molecules of the orthorhombic system begin to rotate around their central axis. This phase transition results in the material exhibiting a soft pliable state [6]. At room temperature, 40#PA and 45#PA exhibit soft plastic characteristics. PTC materials display notable flexibility. Conversely, at the same temperature, 48# PA does not fully transition into this soft phase, thereby potentially reducing the material's overall flexibility. Figure 5 shows the flexible characterization of samples No. 1, No. 4, and No. 5. It can be found that the ratio of PA to SM also affects the flexibility of the material. In this process formulation, SM provides a flexible support frame for the material as a whole. Given the substantial presence of SM in the formula, this series of compounds exhibits a certain degree of flexibility. If the content of PA distributed in the frame is too high, the coating capacity of the frame will reduce the overall flexibility.

3. Adaptive temperature control performance

There is a special change rule between the resistivity and temperature of PTC material, which makes it have adaptive temperature control performance. The principle is as follows: At the commencement of the power supply, the PTC heater's initial resistivity is small due to the system's low ambient temperature. This results in a substantial initial heating power, causing a rapid temperature rise. As temperatures rise, resistance increases, and heating power diminishes. When the heat exchange with the outside world is equal to the heat of the element, the thermal balance is reached. Thermostatic control is achieved by adjusting the heating power through the sensitivity of the resistivity to temperature.

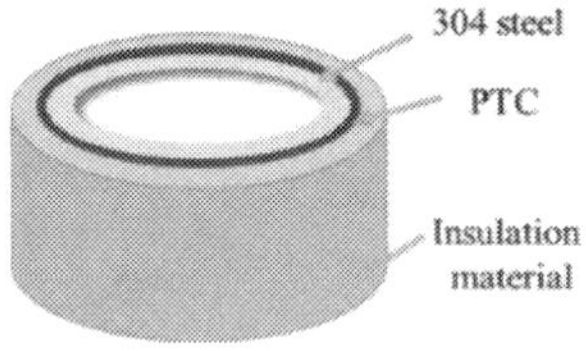

Figure 6. Adaptive temperature control experimental components.

In order to simulate the actual temperature rise of the controlled parts of the surface, the simulated components are shown in Figure 6. Controlled parts selected 304 stainless steel ring column, thickness 4 mm. The heating plate size is 80 mm × 20 mm × 1 mm. To make the heat generated by the heating sheet as much as possible in the form of heat conduction to the controlled parts and reduce the loss of heat to the environment, the heating sheet is coated with a layer of insulation cotton. Constant voltage is applied to the heating sheet. The variation in the inner surface temperature of the ring column over time is used to evaluate the heating effect and temperature control capability of the sample.

Figure 7 (a) presents the temperature profiles of a controlled toroidal column heated by three PTC plates held near 0°C. A common pattern emerges rapid initial temperature rise, a plateau in the transition region, and stabilization. The equilibrium temperatures for 40#PA, 45#PA, and 48#PA PTC heaters are 24.4°C, 26.3°C, and 27.2°C, respectively. Initially, similar heating rates are observed due to comparable resistivities at low temperatures. As heater temperatures exceed the Curie point, resistivity surges, diminishing heating power and halting further temperature ascent. The final equilibrium temperatures may fall below the Curie point due to environmental heat exchange. Lower melting point PA matrices lead to lower equilibria, narrowing the temperature differential for heat transfer, thereby shortening equilibration times. Thus, PA properties notably modulate the adaptive temperature control behavior of PTC materials under identical operational conditions.

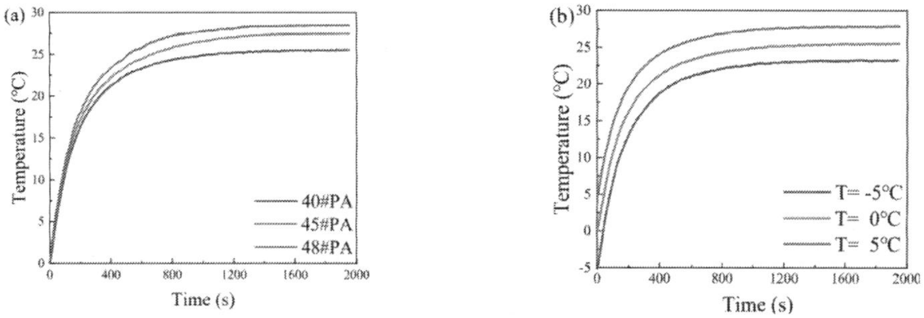

Figure 7. (a) The temperature rise of the controlled component under different heating sheets; (b) Temperature rise of the controlled component at different ambient temperatures.

To investigate the impact of ambient temperature on the temperature control process, a 40# PA matrix PTC was utilized as a heating sheet. Set the ambient temperature at -5°C, 0°C, and 5°C, and compare the differences between the temperature control curves, as shown in Figure 7 (b). The equilibrium temperature and temperature rise are shown in Table 3. As the ambient temperature rises, so does the equilibrium temperature of the controlled component. However, this temperature increase tends to diminish gradually. For instance, when the ambient temperature increases from -5°C to 5°C, the equilibrium temperature of the controlled cylindrical ring only experiences a change of 4.7°C. This magnitude of change represents merely one-third of the environmental temperature change. These results suggest that without any external control mechanisms, the PTC heating sheet can effectively mitigate the impact of external temperature variations on temperature control.

Table 3. Equilibrium temperature and temperature rise at different ambient temperatures.

ambient temperature	equilibrium temperature	Temperature rise amplitude
-5°C	22.4°C	27.2°C
0°C	24.4°C	24.4°C
5°C	27.2°C	22.2°C

4. Conclusion

In this paper, a new type of highly flexible polymer-based PTC material was prepared by experiments. The effects of PA with different thermal properties on the electrical properties and flexibility of the materials were studied. The adaptive temperature control ability of the material is investigated by experiments. The conclusions derived from this study are as follows:

(1) Polymer-based PTC materials were prepared using PA, support material, and conductive filler. When the temperature falls within the phase transition range of PA, the thermal properties of PA significantly impact the temperature resistance of PTC materials. The Curie temperature is determined by this PA phase transition temperature. However, the resistivity was not affected by the type of PA before the phase transition. When the temperature reaches 50°C, the PTC material has reached the insulation state, so the PTC strength can reach 6.3. On the whole, the resistivity-temperature characteristic trend of the material is not affected by the PA type.

(2) The phase composition and proportion of PA significantly impact the flexibility of PTC materials. The SM network structure imparts flexible support to the PTCs. Given the presence of two solid wax types, the macrostructure can manifest as either a brittle or a soft configuration. The PTC material is in a soft phase state at room temperature due to its low solid-solid transition temperature. Furthermore, reducing PA content minimizes disruption to the overall flexible grid structure, thereby further improving material flexibility. Highly flexible PTC materials have broad application prospects in wearable fields.

(3) The novel high-flexibility polymer-based PTC exhibits excellent temperature responsiveness and adaptive control capabilities. Its range of thermal equilibrium temperature variation is merely one-third that of the ambient environment. Varying PA formulations yield distinct Curie temperatures for the material, leading to different stable temperatures under identical operating conditions. Thus, PTC materials tailored to specific thermal control needs can be designed by selecting PA with appropriate thermal properties.

Acknowledgment

This research is sponsored by the National Key R&D Program of China for financial support (Grant No. 2022YFC2204403) and Major Science and Technology Projects in Anhui Province (Grant No. 202203a07020023).

Journal of Physics: Conference Series **2879** (2024) 012010 doi:10.1088/1742-6596/2879/1/012010

References

[1] W. Zhang, R. Zhao, W. Cheng, "Temperature control performance of a spaceborne PTC heating system: Dynamic modeling and parametric analysis," Thermal Science and Engineering Progress, 44(2023) 102062.
[2] D. Ren, Z. Ding, Y. Han, "Study on adaptive infrared camouflage of novel positive temperature coefficient (PTC) materials in space," Infrared Physics & Technology, 126(2022) 104346.
[3] R. Wang, W. Cheng, "A novel flexible room temperature positive temperature coefficient material for thermal management," Advanced Composites and Hybrid Materials, 2(2019) 83-92.
[4] J. Hao, Y. Yang, Y. Shen, R. Zhou, W. Zhang, H. Chen, W. Cheng, Positive temperature coefficient material based on silicone rubber/ paraffin/ graphite/ carbon nanotubes for wearable thermal management devices, Chemical Engineering Journal, 493(2024) 152427.
[5] Go Gwang Myeong, Park Siwoo, Lim Minseob, Jang Byungkwon, Park Ji Young, Cho Hong Baek, Choa Yong Ho, "Enhanced positive temperature coefficient intensity and reproducibility with synergistic effect of 0-D and 2-D filler composites," Journal of Materials Science, 2022, 57 (38): 18037-18050.
[6] S.P. Srivastava, Jyoti Handoo, K.M. Agrawal, G.C. Joshi, "Phase-transition studies in n-alkanes and petroleum-related waxes—A review," Journal of Physics and Chemistry of Solids,1993 0022-3697.

Journal of Physics: Conference Series **2879** (2024) 012011 doi:10.1088/1742-6596/2879/1/012011

A novel output window for dual-band gyrotron amplifier

Xiaoyan Wang[1*], Jinyong Fang[1], Jiangniu Wu[1] and Kai Peng[1]

[1]Xi'an Branch of China Academy of Space Technology, Xi'an 710100, China

[*]E-mail: wangxiaoyanns@163.com

Abstract: As a high-frequency, high-power, and high-bandwidth microwave source, the gyrotron amplifier has good application prospects in the aerospace field. The microwave window is of great significance for the high-frequency characteristics, power capacity, lifespan, and reliability of the gyrotron amplifier. This paper introduces a novel double-disk sapphire output window, which has excellent mechanical strength, high-temperature resistance, and high power capacity. The transmission coefficient S_{21} of TE_{01} mode is greater than -1 dB in 8.29 GHz~9.73 GHz, and the S_{21} of TE_{02} mode is greater than -1 dB in 16.33 GHz~17.8 GHz, which can ensure wide bandwidth operation of high microwave transmission in X and Ku bands. It solves the problem of using one port to achieve microwave output in two bands for dual-band gyrotron amplifiers.

1. Introduction

Gyrotron amplifiers have great application prospects in the aerospace field. In recent years, some advanced radar, electronic countermeasures, and communication systems require dual-band operation. To reduce the size, quality, and cost of the entire transmitter system, studying dual-band gyrotron amplifiers has great practical value [1] [2]. The microwave window isolates the high vacuum environment in the tube from the external atmosphere and seals it. Simultaneously, realizing the input and output of the microwave and achieving a dual-band output window is one of the technical difficulties in stabilizing dual-band gyrotron amplifiers. Sapphire material is selected as the window material due to its low microwave loss, high mechanical strength, high-temperature resistance, and high power capacity. Compared with single-disk and triple-disk window structures, the double-disk window can ensure high microwave transmission in a wider operating frequency. This novel dual-disk sapphire output window has achieved excellent performance in all aspects, laying the foundation for the development of dual-band gyrotron amplifiers.

2. Design of output window

For a dual-band gyrotron amplifier, the amplified electromagnetic waves in both bands are output from the same circular waveguide port. Therefore, the output window should satisfy the good microwave transmission performance in both X/Ku bands. Secondly, the dual-band gyrotron amplifier operates in TE_{01} mode in X-band and TE_{02} mode in Ku-band, which are all non-fundamental modes, making the design of the dual-band output window more difficult [3][4].

Output window also needs to ensure high mechanical strength, high power capacity, and good thermal stress strength, so the material of the window can generally be ceramic, diamond, sapphire, etc. Diamond has a low dielectric loss, high hardness, and good performance, but it is too expensive. Sapphire has a low dielectric loss, and high mechanical strength, and can be processed into smaller thin slices. Compared with alumina ceramics, sapphire has higher thermal conductivity, high-temperature resistance, and higher power capacity [5]. By contrast, sapphire is chosen as the window material for the dual-band output window. The tangent loss angle and relative dielectric constant are $\delta_l = 0.0005$, $\varepsilon_r = 9.4$.

The thickness of the cylindrical output window t is an integer multiple of the half-wavelength [6]:

$$t = N \cdot \lambda_d / 2 \quad (1)$$

where $\lambda_d = 2\pi / \sqrt{(2\pi/2)^2 \varepsilon_r - (\chi_{mn}/R)^2}$, χ_{mn} is the characteristic roots of the corresponding pattern, and N is an integer. The thickness of the TE_{01} mode window at a frequency of 9.35GHz is calculated as $t = N \cdot 5.4mm$, and that of the TE_{02} mode window at a frequency of 16.77GHz is $t = N \cdot 2.8mm$. In order to adapt to the transmission of the two bands, the reference value of the window thickness is around 5.5 mm, and the specific values are calculated by simulation.

After the initial selection of the window structure size, the transmission performance and electric field distribution of the single-disk, double-disk, and triple-disk output window structures in the X-band TE_{01} mode and in the Ku-band TE_{02} mode are calculated by using HFSS. It is found that the operating frequency band of the single-disk window is very narrow, and the center frequency points of the two bands are completely doubled, which cannot meet the X/Ku band microwave transmission with 9.35 GHz and 16.77 GHz as the center frequency at the same time [3]. With the increase in the number of window disks, the operating frequency band will be significantly broadened, which can ensure the transmission of microwaves in the required frequency band. However, increasing the number of disks will also result in more microwave losses. The microwave transmission performance of the triple-disk window structure is not as good as that of the double-disk structure. The final output window is selected as a double-disk window structure, as shown in Figure 1.

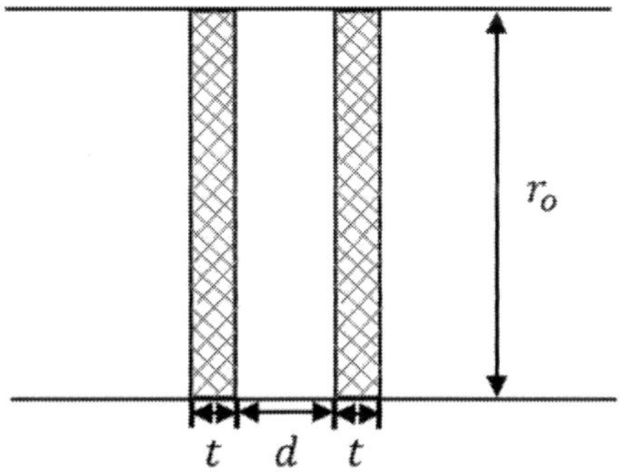

Figure 1. Schematic diagram of the double-disk output window structure.

3. Simulated results

Figure 2 (a) (b) shows the S_{11} on two bands when the thickness of the output window changes from 5.65 mm to 5.85 mm. It can be seen that when the window thickness changes by 0.2 mm, the output frequency band of the X-band shifts by 300 MHz, and the Ku-band shifts by 400 MHz. The operating frequency band of the output window is very sensitive to the disk thickness. When the output window operates in the X-band, the thinner the disk is, the smaller the S_{11} is, and the overall band shifts to the right; When the Ku-band is operating, the thicker the disk is, the smaller the S_{11} is, and the overall band shifts to the left. The optimal disk thickness is not consistent when operating on the X/Ku band. The thinner the window is, the better the transmission performance of the X-band is. The thicker the window is, the better the transmission performance of the Ku-band is. In order to ensure that the output window can meet the transmission within the required frequency band on both X/Ku bands without changing the structural parameters, there is a certain compromise in transmission. The blue line is the best choice for integrating the operating of two bands in Figure 2 (a) (b).

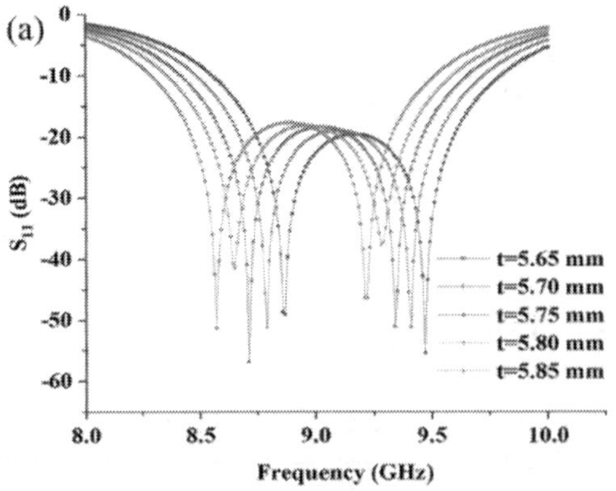

(a) X-band

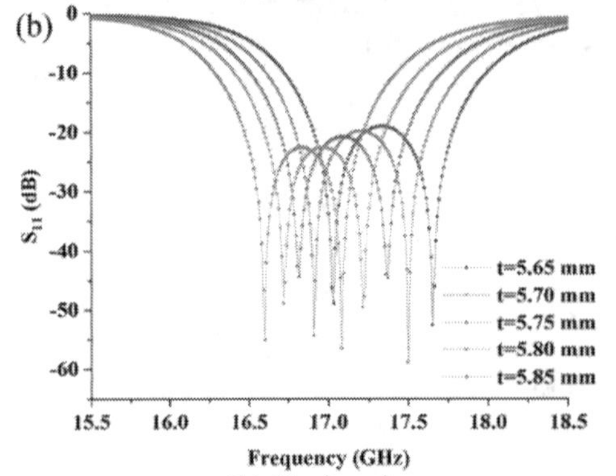

(b) Ku-band

Figure 2. S_{11} of output window varies with disk thickness t.

Finally, the disk thickness is 5.76 mm and the disk spacing is 7.4 mm. Figure 3 (a) (b) shows the S parameters calculated by the double-disk output window in X-band and Ku-band respectively. It can be seen that the transmission coefficient $S_{21} > -1$ dB in 8.29 GHz~9.73 GHz and the reflection coefficient $S_{11} < -20$ dB in 8.63 GHz~9.5 GHz, while the transmission coefficient $S_{21} > -1$ dB in 16.33 GHz~17.8 GHz, and the reflection coefficient $S_{11} < -20$ dB in 16.56 GHz~17.48 GHz. Figure 3 (a) and (b) also show the mode field distribution of 9.35 GHz and 16.77 GHz, respectively. It can be seen that the input and output electric field intensity are basically the same and the mode purity is high. The double-disk window output window can ensure good microwave transmission in X-band TE_{01} mode and Ku-band TE_{02} mode at the same time, meeting the working requirements of the X/Ku dual-band gyrotron amplifier.

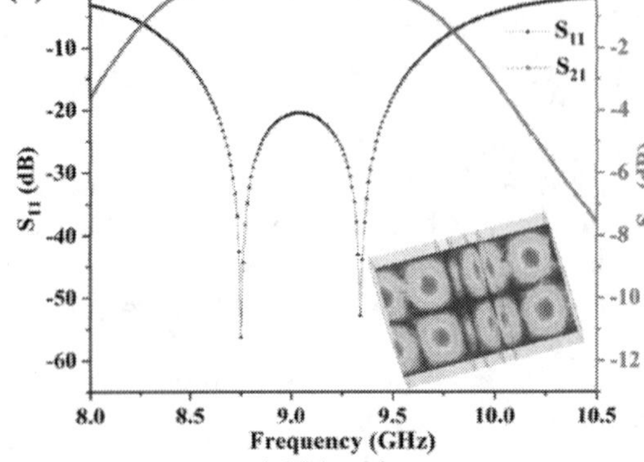

(a) X-band

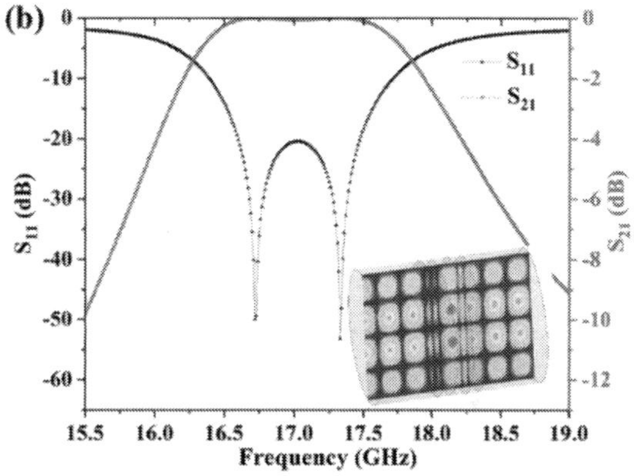

(b) Ku-band

Figure 3. S-parameter and field distribution of dual-band output window.

4. Conclusion

In this paper, a double-disk sapphire output window is designed for a dual-band gyrotron amplifier. Simulation results show that the thinner the window is, the better the transmission performance of the X-band is. The thicker the window is, the better the transmission performance of the Ku-band is. The final designed novel double-disk sapphire output can operate in X-band (8.29 GHz~9.73 GHz) TE_{01} mode and Ku-band (16.33 GHz~17.8 GHz) TE_{02} mode respectively, which contributes to the development of dual-band gyrotron amplifiers and promotes the development of dual-band radar, electronic countermeasures, and other technologies. In the future, it is necessary to continue exploring new materials and structures as window panels.

References

[1] Yong XU, Luo Y, Hong-Fu LI, et al. Design of a new type of broadband output window for high-power gyro-TWT [J]. Journal of Infrared and Millimeter Waves: 2013, 12:45-64.
[2] Samsonov S, Denisov G, Bogdashov A,et al.Recent Activity and Results on Gyro-TWT at the IAP [J].2023 24th International Vacuum Electronics Conference (IVEC), 2023:1-2. DOI:10.1109/IVEC56627.2023.10157649.
[3] Wang XY, Gao DP, Wang Y, et al. Design and Thermal Analysis of Magnetron Injection Gun for Dual-Band Gyroklystron [J]. IEEE Transactions on Plasma Science: 2022, 50(3): 670-677.
[4] Ma M, Zhao Q, Mo K, et al. Design of a dual-mode input structure for K/Ka-band gyrotron TWT [J]. Electronics, 2022, 11(3): 432.
[5] Tang Y, Luo Y, Xu Y, et al. Design of a Novel Dual-Band Gyro-TWT [J]. IEEE Transactions on Electron Devices, 2014, 61(11):3858-3863.
[6] Li S, Xu Y, Luo Y. Design of broad-band output window for Ku-band TE11 mode gyro-traveling wave tube [J]. Chinese Journal of Vacuum Science and Technology, 2014. DOI:10.3969/j.issn.1672-7126.2014.02.08.

Design optimization of anti-spin parachute for fighter-class aircraft

Qi Zhang* **and Jiaao Cheng**

Aircraft Flight Test Technology Institute, Chinese Flight Test Establishment, Xi'an 710089, China

*Corresponding author: qizhang0818@126.com

Abstract: In the absence of spin wind tunnel test and model free flight test support, the traditional empirical formulas had low accuracy and large dispersion, which made it difficult to effectively support the anti-spin parachute design of fighter planes. Based on the principle of spin motion, the empirical formula is modified from two aspects: aircraft aerodynamic layout and mass characteristics. At the same time, the relationship between the parachute area and the flight weight is summarized, which makes the estimated result closer to the actual value and more applicable. In addition, aiming at the problem of excessive load in the design of anti-spin parachutes, the causes of the problem are analyzed in detail. The design criterion of the anti-spin parachute is optimized. Under the premise of ensuring the recovery effect, the need to strengthen the aircraft body structure is effectively reduced. The results show that the engineering estimation methods and design criteria of the anti-spin parachute proposed in this paper are effective after simulation and actual application and can provide technical support for subsequent anti-spin parachute design of new fighter planes.

1. Introduction

In order to fully tap the combat potential of the aircraft on the left boundary, improve operational safety, and maximize the combat capability of the troops, currently, most combat aircraft at home and abroad are required to carry out high angle of attack stall and tailspin flight tests. Stall and tailspin are typical complex and dangerous conditions of aircraft. In order to ensure the safety of flight tests, GJB 3814-99 stipulates that an approved emergency recovery system should be installed for each type of QX and JQ spin-testing aircraft [1]. At present, the high angle of attack emergency recovery system mainly includes two kinds: anti-spin parachute and anti-spin rocket. The anti-spin parachute is widely used in domestic military aircraft flight tests because of its advantages, such as simple operation, high reliability, convenient operation, and maintenance.

The design of anti-spin parachute systems involves knowledge from multiple disciplines, including aerodynamics, flight mechanics, structural mechanics, drag parachute design, etc. This process is a complex and systematic project. The main work includes the selection of parachute type and the determination of several key design parameters, such as the length of the connecting rope, the length of the parachute rope, the parachute resistance area, the working load, the inflating time, and the porosity. At the same time, the design needs to take into account the stability, packaging volume, system quality, and other limiting factors [2].

Based on the experience and lessons of anti-spin parachute design for multi-type fighter aircraft in China, this paper analyzed in detail the two main problems in the design process of anti-tailspin parachutes for different types of aircraft and put forward the corresponding solutions.

2. Optimization of engineering estimation method

Among numerous design indicators, the resistance area of the anti-spin parachute is the most critical parameter in system design. If the area is too small, the parachute cannot provide enough downward pitching moment and stopping yaw moment after release. The aircraft cannot effectively recover from a high angle of attack state. If the area is too large, it will increase the system quality and packaging volume, elevating the installation difficulty and impacting aircraft aerodynamic characteristics. In addition, the increase of the parachute area, in the same state, although it can produce a greater recovery moment, will also increase the parachute load, resulting in higher structural strength requirements for the rear fuselage of the aircraft, which is often difficult to achieve. Therefore, it is necessary to reasonably design the parachute resistance area under the balanced consideration of various factors.

In order to ensure the rationality of the design of the umbrella resistance area, the preliminary scheme design is generally completed by engineering estimation according to the aerodynamic layout, mass/inertia characteristics, and body structure of the aircraft. Then, the model-free flight test or vertical wind tunnel test will be carried out to verify the effect of the anti-spin parachute, and then the final scheme will be designed. The specific process is shown in Figure 1.

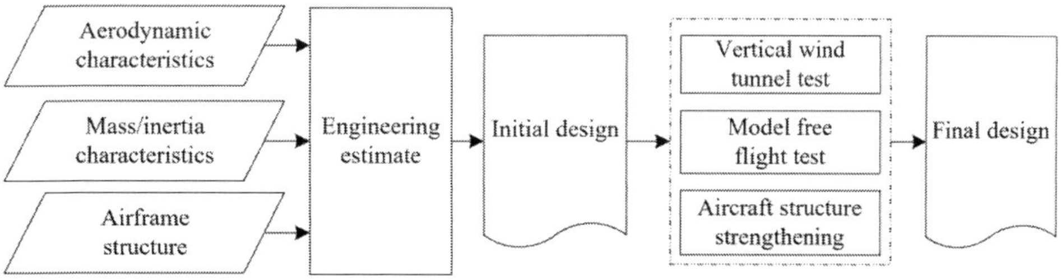

Figure 1. Design flow of anti-spin parachute.

However, in fact, in many aircraft flight tests, due to the tight project cycle, the anti-tailspin parachute design must be completed before the vertical wind tunnel test and model free flight test to ensure the test node. In this case, a reliable engineering estimation method is essential. However, according to the previous model design experience, the existing engineering estimation methods have large dispersion and low accuracy. Therefore, it cannot be directly used to guide the design of new aircraft anti-tailspin parachutes.

In this paper, the research on optimizing the anti-spin parachute engineering estimation method was carried out using the two aspects of engineering empirical formula modification and actual aircraft design data statistics.

2.1 Modification of engineering empirical formulas

2.1.1 Introduction of engineering empirical formulas. At present, there are four kinds of engineering empirical formulas commonly used in estimating the resistance area of the anti-spin parachute[3], [4], as shown in Table 1 for details.

Table 1. Engineering empirical formulas.

Formula	Remark
$C_D S_p = 0.0914 \dfrac{Sb}{l_p}$	(1)
$C_D S_p = 0.0325 \dfrac{W}{V_s}$	(2)

Journal of Physics: Conference Series **2879** (2024) 012012 doi:10.1088/1742-6596/2879/1/012012

$$C_D S_p = \frac{2\left(I_x \cos^2 \alpha_w + I_z \sin^2 \alpha_w\right)}{\rho V_w l_p \sin \alpha_w \left(l_p \sin \alpha_w + R_w\right) T_w} \quad (3)$$

$$C_D S_p = 4 \frac{(I_z - I_x) \cdot S}{W \cdot l_p^2} \quad (4)$$

In the formula:
C_D: Parachute resistance coefficient,
S_p: Parachute area (m^2),
S: Wing area (m^2),
B: Wing span (m),
L_p: Distance from hook to center of gravity (m),
W: The weight of the aircraft (kg),
V_s: Stall speed of clean configuration without power (km/h),
I_x、I_z: The moment of inertia of the roll and yaw axes, respectively (kg.m^2),
A_w: Angle of attack during spin (°),
ρ: Atmospheric density corresponding to the height at which the parachute is released (kg/m^3),
l_p: The distance from the hook to the center of gravity of the aircraft (m),
V_W: True speed in the spin (m/s),
R_W: Spin radius (m),
T_W: Expected or specification required spin-out delay time (s).

2.1.2 Analysis of engineering empirical formulas. At present, there is no clear literature on the validity of the above empirical formula. In this paper, the accuracy of the above four empirical formulas has been analyzed according to the design parameters of five different types of aircraft tailspin parachutes. The effect of these anti-spin parachutes has been verified by model-free flight test. It has been shown to be effective in getting the aircraft to recover from spin. The actual resistance area of the anti-spin parachutes and the estimated results based on the empirical formula are shown in Figures 2-5.

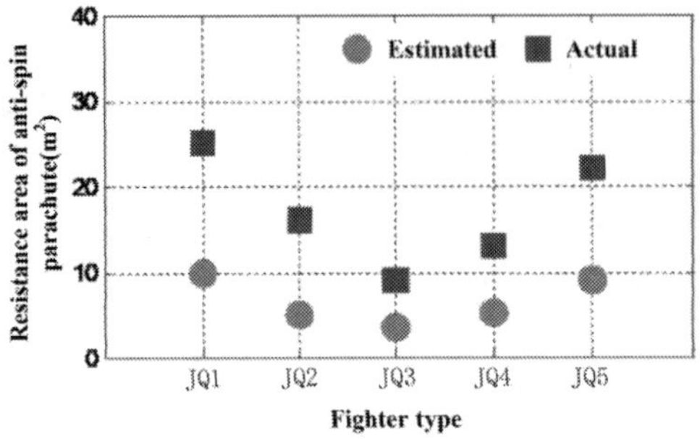

Figure 2. Comparison between the estimated value and actual value (Formula 1).

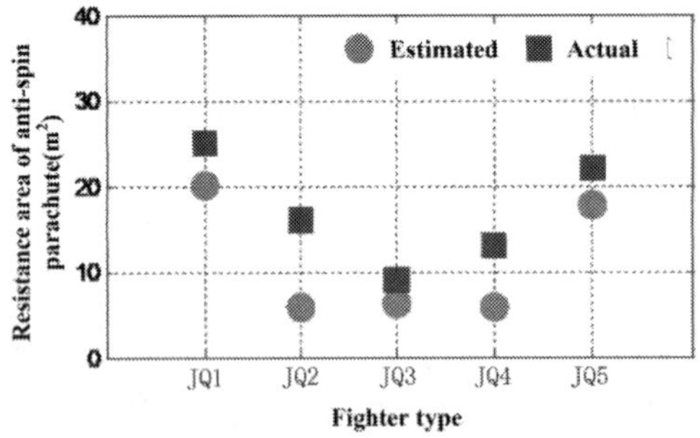

Figure 3. Comparison between the estimated value and actual value (Formula 2).

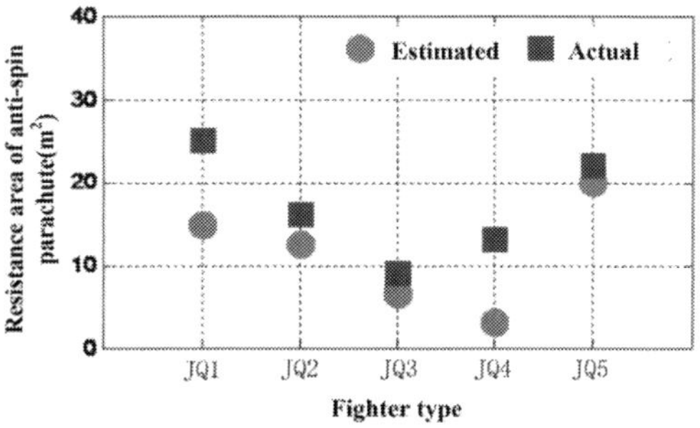

Figure 4. Comparison between the estimated value and actual value (Formula 3).

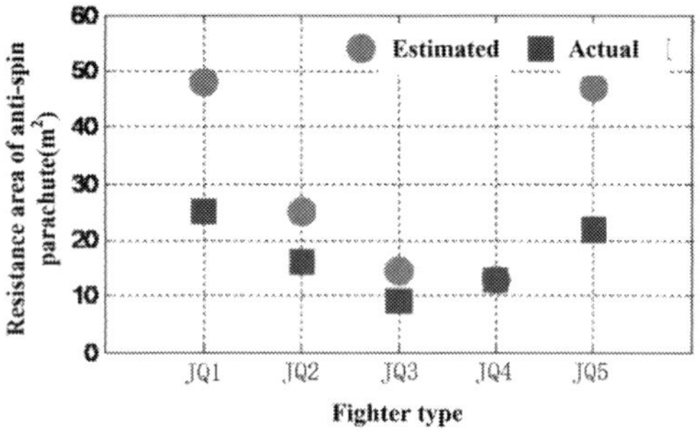

Figure 5. Comparison between the estimated value and actual value (Formula 4).

It can be seen that the estimated results of the four empirical formulas are only consistent with the actual values in some cases. In most cases, the D-value is large. Therefore, it cannot be directly used to guide the design of new aircraft anti-tailspin parachutes.

The analysis shows that the reason why the estimated results are poor is mainly because the above empirical formula was put forward earlier. In recent years, with the transformation of aircraft combat mode and the development of aviation technology, the aerodynamic layout and design ideas of aircraft have undergone tremendous changes. The aerodynamic characteristics of high angle of attack and mass inertia characteristics also change considerably[5]. Therefore, it is difficult to apply the previous empirical formula.

2.1.3 Modification of empirical formulas. Further analysis of the data in Figures 2-5, as well as Formulas (2), (3), and (4) shows no apparent correlation between the estimated value and the actual value. Reliable estimates are available only for individual aircraft. Therefore, it is difficult to correct. Although the estimated result of formula 1 is generally small, it is easy to correct, because there is an apparent linear relationship with the actual value of the resistance area of the anti-tailspin parachute. Accordingly, the proportional coefficient in Formulas (1) is modified by linear fitting. The revised estimation formula is as follows:

$$C_D S_p = 0.238 \frac{Sb}{l_p} \tag{5}$$

According to Formula (5), the anti-spin parachute resistance area of the above five types of aircraft was re-estimated. The comparative analysis results are shown in Figure 6.

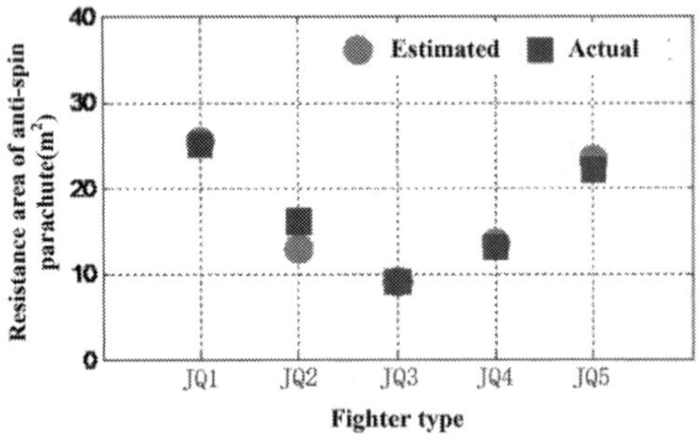

Figure 6. Comparison between the estimated value and actual value (Formula 5).

It can be seen from Figure 6 that the resistance area of the anti-spin parachute calculated according to the revised estimation formula is basically consistent with the actual value.

Further analysis shows that in the past, the tailspin characteristics of aircraft with wasp-waisted fuselage and large swept-wing layouts were usually buffeted tailspin with low steady-state spin Angle of attack. However, modern aircraft generally adopt the aerodynamic layout of the side wing and wing-body fusion. According to the results of model free flight tests, the tailspin modes of modern aircraft are mostly flat tailspin with high steady-state spin Angle of attack. It's harder to get the plane to recover from spin. Therefore, a larger anti-spin parachute is required [6], [7].

2.2 Statistical rules of different aircraft design data

Formula (5) is mainly based on the aerodynamic layout parameters of the aircraft. However, according to the principle of spin motion, the spin characteristics of aircraft are affected by both aerodynamic layout and mass/inertia characteristics [8] [9]. Therefore, in order to estimate the parachute area more comprehensively and accurately, the influence of mass/inertia characteristics must be considered.

For this reason, data on more than ten types of aircraft at home and abroad are collected and sorted out. Since moment of inertia data is challenging to obtain, it is also considered that the mass of modern aircraft is usually concentrated in the fuselage. The inertia characteristics have similar characteristics. That is, the moment of inertia of the pitch axis is equivalent to that of the yaw axis and much larger than that of the rolling axis. Therefore, the influence of quality characteristics is mainly considered in the analysis.

Figure 7 shows the relationship between the anti-spin parachute resistance area and the flight weight of more than ten different types of aircraft at home and abroad. It can be seen from the figure that there is a good linear relationship between the anti-spin parachute resistance area and the flight weight of the aircraft. That is, as the flight weight of the aircraft increases, a larger anti-spin parachute is required to recover from the spin.

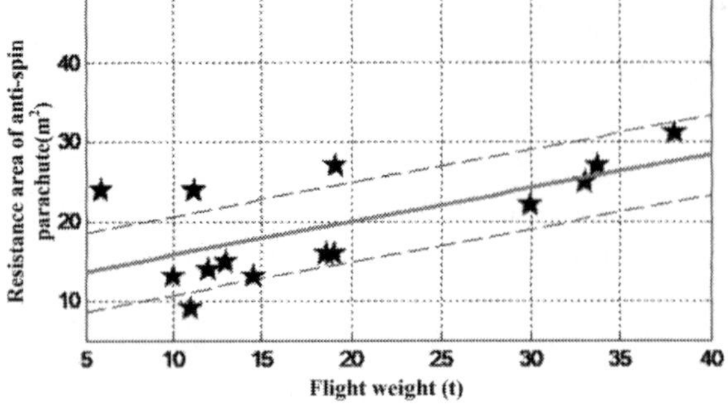

Figure 7. The resistance area of the anti-spin parachute varies with the flight weight.

Through linear fitting, the relationship between the resistance area of the anti-spin parachute and the flight weight is obtained, as shown in Formula (6).

$$C_D S_p = 4.2 \times W \times 10^{-4} + 11.5 \tag{6}$$

In addition, it can be seen from the figure that the anti-spin parachute resistance area of different aircraft is basically within the range of ±5 m² of the fitting result. That is within the two dotted lines in Figure 7. This indicates that due to the different aerodynamic layouts of the aircraft with similar weights, the anti-spin parachute resistance area required to recover from the tailspin will also be different. Therefore, Formula (5) should be applied comprehensively in the estimation. In the case that the structural strength of the aircraft's rear fuselage is sufficient, it is recommended to take the maximum value of the estimated results of Formulas (5) and (6) as the design value of the anti-spin parachute resistance area.

2.3 Validity verification

2.3.1 Comparison and verification of estimation results of foreign fighter aircraft. In order to verify the validity of the revised formula, the anti-spin parachute resistance area of 8 different types of foreign aircraft, such as F22, F16, and Su-27, was estimated according to Formulas (5) and (6). The results are shown in Table 2.

Table 2. Estimation results of anti-spin parachute area of typical foreign aircraft.

Aircraft type	Estimated result (m²)		Comprehensive estimation result (m²)	Actual Value (m²)
	Formula (5)	Formula (6)		
Kfir	13.6	17.6	17.6	13
X-29	9.9	14.6	14.6	14
AMX	8.9	16.9	16.9	15
SSAB37	15.5	19.3	19.3	16
Su27	25.5	24.1	25.5	26
F16	10.4	19.5	19.5	27
F14	27.1	25.7	27.1	27
F22	28.1	27.5	28.1	31

Compare and analyze the comprehensive estimate results (The maximum value in the estimation results of Formulas (5) and (6)) and the actual values in Table 2, as shown in Figure 8 for details. It can be seen that the estimated results of different aircraft agree well with the actual values. It shows that the formula for estimating the area of the anti-spin parachute proposed in this paper is effective and universal.

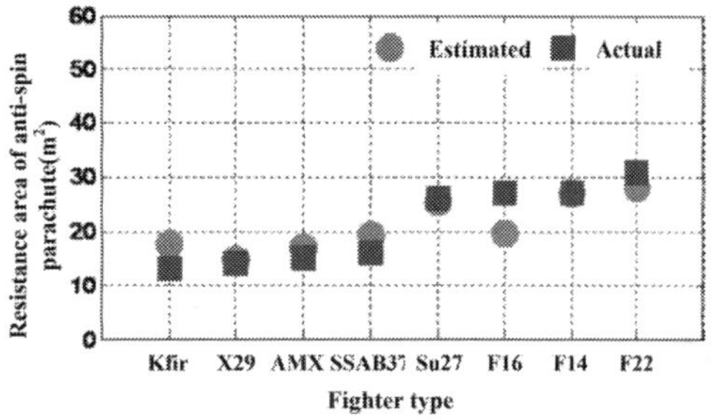

Figure 8. Comparative analysis results of anti-spin parachute area of typical foreign aircraft.

2.3.2 Simulation Verification. In order to further verify the validity of Formulas (5) and (6), a trainer aircraft is taken as an example, and the resistance area of the anti-spin parachute is estimated to be between 13m^2 and 16m^2. To be conservative, 16m^2 is taken as the design value. At the same time, based on MATLAB/SIMULINK, the high angle of attack dynamics and kinematics simulation model [10] and the anti-spin parachute dynamics model [11] were established.

The simulation analysis results are shown in Figure 9. Initially, the aircraft flew at a stable level with a basic altitude of H_p=10000m and an indicated airspeed of V_i=245km/h. t=10s, the pilot began to slowly pull the lever. When the aircraft stalls, we quickly pull the rod to the bottom while pushing the left rudder to the bottom and hold. The aircraft entered a stable tailspin with an angle of attack of 45° to 50° and t=80s, releasing the anti-spin parachute. Subsequently, the aircraft's yaw rate rapidly decreased. Meanwhile, the angle of attack rapidly decreases below the stall angle of attack. This indicates that the designed anti-spin parachute can provide sufficient pitch and yaw moment to recover the aircraft from a spin state.

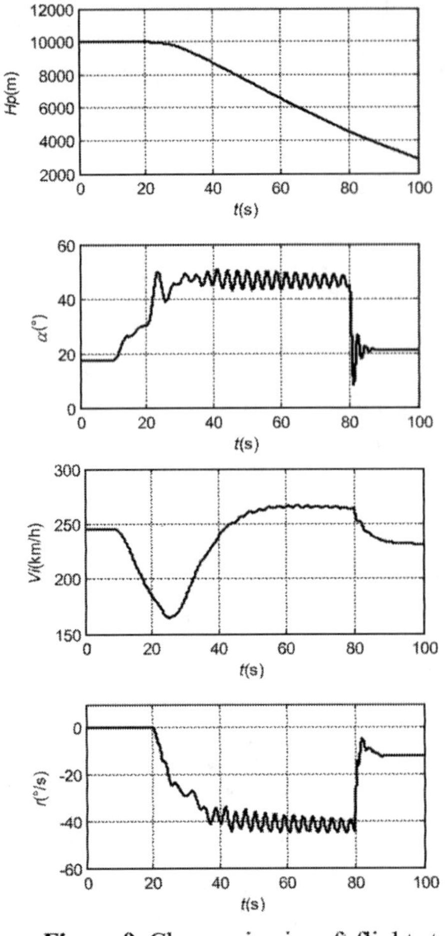

Figure 9. Changes in aircraft flight status.

2.4 Brief summary

Based on data from more than ten types of aircraft at home and abroad, this chapter revises and optimizes the resistance area estimation formula for fighter aircraft anti-spin parachutes from two aspects of aircraft aerodynamic layout and mass inertia characteristics. The results of comparative verification and simulation analysis show that the estimation formulas presented in this chapter are effective. The estimated anti-spin parachute can effectively recover the aircraft from the spin state.

In the absence of sufficient wind tunnel test data and model-free flight test data, it is suggested that the estimation formulas proposed in this chapter be used for the anti-spin parachute design.

3. Revision of design criteria for anti-spin parachute

3.1 Problem Description

Another prominent problem often faced in the design of anti-spin parachutes is that the strength of the aircraft structure makes it challenging to meet the requirements of parachute load.

For aircraft flight tests, in order to ensure the effectiveness of the anti-spin parachute, it is often expected that the larger the parachute, the better, and it can be used in various flight conditions. However, this also leads to a large parachute load, which sharply increases the difficulty of strengthening the rear fuselage bearing structure of the aircraft, especially for mass-produced aircraft. It is difficult to make significant changes to the aircraft's main load-bearing mechanism, which may result in the problem of not being able to meet the requirements for anti-spin parachutes.

3.2 Problem Analysis

To solve this problem, first analyze the umbrella load. According to the force analysis, we decompose the parachute loads into the body axis system. The parachute loads under the body shafting in different directions are calculated as follows:

$$\begin{aligned} F_x &= \bar{q} \cdot C_D \cdot S_p \cdot K_p \cdot \cos\beta_w \cdot \cos\alpha_w \\ F_y &= \bar{q} \cdot C_D \cdot S_p \cdot K_p \cdot \sin\beta_w \\ F_z &= \bar{q} \cdot C_D \cdot S_p \cdot K_p \cdot \cos\beta_w \cdot \sin\alpha_w \end{aligned} \tag{7}$$

In the formula:

F_x, F_y, F_z: Parachute loads in different directions under the body shafting (N),

$\bar{q}$: Dynamic pressure (pa),

K_p: Impact coefficient when releasing the parachute,

β_w: Sideslip angle (°).

From the formula, it can be seen that the values of parachute loads in different directions are mainly related to the parachute resistance area and operating conditions. Among them, the operating conditions refer to the flight states of the aircraft when the anti-spin parachute is used. In the design process, it is required that the anti-spin parachute can normally be used under various motion modes that may exist after the aircraft enters a high angle of attack state, including deep stall, post-stall rotation, spin, etc., as well as during the aircraft leveling process. However, the motion modes of the aircraft at high angles of attack belong to the inherent characteristics of the aircraft, which are difficult to change.

Accordingly, to solve the problem of excessive load, solutions can only be analyzed and studied from the perspective of the parachute resistance area.

3.3 Rationality analysis of design criteria

Currently, the determination of the resistance area of the anti-spin parachute is mainly based on GJB3814-99 while referring to the relevant provisions on spin recovery in GJB 185-86.

GJB3814-99 specifies that the anti-spin parachute should be able to reverse the tailspin at the most adverse possible maneuvers within the appropriate altitude loss. Reference to the requirements of GJB 185-86: fighter class aircraft should be recovered within a delay of 2 turns of tail rotation. In summary, it can be concluded that the design criteria of the anti-spin parachute are as follows: the aircraft should be recovered out of the tailspin in the most unfavorable maneuvering position possible within the delay of 2 turns of the tailspin.

In the past design of anti-tailspin parachutes, the most unfavorable control position was usually defined as the assisted spin control position, which is to maintain the control during spin entry and the control opposite to changing out spin. This requirement can be met for some aircraft. However, with the increase of aircraft types, it is found that in many aircraft meeting the requirements of this article, the design of the parachute will be unusually large, resulting in the parachute load far exceeding the structural strength limit of the aircraft rear fuselage. The analysis shows that the reason why the area of the parachute is abnormally large in this case is mainly because the longitudinal control surface and transverse direction control surface of some aircraft still have high control efficiency under the condition of a high angle of attack. That is, the pull rod can still generate a lift moment, the pressure plate can still generate a roll moment, and the rudder can still generate a yaw moment for the aircraft.

According to the spin motion equation, the key to correcting the spin is to prevent rotation[12] [13]. However, under assisted tailspin control, in order to stop spinning, the anti-spin parachute must overcome the additional aerodynamic torque caused by assisted tailspin control. At this point, if the aircraft has high control efficiency, greater torque is required to stop spinning, which results in a large anti-spin parachute area.

In fact, generally speaking, aircraft with higher operational efficiency at high angles of attack are more likely to recover from spin. Therefore, there is no need for a large anti-spin parachute to provide additional pull-out moments. This is contrary to the design result. Consequently, it is obviously unreasonable to design an anti-spin parachute based on the above criteria.

3.4 Revision of design criteria

Further analysis of the national military standard requirements, the reason why it is required to change the spin in the most unfavorable maneuvering position is mainly to consider that the wrong maneuvering mode may be implemented in the process of spin change due to the unreasonable change mode determined by the previous theoretical research, the wrong judgment of the spin state (left and right) or the pilot's orientation loss. It is mainly reflected in transverse and heading control[14].

Through research and analysis, in order to avoid such erroneous manipulation during spin recovery and reduce the parachute resistance area, considering a compromise, the pilot's control when using an anti-spin parachute to recover from the spin will be clearly defined as three neutral positions of the stick and rudder. In this way, under a clear form of control, the pilot will not make mistakes and can try to avoid adverse torque changes caused by incorrect push or pull rods, pressure plates, and rudder pedals. Therefore, the demand for anti-spin parachutes can be significantly reduced, and the same applies to anti-spin parachute loads. This alleviates the need for structural strengthening of the aircraft. Simulation analysis shows that for some aircraft, this method can reduce the anti-spin parachute loads by about 50%.

3.5 Brief summary

To balance the structural strength limitations of the aircraft body and ensure the recovery effect, the design criteria for anti-spin parachutes should be revised to: the aircraft should be recovered from spin within a delay of 2 turns, while maintaining the three center positions of the rudder and stick.

4. Conclusions

This article proposes a more applicable and accurate method for estimating the resistance area of the anti-spin parachute in fighter-class aircraft. At the same time, by further interpreting the requirements of the national military standards, the design criteria for anti-spin parachutes have been optimized. On the premise of ensuring the recovery effect, the demand for strengthening the aircraft body structure has been effectively reduced. The relevant research results can provide technical support for the design of anti-spin parachutes for future new aircraft.

References

[1] Li Shuyou, et al. GJB3814-99 Stall/post-stall/spin flight test demonstration requirements for military airplanes [s]. Beijing: Chinese Aviation Ministry of Industry Thirty-first Research Institute,1999.

[2] Li Shuyou, ZHANG Pei tian. The Design Criteria of Anti-Spin Parachute System [J]. Flight Dynamics, 2002, 20(1): 46-50.

[3] Wang Qi. A Study of Anti-spin Parachute System Designing and Testing [D]. Xi'an: Northwestern Polytechnical University, 2007

[4] Zhao Tao, WANG Qi, WANG Jianpei, et al. Research on simulation of anti-spin parachute system [J]. Flight Dynamics, 2001, 19(4): 19-22.

[5] Zhang Jigao. The partition to the fighter configuration[J] Acta Aerodynamica Sinica, 2009, 27(5): 616-622.

[6] Sang Yusheng, TIAN Peiyan, WEI Yusheng, et al. Analysis of Departure/Spin Characteristics of F-16 at High Angle of Attack [J]. Flight Dynamics, 2002,20(1):58-62.

[7] Li Xue qin, Gong Xi qing, Jia Xiaopeng, et al. Research on JL8 aircraft stall and spin flight test [J]. Flight Dynamics,2001,19(2):57-61.

[8] Duan Yiqian, Han Taofeng,Yang guang, et al. Study about the influence of inertia characteristics on Aircraft Spin Characteristics and Recovery[J].Trainer, 2019,(4):24-28.

[9] Xu Yanjun, Wang Rui. Analysis on inertial moment of spinning aircraft based on data from flight parameter recorder[J]. aircraft design,2023,43(01):29-32.

[10] Rao Qiulei, Zhang Qi. Solving method of steady state spin based on optimization theory[J].Chinese Journal of Applied Mechanics,2019,(4): 759-764.

[11] Rao Qiulei, Wang Qi. Analyse and Numerical Simulation about Deep-Stall Character of High Horizontal Tail Plane and Stall Recovery Parachute [J]. Advances in Aeronautical Science and Engineering, 2010,(3):214-218.

[12] Tan Jianmei, Wang Junqiu. Aircraft spin recovery technique based on deep r enforcement learning [J]. Aero weaponry, 2024,31(01):77-88.

[13] Zhang Zijun, Zhao Tong, Sun Ye, et al. Review of the Study on High-angle-of-attack Flight Problems of Aircraft[J].Advances in Aeronautical Science and Engineering, 2022,13(03):74-85.

[14] Fang Zili, He Tai, Li Jikuan, et al. A Flight Test Method for Departure Sensitivity at High Angle of Attack Based on Decelerated Climbing Maneuver[J]. Jounal of Air Force Engineering University, 2023,24(01):76-81.

Journal of Physics: Conference Series **2879** (2024) 012013 doi:10.1088/1742-6596/2879/1/012013

Study on the mechanical properties and microstructures of Al-Cu-Mg alloy thick plates obtained by different hot rolling processes

Kai Zhu[1,2,3], Qilong Liu[1,2,3], Jinlai Fu[4], Jing Lv[1,2,3], Lizhen Yan[1,2,3], Yanan Li[1,2,3] and Xiwu Li[1,2,3,*]

[1]State Key Laboratory of Nonferrous Metals and Processes, China GRINM Group Co., LTD., Beijing 100088, China
[2]GRIMAT Engineering Institute Co., LTD., Beijing 101407, China
[3]General Research Institute for Nonferrous Metals, Beijing 100088, China
[4]Northeast Light Alloy Co., Ltd, Harbin 150060, China

[*]Correspondence: lixiwu@grinm.com

Abstract: In the research, the microstructures of large-scale 2324 aluminum alloy thick plates obtained through two different hot rolling processes are investigated by using Scanning Electron Microscopy (SEM), Energy Dispersive Spectroscopy (EDS), and Electron Backscatter Diffraction (EBSD). Additionally, the tensile properties of the plates are evaluated at room temperature. The results indicate that the thick plates produced by Process A exhibit higher tensile and yield strengths along the L (longitudinal) and LT (long transverse) directions. The higher proportion of deformed microstructures is an important factor contributing to their enhanced tensile and yield strengths. Meanwhile, the different rolling processes have minimal impact on the kind and distribution of the second phase in the hot-rolled thick plates, with the second phase primarily consisting of Al_2Cu and Al_2CuMg phases, which are distributed in a streamlined way along the rolling direction of the thick plates. The microstructures of the hot-rolled thick plates obtained from both hot-rolling processes are dominated by deformed structures. The thick plates produced by Process B have relatively smaller grain sizes and a higher degree of fragmentation, making their microstructure more susceptible to recrystallization transformation during subsequent solution heat treatment.

1. Introduction

Aluminum alloys, often favored for their lightweight properties, offer numerous advantages such as exceptional conductivity, low density, high strength, robust corrosion resistance, malleability, and the ability to be recycled. They have been utilized in various sectors, including as electronics, transportation, aircraft, and architectural design [1, 2]. Especially in the aerospace industry, aluminum alloys are treated as a critical role. Today, aluminum alloys have become the primary material choice in large aircraft design and manufacturing processes. For instance, the 2xxx series (Al-Cu-Mg) hard aluminum alloys represented by 2024-T3 and 2524-T3 are mainly used in positions such as fuselage skin and wing cover [1, 3]. The 7xxx series (Al-Zn-Mg-Cu) high-strength aluminum alloys represented by 7050-T7451 and 7085-T7451 are mainly applied in positions like fuselage frames, longerons, and wing panels [4]. Additionally, the novel Al-Li alloys represented by 2050-T8 and 2196-T8 are mainly used in positions such as bulkhead support structures, floor beams, or slide rails [5, 6].

In recent years, higher performance standards have been set for aluminum alloys in response to the aviation industry's need for lightweight structural materials with improved strength, hardness, and reliability. Based on the conventional AA2024 alloy, Alcoa successfully created a new kind of age-hardenable high-purity aluminum alloy known as 2324 in the 1980s by lowering the proportion of the primary alloying element Cu and tightly regulating the content of impurity elements like Fe and Si [7-9]. Because there are fewer Cu-containing second phases in the AA2324 alloy than in the AA2024 alloy, the alloy has better strength, fracture toughness, and a lower density than the AA2024 alloy. Additionally, the alloy's fatigue resistance is enhanced.

The primary form of the 2324-T39 aluminum alloy is a thick plate, which is primarily utilized for manufacturing wing integral structural components that have stringent demands for fracture toughness and fatigue performance. This is mainly because, by integrating components, the overall number of aircraft body parts is greatly decreased, resulting in a considerable improvement in the fatigue resistance of the aircraft body. This, in turn, leads to a substantial increase in the service life and reliability of the aircraft body. The advancement of thick plate billet preparation technology and improved machining technology has led to the emergence of large-scale, integrated, and complicated aluminum alloy components for aircraft bodies as an unavoidable trend. The hot rolling procedure is crucial in the production of thick plates made from aluminum alloys. It has a direct impact on the homogeneity of the structure, the distribution of secondary phases, and the solid solution heat treatment system of the alloy. This procedure is essential in defining the ultimate microstructure and mechanical properties of the alloy. However, there is a scarcity of research publications on this topic [10-12]. Therefore, this paper focuses on studying the mechanical properties of thick plates made from the 2324 aluminum alloy obtained by two different hot-rolling processes. Moreover, the mechanical characteristics of the aluminum alloy were explored by characterizing and analyzing its microscopic structures under T39 and hot-rolling circumstances in order to investigate the influence of different hot-rolling processes. This research can be conducive to guiding the manufacturing of thick plates made from 2324-T39 aluminum alloy.

2. Experimental

The 2324 aluminum alloy thick plates studied in our work were manufactured by Northeast Light Alloy Co., Ltd. The semi-continuous casting technique was used to make aluminum alloy ingots, which were then subjected to homogenization heat treatment. Following that, two processes named A (controlled cooling and rolling) and B (traditional hot rolling) were employed for the hot rolling procedure, respectively. Subsequently, the process involved solution heat treatment, cold rolling, and a 1.5% pre-stretching treatment (T39), resulting in the production of thick plates with a thickness of 25.4mm. The tensile properties of the alloy at room temperature were evaluated using a SANS CMT 4304 microcomputer-controlled universal testing machine with a tensile rate of 2mm/min. The LS surface of the samples was examined and analyzed using EBSD to investigate and characterize its microstructure. Following the grinding and polishing of the samples, an electrochemical polishing process was carried out using a high-chloric acid electrolytic polishing solution consisting of 10% ethanol. Throughout the electrochemical polishing procedure, the voltage was regulated within the range of 20V to 25V, the current was maintained below 0.5A, and the duration of the polishing operation was roughly 10 seconds. The microstructures and textures of the alloys were observed and characterized by using a JEOL JSM 7001F scanning electron microscope (SEM, JEOL Corporation, Tokyo, Japan) equipped with an electron backscatter diffraction system (EBSD, EDAX Inc., America). Additionally, an EDS (Oxford Instruments, Oxford, England) analysis was utilized to identify the types of second-phase particles present in the plates. The microstructure observation samples were polished in a sequential manner using water sandpapers with grit sizes of 240#, 600#, 1000#, 2000#, and 5000# on the surface being observed. Following the first grinding, a 1μm diamond polishing paste was then applied to the observation surface until all apparent scratches were eliminated. Figure 1 shows the locations where the room temperature tensile properties evaluation and the microstructure observation of the alloys are concerned. It should be noted that the specimens used for both tensile testing and microstructure investigation were extracted from the middle section of the plate.

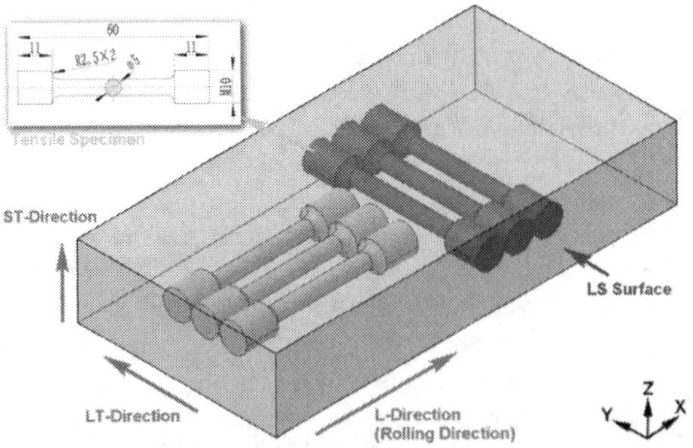

Figure 1. Sampling schematic of the specimens for room temperature tensile test, microstructure observation, and chemical composition analysis.

3. Results and discussion

3.1 Room temperature tensile properties

Figure 2 exhibits the outcomes of the tensile test performed on thick plates composed of 2324-T39 aluminum alloy at room temperature. The plates were produced using two distinct hot rolling methods, referred to as A and B. The tests were carried out in both the L and LT directions. Firstly, the figure shows that the alloy produced using hot rolling process A has better tensile strength and yield strength along the L direction at room temperature compared to the alloy samples produced by hot rolling process B. The alloy produced through hot rolling process A exhibits a tensile strength that is about 21 MPa more and a yield strength that is approximately 88 MPa greater than the alloy produced through process B. Nevertheless, the elongation of the alloy produced using hot rolling method A is comparatively reduced. Furthermore, in line with the room temperature tensile test results observed in the L direction, the mechanical properties of the alloy manufactured using hot rolling process A in the LT direction likewise surpass those of the alloy generated by process B. The alloy from procedure A has a tensile strength of 509 MPa and a yield strength of 393 MPa along the LT direction. Similarly, elongation in this direction is not an advantage of the alloy produced from hot rolling process A compared to that fabricated by using process B. Furthermore, a comprehensive examination of the tensile characteristics of the alloy fabricated through hot rolling process A at ambient temperature reveals that, apart from the increased tensile strength in the LT orientation, the yield strength and elongation in the LT orientation are comparatively lower than those in the L orientation.

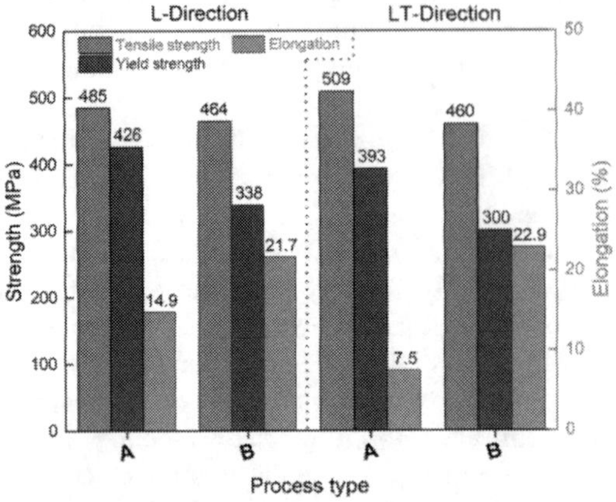

Figure 2. The mechanical characteristics of the alloys produced using two distinct hot-rolling methods.

3.2 EBSD Characterizations of the as-finished 2324 aluminum alloy

Figure 3 displays the EBSD characterization data of the LS surface of the 2324-T39 aluminum alloy, corresponding to two distinct hot rolling methods. Figures 3(a_1) and (a_2) demonstrate that the alloy produced by procedure A has a significant proportion of grains with an average misorientation ranging from 0.7 to 5 and a substantial proportion of low-angle grain boundaries (<15°). This suggests that the microstructure of the alloy produced by process A consists mainly of deformed structures. In contrast, the alloy produced through process B exhibits a higher proportion of grains with an average misorientation ranging from 0 to 0.7 and a more significant presence of high-angle grain boundaries (>15°). This indicates that the microstructure of the alloy produced by process B is predominantly composed of recrystallized structures. Thick plates or forgings that have dominant deformation structures in their microstructure are likely to demonstrate increased tensile strength and yield strength. In other words, if the alloy is mainly composed of recrystallized structures, its mechanical qualities will be comparatively diminished [13]. Furthermore, the statistical findings presented in Figure 3(b_3) demonstrate that the alloy manufactured using process B exhibits comparatively reduced grain sizes, leading to a more excellent elongation in comparison to the alloy created by process A. To clarify the connection between micro-morphologies and mechanical properties, previous studies have shown that the shape of grains affects the elastic responses [14]. This phenomenon occurs as a result of the synchronized distortion generated by either the rolling process or heat treatment, resulting in distinct elastic characteristics. These disparities are thought to be accountable for the emergence of inherent tension at the interfaces between grains. In addition, when the externally applied force is paired with it, it leads to a higher level of shear stress at the boundaries between the grains rather than within the grains themselves. Hence, plastic deformation, which refers to the displacement of dislocations, predominantly occurs in proximity to grain boundaries at the onset. When external stress is applied in the direction of the long axes at a 45° angle, elongated grains will undergo plastic deformation more efficiently. This also leads to the propagation of dislocations into the matrix, resulting in a decreased yield strength of the material. The 2324-T39 alloy thick plate produced by hot rolling method A has a higher yield strength due to its internal structure consisting primarily of elongated grains aligned along the rolling direction. On the other hand, the alloy thick plate produced using hot rolling method B exhibits a comparatively reduced yield strength due to its relatively tiny internal grain size. The preferred orientation along the rolling direction exhibits a low degree, resulting in the rotation of the grains' stretching axes towards the 45° direction when subjected to tensile loading. The alloy's yield strength is weakened by this rotation.

Additionally, the impact of grain boundary strengthening on the yield strength of alloys is also regarded as a significant component. The grain boundary strengthening effect can be mathematically represented by Equation (1) [14, 15]:

$$\sigma_{gb} = \alpha_2 Gb[(1 - f_{rex})(1/\delta) + f_{rex}(1/D)] \tag{1}$$

In the equation, G represents the alloy's shear modulus, b is the Burgers vector, f_{rex} denotes the recrystallization fraction of grains, δ is the size of the sub-grains (within the deformed grains), D is the size of recrystallized grains, and α_2 is a constant. Since the size of recrystallized grains D is much larger than the size of the substructure δ, the above equation can be simplified to the following Equation (2):

$$\sigma_{gb} = \alpha_2 Gb(1 - f_{rex})(1/\delta) \tag{2}$$

Based on the previous analysis, it indicates that the alloy produced by using hot rolling process A has a large fraction of low-angle grain boundaries (<15°) in the grain boundary reconstruction diagram, which means that there is a higher proportion of sub-grains in the microstructure of the alloy. Equation (2) is another reason for the alloy, which was produced through hot rolling process A, having a greater yield strength. On the other hand, the alloy produced by hot rolling process B has a lower yield strength due to a larger degree of microstructure recrystallization that limits the effect of reinforcing grain boundaries.

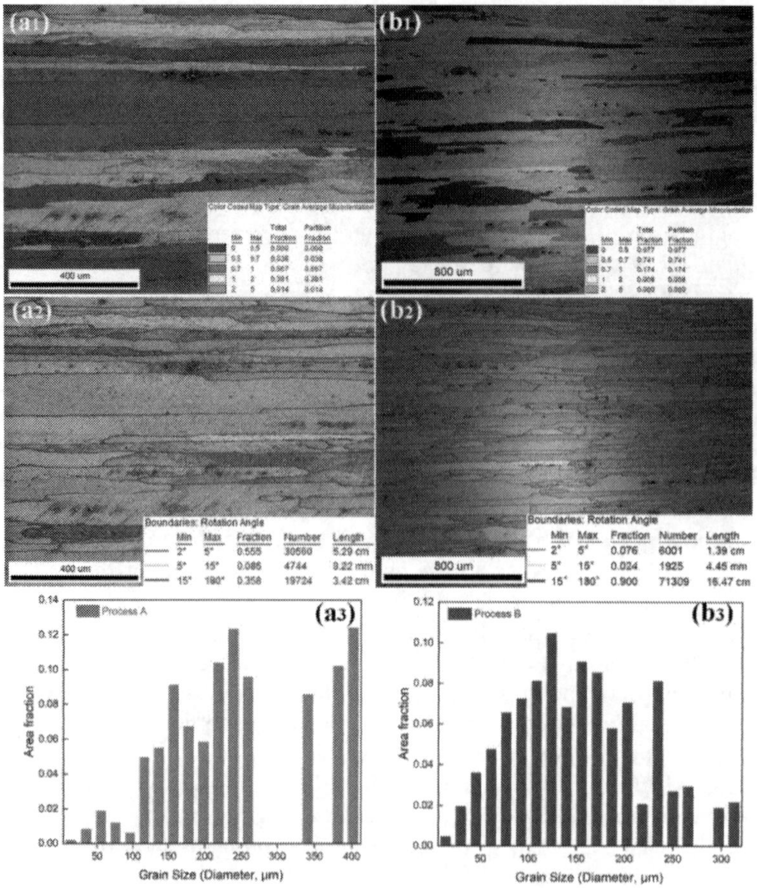

Figure 3. EBSD characterization results of the two different types of 2324 aluminum alloys: (a_1), (a_2), and (a_3), respectively, represent the grain average misorientation map, grain boundary reconstruction map, and the statistical results of grain size for the alloy produced by process A, as well as (b_1), (b_2), and (b_3) for the alloy produced by process B.

3.3 SEM observations and DSC analysis of the as-hot-rolled 2324 aluminum alloy

Figure 4 illustrates the SEM images and the DSC curves of the hot-rolled alloys, which were processed using two different hot-rolling techniques. The scanning electron microscopy (SEM) data depicted in Figures 4(a) and (b) demonstrate that the second phases in both types of alloys display a streamlined alignment that is parallel to the direction of rolling. The EDS investigation indicates that the main secondary phases present are Al_2Cu and Al_2CuMg. It is worth mentioning that the Al_2CuMg phase is often observed together with the Al_2Cu phase. This indicates that some Al_2CuMg phases with low melting points have formed surrounding the Al_2Cu phases during the air-cooling process after the massive ingots have been treated with homogenization. The distribution and kinds of second phases inside the alloys are minimally affected by the varied hot-rolling methods, as observed in the SEM characterization data. In addition to the scanning electron microscope (SEM) results, the differential scanning calorimetry (DSC) analysis, shown in Figure 4(c), reveals a significant increase in temperature near 510°C for both hot-rolled alloys. The temperature is marginally below the distinctive endothermic peak temperature of the Al_2Cu phase, indicating that the presence of the Al_2CuMg phase, which has a lower melting point, might be responsible for the observed change in the peak temperature.

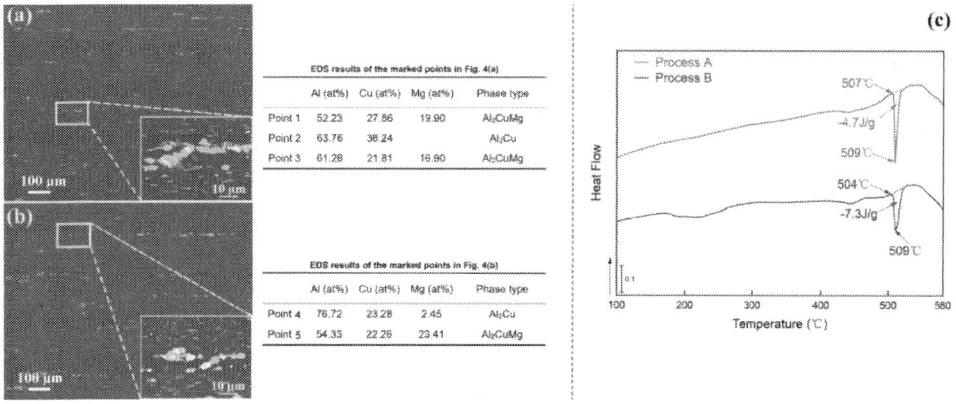

EDS results of the marked points in Fig. 4(a)

	Al (at%)	Cu (at%)	Mg (at%)	Phase type
Point 1	52.23	27.86	19.90	Al_2CuMg
Point 2	63.76	36.24		Al_2Cu
Point 3	61.28	21.81	16.90	Al_2CuMg

EDS results of the marked points in Fig. 4(b)

	Al (at%)	Cu (at%)	Mg (at%)	Phase type
Point 4	76.72	23.28	2.45	Al_2Cu
Point 5	54.33	22.26	23.41	Al_2CuMg

Figure 4. The SEM images and EDS point analysis results of the two alloys are shown in (a) and (b), and the corresponding EDS analysis curves are plotted in (c).

3.4 EBSD Characterizations of the as-hot-rolled 2324 aluminum alloy

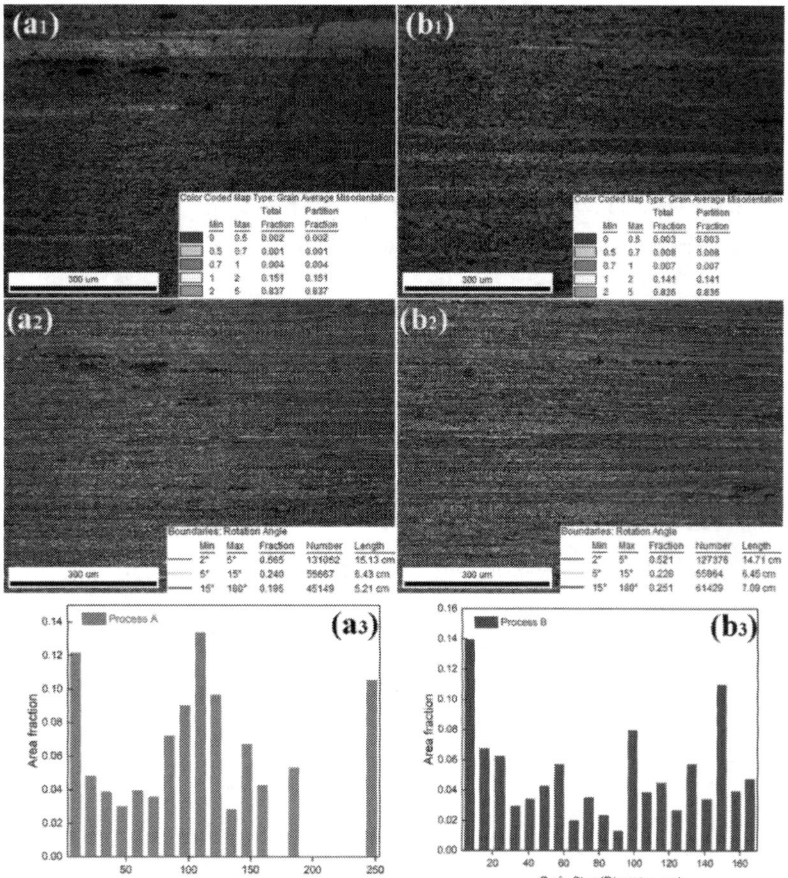

Figure 5. EBSD characterization results of the two different types of as-hot-rolled 2324 aluminum alloys: (a_1), (a_2), and (a_3), respectively, represent the grain average misorientation map, grain boundary reconstruction map, and the statistical results of grain size for the alloy produced by process A, as well as (b_1), (b_2), and (b_3) for the alloy produced by process B.

Figure 5 shows the Electron Backscatter Diffraction (EBSD) outcomes of the microstructures produced by two distinct hot-rolling methods. The figure displays grain average misorientation and grain boundary reconstruction maps. It is evident from the maps that the proportion of grains with an average misorientation ranging from 0.7 to 5 and the percentage of small angle grain boundaries below 15° are both considerably high in the hot-rolled thick plates of the two processes. This suggests that the prevailing microstructure type in the hot-rolled thick plates is a deformed structure. Furthermore, as can be seen from the statistical results shown in Figures 5(a_3) and (b_3), the alloy made with Process A has a relatively larger grain size in its hot-rolled microstructure. In comparison, the hot-rolled microstructure of the alloy produced by Process B has a relatively lower grain size. The statistical analysis of grain size indicates that the alloy thick plates produced through Process B undergo a greater level of grain fragmentation during the hot rolling procedure, resulting in increased energy storage. As a result, the microstructure of the alloy becomes more prone to undergoing recrystallization transition during successive solution heat treatment processes, which ultimately causes a decrease in the mechanical characteristics of the alloy.

4. Conclusions

The current study focused on the main results of 2324 aluminum alloy thick plates manufactured utilizing two different hot rolling methods. To summarize, the results can be succinctly stated as:

(1) Compared to Process B, the thick plates made of 2324 aluminum alloy produced by Process A exhibit better yield and tensile strengths in the L and LT directions at room temperature. This is mainly because the microstructure of the plates is mostly made up of deformed structures. Meanwhile, the 2324 aluminum alloy plates produced using Process B, which have smaller grain sizes, exhibit a greater elongation.

(2) The morphology and arrangement of the grains are crucial factors in determining the mechanical characteristics of alloys. The alloy formed through the hot rolling method A exhibits an internal structure primarily consisting of a deformation structure characterized by larger grain size and straight alignment along the rolling direction. As a consequence, the alloy demonstrates enhanced yield strength. In contrast, the alloy produced through process B exhibits a greater proportion of recrystallized structure inside its internal arrangement, characterized by smaller grain size. This facilitates grain having more opportunities to rotate under external loads, consequently reducing the alloy's yield strength.

(3) The impact of different hot rolling procedures on the distribution and composition of second phases in thick plates manufactured of the 2324 aluminum alloy is modest. The second phase, which consists primarily of Al_2Cu and Al_2CuMg phases, displays a uniform distribution along the rolling direction in the thick hot-rolled plates. Both hot-rolled thick plates produced using the two methods exhibit a substantial amount of deformed structure. However, the hot-rolled thick plates obtained by Process B have a comparatively smaller grain size and a higher degree of grain fragmentation. This makes their microstructure more susceptible to recrystallization transformation during the following heat treatment procedures.

(4) The solution heat treatment process after hot rolling also has a significant impact on the microstructure and mechanical properties of the thick plate. Further studies will be carried out on it shortly.

Acknowledgments

The National Key R&D Program of China (No. 2020YFF0218200), the GRINM Innovation Fund Project, and other relevant programs provided financial assistance for this work.

References

[1] J. Immarigeon, R. Holt, A. Koul, L. Zhao, W. Wallace, J. Beddoes, Lightweight materials for aircraft applications, Materials characterization 35(1) (1995) 41-67.

[2] X. Zhang, Y. Chen, J. Hu, Recent advances in the development of aerospace materials, Progress in Aerospace Sciences 97 (2018) 22-34.

[3] X. Luo, Q. Zha, L. Xia, Research progress of 2 ××× series aviation aluminum alloys, Light Alloy Favrication Technology (09) (2018) 17-25.
[4] M. Zhang, D. Liu, Aerocraft Aluminum Alloys and Their Materials Processing, Materials China (01) (2013) 39-55.
[5] P. Li, Y. Ye, L. Deng, M. Zhang, Progress of Aerocraft Al-Li Alloys, Materials China (11) (2016) 856-862.
[6] J. Yang, J. Lu, H. Feng, A. Li, Z. Lu, L. Dai, The History of AlLi Alloys and the Research Development in China, Materials Reports (S2) (2014) 430-435.
[7] J.C. Williams, E.A. Starke Jr, Progress in structural materials for aerospace systems, Acta materialia 51(19) (2003) 5775-5799.
[8] T. Zhang, R. Bao, S. Lu, B. Fei, Investigation of fatigue crack propagation mechanisms of branching crack in 2324-T39 aluminum alloy thin plates under cyclic loading spectrum, International Journal of Fatigue 82 (2016) 602-615.
[9] L. Songsong, B. Rui, Z. Ting, F. Binjun, Mechanism of crack branching in the fatigue crack growth path of 2324-T39 Aluminium alloy, Frattura ed Integrità Strutturale 10(35) (2016) 73-81.
[10] T. Zhang, R. Bao, B. Fei, Load effects on macroscopic scale fatigue crack growth path in 2324-T39 aluminum alloy thin plates, International Journal of Fatigue 58 (2014) 193-201.
[11] C. Cepeda-Jiménez, O.A. Ruano, M. Carsí, F. Carreño, Study of hot deformation of an Al–Cu–Mg alloy using processing maps and microstructural characterization, Materials Science and Engineering: A 552 (2012) 530-539.
[12] T. Liu, Y. Zhang, C. Liu, Q. Ran, L. Cui, A Brief Analysis of the As-cast Microstructure and Homogenization Treatment of 2324 Aluminum Alloy, Chinese High-tech Enterprise (24) (2016) 64-65.
[13] K. Zhu, H.W. Yan, Correlation among Microstructure, Texture, and Properties along the Thickness Direction in As-Hot-Rolled and As-Solution-Quenched Al7055 Thick Plates, Materials Science Forum, Trans Tech Publ, 2021, pp. 65-73.
[14] J. Chen, L. Zhen, S. Dai, W. Shao, B. Zhang, Effects of grain shape and texture on the through-thickness yield strength of AA7055 aluminum alloy plate, Rare Metal Materials and Engineering 37 (11) (2008)
[15] M. Starink, S. Wang, A model for the yield strength of overaged Al–Zn–Mg–Cu alloys, Acta Materialia 51(17) (2003) 5131-5150.

Simulation-based analysis of welding process parameters for TC4ELI titanium alloy thick-walled cylinders

Guanghui Gao, Xiaowei Cheng and Yue Gao*

School of Mechanical and Electrical Engineering, Soochow University, Suzhou, China

*Corresponding author: yuegao@suda.edu.cn

Abstract: In this paper, ABAQUS is used to numerically simulate the residual stresses during welding of titanium alloy thick-walled cylinders. A detailed simulation analysis of the stress field during the welding process is performed by establishing an accurate finite element model, defining the thermal physical properties, boundary conditions, and heat source model, and using the indirect heat-force coupling method. The results of the study show that the distribution of the weld stress field has an important influence on weld quality, material properties, and structural strength. The movement of the welding heat source results in a dynamic change in the stress field, which gives rise to a significant stress concentration at and in the vicinity of the welding heat source.

1. Introduction

The titanium alloy pressure-resistant thick shell material used for deep-sea resource development has been increasingly emphasized in the marine field. Welding is the key technology used in its processing and manufacturing processes. Titanium metal is classified as an active metal. If the welding process is not properly protected, it is very easy to react with hydrogen, oxygen, and nitrogen in the air and generate a brittle phase, reducing the plastic toughness of welded joints. In addition, titanium alloys have about half the modulus of elasticity of steel and are more susceptible to deformation at elevated temperatures. The majority of titanium alloy thick-walled cylinders are produced using the narrow gap welding method, which involves reducing the amount of molten metal filler and the heat input. This approach is employed to achieve the desired reduction in welding residual stress and deformation. However, due to the complexity of the welding process, the number of welding layers within the thick plate is susceptible to the formation of three-way stress, which can be exacerbated by the superposition of working stress during service. Once the yield strength of the material is exceeded, plastic deformation will occur, which may result in structural failure and subsequent harm to the reliability and safety of the structure.

Currently, there is a paucity of research investigating welding residual stresses in titanium alloy thick-walled cylinders. Titanium alloys are chemically active and readily react with elements such as nitrogen, hydrogen, and oxygen in the air under high temperatures, which presents a challenge to welding. It is of paramount importance to elucidate and master the distribution pattern of residual stresses under narrow-gap welding of thick-walled cylinders, as this knowledge will be instrumental in ensuring the serviceability and safety of the structure in the later stages [1-3]. Finite element analysis of the temperature field and stress field during welding represents a significant research area within the field of welding. This analytical approach is of great importance for the prediction and control of temperature distribution, welding deformation, and residual stress during the welding process. This paper is devoted to the numerical simulation of the stress field in the thick-walled cylindrical weld of TC4ELI titanium alloy.

The objective is to investigate the effects of welding parameters on weld quality, material properties, and structural strength.

2. Simulation analysis of titanium alloy thick-walled cylinder welding

2.1 Establishment of a finite element model for a titanium alloy thick-walled cylinder

This chapter presents a numerical investigation of the temperature field and post-weld residual stress field of titanium alloy thick-walled cylinders using finite element software ABAQUS [4]. By modeling and analyzing the distribution and change law of the residual stress formed by welding speed on the welded parts, we can reduce the welding deformation and improve the service life of the welded structure to provide reference significance.

2.2 Setting of material property parameters

The performance parameters of titanium alloy thick-walled cylinder material are subject to change due to the influence of temperature during welding. The density of the material remains relatively constant with changes in temperature, with a value of 4.51×10^{-9} kg/mm³. The relationship between other parameters and temperature is illustrated in Figures 1(a)-(d).

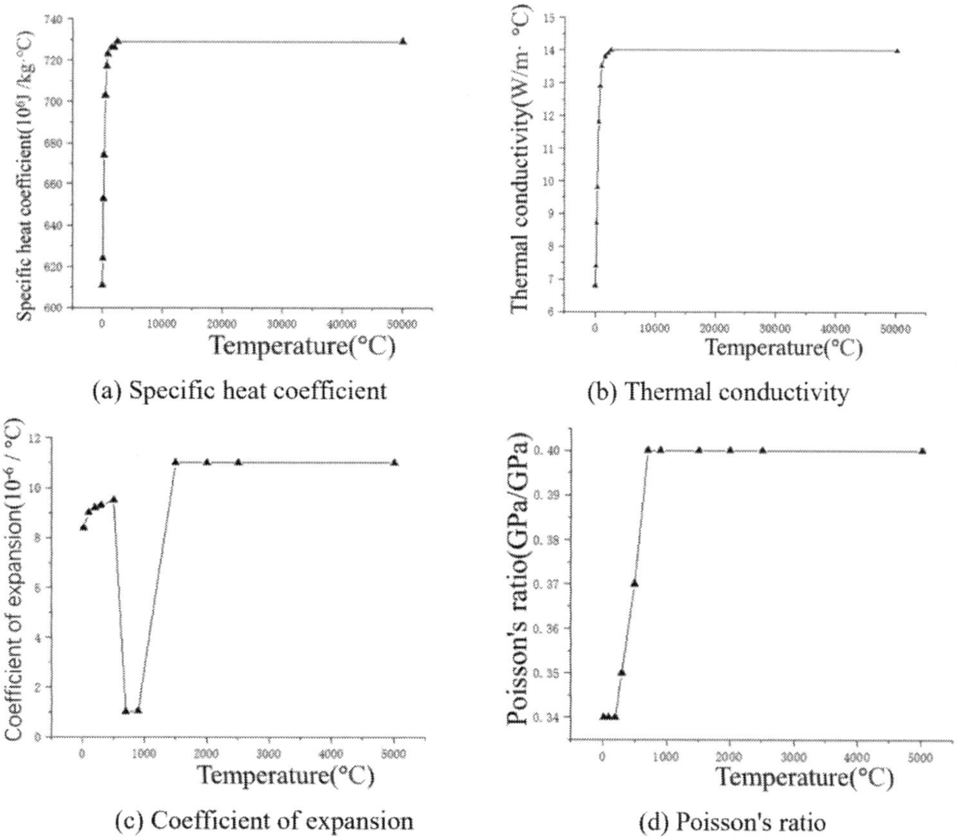

(a) Specific heat coefficient (b) Thermal conductivity

(c) Coefficient of expansion (d) Poisson's ratio

Figure 1. Variation curve of each parameter with temperature.

2.3 Calculation steps for welding heat stress

The welding heat stress is calculated using the indirect coupling method [5]. The flow chart of the analysis is shown in Figure 2.

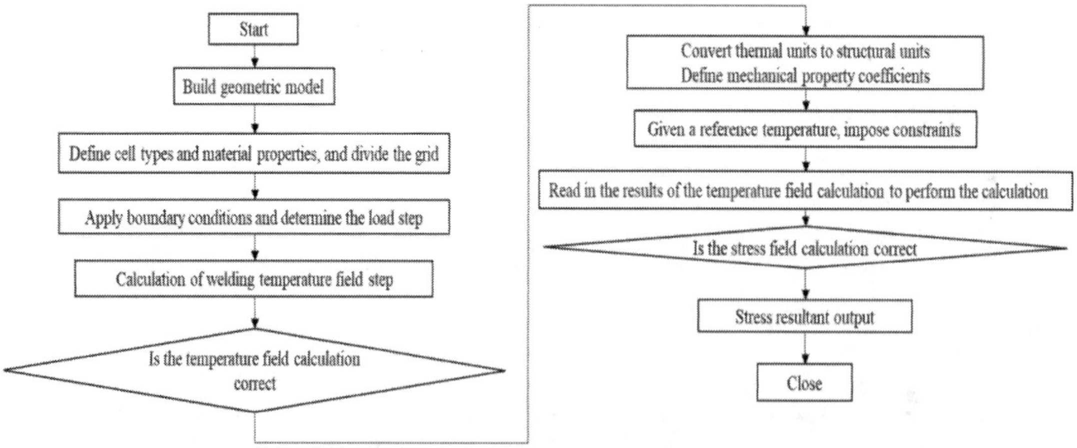

Figure 2. Chart of the analysis.

2.4 The setting of initial and boundary conditions

In the initial condition definition, the temperature of the node is set to 25°C. In the boundary condition settings, the convective heat transfer coefficient describes the transfer of heat from a fluid to a solid surface by convection and is set to a constant value of 0.02 W/mm²-K. The radiant heat transfer coefficient describes the exchange of heat between the surface of an object and the environment due to radiation. The radiative emissivity is set to a constant value of 0.8. The positional boundary conditions are set based on the correctly solved temperature field. This essentially simulates the situation where the weldment is fixed and constrained to have degrees of freedom in a certain direction. This is shown in Figures 3(a)-(c).

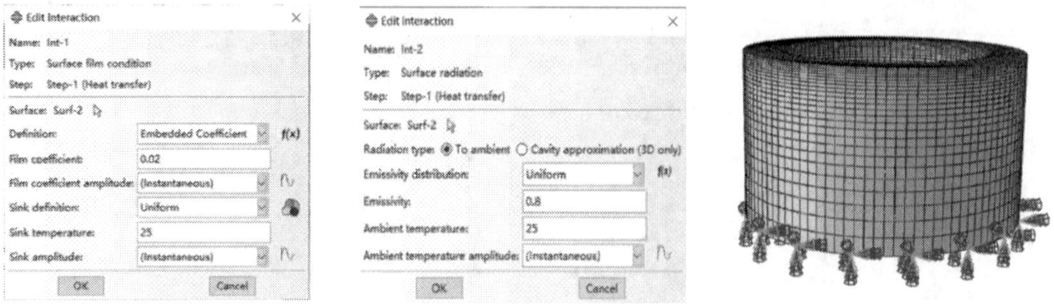

(a) Convection heat transfer coefficient (b) Radiant heat transfer coefficient (c) Displacement boundary condition

Figure 3. Boundary conditions.

2.5 Selection of heat source model

In ABAQUS, the choice of heat source models is critical to simulating the welding process, as they determine the manner and efficiency of heat transfer. In this simulation, a dual ellipsoid heat source model is selected because it is capable of accurately reproducing the traveling melt pool [6].

3. Analysis of stress field results for titanium alloy thick-walled cylinders

3.1 Stress field

The cloud diagrams of the stress field at different moments during the welding process in the transverse and longitudinal directions are presented in Figures 4(a)-4(d), 4(e)-4(h), respectively. The stress field cloud diagram reveals that the location of the welding heat source and its surrounding area exhibits a pronounced stress concentration phenomenon. At the beginning of the cooling phase, there is a significant concentration of tensile stresses at the welded joints. As the cooling time increases, the tensile stress gradually increases and extends to the surrounding area. Later in the cooling phase, the stress field distribution gradually stabilizes.

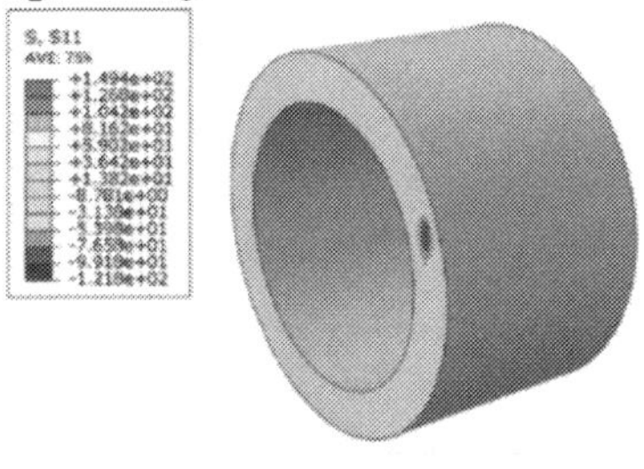

(a) Starting moment of welding in X direction

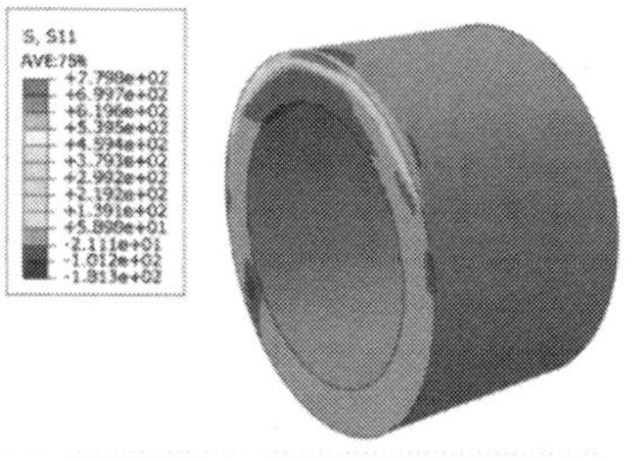

(b) Middle moment of welding in X direction

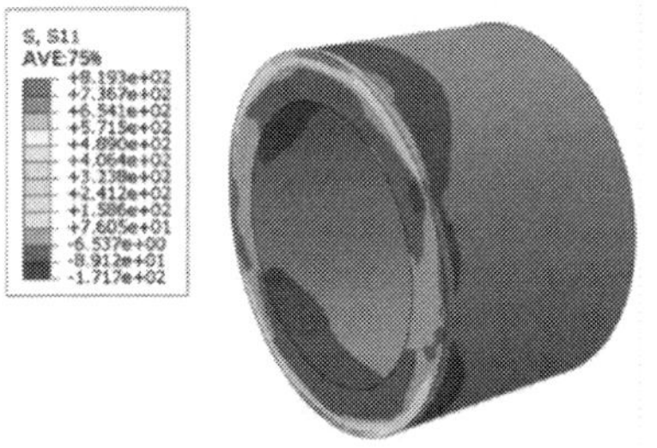

(c) End moment of welding in X direction

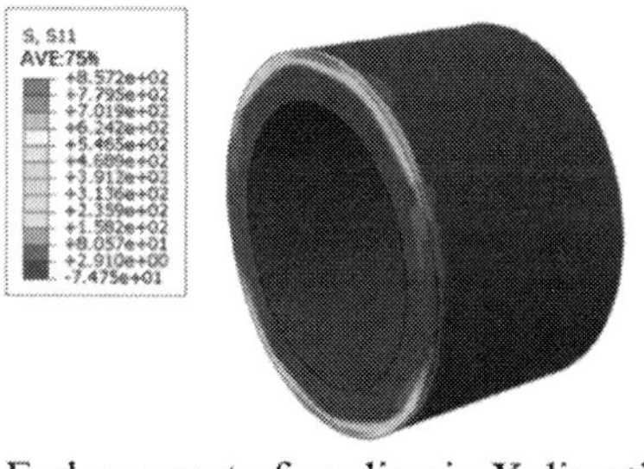

(d) End moment of cooling in X direction

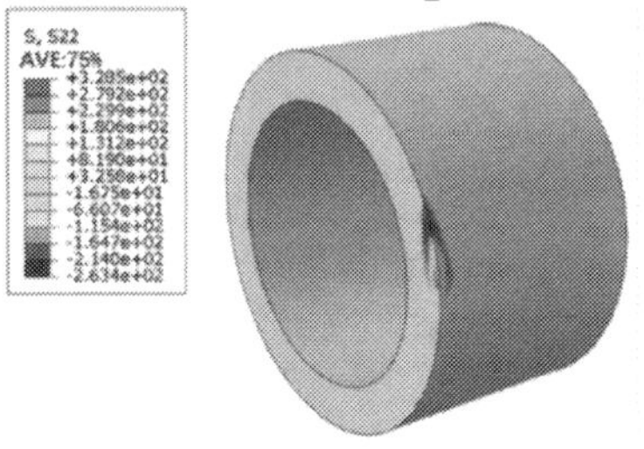

(e) Starting moment of welding in Y direction

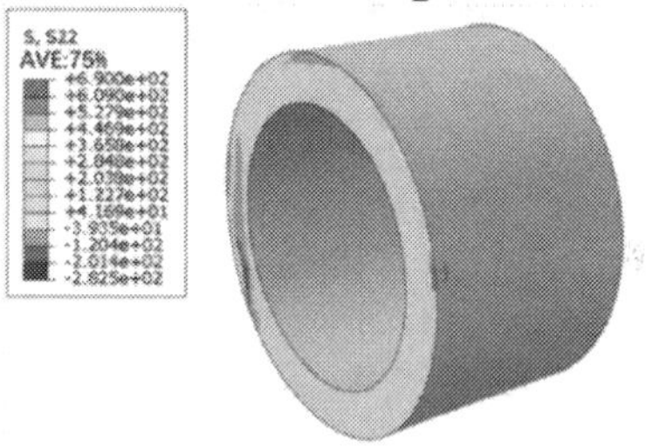

(f) Middle moment of welding in Y direction

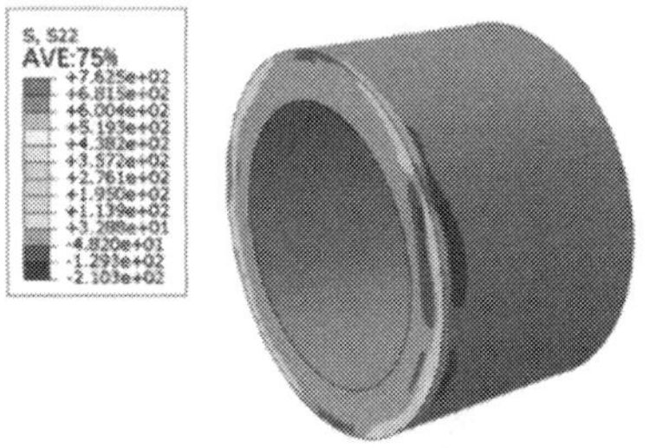

(g) End moment of welding in Y direction

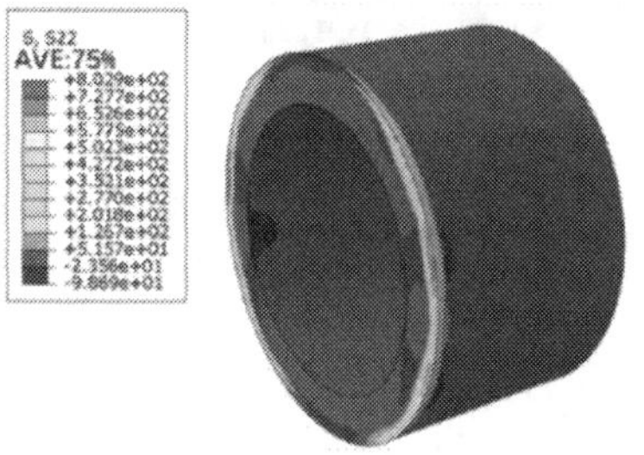

(h) End moment of cooling in Y direction

Figure 4. Cloud plots of stress field at different moments during welding in transverse and longitudinal directions.

3.2 Variation curves of X and Y direction stress with temperature at different locations

The cloud plots of stress variation with time in X and Y directions at four selected points, A, B, C, and D, are shown in Figures 5(a)-(b). At the weld location, the stresses in the X-direction are manifested as

Journal of Physics: Conference Series **2879** (2024) 012014 doi:10.1088/1742-6596/2879/1/012014

tensile stresses due to thermal expansion and cooling contraction during the welding process. In the Y-direction, thermal expansion and cooling contraction of the material in the weld region also occurs, which gives rise to the phenomenon of stress concentration. The restraining effect of the material on both sides of the weld makes it possible to have an alternating distribution of tensile and compressive stresses in the Y direction. In the X-direction, the heat-affected zone is affected by the tensile stresses in the weld region and the restraining effect of the surrounding base material. The stress distribution is more complicated. In the Y-direction, the material in the heat-affected zone experiences less thermal expansion and cooling contraction than in the weld region, resulting in a relatively low-stress level in the Y-direction. The base material region is situated at a considerable distance from the weld and the heat-affected zone. The stress level in the X-direction is relatively low. In the Y-direction, the stress distribution in the base material region is relatively uniform and low.

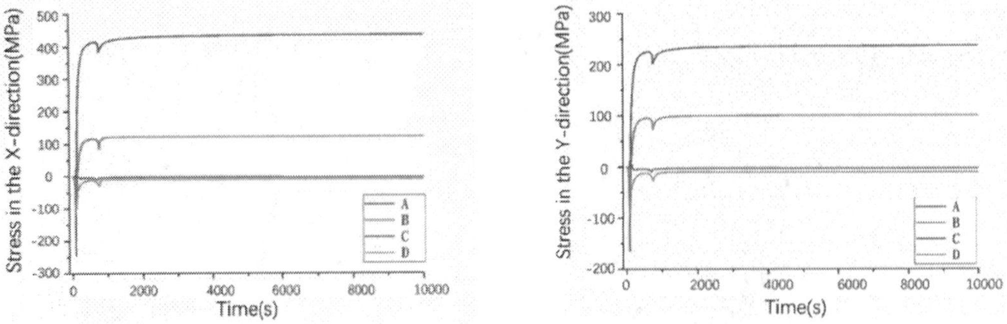

(a) X-direction (b) Y-direction

Figure 5. Cloud plots of stress variation with time in X and Y directions at four points: A, B, C, and D.

3.3 Effect of different speeds on the variation of the stress field

The following section investigates the effect of varying the weld line speed on the variation of the weld temperature and stress fields. The line speeds of 1.25 mm/s and 1.5 mm/s are taken as examples.

1) Stress in the X-direction

As the welding speed increases, it can be observed from Figure 6(a) that the peak stress in the X-direction at the weld location gradually decreases while the rate of stress change increases. From Figure 6(b), it can be observed that the peak stress in the heat-affected zone in the X-direction decreases, and the rate of stress change increases. Figure 6(c) illustrates that the stress change in the base material region is relatively flat. As the welding speed increases, the rate of stress change increases, with little change in the peak value.

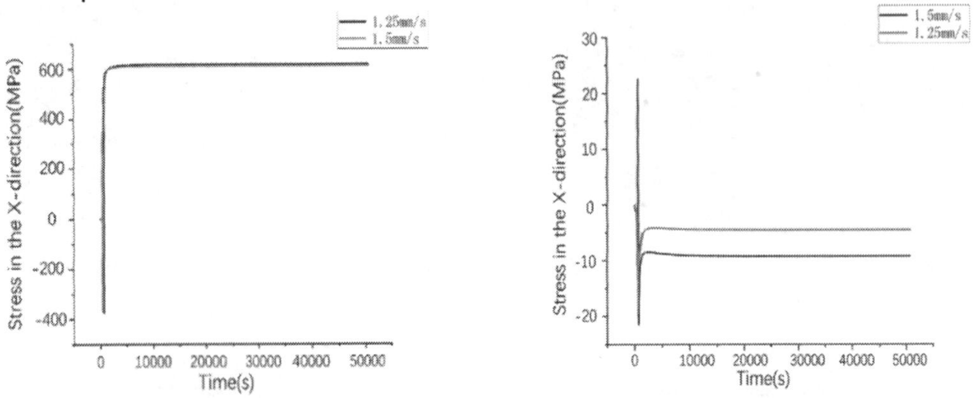

(a) Point A (b) Point B

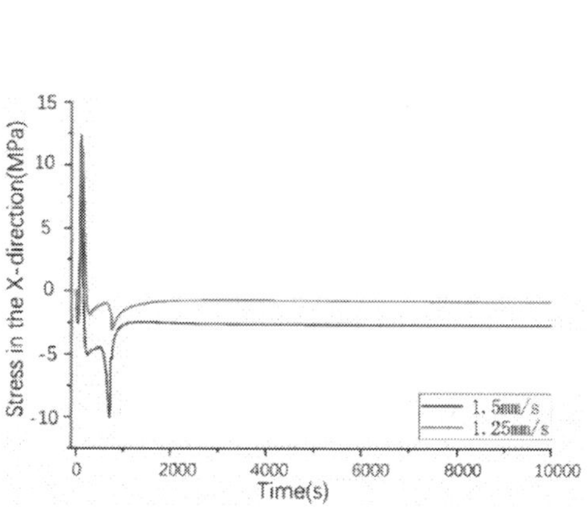

(c) Point C

Figure 6. Cloud plots of stress variation with time for different velocities in the X-direction at points A, B, and C.

2) Stress in the Y-direction

As the welding speed increases, it can be observed from Figures 7(a)-(c) that the Y-direction stress in the weld region exhibits a gradual accumulation at low welding speeds, reaching a peak value and subsequently stabilizing. At high welding speeds, the stress changes in the Y direction in the weld region are more dramatic, with peaks occurring at an earlier stage and overall stress levels being relatively low. The peak stress in the Y direction in the heat-affected zone decreases while the rate of stress change increases. The change in stress in the Y direction in the base material region is relatively minor, with a slight increase in the rate of change.

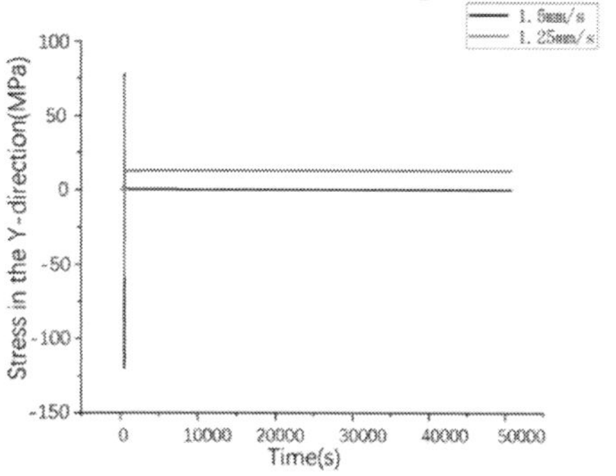

(a) Point A

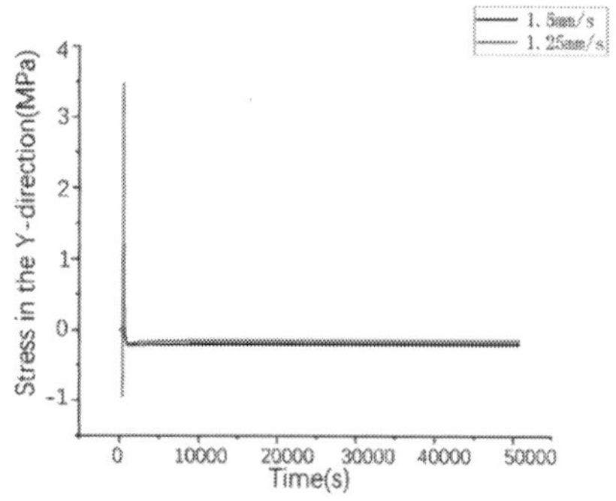

(b) Point B

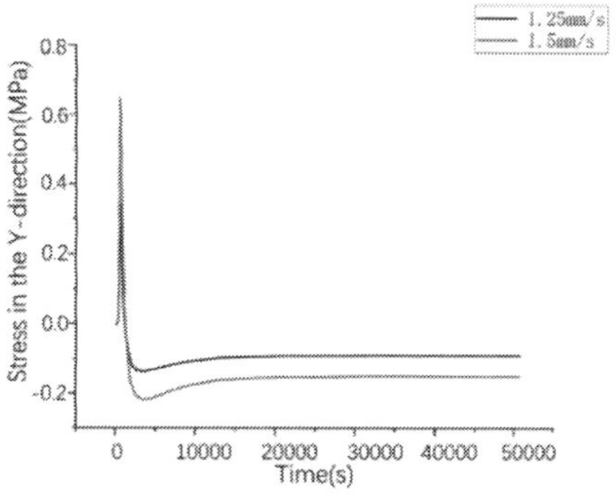

(c) Point C

Figure 7. Cloud plots of stress variation with time for different velocities in the Y-direction at points A, B, and C.

4. Conclusion

The present study employs an in-depth numerical simulation to investigate the residual stresses that arise during the welding of titanium alloy thick-walled cylinders. By establishing an accurate finite element model using ABAQUS, defining the thermal physical properties, boundary conditions, and heat source model, and applying the indirect coupling method of heat-force, a detailed simulation analysis of the stress field during the welding process is carried out. The results of the study show that:

1) The dynamic changes in temperature and stress fields result from the movement of the welding heat source.

2) An increase in welding speed leads to a decrease in peak stress and an increase in the rate of stress change.

3) The study deepens the understanding of the temperature and stress distribution law during the welding process of titanium alloy thick-walled cylinders and provides a scientific basis for the reliability assessment and life prediction of welded structures.

Future research could further explore the microstructural changes and property evolution of welded joints and the effect of these changes on the performance of welded joints.

Acknowledgment

This work is supported in part by the National Natural Science Foundation of China (Grant No. 72201186), the Natural Science Foundation of Jiangsu Province (Grant No. BK20220481), the Natural Science Foundation of the Jiangsu Higher Education Institutions of China (Grant No. 22KJB410002), and the China Postdoctoral Science Foundation (Grant No. 2023M731450).

References

[1] Wisdom Li, Yulong Li, and Anmin Tang, Plastic deformation conditions and brittle fracture analysis of thick-walled cylinders subjected to internal pressure. Journal of Applied Mechanics, 2020. 37(04): 1515-1520+1857.

[2] A Keane, A. Bhaskar, T M Olsen. Weld sequence optimization: the use of surrogate models for solving sequential combinatorial problems [J]. Computer methods in applied mechanics and engineering, 2005(30): 3535-3551.

[3] Li Shangqing et al., Research on hot assembly process of combined thick-walled cylinders. Machinery Manufacturing, 2021. 59(01): 63-65.

[4] Zhang, L. X., et al., Effect of quenching on residual stresses in 6061-T6 aluminum alloy castings based on ABAQUS simulation. Shanghai Metals, 2023. 45(05): 75-80+87.

[5] Welding Society of the Chinese Mechanical Engineering Society, ed. Welding Handbook Welding Methods and Equipment Volume 1. Beijing: machinery industry press 2016

[6] Bjelic, M., et al. Multi-objective calibration of the double-ellipsoid heat source model for gmaw process simulation. Thermal Science, 2022. 26(3): 2081-2092.

Journal of Physics: Conference Series **2879** (2024) 012015 doi:10.1088/1742-6596/2879/1/012015

The effect of functional group of Graphene-based woven filter on the desalination performance

Ziqi Huang[1] and Lijun Liang[1,*]

[1]College of Automation, Hangzhou Dianzi University, Hangzhou, 310018, China

[*]Corresponding author E-mail address: llj@hdu.edu.cn

Abstracts. Graphene-based braided filter membranes (GWFM) are becoming increasingly crucial in seawater desalination because of their excellent performance, moving from research to practical use and helping to tackle water scarcity. However, the mechanisms associated with GWFMs functionalized by functional groups are unclear. This paper demonstrates that graphene strips are intricately integrated into a filtration membrane characterized by X-functional groups (GWFM-X). Molecular dynamics (MD) simulations are effective for calculating water flux, assessing salt discharge effects, and understanding the physical mechanisms in seawater desalination, making them crucial for optimizing and advancing desalination technology and salt rejection rate of four types of GWFM-X, namely GWFM-1.2nm-COOH, GWFM-1.2nm-NH_2, GWFM-1nm-COOH, and GWFM-1nm-NH_2, which were designed with different functional groups and nanopore diameters. These results demonstrate the superior performance of GWFM-X in water filtration applications. Salt rejection efficiency of four distinct variants of GWFM-X, namely GWFM-1.2nm-COOH, GWFM-1.2nm-NH_2, GWFM-1nm-COOH, and GWFM-1nm-NH_2, is designed with different functional groups and nanopore diameters. The test membrane shows exceptional water flux, significantly outperforming current commercial reverse osmosis membranes, with measurements of 30.00±0.49 L cm^{-2} day^{-1} MPa^{-1}, 35.41±0.71 L cm^{-2} day^{-1} MPa^{-1}, 9.31±0.31 L cm^{-2} day^{-1} MPa^{-1}, and 20.66±0.21 L cm^{-2} day^{-1} MPa^{-1}, respectively. The free energy profiles of GWFM-X reveal a pronounced disparity in free energy barriers between water molecules and Na+/Cl- ions, with a difference of up to 14.26 k_BT. The simulation results show that the seawater desalination system maintains over 75% efficiency at pressures between 10 and 500 MPa, highlighting its effectiveness and versatility. Moreover, advancements in nanofabrication technology are paving the way for new membrane materials that could improve desalination methods and help address water scarcity.

1. Introduction

Due to swift population expansion and nuclear leaks in recent years [1-4], clean freshwater resources have become a resource that countries must strive for [3-5]. Therefore, obtaining freshwater resources by developing desalination technologies can effectively alleviate the plight of freshwater resources. So far, desalination technologies include MSF, MED, and RO [6-7]. However, despite significant advances in reverse osmosis technology, the core filtration membrane is still based on TFC membranes invented 40 years ago [10]. These membranes are very sensitive to chemical chlorine and are susceptible to biological contamination, and their membrane structure and degradation issues remain insurmountable constraints [6]. Improving desalination efficiency has been a long-standing objective in seawater desalination, water treatment, and chemical separation. The aim is to achieve efficient and pure separation effects through innovative designs and the optimization of membrane materials. Its remarkable permeability and exceptional selectivity create an effective pathway for water molecules to swiftly and accurately navigate through a designated medium or membrane. This characteristic facilitates the unhindered passage of water molecules and effectively prevents the ingress of non-target substances, resulting in optimal concentration and precise isolation of water molecules. In the vital domains of water resource management, seawater desalination technologies and biofilm separation processes, this attribute demonstrates outstanding potential and significance, contributing significantly to the technological advancement and evolution of related industries and effective barrier of impurities, thereby demonstrating an extraordinary capacity for seawater desalination. Concurrently, through a multitude of preparation strategies and technical optimizations, the stability and durability of the material are further consolidated, indicating that the material has substantial potential for broad application in seawater desalination and water treatment [11-12]. Subsequently, some researchers investigated carbon nano-based fillers (graphene oxide [8-9] and carbon nanotubes [13-14]), nanofillers composed of metals and metal oxides, such as zinc oxide [15,16], silver [17-18], etc., and inorganic salt-based nano-based fillers (boron nitride [19-20], MoS_2 [21-22], etc.) for reverse osmosis desalination applications. Therefore, the thin-film nanocomposite membranes replace the thin-film composite membranes as an excellent alternative to next-generation RO membranes in desalination [10].

The distinctive microscopic structure and chemical properties of NPG markedly enhance the rate of water molecular transport and achieve a substantial increase in water flux while maintaining excellent solute screening capabilities. Its exceptional plasticity and adaptability offer a novel avenue for in-depth optimization of membrane properties, suggesting that the material will have a significant and promising future in core fields such as seawater desalination and water treatment [8,9,11,12]. In fact, the greater the porosity of NPG, the higher the water flux. Graphene woven filtration membrane (GWFM) with good mechanical properties has also been noted [23]. This membrane has high water flux while possessing stronger mechanical properties than NPG, and due to the braided structure, the appropriate pore size can be achieved by modifying the braiding distance in practical applications. However, there is a limit to the size of NPG in practical applications. Functional graphene (FG) is a distinctive material in the domain of filtration membranes, having emerged as a prominent subject of scientific inquiry. Following ingenious functional modification, the FG membrane has demonstrated the ability to expand the aperture while simultaneously preserving exceptional water permeability and salt-blocking efficacy, markedly enhancing the efficiency of water treatment and the quality of effluent [24]. Its extensive potential for deployment in cutting-edge applications such as seawater desalination, water purification, and environmental protection suggests that the FG membrane will assume a pivotal role in future technological advancement.

Recently, some researchers have mimicked the weaving technique by simulating and studying the interwoven composite technology of carbon nanotubes and graphene, which aims to develop a new material that will distinguish itself from nanostructured materials due to its exceptional resilience and strength. This material has become a prominent emerging force, indicating a vast range of potential applications and future development prospects [22,25-27]. Meanwhile, some researchers have also found that adding carboxyl and amino groups to the graphene surface to achieve fast water transport channels during experiments [28,29].

There are no reports on the structural modification of graphene strips with different functional groups for desalination applications. This study presents a new approach to modifying graphene strips for use in filtration membranes. The fine woven filter membrane exhibits excellent net hydraulic properties and is modified with amino and carboxyl groups (GWFM-COOH and GWFM-NH_2) around the pore sizes. Previous studies have shown that the nanopore sizes of GWFMs have significantly affected water fluxes and desalination performance [23]. This work modulates the pore size of GWFM by changing the functional groups near the braided structure. Our molecular dynamics simulations comprehensively calculated and analyzed the water flux and desalination rate of GWFM-X at different pressures.

Journal of Physics: Conference Series **2879** (2024) 012015 doi:10.1088/1742-6596/2879/1/012015

2. Modeling and methodology

2.1. System setup

This paper presents an innovative generalized water framework model based on functional group-modified graphene bands to optimize the performance of water filtration systems. This paper employs the MD simulation technique to examine the profound impact of diverse functional groups and their modification densities on the surface characteristics of graphene bands. The study demonstrated that functional groups regulated essential properties of graphene bands, such as surface energy and wettability. It also revealed that alterations in modification density significantly impacted these properties. By precisely regulating these properties, the functionality of graphene bands can be optimized, providing a theoretical foundation for potential applications in energy, catalysis, and sensing. This study enhances the comprehension of graphene surface modification and facilitates its advancement. The simulation results show that functional group modification can significantly regulate the permeation path and the behavior of water molecules in the graphene band, thereby improving efficiency. In addition, the potential applications of the GWFM-X model in water treatment, seawater desalination, and biomolecular separation are also discussed, demonstrating its broad application prospect and essential research value. In the system, the membrane element is precisely arranged in the center, with the salt solution and pure water distributed on both sides. This simulation aims to provide a detailed analysis of membrane materials' ion separation mechanism and desalting ability, thereby offering a robust theoretical foundation for optimizing water treatment technology. Reducing the pore size of the nanopore or modifying the salinophilic groups will increase salt rejection, but this will inevitably reduce the water flux through the GWFM-X. Table 1 provides a detailed account of the precise size parameters of the system, which have been meticulously crafted to mirror the physical attributes and behavior of the box. The GWFM-X system, with its highly customized and precise simulation capabilities, ensures the reliability of simulation results and offers substantial support for box performance analysis. Through this system, we can comprehensively investigate the influence of box parameter alterations on the overall performance, thereby providing a scientific foundation for optimal design and application.

2.2. Simulation details

The study utilized Gromacs 2019.6 with a 2-fs time step to create an effective MD simulation platform, ensuring accurate simulations for investigating membrane materials' ion screening and desalination properties [30]. In this study, we utilized the Nose-Hoover technique to uphold a consistent system temperature of 298 K, ensuring the simulation results' accuracy, and conducted an isothermal isobaric system (NPT) to simulate a realistic seawater environment. To eliminate the simulation boundary effect and ensure a seamless system connection, the fixed GWFM-X model remains static, preserving its structure and properties and accurately reflecting ion sieving and water transport performance under specific conditions [31]. The generic force field used in this paper is CHARMM, which is applied to the membrane model GWFM-X, and the central component of the push plate is constituted by carbon atoms embedded between graphene sheets, combined with an ionic solution of sodium and chloride to simulate the actual aqueous environment. This composition is highly compatible with the TIP3P water model, which ensures that the simulation accurately reflects the behavior of water molecules [32-33]. The interactions between various particles governed by the Lennard-Jones potential are handled utilizing the Lorentz-Berthelot mixing rule to ensure the most accurate modeling of these interactions' energy and distance parameters [34]. In the simulation, the L-J action cutoff is set at 1.3 nm, the short-range force range is defined, the resource utilization is optimized, and the accuracy and physical rationality are ensured. The potential molecular electrostatic (PME) method is employed to segment the electrostatic field, and a 1.3 nm separation threshold is set to actualize electrostatic action, reduce error, improve efficiency, and facilitate an in-depth analysis of the electrostatic effect of the molecular system [35].

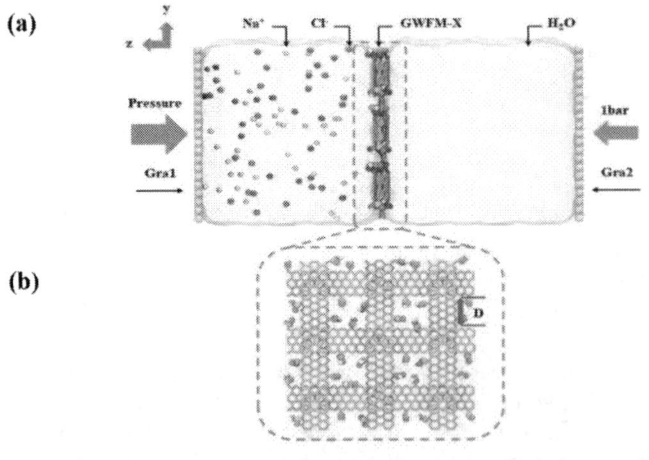

Figure 1. MD framework analyzes GWFM-X to optimize water filtration performance, Na^+ ions (yellow), Cl^- ions (pink), and water molecules (white fluid). (b) GWFM-X model. There are two types according to the pore diameter (D) of the nanopores: the GWFM-1.2 nm-X and the GWFM-1 nm-X, which refer to graphene-based filtration membranes with nanopore diameters of 1.2 nm and 1 nm, respectively.

Table 1. Details of the MD simulation systems.

System	Size of box (nm^3)	Amount of water molecules	Area of the GWFM-X (nm^2)
GWFM-1.2nm-COOH	5.88×5.95×15.14	10,427	35.00
GWFM-1.2nm-NH_2	5.88×5.95×15.14	10,451	35.00
GWFM-1nm-COOH	5.00×5.08×15.14	7,371	25.40
GWFM-1nm-NH_2	5.00×5.08×15.14	7,402	25.40

2.2.1. Calculation of water flow at different pressures. The graphene sheets, Gra1 and Gra2, were positioned in parallelism at z = 14.986 nm and z = 4.986 nm on either side of the membrane model GWFM-X [30]. Molecular simulation is a crucial technical measure in studying complex systems involving water. Its purpose is to guarantee the precision and reliability of the simulation process, prevent the simulated water molecules from inadvertently escaping from the predetermined baffle boundary, and maintain the stability of the pressure within the system. In particular, a constant pressure differential of 1 bar is applied between two simulated regions, designated Gra2 and Gra1, in opposite directions. This pressure difference is designed to create an effective pressure barrier that not only effectively prevents water molecules from crossing the boundary of the baffle but also maintains the number and distribution of water molecules in the simulated area. It ensures that the internal pressure of the entire simulated system is maintained at a stable and controllable level. Through this setting, the behavior characteristics of water molecules under specific conditions can be more accurately simulated and analyzed, and more reliable and accurate experimental data can be provided for research in related fields. This strategy enhances the accuracy of the simulation, accurately reflects the water filtration characteristics of the GWFM-X model under pressure, and provides reliable simulation data to support the optimization of the transport efficiency and selectivity of water molecules. In Figure 1a, the GR1 is designed as a dynamically rigid piston that is vertically pressed against the GWFM-X surface to simulate the actual pressure conditions. The pressure calculation formula, based on total pressure and contact area, ensures the accuracy of pressure simulation, provides reliable data for ion screening, water transport, and other studies, and enhances the authenticity of simulation and the accuracy of research [37]:

$$\Delta P = P - P_0 \tag{1}$$

In this expression, the reference pressure P_0=1 bar for Gra2, while P denotes the pressure on gr1. Since P is much greater than P0, the system's actual pressure change ΔP can be approximated as ΔP=P-P_0. This simplification shows that the transmembrane pressure mainly comes from P on gr1 [37-38]. In order to accurately calculate the described performance parameters, we only considered data from the seawater side before half of the solution passed through the GWFM-X. To enhance the precision and dependability of MD simulation outcomes, we implemented a meticulous repeated simulation approach, conducting five independent filtration simulation experiments for each system. This technique effectively mitigates the impact of random fluctuations. Further, it reduces the fluctuation range of simulation outcomes through averaging, thereby ensuring the robustness and authenticity of the final conclusions and providing a robust foundation for the accuracy of MD simulation results.

2.2.2. Free energy barrier calculation: umbrella sampling. Umbrella sampling was employed to determine the distributions of free energy barriers for H_2O and Na^+/Cl^- during their traversal through the GWFM-X. The PMF profiles were derived from the reaction coordinates [36]. Therefore, the umbrella samples were taken, and only the reaction coordinates within the -0.5 nm to 2 nm range were considered. For the accuracy of the measurements, 1.5 ns equilibration is performed before each window simulation. Finally, the free energy barriers of individual atoms are characterized by counting the highest PMF peaks.

3. Results and discussion

3.1. Water flux and desalination rate

To enhance clarity, the water fluxes corresponding to ΔP of 10 MPa, 100 MPa, and 500 MPa are illustrated in Figure 2a. To quantitatively analyze the water flux of GWFM-X, the diffusion velocity of water molecules in the membrane was first calculated using $D_i = \lim_{t\to\infty} \frac{\langle|r(t)-r(0)|^2\rangle}{2t}$ (Figure 2b), where r(t) represents the distance travelled by the water molecules present in the nanopore, and the efficiency of water flux is defined in this context as:

$$V_e = \frac{D_i c_m}{l}\left\{1 - \alpha exp\left[\frac{-v_i \Delta p}{RT}\right]\right\} \quad (2)$$

This paper defines several key parameters that characterize the movement of water molecules in the membrane, including rate, film thickness, diffusion coefficient, concentration, and density ratio. Constant temperature conditions are assumed. The accuracy of the theoretical model was verified, and the experimental deviation was revealed by calculating the formula and comparing the moisture content directly with the experiment. This provided a reference for the optimization of membrane filtration technology.

The findings indicate that the pore size is the primary determinant of the water flux. At small pore sizes, the movement of water molecules is constrained, resulting in a notable reduction in flux. As pore size increases, the freedom of activity of water molecules rises, resulting in a notable enhancement in flux. This finding enhances our comprehension of nanoscale fluid transport mechanisms and offers a novel perspective for the design and application of nano-porous materials. By meticulously adjusting the size of nanopores, material properties can be optimized, water treatment efficiency can be enhanced, and water resources management and purification technology can be transformed. The quantity of water molecules traversing the nanopore increases linearly over time and is proportional to the magnitude of ΔP. When the functional group is held constant, the pore diameter is increased to accelerate the velocity. The water flux of GWFM-NH2 is observed to exceed that of GWFM-COOH when the aperture is held constant. In order to provide a comprehensive evaluation, the filter membrane water flux per unit area was introduced, and the results demonstrated that the effective water flux varies linearly with ΔP. At the same ΔP, the larger the nanopore size, the larger the V_e among the same functional groups; at the same ΔP, the modified NH_2 around the nanopore has a larger V_e than COOH among the same size nanopores. Further calculations showed that the water fluxes of GWFM-1.2nm-COOH, GWFM-1.2nm-NH_2, GWFM-1nm-COOH, GWFM-1nm-NH_2, respectively, were as high as 30.00±0.49 L cm^{-2} day^{-1} MPa^{-1}, 35.41±0.71 L cm^{-2} day^{-1} MPa^{-1}, 9.31±0.31 L cm^{-2} day^{-1} MPa^{-1}, and 20.66±0.21 L cm^{-2} day^{-1} MPa^{-1}.

In order to quantitatively evaluate the salt removal efficiency of the GWFM-X, this paper introduces the concept of salt removal efficiency to characterize its performance accurately:

$$E = 1 - \frac{N_{\frac{1}{2}}}{N_0} \quad (3)$$

In the given formula, N_0 is the initial number of ions in the solution, and $N_{1/2}$ is the number of ions remaining after the solution volume is halved through the GWFM-X nanopore. Experiments show that GWFM-X with different particle sizes has excellent salt inhibition ability (Figure 2b). In particular, under the same pressure, the desalting efficiency of GWFM-X for GWFM-COOH-modified solution is higher than that of GWFM-NH2. GWFM-1nm-X can reach a 100% desalination rate under low P, and the desalination rate starts to decrease when the pressure increases to 300MPa. In contrast, the salt rejection rate of GWFM-1.2nm-X firstly decreases with the increase of ΔP, then increases and then decreases, and reaches the highest salt rejection rate near 300MPa but can still reach more than 75%. The counter-intuitive phenomena of the salt rejection of GWFM-1.2nm-X will be discussed later in the article.

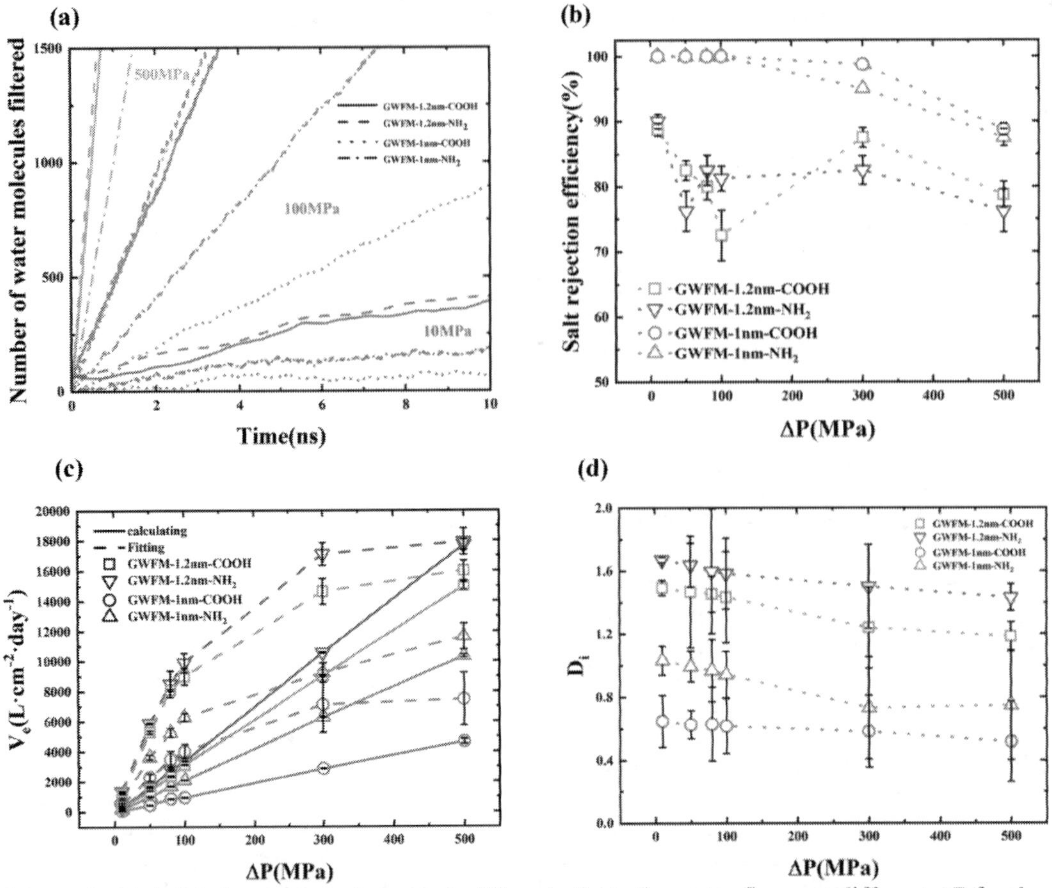

Figure 2. (a) Solid, drawn, dotted, and dotted lines indicate the water fluxes at different ΔP for the GWFM-1.2nm-COOH, GWFM-1.2nm-NH_2, and GWFM-1nm-COOH, GWFM-1nm-NH_2 systems, respectively. The lines of identical coloration in each diagram illustrate the system under consistent ΔP conditions. Section (b) demonstrates the variation in desalination efficiency with ΔP. (c) elucidates the relationship between the diffusion coefficient of water molecules in the pore along the Z-axis and ΔP. Figure (d) depicts the dynamic performance of effective flow rate Ve as a function of ΔP.

Journal of Physics: Conference Series **2879** (2024) 012015 doi:10.1088/1742-6596/2879/1/012015

3.2. Microstructure of water flow in GWFM-X nanopores

Figure 3 demonstrates that for the smaller nanopore size GWFM-1nm-X, water molecules tend to pass through the middle of the nanopore, whereas, for GWFM-1.2nm-X, they prefer to pass near the functional groups located in the nanopore, which results in an approximate 2-fold water channel. The low potential energy of water molecules in proximity to the GWFM-1.2 nm-X functional group and at the center of the GWFM-1 nm-X cavity can be attributed to the functional group-water interaction within the nanopore, which is subject to regulation by the dimensions of the graphene pore.

On the other hand, with the same size of the nanopore, for example, at 1.2nm pore size, the water molecules of GWFM-1.2nm-COOH are more compact and regular in the pore channel than those of GWFM-1.2nm-NH_2, showing a "windmill shape". This may be because the diffusivity of GWFM-1.2nm-COOH has higher diffusivity Di (Figure 2c) than that of GWFM-1.2nm-NH_2, and the water molecules in the pore channel of GWFM-1.2nm-COOH are less likely to pass through it and have higher hydrophobicity. This result is mainly related to the functional groups of graphene pore functionalization. The above results lead to a completely different flow pattern, and the water fluxes of GWFM-1.2 nm-COOH (30.00±0.49 L cm^{-2} day^{-1} MPa^{-1}) and GWFM-1.2 nm-NH_2 (35.41±0.71 L cm^{-2} day^{-1} MPa^{-1}) are 1.5-2 times the water fluxes of GWFM-1nm-COOH (9.31±0.31 L cm^{-2} day^{-1} MPa^{-1}) and GWFM-1nm-NH_2 (9.31±0.31 L cm^{-2} day^{-1} MPa^{-1}), respectively. Conclusions are consistent.

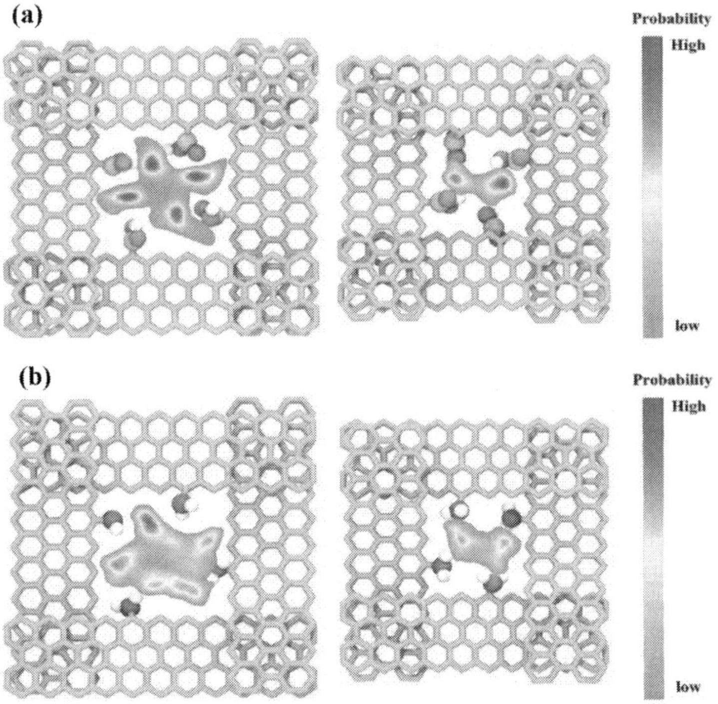

Figure 3. Distribution of water molecules within the aperture. (a) Positional density distribution of GWFM-1.2nm-COOH vs. GWFM-1nm-COOH filtered water molecules. (b) Positional density distribution of GWFM-1.2nm-NH_2 vs. GWFM-1nm-NH_2 filtered water molecules.

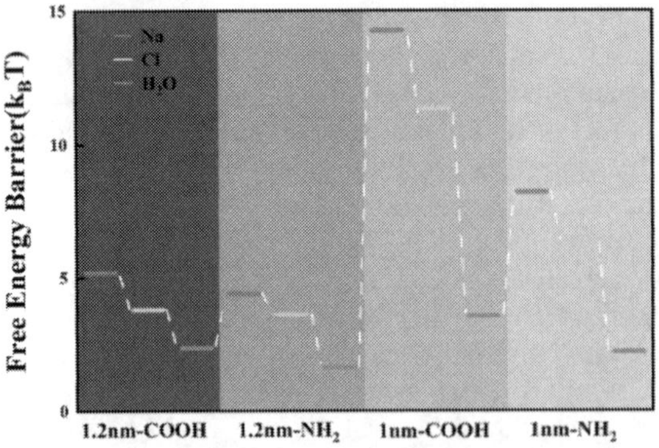

Figure 4. Energy barriers for Na^+, Cl^- and water molecules traversing GWFM-X nanopores.

3.3. Energy-barrier profiles of water/ions through the GWFM

The free energy barrier of H_2O and Na^+/Cl^- across the membrane material was calculated with great precision, and the results are presented in Figure 4. In the case of the GWFM-1nm-COOH model, the maximum energy barriers for Na^+ and Cl^- in nanopore channels are 14.26 k_BT and 11.35 k_BT, respectively. In contrast, H_2O molecules exhibit a lower energy barrier, with a maximum value of only 3.54 k_BT. For GWFM-1nm-NH_2, the highest free energy barriers were 8.21 k_BT and 6.45 k_BT, respectively, while the highest free energy barrier was only 2.20 k_BT for H_2O. Therefore, salt ions are more difficult than water molecules to penetrate the nanopore of GWFM-1nm-COOH due to the large difference in the free energy barriers of Na^+/Cl^- and H_2O. There are no ions passed through the nanopore at $\Delta P<100$ MPa.

For the larger nanopore size GWFM-1.2nm-COOH, the highest free energy barriers for Na^+, Cl^-, and H_2O through the nanopore are 5.20 k_BT, 3.81 k_BT, and 2.35 k_BT, respectively, and for GWFM-1.2nm-NH_2, they are 4.39 kB T, 3.62 k_BT, and 1.63 k_BT, respectively. The results mentioned above are conducted when ΔP=0. When ΔP is greater than zero, the enhancement of the system's driving force has been demonstrated to significantly promote the ion transport process, thereby facilitating ion migration in the medium. This is achieved by reducing the energy barrier between ions. This discovery provides a new perspective for understanding the mechanism of ion transport. It is expected to be applied to batteries, electrolytes, and other fields to improve the efficiency of energy conversion and storage. By regulating the driving force of the system, the ion conductivity can be optimized, thereby promoting the progress and development of related technologies and water crossing the GWFM-X membrane, accelerating particle crossing following transition state theory, improving material transport efficiency, and providing a theoretical foundation for process optimization, including membrane separation. To determine the rate at which water molecules or ions traverse a medium (such as a specific membrane), it is possible to employ several strategies that will facilitate the acquisition of accurate data:

$$R_{(particle)} \propto e^{-\frac{\Delta G(particle)}{k_B T}} \quad (4)$$

This R rate is influenced by three primary factors: the free energy barrier ΔG, temperature T, and the Boltzmann constant k_B. Furthermore, R is also influenced by other variables, such as the particles' shape, size, and concentration. Together, these elements dictate the crossing process's speed and efficacy. Ignoring the difference between ions and water molecules during the calculation process allows a quick assessment of the model's reasonableness. The abovementioned formula approximates the flux ratio of salt ions and water molecules across the GWFM-X membrane.

The relationship between the reaction rates for water and ions can be approximated based on the above equation. For GWFM-1nm-X, GWFM-1nm-COOH, and GWFM-1nm-NH_2, the $R(H_2O):R(Na^+)$, $R(H_2O):R(Cl^-)$ are >105, > 104 and >103, >102 , respectively.

For GWFM-1.2nm-X, GWFM-1.2nm-COOH and GWFM-1.2nm-NH_2, the $R(H_2O):R(Na^+)$, $R(H_2O):R(Cl^-)$ were about 17.30,4.32 and 15.79,7.31, respectively. Therefore, even low ΔP will still pass Na^+/Cl^- in these two systems.

Using the above estimation method, the water flux of GWFM-1.2nm-X should be 2 times that of GWFM-1nm-X. In fact, the proportion of water flow resulting from the influence of ΔP is about 1.5-2 times, almost the same as the water flux calculated in the previous simulation.

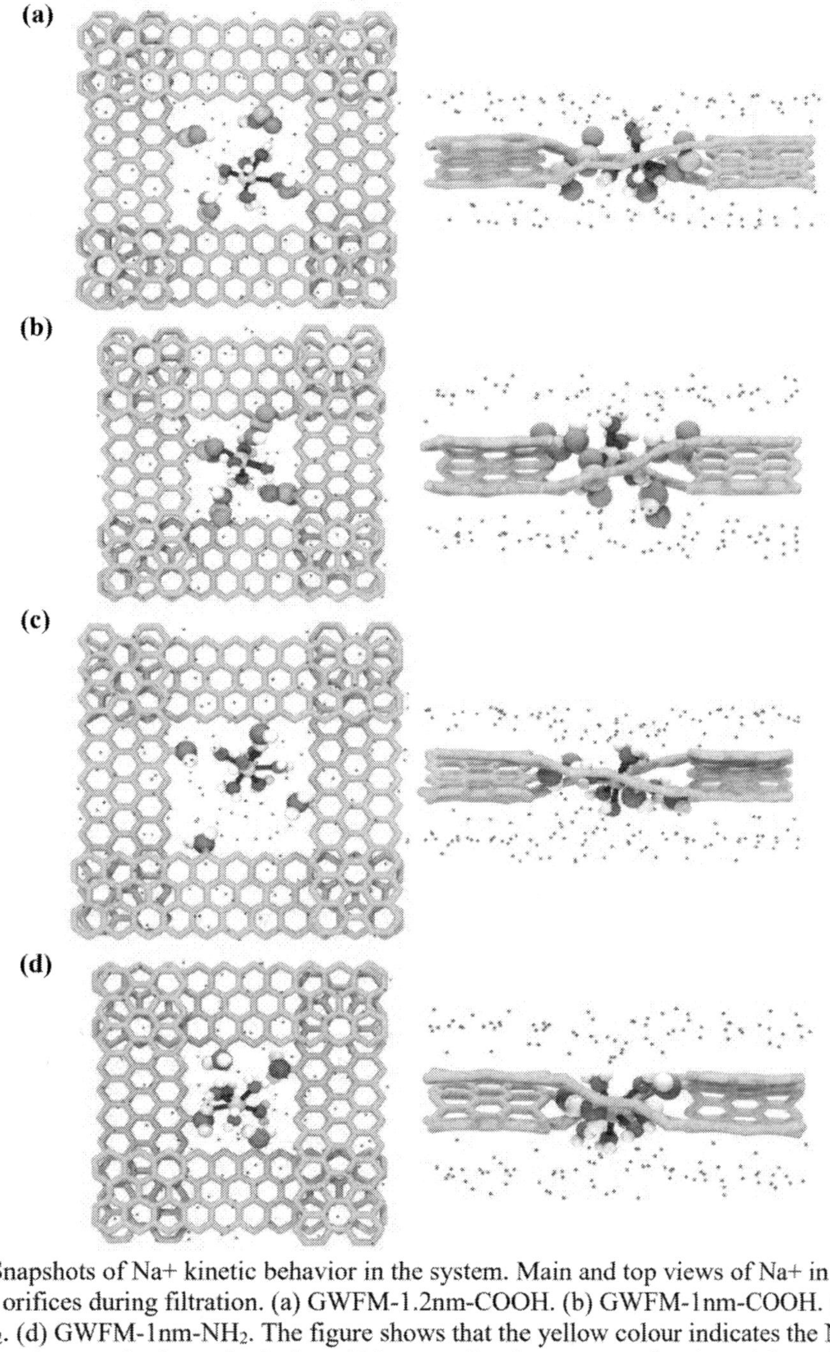

Figure 5. Snapshots of Na+ kinetic behavior in the system. Main and top views of Na+ in the range of GWFM-X orifices during filtration. (a) GWFM-1.2nm-COOH. (b) GWFM-1nm-COOH. (c) GWFM-1.2nm-NH_2. (d) GWFM-1nm-NH_2. The figure shows that the yellow colour indicates the Na+ ion, and the diagram shows red spheres for hydrated Na+ coordination water molecules, pink marks for other water molecules, and blue lines for interatomic connections. Green and orange represent carbon and oxygen in the carboxyl group, while purple. Cerulean highlights the carbon atoms of the GWFM-X film material and their connected structures, forming the structural and functional basis of the film.

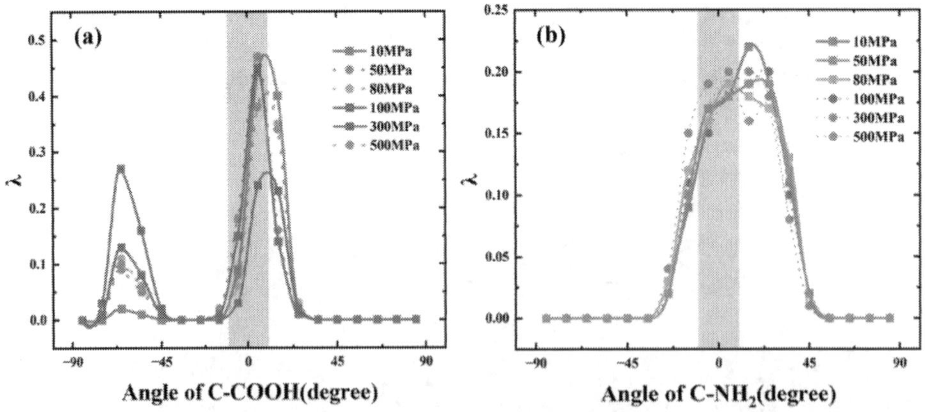

Figure 6. Angular probability distribution of nanopore C-X versus xOy in different ΔP. 0° indicates that the functional group is parallel to the membrane, positive numbers indicate that the functional group is biased towards the permeate side, and negative numbers indicate that the functional group is biased towards the feed side. (a) Angular distribution of C-COOH with xOy for GWFM-1.2nm-COOH, (b) Angular distribution of $C-NH_2$ with xOy for GWFM-1.2nm-NH_2.

3.4. Desalination kinetic processes

The internal mechanism of achieving excellent desalting performance is thoroughly examined from the perspective of the dynamics. Due to the attraction of the functional groups of GWFM-1nm-COOH to salt ions to form clusters, Na^+ is difficult to pass through easily (Figure 5).

Na^+ will undoubtedly exhibit specific penetration behavior as it passes through the membrane. When Na^+ is pushed near the nanopore by the pressure of water, it must first pass through the dehydration process and then bind to the functional groups. It is done by breaking the hydration and overcoming the functional groups' binding effect on Na^+. Therefore, even at large ΔP up to 500 MPa, small-sized nanopores such as GWFM-1nm-COOH can and will maintain high salt-exclusion efficiency. In contrast, Cl^- has a similar desalination mechanism despite its different hydride number and hydration radius.

Different GWFM-X still has high salt removal efficiency at 500MPa. Na+ permeable film trajectory analysis (Figure 7a) showed that their X-Na^+-OW angles in the nanopore were differently distributed under different functional groups and the length and average binding time of Na^+-X. α, l denote the X-Na^+-OW angle and the average length of X-Na^+, respectively. l lengths of GWFM-NH_2 were almost the same. At the same time, those of GWFM- COOH have l lengths of about, respectively, which are clearly more tightly bound (Figure 7b). Through the average time of average binding (Figure 7c), GWFM-1nm-COOH has significantly more time than the other GWFM-X, indicating that Na^+ will stay in the nanopore due to the attraction of the functional group. In the X-Na^+-OW angle, the GWFM-1nm-COOH is almost not present from 150 degrees to 180 degrees. Still, the GWFM-1.2nm-NH_2 is also almost not present from 60 degrees to 90 degrees (Figure 7d). Observing the change in hydration molecules (Figure 7e), Na^+ requires the most water molecules to be shed in GWFM-1nm-COOH, nearly two, which agrees with the conclusion in Figure 4.

For the large-size aperture GWFM-1.2nm-X, there is a first decrease and then an increase in the salt rejection rate at low ΔP. As illustrated in Figure 6a, for GWFM-1.2nm-COOH from 10 MPa to 100 MPa, the angle of C-COOH with the xOy plane becomes more and more diffuse, suggesting that the position of C-COOH becomes more unstable with the change of pressure, resulting in the salt ions passing through the pore more easily. However, as the pressure continues to increase, the angle of C-COOH tends to parallel the membrane, which narrows the pore size and increases the salt rejection rate. As shown in Figure 6b, GWFM-1.2nm-NH2 will be more stable than GWFM-1.2nm-COOH, but it will also trough near 50 MPa. Compared to 10 MPa, $C-NH_2$ oscillates more drastically at 50 MPa, leading to larger pore size and lower salt rejection. This is the reason that causes the salt rejection rate of GWFM-1.2nm-X in Figure 2b to decrease first and then increase at low ΔP.

By modifying the angle of the functional group and the Na+ cluster, the GWFM-1.2nm-COOH membrane surmounts the energy barrier, thereby achieving a high water flux and a high salt discharge rate. This demonstrates the membrane's considerable potential for use in water treatment. At the same size, the cluster X-Na^+ length of GWFM-COOH with Na+ is shorter and has a longer average lifetime than that of GWFM-NH_2, suggesting that carboxylate groups are more tightly bound to NA^+. Through the large pore size of GWFM-X, less structural deformation in the passage through GWFM-X requires only a dehydration operation. In contrast, the GWFM-X membrane's small aperture design increases the free energy barrier for the passage of hydrated ions, improves selective ion retention efficiency, and enhances the membrane's separation performance. This makes it a promising candidate for high-precision ion separation applications such as seawater desalination and ion exchange. The passage of ions requires dehydration deformation and freedom from the attraction of functional groups to NA^+. This explains the excellent efficiency of GWFM-1nm-COOH in small pore sizes with very high salt rejection in low ΔP.

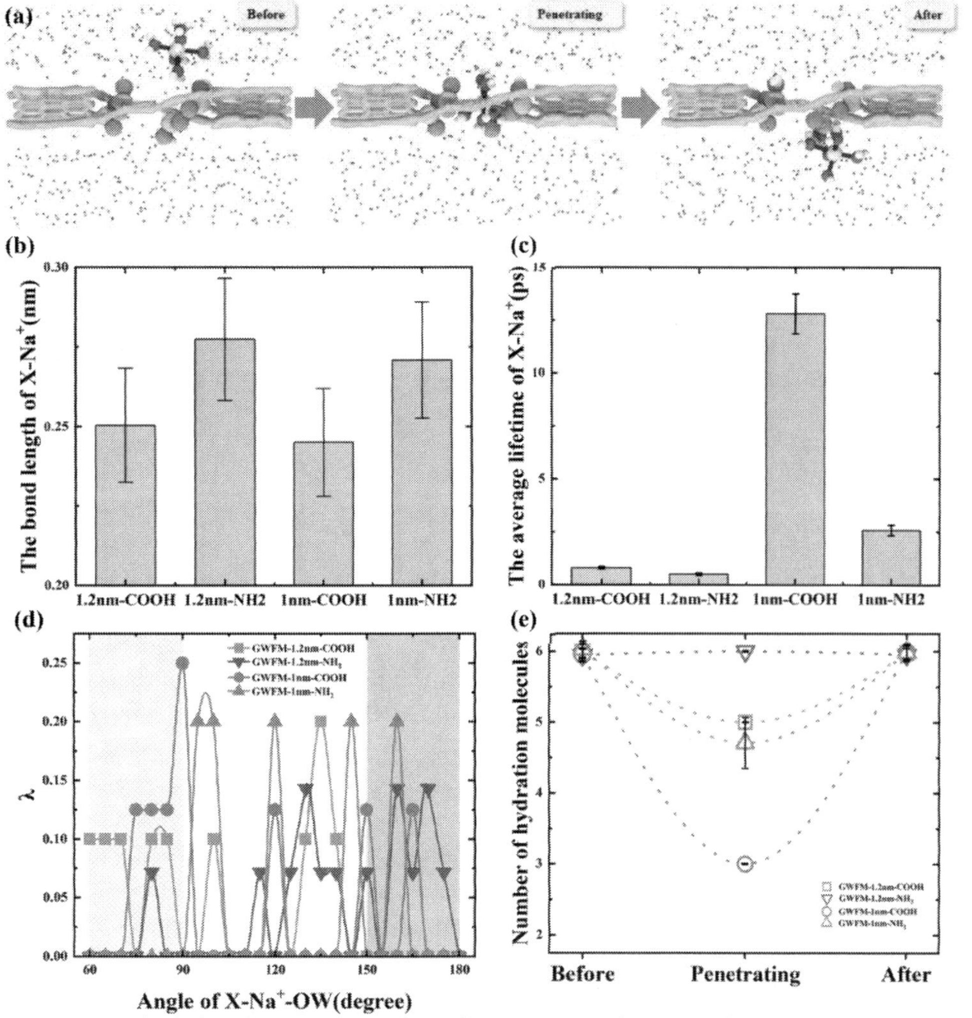

Figure 7. Images depicting the transformation of X-Na+-OW clusters as they traverse GWFM-X at a pressure of 500 MPa. (a) The kinetic mechanism of X-Na+-OW hydrated cluster binding length and angle changes. α, l denotes the penetrating X-Na+ binding length and OX-Na+-OW angle, respectively. (b) Mean binding length of X-Na+. (c) Mean duration of X-Na+ binding interactions. (d) Angular distribution of X-Na+-OW, where λ represents the frequency fraction of the angular distribution. (e) Distribution of the number of hydrated molecules.

4. Summary

This work demonstrates the potential of a woven filter composed of graphene that is founded on an X-functional group membrane (GWFM-X) for desalination purposes through molecular dynamics simulations. The study includes screening for suitable nanopore size and functional groups and a detailed analysis of water flux, flow characteristics, and desalination performance. The analysis of water fluxes of GWFM-X confirms their effectiveness for desalination. The permeability coefficients of GWFM-1.2nm-COOH, GWFM-1.2nm-NH_2, GWFM-1nm-COOH, and GWFM-1nm-NH_2 were found to be high, ranging from 9.31 ± 0.31 L cm^{-2} day^{-1} MPa^{-1} to 35.41 ± 0.71 L cm^{-2} day^{-1} MPa^{-1}. The results show that GWFM-1nm-COOH and GWFM-1nm-NH_2 achieved salt removal efficiencies of 88.75% and 87.5%, respectively, at 500 MPa during the simulation period. Similarly, GWFM-1.2nm-COOH and GWFM-1.2nm-NH_2 exhibited salt removal efficiencies of 87.5% and 82.5%, respectively, at 300 MPa. The GWFM-1.2nm-COOH membrane represents a promising new development in filtration technology, exhibiting excellent performance across a broad range of pressure variations. The combination of a meticulously crafted nanoscale pore structure and a surface carboxyl chemical modification result in a highly efficient mode of water molecule permeation, akin to "windmill rotation," while salt ions are effectively intercepted due to the formation of stable clusters within the pores and the robust interactions with the membrane surface. This enables a high salt removal rate while maintaining a high water flux. In terms of production application, this study can change the pore size by modifying the gap between graphene braided strips and achieve the filtration effect of different salts by adding different functional groups to graphene. This not only brings a certain degree of development for life and production but also a certain degree of convenience, making the filter membrane suitable for various scenarios. In the future membrane design, Ag^+ and O_3^- can be added to the membrane to achieve seawater desalination and purify water resources, perhaps more pure freshwater resources. Therefore, our work will not only provide a direction to explore the application of nano-woven modified functional group technology but also lay the foundation for future research on designing better desalination membranes.

Acknowledgment

This research was funded by the National Natural Science Foundation of China (Grant No. 22276045), the Key Research and Development Project of Zhejiang Province (Project No. 2024C03258), the Collaborative Research and Development Initiative of the Yangtze River Delta Science and Technology Innovation Community (2023CSJGG1000), and the Key R&D Project of Anhui Province (Grant No. 202304a05020022).

Author Contributions

Ziqi Huang: Research, Data Management, Original Draft Writing. Lijun Liang: Concept Development, Methodological Framework, Supervision, Resource Allocation, Manuscript Review and Editing.

References

[1] K. Elsaid, et al. Environmental impact of desalination technologies: a review, Sci. Total Environ. 748 (2020) 141528–141547.
[2] M. E. a. W. A. Phillip, The future of seawater desalination energy, technology, and the environment, †, Science 333 (2011) 712–717.
[3] Metsämuuronen et al. (2014). Natural Organic Matter Removal from Drinking Water by Membrane Technology. Separation & Purification Reviews, 43(1) 1–61.
[4] Y. G. Yan et al. A graphene-like membrane with an ultrahigh water flux for desalination, Nanoscale 9 (2017) 18951–18958.
[5] T. Mezher et al. Techno-economic assessment and environmental impacts of desalination technologies, Desalination, 266 (2011) 263-273.
[6] J. Pei et al. Fabrication of reduced graphene oxide membranes for highly efficient water desalination, RSC Advances, 6 (2016) 101948-101952.
[7] X. Li, B. Zhu, J. Zhu, Graphene oxide based materials for desalination, Carbon, 146 (2019) 320-328.

[8] D. Rana et al. Development of antifouling thin-film-composite membranes for seawater desalination, Journal of Membrane Science, 367 (2011) 110-118.
[9] T. Li, Q. Tu, S. Li, Molecular dynamics modeling of nano-porous centrifuge for reverse osmosis desalination, Desalination, 451 (2019) 182-191.
[10] T. Yang et al. Tailoring pores in graphene-based materials: from generation to applications, Journal of Materials Chemistry A, 5 (2017) 16537-16558.
[11] Hebbar et al. Carbon nanotube- and graphene-based advanced membrane materials for desalination. Environ Chem Lett, 15 (2017) 643–671.
[12] Wang R et al. Recent Advances in Applications of Carbon Nanotubes for Desalination: A Review. Nanomaterials, 10 (2020) 1203.
[13] Alabi et al. Review of nanomaterials-assisted ion exchange membranes for electromembrane desalination. npj Clean Water, 1 (2018) 10.
[14] Shim, H. E et al. Silver Nanomaterial-Immobilized Desalination Systems for Efficient Removal of Radioactive Iodine Species in Water. Nanomaterials, 8 (2018) 660.
[15] P. Srimuk, S. Husmann, V. Presser, Low voltage operation of a silver/silver chloride battery with high desalination capacity in seawater, RSC Advances, 9 (2019) 14849-14858.
[16] H. Gao et al. Rational Design and Strain Engineering of Nano-porous Boron Nitride Nanosheet Membranes for Water Desalination, The Journal of Physical Chemistry C, 121 (2017) 22105-22113.
[17] J. H. Lee et al. Vertically Aligned Carbon Nanotube Membranes: Water Purification and Beyond, Membranes, 10 (2020) 273.
[18] Y. Liu et al. MoS2 nanoflakes-coated electrospun carbon nanofibers for "rocking-chair" capacitive deionization, Desalination, 520 (2021) 115376.
[19] T. Liu et al. Graphene-based woven filter membrane with excellent strength and efficiency for water desalination, Desalination, 533 (2022) 115775.
[20] F. Malekian et al. Recent progress in gas separation using functionalized graphene nanopores and nano-porous graphene oxide membranes, The European Physical Journal Plus, 134 (2019) 212.
[21] L. Zhang, M. Becton, X. Wang, Mechanical analysis of graphene-based woven nano-fabric, Mater. Sci. Eng. A 620 (2015) 367–374.
[22] B. Liu et al. Atomic-scale finite element method in multiscale computation with applications to carbon nanotubes, Phys. Rev. B 72 (2005) 035435-035143.
[23] D. A. Leigh et al. A molecular endless (74) knot, Nat. Chem. 584 (2020) 562–568.
[24] H. Khorramdel et al. Synthesis and characterization of graphene acid membrane with ultrafast and selective water transport channels, Separation and Purification Technology, 212 (2019) 497-504.
[25] Y. Yuan et al. Enhanced desalination performance of carboxyl functionalized graphene oxide nanofiltration membranes, Desalination, 405 (2017) 29-39.
[26] H. J. C. B et al. GROMACS: a message-passing parallel molecular dynamics implementation – ScienceDirect [J], Comput. Phys. Commun. 91 (1995) 43–56.
[27] N. L., P. Mark, Structure and dynamics of the TIP3P, SPC, and SPC/E water models at 298 K [J], J. Phys. Chem. A 105 (2001) 9954–9960.
[28] J. O. Hirschfelder, C. F. Curtiss, R. B. Bird, Molecular Theory of Gases and Liquids, in: John Wiley and Sons, New York, 1954.
[29] Y. D. M., T. A. Darden, L. G. Pedersen, Particle mesh Ewald: an N□log(N) method for Ewald sums in large systems, J. Comput. Chem. 18 (1993) 1463–1472.
[30] Z. Hu, Y. Chen, J. Jiang, Zeolitic imidazolate framework-8 as a reverse osmosis membrane for water desalination: insight from molecular simulation, J. Chem. Phys. 134 (2011) 134705.
[31] B. Roux, The calculation of the potential of mean force using computer simulations, Comput. Phys. Commun. 135 (1995) 275–282.
[32] D. Cohen-Tanugi, J. C. Grossman, Water permeability of nano-porous graphene at realistic pressures for reverse osmosis desalination, J. Chem. Phys. 141 (2014) 074704–074709.

[33] M. Souaille, B. t. Roux, <Extension to the weighted histogram analysis method. pdf>, Comput. Phys. Commun. 135 (2001) 40–57.
[34] S. Kumar et al. The weighted histogram analysis method for free-energy calculations on biomolecules. I. The method, J. Comput. Chem. 13 (1992) 1011–1021.
[35] Werber et al. Permselectivity limits of biomimetic desalination membranes. Science advances, 4 (2018) 8266.
[36] R. W. Baker, Membrane Technology and Applications, in: John Wiley & Sons, Ltd, 2012.
[37] D. G. Truhlar et al. Current status of transition-state theory, J. Phys. Chem. 100 (1996) 12771–12880.
[38] L. Liang et al. Direct proof of soft knock-on mechanism of ion permeation in a voltage gated sodium channel, Int. J. Biol. Macromol. 188 (2021) 369–374.

Journal of Physics: Conference Series **2879** (2024) 012016 doi:10.1088/1742-6596/2879/1/012016

Study on tensile damage evolution of Ti6Al4V alloy based on DIC technique

Dexin Mao, Lei Li*, Xiaodong Li, Zhankun Sun, Li Zhang and Qiao Liu

Inner Mongolia University of Technology, Hohhot, 010051, PR China

*Corresponding author: leillt@imut.edu.cn

Abstract. Both optical microscopy (OM) and scanning electron microscopy (SEM) were used to analyze the microstructure of the Ti6Al4V alloy and evaluate its tensile damage behavior. Using Lemaitre's definition of damage, the variation rule of damage during the reduction of elastic modulus was obtained. Combining the digital image correlation (DIC) analysis approach allowed for an experimental investigation of the alloys' tensile mechanical characteristics. The strains at 500 independent points were extracted using VIC-2D software. The damage factor of the whole strain field was established by equations and 3D graphics, and the damage evolution process was visualized and analyzed. The results show that the Ti6Al4V alloy has good plasticity, and its fracture mode is a ductile fracture. The sample's modulus of elasticity falls, and the extent of damage increases as the stretching continues. The sample undergoes a localized violent deformation during the tensile process, as opposed to a homogenous deformation state. The harm progressively concentrates and increases at an increased rate.

1. Introduction

Titanium alloys are essential in high-end manufacturing industries such as aerospace, biomedical, and automotive because of their superior mechanical qualities, high biocompatibility, and corrosion resistance. [1-3]. Because of its distinctive mechanical characteristics and numerous application opportunities, the Ti6Al4V alloy—a representative of titanium alloys—has drawn the attention of scientists and engineers. However, with the complexity of the application environment and the harshness of the working conditions, the mechanical properties of Ti6Al4V alloys show certain limitations. Their damage and fracture behavior need to be explored in depth.

The goal of damage mechanics, a crucial subfield in the study of material behavior under stress, is to understand how materials deteriorate under complicated loads and how this affects their mechanical characteristics [4]. The in-depth understanding of the damage behavior of materials can guide the design and optimization of materials and improve their reliability and service life under extreme working conditions [5]. Therefore, a systematic study of the tensile damage behavior of Ti6Al4V alloys is of great theoretical significance and practical value for revealing the microscopic mechanism, optimizing the macroscopic properties, and expanding the application areas.

There are many methods to characterize material damage. Zhang and Zheng [6] developed a full Gurson model by combining the Gurson [7] model with the microporous aggregation criterion. The model indicates that the ductile damage of the material is related to the microporous kernel parameters using the model to simulate the crack expansion of the material, but this method has its limitations. The simulated crack expansion is zero when the strain in the specimen is small, which does not characterize the damage throughout the specimen. Through thermostatic loading, Zhai et al. [8] examined the progression of grain boundary damage and the beginning of cracks in 690 alloy, and the results show that the grain boundary damage evolution clearly indicates creep deformation of the material The homogeneous deformation of the grain boundaries is one of the major drawbacks in characterizing the damage of materials based on the grain boundaries. Similarly, damage studies based on grain boundaries have shortcomings. Since grain boundaries are not uniformly distributed in a material, this leads to varying grain boundary properties and their effect on material properties. Grain boundaries may lead to stress concentration due to their specific orientation and structure, which increases the weakness of the material and leads to crack initiation and extension. In contrast to the previously mentioned methods, digital image correlation (DIC) can non-destructively obtain information on the displacement and strain distribution of material during loading by analyzing digital images of the material's surface, providing a completely new perspective on the study of the damage behavior of materials [9, 10].

In this study, the tensile damage behavior of Ti6Al4V alloys will be investigated in depth using the DIC technique. The analysis is carried out in combination with the tissue and fracture morphology. According to Lemaitre's definition of damage [11], the variation rule of damage based on elastic modulus is explored. The deformation behavior of Ti6Al4V alloy in the tensile process was recorded in real time by using DIC technology, and highly accurate displacement and strain data was obtained. Then, the global strain field is established, and the damage factor is defined based on the scatter plots taken by DIC, which provides a scientific basis for further optimizing the properties of this material. Through this study, it is expected to have a new understanding of the damage mechanism of Ti6Al4V alloy, provide theoretical support and technical guidance for its application in extreme working conditions, and promote the development of scientific research and engineering applications of titanium alloy materials.

2. Materials and methods

In this work, vacuum non-consumptive electrode arc melting has been applied for material preparation. The samples were turned over and melted ten times, each lasting around three minutes, and electromagnetic stirring was activated to guarantee chemical homogeneity. The alloy ingots were obtained by cooling the samples to room temperature following the completion of the melting process. The nominal composition of the Ti6Al4V alloy is shown in Table 1.

Table 1. Chemical composition of TC4 titanium alloy (mass fraction, %).

Al	V	Fe	O	C	N	Ti
6.0	4.0	0.05	0.02	0.02	0.03	Bal.

They were machined to uniform dimensions using a wire-cutting machine (DK7740). The final experimental specimens are shown in Figure 1(a) for the 3D schematic and Figure 1(b) for the planar schematic and size.

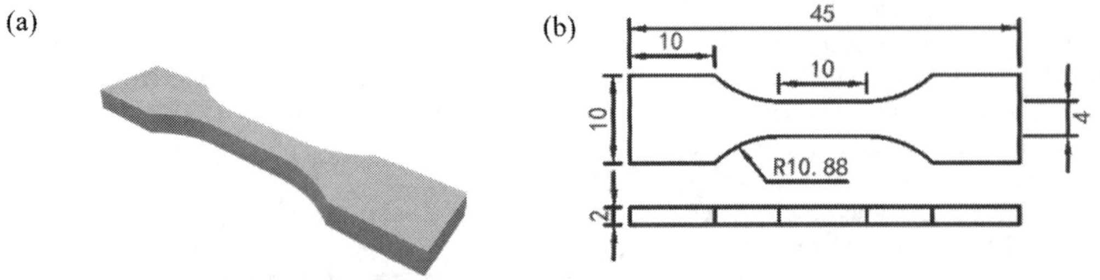

Figure 1. (a) 3D Schematic. (b) Specimen size diagram(mm).

Scanning electron microscopy (SEM) and optical microscopy (OM) were used to examine the materials' microstructure. To determine the alloy's tensile strength and elongation, the tensile characteristics of a specimen made of Ti6Al4V alloy were examined. The tensile specimens were finely ground with sandpaper and cleaned by ultrasonic vibration to obtain clean and smooth specimens. The

MTS Landmark fatigue tester with a tensile rate of 0.1 mm/min was used to perform tensile testing. The experimental findings were determined by taking the average of the tests performed on the three specimens.

DIC is an optical measuring technique that combines numerical analysis and image processing. The DIC technique starts with random texture marking of scattered spots on the surface of an object and then changes by tracking the correlation of these markings between images before and after deformation. By calculating the best matching position of each small area in the image sequence, the displacement field can be obtained, and then the strain and deformation of the object can be analyzed. DIC can be applied locally, is user-friendly, environmentally gentle, and suitable for various materials. It works by using a CCD camera to capture images of the speckle pattern on the specimen's surface during the experiment. Then, using displacement analysis of the preset area, the target image's scattering field is connected to the most relevant point in the reference image. The offset between the target image's scattering field and the experiment's reference image is calculated to get the displacement value, which is then used to determine the relevant strain value using the theory of strain. Figure 2 displays the DIC measuring system's functioning schematic diagram. In the test experiment, the CCD camera lens's maximum resolution is 1624(H)×1224(V) pixels, its maximum frame rate is 30 fps, and its strain accuracy is≤50 $\mu\varepsilon$.

Through the DIC method, not only the full-field strain distribution of Ti6Al4V alloy in different tensile stages can be obtained, but also the deformation characteristics of local regions can be obtained by processing the full-field experimental data. The tensile damage progression of Ti6Al4V alloy is observed in this experiment using the DIC system, which is set up as follows:

(1) To increase the experiment's accuracy, the Ti6Al4V alloy was first sandpapered and then physically polished to lessen the alloy's local stress concentration caused by surface roughness. Next, surface scattering spots were sprayed on the polished metal specimens.

(2) Clamp the alloy specimen on the MTS fatigue tester, make sure that the sprayed side is facing the CCD camera, and pre-tighten the specimen to ensure that it is fixed.

(3) Switch on the image acquisition system and make sure the image display is normal. Adjust the parameters of the CCD camera to ensure a clear image.

(4) After completing the debugging of the acquisition system and the MTS fatigue tester, at the beginning of the tensile test, make the CCD camera record a photo every second.

(5) Perform the relevant data processing on the VIC-2D software with the captured pictures.

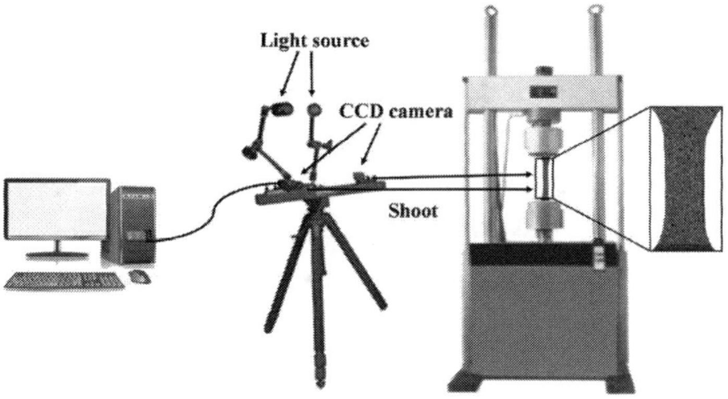

Figure 2. Working diagram of the DIC measurement system.

3. Results

3.1. Microstructure morphology

The net basket organization can be seen in Figure 3(a), which is formed by a staggered arrangement of smaller sized α-collections. This organization indicates that Ti6Al4V alloy has good plasticity and toughness. From Figure 3(b), small and shallow tough nests can be seen distributed at the fracture of the specimen, which indicates that the fracture mode of the specimen is ductile fracture [13-15].

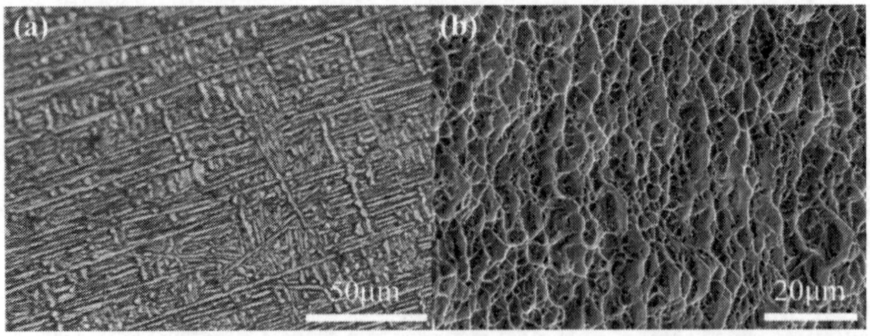

Figure 3. (a) OM pattern of Ti6Al4V. (b) SEM of Ti6Al4V fracture morphology.

3.2. Contour analysis using digital image correlation (DIC)

The specimen's DIC strain cloud at various tensile process phases is displayed in Figure 4. The strain contours at the elastic stage, uniform plastic deformation stage, necking beginning stage, and necking stage—typical locations on the tensile stress-strain curve—were chosen. The part of the tensile curve that contains the DIC contours for this kind of location is represented by the letters A, B, C, D, and E in Figure 4. The image illustrates how the material is in the elastic stage at the start of the specimen's loading and exhibits a slight deformation (A in Figure 4); As the load is increased gradually, the material moves into the plastic deformation stage (B–C in Figure 4), where plastic deformation extends out and the strain cloud graphs exhibit various color variations. The deformation increases until it reaches the necking selection state, at which point the tiny deformation zone's deformation slows down and the big deformation region starts to emerge. The distortion in the big deformation zone is substantial, whereas the tiny deformation region progressively stops deforming. At this point, the less deformation region ceases to cause damaging deformation, and the larger deformation region is primarily where damage occurs (D-E in Figure 4). [16, 17].

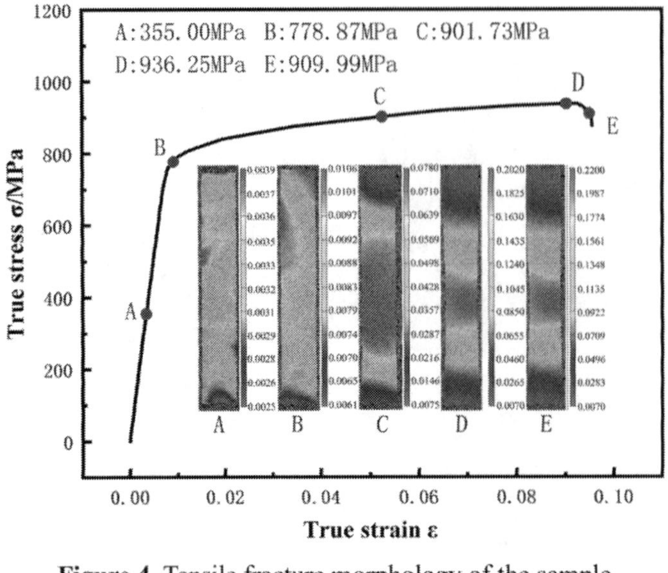

Figure 4. Tensile fracture morphology of the sample.

3.3. Damage evolution

According to the damage equation defined by Lemaitre [11]:

$$D = 1 - \tilde{E}/E \tag{1}$$

E is the material's modulus of elasticity in its undamaged state, $\tilde{E}$ is the material's modulus of elasticity after damage.

$\tilde{E}$ is obtained by using the "Secant Method" [12] and plotted in Figure 5(b) according to equation (1). From Figure 5(a), it is evident that when strain increases, the modulus of elasticity falls. Figure 5(b) illustrates the extent of material degradation that occurs when the elastic modulus is reduced, with large damage produced at the beginning of the deformation of the material and the growth of the damage becoming slow afterward.

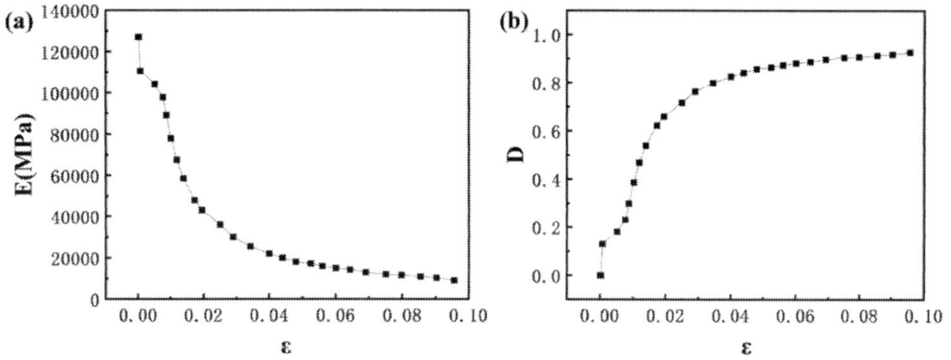

Figure 5. (a) E modulus variations with strain. (b) Evolution of damage D with strain.

It is evident that the pattern shown in Figure 5 does not adequately adapt to the specifics of the specimen damage development. 500 locations were randomly chosen from the strain field to quantitatively examine the damage process. The Ti6Al4V alloy's tensile damage process was characterized using the damage factor D, and the evolution of the damage was described according to the variation rule of the obtained data [18, 19]. The difference between the longitudinal point strain ε_{yy} and the minimum strain ε_{min} in each state is used to represent the relative strain factor ε_r.

$$\varepsilon_r = \varepsilon_{yy} - \varepsilon_{min} \tag{2}$$

ε_{yy} is the longitudinal strain at a single point; ε_{min} is the minimum strain in each state.

From J. Lemaitre's damage definition $D = \delta S_D / \delta S$ [20]. In order to fit the data obtained by DIC in this experiment, its δS_D is replaced by ε_r, and δS is replaced by $\varepsilon_{r_{max}}$. The following is the definition of the damage factor:

$$D = \varepsilon_r / \varepsilon_{r_{max}} \tag{3}$$

$\varepsilon_{r_{max}}$ is the maximum value of ε_r; D is the damage factor.

Figure 6 shows the three-dimensional variation of the damage factor for 500 individual points on the surface of Ti6Al4V alloy. According to the figure, the damage factor grows gradually as strain increases during the homogeneous deformation stage, and it does not significantly change when the number of point sequences increases for each point. However, the damage grows much more quickly during the local deformation stage. High in the center and low at both ends is a phenomenon that is demonstrated by the damage factor's change with the number of point sequences. And as the tension grows, this tendency becomes more and more apparent. Conversely, because the strain field's points are consistently chosen in a certain order, the point sequence may be used to associate the damage with its location. The final fracture of the specimen occurs in this region where the point sequence is from 200 to 350. The corresponding change that occurs in the Ti6Al4V alloy specimen is the increasing concentration of stress, which eventually produces a localized large deformation and then fracture.

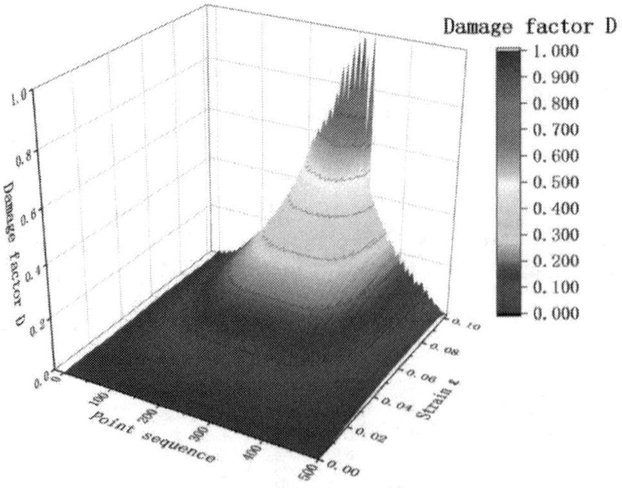

Figure 6. 3D graphic of damage factor evolution.

4. Conclusion

In this paper, the tensile process of cast Ti6Al4V alloys has been analyzed in conjunction with the DIC technique. The alloy's microstructure is explained by both SEM and OM. Using Lemaitre's definition of damage, a damage evolution law based on the change in elastic modulus is obtained. The tensile damage behavior of Ti6Al4V alloy was studied by defining the damage factor and based on the strain data collected by DIC. The damage factors for the whole strain field were established, and the damage law was analyzed. The following conclusions were drawn:

(1) The Ti6Al4V alloy experiences necking when the plastic strain exceeds a particular value, and early plastic deformation is primarily homogenous.

(2) According to the microstructure and fracture morphology of the alloy, the fracture mode of the specimen was determined to be a ductile fracture.

(3) As stretching proceeds, the modulus of elasticity of the specimen becomes smaller, and the degree of damage increases.

(4) As the strain increases, the damage tends to become more concentrated, and the damage factor increases rapidly.

Through this experiment, the connection between the strain and the damage factor of the Ti6Al4V alloy was established, allowing for the prediction of the alloy's damage degree at various strains. By analyzing the damage factor of Ti6Al4V alloy under specific strains and the damage site of the specimen, the correlation between the damage factor and the damage site can be found, which provides key information for the damage prediction and performance optimization of the material. In practical applications, targeted improvement of the material's weak points can optimize the material design and increase damage resistance, hence improving the material's service life and engineering application value.

Acknowledgments

This work was supported by the National Natural Science Foundation [grant number 12262029]; the Inner Mongolia Fund [grant number 2023MS01007]; and the Inner Mongolia Doctoral Fund [grant number BS2021053].

References

[1] Sofinowski K, Šmíd M, Kuběna I, Vives S and Swygenhoven HV (2019) In situ characterization of a high work hardening Ti-6Al-4V prepared by electron beam melting. *Acta Materialia* 179224-236.

[2] Ren Y, Lin X, Fu X, Tan H, Chen J, and Huang W (2017) Microstructure and deformation behavior of Ti-6Al-4V alloy by high-power laser solid forming. *Acta Materialia* 13282-95.
[3] Wang L, Luo J, Xu J and Zhang J (2021) Study on the effect of SPS sintering temperature on the organization and properties of (GNPs-Cu)/Ti6Al4V composites. *Rare Metals* 45(12): 1429-1437.
[4] Pei C, Zeng W and Yuan H (2020) A damage evolution model based on micro-structural characteristics for an additive manufactured superalloy under monotonic and cyclic loading conditions. *International Journal of Fatigue* 131105279-105279.
[5] Xu L, Chong K, Su Y Chang C and Zhao L (2023) Meso-scale low-cycle fatigue damage of polycrystalline nickel-based alloy by crystal plasticity finite element method. *International Journal of Fatigue* 175.
[6] Zhang K and Zheng C (1992) A simplified multilevel nucleus model for cavities. *Journal of Solid Mechanics* (02):187-190.
[7] Zhang Z and Thaulow C (2001) A Complete Gurson model approach for ductile fracture. *Engineering Fracture Mechanics* 67(2):155-168.
[8] Zhai Z, Toloczko M, Kruska K, Schreiber D and Bruemmer S (2019) Grain Boundary Damage Evolution and SCC Initiation of Cold-Worked Alloy 690 in Simulated PWR Primary Water. In: *Proceedings of the 18th International Conference on Environmental Degradation of Materials in Nuclear Power Systems–Water Reactors*. Berlin. pp. 457-483.
[9] Liu H, Sun R, Zhou Y, Bai R, Xia M, Cai X, Wang F, and Zhang W (2023) Study on the crack extension mechanism of SLM-IN718 fine surface cracks based on digital image correlation technology. *Rare Metals Materials and Engineering* 52(10):3433-3441.
[10] Peng J, Xue Z, Dai Q, Miu X, and Liu X (2022) Study on the local material structure of welded joints of nickel-based welding consumables based on DIC. *Rare metal materials and engineering* 51(05):1667-1673.
[11] Lemaitre J (1985) A Continuous Damage Mechanics Model for Ductile Fracture. *Eng. Mater. Technol* 107(1): 83–89.
[12] Xie H, Ju Y and Dong Y (1997) Discussion on the "modulus of elasticity method" in the definition of classical damage. *Mechanics and Practice* (02): 2-6.
[13] Liu S, Cai J, Zhou Y, Cao J, Zhang W, Dai S, Huang X and Cao C (2024) Effect of annealing temperature on microstructure evolution and tensile behavior of multidirectional forged Ti-6Al-4V alloy. *Transactions of Nonferrous Metals Society of China* 34(06): 1864-1877.
[14] Shi G, Wang X, Chen Y, and Du Z (2022) Effect of laser power on the organization and properties of laser fusion deposited Ti-6Al-4V-0.25C alloy. *Rare Metal Materials and Engineering* 51(12):4570-4578.
[15] Ma J, Li J, Wang Z, Wang Y, and Wang J (2021) Microstructure and mechanical properties of the bonding zone of Ti-6Al-4V alloy fabricated by forging-additive composite. *Journal of Metals* 57(10):1246-1257.
[16] Wei L, Li L, Cui X, and Liu D (2020) Study on damage evolution of Cu-Ni19 alloy at different annealing temperatures based on DIC method. *Engineering Mechanics* 37 (04): 227-235.
[17] Zhao G, Zhang R, Liu J, Li H, Ma L, and Li Y (2022) Study on microstructure and tensile damage evolution of 2205 duplex steel at different solution temperatures. *Materials Today Communications* 33.
[18] Zhao Y, Wang L, Lei Z, Han X, and Shi J (2018) Study on bending damage and failure of basalt fiber reinforced concrete under freeze-thaw cycles. *Construction and Building Materials* 163460-470.
[19] Li L, Li X, Liu Q and Mao D (2024) Effect of heat treatment on the microstructure and surface damage evolution of selective-laser-melted IN718 alloy. *Matéria (Rio de Janeiro)* 29: e20230311.
[20] Lemaitre J (1996) Tutorial on damage mechanics. *Science Press*, Beijing.

Journal of Physics: Conference Series **2879** (2024) 012017 doi:10.1088/1742-6596/2879/1/012017

The integrated performance of adaptive temperature control and temperature measurement of PTC materials with high thermal cycle repeatability

Yinfa Yang, Jianmin Hao, Hua Chen and Wenlong Cheng*

Department of Thermal Science and Energy Engineering, University of Science and Technology of China, Hefei, Anhui, 230027, PR China

*Corresponding author's email: wlcheng@ustc.edu.cn

Abstract: As the temperature increases, the resistance of the positive temperature coefficient (PTC) material becomes larger. Using this material can reduce the complexity of the temperature control system. In this paper, a new PTC material is developed by means of melt blending, which still has operable PTC characteristics after 60 high and low temperature cycles. The data points collected in multiple thermal cycles were fitted to obtain the relationship between the sample's temperature and resistance. In addition, the integrated performance of adaptive temperature control and temperature measurement of PTC materials is proposed for the first time. Based on the prepared samples, the temperature control characteristics of PTC materials on the controlled object under different environmental conditions are studied experimentally, and the temperature value converted from the resistance value is compared with the temperature value measured by the platinum resistance. The results show that the prepared sample has good temperature control performance, and the material has an operable temperature measurement function in the PTC effect range.

1. Introduction

Electric heating is a common means of heating objects, which is extensively used in varied occasions requiring low temperature heating. The traditional electric heating temperature control technology mainly provides additional heat to the controlled device through the resistance of the fixed resistance value, collects the temperature information and feeds it back to the heater, and then adjusts the power of the heater according to the preset temperature value, thus the temperature of the equipment can be maintained within a reasonable range [1]. However, this thermal control technology requires the use of temperature measuring instruments and corresponding control components, which will increase the complexity and quality of the system [2].

For PTC material, its temperature will affect its resistance value. The resistance value of PTC material is in a stable state when the temperature is less than a certain value, which is called the fixed resistance area. When it exceeds this value (Curie point), the resistance value will increase sharply with the change in temperature, which is defined as the PTC effect area. When the temperature continues to rise to another point (termination point), its resistance value remains at a fixed value, and the region after the termination temperature is called the PTC effect termination region [3]. With this unique resistance-temperature characteristic, when a resistor made of PTC material is used as a heating element, a fixed voltage is applied to it without using a temperature measuring instrument and a control element.

It can realize self-regulation of heating power according to its temperature change, thereby achieving adaptive temperature control [4]. When the heating element made of PTC material is applied in thermal control, its electrical resistance is closely related to its temperature in the PTC effect region. Once the material temperature reaches the range that can trigger the PTC effect, a specific relationship can be used to convert the resistance value of the material into the corresponding temperature value in real time. This method makes it possible to measure the temperature of the system in real time without the need for additional temperature measuring elements.

When the PTC material is made into a heating source, its good PTC electric cycle reproducibility is the basis for the temperature control system to have operable self-regulating thermal control performance and high-precision temperature measurement [1]. At present, researchers have explored many ways to improve the PTC repeatability of materials through experiments. Gwang-Myeong et al. [5] proposed PTC composites based on ethylene-vinyl acetate (EVA), which makes the prepared samples have operable repeatability through the cooperative effect of 0-D and 2-D carbon fillers. Chen et al. [6] developed a PTC composite based on high density polyethylene (HDPE) / polyvinylidene fluoride (PVDF). By adding ionic liquid to the composite, the interfacial gap between HDPE and PVDF is reduced, which effectively improves the PTC repeatability of the prepared sample. Xu et al. [7] improved the stability of the developed material by filling CB and TiC into the HDPE matrix, making CB particles serve as a bridge between TiC particles.

In this paper, a new type of PTC material was developed by melt mixing with GP and CNTs as conductive objects, EVA and lauric acid (LA) as a matrix, and the electrical properties and thermal cycle stability of the prepared samples were measured. Finally, based on the prepared PTC materials, a PTC self-regulating temperature control experimental device was built to study the temperature regulation and temperature measurement effect of the prepared samples.

2. Preparation and electrical properties of materials

2.1 Materials

EVA is used as the supporting material, and LA is used as the phase change material. The two are mixed to form the matrix of the prepared material. EVA (VA composition of 32%) was purchased from DuPont and LA (density of 880 kg/m^3) was purchased from Xilong Science Co., Ltd. CNTs (purity > 95%, outer diameter: 0.008-0.015 nm, length: ~ 0.05 mm) and GP (carbon composition: ≥ 99.85 wt %, Standard size value (30 mm): ≥ 95%) are filled into the matrix as conductive fillers. CNTs were purchased from Aladdin Biochemical Technology Co., Ltd. GP was purchased from Sinopharm Chemical Reagent Co., Ltd.

2.2 Sample preparation process

The process required to prepare the sample is shown in Figure 1. The temperature in the kettle was heated to 80°C. The LA was placed in the stirred kettle and after it was completely melted, a certain amount of EVA was added to the stirred tank. The mass ratio of dodecanoic acid to EVA was 3:7. At this time, EVA and dodecanoic acid were mechanically mixed in a stirred tank for 30 min, and then carbon nanotubes and graphite were added, accounting for 2 % and 40 % of the mass of the matrix material, respectively. Start the stirring tank, mechanically stir and mix for 2 h, and place the molten material in the mold to press into a 1 mm sheet. Finally, the prepared material was cooled and cut into sheets with a size of 40 mm * 40 mm * 2 mm.

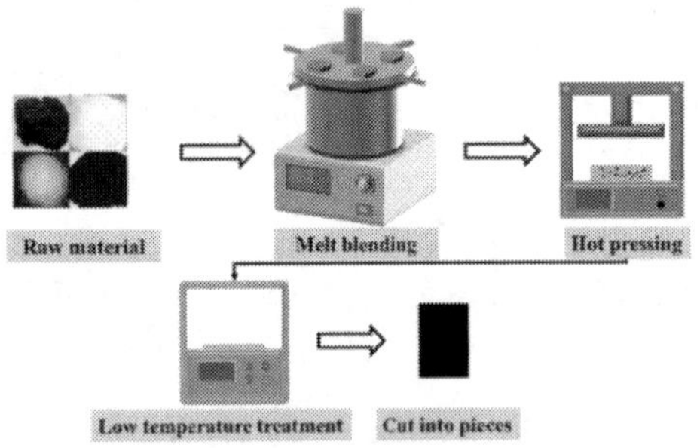

Figure 1. Preparation process of materials.

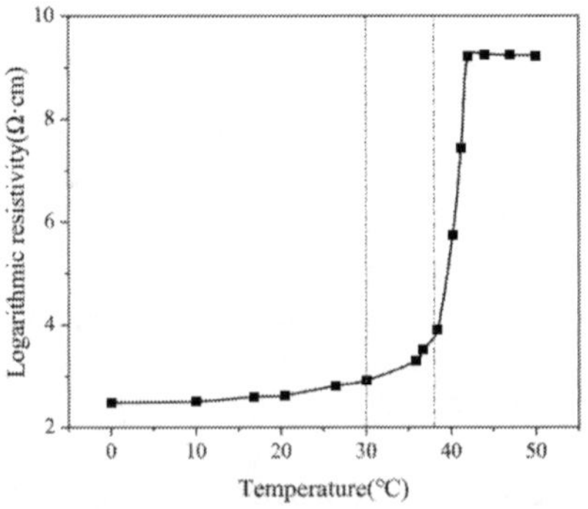

Figure 2. The curve of log ρ and temperature of the sample.

2.3 Electrical properties of materials

The two surfaces of the PTC sample prepared in 2.2 were coated with a flat conductive film, and then the electrode was attached to the surface of the sample coated with a conductive film. A fixed voltage is provided to both ends of the self-made sample. The temperature value of the sample surface is collected by the collector, and the resistivity of the material is measured in the temperature range of 0°C ~50°C. The measured resistance-temperature value is shown in Figure 2. Before the material temperature reaches 30°C, the resistivity variation range of the material is small. When the temperature value is within 30 ~ 38°C, the resistivity value of the self-made sample tends to increase continuously, but the change range is small, only one order of magnitude, so we call it the weak PTC characteristic interval. When the temperature of the sample itself rises to 38°C, the resistivity value of the sample increases by about 5 orders of magnitude in the numerical range of 38 ~ 42°C, so we call it the strong PTC effect interval. For PTC materials, PTC intensity (I_{PTC}) is usually used to represent the strength of PTC characteristics that can be denoted as:

$$I_{PTC} = lg \cdot \frac{\rho_t}{\rho_0} = \lg\rho_t - lg\rho_0 \quad (1)$$

where ρ_t represents the resistivity of the sample at the termination temperature, ρ_0 represents the resistivity of the material at the Curie temperature.

According to Eq. (1), the PTC intensity of the sample in the first PTC characteristic range is 1, the Curie temperature in the first PTC characteristic range is 30°C, the termination temperature is 38°C, the PTC intensity in the second PTC effect range is 5, the Curie temperature is 38°C, and the termination temperature is 42°C. The material has ultra-high PTC intensity in the second PTC effect range.

2.4 The cyclic repeatability of the PTC characteristics of the samples.

When using the prepared PTC sample as a temperature control element, it is necessary to improve the PTC cycle repeatability of the material itself, which is an important condition for the temperature control system based on it to have operable self-regulating temperature control characteristics and high temperature measurement accuracy. The prepared samples were placed in an environmental incubator, and the resistivity of the samples was measured under the environmental conditions of 0-50°C. As shown in Figure 3, the relationship curve between the resistance and temperature of the sample did not change decidedly after 60 high and low temperature cycles, indicating that the developed PTC material has excellent PTC cycle repeatability.

2.5 Microstructure of PTC materials

The microscopic image of the sample under the scanning electron microscope is shown in Figure 4(a), and the graphite peaks composed of agglomerated graphite stacks are uniformly dispersed in the prepared sample. Figure 4(b) shows a further enlarged picture of the microstructure of PTC material, showing that the graphite peak is connected by CNTs to construct a unique island bridge structure. This island bridge structure makes the conductive network inside the PTC material more stable [8], which further explains the reason why the prepared sample has operable PTC cyclic repeatability.

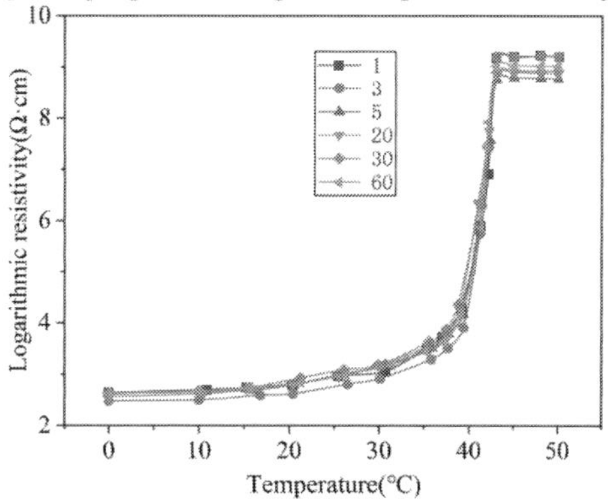

Figure 3. The resistance-temperature characteristics of the material under different thermal cycles.

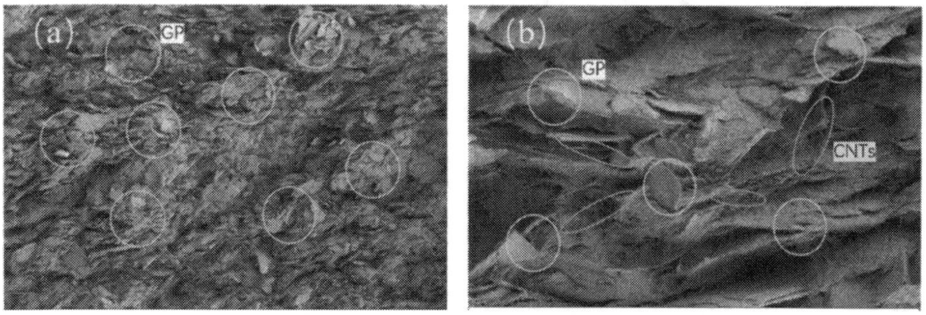

Figure 4. The microstructure diagram of the material.

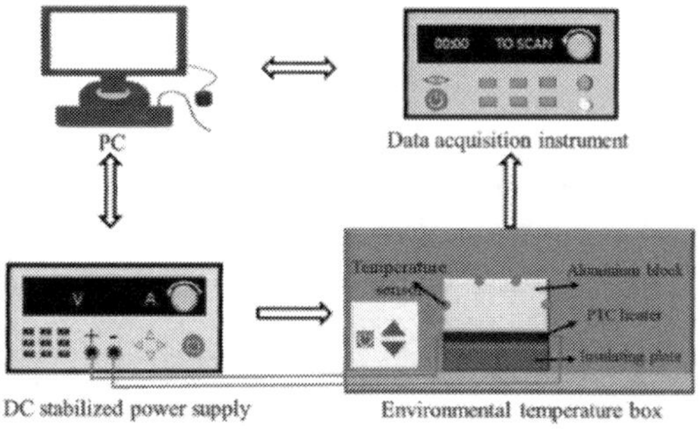

Figure 5. Temperature control test system.

3. Adaptive temperature control and temperature measurement integration experiment

3.1 Instrument and instrument error value

Figure 5 is a self-adjusting temperature control test system based on the prepared sample. The electrode is fixed on both sides of the prepared sample, and the two are fixed by insulating tape so that it can perform temperature measurements and provide heat to heat the object in the system. A fixed voltage is applied to both ends of the PTC heating device through a power supply with minimal voltage fluctuation. Taking the aluminum block (40 mm × 40 mm × 2 mm) as the controlled object, a small hole is drilled at the bottom of the aluminum block, and a platinum resistance is placed. The temperature measurement accuracy of the temperature measuring resistor used in the experiment is 0.2°C. A layer of dense thermal conductive adhesive is coated between the aluminum block and the self-made heating element, and the pressure is applied through the tape to make the two in close contact, to reduce the contact thermal resistance between the two and transfer the generated heat to the aluminum block as much as possible. The self-regulating temperature control and temperature measurement experimental platform is placed in the environmental simulation device to adjust the temperature of the surrounding environment, and the voltage, current, and temperature data of the controlled object in the circuit are collected by the data acquisition instrument. The error of the instruments used above is obtained from the instrument manufacturer, and the error of each instrument is shown in Table 1.

3.2 Fitting of temperature-resistance relationship

We fit the values collected in the thermal cycle repeatability experiment in Figure 6(a), and the temperature-resistance relationship obtained by fitting is shown in Eq (2). The deviation between the fitted results and the experimental data is shown in **Figure 6(b)**. In the region with low resistivity, the resistivity value does not fluctuate significantly with the increase of temperature, and the low temperature resistivity measured under each high and low temperature cycle is different, resulting in a large deviation of the fitting curve when the temperature of the sample itself is low, indicating that the prepared sample only has a heating function in this region. In the weak PTC characteristic region, although the deviation of the fitting curve temperature decreases, the maximum deviation still reaches ±4°C. In the strong PTC characteristic region, the temperature deviation of the fitting curve is significantly smaller than that in the weak PTC characteristic region, and the maximum temperature deviation is only ±0.35°C. At this time, the prepared sample has both heating and temperature measurement functions. When the temperature of the controlled object is self-regulated by the prepared sample, the resistance value can be converted into a temperature value to perform the temperature measurement function.

Table 1. An error in the instrument.

Metrical data	Device	Error
Ambient temperature	Environmental temperature box	±0.5 K
The temperature of the controlled object	platinum-cobalt method resistance temperature sensor	±0.15 K
Heating sheet temperature	platinum-cobalt method resistance temperature sensor	±0.15 K
Voltage	DC stabilized power supply	±0.3 mv

$$T_P = \begin{cases} 63lg\rho - 163 \\ 39 - 809e^{-lg\rho/0.7} \\ 42.5 - 26.6e^{-lg\rho/2.4} \end{cases} \tag{2}$$

where T_P is the fitted temperature, ρ is the resistivity of PTC material.

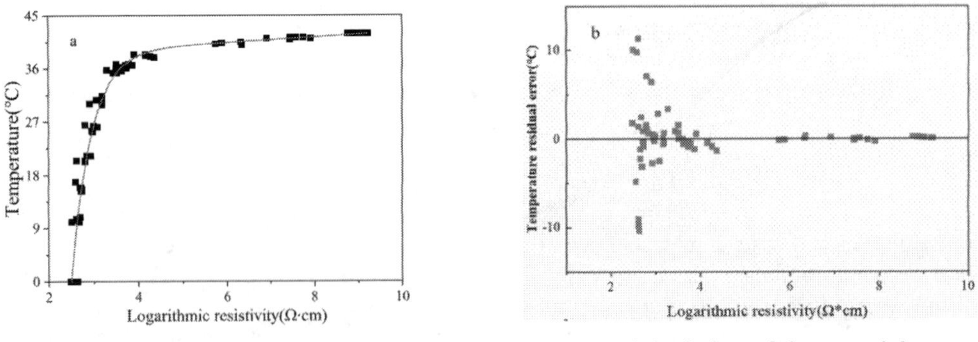

Figure 6. The resistance-temperature fitting curve and deviation of the material.

3.3 The temperature management and temperature measurement characteristics of the prepared samples at different ambient temperatures

Through the ambient temperature box, the environmental conditions of the experimental system were adjusted to 0, 5, 10, and 15 °C in turn. By comparing the temperature curve measured by the platinum resistance with the temperature curve converted by the PTC resistance fitting formula, the adaptive temperature control and temperature measurement functions of the controlled object under different conditions are explored respectively. The initial temperature of the whole system is the same as the environmental temperature, and a fixed voltage of 15 V is provided to the PTC sample through the DC regulated power supply. The temperature control and temperature measurement curves of the system under different environmental conditions are shown in Figure 7. There are three main processes in the temperature control process of the system composed of samples under different conditions.

The system heats up rapidly in the first process. This is because the temperature of the sample itself is less than the Curie temperature, and the resistance value is low, so the heating power of the PTC heating plate is higher.

In the second process, the system heats up slowly. This is because the temperature of the sample is in the range of PTC characteristics, and the resistance value of the sample increases with the increase in temperature. Therefore, the heating power of the PTC heating plate is reduced, and the controlled object is slowly heating up.

In the third stage, the sample temperature reaches the equilibrium temperature, and the system is in thermal equilibrium with the surrounding environment. In addition, in the whole process of self-regulating temperature control, the temperature measurement process of the sample under different environmental conditions also has three main processes. The first process has no temperature measurement function. At this time, the temperature of the material itself can only be obtained according to the temperature measurement curve. The second stage is the transition zone to realize the temperature measurement function. At this time, the temperature of the sample is in the weak PTC characteristic region, and the deviation between the temperature measurement curve and the actual temperature gradually decreases. The third process is the implementation of the sample temperature measurement function. At this time, the temperature of the sample is in the strong PTC characteristic region. At this time, the deviation between the temperature measurement curve and the actual temperature is kept within ±0.35°C, which has a good temperature measurement function. When the environmental temperature increases from 0 to 15°C, the final temperature of the controlled object increases from 38 to 38.5°C, and the change range is just 1 % of the ambient temperature, which weakens the incidence of environmental conditions on the system.

The above shows that the sample has excellent self-regulating thermal control characteristics. At the same time, the prepared material can perform heating functions in the PTC effect zone while performing temperature measurement, and finally realize the integration of thermal management and temperature measurement of the system.

4. Conclusion

In this article, a new material with a low Curie point was prepared and the electrical performance of the sample itself was tested. Based on the prepared material, an experimental system was built to study the self-regulating heating and temperature measurement functions of PTC materials:

(1) The prepared PTC material has two PTC effect intervals, and the Curie temperatures are 30°C and 38°C, respectively. The PTC intensities of the two PTC effect intervals are 1 and 5, respectively.

(2) PTC material has no obvious change in PTC characteristics after 60 thermal cycles and has excellent thermal cycle stability.

(3) The prepared samples have excellent self-regulating heating and temperature measurement functions. When the ambient temperature changes from 0 to 15°C, the temperature of the controlled object changes by 0.5°C. At the same time, in the range of PTC effect, the temperature measurement error of the material is ±0.35°C.

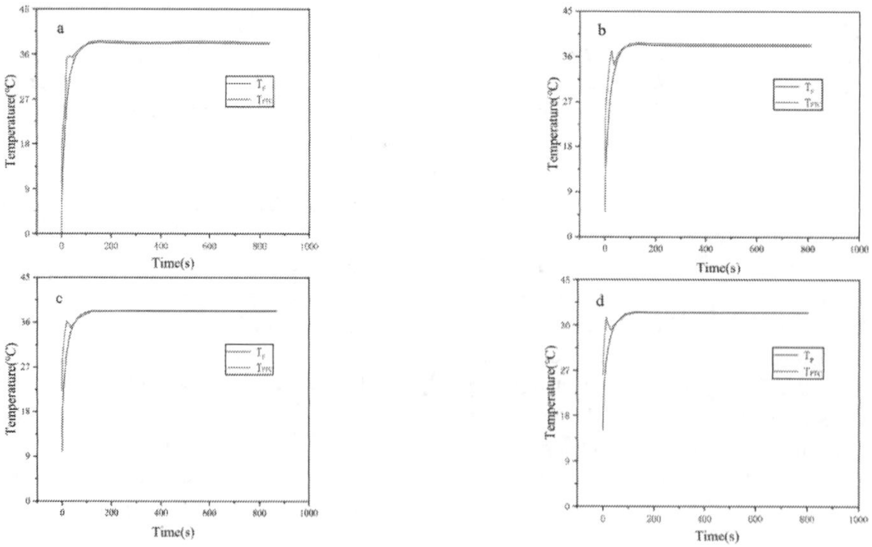

Figure 7. The characteristics of temperature control and temperature measurement of materials under various environmental conditions (a) 0°C; (b) 5°C; (c) 10°C; (d) 15°C.

Acknowledgements
This work is supported by the National Key R&D Program of China (Grant No. 2022YFC22010203) and the National Science Foundation of China (Grant No. 52376153).

References

[1] Y.-F. Yang *et al.*, "High-precision adaptive temperature control performance of thin positive temperature coefficient materials with ultra-high resistance-temperature coefficient," *Applied Thermal Engineering,* vol. 253, 2024.

[2] J.-M. Hao *et al.*, "Positive temperature coefficient material based on silicone rubber/ paraffin/ graphite/ carbon nanotubes for wearable thermal management devices," *Chemical Engineering Journal,* vol. 493, 2024.

[3] W.-l. Cheng, J.-l. Song, Y. Liu, S. Yuan, W.-f. Wu, and Z.-m. Xu, "Theoretical and experimental studies on thermal control by using a novel PTC material with room temperature Curie point," *International Journal of Heat and Mass Transfer,* vol. 74, pp. 441-447, 2014.

[4] W.-l. Cheng, S. Yuan, and J.-l. Song, "Studies on preparation and adaptive thermal control performance of novel PTC (positive temperature coefficient) materials with controllable Curie temperatures," *Energy,* vol. 74, pp. 447-454, 2014.

[5] G.-M. Go *et al.*, "Enhanced positive temperature coefficient intensity and reproducibility with synergistic effect of 0-D and 2-D filler composites," *Journal of Materials Science,* vol. 57, no. 38, pp. 18037-18050, 2022.

[6] L. Chen, X. Wu, X.-F. Zhang, and J.-M. Zhang, "Enhanced Reproducibility of Positive Temperature Coefficient Effect of CB/HDPE/PVDF Composites with the Addition of Ionic Liquid," *Chinese Journal of Polymer Science,* vol. 39, no. 2, pp. 228-236, 2020.

[7] H.-P. Xu, H.-J. Liu, H.-Q. Xie, W. Yu, and D.-D. Yang, "Excellent positive temperature coefficient behavior and electrical reproducibility of HDPE/(TiC-CB) composites," *European Polymer Journal,* vol. 175, 2022.

[8] X. Zhang *et al.*, "Two-step positive temperature coefficient effect with favorable reproducibility achieved by specific "island-bridge" electrical conductive networks in HDPE/PVDF/CNF composite," *Composites Part A: Applied Science and Manufacturing,* vol. 94, pp. 21-31, 2017.

Journal of Physics: Conference Series **2879** (2024) 012018 doi:10.1088/1742-6596/2879/1/012018

Study on the influence of trace elements H on the properties of A356 aluminum alloy

Jianjun Lv[1*], Jiang Zhang[1], Xiaolu Hong[2], Peixu Zhou[2], Dahui Chen[2] and Yingjiang Peng[2]

[1]Inner Mongolia Metal Materials Research Institute, Baotou 014000, Inner Mongolia, China

[2]Ningbo Branch of China Academy of Ordnance Science, Ningbo 315000, Zhejiang, China

[*]Corresponding author's e-mail: zhangjiangjixie062@126.com

Abstract: The vacuum melt purification refining technology can significantly reduce the content of hydrogen elements in alloy melts and prevent defects such as hydrogen-induced pinholes. This article investigates the effect of hydrogen content on the chemical composition, mechanical properties, and structural defects of A356 alloy melt by using vacuum melt purification refining technology to reduce the hydrogen content in A356 alloy melt. The experimental results show that: 1) there is basically no loss of chemical elements during the vacuum melting process; 2) The use of vacuum refining and purification control technology can control the hydrogen content inside the melt to be ≤ 0.10ml/100g; 3) The effect of H content on the metallographic structure of A356 alloy is not significant, but it has a certain improvement on the tensile strength at room temperature, but the increase is ≤ 5%.

1. Introduction

In terms of hydrogen removal and purification of high-performance aluminum alloy melts, due to the toxicity and environmental pollution of refining agents such as hexachloroethane and chlorine gas, non-adsorption purification technology using rotary nozzles is widely used abroad. This method has the advantages of high efficiency, stability, and easy operation. It also has unique advantages in terms of bubble dispersion, sufficient contact with aluminum melt, and selection of gas refining agents [1,2].

The SNIF method developed by the Union Carbide Corporation in the United States [3] uses a purification principle of injecting inert gas into the melt from a high-speed rotating nozzle, thereby achieving the purpose of degassing and slag removal. The purification process involves the aluminum alloy melt entering the initial zone through the inlet of the device, being purified by the first rotating nozzle, and then entering the rear end through the partition for a second purification. The melt flows back to the storage room at the front of the device through the graphite or silicon nitride conduit at the bottom of the furnace, and finally flows out through the outlet at the upper end. This process can reduce the hydrogen content in the aluminum liquid to 0.105ml/100g, achieve a removal rate of over 80% for inclusions above 50 μ m, and reduce the alkali metal content to below 10-4. MINT method is an online aluminum liquid treatment method developed by United Aluminum Corporation in the United States, which is suitable for continuous aluminum alloy processing. It has good removal effects on hydrogen, inclusions, and alkali metals in aluminum melt. The American company ALmex has developed LARS technology, which uses inert gas for on-site preheating, friction stirring of aluminum melt with inert gas, and various measures to prevent bubble aggregation. This can fundamentally eliminate physical inclusions, metallurgical inclusions, and chemical inclusions in aluminum melt, thereby achieving a purification effect [4].

China's Yang Changhe et al. have developed a new method of rare earth inert gas (argon) combined degassing [5], and Ding Wenjiang et al. have developed JDN-1 multifunctional rare earth flux, which has significant degassing and purification effects on ZL114A aluminum alloy and has a certain degree of modification and refinement [6]. On the basis of introducing some advanced purification technologies and equipment from abroad, research on aluminum alloy melt purification technology has also begun domestically. For example, the Dalian University of Technology and East China Aluminum Processing Plant designed and developed DUT-89 aluminum liquid purification equipment, the Southwest Aluminum Processing Plant developed DFU, Northeastern University and Shijiazhuang Aluminum Plant developed the DGLJ refining device, Zhuoshen Company developed the CQ500 refining device, etc[7]. Domestic invention patents: CN00125770.6 for hydrogen removal from aluminum alloy melt by rotary pulse injection, CN201510688226. X for an aluminum alloy refining agent and its preparation method, mainly focused on refining agents and rotary injection. Although the same degassing methods and technologies as foreign countries have been adopted, and improvements have been made while introducing and digesting them, there is still a certain gap between China's aluminum alloy melt treatment level and the international advanced level [8].

High-quality aluminum alloy melt is a prerequisite for obtaining high-quality castings. The vacuum refining and purification control technology of molten metal is the main factor affecting the quality of aluminum alloy melt. Addressing the issues of complex melt processing, low mechanical properties of components, and low automation of casting processes and equipment for high-quality automotive aluminum alloy A356 material. This article adopts the integrated technology of rotary blowing, stirring, and rapid vacuum extraction to design and manufacture vacuum refining equipment for A356 aluminum alloy melt. It overcomes the technical bottleneck that the hydrogen content inside the melt cannot exceed 0.10ml/100g under conventional atmospheric pressure, fundamentally solves the problems of high hydrogen and inclusion content in the aluminum melt, realizes the industrialization of melt vacuum refining treatment technology, and provides a material basis for the development of high-quality aluminum-silicon alloys.

2. Design and manufacturing scheme for vacuum refining hydrogen removal device

We carry out the structural design, preparation, and commissioning of vacuum melt refining equipment in a 300kg resistance insulation furnace, and verify the effectiveness of vacuum melting technology under industrial production conditions. The specific plan is as follows:

1) The device includes a 300kg aluminum alloy resistance insulation furnace with a power requirement of $\geq$ 45 KW. The control cabinet of the vacuum hydrogen removal device, vacuum furnace cover, and lifting and moving device module are integrated into one. The vacuum pump is arranged separately and can be quickly connected to the vacuum furnace cover;

2) We design and manufacture specialized 310s heat-resistant stainless steel intermediate connectors, connecting resistance furnaces, vacuum furnace covers, and graphite SiC crucibles;

3) The crucible used requires smooth upper and lower-end faces to ensure vacuum sealing requirements;

4) The crucible and intermediate connecting body, as well as the empty furnace cover and intermediate connecting body, are sealed with graphite packing seals;

5) Within 15 minutes, the vacuum degree requirement is $\leq$ 100mbar. Choose a suitable vacuum pump based on this technical requirement;

6) When vacuuming, the small diameter should be pre-evacuated first, and after reaching a certain degree of vacuum, it should be switched to large diameter vacuuming until the specified vacuum requirements are met;

7) We add a set of inflation valves to the vacuum pipeline, which can be filled with protective gas (argon or nitrogen);

8) Furnace cover handling and vacuum pipeline connection time<10 minutes;

9) The vacuum hydrogen removal device adopts the commonly used graphite rotor and graphite baffle structural components in the market, and the position of the graphite baffle can be manually adjusted up and down.

The 300kg vacuum melting device developed according to the above plan, as shown in Figure 1, meets the expected design requirements for all functions.

(a) Vacuum melting device placed on the ground (b) and submerged underground

Figure 1. Vacuum melting device.

The vacuum melting device is placed above the ground, as shown in Figure 1 (a). During use, it is necessary to climb a step (1 meter high) to complete the scooping and pouring of aluminum liquid, which poses a safety hazard of burns and falls. To eliminate safety hazards, the vacuum melting device is moved underground, as shown in Figure 1 (b).

3. Research scheme for vacuum hydrogen removal and purification technology of aluminum alloy melt

Using a YB-500 four-column hydraulic press and vacuum melt refining device, after melting Al Si Mg alloy, a refining agent is used to remove gas at 720 °C~730 °C. AlSr10 is modified at 730 °C~740 °C (added at 0.5-0.6% AlSr10), and AlTiCB is added at 730 °C~740 °C for refinement treatment. Finally, rotary blowing is used for secondary degassing. Using aluminum alloy metal standard sample molds and 110 series extrusion casting sample molds, Al Si Mg alloy was refined and degassed under different vacuum pumping process parameters. We conducted the following research plan: (1) The performance changes were tested under different vacuum degassing times and settling times (0-90 minutes), and the "settling time tensile performance" change curve was drawn; (2) The changes in hydrogen content at different vacuum degassing times and settling times (0-90 minutes) were measured, and a "settling time hydrogen content" curve is drawn; (3) At different vacuum degassing times and settling times (0-90 minutes), the metallographic structure was examined to determine the "settling time metallographic structure evolution" pattern. The specific sample plan table is shown in Table 1.

Table 1. Correspondence table of the relationship between sample number and test method.

Number	Process status	Metal type specimen	Squeeze casting cake sample	Notes
0	Vacuum Degassing(0min)	3 pieces	3 pieces	When pouring in the same state, pour the metal mold and the squeeze casting
1		Pour 3 samples after vacuum degassing	Pour 3 samples after vacuum degassing	
2	Vacuum Degassing(15min)	After standing for 30 minutes, pour 3 samples	After standing for 30 minutes, pour 3 samples	
3		After standing for 60	After standing for 60	

		minutes, pour 3 samples	minutes, pour 3 samples	sample successively.
4		Pour 3 samples after vacuum degassing	Pour 3 samples after vacuum degassing	
5	Vacuum Degassing(30min)	After standing for 30 minutes, pour 3 samples	After standing for 30 minutes, pour 3 samples	
6		After standing for 60 minutes, pour 3 samples	After standing for 60 minutes, pour 3 samples	
7		After standing for 90 minutes, pour 3 samples	After standing for 90 minutes, pour 3 samples	

Heat treatment process: The sample is subjected to T6 heat treatment process (quenching process 535 ± 5 °C, holding time 8h, water quenching, water temperature 60-80 °C, intermediate retention 8h, aging treatment 180 ± 5 °C, holding time 8h, air cooling);

Experimental equipment: The microstructure of the sample was observed using a metallographic microscope (ZEISS Axio observer), and the mechanical properties of the sample were tested using a CMT-4204 microcomputer-controlled electronic universal material testing machine.

4. Test results and analysis

4.1. Evaluation of H content in A356 aluminum alloy in different states

Stealing a diameter of 110mm × 10mm from samples cast in different states for hydrogen content testing, the test results are shown in Table 2.

Table 2. Hydrogen content of samples.

	Sample number	H content Hml/100g	Sample number	H Hml/100g	Notes
Squeezing type specimen	0	0.12	20	0.09	Not subjected to vacuum treatment Not subjected to vacuum treatment
	1	0.11	21	0.08	Immediately pour after vacuum treatment
	4	0.09	24	0.09	Secondary vacuum treatment, immediate pouring
	Sample number	H content Hml/100g	Sample number	H content Hml/100g	notes
Metal type specimen	0	0.12	20	0.08	Not subjected to vacuum treatment Not subjected to vacuum treatment
	1	0.10	21	0.08	Immediately pour after vacuum treatment
	4	0.10	24	0.08	Secondary vacuum treatment, immediate pouring

From the test results of hydrogen content, it can be seen that after refining, degassing, metamorphism, and refining, the hydrogen content of A356 alloy in ordinary ingots has been reduced to 0.12ml/100g, and in semi-continuous ingots, it has been reduced to 0.08-0.09ml/100g by using rotary blowing for secondary degassing, without vacuum refining degassing. After vacuum treatment, the hydrogen content in the aluminum alloy melt of ordinary ingots with higher hydrogen content was significantly reduced by 0.03-0.04 ml/100g. However, the semi-continuous ingots with lower hydrogen content did not further reduce the hydrogen content in the melt.

Under the same process conditions, the purity of semi-continuous cast aluminum alloy is better than that of ordinary aluminum alloy ingots after refining and degassing. After vacuum treatment, the hydrogen content of the two alloys can reach the lowest limit equilibrium value, and after secondary vacuum refining, the hydrogen content will stabilize around this limit equilibrium value.

From this, it can be seen that when the hydrogen content in aluminum alloy decreases to a certain extent, it will reach a minimum limit equilibrium value, and then vacuum degassing will be used, and the hydrogen content in the aluminum alloy will not further decrease. Therefore, vacuum refining of A356 alloy only needs to be carried out once.

4.2. The effect of trace elements H on the chemical composition of A356 aluminum alloy

The chemical composition changes of A356 aluminum alloy ingots after remelting, refining, degassing, and vacuum treatment for 15 minutes and 30 minutes are shown in Table 3. It can be seen that under the current method of using stainless steel crucibles with graphite crucibles, chemical elements are basically not burned during vacuum treatment, and the composition content remains unchanged.

Table 3. Composition of Al Si Mg alloy ingots.

Status	Composition				Notes
	Si	Mg	Ti	Fe	
Technical requirement	6.5～7.5	0.30～0.45	0.08～0.20	0.12	
After remelting	6.61	0.306	0.133	0.125	
15min vacuum	6.41	0.300	0.139	0.120	
30min vacuum	6.24	0.294	0.170	0.118	

4.3. The effect of trace element H on the metallographic structure of A356 aluminum alloy

The mechanical properties of castings are closely related to their microstructure. The equilibrium microstructure of A356 alloy is composed of α - Al+(α - Al+Si) eutectic+Mg2Si phase. A356 alloy is a typical Al Si Mg ternary alloy, which is a strengthening phase Mg2Si formed by adding magnesium to Al Si binary alloy. Heat treatment can significantly enhance the aging strengthening ability of the alloy and improve its mechanical properties.

After remelting A356 aluminum alloy ingots, they were subjected to vacuum refining treatment, poured with metal samples, and then subjected to T6 heat treatment to compare their microstructures. It was found that there was little change in the microstructure of the metal samples poured before and after vacuum refining treatment for ordinary ingots. The organization is as follows: the solid solution is fine, and the eutectic silicon is in the form of small pieces or dots. The alloy phase distribution is relatively uniform, and no defects such as cracks, overburning, pores, or oxide inclusions are observed. The morphology is shown in Figures 2-4.

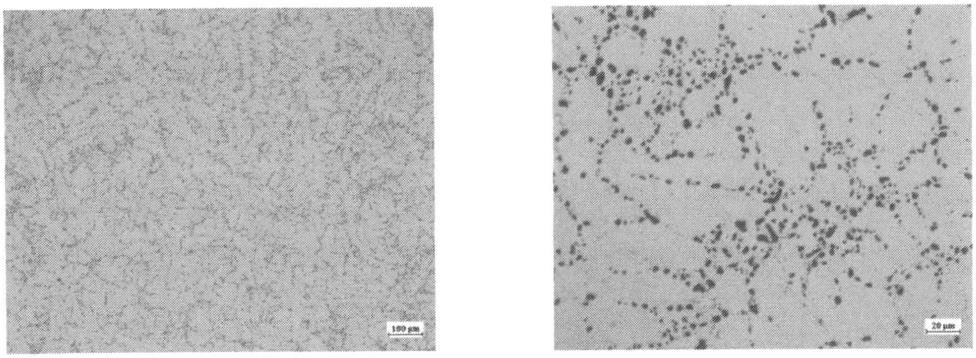

Figure 2. A356 ordinary ingot remelting+no vacuum treatment+T6 metallographic structure.

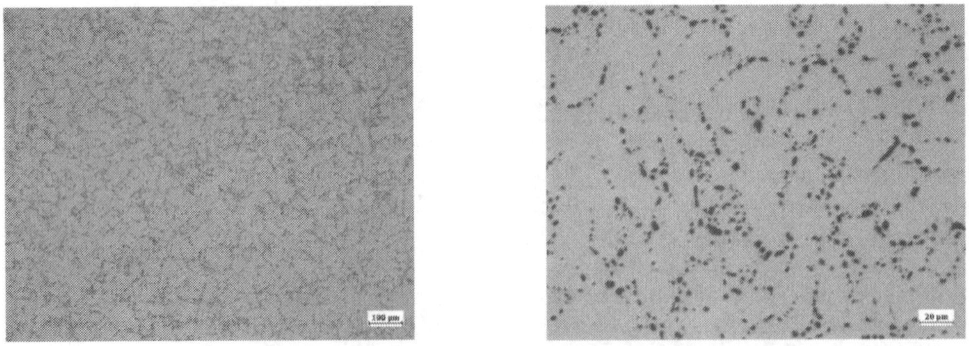

Figure 3. A356 ordinary ingot remelting+vacuum treatment for 30 minutes+T6.

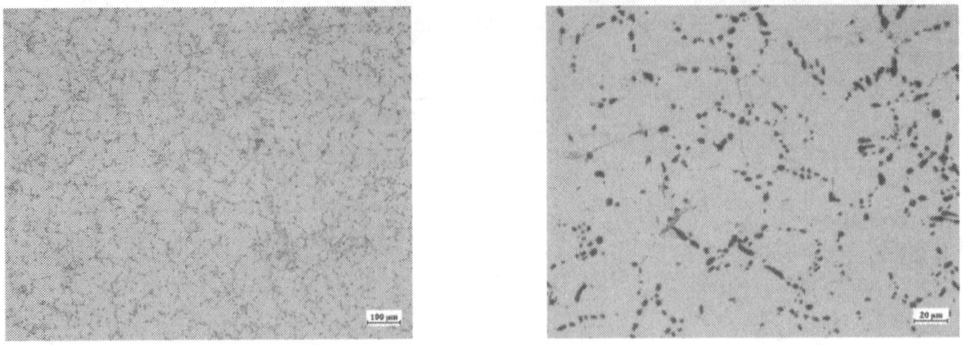

Figure 4. A356 ordinary ingot remelting+secondary vacuum treatment for 30 minutes+T6.

4.4. The effect of trace element H on the mechanical properties of Al Si Mg aluminum alloy

We test A356 ordinary ingots in different states, pour metal and extruded samples separately, and test the mechanical properties of the samples in different states after T6 heat treatment. Five or more test bars are taken from each group of samples for testing, and the average value is taken. The test results are detailed in Table 4.

Table 4. Mechanical properties of A356 alloy with different test schemes.

Sample number	Tensile strength (MPa)	Yield strength (MPa)	Elongation after fracture A (%)	Notes
0	298	244	14.0	Not subjected to vacuum treatment
1	306	251	12.0	Immediately pour after vacuum treatment
2	305	247	15.5	Vacuum treatment, let it stand for 30 minutes
3	314	245	12.5	Vacuum treatment, let stand for 60 minutes
4	312	248	10.0	Not subjected to vacuum treatment
5	303	241	9.0	Immediately pour after vacuum treatment
6	292	244	4.0	Vacuum treatment, let it stand for 30 minutes
7	300	243	4.0	Vacuum treatment, let stand for 60 minutes

The mechanical properties of ordinary ingot extruded specimens are shown in Table 5.

Table 5. Mechanical properties of ordinary ingot extrusion samples.

Sample number	Tensile strength (MPa)	Yield strength (MPa)	Elongation after fracture A(%)	Notes
0	296	240	9.5	Not subjected to vacuum treatment
1	303	248	12.0	Immediately pour after vacuum treatment
2	299	230	11.5	Vacuum treatment, let it stand for 30 minutes
3	302	239	10.5	Vacuum treatment, let stand for 60 minutes
4	298	240	10.0	Not subjected to vacuum treatment
5	296	241	8.0	Immediately pour after vacuum treatment
6	297	239	7.0	Vacuum treatment, let it stand for 30 minutes
7	297	226	7.5	Vacuum treatment, let stand for 60 minutes

From the above data, it can be seen that the effects of the vacuum degassing process and static time on tensile strength and specified plastic elongation strength Rp0.2 performance are not significant. The difference between the data before and after vacuum treatment is not significant, and the effect of the settling time after vacuum treatment on strength is not apparent. However, the effects of vacuum degassing and settling time on plasticity are more prominent. As the settling time after vacuum degassing increases, the elongation at break gradually decreases.

5. Conclusion

This article adopts the integrated technology of rotary blowing, stirring, and rapid vacuum pumping to design and manufacture vacuum refining equipment for the molten aluminum alloy A356 material. It overcomes the technical bottleneck that the hydrogen content inside the melt cannot exceed 0.10ml/100g under conventional atmospheric pressure, and masters the influence of vacuum refining treatment on the hydrogen content of different alloys.

1) During the vacuum melting process, chemical elements are basically not burned, and the use of vacuum refining and purification control technology can control the hydrogen content inside the melt to be $\leq$ 0.10ml/100g;

2) When the hydrogen content in aluminum alloy drops to a certain level, it will reach a minimum equilibrium value. Then, vacuum degassing is used, and the hydrogen content in the aluminum alloy will not further decrease. That is, the vacuum refining of A356 alloy only needs to be carried out once.

3) The effect of H content on the metallographic structure of A356 alloy is not significant, but it has a certain improvement on the tensile strength at room temperature, but the increase is $\leq$ 5%.

Acknowledgments

This study was supported by the Inner Mongolia Autonomous Region Science and Technology Plan Project, china, Nos.2023 YFHH0050.

Reference

[1] Du M, Chen Y, Zhang H X et al. Development status of molten processing technology for casting aluminum alloys[J]. CASTING. 2024,73(06):738~745.

[2] Zhang J, Peng Y J, Zhu H L et al. The Effect of Al-Ti-C-B Refiner on the Microstructure and Properties of A356 Aluminum Alloy[J]. Special Casting & Nonferrous Alloys. 2021,41(11) 1416-1419
[3] Zhang F M, Peng X S. Several issues to be noted when using SNIF purification device[J]. Light alloy fabrication technology. 1997(12): 1~3.
[4] Li J H, Hao Q T. The current situation and development trend of purification technology for aluminum alloy melt[J].Chinese Foundry Equipment and Technology. 2005 (06):1~4
[5] Li D X, Han M X, Zhang J. Hereditary microstructure and high yield strength design of fine-grained Al Si Mg alloy[J]. Materials Reports. 2021,35(9):09003~09008.
[6] Ding W J,Wu G H, Tong X etal. Research Status and Prospects of Magnesium Rare Earth Alloy Melt Treatment[J]. CASTING. 2021,70(01):2-14.
[7] Zhou Q, Niu D C, Huang P. Development of DFU Aluminum Melt Purification Device[J]. Light alloy fabrication technology. 2000(03)):5~8.
[8] Qin M, Yang T C. Purification Treatment and Mechanism Study of A356 Aluminum Alloy Melt[J].Modern Manufacturing Technology and Equipment,2016(09):52~54.

Journal of Physics: Conference Series **2879** (2024) 012019 doi:10.1088/1742-6596/2879/1/012019

Study on warm formability of 2219 aluminum alloy with pre-deformation

Qi Gao[1], Zhichao Zhang[2*], Jinlong Meng[2] and Xiaobo Fan[1]

[1]School of Mechanical Engineering, Dalian University of Technology, Dalian, Liaoning, 116000, China

[2]Shanghai Aerospace Equipments Manufacturer Co., Ltd., Shanghai, 200000, China

[*]zczhang_hit@163.com

Abstract. The manufacturing process of launch vehicle storage tanks requires urgent simplification, which can be improved by forming the local features after artificial aging. Therefore, a new process of warm flanging aluminum alloys with pre-deformation was proposed to meet this requirement. The mechanical properties of 2219 aluminum alloy with pre-deformation were investigated at different temperatures using thermomechanical property testing. The deformation law of warm flanging was analyzed through simulation. The flanging limit of 2219 aluminum alloy with pre-deformation was obtained through experiments. Results showed that the ductility of 2219 aluminum alloy with pre-deformation increased then decreased with the increasing temperature, a maximum strain of 0.65 was obtained at 200°C. The equivalent strain gradually rose from the corner zone to the straight wall zone during flanging. The maximum equivalent strain is located at the edge of the straight wall, which gradually decreases with the increasing flanging coefficient. A flanging coefficient of 0.60 can be obtained at 200°C without splitting or seriously thinning. These results guide the local flanging of the integral bottom of 2219 aluminum alloy.

1. Introduction

Aluminum alloys have been widely used in the aerospace field, such as carrier rockets and aircraft, owing to their advantages of high strength, lightweight, and corrosion resistance [1]. With the rapid development of the aerospace industry, the requirements of transportation equipment are towards high reliability, long life cycles and lightweight materials. These trends promote the development of integral structures to replace the traditional split-welded structures, leading to the emergence of large-size thin-walled aluminum alloy integral components [2]. These integral components usually require local features for connection, which are difficult to form with the major structure simultaneously. For example, the flange of the launch vehicle storage tank's bottom is necessary to connect and transport media, which is usually formed through local flanging [3]. However, the materials in the pre-formed bottom have low ductility and high strength due to work hardening, which exhibits an uneven distribution. These lead to differences in the formability of the post-forming flanging.

The development of the aerospace industry has also put forward higher requirements for the production efficiency of transportation equipment. Forming local features after aging can effectively simplify the process and shorten the manufacturing cycle. However, current researches on post-forming flanging are mainly focused on the materials in the W temper [3] and T6 temper [4]. The deformation law of local flanging of aluminum alloy with pre-deformation needs further research. The aluminum alloys with pre-deformation have limited ductility at room temperature, making it difficult to form local features by cold stamping. To improve the formability and control the strength reduction, the warm forming process has been studied and applied [5-7]. Complex aluminum alloy components can be produced through warm forming due to the enhancement of formability. Additionally, the rebound could be reduced in the warm forming process, making the components have better accuracy and quality. Warm forming is widely used in many fields [8], and it is suitable to solve the problem of limited formability of aluminum alloy with pre-deformation.

In this study, a new warm flanging process for aluminum alloys with pre-deformation was proposed to form local features in the integral bottom, thereby reducing the manufacturing cycle time. 2219 aluminum alloy was selected as the experimental material, the major material of carrier rocket fuel tanks. The true stress-strain curves of the materials at different temperatures were obtained with the aid of DIC. The forming laws of warm flanging under different flanging coefficients are analyzed by simulation. The forming limit of warm flanging is obtained through the process experiments. These guide the local flanging of the integral bottom of 2219 aluminum alloy.

2. Experimental materials and method

2.1. Thermo-mechanical properties test

Figure 1 shows the test principle of thermo-mechanical properties. The 2219-O aluminum alloy sheets were chosen as the original materials, which have a thickness of 4.1 mm. Uniaxial tensile specimens were fabricated from the original sheets, with a parallel section size of 30 mm × 5 mm. Because the flanging zone of large-size integral components has a pre-formed degree of approximately 7%, a room-temperature pre-forming of original sheets (with an elongation of 7%) was proceeded between solution treatment (heating 535°C × 1 h and water quenching) and artificial aging (175°C × 18 h). The effect of temperature on the mechanical properties and deformation behavior of 2219 aluminum alloy with pre-deformation was investigated through uniaxial tensile tests and DIC analysis, with an initial strain rate of 0.001 s^{-1}. The test temperatures were set to room temperature (25°C), 150°C, 175°C, 200°C, 225°C, 250°C, 275°C, and 300°C.

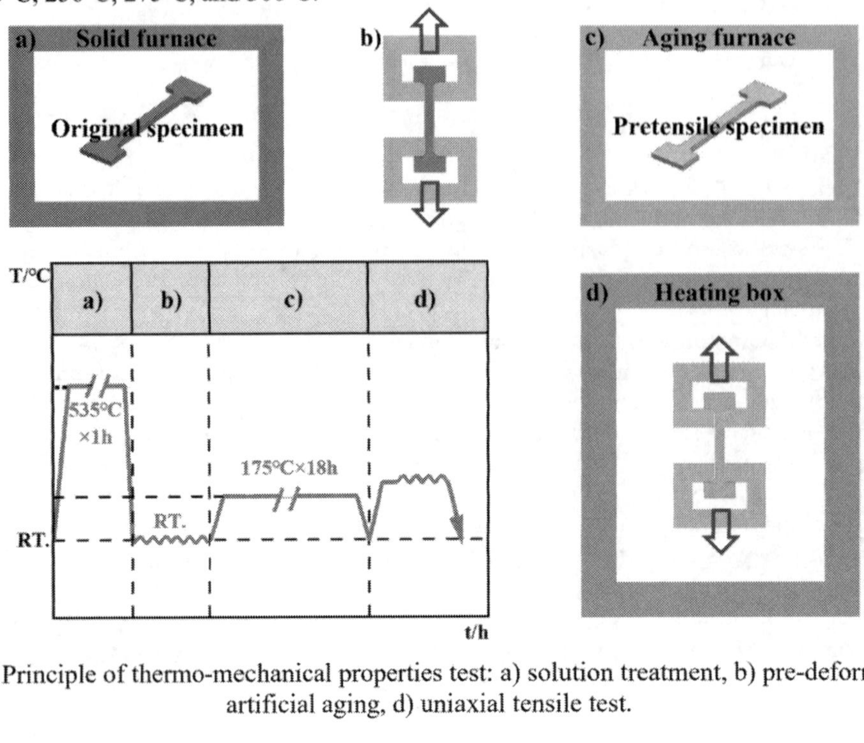

Figure 1. Principle of thermo-mechanical properties test: a) solution treatment, b) pre-deformation, c) artificial aging, d) uniaxial tensile test.

2.2. Numerical simulation
A simulation model was established to investigate the forming law of warm flanging. The density of the material was set to 2.7 g/cm^3, with Young's modulus of 79000 MPa and Poisson's ratio of 0.33. The flanging dies were defined as discrete rigid body units, with convex and concave shapes, while the sheets were defined as solid units. The mesh size of the dies was set to 2 mm, and the mesh size of the sheet was set to 1 mm. The unit types of sheets were set to C3D8R, which were divided into 6 layers in thickness direction to ensure the accuracy of results. The contact between dies and sheet was set as surface-to-surface. The stress-strain curves were fitted using the Swift function. The material property at 200°C was imported into the simulation model. To investigate the strain distribution and evolution during warm flanging, simulations were conducted with different flanging coefficients (0.50, 0.55, 0.60). The friction coefficient between the die and sheet was set as 0.4.

2.3. Warm flanging process experiment
Experiments were conducted to investigate warm flanging deformation behaviors with different flanging coefficients (flanging aperture × flanging coefficient = pre-opening diameter) at 200°C, using flanging specimens with an aperture of 60 mm. The warm flanging forming device schematic is shown in Figure 2, including three main parts: warm flanging dies, heating devices, and a temperature control device. The warm flanging die was fixed to a bracket, and the punch was driven by the connecting rod and cylinder. The use of dies ensured coaxiality during the flanging process. The dies were directly heated through the heating device, while the sheet was indirectly heated by the conduction with heat dies. To reduce the energy loss caused by thermal radiation, heat preservation treatment is applied around the outside of the device. Heat shields were added between the die and bracket to prevent the heat dissipation. The temperature control device includes temperature sensors (K-type thermocouples) and a PID control unit. The temperature sensors could provide an accuracy of ±0.1°C. The temperature data is transmitted to the PID control unit for adjustment and control, with a control accuracy of ±1.0°C.

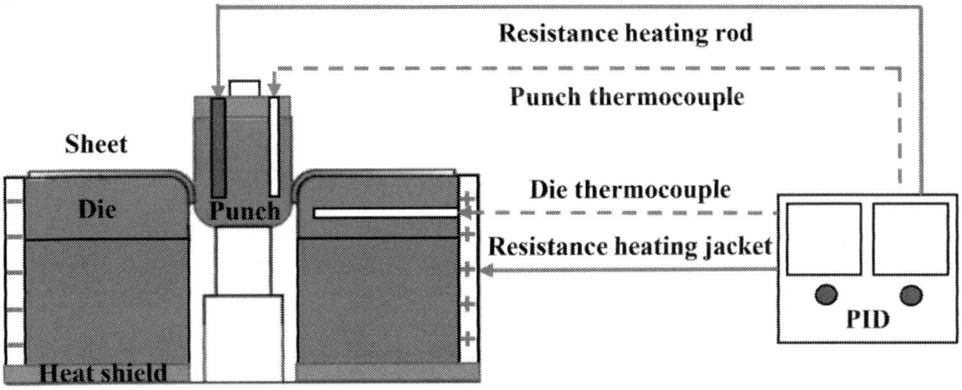

Figure 2. Schematic diagram of warm flanging forming device.

3. Results and discussion

3.1. Thermal tensile deformation mechanism of materials

3.1.1. Strain distribution under different deformation stages. Figure 3 shows the DIC analysis of a uniaxial tensile specimen at different deformation stages at 200°C, with a maximum strain of 0.70. The strain in the parallel section gradually increases with the continuous increase of the tensile displacement. In the early stage of tensile deformation, the strain distribution of the parallel section is relatively uniform. As deformation progresses, the strain distribution in the parallel section becomes increasingly uneven, leading to a more pronounced concentration of strain. Significant necking was observed in the final deformation stage, leading to a notable increase in the difference in strain distribution among different areas. The maximum difference of strain exceeded 0.60, indicating serious inhomogeneous deformation during the test at 200°C. It is clear that with the increasing deformation, the deformation uniformity of sheets decreased significantly, leading to local instability.

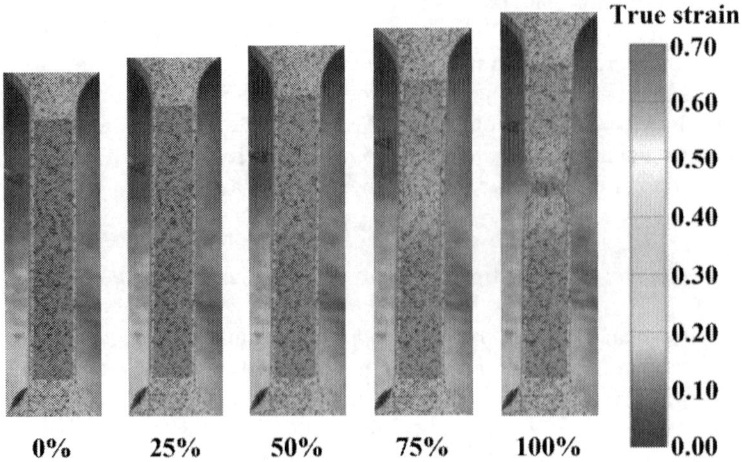

Figure 3. Strain distribution under different deformation stages.

3.1.2. True stress-strain curve at 1 mm interval. The strain of the specimen exhibited significantly different values with different selected zones due to the inhomogeneous deformation at 200°C. The true stress-strain curve obtained through traditional methods significantly differed in strain from the deformation zone, which cannot describe the deformation behaviors accurately. It is obvious that the different of strain reduced significantly with the selected zone around the necking zone. The strain calculated in a short distance in the necking zone can be closer to the flow strain of the material, resulting in a more accurate stress-strain curve. Therefore, a zone of 1 mm in length was selected to analyze the stress-strain curves of the material at different temperature conditions.

Figure 4 displays the stress-strain curves of the selected zone with 1 mm length at various temperatures. The flow stress of sheets decreased with higher temperature, while the ductility increased then decreased with the increasing temperature. The maximum true strain value of 0.65 was obtained at 200°C. The ultimate strain also exceeded 0.60 at 175°C. However, from 200°C to 300°C, the maximum true strain decreased and the plasticity worsened with the increasing temperature. Therefore, 200°C is chosen as the experimental temperature for the warm flanging process to prevent specimen splitting.

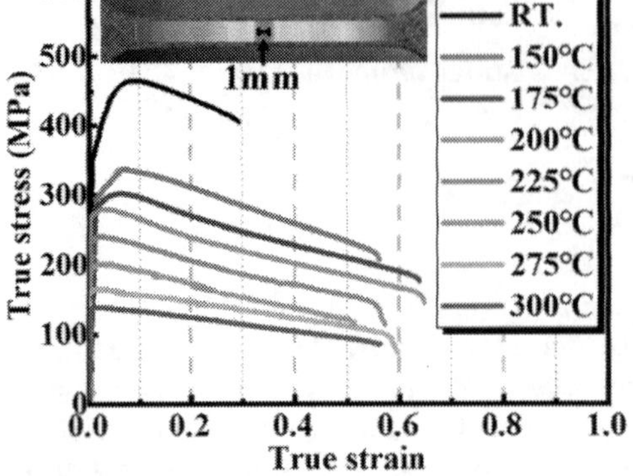

Figure 4. True stress-strain curve of selected zone at various temperatures.

3.2. Forming limit of warm flanging

3.2.1. Strain distribution at different flanging deformation stages. Figure 5 illustrates the equivalent strain distribution of specimens at various deformation stages with a flanging coefficient of 0.50 at 200°C. The distribution of equivalent strain is similar under different stages, with circumferential strain becoming the primary factor affecting the process of flanging. The equivalent strain gradually increased from the corner zone to the straight wall zone, with the maximum equivalent strain located at the edge of the straight wall. The maximum strain was approximately 0.80 in this condition, making the edge a risk region during flanging, Therefore, it is crucial to focus on the maximum strain at the edge under different flanging conditions.

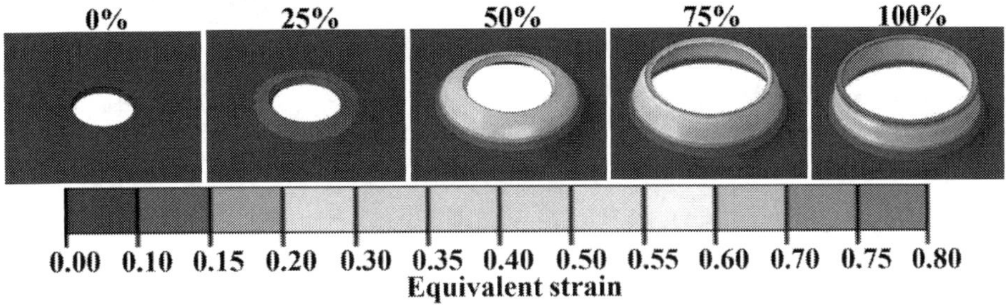

Figure 5. Equivalent strain distribution at different deformation stages.

3.2.2. Circumferential strain evolution at the edge of the straight wall. Figure 6 displays the circumferential strain evolution curves at the edge of flanging specimens with different flanging coefficients at 200°C. The evolution laws of circumferential strain at the straight wall edge were similar. During the flanging process, the displacement and circumferential strain at the straight wall edge increased gradually. The maximum circumferential strain decreased with the increasing flanging coefficients, with corresponding values of 0.74, 0.66, and 0.58. At the edge of the straight wall, the material experienced a unidirectional tensile strain. The flange process could be smooth under the flanging coefficient of 0.60 because the maximum circumferential strain of 0.58 is lower than the maximum true strain of 0.65 obtained at 1 mm intervals. However, if the flanging coefficient was reduced to 0.50 or 0.55, the specimen may fail during flanging for large circumferential strain.

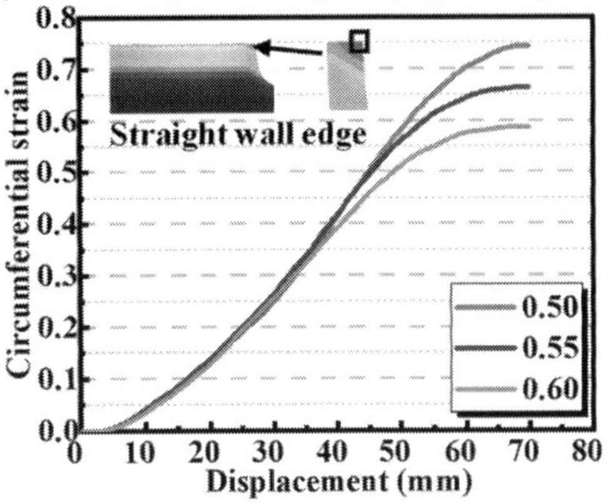

Figure 6. Circumferential strain of the straight wall edge of flanged specimens with different flanging coefficients.

3.3. Warm flanging formability

Figure 7 shows the specimens with different flanging coefficients (0.50, 0.55, 0.60) formed at 200°C. The specimen with a flanging coefficient of 0.6 can be formed successfully, as shown in Figure 7 c) respectively. As the flanging coefficient further decreases, the straight wall edge will experience local necking and splitting, as shown in Figure 7 b) and a) respectively. The limit coefficient of the flanging with an aperture of 60 mm at 200°C is about 0.6. The experimental results are consistent with the warm tensile test and numerical simulation.

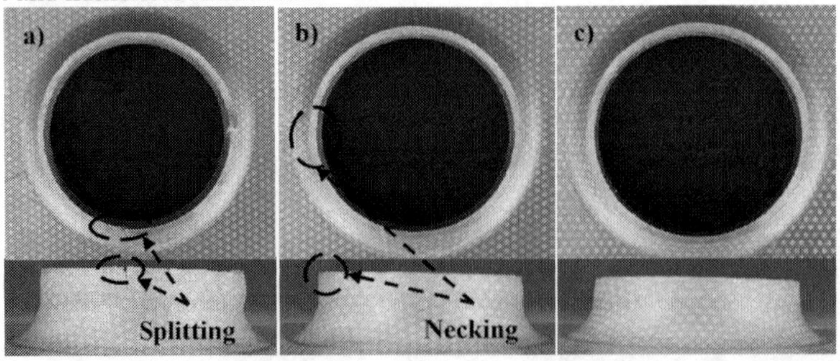

Figure 7. Shaped specimens with different flanging coefficients: a) 0.50, b) 0.55, and c) 0.60.

Figure 8 shows the prediction flow chart of the flanging forming limit. To form local features successfully, the maximum circumferential strain at the edge of the specimen during the warm flanging process for aluminum alloys with pre-deformation should not be greater than the limit strain under uniaxial tensile. As the flanging coefficient is further reduced, the maximum strain of the specimen will exceed the material's forming limit, leading to excessive splitting and even splitting. Therefore, the forming temperature and flange coefficient should be comprehensively considered in the warm flanging process. This can be determined through the limit strain of 1 mm intervals under uniaxial tensile test and numerical simulation.

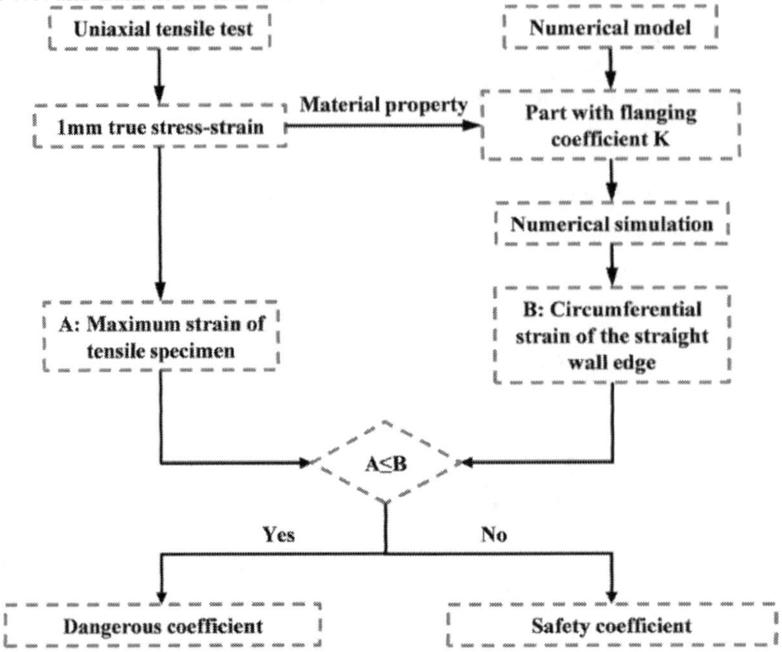

Figure 8. Prediction flow chart of flanging forming limit.

4. Conclusions

In this study, the warm deformation behavior and flanging formability of the 2219 aluminum alloy with pre-deformation were discussed. The flanging deformation behaviors were analyzed through numerical simulation. The warm uniaxial tension behaviors and flanging formability were determined by experiments. These results could guide the warm flanging process of the 2219 aluminum alloy with pre-deformation. The main conclusions are as follows:

(1) The tensile properties of the 2219 aluminum alloy with pre-deformation can be significantly enhanced by increasing the temperature. At 200°C, the limit strain reaches 0.65, which is 2.2 times higher than that at room temperature (0.30). Meanwhile, an increase in temperature can also lead to strain softening and localized concentrated deformation.

(2) During the flanging process, the deformation is concentrated at the edge of the straight wall, while the deformation in other areas is minimal. Circumferential tensile deformation mainly occurs at the edges of straight walls, similar to the process of uniaxial stretching.

(3) The appropriate flanging coefficient can be determined by comparing the limit strain under uniaxial tension with the circumferential strain at the straight wall edge simulated during flanging forming. The specimen with a flanging coefficient of 0.6 can be smoothly formed. However, local necking and splitting will occur as the flange coefficient decreases to 0.55 and 0.50. The results are consistent with the warm tensile test and numerical simulation of warm flanging.

Acknowledgments
This work was financially supported by the National Key Research and Development Program of China (2019YFA0708804). National Natural Science Foundation of China [Grant No. 52375311]; and Shanghai Rising-Star Program.

References
[1] Tolazzi M (2010) Hydroforming applications in automotive: A review. J. International Journal of Materials Forming, 3: 307-310. [10.1007/s12289-010-0768-2]
[2] Yuan S J (2021) Fundamentals and processes of fluid pressure forming technology for complex thin-walled components. J. Engineering, 7: 358-366. [10.1016/j.eng.2020.08.014]
[3] Zhang X L, Guo H and Zhao H F (2021) Heat flanging process of quenched 2219 aluminum alloy plate. J. Forging Technology, 46: 136-140. [10.13330/j.issn.1000-3940.2021.10.022]
[4] Gong J X, Yang Y and Fan H M (2022) Experimental study on the hole-forming of the bottom of the rigid tank. J. Aerospace Manufacturing Technology, 231: 11-15. [10.3969/j.issn.1674-5108.2022.01.003]
[5] Bariani, P F, Bruschi S, Ghiotti A and Michieletto F (2013) Hot stamping of AA5083 aluminium alloy sheets. J. Cirp Annals-manufacturing Technology, 62: 251-254. [10.1016/j.cirp.2013.03.050]
[6] Bolt P, Lamboo N and Rozier P (2001) Feasibility of warm drawing of aluminium products. J. Journal of Materials Processing Technology, 115: 118-121. [10.1016/S0924-0136(01)00743-9]
[7] Serkan T, Fahrettin O and Ilyas K (2008) Review of warm forming of aluminium-magnesium alloys. J. Journal of Materials Processing Technology, 207: 1-12. [10.1016/j.jmapro.2020.10.016]
[8] Wang P Y, Zhuang W M, Gao R J and Xie D X (2021) Effect of hot forming on static mechanical properties of AA5754 aluminum alloy. J. Journal of Jilin University, 51: 847-854. [10.13229/j.cnki.jdxbgxb20200180]

Journal of Physics: Conference Series **2879** (2024) 012020 doi:10.1088/1742-6596/2879/1/012020

Preparation and performance study of a new solvent-free insulation coating for low-voltage busbar

Xingyue Deng

School of Mechatronic Engineering and Automation, Shanghai University, Shanghai, 200444, China

487005822@qq.com

Abstract: This study focuses on low-voltage busbar insulation and is committed to developing a new solvent-free insulation coating. A coating formula with specific properties was successfully synthesized by screening and optimizing various raw materials. A comprehensive and in-depth testing and analysis were conducted on the developed coatings' physical properties, electrical insulation performance, heat resistance, and chemical corrosion resistance. The results show that the new solvent-free insulation coating has excellent heat resistance, corrosion resistance, and mechanical strength while maintaining good insulation performance. Compared with traditional insulation coatings, its solvent-free properties effectively reduce the emission of volatile organic compounds, meeting environmental requirements. This study provides a more efficient, reliable, and environmentally friendly solution for the insulation protection of the low-voltage busbar, which has significant practical application value and broad market prospects.

1. Introduction

In modern power systems, the low-voltage busbar is a critical component of power distribution and transmission, and its safe and reliable operation is essential. With the increasing demand for power and the increasingly complex operating environment of power equipment, the insulation performance of low-voltage busbar is required [1-2] to be higher. As a crucial insulating protection material, insulation coating can effectively improve the low-voltage busbar's insulation performance and ensure the power system's stable operation. The low-voltage busbar efficiently transmits the electric energy from the incoming line end to each outgoing line end and undertakes the transmission task of a large current. The insulation performance of the low-voltage busbar is one of the key factors in ensuring the safe and stable operation of the power system. Still, it also faces a series of problems and challenges. The busbar is usually covered with an insulating sheath or RTV insulating coating to improve its insulation type, ensure safety, and prevent short circuits. The insulation sheath [3-6] is located outdoors for a long time and is susceptible to extreme weather, pollution, salt spray corrosion, and other causes of abnormal aging. In addition to bird damage, bird nests, and animals, the insulation sheath grounding point is not misdistributed, or the distance is less than 1m. Therefore, its service life often does not reach six years after the occurrence of falling off, cracking, and other phenomena, resulting in line trips all year round, which accounted for about 40% of the total number of trips. RTV insulating coating [7-8] generally refers to room-temperature vulcanized silicone rubber. It has a certain electrical insulation performance and bonding strength. By spraying the RTV covered in the low-voltage busbar, a single operation spray thickness is low, usually 0.3mm thickness. Coating too thick can easily cause cracking and falling off. Although it can achieve surface antifouling flash to a certain extent, due to low insulation strength, the electric strength is only 8kV/mm, and the volume resistivity is about $1\times10^{12}\Omega\cdot m$ (thickness is 0.3mm). Usually, after one year of use, its insulation performance will be reduced to a certain extent. In addition to its high cost, strength and hardness are relatively low. Achieving more than six years of service life is difficult. Most of the traditional insulating coatings contain solvents, which will release volatile organic compounds (VOCs) during construction and use, causing harm to the environment and human health. At the same time, the presence of solvents will also lead to a reduction in the solid content of the coating, affecting its performance and service life. Therefore, developing a new solvent-free insulating coating is of great practical significance [9-10].

Given the above two solutions and the technical bottleneck, we optimize the insulation design and use a new solvent-free epoxy insulation coating [11-13] as the main insulation material to replace the insulation jacket and RTV, which is coated on the low-voltage busbar for packing. It can achieve excellent insulation performance, good moisture and heat resistance, excellent weather resistance, low VOC emission, good mechanical properties, and corrosion resistance [14-16]. Even in high temperature and high humidity environments, it will not reduce its insulation performance [17-19], and its durability reaches more than ten years, which has good scalability and technical prospects [20-23].

In recent years, with the continuous development of material science and chemical technology, solvent-free insulating coatings have gradually become the focus [14] of research. Solvent-free insulating coatings have the advantages of high solid content, low VOC emission, excellent insulation performance, and good construction performance. They can meet the requirements of modern power systems for low-voltage busbar insulation protection. However, there are still some shortcomings in the performance of solvent-free insulating coatings on the market at present, such as poor weather resistance, weak adhesion, insufficient mechanical strength, etc., which limits their wide application in the field of power.

2. Experimental part

2.1 Experimental materials

Epoxy-modified silicone resin: Jiangsu Sanmuchem; Liquid petroleum resin LA300: Ruetger; Modified polyamides and modified phenolic amines: Guolong Chemical; Cashew phenol glycidyl ether: Cadle Chemical; Defoamer BYK1790: BYke Chemical; Dispersant BYK2055: BYK Chemical; Leveling agent BYK361N: Silane coupling agent A-187: American Metu; Fumed silica 972: Evonik; Byke Chemical; Titanium dioxide 800 mesh: commercially available; Talc powder 800 mesh: commercially available; Mica powder 800 mesh: commercially available; 1200 mesh aluminum tripolyphosphate: commercially available; Corundum 500 mesh: commercially available; Flame retardants: modified magnesium hydroxide and modified aluminum hydroxide 1200 mesh: commercially available.

2.2 Test Instruments

Electric strength testing equipment: Merrick; Volume resistivity test: Merrick; Salt spray box: Q-LAB; Wear test machine: Biuged; Oven: Biuged; Adhesion tester: Elcometer; Aging test box: Biuged.

2.3 Experimentation

2.3.1 Preparation of insulating coating. Epoxy-modified silicone resin, liquid petroleum toughening resin, cashew phenol glycidyl ether, and dispersant were dispersed at low speed for 15 minutes. Then, we dispersed fumed silica, defoamer, dispersant, levelling agent, and silane coupling agent at high speed for 20min. Finally, titanium dioxide, talc, mica powder, aluminum tripolyphosphate, corundum, modified magnesium hydroxide and modified aluminum hydroxide were dispersed at high speed for one hour to a fineness of 60μm. The main agent was obtained after filtration. The main agent and curing agent are mixed evenly according to the mass ratio of 5:1, that is, solvent-free insulating coating.

2.3.2 Preparation of insulating coating sample. We prepare several samples of low-voltage busbars, polish their surfaces with 60-mesh sandpaper, clean them until no dust is visible to the naked eye, and then spray insulating paint on the low-voltage busbar through high-pressure airless spraying equipment. Finally, we maintain them at a humidity of 50%±5% and a temperature of 23±2°C for at least 14 days and then perform various performance tests.

2.3.3 Performance test. We test the electric strength according to GB/T 1408.1-2016, test the volume resistivity according to GB/T 1410-2006, test 0 distance long-term voltage test according to GB/T 14049-2008, and test leakage marks according to GB/T; We also test humidity and heat resistance according to GB/T 1740, test flame retardant according to GB/T 10707, and test resistance to neutral salt spray according to GB/T 1771-2007; The aging of coating is tested according to GB/T 1766-2008; The dielectric resistance was tested according to GB/T 9274-1988, and the media were 10%H_2SO_4, 10%NaOH solution, 5%NaCl solution and 80°C diesel oil (0#). The drying time was tested according to GB/T 1728-2020; Then, we test non-volatile matter content according to GB/T 31361-2015, test wear resistance (1000g/1000r, CS-17) according to GB/T 1768-2006, test hardness according to ASTM D2240; and test adhesion (pulling method) according to GB/T 5210-2006.

3. Results and discussion

3.1 The proportion of resin quality selection

Epoxy-modified silicone resin is prepared by introducing epoxy groups into the body of the silicone resin through a condensation reaction. It has various excellent properties, such as excellent electrical insulation properties, high dielectric strength, and large-volume resistivity. The paint film coated on the busbar has good aging resistance, thermal stability, and high bonding strength. Epoxy-modified silicone resin synthesizes the advantages of epoxy resin and silicone resin. It overcomes some shortcomings of pure silicone resin, such as high-temperature drying, long curing time, inconvenience in large-area construction, poor adhesion and resistance to organic solvents, and poor mechanical strength of paint film when the temperature is high. At the same time, it improves the shortcomings of epoxy resin, such as enormous internal stress, brittle quality, fatigue resistance, poor heat resistance and impact resistance, and poor weather resistance. The high-temperature resistance of epoxy-modified silicone resin is usually above 200°C, and even at 500°C, but it can still maintain the stability of its performance. However, the different mass ratios of epoxy-modified silicone resin in the main agent will have a certain impact on the insulation performance, flame retardant performance, and corrosion resistance of the coating, as shown in Table 1.

Table 1. Effect of mass ratio of epoxy-modified silicone resin on coating properties.

Items	Resin mass ratio			
	25%	35%	45%	55%
Electric strength (kV/mm)	18	42	45	39
Volume resistivity (Ω·m)	1.39×10^{14}	1.05×10^{15}	1.23×10^{15}	1.05×10^{14}
Flame retardant	FV-1	FV-0	FV-0	FV-1
Salt spray resistance (h)	2000	3000	3000	2000

Epoxy-modified silicone resin forms a three-dimensional crosslinked network during curing. Higher epoxy resin content means higher crosslinking density, which helps to improve the mechanical strength and electrical insulation properties of the coating. The tightness of the crosslinked network is directly related to the insulation strength of the coating because the more crosslinked points, the less free movement between molecules, thereby reducing the migration of charge carriers and enhancing insulation performance. A higher epoxy content provides a denser coating, reducing the chance of moisture and salt penetration, thereby improving salt spray resistance. However, other components in the formulation also affect the coating's compactness and salt spray resistance.

It can be seen from the relevant data in Table 1 that when the mass ratio of epoxy-modified silicone resin is 25 %, the electric strength is only 18kV. This is because the mass ratio of resin is too low, and the corresponding mass ratio of pigment and filler is too high, exceeding the critical value. Then, the resin cannot wrap the pigment and filler, and there are voids inside the coating, which is not dense, resulting in the coating being easily broken down by high voltage. Its electric strength is also reduced. When the mass ratio of epoxy-modified silicone resin is 55 %, its electric strength is 39 kV, and the flame retardancy is FV-1. The electric strength and flame retardancy of epoxy-modified silicone resin are decreased. This is because the resin mass ratio is too high, and the corresponding pigment and filler mass ratio are relatively small, which is limited to improving epoxy-modified silicone resin's insulation performance and flame retardancy. When the mass ratio of epoxy-modified silicone resin is 35 % ~ 45 %, the resin tightly wraps the pigments and fillers, the coating is dense, and the impermeability and pressure resistance are strong, which not only improves the inherent insulation and flame retardancy of epoxy-modified silicone resin but also improves its corrosion resistance. Therefore, when the mass ratio of epoxy-modified silicone resin is 35% ~ 45%, its electric strength, volume resistivity, flame retardancy, and salt fog resistance are the best and meet the relevant performance requirements of insulation coating.

3.2 Selection of mass ratio of active diluent

The active diluent usually has an epoxy group, which can crosslink with the resin and become part of the paint film, achieving the environmental protection requirements of no VOC volatilization. In addition, the active diluent can reduce the viscosity of the coating, which is convenient for field operation and construction, and improve the insulation coating's flexibility, impact resistance, wear resistance, and acid and alkali resistance. Cardanol glycidyl ether is a flexible bifunctional glycidyl ether. Its carbon eight-side chain separates the aromatic ring groups, which can be well compatible with the resin, increasing its flexibility, impact resistance, wear resistance, and acid and alkali resistance. However, the main agent's different mass ratios of cardanol glycidyl ether will impact the coating and coating performance.

Table 2. The effect of different mass ratios of cardanol glycidyl ether on the properties of coatings and coatings.

Items	The mass ratio of cardanol glycidyl ether			
	5%	10%	15%	20%
Viscosity (mpa.s)	10250	6320	3440	1350
Pliability (mm)	10	5	5	5
Impact resistance (cm)	40	50	50	40
Acid resistance (10%H_2SO_4), day	20	30	30	20
Alkali resistance (10%NaOH), day	20	30	30	20
Abrasion resistance (mg)	86	53	41	112

It can be seen from the relevant data in Table 2 that as the proportion of cardanol glycidyl ether continues to increase from 5% to 20%, the viscosity of the coating gradually decreases, and the flexibility of the coating is improved to a certain extent, reaching 5mm. The impact, acid, alkali, and wear resistance of the coating increase first and then decrease. When the mass ratio of cardanol glycidyl ether is 5%, its viscosity is large, which is not conducive to on-site spraying construction. When the mass ratio is 20%, the viscosity is too low, making it easy to produce sagging and settle. This is not conducive to coating storage, and its wear resistance is poor. Considering comprehensively, the cardanol glycidyl ether mass ratio of 10%~15% is the best choice. Its viscosity is moderate, and its flexibility, impact resistance, acid and alkali resistance, and wear resistance meet the relevant requirements.

3.3 Selection of curing agent

The insulating coating should have a dense film, good corrosion resistance, long service life after mixing, and the advantages of high electric strength, large volume resistivity, good flame retardancy, strong corrosion resistance, and high adhesion. These properties are primarily affected by the curing agent. Based on determining the mass ratio of the main resin and the proportion of the active diluent, the use of a modified polyamide curing agent alone, the use of a modified phenolic amine curing agent alone, and the mixing of the two in different mass ratios will have a significant impact on the performance of the coating.

Table 3. The effect of the ratio of modified polyamide and modified phenolic amine curing agent on the insulation properties of the coating.

Items	Curing agent ratio (polyamide: phenolic)			
	1:0	0:1	1:1	3:1
Electric strength (kV/mm)	41	42	44	42
Volume resistivity (Ω·m)	1.08×10^{15}	1.51×10^{15}	1.24×10^{15}	1.92×10^{15}
Flame retardant	FV-1	FV-1	FV-1	FV-1
Salt spray resistance (h)	2000	2000	3000	3000
Adhesion (MPa)	11.3	9.7	20.8	15.6
The use period after mixing	1h	1h	3h	3h

The coating formed by the modified phenolic amine curing agent generally has a higher crosslinking density, which can improve the coating's electrical strength and volume resistivity because it reduces the free path of charge and the risk of dielectric breakdown. On the contrary, a polyamide curing agent may provide better flexibility, but lower crosslinking density may negatively impact the electrical strength. Therefore, selecting the appropriate curing agent ratio can optimize the insulation of the coating while maintaining good mechanical properties.

It can be seen from the relevant data in Table 3 that when the ratio of modified polyamide and modified phenolic amine curing agent is used separately with the main agent, the data of electric strength, volume resistivity, and flame retardancy are basically the same, which meet the requirements of insulation performance. When the mixing ratio of modified polyamide and modified phenolic amine curing agent is 1:1 and 3:1, its salt spray resistance, adhesion, and service life after mixing are better. This is because the modified polyamide curing agent molecules usually contain long aliphatic hydrocarbon carbon chains and polar amide groups, which can make the epoxy-modified silicone resin crosslink at room temperature. The aliphatic hydrocarbon chain can isolate the rigid benzene ring in the epoxy-modified silicone resin molecule. It can give the bond greater freedom when subjected to load or impact, thus endowing the cured epoxy resin with excellent flexibility, insulation, water resistance, wear resistance, and chemical stability. The modified phenolic amine curing agent has a long aliphatic side chain, long operating period, toughness, adhesion, water resistance, and corrosion resistance. The two synergize after mixing; the coating is the densest, and the salt spray resistance and adhesion are better. In addition, the service life after mixing reaches three hours, which gives the field construction more operation time and meets the coating work of insulating coatings in various environments. Generally, modified polyamide's procurement cost is lower than modified phenolic amines. Considering comprehensively, the cost performance is higher when the modified polyamide and phenolic amine curing agent are mixed at 3:1.

3.4 The choice of pigment base ratio

Pigments can fill the pores in the coating and improve its compactness, thereby enhancing its insulation. However, a pigment-to-binder ratio that is too high may lead to the aggregation of pigment particles, the formation of conductive pathways, and the reduction of the insulation properties of the coating. Increasing the content of pigments and fillers can improve the flame retardancy of the coating. Still, excessive pigments may reduce the flexibility and mechanical strength of the coating and affect the overall performance. The salt spray resistance is related to the compactness and chemical stability of the coating. The appropriate pigment-to-binder ratio can fill the tiny pores in the coating and improve the compactness of the coating, thereby enhancing salt spray resistance. However, a pigment-to-binder ratio that is too high may lead to an increase in the porosity of the coating and a decrease in salt spray resistance.

Under the premise of determining the mass proportion of the main agent and the curing agent in the main agent, the influence of different face base ratios on the insulation performance of the coating is shown in Table 4.

Table 4. The effect of different pigment-to-binder ratios on the insulation properties of the coating.

Items	Pigment-to-binder ratio			
	0.3	0.7	1.0	1.3
Electric strength (kV/mm)	14	34	45	40
Volume resistivity (Ω·m)	1.97×10^{13}	1.14×10^{15}	1.87×10^{15}	1.01×10^{14}
Flame retardant	FV-1	FV-1	FV-0	FV-0
Salt spray resistance (h)	1000	2500	3000	2000

It can be seen from the relevant data in Table 4 that when the face base ratio is 1.0, the electric strength of the insulation coating is about 45kV/mm, the volume resistivity is above $1\times10^{15}\Omega\cdot m$, and its flame retardant reaches FV-0 level. In addition, its salt spray resistance is not abnormal for 3000h. Choosing a pigment-to-binder ratio of 1.0 is more appropriate based on its performance.

From the relevant data in Table 4, it can be concluded that when the pigment-to-binder ratio is 1.0, the breakdown strength of the insulating coating is about 45kV/mm, the volume resistivity is above $1\times10^{15}\Omega\cdot m$, and its flame retardancy reaches FV-0 level. In addition, its salt spray resistance is not abnormal for 3000h. This is because a reasonable pigment-to-binder ratio is crucial to the coating's electric strength and corrosion resistance. When the pigment-to-binder ratio is 1.0, the resin can tightly wrap the pigment. Many flake structures in the pigment make the coating more compact and form a dense microstructure, effectively hindering current conduction. In addition, the internal resistance of the coating is high, which can not only isolate the free flow of internal electrons but also effectively prevent water molecules and salt mist molecules from penetrating the paint film to the surface of the substrate and causing corrosion, which greatly improves the insulation performance and anti-permeability of the coating. When the pigment-to-binder ratio is 0.3, the filler in the coating is insufficient to form a dense microstructure. When the pigment-to-binder ratio is 0.3, the filler in the coating is insufficient to form a dense microstructure. When the pigment-to-binder ratio is 1.3, the filler in the coating is too high to be wrapped by the resin, and there are voids in the coating, which all lead to poor shielding performance of the coating and cannot block the free flow of electrons inside the coating. Considering its performance, it is more appropriate to select the pigment-to-binder ratio of 1.0, which not only can limit the free flow and conduction of electrons but also have the characteristics of high resistance and dense structure to achieve the purpose of insulation, increasing the corrosion resistance of the coating.

3.5 Selection of coating thickness

After determining the resin proportion of the insulating coating, the mixing ratio of the curing agent, and the face base ratio, the minimum electric strength is at least 10kV/mm. The maximum electric strength is about 42kV/mm for the low-pressure busbar. The thickness of different insulating coatings will have a certain impact on the electric strength. Table 4 lists the electric strength of different insulating coatings.

Table 5. The electric strength of different insulation coating thicknesses.

Coating thickness (mm)	0.5	1.0	1.5	2.0	2.5	3.0
Electric strength (kV/mm)	15	38	45	40	34	31

As the coating thickness increases, the breakdown voltage usually increases because a thicker coating can provide a higher resistance to prevent the current from passing through. However, if the coating is too thick, it may lead to an increase in internal stress, thereby reducing the overall mechanical strength of the material, which may eventually lead to cracks and breakdown. The thickness may also affect the coating's dielectric constant and dielectric loss factor. Generally, a thinner coating may have a lower dielectric constant and dielectric loss. Still, the premise is that the coating must be sufficiently uniform and defect-free to maintain good electrical properties.

It can be seen from Table 5 that with the increase of the thickness of the insulating coating, the electric strength increases first and then decreases. Excessive thickness does not bring higher electric strength. This may be because the coating is too thick, and cracks and discontinuities are inside the paint film, leading to a decrease in density. A dense microstructure cannot be formed, which cannot effectively hinder the conduction of the current. Therefore, the thickness of the insulation coating is recommended to be about 1.5mm, and its economic cost is also low. Its withstand voltage strength reaches 45kV/mm, which meets the requirements of the withstand voltage strength of the low-voltage busbar.

3.6 Comprehensive performance index

After determining the insulation coating formula and thickness, the insulation coating is sprayed on the busbar by high-pressure airless spraying equipment. After sampling and sampling, the comprehensive performance of the coating and coating is tested. The test data is listed in Table 6, Table 7 and Table 8.

Table 6. Test data on solvent-free insulation coatings' physical and chemical properties.

Serial number	Test item	Test criteria	Technical indicators	Test results
1	No volatile matter content /%	GB/T 1725-2007	≥95%	99.7
2	Watch dry time /h	GB/T 1728-1989	8	8
3	Work time/h	GB/T 1728-1989	24	24
4	VOC (g/L)	GB/T 23985-2006	20	7
5	Wear resistance/mg	GB/T 1768-2006	≤100	26
6	Hardness	ASTM D2240	≥50	60
7	Adhesion/MPa	GB/T 5210-2006	≥5	20.5
8	Impact resistance/J	GB/T 1732-1993	≥5	5
9	Resistance to bending (mm)	GB/T 6742-2006	≤10	10
10	Flame retardant	GB/T 10707	≥FV-1	FV-0

Table 7. Corrosion resistance test data of solvent-free insulation coatings.

Serial number	Test item	Test criteria	Technical indicators	Test results
1	Acid resistance (10% H_2SO_4)	GB/T 9274—1988	30 days normal	30 days, no abnormalities
2	Alkaline resistance (10% NaOH)	GB/T 9274—1988	30 days normal	30 days, no abnormalities
3	Salt-resistant water (5% NaCl solution)	GB/T 9274—1988	30 days normal	30 days, no abnormalities
4	Oil resistance (0# diesel)	GB/T 9274—1988	30 days normal	30 days, no abnormalities
5	Resistance to neutral salt spray	GB/T 1771—2007	1000h No exception	There is no exception in 3000h
6	Aging resistance	GB/T 1766—2008	1000h Nothing abnormal	There is no exception in 1000h

The electrical strength test of the busbar insulation coating layer after 14 days of curing at room temperature (23±2°C) is shown in Figure 1.

Figure 1. Insulation coating withstands voltage tests.

Table 8. Insulation performance test data of solvent-free insulation coating at room temperature.

Serial number	Test item	Test criteria	Technical indicators	Test result
1	Electric strength (14 days at room temperature) kV/mm	GB/T 1408.1-2016	≥10	45
2	Volume resistivity (Ω·m)	GB/T 1410-2006	$\geq 1\times 10^{12}$	1.88×10^{15}

3	0 Distance long-term withstand voltage test	GB/T 14049-2008	12kV power frequency withstand voltage, 4h no breakdown, no discharge	12kV power frequency voltage resistance, 4h no breakdown, no discharge
4	Leakage tracking (kV)	GB/T 6553	10 or higher	50

Under the extremely harsh environmental conditions of high temperature (200°C) and high humidity (100%), the electric strength of the busbar insulation coating layer after 14 days of phased cumulative curing is shown in Figure 2.

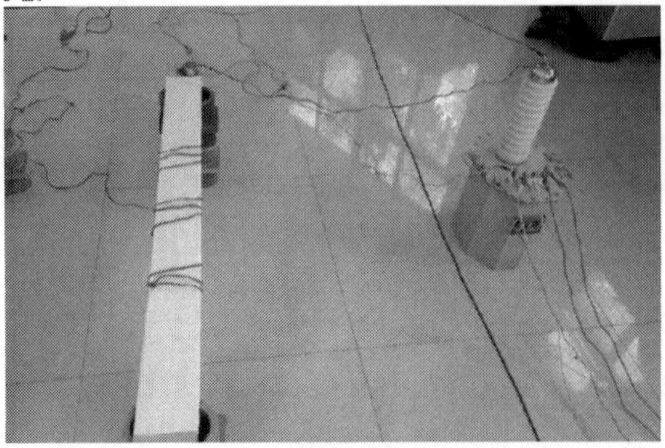

Figure 2. Electrical strength test of insulating coating in high temperature and high humidity environment.

Table 9. Insulation performance test data of solvent-free.

Serial numbe r	Test item	Test criteria	Technical indicators	Test result
1	Electrical strength (temperature 200°C, humidity 100%, for 3 days) kV/mm	GB/T 1408.1-2016	≥10	45
2	Electrical strength (temperature 200°C, humidity 100%, for 6 days) kV/mm	GB/T 1408.1-2016	≥10	43
3	Electrical strength (temperature 200°C, humidity 100%, for 9 days) kV/mm	GB/T 1408.1-2016	≥10	39
4	Electrical strength (temperature 200°C, humidity 100%, for 12 days) kV/mm	GB/T 1408.1-2016	≥10	35
5	Electrical strength (temperature 200°C, humidity 100%, for 14 days) kV/mm	GB/T 1408.1-2016	≥10	31

From the relevant data in Table 6, Table 7, and Table 8, it can be concluded that the indicators of the new solvent-free insulation coating meet the relevant standards of insulation performance. Coating it to the low-voltage busbar can achieve excellent insulation performance, good heat and humidity resistance, excellent weather resistance, low VOC emissions, good mechanical properties, and corrosion resistance. It can be seen from the data in Table 9 that the solvent-free insulation coating is placed in an extremely harsh environment of high temperature and high humidity for 14 days, and the pressure strength data gradually decreases. This is because, in a high-temperature environment, the coating is exposed for a long time and then oxidized. The reaction accelerates the aging process. In addition, the internal structure of the coating begins to become unstable, and the internal chemical bonds begin to break and reorganize, resulting in a decrease in the compactness of the coating. In addition, water molecules have a certain conductivity in the high humidity. When they continue to penetrate the coating, they form tiny water channels. These waters also provide ion transport media, making the ions in the coating easier to migrate, thereby increasing the conductivity of the coating and resulting in a decrease in the insulation performance of the coating. Even though the insulation coating is placed in an extremely harsh environment of high temperature and high humidity for 14 days, its withstand voltage strength is reduced to a certain extent. However, it still meets the requirements of the low-voltage busbar to withstand a voltage strength of more than 10kV/mm. It can be speculated that the durability of the new solvent-free insulation coating can reach more than ten years, and its durability can reach more than ten years, which has good popularization and technical prospects.

4. Conclusion

(1) The main resin of solvent-free insulation coating is high-temperature epoxy-modified silicone resin, which accounts for 35%~45% of the main agent. When the ratio of polyamide and modified phenolic amine curing agent is 3:1, the coating density is the best, the salt spray resistance is the highest, and the adhesion is the best. When the ratio of pigment to binder is 1.0, it has the highest electric strength, the highest electric strength, and the best corrosion resistance, and it meets the insulation requirements of a low-voltage busbar.

(2) The thickness of solvent-free insulation coating is not linear with the electric strength and does not increase with the increase of thickness. The coating thickness of 1.5mm has the highest electric strength and low economic cost.

(3) Even if the new solvent-free insulation coating is placed in an extremely harsh environment of high temperature (200°C) and high humidity (100%) for 14 days, it can still meet the requirements of low-voltage busbar electric strength above 10kV/mm although its electric strength has decreased. It can achieve more than ten years of durability.

References

[1] State grid corporation announces eighteen major grid anti-accident measures (Revised edition) [J]. East China E-lectric Power, 2012, 40(04):607.
[2] ZHANG Ning, HUANG Chao, ZHANG Xiaohai, et al. Discussion on electrical faults [J]. Electric Age, 2002(06):79-80.
[3] LI Yi. Insulation testing and usage suggestions for outdoor busbar heat shrink tubing [J]. Electric Safety Technology, 2010(11):52-54.
[4] Claudia Buerhop-Lutz, Tobias Pickel, Oleksandr Stroyuk, et al. Insulation resistance in relation to distribution of backsheet types in strings and inverters [J]. Solar Energy Materials and Solar Cells, 2022, (246):111913.
[5] LAN Guigang. Application of cross-ink thermal shrinkage materialfor protection on transmission insulations [J]. Electric Power, 1998(08):69-71.
[6] Mohammad IZADINIA, Reza SOLTANI, Mahmoud HEYDARZADEH SOHI, et al. Thermal insulation capability and adhesion strength of unpyrolyzed thick thermal barrier coatings [J]. Transactions of Nonferrous Metals Society of China, 2024, 34(4):1226-1236.

Journal of Physics: Conference Series **2879** (2024) 012020 doi:10.1088/1742-6596/2879/1/012020

[7] Tao Deng, Qian Wang, Shuming Liu, et al. Breakdown mechanism of RTV silicone rubber coated insulators under steep-front impulse voltage [J]. Electric Power Systems Research, 2023 (224):109772.
[8] Chengyu Zheng, Dechao Hu, Zhengbiao Liu, et al. A simple and efficient in situ polymerization of silica xerogel-acrylic thermal insulation coatings [J]. Progress in Organic Coatings, 2024, 187:108142.
[9] WANG Yafei, CHENG Hao, WANG Xiaomei, et al. Preparation and properties of odorless and environmental insulating varnish with low volatility [J]. Insulating Materials, 2018, 51(06):32-34.
[10] Guo Wei, Hui Liu, Yufan Zhou, et al. Solvent-free high-performance insulating coatings from photocurable polyimide resin for electronic circuit boards [J]. Progress in Organic Coatings, 2024, 189:108260.
[11] Tengfei Liu, Yuanshaung Liu, Dingrong Qu, et al. Multifunctional epoxy coatings reinforced with LDHs-modified silica microspheres: Synergistic effects on corrosion resistance, thermal insulation, and flame retardancy [J]. Colloids and Surfaces A: Physicochemical and Engineering Aspects, 2024, (698):134578.
[12] Guorong Wang, Zhiyuan Mei, Yu Li. Zinc-containing metal-organic frameworks nanospheres for flame retardation and thermal insulation performance in epoxy resin-based coatings [J]. Polymer Testing, 2023 (128):108209.
[13] Feiyang Li, Jianpeng Zou, Wenyi Zhan, et al. The influence of sodium silicate water glass content on the structure and properties of silicate thermal insulation coating [J]. Materials Letters, 2024, 363(14):136301.
[14] Emilio Bastidas-Arteaga , Mark G. Stewart. Damage risks and economic assessment of climate adaptation strategies for design of new concrete structures subject to chloride-induced corrosion [J]. Structural Safety, 2015(52): 40-53.
[15] Ehsan Bakhshandeh, Ali Jannesari, Zahra Ranjbar, et al. Anti-corrosion hybrid coatings based on epoxy–silica nano-composites: Toward relationship between the morphology and EIS data [J]. Progress in Organic Coatings, 2014, 77(07): 1169-1183.
[16] M. Behzadnasab , S. M. Mirabedini , K. Kabiri , S. Jamali. Corrosion performance of epoxy coatings containing silane treated ZrO_2 nanoparticles on mild steel in 3.5% NaCl solution [J]. Corrosion Science, 2011, 53(01): 89-98.
[17] Jianyu Liu, Yongkai Quan, Guoqiang Xu, et al. An online multi-spectrum-based method for evaluating thermal insulation performance of smart thermal barrier coatings [J]. Cell Reports Physical Science, 2024, 5(5):101958.
[18] ZHANG Pengfei, WU Jin. Development of aging and acid alkali resistant power cables for chemical industry [J]. New Technology & New Products of China, 2019(13):49-50.
[19] Wenjuan Wang, Yan Gao, Shanming Zhang, et al. A novel thermal insulation coating with layer-by-layer stacked nanosheets towards ultralow thermal conductivity [J]. Journal of the European Ceramic Society, 2024, 44(13):7883-7890.
[20] YU Guoling, CHEN Wanyao, WANG Xueke, et al. Progress of research on functional paints in China [J]. Electroplating & Finishing, 2020(02):111-115.
[21] Zhiyi Zhang, Xiaofei Han, Lan Jia, et al. Highly water-resistant transparent waterborne polyurethane thermal-insulation coating material with multiple self-crosslinking network based on controllably activated end-capping reagent [J]. Progress in Organic Coatings, 2024, 187:108104.
[22] Hui Liu, Long Yan, Feiyue Wang. Synthesis of antioxidant functionalized nano-TiO_2 for improving fire resistance and UV-shielding behavior of intumescent fire-retardant coatings for cable [J]. Progress in Organic Coatings, 2024(194):108622.
[23] D. Uglietti, A. Testino, R. Sobota. Examination of commercial ceramic coatings for the electrical insulation of Bi-2212 wire [J]. Cryogenics, 2023(129): 103626.

Journal of Physics: Conference Series **2879** (2024) 012021 doi:10.1088/1742-6596/2879/1/012021

Material selection and design of intake blades for aircraft engines based on additive manufacturing

Yibing Hou*, Haocheng Luo and Siying Dong

Beihang University, Beijing, 100083, China

*Corresponding author's e-mail: houyibingyeye@126.com

Abstract. The intake blades of aircraft engines are mainly used to capture and guide external air into the engine and are the foundation of engine operation. Their materials and design are related to the performance level and reliability of the entire engine. The "additive" 3D printing manufacturing method has the advantages of high precision, low cost, less material waste, strong design, and a short research and development cycle. This article proposes the use of GH4169 high-temperature alloy as the 3D printing material for the service requirements of aircraft engine intake blades, constructs a 3D printing model, and designs and realizes the 3D printing of engine intake blades.

1. Introduction

The engine is the "heart" of aviation equipment, and its performance determines a series of performance indicators such as flight speed, maneuverability, range, and effective load capacity [1]. In gas turbine engines, the turbine is an extremely critical component, and the turbine inlet blades are essential components that convert the engine's thermal energy into mechanical energy. Its working environment is the most demanding. During the working process, rotor blades need to withstand the combined effects of high temperature and high pressure, centrifugal load, vibration load, thermal stress, gas corrosion, and high-temperature oxidation [2,3]. The harsh service environment results in the highest probability of inlet blade failure. Statistically, intake blades account for over 70% of engine component failure events [1]. There are three key service environment characteristics of the turbine blades: high working temperature, complex working stress, and high corrosiveness of the working environment [1,3]. In addition, the design of the intake blade structure is complex to minimize the resistance and turbulence during airflow, improve intake efficiency, and prevent foreign objects, such as birds, from directly impacting the engine core components during flight. The design, manufacturing, and material level of turbine inlet blades will significantly affect the performance and reliability of the entire engine [4] and are one of the research difficulties and hot topics.

3D printing, also known as additive manufacturing, is a technology based on the "discrete stacking" principle, which uses digital models as the basis to form materials in three-dimensional space through layer-by-layer stacking. Unlike traditional "equal material" and "subtractive" manufacturing methods, the "additive" 3D printing manufacturing method can substantially simplify the manufacturing process, considerably shorten the development cycle, and also has advantages such as high precision, low cost, less material waste, and strong design. The application of 3D printing technology provides flexibility for blade design, allowing researchers to introduce more advanced aerodynamic characteristics, conduct innovative design research, and optimize engine performance. The material selection and design of intake blades for aircraft engines based on additive manufacturing reduce the blades' manufacturing cost

Journal of Physics: Conference Series **2879** (2024) 012021 doi:10.1088/1742-6596/2879/1/012021

and promote innovation in engine blade design, providing a new and convenient way to transform the engine manufacturing field. This technology has gradually become one of the most promising key technologies for designing intake blades for aircraft engines.

2. Requirements for material selection of intake blades for aircraft engines

2.1. Characteristics of service environment

The service environment characteristics of the intake blades of aircraft engines are:

(1) The working temperature is high and unevenly distributed. The known temperature before the turbine of the fourth generation military engine has reached 1850-2000K [5]. The temperature distribution of different blade parts is uneven, with the lowest temperature at the ten on and root in the longitudinal direction of the blade and the high-temperature zone from the middle section to the blade tip. The inlet and outlet edges of the blade are both high-temperature zones in the cross-sectional direction, with the outlet edge having a higher temperature.

(2) High speed and high stress. The rotational speed of turbine rotor blades during operation is usually around 12000r/min. The stress distribution of the service blades is uneven, and during the service process, the blade body may bear a centrifugal tensile stress of about 140MPa. In contrast, the average centrifugal tensile stress borne by the blade root reaches 280~560MPa [6].

(3) The working environment is highly corrosive. Turbine blades are subjected to oxidation and thermal corrosion caused by high-temperature gas [3,7]. Even if high-quality aviation kerosene is used for aircraft engines, the fuel still contains trace amounts of sulphur. When the aircraft is in coastal areas, the engine inhales air containing NaCl, which forms Na2SO4 and deposits on the surface of the turbine blades. The molten salt film causes thermal corrosion of turbine blades, which is more severe than pure oxidation.

(4) The structure of turbine blades is complex. At present, turbine blades are generally hollow structures with many small pipes inside, which play a role in forced cooling. In order to improve cooling efficiency, the internal cavity structure of turbine blades is becoming increasingly complex, and the wall thickness is becoming thinner, resulting in a more uneven distribution of temperature and stress in the blades in service. The hollow structure also reduces the effective load-bearing area, increasing the blade failure risk.

2.2. Material selection requirements

Due to the extremely harsh service environment of aircraft engine turbine blades, the requirements for the materials used are also highly demanding. The material of turbine working blades mainly needs to meet the following requirements [8]:

(1) Turbine blade materials should have good mechanical properties, including high-temperature creep performance, mechanical fatigue performance, thermal fatigue performance, impact resistance, and good high-temperature plasticity.

(2) Turbine blade materials should have good thermal corrosion and oxidation resistance, and their surfaces should be suitable for coating various protective coatings, such as Al-Si coatings, MCrAlY coatings, and thermal barrier coatings.

(3) Turbine blades should have good process performance, such as high thermal conductivity, low thermal expansion coefficient, density, and good casting and welding performance.

(4) The turbine blade material should have a high initial melting temperature and be able to withstand short-term overheating. The material of turbine blades should have good tissue stability, which can maintain relatively stable tissue during long-term use and avoid the precipitation of harmful phases such as TCP.

3. 3D printing material selection and design of intake blades for aircraft engines

3.1. Material selection

Super Alloy, literally translated as Super Alloy, refers to a type of metal material based on iron, nickel, and cobalt that can work for a long time under high temperatures and certain stress above 600 °C. It has excellent high-temperature strength, good oxidation and thermal corrosion resistance, good fatigue performance, fracture toughness, and other comprehensive properties. It is mainly used in the aerospace and energy fields. Superalloys can withstand extreme conditions. Nickel iron-based superalloys are particularly suitable for high-temperature environments requiring resistance to creep, corrosion, and thermal shock. GH4169 has the following characteristics:

(1) GH4169 is the mainstay of high-temperature alloys and the cornerstone of thermal structural aerospace applications. It can be permanently used above 600 ° C, with good creep, fracture strength, high fatigue resistance, and corrosion resistance at high temperatures. Its Young's modulus is almost twice that of titanium alloy Ti6Al4V (TC4), similar to nonalloy hard enable carbon steel (45 steel). GH4169 accounts for over 50% of the weight of aircraft turbojet engines and is the main material for the disks, blades, casing, and turbine blades in the high-pressure section of the compressor. GH4169 has excellent high-temperature performance and good toughness at low temperatures (preventing brittle fracture of parts), so it has found various applications in rocket engines and other low-temperature working conditions.

(2) GH4169 has high iron content and low alloy cost. GH4169 contains a large amount of iron, which reduces the price per kilogram and gives it a precipitation-hardening effect. The low migration rate of iron in the matrix endows the main strengthening phase (γ ") with slow precipitation kinetics, thereby reducing the sensitivity to post-weld cracking. If the migration/diffusion rate is high, it is easy to form alloy element segregation in the weld and heat-affected zone, leading to cracks. GH4169 has good welding performance. Specific alloying elements endow GH4169 with excellent corrosion resistance even at temperatures up to 1000 ° C. Nickel can be used to resist chloride ion stress corrosion cracking and prevent corrosion of many inorganic and organic oxidizing compounds in various acidity and alkalinity ranges. Chromium endows it with the ability to resist oxidative media and sulfur compounds, while molybdenum can enhance its resistance to pitting corrosion.

Room temperature tensile tests are conducted according to reference [9] standards to test the tensile properties of laser cladding manufactured specimens. The tensile specimen was subjected to direct aging heat treatment, with a 2 mm/min loading tensile rate in the indoor tensile test. The stress-strain curve of the GH4169 alloy tensile specimen at room temperature after heat treatment is shown in Figure 1 [10]. The figure shows that the yield strength after heat treatment is 1052MPa, the tensile strength is 1314MPa, and the tensile limit is 12.5%. According to the China Aerospace Materials Industry Manual, the tensile strength of GH4169 after direct aging heat treatment is 1240MPa. After manufacturing laser cladding, the material's tensile strength is 105.9% of the original material. Figure 2 shows the trend of corrosion weight gain of the sample after the salt spray heat exposure test with increasing heat exposure temperature. At 550 and 650 °C, a complete and dense protective Al_2O_3 oxide film was formed on the sample's surface after a high-temperature oxidation test, which protected the substrate from further oxidation [11]. Therefore, GH4169 high-temperature alloy has excellent performance as a 3D-printed material for aircraft engine blades.

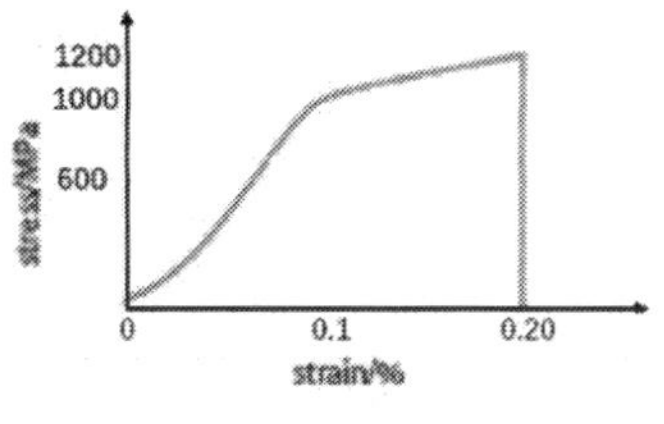

Figure 1. The stress-strain curve.

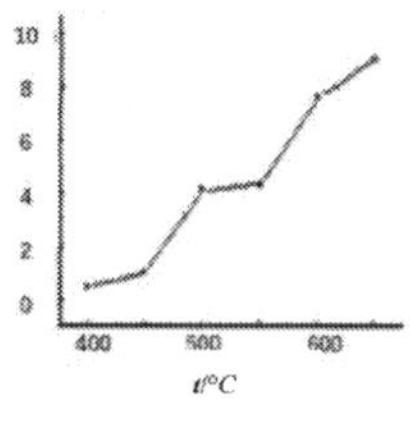

Figure 2. The trend of corrosion weight gain.

3.2. GH4169 high-temperature alloy 3D printing process parameter setting

The 3D printing equipment of laser metal powder melting instruments uses an infrared laser to melt and form a metal powder directly. The crucial parameters that affect the comprehensive performance of SLM technology formed parts include scanning speed, scanning power, and scanning path.

An in-depth study was conducted on the performance of single-walled walls with different printing parameters [8]. It was found that single-walled walls formed with a scanning power of 140W and a scanning speed of 400mm/s had no collapse, cracks, or warping, and the melt pool was superior to those formed with other parameters. This is also the basis for selecting scanning power and speed in this paper.

For the parameter selection of scanning paths, optical microscopy (OM), scanning electron microscopy (SEM), X-ray diffraction (XRD), transmission electron microscopy (TEM), electron backscatter diffraction (EBSD), etc., are used to study the microstructure of samples formed by three scanning paths (as shown in Figure 3) [8]. It was found that the microstructure of the samples formed by the upper and lower scanning paths at an angle of 45 ° was uniform, without generating small cracks, and the scanning electron microscopy showed the precipitation of the Laves phase. By phase analysis and transmission electron microscopy observation, a strengthening phase γ "(Ni_3Nb) was produced, and the matrix grains displayed by EBSD were relatively small. The mechanical properties of the samples formed by three different scanning methods showed that the tensile strength of the samples formed by the scanning path angle of 45 ° between the upper and lower layers reached 1115MPa. The elongation rate of the samples formed by the scanning path angle of 67 ° between the upper and lower layers reached 37.8%, slightly higher than that of the samples formed by the scanning path angle of 45 ° between the upper and lower layers (as shown in Figure 4). By observing the fracture morphology, it was found that the size of the tough dimples in the sample formed by the 45 ° angle between the scanning paths of the upper and lower layers was relatively large. The micro-hardness of samples formed by three different scanning methods was tested. It was found that the micro-hardness of samples formed by the 45 ° scanning path angle between the upper and lower layers was 320HV higher than that of samples formed by the other two forming methods but lower than the 350HV of forgings [12]. Based on this, this article adopts a 45 ° scanning path.

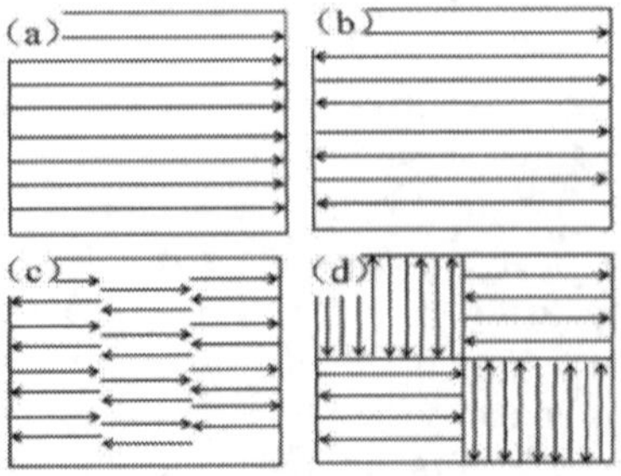

Figure 3. Single layer laser scanning path: (a) Single layer laser unidirectional scanning; (b) Single layer laser reverse scanning; (c) Single layer laser short distance reverse scanning; (d) Single layer laser segmented reverse scanning [9].

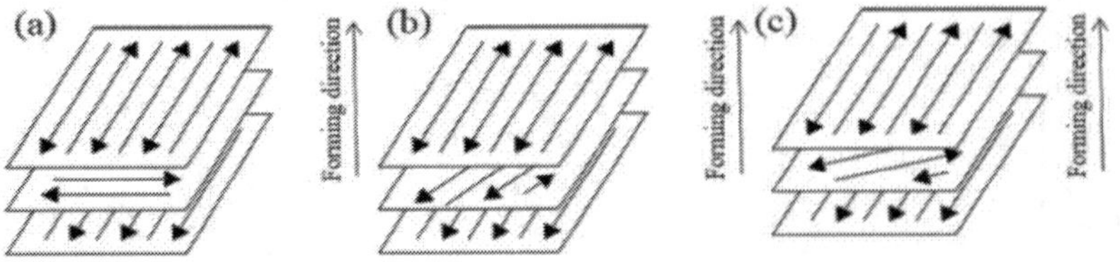

Figure 4. Three scanning methods: (a) The angle between the scanning paths of the upper and lower layers is 90 °; (b) The angle between the scanning paths of the upper and lower layers is 67 °; (c) The angle between the scanning paths of the upper and lower layers is 45 ° [10].

We convert the pre-generated 3D image into STL format and input it into the device before printing. According to the technical standards proposed by Zhang et al. [13,14], a powder layer thickness of about 40 μ m and a laser spot diameter of Φ0.1mm are selected. After the preparation is completed, the residual powder on the surface of the 3D-printed GH4169 alloy sample is swept away, and then the substrate is removed. The entire selective laser melting forming experiment ends here. Table 1 presents the chemical composition of 3D-printed GH4169 alloy samples.

Table 1. Chemical composition of GH4169 alloy samples for 3D printing.

C	Si	Mn	Cr	Ni	Co	Ti	Al	Mo	P	S	Nb
0.03	0.2	0.2	18.0	54.5	0.8	1.0	0.5	3.1	0.004	0.0022	5.1

3.3. 3D printing modeling and implementation

3.3.1 Modelling. SolidWorks software is used to create a 3D model of the rotating blades at the intake of an aircraft engine and it is converted to STL format, as shown in Figure 5.

Figure 5. 3D modeling of rotating blades at the intake of an aircraft engine.

3.3.2 Printing implementation. We use the Renishaw AM400 additive manufacturing platform (Figure 6), equipment-specific software Renishaw QuantAM with file 3D model processing tool, Renishaw Material Editor Modelling, and implementation using a material editor. The implementation steps are as follows:

Figure 6. RENISHAW AM4003D printing equipment diagram.

(1) Parts placement

We import the formed STL format model file into the specialized 3D model processing tool of the AM400 system and determine the reasonable placement direction of the 3D model in advance, as shown in Figure 7, to reduce processing time and required support, thereby optimizing the forming process of the component.

(a) Zoom to fit the base size (b) Image before the model laid flat (c) Image after the model laid flat

Figure 7. Design diagram for part placement.

(2) Support structure generation

We set up a support structure below the hollow area of the component to enable metallurgical bonding between the hollow structure and the support structure, ensuring that the component can be fully formed. The system will automatically generate a support structure, as shown in Figure 8.

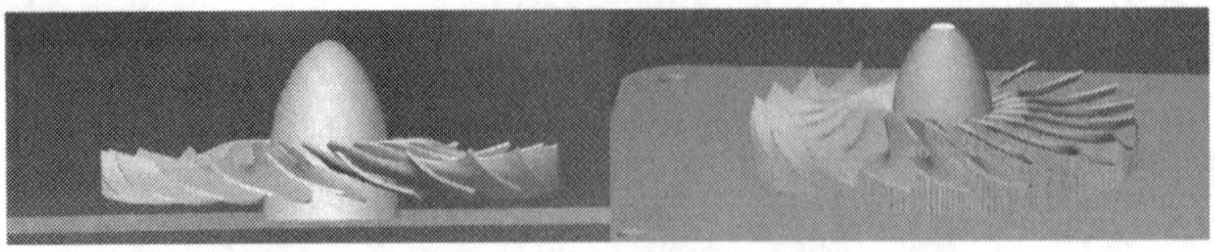

Figure 8. Support structure design diagram.

(3) Material selection and parameter setting

We set the selected materials in the material editor, imported the file processing tool, and set the parameters and scanning strategy of the selected materials. This article selects a power of 160W, a layer thickness of 0.09mm, and a rotation increment angle of 45 °, as shown in Figure 9.

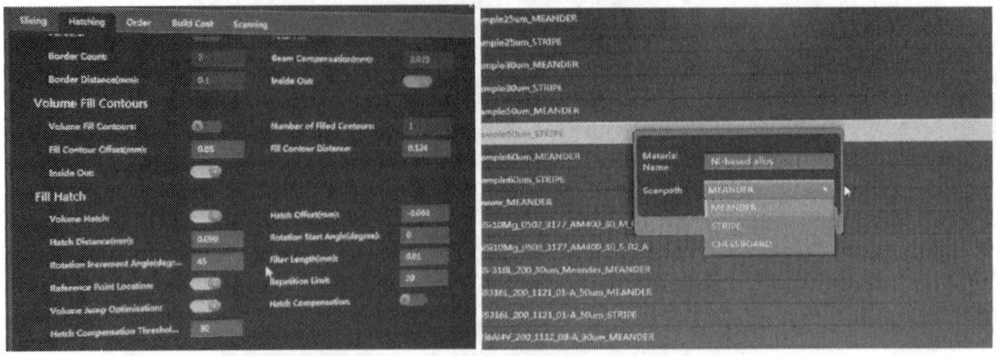

Figure 9. Display of 3D printing parameter settings.

(4) Slice scanning

We plan the scanning path of the laser at each layer, calculate the processing time, export the MTT format, and complete the final inspection， as shown in Figure 10.

Figure 10. Scanning slice image.

It takes a total of 33 hours to complete the overall 3D printing design, as shown in Figure 11. After completing the above software operations, it is converted to the MTT format that the printing device can recognize. The resulting image is exported, as shown in Figure 12.

Figure 11. 33 hours to complete the overall 3D printing design.

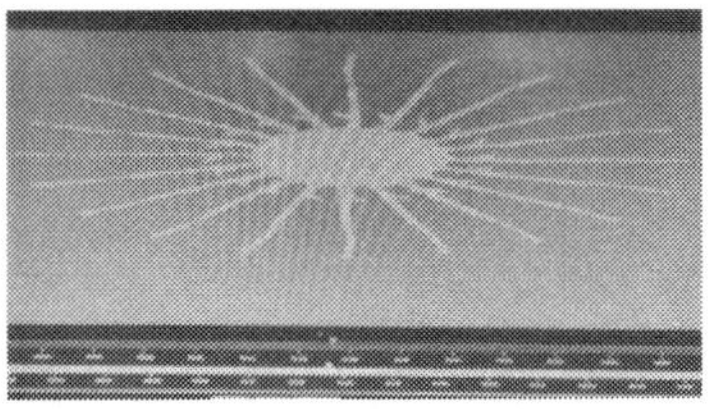

Figure 12. Export the resulting image.

4. Conclusion

3D printing has the advantages of fast molding speed, high design freedom, and minimal material waste, making it highly advantageous in preparing complex structural products. In recent years, it has received increasing attention and research. This article focuses on the 3D printing design of aircraft engine intake based on GH4169 high-temperature alloy and implements it using RENISHAW AM 400 3D printing equipment. Further research can be carried out from the following two aspects: 1) Given the current scattered and small sample size of additive manufacturing material research, it is necessary to expand further the experimental data samples of materials selected for aircraft engines to facilitate in-depth study of material mechanical properties. 2) The performance, process parameters, and service environment of additive manufacturing structures for aircraft engines are closely related. Therefore, how to mine process parameters, defect morphology, microstructure, and other data, integrate them with engine service data, and introduce artificial intelligence technology is the development trend of aircraft engine additive manufacturing structural design.

References

[1] Tao Chunhu, Zhong Peidao, Wang Renzhi. Failure and Prevention of Rotating Components in Aircraft Engines [M]. Beijing: National Defense Industry Press, 2000.

[2] Carter T J. Common failures in gas turbine blades [J]. Engineering Failure Analysis, 2005, 12: 237-247.

[3] Ali L, Ghauri K M, Hasan F. Cracking failure in gas-turbineblades [J]. Pakistani Journal of Engineering & Applied Sciences, 2011, 9:98-102.

[4] Feng Qiang, Tong Jinyan, Zheng Yunrong, et al. Service damage and repair of gas turbine blades [J]. China Materials Progress, 2012, 31 (12): 21-34.

[5] Liu Daxiang, Cheng Ronghui. The current situation and development trend of world aviation power technology [J]. Journal of Beihang University, 2002, 28 (5): 490-496.

[6] Kargarnejad S, Djavanroodi F. Failure assessment of Nimonic80A gas turbine blade [J]. Engineering Failure Analysis, 2012, 26: 211-219.

[7] Xie Y J, Wang M C, Zhang G, et al. Analysis of superalloyturbine blade tip cracking during service [J]. Engineering Failure Analysis,2006, 13(8): 1429-1436.

[8] Autumn Adventure Study on the Properties of GH4169 Alloy Formed by SLM [D]. Nanchang University of Aeronautics and Astronautics, 2020.DOI: 10.27233/d.cnki.gnchc.2020.000078.

[9] Gao Yifei, Liang Xinbang, Deng Xinglin. GB/T 228.1-2010, Tensile Testing of Metallic Materials Part 1: Room Temperature Test Method-Implementation Guide [M]. Beijing: China Quality Inspection Press, 2012.

[10] Zhang Sheng, Wang Changhao, Lv Shuai, et al. Research on the Process and Properties of GH4169 Material Prepared by Laser Cladding [J]. Mechanical and Electrical Engineering Technology, 2022, 51 (10): 65-67.

[11] Zhao Honglu, Wang Li, Tong Wenwei, et al. Experimental study on salt spray heat exposure cyclic corrosion resistance of GH4169 alloy [J]. Aero Engine, 2009, 35 (06): 36-39.

[12] Gu Dongdong and Shen Yigo. The current status and technological prospects of rapid prototyping of metal parts based on selective laser melting [J]. Aviation Manufacturing Technology, 2012, 404 (8): 32-37.

[13] Zhang Xuefeng, Li Huaixue, Hu Quandong, et al. Effect of Heat Treatment on the Microstructure and Tensile Properties of Laser Selective Melting GH4169 High Temperature Alloy [J]. Aerospace Manufacturing Technology, 2019, 62 (19): 78-85.
[14] Zhang Xuefeng, Li Huaixue, Hu Quandong, et al. Effect of heattreatment on the microstructure and tensile properties of GH4169superalloy fabricated by selective laser melting [J]. Aeronautical Manufacturing Technology, 2019, 62(19): 78-85.

Journal of Physics: Conference Series **2879** (2024) 012022 doi:10.1088/1742-6596/2879/1/012022

Quality evaluation and crack cause analysis of a type B sleeve service welding

Yan Xu[1*], Qiang Bai[1], Fengping Yang[1], Zhenjun Feng[1], Yonggang Chang[1] and Liang Li[1]

[1]CNPC Tubular Goods Research Institute, Xi'an, Shaanxi, 710077, China

[*]Corresponding author's e-mail: xuyan008@cnpc.com.cn

Abstract. Type B sleeve repair technology is important in service welding repair because the quality of fillet weld directly affects the sealing and bearing capacity of Type B sleeve repair. In this paper, a cover weld crack at 0 o 'clocks of fillet weld found by non-destructive testing is analyzed. The tensile strength of fillet weld and longitudinal weld is much higher than the technical requirements of X80 steel grade pipe, the welding material belongs to a strong matching state, and its impact toughness value is low. At the same time, the hardness test was carried out at the fillet weld, and the hardness of the tube body at the welding position of the cover was higher than that at other positions, and there was a risk of cracking. The defects were dissected as hot cracks, and the reason was related to the strong matching of the selected welding material. It is suggested that when selecting welding materials for lap fillet welding, it is possible to try to reduce the matching strength level of the weld without affecting the sealing and bearing capacity.

1. Introduction

Type B sleeve repair technology, as an important in-service welding repair means, can ensure continuous pipeline transportation, reduce environmental damage, shorten repair cycles, reduce repair costs, and have broad prospects for development[1-4]. Two longitudinal welds and two annular lap fillet welds must be welded for Type B sleeve installation. Among them, the fillet welds overlap the steel pipe and Type B sleeve, and the welding quality directly affects the sealing and bearing capacity of Type B sleeve repair. At present, the wall thickness of the low-strength type B sleeve used in the field reaches 50~60 mm, resulting in excessive labor intensity in sleeve installation and welding, and it is not easy to guarantee welding quality. To achieve effective repair of high-steel pipe girth welds, a selection of welding materials and welding processes for type B sleeve welding are in urgent need of improvement. In particular, the strength mismatch and degree of welding material selection will inevitably affect the mechanical properties and bearing capacity of welded joints. At present, there are more studies on the strength matching of butt girth welds, but few reports on the strength matching of lap fillet welds [5-10]. In this paper, a field fillet weld cracking problem was studied, the mechanical properties of the welded joint with the matching lap fillet weld were mastered, and the selection of welding materials and welding quality control were guided. A magnetic particle flaw detector was used to detect the sleeve fillet weld. The type of magnetic powder was black magnetic powder, the current type was AC, the lifting force was ≥45N, and the surface illumination was >1000lx. A 35.7 mm length crack was found on the upstream fillet weld, located near 0 o'clock, as shown in Figure 1 (a). A crack with a crack length of 4.9 mm was found on the downstream fillet weld, also located near the position of 0 o'clock, as shown in Figure 1 (b).

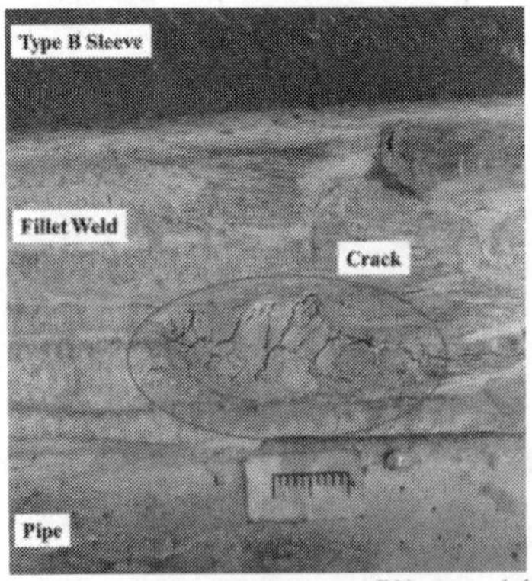

(a) Cracks in upstream fillet welds

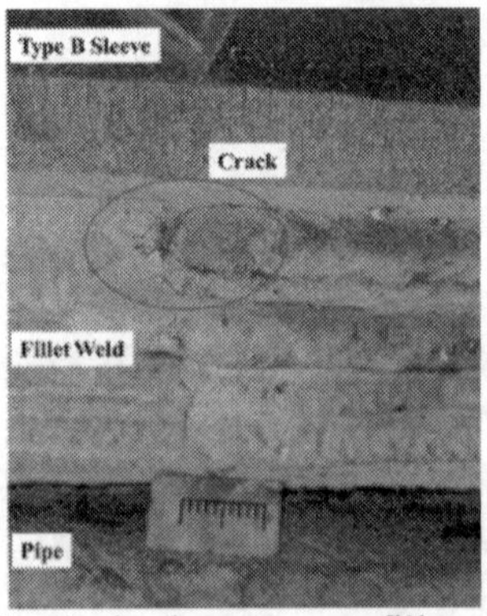

(b) Cracks in downstream fillet welds

Figure 1. Magnetic particle detection result.

2. Test methods and results

2.1. Chemical composition analysis

The sleeve body, fillet weld, and longitudinal weld were sampled, respectively, and the samples were 30×20×t. The chemical composition was analyzed by the direct reading spectrometer, and the results are shown in Table 1.

Table 1. Chemical compositions (Wt.%).

Sample	C	Si	Mn	Cr	Mo	Ni	Nb	V	Ti	Cu
Sleeve body	0.13	0.3	1.43	0.032	0.0046	0.025	0.018	0.0036	0.0027	0.07
Fillet weld	0.066	0.51	1.38	0.032	0.061	1.06	0.014	0.017	0.04	0.094
Longitudinal weld	0.068	0.51	1.41	0.03	0.055	1.00	0.014	0.017	0.043	0.092

2.2. Tensile test

Rod-like samples were taken from the sleeve body, fillet weld, and longitudinal weld, respectively for the tensile test. The diagram of the tensile specimen is shown in Figure 2. The diameter of the standard distance segment was 6.25 mm, and the length of the standard distance segment was 25 mm. The results are shown in Table 2. The mechanical tensile properties of the Type B sleeve meet the requirements of Q345R material. However, the yield strength and tensile strength of the cladding metal of fillet weld are much higher than the requirements of X80 steel grade, and the yield ratio is not less than 0.94. The strength of the root welding position of the cladding metal in the longitudinal weld is slightly close to the requirements of the X80 steel grade, and the strength of other sampling positions is much higher than the requirements of the X80 steel grade. The matching of welding material is higher than that of steel pipe and Type B sleeve.

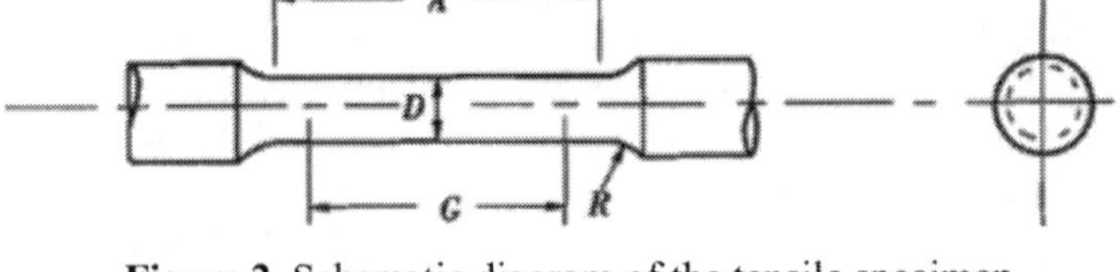

Figure 2. Schematic diagram of the tensile specimen.

Journal of Physics: Conference Series **2879** (2024) 012022 doi:10.1088/1742-6596/2879/1/012022

Table 2. Tensile test result.

No.	Sample position		Tensile strength (MPa)	Yield strength (MPa)	Yield ratio
1	Sleeve body		517	372	0.72
2	Upstream fillet weld		742	703	0.95
3	Downstream fillet weld		746	704	0.94
4	Longitudinal weld	Near outer wall	732	687	0.94
		Thickness center	754	715	0.95
		Near inner wall	641	571	0.89

2.3. Charpy impact test

Samples were taken from the sleeve, fillet weld, and longitudinal weld, respectively, and Charpy V-notch impact samples were processed. The impact sample diagram is shown in Figure 3. The sample size was 10mm×10mm×55mm, and the test temperature was -10°C. The test results are shown in Table 3.

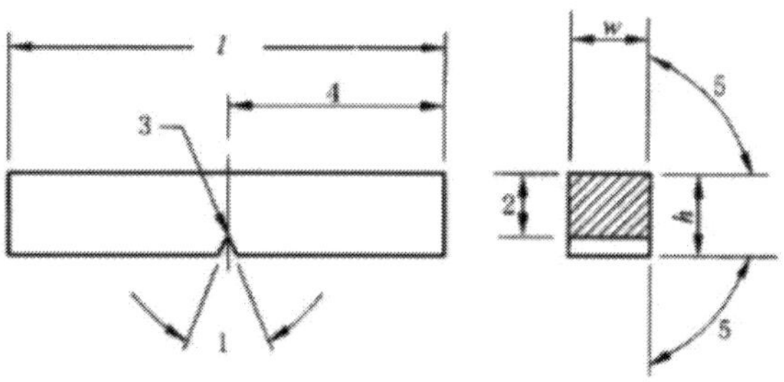

Figure 3. Schematic diagram of impact specimen.

Table 3. Impact test result.

No.	Sample position		Absorbed energy (J)				Shear section ratio (%)			
			Single			Mean	Single			Mean
1	Sleeve body		194	239	240	224	100	100	100	100
2	Upstream fillet weld		80	69	104	84	90	90	95	92
3	Downstream fillet weld		95	68	47	70	90	80	60	77
4	Longitudinal weld	Near outer wall	81	77	83	80	99	90	100	96
		Thickness center	101	100	94	98	98	97	95	97
		Near inner wall	120	85	95	100	75	70	80	75

2.4. Hardness test

The fillet weld was sampled, and a 10kg load Vickers hardness test was carried out. The test results of fillet weld hardness are shown in Figure 4. The average hardness of the upstream fillet weld was 242HV, the minimum was 198HV, and the maximum was 285HV. The maximum hardness appeared in the sealing position and the covering position of the root welding. The average hardness of the downstream fillet weld is 268HV, the minimum is 241HV, the maximum is 306HV, and the maximum appears in the cover welding position.

Journal of Physics: Conference Series **2879** (2024) 012022 doi:10.1088/1742-6596/2879/1/012022

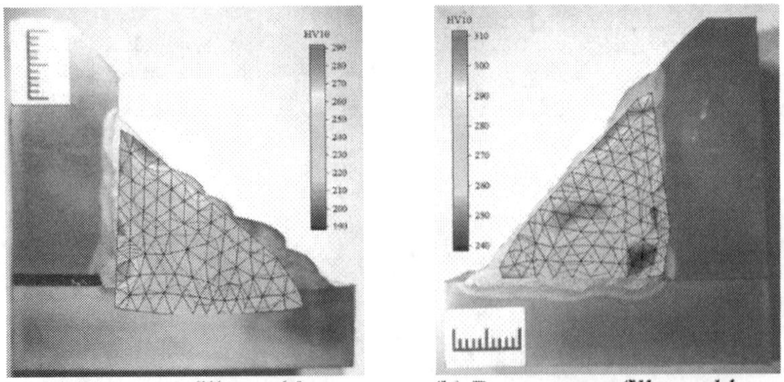

(a) Upstream fillet welds (b) Downstream fillet welds

Figure 4. Hardness distribution in the direction of wall thickness in different areas.

2.5. Guided bending test

The lateral bending sample was taken at the longitudinal weld with a width of 13 mm and a length of 230 mm. The face-bending sample is taken at the fillet weld with a width of 25 mm and a length of 230 mm. The diameter of the bending shaft is 90 mm. In the lateral bending test of the longitudinal weld, there are small non-fusion defects and 3 mm cracks in the weld, as shown in Figure 5.

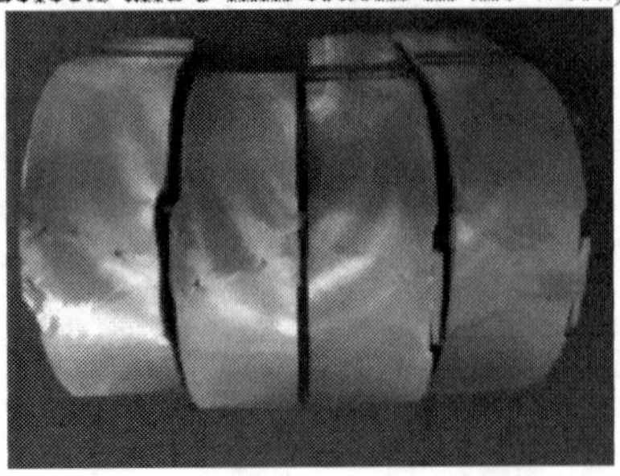

Figure 5. Bending test crack of the longitudinal weld.

2.6. Metallographic analysis

The defects found in magnetic particle detection were dissected and analyzed, and the metallographic sampling position and macroscopic morphology of the upstream fillet weld cracks were obtained, as shown in Figure 6. The anatomical results of crack defects in the welding area of the cover are shown in Figure 7. The structure of the crack area is loose, and the local crack has intergranular characteristics, which is called a welding hot crack. In addition, the upstream fillet weld has continuous non-fusion defects, as shown in Figure 8, and these defects are found in the same position of two different cross sections.

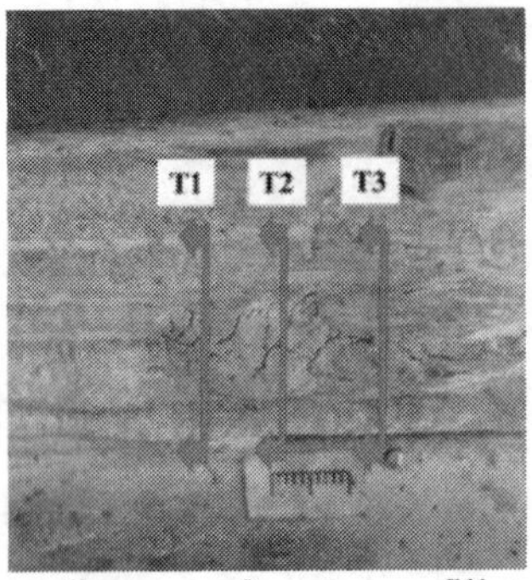

(a) Sampling diagram of upstream fillet weld crack

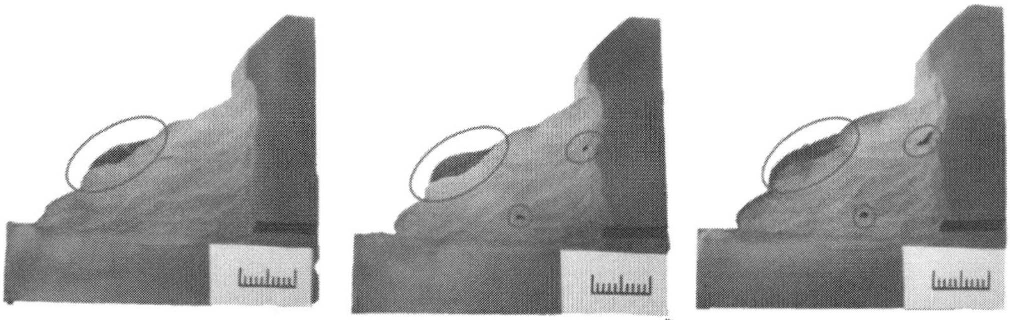

(b) Macro morphology of T1 (c) Macro morphology of T2 (d) Macro morphology of T3

Figure 6. Metallographic sampling position and macroscopic morphology.

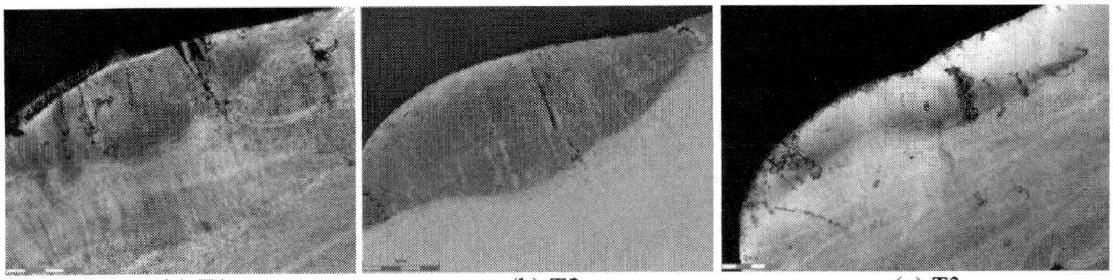

(a) T1 (b) T2 (c) T3

Figure 7. Crack appearance of the cover welding.

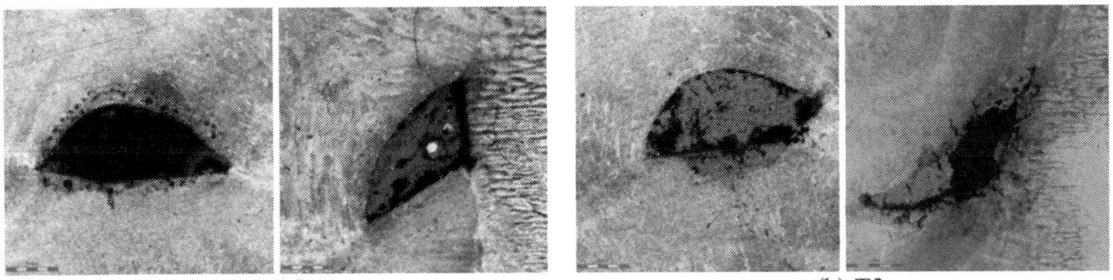

(a) T2 (b) T3

Figure 8. Non-fusion defect.

3. Discussion and analysis

In this case, surface cracks appeared in the fillet weld position when the Q345 material Type B sleeve was used to repair the X80 steel pipe by automatic welding. It was confirmed by magnetic particle detection that there was a crack in the cover weld at 0 o'clock of the fillet weld, and the maximum length was 35.7 mm. No obvious abnormality was found in the chemical composition of the Type B sleeve tube body, fillet weld, and longitudinal weld. The tensile strength of fillet weld and longitudinal weld is much higher than the technical requirements of X80 steel grade pipe, the welding material belongs to a strong matching state, and its impact toughness value is low. At the same time, the hardness test was carried out at the fillet weld, and the hardness of the tube body at the welding position of the cover was higher than that at other positions, and there was a risk of cracking. Through the crack location sampling and metallographic inspection, no obvious abnormality was found. The structure of the crack area was loose, and the local crack had the characteristics of intergranular, which is a welding hot crack. Therefore, the cause of crack = is related to the cooling rate of molten metal in different parts of the welding process, especially as the cooling rate of the outer surface cover welding is faster, resulting in an abnormally high hardness value of the outer wall layer, and further causing cracks. Therefore, it is suggested that when selecting welding materials for lap fillet welds, it is possible to try to reduce the matching strength level of weld seams without affecting sealing and bearing capacity. The application of high-strength welding materials makes it necessary to solve the problem through research, such as preventing hydrogen-induced cracking and increasing preheating temperature, which may affect pipe performance, decreasing weld toughness caused by increasing weld metal strength, and increasing defect control difficulty, so as to avoid the problem of system reliability reduction caused by only considering the improvement of a single factor.

4. Conclusions

The surface weld crack at 0 o'clock of fillet weld found by non-destructive testing is an anatomically hot crack, which is related to the strong matching of the selected welding material. It is suggested that when selecting welding materials for lap fillet welding, it is possible to try to reduce the matching strength level of the weld without affecting the sealing and bearing capacity.

References

[1] Crapps J M, Yue X, Berlin R A, et al. Strain-based pipeline repair via type B sleeve[J]. International Journal of Offshore and Polar Engineering, 2018, 28(3):280-286.
[2] Amori K E, Hussain M N, Hilal H B. Thermal analysis of in-service welding process for pipeline [J]. Journal of Petroleum Research and Studies, 2019, 9(1): 1-20.
[3] Amori K E, Hussain M N, Hilal H B. Experimental investigation of pipeline in-service welding process[J]. Journal of Petroleum Research and Studies, 2018, 8(1): 18-28.
[4] Cheng Z, Long W, Xue S, et al. Research on the influence mechanism of brazing seam geometry on gas pores in brazed joints[J].China Welding,2020,29(04):13-18.
[5] Farzadi A, Sanaei S. Analysis of failure caused by in-service welding in an X52 gas pipeline[J]. Journal of Welding Science and Technology of Iran, 2018, 3(2): 9-19.
[6] Teng Bin, Li Xiaoyu, Lei Zhen, et al. Analysis on cold crack sensitivity of low alloy high strength steel weld by laser-arc hybrid welding[J]. Transactions of the China Welding Institution, 2010, 31(11): 61 – 64. [10]
[7] Xu Yan, Liu Yinglai, Nie Xianghui , et al. Analysis of mechanical properties and application effect of over-matching lap angle weld joint[J]. Transactions Of The China Welding Institution, 2022, 43(10): 101-105.
[8] Dunđer M, Vuherer T, Samardžić I, et al. Analysis of heat-affected zone microstructures of steel P92 after welding and after post-weld heat treatment[J]. The International Journal of Advanced Manufacturing Technology, 2019, 102(9): 3801-3812.
[9] Song W, Sun X, Liu X, et al. Strength mismatch effect on residual stress of high-strength steel butt-welded joints[J]. Procedia Structural Integrity, 2021, 33: 795-801.
[10] Ran M M, Sun F F, Li G Q, et al. Experimental study on the behavior of mismatched butt welded joints of high-strength steel[J]. Journal of Constructional Steel Research, 2019, 153: 196-208.

Large amplitude and small size ultrasonic milling for BT40 tool holders

Yu Zhao[1*], Xuewei Li[2], Shengrong Shu[2] and Chenjun Wu[1]

[1]The 36th Research Institute of China Electronics Technology Group Corporation, Jiaxing 314001, Zhejiang, China
[2]School of Mechatronics & Vehicle Engineering, East China Jiaotong University, Nanchang 330013, Jiangxi, China

[*]Corresponding author's e-mail: 314425426@qq.com

Abstract: The current ultrasonic milling tool holders are relatively long in structure, and the tool is susceptible to generating detrimental radial vibrations during high-speed rotation. To achieve high-quality milling of difficult-to-machine materials, a longitudinal-torsional ultrasonic oscillator suitable for BT40 tool holders has been proposed, which can generate a large amplitude while maintaining a compact length. A longitudinal vibration ultrasonic transducer with an exponential front cover is designed using the analytical method. Based on the principle of acoustic wave reflection, four spiral grooves are introduced on the exponential variable cross-section bar to convert the longitudinal vibration into longitudinal-torsional composite vibration. The effects of groove width, groove depth, and spiral angle on the longitudinal-torsional resonance were analyzed using the single-factor method in finite element software, and the optimal spiral groove parameter combination was determined. The longitudinal-torsional ultrasonic transducer is assembled with the BT40 tool holder to construct a longitudinal-torsional ultrasonic tool holder model, and modal analysis was performed. The resonant frequency was determined to be 22,845 Hz. An experimental platform was established to carry out impedance analysis and amplitude testing on the longitudinal-torsional ultrasonic tool holder. The results show that its resonant frequency is 22,152 Hz, which is slightly different from the simulation value. The tool holder's maximum longitudinal amplitude is 11.2 μm, and the maximum torsional amplitude is 4.3 μm, with a torsional-to-longitudinal ratio of 0.38. The performance of the longitudinal-torsional ultrasonic tool holder meets the processing requirements of most ultrasonic milling applications, and the correctness of the design method is verified.

1. Introduction

High-performance composite materials are widely used in aerospace, marine, and other fields. Although these materials exhibit excellent performance, traditional machining methods fail to meet the requirements for high-efficiency, high-precision, and low-damage machining. This led to the development of ultrasonic vibration machining technology [1]. Ultrasonic vibration machining is known for its high precision, low surface roughness, and low cutting force, making it particularly suitable for hard-to-machine materials such as titanium alloys, carbon fiber composites, and optical glass [2]. With further research, composite vibration machining modes such as longitudinal-bending composite and longitudinal-torsional composite have been developed [3], showing better machining effects compared to one-dimensional vibration for difficult-to-machine materials [4]. These have become popular research topics in recent years.

The longitudinal-torsional composite ultrasonic transducer is a crucial component of the longitudinal-torsional ultrasonic machining system. The Langevin-type transducer generates vibration based on the inverse piezoelectric effect principle, with two primary methods for achieving longitudinal-torsional vibration [5]. The first method is to achieve longitudinal-torsional resonance by coupling the longitudinal vibration generated by axially polarized piezoelectric ceramics with the torsional vibration generated by tangentially polarized ceramics through the proper assembly of two ceramics with different polarization directions. However, torsional ceramics are difficult to manufacture, and devices made using this principle are not stable during high-speed operation [6]. The second method involves special structural designs on the front cover plate, such as inclined slits or spiral grooves, to convert part of the axial vibration generated by axially polarized piezoelectric ceramics into torsional vibration, achieving longitudinal-torsional vibration [7]. This method has been extensively researched by scholars both domestically and internationally. Pi [8] designed uniform inclined grooves on the transducer's vibration rod based on the propagation principle of sound waves, causing partial reflection of longitudinal waves into transverse waves. These studies analyzed the design methods of longitudinal-torsional ultrasonic transducers, and researchers have applied longitudinal-torsional ultrasonic vibration characteristics to ultrasonic welding, ultrasonic grinding, and ultrasonic drilling, but research on longitudinal-torsional composite ultrasonic milling tool holders is relatively scarce.

This dissertation combines theoretical derivation, finite element analysis, and experimental validation to design a sandwich-type longitudinal-torsional composite ultrasonic transducer suitable for BT40 tool holder milling. Based on the principles of modal conversion, spiral grooves were machined onto the exponential horn to achieve synchronous longitudinal-torsional vibration. The effects of spiral groove parameters on the resonant frequency of the transducer and the relationship between spiral groove parameters and the torsional-to-longitudinal ratio were explored through finite element analysis, selecting suitable spiral groove parameters to achieve ideal results. The longitudinal-torsional ultrasonic transducer was assembled with the BT40 tool holder, and impedance analysis and amplitude testing were conducted to verify the theoretical design and simulation results.

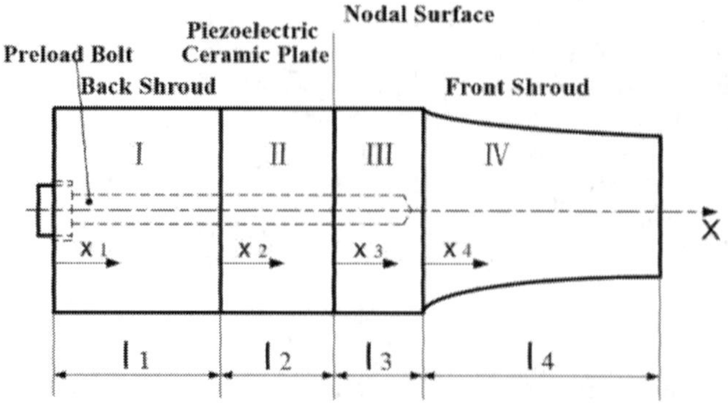

Figure 1. Schematic diagram of transducer.

2. Design of sandwich-type longitudinal vibration ultrasonic transducer

2.1 Design theoretical derivation

Based on different energy conversion principles, ultrasonic transducers are divided into piezoelectric transducers and magnetostrictive transducers. Compared to magnetostrictive transducers, piezoelectric transducers are smaller, easier to process, and have higher energy conversion efficiency. Therefore, this design adopts a sandwich-type piezoelectric transducer. According to wave theory, the overall length of the transducer should be an integer multiple of half the wavelength. The designed transducer length is half a wavelength. To reduce performance degradation due to looseness or gaps and further reduce the axial size of the transducer, the front cover plate and horn are combined into one integral piece, referred to as the front cover plate. The transducer structure is shown in Figure 1. Section l_1 represents the rear cover plate, Section l_2 represents the piezoelectric ceramics, Section l_3 represents the cylindrical portion of the front cover plate, and Section l_4 represents the exponential horn. The node is assumed to be at the contact surface between the front cover plate and the piezoelectric ceramics, and serves as the reference point for setting the flange.

Due to the complex load of the transducer and difficulty in measurement, to simplify the analysis, the front and rear cover plates of the transducer are assumed to be unloaded, and the effect of the preload bolt size on the transducer is ignored in the following derivation. It is assumed that the materials used for the front and rear cover plates of the transducer are uniform and isotropic. Under harmonic vibration conditions, the vibration velocity equation for any variable-section rod undergoing longitudinal vibration is:

$$\frac{\partial^2 v}{\partial x^2}+\frac{1}{S}\cdot\frac{\partial S}{\partial x}\cdot\frac{\partial v}{\partial x}+k^2 v=0 \tag{1}$$

where $v=v(x)$ represents the point vibration velocity; $S=S(x)$ is the cross-sectional area coefficient of the rod; k is the circular wavenumber, and $k=\frac{\omega}{c}=\frac{2\pi f}{c}$.

The frequency equations are derived separately for both sides of the nodal surface. The rear cover plate and the piezoelectric ceramic on the left side of the nodal surface are uniform rods with constant cross-sections, and the cylindrical section of the front cover plate on the right side of the nodal surface is also a uniform rod with a constant cross-section, while the exponential section of the front cover plate is a variable-section rod.

2.1.1 Derivation of the frequency equation for the left side of the nodal surface. When the cross-section is uniform, $S(x)$ is a constant, and Equation (1) simplifies to:

$$\frac{\partial^2 v}{\partial x^2}+k^2 v=0 \tag{2}$$

Solving this equation gives the general solution for the vibration velocity and longitudinal stress of the uniform rod:

$$v(x)=A\sin kx+B\cos kx \tag{3}$$

$$\partial(x)=\frac{E}{j\omega}\frac{\partial v}{\partial x}=-j\rho c\left(A\cos kx-B\sin kx\right) \tag{4}$$

where A and B are undetermined coefficients.

The boundary conditions for the velocity and force on the left side of the nodal surface of the transducer are:

$$\begin{cases} v_1(l_1)=v_2(0), \\ v_2(l_2)=0, \\ \partial_1(l_3)=0, \\ \partial_1(l_1)=\partial_2(0). \end{cases} \tag{5}$$

Substituting the boundary conditions into Equations (3) and (4), and solving the system of equations, the frequency equation for the left side of the nodal surface is obtained:

$$\tan k_1 l_1 \tan k_2 l_2=\frac{z_2}{z_1} \tag{6}$$

2.1.2 Derivation of the frequency equation for the right side of the nodal surface. The l_3 section on the right side of the transducer nodal surface is a cylindrical body with a constant cross-section, and its general solution is similar to Equations (3) and (4). The l_4 section is an exponential variable cross-sectional body, with the large end radius R_3 and the corresponding cross-sectional area S_3, and the small end radius R_4 and the corresponding cross-sectional area S_4. The radius and area variation functions of the cross-sections are respectively:

$$R_4(x) = R_3 e^{-\beta x} \tag{7}$$

$$S_4(x) = S_3 e^{-2\beta x} \tag{8}$$

where β is the taper coefficient of the exponential horn, $\beta = \frac{\ln N}{l_4}$, and $N = \frac{R_3}{R_4}$.

Substituting Equation (8) into Equation (1) yields the vibration velocity equation for the exponential rod:

$$\frac{\partial^2 v}{\partial x^2} - 2\beta\frac{\partial v}{\partial x} + k^2 v = 0 \tag{9}$$

The vibration velocity solution and longitudinal stress are:

$$v_4(x_4) = e^{\beta x}\left(A_4 \sin k_4' x_4 + B_4 \cos k_4' x_4\right) \tag{10}$$

$$\partial_4(x) = -j\frac{\rho c}{k}\left[A_4 e^{\beta x}(\beta \sin k_4' x_4 + k_4' \cos k_4' x_4) + B_4 e^{\beta x}(\beta \cos k_4' x_4 - k_4' \sin k_4' x_4)\right], \tag{11}$$

where $k_4' = \sqrt{k_4^2 - \beta^2}$.

The boundary conditions for the velocity and force on the right side of the nodal surface are:

$$\begin{cases} v_3(0) = 0, \\ v_3(l_3) = v_4(0), \\ \partial_3(l_3) = \partial_4(0), \\ \partial_4(l_4) = 0. \end{cases} \tag{12}$$

Substituting the boundary conditions into Equations (10) and (11), and solving the system of equations, the frequency equation for the right side of the nodal surface is obtained:

$$k_4 \tan k_3 l_3 \tan k_4' l_4 = k_4' + \beta \tan k_4' l_4 \tag{13}$$

2.2 Determination of the material and dimensions of the longitudinal vibration transducer

2.2.1 Selection of transducer materials. 3Cr13 stainless steel has low acoustic impedance, high fatigue strength and low cost, and is easy to machine. Therefore, 3Cr13 steel is used for both the front and rear cover plates. PZT-8 emission-type material is selected for the piezoelectric ceramic due to its low dielectric loss and high mechanical quality factor. The material properties are shown in Table 1.

Table 1. Material properties of the transducer.

Material	Density $\rho / (kg/m^3)$	Sound Speed $c / (m/s)$
PZT-8	7600	3120
3Cr13	7750	5170

2.2.2 Modal analysis of the longitudinal vibration ultrasonic transducer. Based on the theoretical calculations, a three-dimensional model of the longitudinal vibration ultrasonic transducer was created in SolidWorks and imported into ANSYS Workbench for modal analysis. The piezoelectric ceramic material was set to the SOLID226 element, which is a coupled-field 20-node hexahedral element used for piezoelectric analysis. An M8 hex socket countersunk bolt was selected for the assembly, and the installation schematic is shown in Figure 2. The distance between the bottom of the bolt and the rear end surface of the rear cover plate is H=42 mm, and the depth of the bolt head is h_1. Keeping H constant and varying h_1, the effect of the bolt head depth on the longitudinal vibration frequency and displacement node of the transducer was analyzed using the Path function in ANSYS Workbench, defining the path from the center of the rear end surface of the rear cover plate to the center of the front end surface of the front cover plate. The search frequency range was set to 20~30 kHz, and the results are shown in Figures 3 and 4.

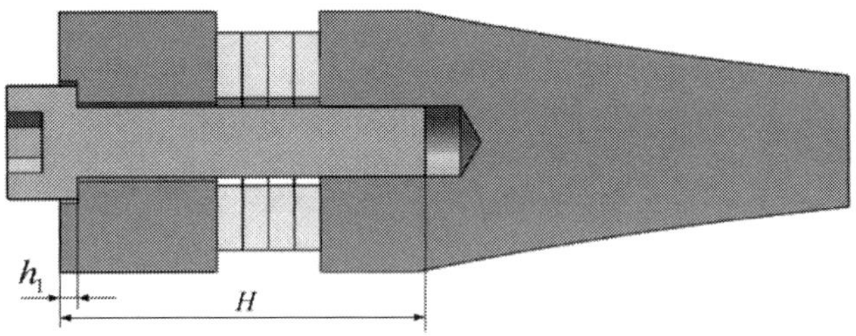

Figure 2. Installation schematic of the countersunk bolt.

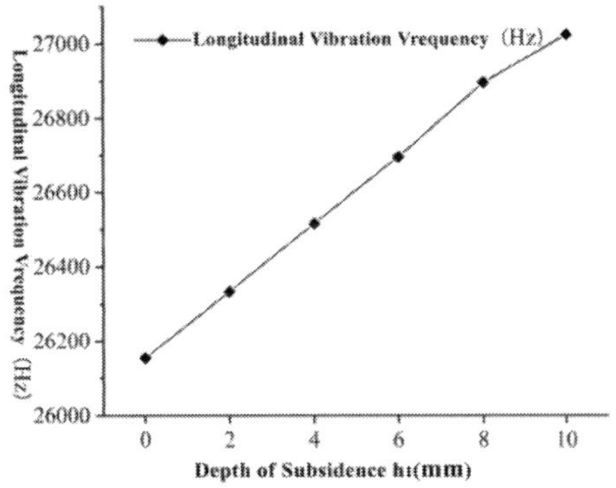

Figure 3. Effect of bolt head depth on transducer frequency.

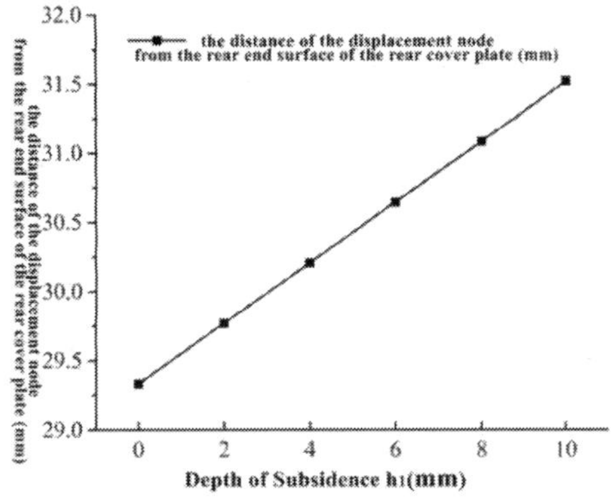

Figure 4. Effect of bolt head depth on displacement node.

From Figures 3 and 4, it can be seen that as the bolt head depth increases from 0 to 10 mm, both the longitudinal vibration frequency of the transducer and the distance of the displacement node from the rear end surface of the rear cover plate gradually increase. The displacement node is near the junction between the piezoelectric ceramic and the front cover plate, confirming the correctness of placing the flange at the displacement nodal surface in the theoretical analysis. With the bolt head depth set to 8 mm, the resonant frequency of the longitudinal vibration transducer is 26,869 Hz, and the vibration mode is shown in Figure 5. The simulation results show an error of 4.04% compared to the design frequency, which is less than 5%. The displacement is greatest at the small end of the front cover plate, and a flange is installed near the junction of the front cover plate to facilitate the installation of the transducer.

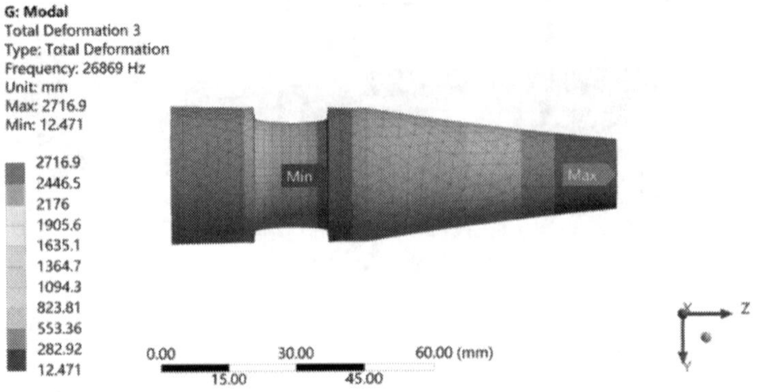

Figure 5. Modal analysis of the longitudinal vibration transducer.

3. Design of the longitudinal-torsional composite ultrasonic transducer

3.1 Principle of the longitudinal-torsional conversion with spiral grooves

According to the principle of acoustic wave propagation, when the longitudinal acoustic wave emitted by the piezoelectric ceramic reflects and superimposes on the spiral groove wall of the exponential front cover plate, the longitudinal vibration is transmitted to the spiral groove section and divided into axial force F_l and tangential force F_τ. Under the influence of the tangential force, part of the longitudinal vibration is converted into torsional vibration. After the two are synthesized, a longitudinal-torsional composite vibration is formed at the end of the transducer. The structural schematic diagram is shown in Figure 6.

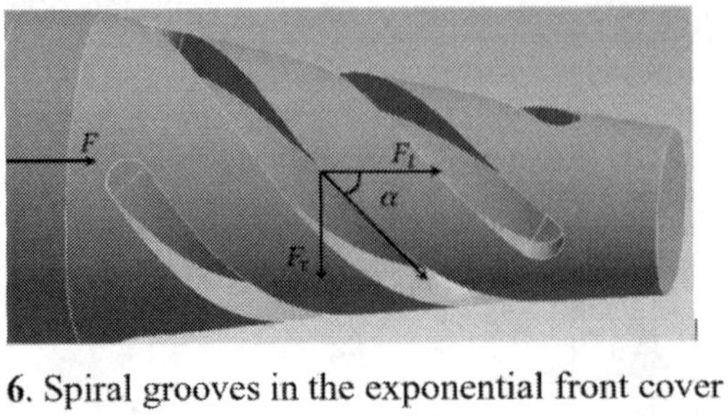

Figure 6. Spiral grooves in the exponential front cover plate.

Based on theoretical analysis, a modal analysis of the transducer with spiral grooves is conducted, applying displacement constraints on the flange. The results are shown in Figure 7, where longitudinal-torsional composite vibration is generated at the end of the front cover plate of the transducer with spiral grooves.

Figure 7. Modal vector diagram of the transducer with spiral grooves.

3.2 Determination of spiral groove parameters and assembly of the longitudinal-torsional ultrasonic tool holder

Based on the simulation analysis results, it is concluded that the larger the groove width and depth are, the larger the torsional-to-longitudinal ratio is. However, the longitudinal-torsional frequency decreases significantly, and when the groove width and depth are too large, the structural rigidity of the transducer becomes insufficient. The front cover plate of the transducer is prone to deformation during use. Considering the influence of the spiral groove structure on the torsional-to-longitudinal ratio, frequency, and structural rigidity of the transducer, the optimal parameters for the spiral groove are determined as a rotation angle of 45°, groove width and depth of 5 mm. At this point, the torsional-to-longitudinal ratio is 0.44, and the longitudinal-torsional frequency is 26,044 Hz. The transducer achieves a large torsional-to-longitudinal ratio, high longitudinal-torsional frequency, and sufficient structural rigidity. The minimum displacement is located near the flange, and the maximum displacement is at the output end of the front cover plate, meeting the usage requirements. The modal analysis is shown in Figure 8.

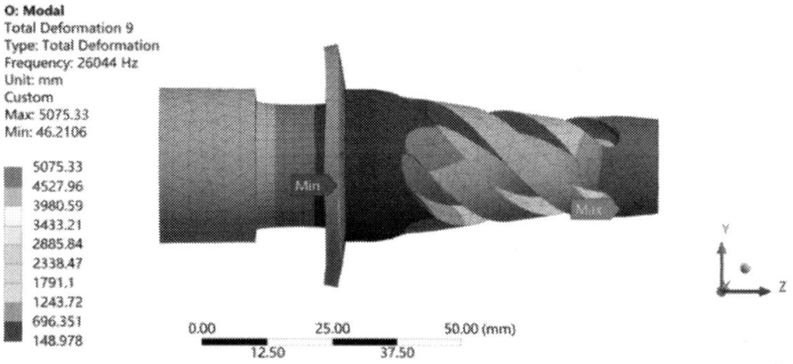

Figure 8. Modal analysis of the exponential longitudinal-torsional composite ultrasonic transducer.

In the milling process, ER collets and ER nuts are commonly used to connect the milling cutter to the horn for convenient clamping and replacement of the milling cutter. Considering the impact of ER collets, ER nuts, and milling cutters on the transducer, the ER collet, nut, and tool are simplified into a cylindrical horn of a certain length using the equal mass method, which is then connected to the original horn to form a new horn. Taking the installation of a $\phi 6*6*50*4F$ milling cutter as an example, the transducer and the BT40 tool holder are assembled according to actual working conditions to form a longitudinal-torsional ultrasonic milling tool holder. The 3D model is shown in Figure 9. The corresponding finite element model is generated and imported into Workbench for mesh generation and modal analysis, with the results shown in Figure 10. The resonance frequency is 22,845 Hz, and longitudinal-torsional composite vibration is generated near the tool tip, meeting the design requirements.

doi:10.1088/1742-6596/2879/1/012023

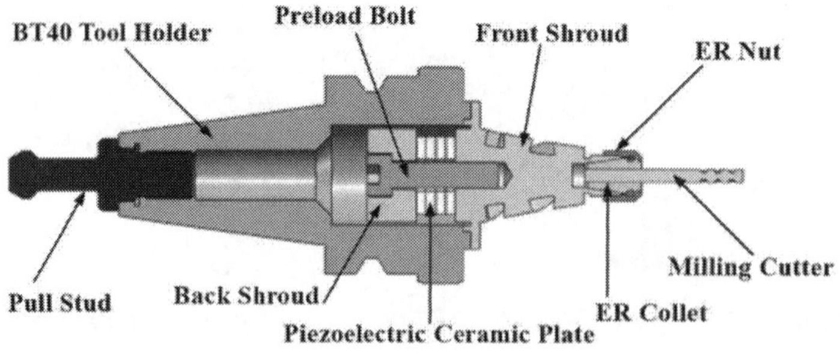

Figure 9. Assembly model of the longitudinal-torsional ultrasonic milling tool holder.

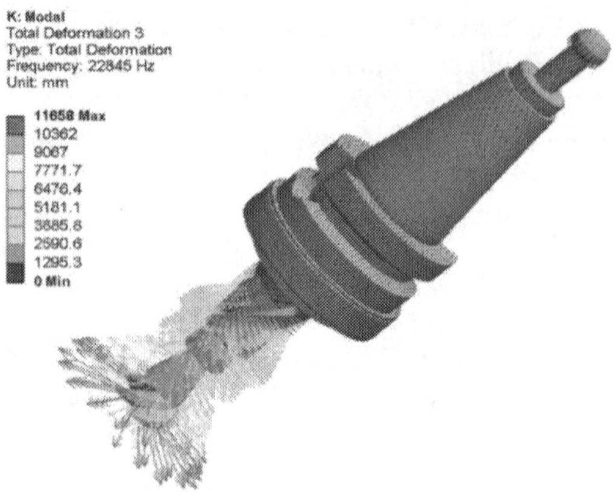

Figure 10. Velocity vector diagram of the longitudinal-torsional ultrasonic milling tool holder.

4. Characteristics testing of longitudinal-torsional ultrasonic tool holder

4.1 Impedance characteristics testing

To comprehensively evaluate the performance and stability of the longitudinal-torsional ultrasonic tool holder system, an impedance test must be conducted. As shown in Figure 11, the PV520A impedance analyzer, produced by Hangzhou Guobiao Ultrasonic Equipment Co., Ltd., was used for the test, and the impedance characteristic curve is shown in Figure 12.

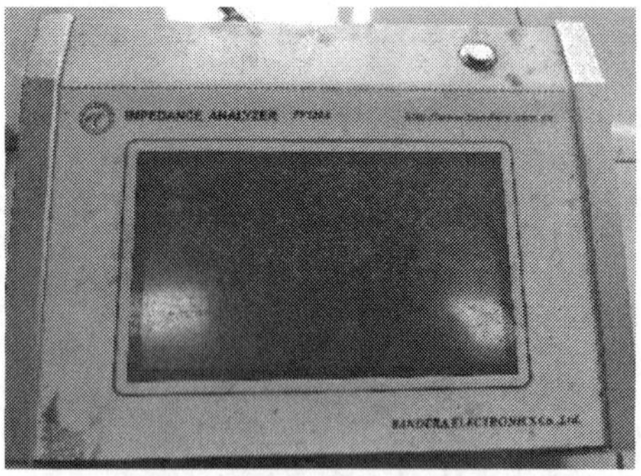

Figure 11. PV520A impedance analyzer.

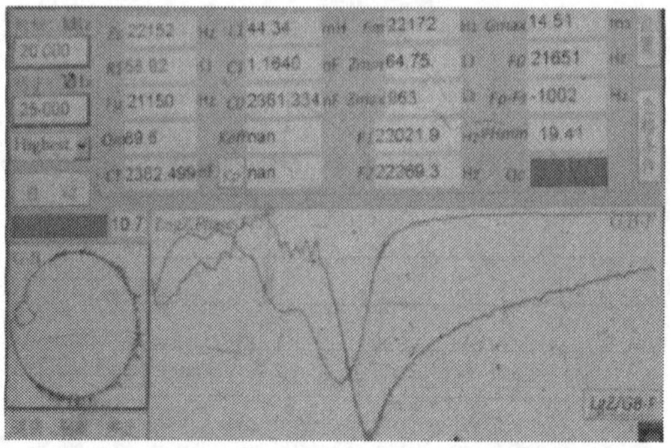

Figure 12. Impedance test results of longitudinal-torsional ultrasonic tool holder.

From Figure 12, it can be seen that the impedance circle of the longitudinal-torsional composite ultrasonic tool holder is nearly a perfect circle, and the phase curve is relatively smooth without noise.

The measured series resonant frequency of the ultrasonic tool holder is 22,152 Hz, compared with the simulation value of 22,845 Hz, with an error of 3.03%, which is within an acceptable range. The main reasons for the error are as follows: firstly, the influence of preloading on the resonant frequency of the tool holder was not considered in the simulation; secondly, the material properties defined in the simulation model were assumed to be uniform, and all contact surfaces were considered to have ideal contact, whereas in actual processing, the material composition and the relationships between contact surfaces are more complex. The equivalent resistance R1 of the longitudinal-torsional ultrasonic tool holder is 68.92 Ω, the dynamic capacitance C1 is 1.16 nF, the static capacitance C0 is 2361.3 nF, and the mechanical quality factor Qm is 89.6. The impedance test data indicate that the device performs well, with minimal mechanical loss, meeting the requirements of actual engineering applications.

4.2 Amplitude characteristics testing

Amplitude is an important parameter for evaluating the performance of the longitudinal-torsional ultrasonic tool holder. To ensure that the tool holder's output amplitude can meet the processing amplitude requirements, amplitude tests were conducted using a laser displacement sensor. In the amplitude test experiment, a 6 mm diameter special tool rod for testing was clamped at the front end of the transducer using a collet and tightening nut, and ultrasonic energy was provided to the transducer in the tool holder by an ultrasonic power supply. The laser displacement sensor was used to sample and analyze the output amplitude of the longitudinal-torsional ultrasonic tool holder. The experiment utilized the CLV 1000 laser displacement sensor produced by Polytec in Germany. When the laser displacement sensor is aligned axially with the tool rod, it measures the longitudinal vibration of the ultrasonic tool holder; when aligned radially, it measures the torsional amplitude of the ultrasonic tool holder. By adjusting the output power of the ultrasonic power supply by changing the power supply voltage, the longitudinal and torsional amplitudes of the tool holder at different excitation voltages were measured, as shown in Figures 13 and 14. The maximum longitudinal amplitude at the end of the transducer was 8.29 μm, the torsional amplitude was 5.38 μm, and the torsional-to-longitudinal ratio was 0.65; the maximum longitudinal amplitude at the end of the tool rod was 11.2 μm, the torsional amplitude was 4.3 μm, and the torsional-to-longitudinal ratio was 0.38. The frequency and voltage were stable, meeting the actual working requirements.

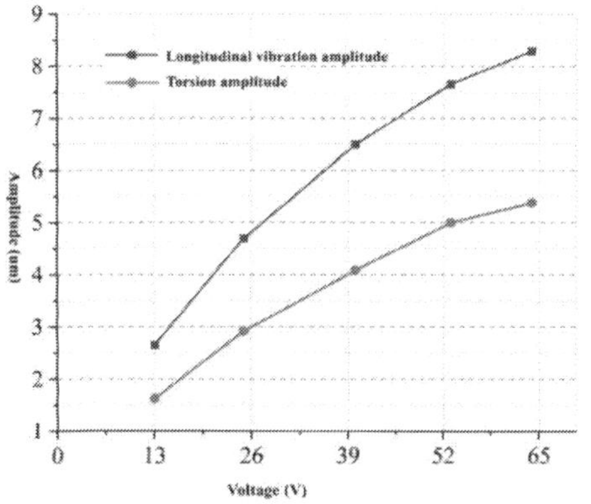

Figure 13. Voltage vs. longitudinal-torsional amplitude at the end of the transducer.

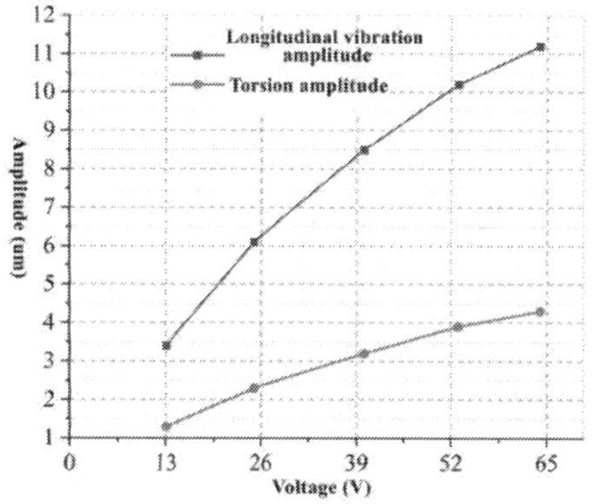

Figure 14. Voltage vs. longitudinal-torsional amplitude at the end of the tool rod.

From Figures 13 and 14, it can be observed that the longitudinal and torsional amplitudes at both the end of the transducer and the end of the tool rod in the longitudinal-torsional ultrasonic tool holder exhibit a clear positive proportional linear relationship with the excitation voltage. Based on this, it is inferred that in milling experiments, the output amplitude of the longitudinal-torsional ultrasonic tool holder can be precisely controlled by adjusting the excitation voltage of the ultrasonic power supply according to processing requirements. This adjustment mechanism is convenient and flexible, making the processing more efficient.

5. Conclusion

(1) A longitudinal vibrating transducer was designed based on the analytical method, and its longitudinal vibration frequency was obtained through finite element analysis. Then, according to the principle of wave transmission and reflection, a helical groove was opened on the front cover plate of the longitudinal vibrating transducer to generate longitudinal-torsional composite vibration.

(2) Using the single-factor method, the influence of helical groove parameters on the torsional-to-longitudinal amplitude ratio γ and resonant frequency f of the longitudinal-torsional transducer was explored in finite element software. Considering the combined effect of the helical groove structure on the torsional-to-longitudinal ratio, frequency, and structural stiffness of the transducer, the optimal parameters for the helical groove were determined.

(3) According to actual working conditions, an assembly model of the longitudinal-torsional ultrasonic transducer, BT40 tool holder, and milling cutter was constructed, and finite element analysis was performed. The results showed that this model could produce longitudinal-torsional composite vibration.

(4) The designed longitudinal-torsional ultrasonic milling tool holder system was manufactured, and the characteristics of the longitudinal-torsional ultrasonic tool holder were verified and evaluated through impedance and amplitude tests. The results showed that the longitudinal-torsional ultrasonic tool holder meets the requirements of actual machining applications.

References

[1] Fang Shanxiang, Zhao Huiling, Zhang Qinjian. The application status and development trends of ultrasonic machining technology [J]. Journal of Mechanical Engineering, 2017, 53(19): 22-32.
[2] Zhang Mingjun, Chen Shuo, Wang Xinbo et al. Study on Surface Quality of SiCp/Al Composites by Longitudinal and Torsional Ultrasonic Assisted Milling [J]. Journal of Henan Polytechnic University (Natural Science Edition)., 2024, 43(05): 45-53.
[3] Han Fengqi, Zhang Deyuan. Experiment of ultrasonic longitudinal torsional and low frequency torsional compound vibration tapping, Journal of Beijing University of Aeronautics and Astronautics. 2024, 50 (04):24-32.
[4] Du P, Han L, Qiu X, et al. Development of a high-precision piezoelectric ultrasonic milling tool using longitudinal-bending hybrid transducer [J]. International Journal of Mechanical Sciences, 2022, 222: 107239.
[5] Zheng Kan, Liao Wenhe, Sun Lianjun, et al. Comparative study on stability of robotic longitudinal vibration and longitudinal-torsional ultrasonic milling [J]. Journal of Mechanical Engineering, 2021, 57(07): 10-17.
[6] Xu S, Shimada K, Mizutani M, et al. Development of a novel 2D rotary ultrasonic texturing technique for fabricating tailored structures [J]. The International Journal of Advanced Manufacturing Technology, 2022, 79: 1161-1172.
[7] Harkness P, Lucas M, Cardoni A. Coupling and degenerating modes in longitudinal–torsional step horns [J]. Ultrasonics, 2012, 52(8): 980-988.
[8] Pi Jun. Longitudinal-torsional vibration converter of cylinder with multiple diagonal slits [J]. Journal of Mechanical Engineering, 2008, (05): 242-248.

Numerical analysis of oil liquid sloshing in fuel tanks of plug-in hybrid electric vehicles under different structure baffles

Enhui Zhang[1], Hao Chen[1*], Lihe Wang[2], Jianbin Huang[3] and Fei Xu[4]

[1]School of Mechanical Engineering, Inner Mongolia University of Science and Technology, Baotou 014000, Inner Mongolia, China
[2]College of Vocational Technology, Inner Mongolia Agricultural University, Baotou 014000, Inner Mongolia, China
[3]Zhejiang Land Rover Automobile Co., Taizhou 318000, Zhejiang, China
[4]Hebei Shichang Automotive Parts Co., Langfang 065000, Hebei, China

[*]Corresponding author's email: chjxgc@163.com

Abstract: To solve a large amount of oil liquid sloshing in the fuel tank and a series of problems caused by oil liquid sloshing in the plug-in hybrid electric vehicle (PHEV) under the condition of electric drive only, this paper analyzes the oil liquid sloshing in fuel tank under different liquid filling rates under complex situations. After determining the representative liquid filling rates, the optimal scheme is finally determined. We conduct the specific analysis of different types, structures, and installation modes of fuel tank baffles under the same working conditions and liquid filling amount. The results show that the less liquid is filled in the fuel tank, the larger the sloshing amplitude is, and the more complicated the sloshing situation is. When the oil liquid sloshes, the non-porous baffle has an obvious zoning effect on the oil liquid and suppresses the oil liquid fluctuations effect, but the oil liquid fluidity is poor. The angle between the linear baffle and the oil liquid flow direction is larger, which has a better attenuation effect on the impact force of oil liquid. In conditions of significant oil liquid sloshing, a high baffle enhances the contact area of the oil liquid, resulting in improved inhibition effects. After considering various advantages and disadvantages, the analysis of the cruciform baffle found that the cruciform baffle limited the multi-directional movement of oil liquid, caused substantial vortex inside the oil liquid, and attenuated the kinetic energy generated by oil liquid sloshing. This greatly reduced the impact of the oil liquid on the fuel tank wall and improved the effect of suppressing oil liquid fluctuations.

1. Introduction

As human awareness of environmental protection and the energy crisis continues to grow, governments and original equipment manufacturers worldwide are imposing increasingly stringent requirements on the environmental performance of automotive components. Plug-in hybrid electric vehicles (PHEV) have garnered significant attention due to their combined advantages of traditional and electric vehicles, positioning them as a key direction in automotive development. However, under the electric driving condition only, the noise is caused by oil liquid sloshing in the fuel tank, the installation of internal parts is not reliable, and vapor volatilization increases, which greatly reduces the performance index of the fuel system. Therefore, it is very necessary to suppress the oil liquid sloshing in the fuel tank.

The research on oil liquid sloshing in fuel tanks has been paid close attention by scholars at home and abroad, and some achievements have been made. In terms of numerical simulation research, in 2010, Luppes et al. [1] used COM FLOW software to simulate the gas-liquid two-phase fluid in the fuel tank and adopted the "uniform gravity" density averaging method, which effectively avoided capturing the free surface acceleration of the liquid and improved the accuracy of the simulation. In 2012, Gao and Shang [2] studied the liquid slosh of tank cars under braking conditions through a fluent simulation system, which provided empirical and theoretical support for the optimization of the liquid slosh suppression scheme. In 2014, Dhaval et al. [3] discussed two CAE methods coupled with a multiphase Euler transient CAE method based on VOF and Lagrange-Euler formula in the study of oil slosh problems in vehicle fuel tanks. The calculation results by the solver show that the VOF method is in good agreement with the experimental test when simulating the oil slosh in the fuel tank. In 2018, Wang et al. [4], established the finite element analysis model of liquid sloshed in a liquid tank car by Fluent, and studied the influence of diameter, shape, and opening position of the hole in the wave plate on liquid sloshed. The results show that the smaller the hole area of the anti-wave baffle is, the better the anti-sloshing effect of the liquid in the tank is. In 2019, Zhang et al. [5-6] only considered the VOF calculation model containing the liquid control unit, and simulated the oil shaking situation of the fuel tank of a certain helicopter under a typical flight attitude, to capture the free liquid surface more accurately. Verma et al. [7] presented a comprehensive design and analysis of a 300L steel fuel tank suitable for heavy-duty off-road vehicles. The performance and durability of the fuel tank under various working conditions were optimized by numerical simulation.

In terms of experimental research, in 2014, Moretti et al. [8] used a self-made transparent rectangular water tank to slide on an inclined track to study liquid slosh, which was compared with numerical simulation and proved the feasibility of the experiment. In 2014, Brar and Singh [9] took an oval tank as an example to establish the liquid slosh model of the model and studied the installation position and shape of the anti-slosh plate. The comprehensive use of the anti-slosh plate can effectively reduce the maximum pressure of the fluid. In 2017, Li et al. [10] used horizontal and vertical damping grilles to optimize the inner cavity structure of the medicine cabinet. The results show that the damping grating can reduce the change of liquid center of gravity and the impact force on the box. In 2022, Zhong et al. [11] studied the test of horizontal baffles in a rectangular water tank. The results show that when the baffle is located at the deep liquid position, the anti-rocking effect is weak and the liquid sloshing is nonlinear. When the baffle is close to the free liquid surface, the anti-sway effect is obvious, the liquid wave height and amplitude are reduced, and the movement is linear. In 2024, Yi et al. [12] took the fuel tank of gasoline vehicles as the research object, introduced the fuel tank assembly from the aspects of structure and function, analyzed the design points of the fuel tank assembly, and provided references and suggestions for the design of the fuel tank assembly.

To sum up, the researchers used numerical simulation and bench test methods to conduct in-depth research on the problem of fuel tank fluid sloshing and improved simulation accuracy through continuous research on the process of oil sloshing in the fuel tank. When studying the design of anti-rocking devices in fuel tanks, researchers have reduced the amplitude of oil sloshing through the application of wave boards, but scholars have focused on simple wave boards, while few scholars have studied the anti-rocking effect of more complex composite wave boards. On this basis, to reduce a large amount of oil slosh to the maximum extent and solve problems such as noise caused by oil slosh, unreliable installation of built-in parts, and increased oil and gas volatilizing, this paper studies the structure of the baffle under different influencing factors through the combination of numerical simulation and bench test and finally determines the optimal scheme to improve the performance index of the fuel system.

2. Basic theoretical equation of oil liquid sloshing

When the vehicle is in motion, the oil surface experiences fluctuations. The theoretical framework for describing this behavior includes the Navier-Stokes equation and the continuity equation from fluid mechanics, along with boundary conditions that characterize the behavior of the liquid's free surface.

For slight sloshing of the oil, the fluid is considered continuous and compressible. The governing equation for the oil in the fuel tank is:

$$\frac{\partial \rho}{\partial t}+\frac{\partial}{\partial x_i}(\rho u_i)=0 \tag{1}$$

Mass conservation equation:

$$\frac{\partial u}{\partial t}+\frac{1}{2}\nabla u^2=-\frac{1}{\rho}\nabla p-\nabla(gZ') \tag{2}$$

Momentum conservation equation:

$$\frac{\partial}{\partial t_i}(\rho u_i)+\frac{\partial}{\partial x_j}(\rho u_i u_j)=-\frac{\partial p}{\partial x_i}+S_i+\left(\mu\frac{\partial u_i}{\partial x_j}+\rho\bar{u}'_i\bar{u}'_j\right) \tag{3}$$

where u_i, x_i, and S_i are the generalized source terms of fluid velocity, displacement, and momentum conservation equations in the direction i, respectively; The values of i and j are (1, 2, 3); p is fluid pressure; ρ is the fluid density; g is the gravitational acceleration; gZ' is the potential function of gravity field; ∇ is the gradient operator; μ is the kinematic viscosity coefficient of fluid; $\bar{u}'_i$ and $\bar{u}'_j$ is the Rello stress.

The Navier-Stokes equation is the basic equation describing the motion of a fluid, taking into account the viscous effects of the fluid, in the form of:

$$\rho\left(\frac{\partial u}{\partial t}+(u\cdot\nabla)u\right)=-\nabla p+\mu\nabla^2 u+\rho g \tag{4}$$

where ρ is the fluid density; u is the fluid velocity vector; p is the fluid pressure; μ is the fluid viscosity; g is the gravitational acceleration.

Vorticity ω is a measure of rotational motion in a fluid, defined as the curl of a velocity field:

$$\omega=\nabla u \tag{5}$$

The vorticity equation can be obtained by calculating the curl of the N-S equation. Curl on both sides:

$$\nabla\times\left(\left(\rho\frac{\partial u}{\partial t}+(u\cdot\nabla)u\right)\right)=\nabla(-\nabla p+\mu\nabla^2 u+\rho g) \tag{6}$$

Due to $\nabla\times(\nabla p)=0$ and $\nabla\times(\rho g)=0$, ρ is a constant, which can be simplified to:

$$\frac{\partial w}{\partial t}+\nabla\times(u\cdot\nabla u)=\mu\nabla^2 w \tag{7}$$

Using the identity in vector analysis:

$$\nabla\times(u\cdot\nabla)=(u\cdot\nabla)w-(w\cdot\nabla)u \tag{8}$$

The vorticity equation can be expressed as:

$$\frac{\partial w}{\partial t}+(u\cdot\nabla)w=(w\cdot\nabla)u+\mu\nabla^2 w \tag{9}$$

The kinetic energy density of a fluid (kinetic energy per unit volume) can be expressed as:

$$e=\frac{1}{2}\rho u^2 \tag{10}$$

Kinetic energy consumption is mainly caused by viscous dissipation. The viscous dissipation rate ε represents the rate at which kinetic energy per unit volume is consumed over time and can be expressed by the following formula:

$$\varepsilon=\mu\left(\frac{\partial u_i}{\partial x_j}+\frac{\partial u_j}{\partial x_i}\right)^2 \tag{11}$$

The presence of vortices increases the velocity gradient and thus the viscous dissipation rate. The specific kinetic energy consumption can be expressed as a function of the velocity gradient:

$$\varepsilon=k\cdot N_v\cdot\mu\left(\frac{\partial u_i}{\partial x_j}+\frac{\partial u_j}{\partial x_i}\right)^2 \tag{12}$$

The rate of change of the fluid kinetic energy with time can be obtained by integrating the viscous dissipation rate:

$$\frac{dE}{dt}=-\int_V\varepsilon dV \tag{13}$$

$$\frac{dE}{dt}=-\int_V(k\cdot N_v\cdot\mu\left(\frac{\partial u_i}{\partial x_j}+\frac{\partial u_j}{\partial x_i}\right)^2)dV \tag{14}$$

where u_i and u_j is the velocity component; μ is the fluid viscosity; N_v is the number of vortices; k is an empirical constant.

3. Research on numerical analysis and experimental method of oil liquid sloshing

3.1 Creation of a three-dimensional fuel tank model

The model creation process follows the principles of computation economy and result accuracy. The local structure that will not affect the calculation result is simplified to calculate the economy. The fuel tank wall is set as a rigid structure, the fluid inside the fuel tank is set as an incompressible fluid, and the gas is ideal. The size of the fuel tank is 666.67 mm×300 mm×200 mm, and the physical model is shown in Figure 1 below.

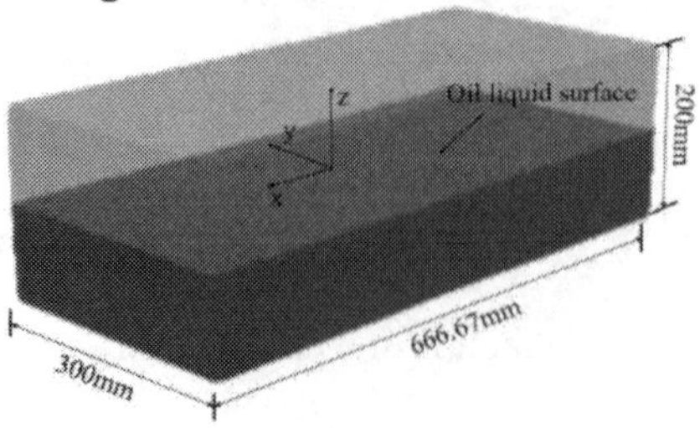

Figure 1. 3D fuel tank model.

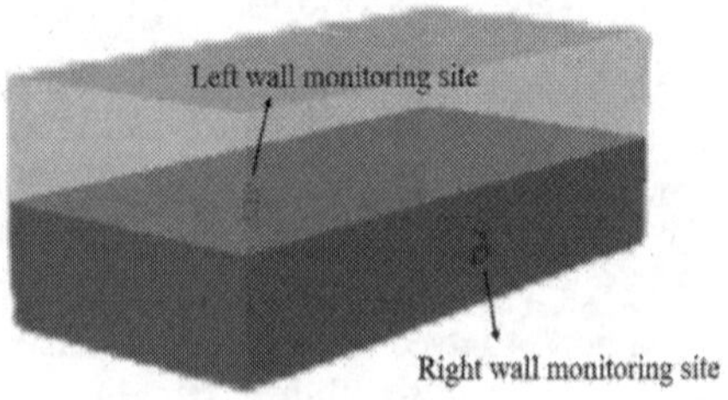

Figure 2. Monitoring points on the left and the right walls of the fuel tank.

3.2 Grid independence test

The convergence and uniformity of the simulation calculation results are the key to the correctness of the results, so it is necessary to verify whether the selection of calculation parameters is reasonable. This verification focuses on two primary factors affecting the simulation results: grid size and time step. The fuel tank is filled to 50%, with the pressure monitoring point located at the center of the liquid level on the side wall, as shown in Figure 2. The simulation conditions are as follows: an initial acceleration of 0.25 g for 1.5 seconds, followed by deceleration at 0.6 g until stopping, and finally allowing the oil to slosh freely for 6 seconds.

3.2.1 Grid size verification. This study evaluates the fuel tank using three different grid sizes: 5 mm, 10 mm, and 15 mm. The pressure curve along the left wall of the fuel tank is simulated, as illustrated in Figure 3. The trend in Figure 3 indicates that the oil slosh pressure history curves for the 5 mm and 10 mm grid models are nearly identical. Considering both economic efficiency and simulation accuracy, this paper selects a grid size of 10 mm.

3.2.2 Time step verification. This article selects three times steps of 0.0005 s, 0.001 s, and 0.005 s, and the pressure curve along the left wall of the fuel tank was simulated in Figure 4. The trend in Figure 4 indicates that the oil slosh pressure history curves for the 5 mm and 10 mm grid models are nearly identical. Considering both economic efficiency and simulation accuracy, this paper selects a grid size of 10 mm. From the perspective of economic efficiency and accuracy in simulation calculations, this paper adopts a time step of 0.001 s.

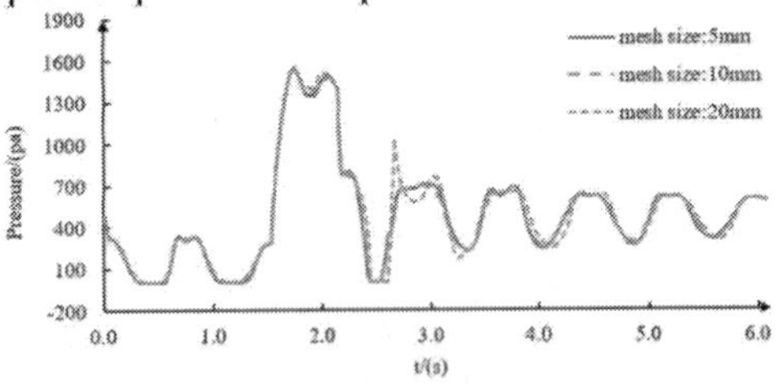

Figure 3. Pressure history curve of liquid level center point on the left wall of the fuel tank with different grid sizes.

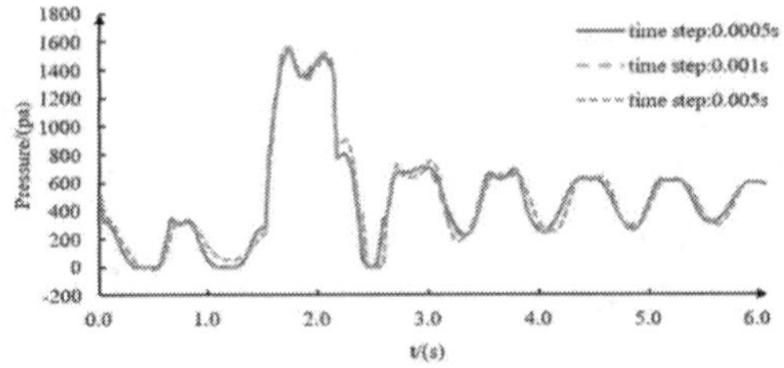

Figure 4. Pressure history curve of the center point of liquid level on the left wall of the fuel tank with different time steps.

3.3 Establishment of experimental bench

To simulate the complex driving conditions of the car and better compare and analyze with the numerical simulation, an experimental platform was built. The experiment is carried out by dragging the horizontally placed fuel tank model along the inclined plane, which is composed of a power system, motion system, braking system, and signal acquisition system. The power system and the braking system drive the movement platform in the movement system to do horizontal acceleration and deceleration, in which the fuel tank is fixed on the movement platform, and finally, the signal acquisition device and computer in the acquisition system complete the collection and processing of pressure signals. The test platform is shown in Figure 5. Figure 6 introduces the specific location of the sensor installation on the fuel tank model. Figure 7 shows the overall diagram of the braking system and the components of the braking system.

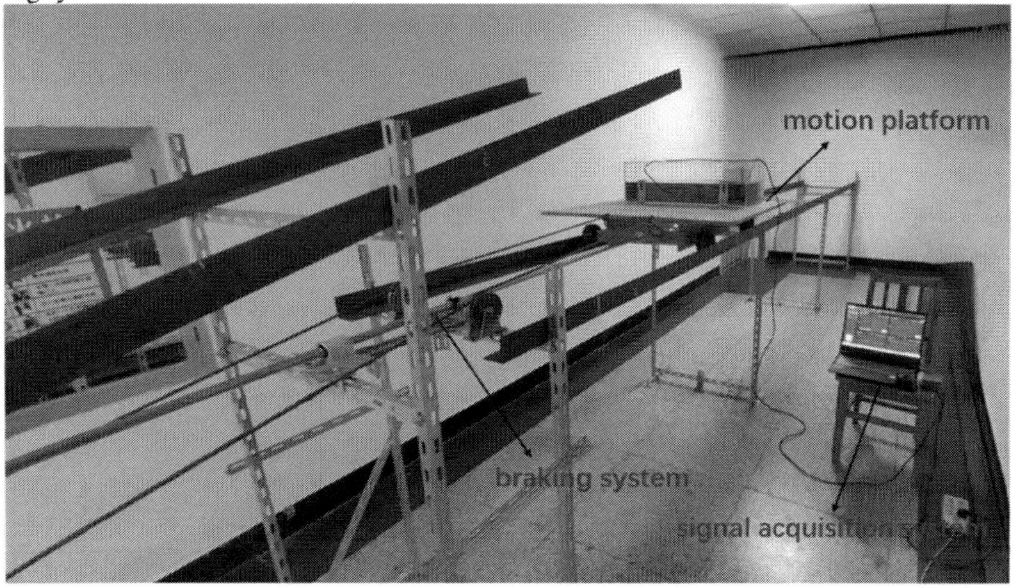

Figure 5. The main structure of the test bench.

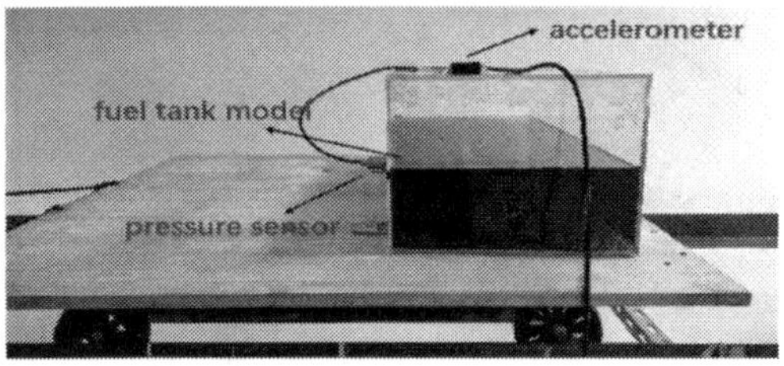

Figure 6. Sensor mounting position.

Figure 7. Detailed drawing of the brake system.

3.4 Experimental procedure

(1) The acceleration of the car in the actual situation is controlled by adjusting the angle of the inclined plane and utilizing the car's weight within the power system. After accelerating to a certain speed, the brake device begins to act, and the movement platform will smoothly slow down to stop the movement.

(2) Considering the test safety, the oil in the tank was replaced with tap water at room temperature. The density: ρ=1.0×103 kg/m^3, and the viscosity: υ=1.005×103 Ns/m^3. The addition of light blue dye in the water can increase the clarity of shooting.

(3) The acceleration sensor is installed on the motion platform of the motion system to verify that the acceleration of the car during operation is under the specified acceleration. The pressure sensor is

positioned at various heights along the midpoint of the fuel tank's side wall, depending on the liquid level, to monitor the pressure at specific points during the oil sloshing process.

(4) The 485 sensor is connected to the computer through the converter, and after Python programming, the signal is picked up, the data is saved to the document, and then the saved data is drawn into a graph to show the overall change process of the signal.

3.5 Comparative analysis of numerical simulation and experimental results

The test conditions are configured to match those of the numerical simulation to ensure accuracy. Figure 8 displays the liquid surface contours from both the numerical simulation and the experiment at three different time points, illustrating the propagation of traveling waves and the climbing of the liquid along the wall during the sloshing process. As shown in Figure 8, the liquid surface profiles from the numerical simulation and the experiment correspond closely at each time point. Figure 9 presents the pressure history curves from both the numerical simulation and the experiment at the same monitoring point. The overall trends of the two sets of curves are quite similar. However, due to experimental errors and sensor accuracy limitations, some peaks and troughs in the experimental data do not align perfectly with the numerical simulation. Despite these discrepancies, the variation in the liquid surface profile and the fluctuations in the pressure history curves confirm the accuracy of the numerical simulation. To save costs and improve research speed, numerical simulation can be used for more complex research.

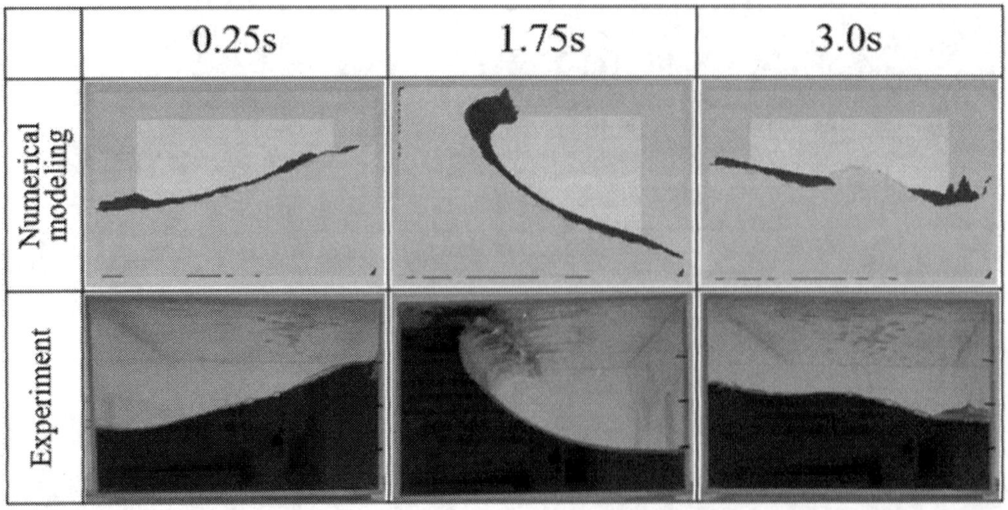

Figure 8. Contrast of liquid surface profile.

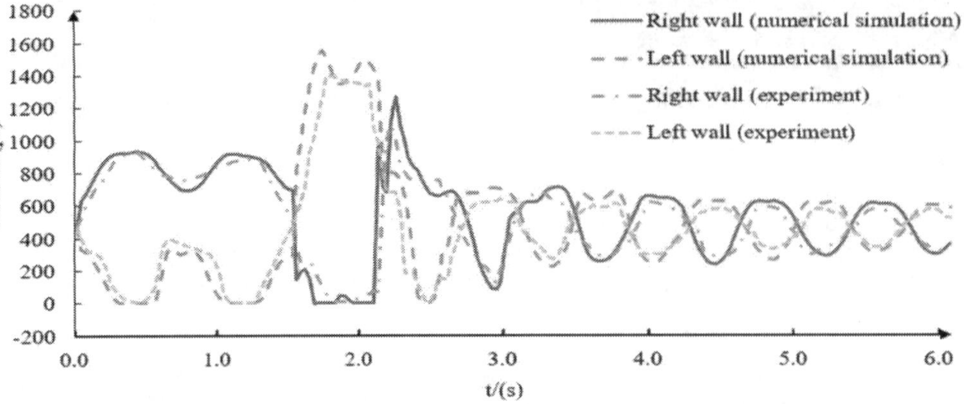

Figure 9. Comparison of pressure history curves.

4. Numerical analysis of oil liquid sloshing in fuel tank under the action of different structure baffles

In this paper, the working condition is set as follows: first, it moves at 0.25 g acceleration for 1.5 s, then slows down to stop at 0.6 g deceleration speed, and finally, we let the oil slosh freely for 6 s.

4.1 Numerical analysis of oil liquid sloshing with different filling quantities

It can be seen from Figure 10, Figure 11, and Figure 12 that the pressure history curve trend of the center point of the liquid level on the left and right walls of the fuel tank with different liquid filling amounts is essentially the same. Under the acceleration condition of the first 1.5 s, the oil liquid accumulates on one side, causing an increase in pressure at the center point of the liquid level on the right wall of the fuel tank. During deceleration, the pressure at the center point of the liquid level on the left wall of the fuel tank surges. After deceleration ends, under the action of oil gravity and inertia force, the oil began to gradually stabilize, making the pressure history curve show a waveform similar to simple harmonic motion, and the double peak phenomenon was more obvious when the liquid content was 25%, and the waveform attenuation period was the longest. Figure 13 illustrates the dynamic pressure history curve of the oil center for different filling amounts after processing the central point pressure values of oil liquid. The figure shows that the dynamic pressure changes more significantly with a filling amount of 25% and is at a minimum with a filling amount of 75%. In general, the less the amount of liquid filled, the greater the change of the pressure history curve waveform is, and the more complicated the oil liquid sloshing situation is.

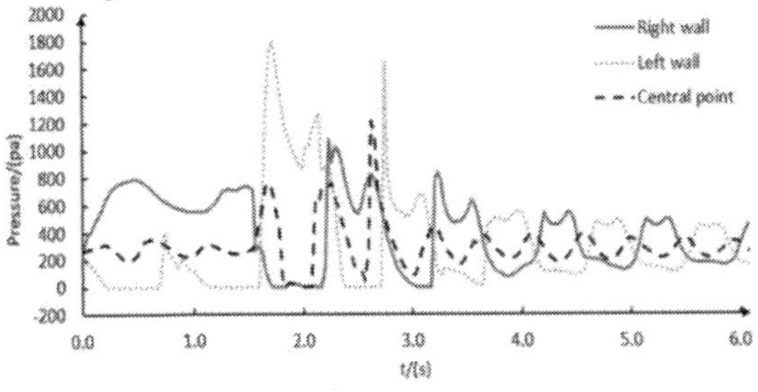

Figure 10. 25% liquid content pressure history curve in the fuel tank.

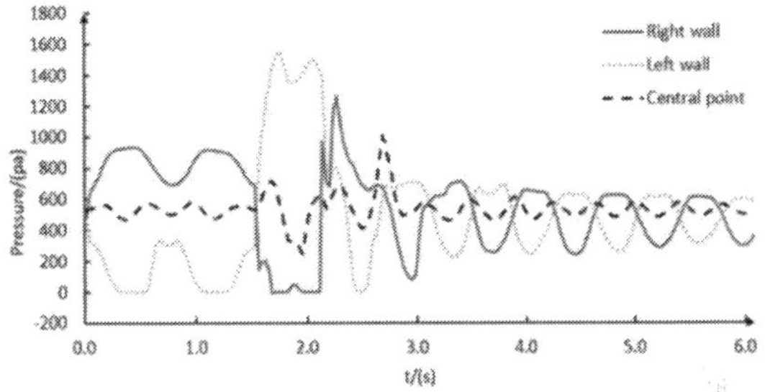

Figure 11. 50% liquid content pressure history curve in the fuel tank.

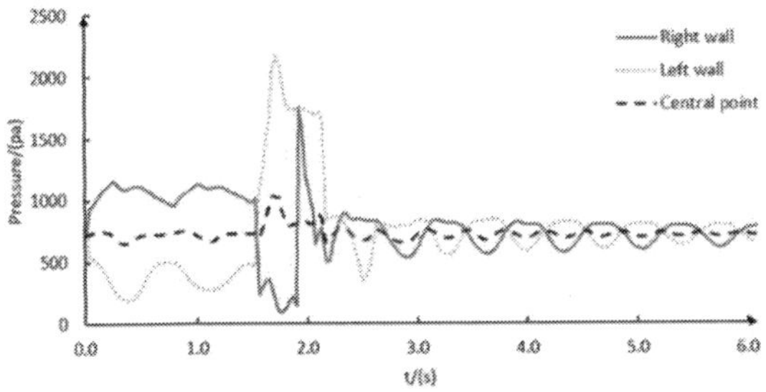

Figure 12. 75% liquid content pressure history curve in fuel tan.

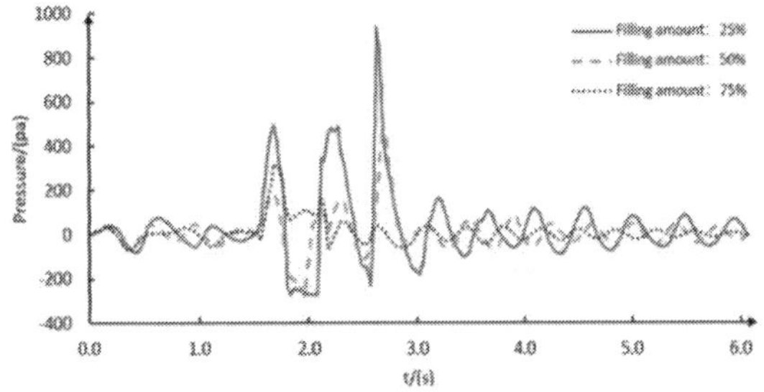

Figure 13. Oil central point pressure history curve with different filling amounts.

4.2 Numerical analysis of oil liquid sloshing under different structure baffles

To minimize the impact of oil liquid on the fuel tank, baffles are typically installed to separate the oil and enhance the tank's structural strength while reducing the oil sloshing amplitude. Based on the actual conditions and the behavior of oil sloshing at different liquid levels, all subsequent studies are conducted with a 50% liquid filling amount.

4.2.1 Different baffle types. Baffles can be divided into two types: non-porous baffles and porous baffles, as shown in Figure 14 and Figure 15.

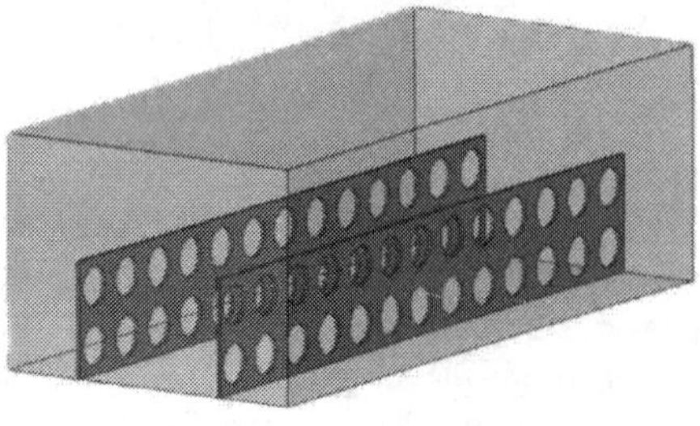

Figure 14. Non-porous baffle fuel tank.

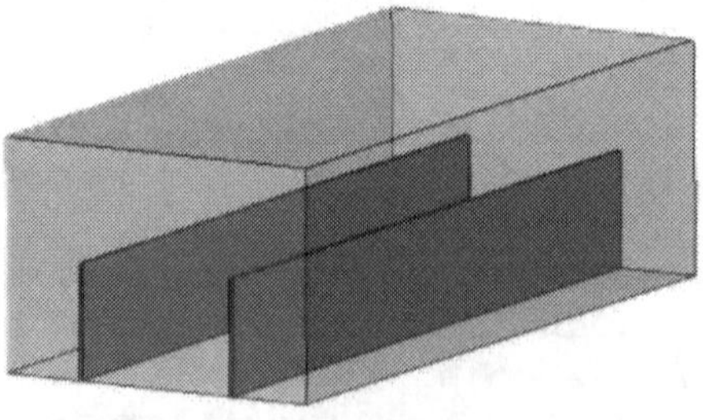

Figure 15. Porous baffle fuel tank.

Figure 16 shows that the pressure values at the center points of the liquid level on the left and right walls of the fuel tank are lower with both perforated and non-perforated baffles compared to the tank without baffles. Additionally, the peak pressure values are reduced and delayed in tanks with baffles. The peak pressure values and fluctuation range are smallest with either type of baffle, indicating that baffles effectively reduce the oil sloshing amplitude and impact pressure on the walls. Among the two, the non-perforated baffle demonstrates a better performance than the perforated one.

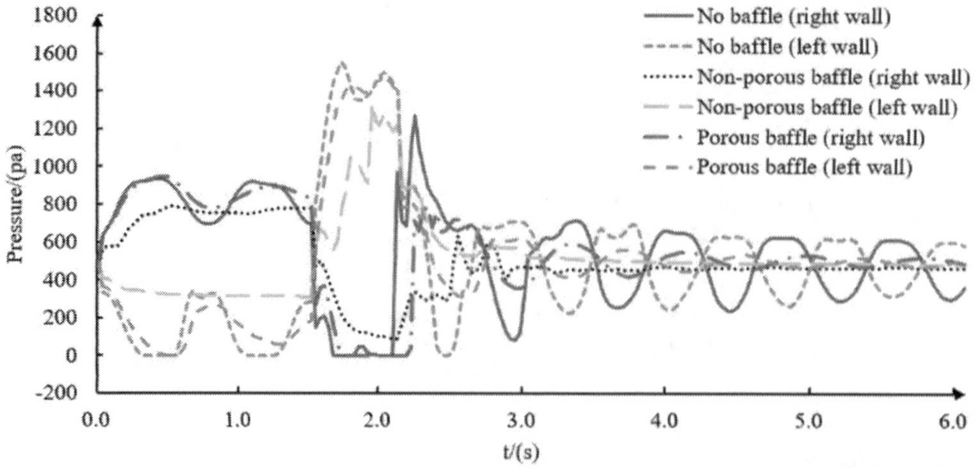

Figure 16. Pressure history curve of the center point of the liquid level on the left and right walls of the fuel tank under the action of different types of baffles with 50% liquid filling.

Figure 17 shows that the addition of baffles reduces the liquid level climbing height caused by inertial forces and clearly defines the oil liquid zone, leading to quicker stabilization. Among the different types of baffles, the non-porous baffle results in the lowest liquid surface climbing height at the same time. The pressure cloud diagrams at three moments indicate that the non-porous baffle achieves the lowest oil liquid pressure compared to the perforated baffle. For instance, at 1.75 s, the oil liquid pressure drops from 1, 986.53 Pa with no baffle to 1, 318.41 Pa with the non-porous baffle, demonstrating the most effective suppression of liquid fluctuations.

Figure 18 shows that in an unbaffled fuel tank, the oil liquid flow is regular and the flow area is relatively uniform. With a perforated baffle, the fuel's movement path is altered, resulting in a multi-directional flow that avoids directional impacts. In tanks with or without perforated baffles, the oil-liquid flow becomes more complex, with a vortex forming between the two baffles at 3.0 s. Overall, the non-porous baffle most significantly alters the direction of oil liquid flow but reduces the oil liquid's fluidity.

4.2.2 Different installation positions. The oil liquid sloshing condition under the action of vertical installation and inclined installation of the baffle is analyzed in Figure 19 and Figure 20.

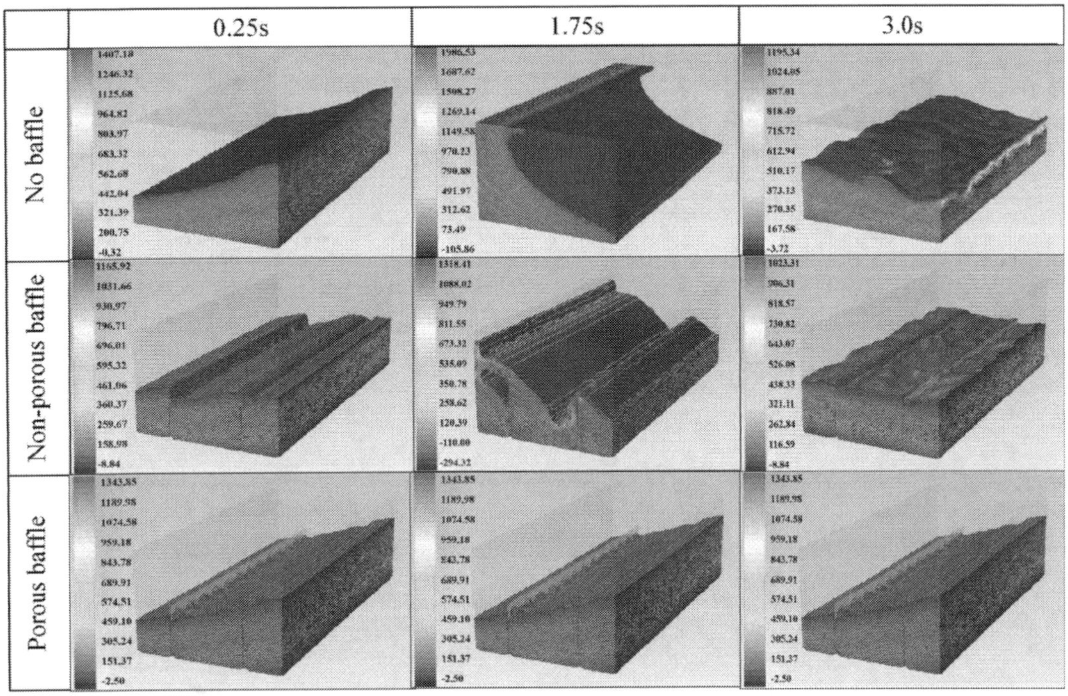

Figure 17. Oil slosh pressure diagram under different types of baffles under the same working.

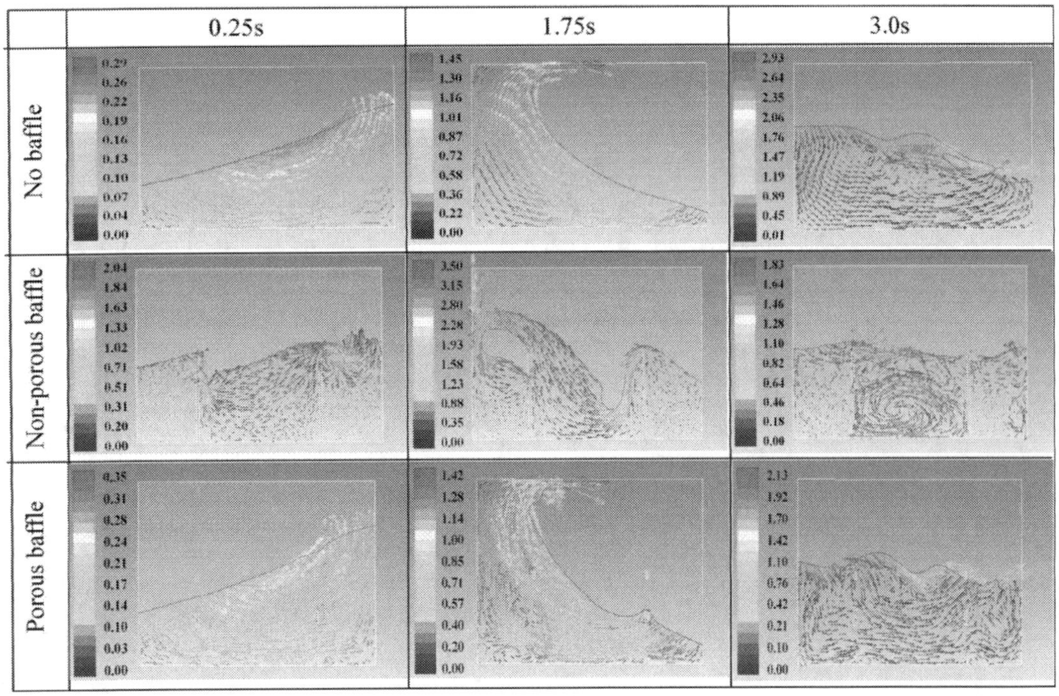

Figure 18. Oil slosh vector diagram under different types of baffles under the same working conditions.

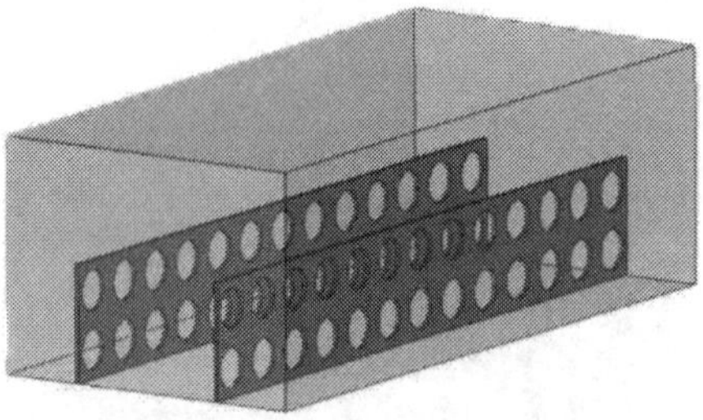

Figure 19. Linear baffle fuel tank.

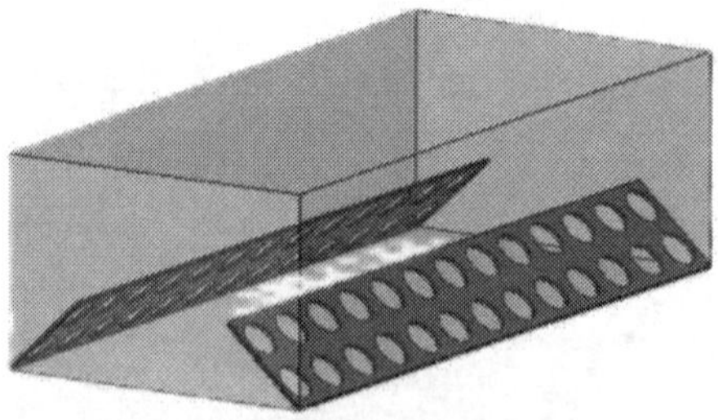

Figure 20. Sloping baffle fuel tank.

It can be seen from Figure 21 that the pressure history curve amplitude of the liquid level center point on the left and right walls of the fuel tank decreases under both installation modes. Under acceleration, the peaks and troughs of the pressure history curve lag. The pressure history curve of the center point of the liquid level on the left and right walls of the fuel tank installed with a linear baffle is more stable and faster, and the attenuation of oil slosh is better.

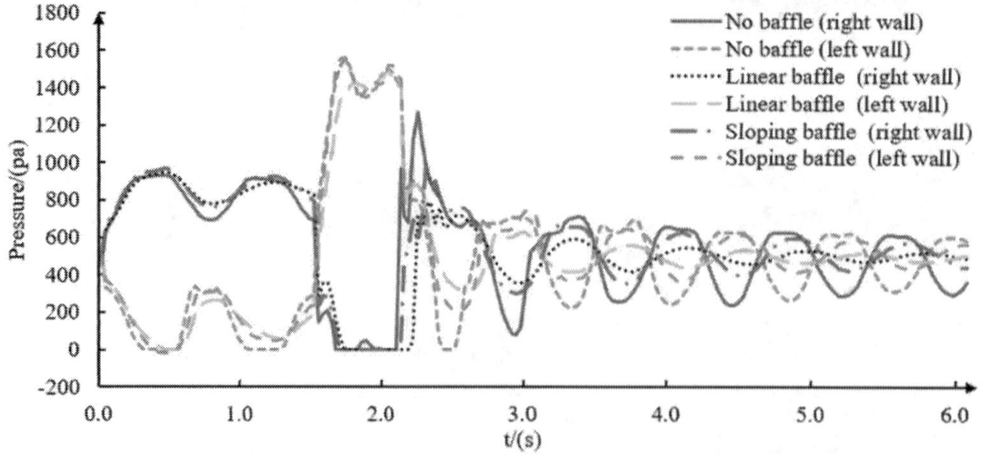

Figure 21. Pressure history curve of the center point of liquid level on the left and right walls of the fuel tank under the action of baffles at different positions with 50% liquid filling.

Figure 22 illustrates that baffles installed at different positions all have an impact on oil liquid slosh. At 1.75 s, the partitioning effect is more pronounced with the linear baffle. By 3.0 s, the oil liquid spattering effect is more evident with the inclined baffle. Comparing the oil liquid pressure values over time, the straight baffle results in the lowest oil liquid pressure, while the inclined baffle results in the second-lowest pressure.

In Figure 23, the oil liquid flow direction inside the tank is altered when equipped with either a linear baffle or a sloping baffle, compared to a tank without a baffle. At 1.75 s, the linear baffle causes the formation of two small eddies within the oil liquid. By 3.0 s, the internal oil liquid flow under the linear baffle becomes more complex.

4.2.3 Different baffles height. This section analyzes flapper fuel tanks of different heights. Figure 24 and Figure 25 show the models of flapper fuel tanks of different heights.

Figure 26 displays the pressure history curves at the center points of the liquid level on the left and right walls of the fuel tank with 50% liquid content and different baffle heights under the same working conditions. The figure shows that, with both high and low baffles, the peak pressure at these points exhibits a noticeable delay and reduced amplitude. During acceleration and deceleration, the two baffles have minimal impact on the pressure values at the center points. However, during the free sloshing stage, the curve with the high baffle is smoother, less abrupt, and exhibits greater stability and quicker recovery.

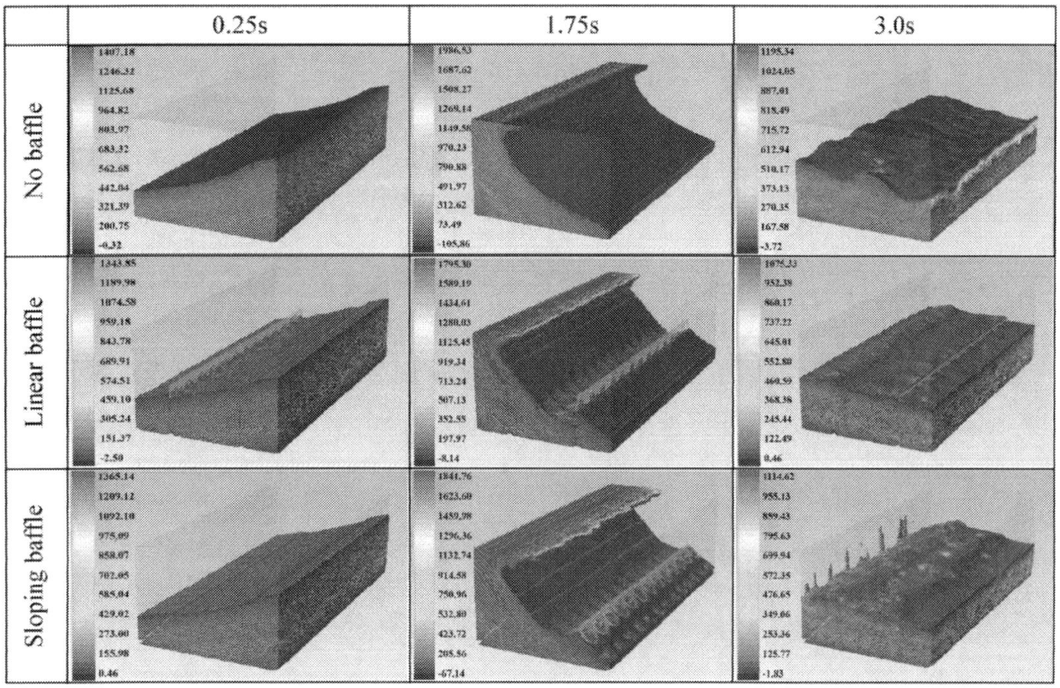

Figure 22. Oil slosh pressure under the action of baffles at different positions under the same working conditions.

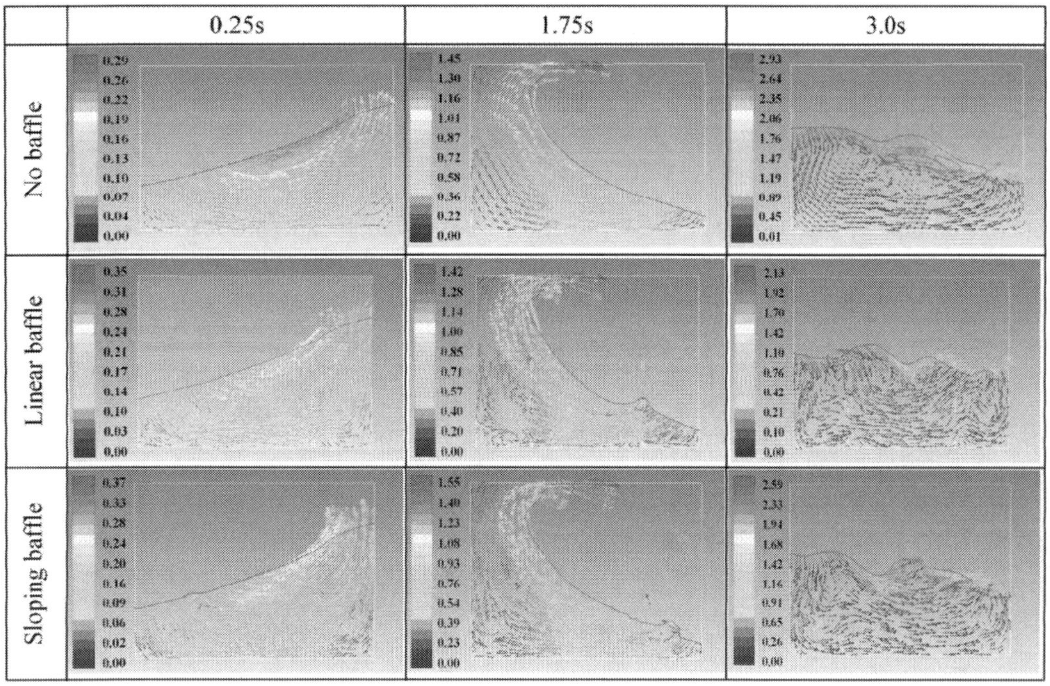

Figure 23. Oil slosh vector diagram under the same working condition and different position baffles.

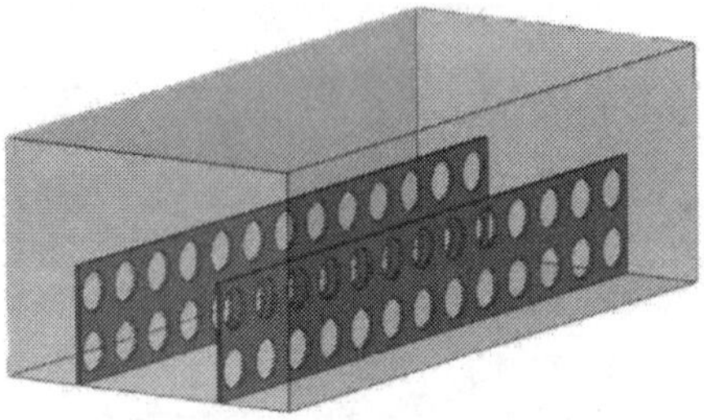

Figure 24. Low baffle fuel tank.

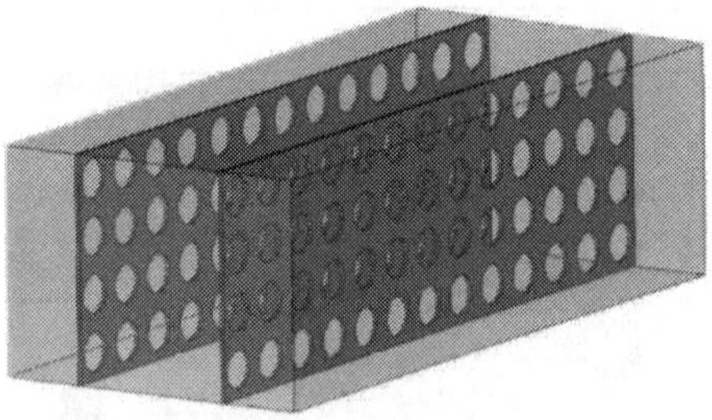

Figure 25. High baffle fuel tank.

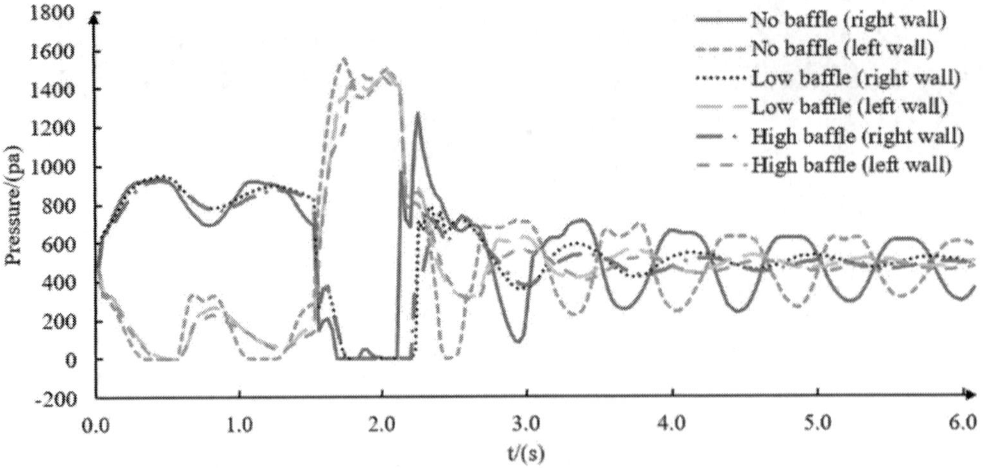

Figure 26. Pressure history curve of the center point of the liquid level on the left and right walls of the fuel tank under the action of baffles of different heights with 50% liquid content.

Figure 27 shows that at 0.25 s, the effects of the low baffle and high baffle are similar. By 1.75 s, the liquid level rise is lower with the high baffle. At 3.0 s, during the free-slosh phase, the oil liquid stabilizes more quickly with the high baffle.

Figure 28 shows that at low liquid levels, the effects of the low and high baffles are similar. However, when the oil sloshing amplitude is significant, the high baffle influences the oil liquid flow direction, directing the oil towards the bottom of the fuel tank. Additionally, the oil liquid flow rate with the high baffle remains relatively stable.

4.2.4 Analysis of cruciform baffle fuel tank. Based on the analysis, the non-porous baffle's strong resistance to oil liquid flow impacts the overall flow pattern, which can be seen in Figure 29 and Figure 30. The straight baffle performs better than the inclined baffle and the high baffle is more effective than the low baffle. Therefore, the following analysis will focus on the design of various-shaped baffles.

Figure 31 displays the pressure history curves at the center points of the liquid level on the left and right walls of the fuel tank with various baffles under the same working conditions. The data show that with the cruciform baffle, the pressure curve experiences the least variation during acceleration and deceleration, and the time required for oil liquid to stabilize during the free sloshing stage is shorter.

Figure 32 shows that the cruciform baffle, compared to the plain baffle, increases the contact area with the oil. At 0.25 s and 1.75 s, the transverse baffles effectively limit the unilateral accumulation of oil liquid, significantly reducing the liquid surface climbing height and concentrating most of the oil liquid at the bottom of the tank. By 3.0 s, the oil liquid recovers more smoothly during the free-sloshing stage. Throughout all three moments depicted in the oil-liquid pressure cloud maps, the pressure inside the tank with the cruciform baffle is consistently lower.

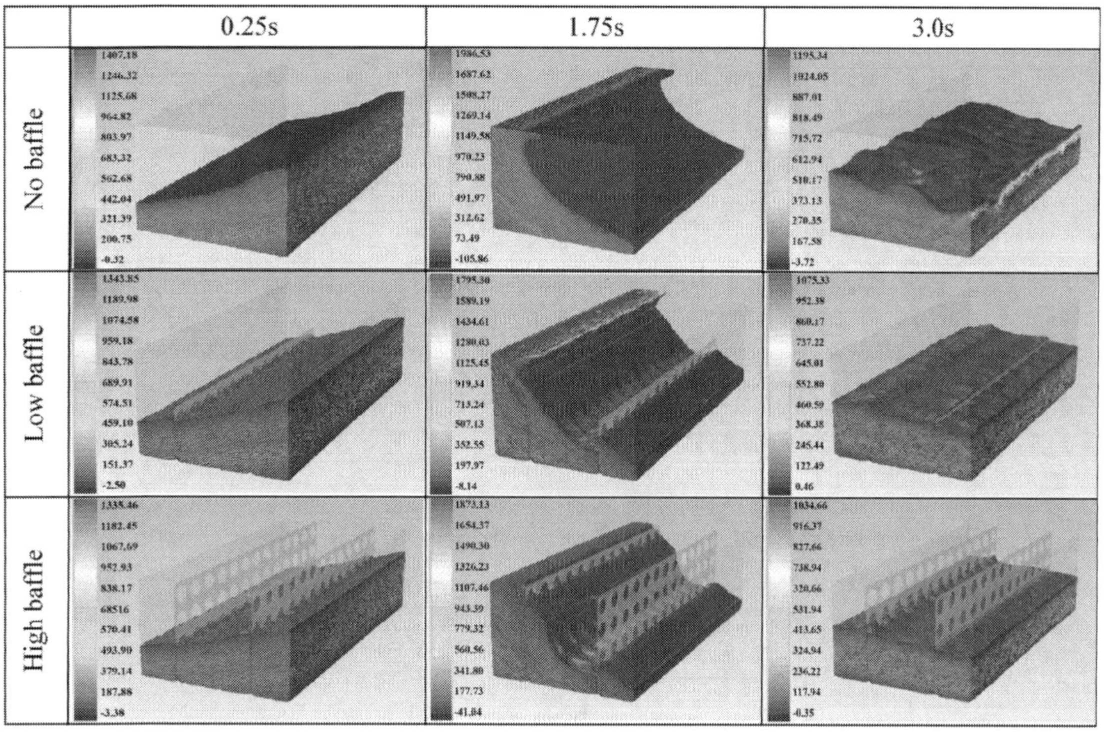

Figure 27. Cloud picture of oil slosh pressure under different height baffles under the same working conditions.

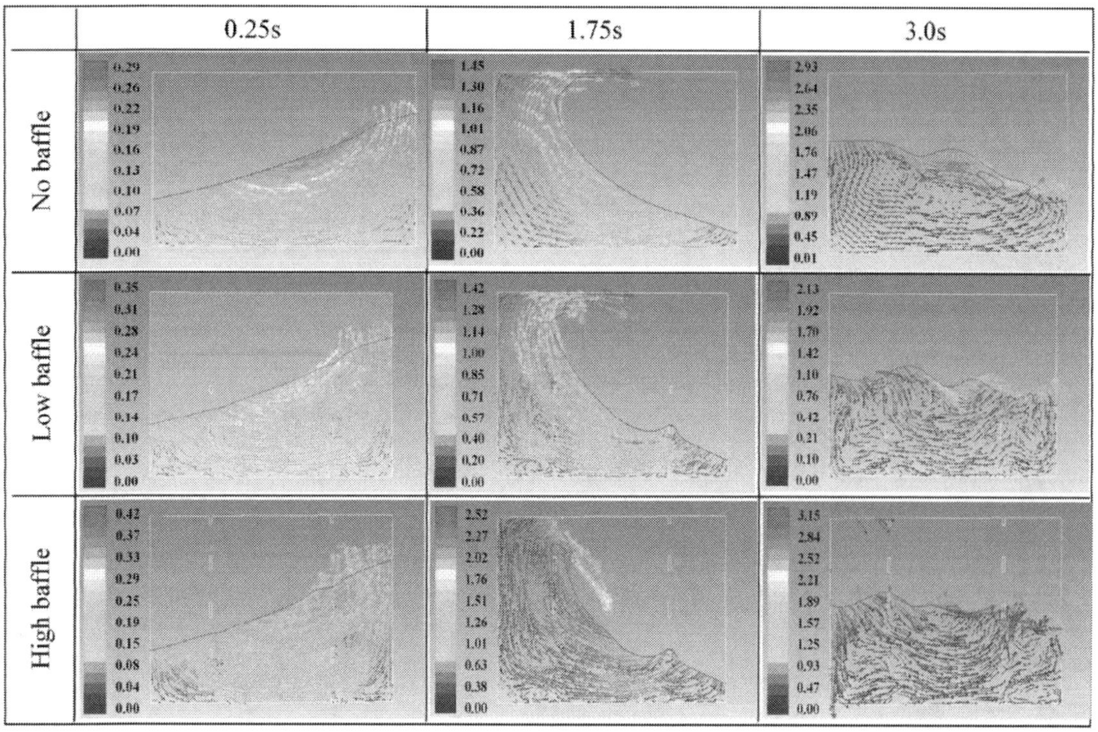

Figure 28. Oil slosh vector diagram under the same working condition and different height baffles.

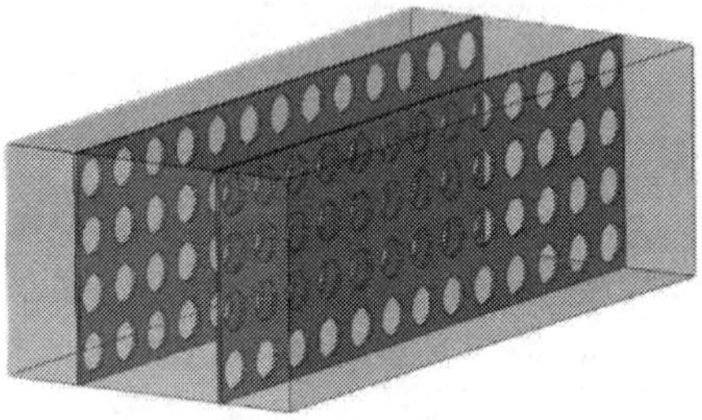

Figure 29. Plain baffle fuel tank.

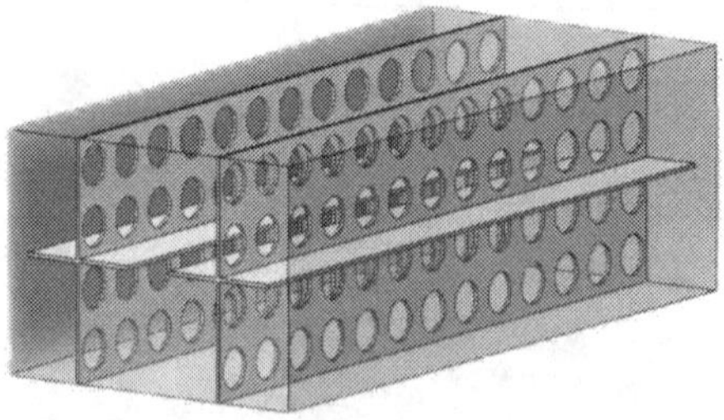

Figure 30. Cruciform baffle fuel tank.

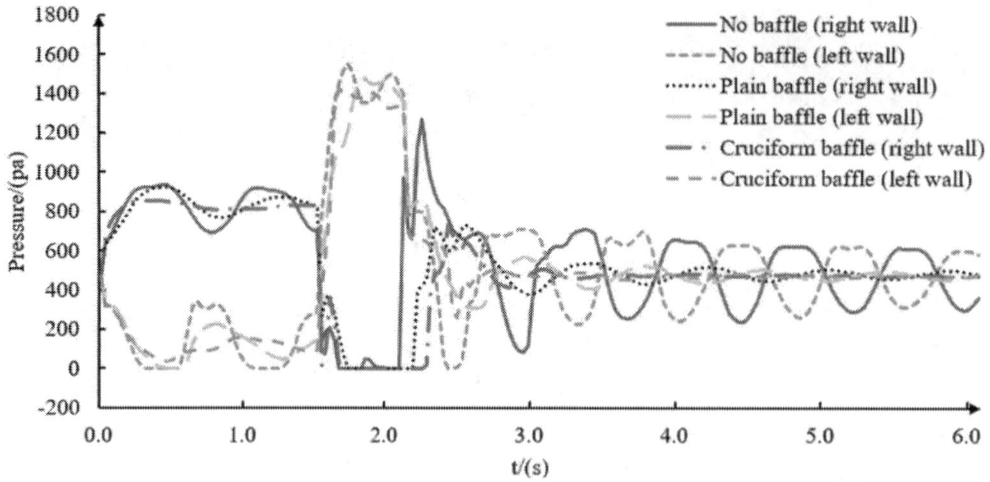

Figure 31. Pressure history curve of the center point of the liquid level on the left and right walls of the fuel tank under the action of different baffles with 50% liquid filling.

Figure 33 reveals that, compared to fuel tanks without baffles or those with plain baffles, the oil liquid movement in a tank with a cruciform baffle is primarily concentrated at the bottom, with overall oil liquid speed remaining relatively stable. At 3.0 s, the cruciform baffle generates numerous vortices during oil movement, resulting in a more complex oil liquid flow direction.

5. Conclusion

(1) Under the same working conditions, the non-porous baffle has a better suppression oil liquid fluctuations effect than the porous baffle, which greatly reduces the impact pressure of oil liquid in the fuel tank, but the oil liquid fluidity under the action of the non-porous baffle is poor.

(2) When the oil liquid sloshes, the angle between the linear baffle and the direction of the oil liquid flow is larger, which attenuates the main impact force on the oil liquid. Compared with the sloping baffle, the linear baffle has a better suppression oil liquid fluctuations effect.

(3) Under conditions of significant oil liquid sloshing, the high baffle increases the contact area with the oil, providing more effective inhibition and accelerating the time needed for the oil to recover smoothly.

(4) The cruciform baffle restricts the multi-directional movement of the oil liquid, causing a large number of vortices inside the oil liquid, attenuates the kinetic energy generated by the oil liquid sloshing, greatly reduces the impact of oil liquid on the fuel tank wall, and improves the fluctuation effect of the oil liquid.

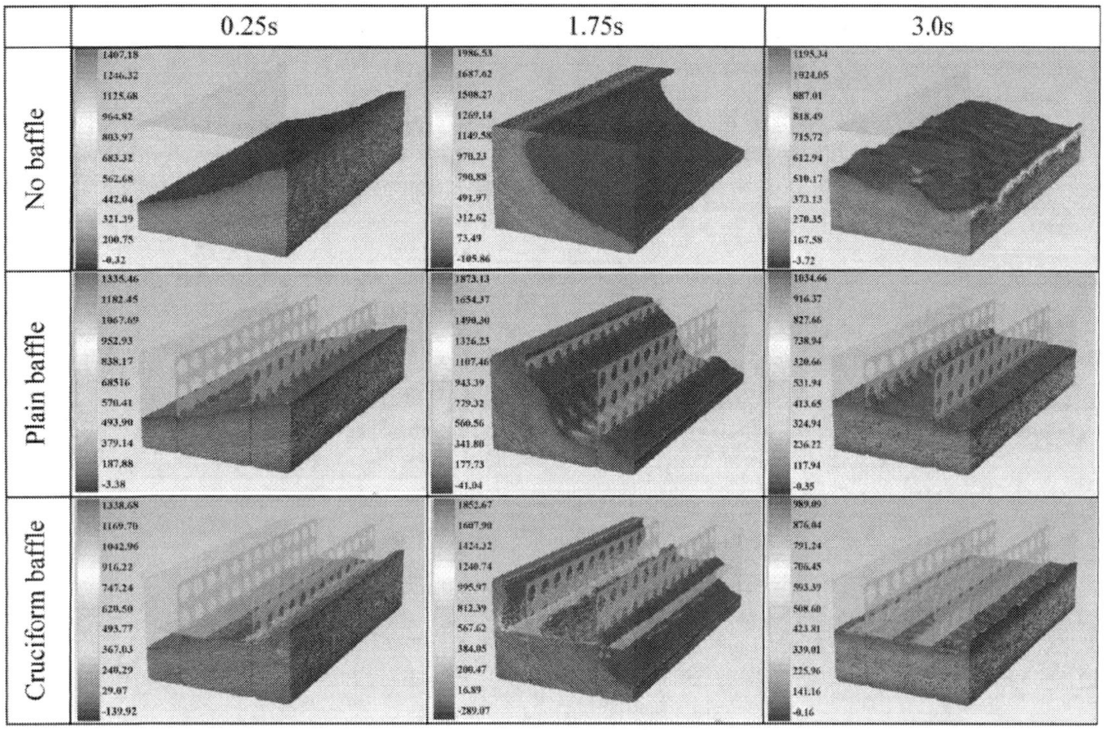

Figure 32. Oil slosh pressure cloud picture under the same working condition and different baffles.

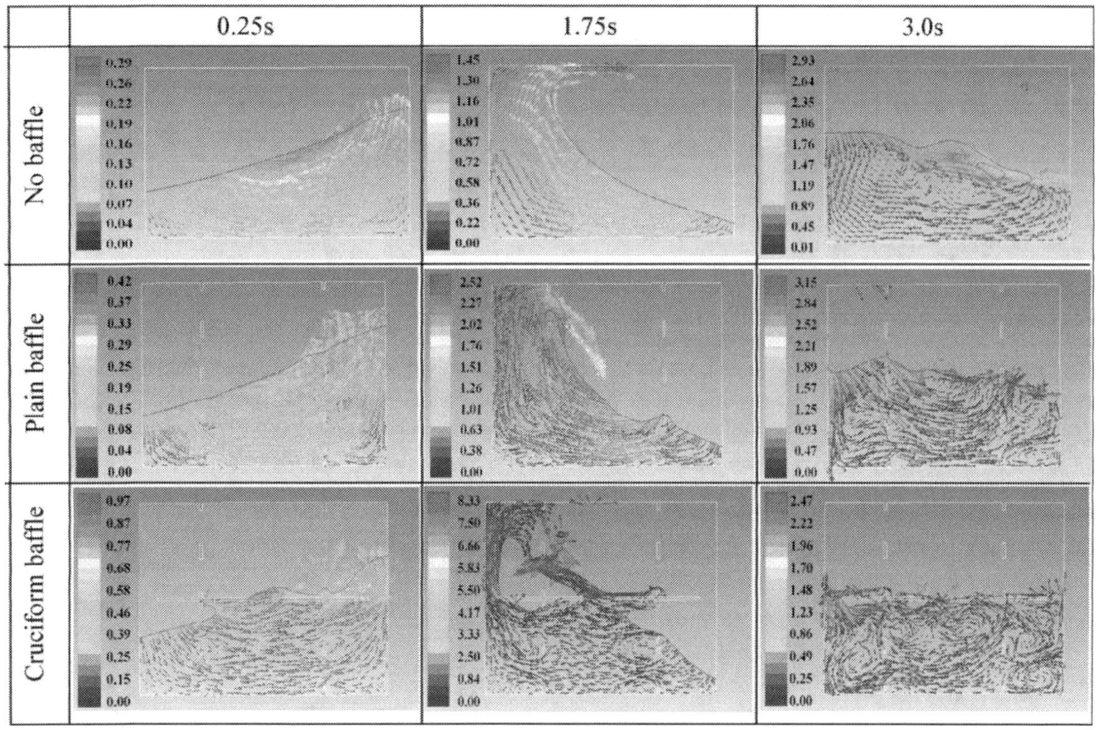

Figure 33. Vector diagram of oil slosh under different baffles under the same working.

Acknowledgments
This paper is one of the stage achievements of the Natural Science Foundation of Inner Mongolia Autonomous Region (2022QN05006), the basic research funds for universities directly under the Inner Mongolia Autonomous Region (2024YXXS022), horizontal topic (NKDHX2024013), and horizontal topic (zyhx2021026).

References

[1] Luppes R., Veldman A., and Wemmenhove R. (2011). Simulation of two-phase flow in sloshing tanks. In Computational Fluid Dynamics 2010: Proceedings of the Sixth International Conference on Computational Fluid Dynamics, ICCFD6, St Petersburg, Russia, on July 12-16, 2010 (pp. 555-561). Berlin, Heidelberg: Springer Berlin Heidelberg.

[2] Gao B. J. and Shang C. W. (2012). The influence of the position of the anti-wave plate on the anti-wave effect of the tanker during braking. Journal of Hebei University of Technology, (03): 46-49 (in Chinese).

[3] Vaishnav D., Dong M., Shah M., Gomez F., and Usman M. (2014). Investigation and development of fuel slosh care methodologies. SAE International Journal of Passenger Cars-Mechanical Systems, 7 (1): 278-288.

[4] Wang Q. Y., Jiang K. H., Rakheja S., et al. Transient response analysis of liquid slosh in a liquid-partially filled tank truck. J. Journal of Vibration and Shock, 2018, 37 (17): 9-16.

[5] Zhang L. N. (2019). Fuel tank slosh simulation for helicopter's typical flight attitudes. Mechanical Engineer, (01): 131-133.3 (in Chinese).

[6] Wu X. D. and He R. (2023). Numerical study on characteristics and mechanism of fuel sloshing in automobile fuel tank. Journal of the Brazilian Society of Mechanical Sciences and Engineering, 46 (1): 40-52.

[7] Verma A., Shankar R., Shaik M. A., et al. (2024). Comprehensive design and analysis of a 300L steel fuel tank for heavy off-road vehicles: numerical and experimental insights. Frontiers in Mechanical Engineering, 5 (2): 158-176.

[8] Moretti F. L., Silveira M. E., Lamin P. C. M. F., et al. (2014). Numerical and experimental evaluation of sloshing in automotive fuel tanks. In: SAE Brazil International Congress and Display. pp. 36-43.

[9] Brar G. S. and Singh S. An Experimental and CFD Analysis of Sloshing in a Tanker (2014). Procedia Technology, 14 (4): 490-496.

[10] Li X., Zhang J. X., Qu F., Zhang W. Q., Wang D. S., and Li W. (2017). Optimal design of antisway inner cavity structure of agricultural UAV pesticide tank. J. Transactions of the Chinese Society of Agricultural Engineering (Transactions of the CSAE), 33 (18): 72-79 (in Chinese).

[11] Zhong W. K., Wu J. R., Fu J. Y., Sun L. Y., and Huang P. (2022). Nonlinear shaking analysis of rectangular reservoirs with built-in horizontal baffles. Engineering Mechanics, (10): 190-199 (in Chinese).

[12] Yi B. B., Sun R. W., Huang J., and Hu B. (2024). Design and research of fuel tank assembly of national six gasoline vehicles. Special Purpose Vehicle, (08): 26-28 (in Chinese).

Initial sizing of a hybrid hydrogen fuel cell commercial aircraft with electric aft-fuselage propulsion

Zhouwei Fan[1], Zhaoguang Tan[1,*], Haiqiang Wang[1] and Yuhan Wang[1]

[1]Shanghai Aircraft Design and Research Institute, Commercial Aircraft Corporation of China, Ltd., Shanghai 201210, China

*Corresponding author's e-mail: tanzhaoguang@comac.cc

Abstract: Hybrid hydrogen fuel cell (HH2FC) aircraft is one of the viable technological pathways to reduce carbon emissions in the aviation industry. For a configuration with aft-fuselage propulsion, a novel method is established for the initial sizing of the HH2FC aircraft in a situation where empirical data is lacking. The carbon emission reduction capabilities and potential for reducing fuel and carbon tax costs can be preliminarily assessed by the proposed initial sizing method. A case study on a narrow-body aircraft indicates that, at the current technological level, the carbon emissions and cost of the HH2FC aircraft will increase to achieve the same payload and range as the conventional aircraft. However, with the improvement of hydrogen technologies, such as fuel cell power density and hydrogen tank gravimetric density, HH2FC aircraft will reduce 6.6% carbon emissions compared to conventional aircraft. As a result, both economic and environmental benefits can be achieved in the future.

1. Introduction

As the economy experiences rapid expansion, the scale of aviation transport continues to grow, concurrently contributing to an increase in the aviation industry's share of global carbon emissions. Conventional aircraft using kerosene fuel show limited potential for significant carbon emissions reduction [1]. Consequently, there is a growing imperative to explore clean energy as a viable alternative in the quest for more sustainable aviation.

Hydrogen stands out with its nearly threefold higher energy density compared to conventional aviation kerosene. Two prospective alternatives of hydrogen propulsion integration are under consideration, including the combustion of hydrogen within the combustion engine and the use of fuel cells to generate the energy for flight [2]. The more sustainable choice for hydrogen utilization employs fuel cells, as the resultant outcome of energy production is limited to water. However, the integration of H2FC requires modifications in aircraft design and consequently poses some challenges in terms of technical feasibility. Hybridizing H2FCs with kerosene as the power source is a more feasible technological approach, recognized as the hybrid hydrogen fuel cell (HH2FC) aircraft.

A fundamental aspect of conceptual design entails the initial sizing of the HH2FC aircraft [3]. For conventional aircraft, establishing the size typically relies on initial sizing methods that reference empirical data. However, the case of HH2FC aircraft features novel components of the power system and a distinct powertrain. Therefore, there is a lack of suitable references, necessitating the development of a new initial sizing method to calculate the overall size of the aircraft in the beginning stage of conceptual design. This paper presents an improved method for the initial sizing of the HH2FC aircraft. The method consists of three parts, including weight analysis, constraint analysis, and sensitivity analysis. The enhancement in the weight analysis model encompasses the introduction of weight models for new components, a model for hydrogen fuel consumption analysis, and a model for analyzing the interdependence of different weights. The constraint analysis integrates the thrust provided by two power sources of the HH2FC aircraft, formulates the dynamic equations of the aircraft, and consequently delimits the performance boundaries of the aircraft under different scenarios to define the feasible design space. Sensitivity analyses are conducted based on weight analysis and constraint analysis to explore the effects of technological parameters on the weight, size, and thrust of the aircraft. A novel HH2FC commercial aircraft with an electric aft-fuselage propulsion configuration is employed to demonstrate the process and results of the improved initial sizing method.

2. Configuration and powertrain

2.1. Configuration

The propulsion system has consistently been a crucial factor influencing the configuration of aircraft. To accommodate H2FC propulsion systems, the configuration of the HH2FC aircraft differs from that of conventional aircraft. NASA's N+4 project [4] adopted a hybrid-powered aircraft with an aft-fuselage fan driven. The study showed that for narrow-body aircraft, powering the aft-fuselage fan through the fuel cell was more efficient than extracting power from the thermal engines.

Based on the above studies, this paper selects the aft-fuselage propulsion configuration for the HH2FC aircraft, which is an environmentally friendly alternative to current narrow-body aircraft. As illustrated in Figure 1, the aircraft configuration features a T-tail design to facilitate the installation of the aft-fuselage fan. Hydrogen storage tanks and fuel cells are strategically placed at the rear of the fuselage to minimize the length of the hydrogen supply pipelines. The aft-fuselage fan is directly driven by the H2FC. The boundary layer ingestion effect created by the aft-fuselage fan involves ingesting the boundary layer airflow from the aircraft's surface, reducing fuselage frictional drag and thus lowering the aircraft's energy consumption.

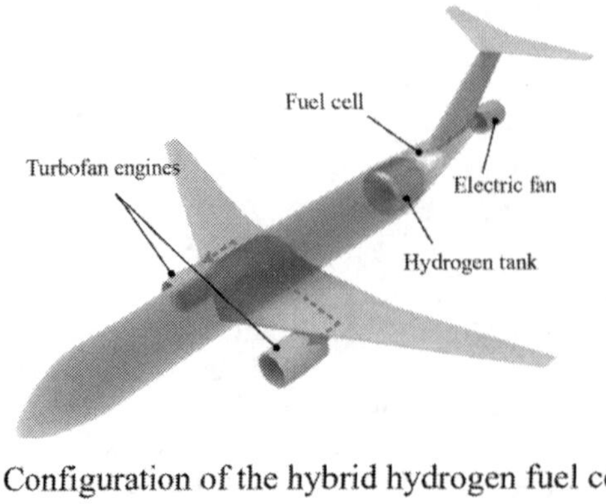

Figure 1. Configuration of the hybrid hydrogen fuel cell aircraft.

2.2. Powertrain

The impact of different powertrains associated with different aircraft configurations should be taken into account in the initial sizing phase. The adoption of a hybrid powertrain introduces a snowball effect on the aircraft weights, wherein the inclusion of individual components may result in a higher weight for the aircraft, consequently diminishing the potential benefits of hybrid power. Therefore, it is essential to simplify the powertrain of the HH2FC aircraft, minimizing the addition of extra components and reducing modifications to the original power system.

The powertrain of the HH2FC aircraft is illustrated in Figure 2. The two power systems operate independently, with minimal alterations to the power system of conventional aircraft. The aft-fuselage propulsion configuration is suitable for retrofitting based on conventional aircraft, making full use of existing conventional aircraft with minimal modifications. The entire new power system, including a hydrogen storage tank, fuel cell, electric motor, and electric fan, is situated at the rear of the aircraft, with minimal alterations to the aircraft's structural framework. The aircraft is propelled primarily by the main thrust generated by the conventional turbofan engine mounted under the wing, complemented by auxiliary thrust from the aft-fuselage electric fan.

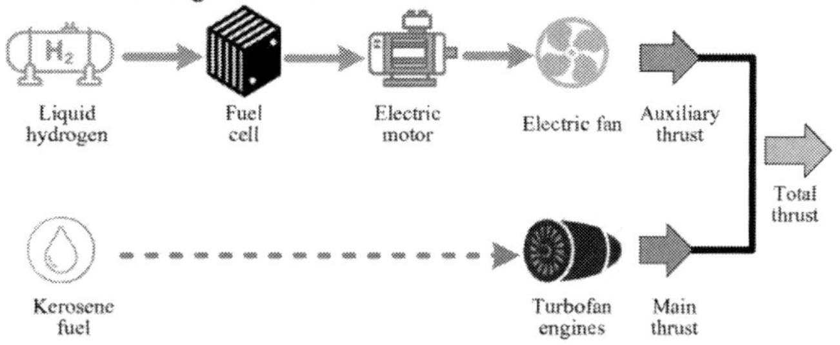

Figure 2. Powertrain of the hybrid hydrogen fuel cell aircraft.

3. Methods

The process of the initial sizing method for the HH2FC aircraft is illustrated in Figure 3. Top-level aircraft requirements, like range, payload, cruise Mach, etc., are utilized as inputs of the process. Then, a component-based weight analysis method, which is suitable for novel aircraft lacking empirical data, is employed to estimate the weights of the aircraft. A constraint analysis in typical scenarios is followed for the transformation of top-level aircraft requirements to feasible design space. After iterations of weight and constraint analyses, requirements are decoupled into fundamental parameters of the sized aircraft. A sensitivity analysis is finally conducted to scrutinize the key technical parameters influencing the aircraft concept. The subsequent sections will furnish a comprehensive elucidation of the improved initial sizing method.

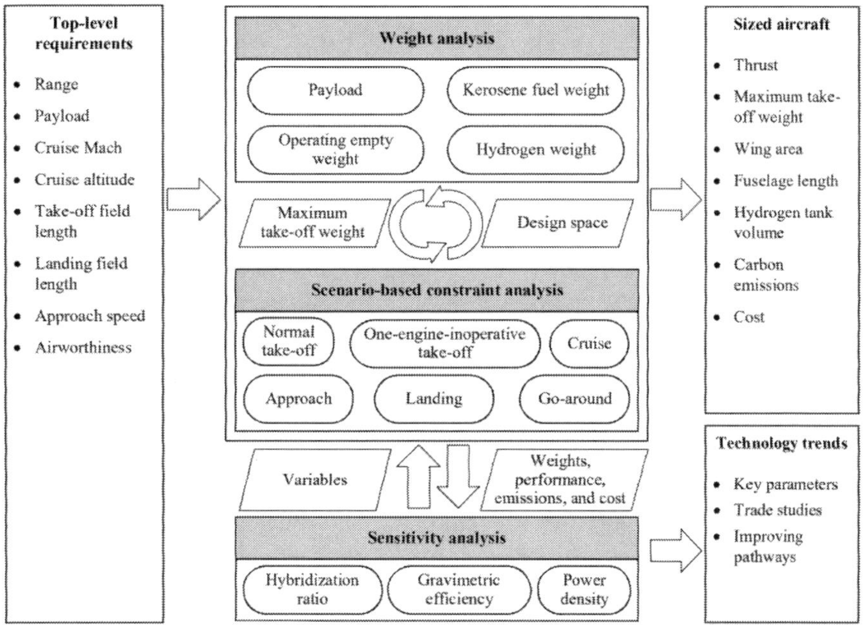

Figure 3. Process of initial sizing for the hybrid hydrogen fuel cell aircraft.

3.1. Weight analysis

For conventional aircraft, the estimation of maximum take-off weight (M_{TO}) can be achieved by empirical formulas. However, for HH2FC aircraft, due to the lack of historical data, it is hard to estimate the aircraft's weights directly. A weight estimation method is proposed for HH2FC aircraft. The analysis process features considerations for the impact of kerosene fuel weight on wing weight and the influence of hydrogen fuel weight on fuselage weight. The auxiliary power unit (APU) system is replaced by the fuel cell in the HH2FC aircraft.

3.1.1. Payload. The payload of commercial aircraft is related to the number of passengers. Each passenger, along with their carry-on luggage, is typically assumed to weigh 75 kg. There is an additional 20 kg for checked luggage per passenger [5].

3.1.2. Fuel weight. Given the distinct ways in which kerosene and hydrogen contribute to aircraft kinetic energy, the consumptions of these two energy sources are considered separately. Subsequently, the mixture-specific fuel consumption ($C_{\text{sfc, mix}}$), which means the total weight of kerosene and hydrogen consumed per unit of thrust generated in a given period, is computed for overall aircraft. Accounting for the efficiency of hybrid kerosene and hydrogen, $C_{\text{sfc, mix}}$ can be calculated using Equation (1).

$$C_{\text{sfc,mix}}=\frac{V}{\Delta h_{\text{ker}}\eta_{\text{ker}}\left(1\text{-}k_{\text{h}}\right)+\Delta h_{\text{H}_2}\eta_{\text{H}_2}k_{\text{h}}} \tag{1}$$

where Δh_{ker} represents the heating value of kerosene; Δh_{H_2} represents the heating value of hydrogen; η_{ker} represents the overall efficiency of converting kerosene fuel energy to aircraft kinetic energy; η_{H_2} represents the overall efficiency of converting hydrogen fuel energy to aircraft kinetic energy; k_{h} represents the fuel hybridization ratio, which is defined in Equation (2).

$$k_{\text{h}}=m_{\text{LH}_2}/\left(m_{\text{ker}}+m_{\text{LH}_2}\right) \tag{2}$$

where m_{LH_2} represents the mass flow of LH_2 and m_{ker} represents the mass flow of kerosene.

Once the $C_{\text{sfc,mix}}$ is obtained, the weight of kerosene and hydrogen consumed by the HH2FC aircraft during the main mission phase can be calculated using the "lost fuel method" proposed by Torenbeek [6]. The reserve fuel weight is specified in the regulation of the Association of European Airlines [6]. Considering both the main mission and reserve mission, the fuel weight (M_{F}) of the HH2FC aircraft can be calculated.

3.1.3. Operating empty weight. The operating empty weight (M_{OE}) of the HH2FC aircraft is obtained by updating corresponding component weights based on the M_{OE} of the conventional aircraft. The method proposed by Loftin [7] is employed in this study for the conventional M_{OE} estimation.

The airframe will be modified for the HH2FC aircraft. The method proposed by Howe [8] is employed to calculate the weight changes of the airframe, which can reflect the structural weight impact of variations in fuselage dimensions, tail type, and inertial relief of kerosene fuel stored in the wing tank.

The bare weight of the turbofan engine is determined using the corrected formula proposed by Jenkinson et al. [9]. Based on the empirical coefficients provided by Howe [8], the turbofan engine system weight is estimated by increasing the bare weight of the engine by a factor of 1.6. The HH2FC aircraft replaces the APU with the H2FC. Therefore, the conventional APU weight should be excluded from the M_{OE} of the HH2FC aircraft. Sforza [10] provided a modified formula for calculating the weight of the conventional aircraft's APU.

The fuel cell system comprises the hydrogen fuel cell stack and balance of plant components. The weight of the fuel cell system is calculated utilizing maximum power and power density. The weight of the hydrogen storage tank is calculated by hydrogen weight and the gravimetric efficiency of the LH_2 storage tank. The electric propulsion system consists of the electric motor, blades, and duct. The weight of the electric motor is obtained through its power density. The weights of the blades and duct are determined based on their dimensions and material densities. The additional weight of the mounting, exhaust system, and pylon is considered through the installation factor of 1.1 [11].

By replacing the airframe weight of conventional aircraft with the new airframe weight, adding the weight of new components, and subtracting the weight of the APU, the M_{OE} of the HH2FC can be updated. After iterations, the weights of the HH2FC aircraft will be estimated.

3.2. Constraint analysis
Constraint analysis is an effective method for the preliminary estimation of sea-level take-off thrust-to-weight ratio (T/W) and take-off wing loading (M/S) for an aircraft. These two parameters are essential for expressing the aircraft's capabilities in various flight scenarios. For constraints of requirements in different scenarios, the relationship between the T/W and M/S can be represented through the "master equation" derived from Newton's second law [12]. Six typical scenarios are considered in this study, including take-off, cruise, approach, landing, and take-off and go-around with one engine inoperative.

3.3. Sensitivity analysis
Sensitivity analysis provides insights into the trends of key technologies influencing the performance of HH2FC aircraft. Key technological parameters studied in this paper include thrust hybridization ratio during cruise ($k_{T,CR}$), power density of fuel cell (σ_{fc}), and gravimetric efficiency of hydrogen storage tank (μ). The influence of variations in key parameters on the HH2FC aircraft will be elucidated through emission analysis and cost analysis.

Carbon emissions resulting from the combustion of kerosene fuel are estimated to be 3.16 times the weight of the fuel itself [13]. The reaction product of hydrogen in a fuel cell is water, leading to the absence of carbon emissions. The cost analysis includes fuel costs and carbon taxes. The economic considerations are closely linked to the consumption of kerosene (M_{ker}) and LH_2 (M_{H_2}), as well as the market trends of these two fuels. The carbon tax cost is associated with the aircraft's carbon emissions and the carbon tax policies.

4. Case study
A case study is employed to demonstrate the application of the initial sizing method for HH2FC aircraft. The selected baseline aircraft is a narrow-body passenger aircraft at the A320neo level, equipped with double underwing turbofan engines. The hybrid propulsion modification is applied to the baseline aircraft using the aft-fuselage propulsion configuration illustrated in Figure 1.

As the inputs of the initial sizing, top-level requirements for both the baseline aircraft and the HH2FC aircraft are specified in Table 1. Considering the inclusion of the aft-fuselage propulsor, the HH2FC aircraft is classified as a three-engine aircraft following airworthiness regulations [14]. Therefore, the climb gradient requirements of the HH2FC aircraft during take-off and go-around with one engine inoperative differ from those of the baseline aircraft.

Table 1. Top-level requirements of the baseline and HH2FC aircraft.

Top-level requirements	Baseline aircraft	HH2FC aircraft
Range (km)	4075	
Passenger number (pax)	158	
Take-off field length (m)	≤ 2200	
Take-off climb gradient with one engine inoperative (%)	≥ 2.4	≥ 2.7
Cruise altitude (m)	11000	
Cruise Mach number	0.785	
Available climb rate at cruise (m s^{-1})	≥ 0.5	
Approach speed (m s^{-1})	≤ 74	
Landing field length (m)	≤ 1800	
Go-around climb gradient with one engine inoperative (%)	≥ 2.1	≥ 2.4

A comparison analysis is carried out, considering the current technological state and the potential technological advancements that could be attained in the future, approximately by 2035. The projections for future technological advancements are derived from existing literature. Pertinent industry reports forecasting technological trends [15–17]. The relevant design parameters of the hybrid propulsion modification are presented in Table 2.

Table 2. Design parameters of the HH2FC aircraft.

Parameter	Current value	Future value
Power density of fuel cell system [17] (kW kg^{-1})	1.5	3.5
Power density of electric motor [18] (kW kg^{-1})	5	10
Efficiency of fuel cell system [17]	0.6	0.75
Efficiency of electric motor [15]	0.9	0.95
Gravimetric efficiency of LH_2 tank [16]	0.6	0.8
Thrust hybrid ratio at take-off	0.2	0.3
Thrust hybrid ratio at cruise	0.1	0.2
Propeller diameter (m)	1.5	

The cost analysis can be conducted based on the current and anticipated future prices of kerosene and hydrogen, as well as the prevailing and projected carbon tax prices. Table 3 presents the parameters used for calculating aircraft fuel costs and carbon tax costs.

Table 3. Fuel price and carbon tax.

Parameter	Current value	Future value
Kerosene price [19] (CNY kg^{-1})	4.2	9.3
Liquid hydrogen price [20] (CNY kg^{-1})	56.9	14.3
Carbon price [20] (CNY kg^{-1})	0.4	1.6

5. Results and discussion

With the aircraft's M/S maintained at 575.08 kg/m^2 and T/W maintained at 0.337, the selected design point for both the HH2FC aircraft and the baseline aircraft falls within the feasible design space. This confirms that the design satisfies the design constraints imposed in the considered scenarios.

Fundamental parameters of the aircraft can be estimated based on the selected value of M/S and T/W through the initial sizing method. Table 4 compares the sizes of the HH2FC aircraft and the baseline aircraft under different technological scenarios. Under current technological conditions, without changing the aircraft's payload and range, the HH2FC aircraft experiences an increase in structural weight, leading to a 29.9% increase in M_{TO} compared to the baseline aircraft. With improvements in fuel cell power density and hydrogen storage ratios in the future, the HH2FC aircraft's M_{TO} is still 14.7% higher than the baseline aircraft. However, there is a decrease in kerosene consumption, contributing to a 6.6% reduction in carbon emissions.

Table 4. Comparisons of fundamental parameters between HH2FC aircraft and baseline aircraft.

Fundamental parameter	Baseline aircraft	HH2FC aircraft (current)	HH2FC aircraft (future)
Wing loading (kg m^{-2})		575.08	
Thrust-to-weight ratio		0.337	
Wing reference area (m^2)	126.07	163.72	144.65
Total take-off thrust (kN)	240	310.95	274.72
Maximum take-off weight (kg)	72500	94153	83183
Operating empty weight (kg)	42100	60670	53017
Kerosene fuel weight (kg)	15390	17925	14406
Liquid hydrogen weight (kg)	-	547	750
Turbofan engine take-off thrust (kN)	120	124.38	96.15
Electric fan take-off thrust (kN)	-	62.29	82.42
Fuselage length (m)	38.90	39.97	40.26
Carbon emissions (kg km-pax^{-1})	0.076	0.088	0.071

Table 5 compares the fuel costs and carbon tax costs of the HH2FC aircraft and the baseline aircraft currently and in the future. According to the results in Table 5, the current HH2FC aircraft does not have an economic advantage compared to the baseline aircraft. However, with improvements in hydrogen fuel cell technology and anticipated increases in kerosene prices and carbon tax rates, alongside the decrease in LH_2 prices in the future, HH2FC aircraft can realize economic benefits compared to conventional aircraft.

Table 5. Comparisons of costs between HH2FC aircraft and baseline aircraft.

Parameter	Current		Future	
	Baseline aircraft	HH2FC aircraft	Baseline aircraft	HH2FC aircraft
Kerosene fuel cost (CNY km-pax^{-1})	0.100	0.117	0.222	0.208
Hydrogen cost (CNY km-pax^{-1})	-	0.048	-	0.017
Total fuel cost (CNY km-pax^{-1})	0.100	0.165	0.222	0.225
Carbon tax cost (CNY km-pax^{-1})	0.030	0.035	0.121	0.113

Sensitivity analysis is conducted to identify the trend of the impact of key technological parameters on the HH2FC aircraft. The gravimetric efficiency of the hydrogen storage tank (μ), fuel cell power density (σ_{fc}), and cruise thrust hybridization ($k_{T,CR}$) have a significant impact on the benefits of the H2FC aircraft. Figure 4, based on the current HH2FC aircraft, uses carpet plots to display the impact of different parameter variations on the M_{TO} and carbon emissions of the HH2FC aircraft.

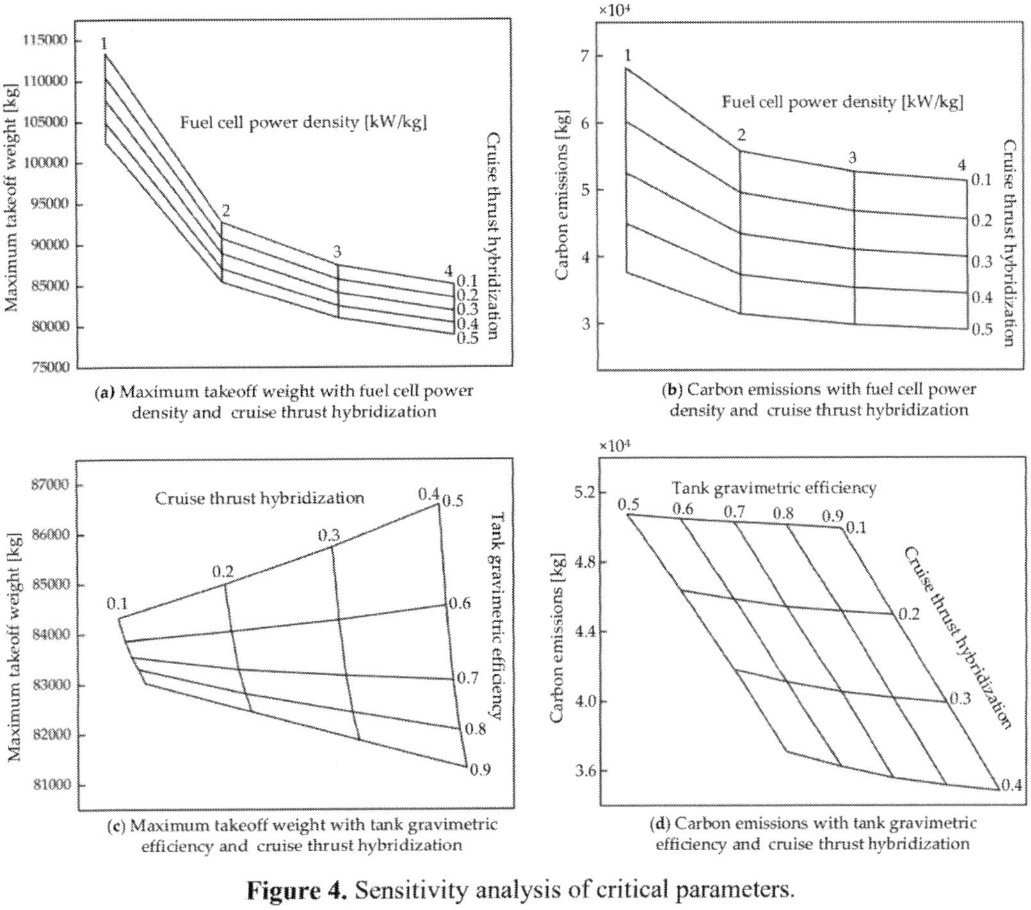

Figure 4. Sensitivity analysis of critical parameters.

As illustrated in Figure 4(a), the increase in the σ_{fc} decreases the M_{TO} of the aircraft. Figure 4(b) shows the effect of σ_{fc} and $k_{T,CR}$ on carbon emissions. The increase of the σ_{fc} and $k_{T,CR}$ both decrease the carbon emissions. The trend of carbon emission reduction due to the increase in σ_{fc} gradually diminishes, while the carbon emission reduction resulting from the increase in $k_{T,CR}$ exhibits an approximate linear relationship. Figure 4(c) indicates that for μ below 0.6, an increase in hydrogen fuel usage as power leads to a higher M_{TO}. Conversely, when the μ exceeds 0.7, the aircraft's M_{TO} decreases as hydrogen fuel usage increases. While the trend of M_{TO} varies with increasing $k_{T,CR}$ at different μ, Figure 4(d) shows that carbon emissions consistently decrease with increasing $k_{T,CR}$ in the μ value range of 0.5 to 0.9. Moreover, the larger the μ is, the faster the reduction in carbon emissions is.

6. Conclusions

An initial sizing method applicable to HH2FC aircraft is introduced in this paper. Considering the technological state of hybrid hydrogen power technology, the sizes of current and future HH2FC aircraft are analyzed separately. Compared with the baseline aircraft with conventional configuration, the economic and carbon emission benefits of current and future HH2FC aircraft are estimated. Based on the assessment of HH2FC aircraft, the following conclusions are drawn:

- At the current technological level, HH2FC aircraft, due to their excessive weight, cannot achieve kerosene fuel savings compared to conventional aircraft. Instead, current HH2FC aircraft may result in higher fuel costs and increased carbon emissions.
- With the improvements in hybrid hydrogen power technology, future HH2FC aircraft can achieve reduced kerosene consumption compared to conventional aircraft, leading to lower carbon emissions. Moreover, as kerosene fuel prices and carbon taxes experience upward trends, coupled with anticipated reductions in LH_2 prices, future HH2FC aircraft can realize certain economic benefits.
- The gravimetric efficiency of hydrogen storage tanks and fuel cell power density significantly impact the benefits of HH2FC aircraft. Future developments in HH2FC aircraft should focus on improving the hydrogen storage gravimetric efficiency and fuel cell power density.

This research provides the initial sizing method for estimating the fundamental parameters of HH2FC aircraft with aft-fuselage propulsion. Subsequent efforts will entail more in-depth analyses employing higher-precision methods, and optimization to enhance the competitiveness of the proposed solutions further.

Acknowledgments

This research was funded by the Shanghai Post-doctoral Excellence Program.

References

[1] Bergero C, Gosnell G, Gielen D, Kang S, Bazilian M, Davis S J (2023). Pathways to net-zero emissions from aviation. Nature Sustainability, 6(4): 404-414. https://doi.org/10.1038/s41893-022-01046-9.

[2] Adler E J, Martins J R R A (2023). Hydrogen-powered aircraft: fundamental concepts, key technologies, and environmental impacts. Progress in Aerospace Sciences, 141: 100922-100952. https://doi.org/10.1016/j.paerosci.2023.100922.

[3] Vonhoff G L M (2021). Conceptual design of hydrogen fuel cell aircraft. Master Thesis, Delft University of Technology, Delft.

[4] Bradley M K, Droney C K (2012). *Subsonic ultra green aircraft research phase II: N+4 advanced concept development*. National Aeronautics and Space Administration, Hampton, Virginia.

[5] Prewitz M, Bardenhagen A, Beck R (2020). Hydrogen as the fuel of the future in aircrafts – challenges and opportunities. International Journal of Hydrogen Energy, 45: 25378-25385. https://doi.org/10.1016/j.ijhydene.2020.06.238.

[6] Torenbeek E (2013). *Advanced aircraft design: conceptual design, analysis and optimization of subsonic civil airplanes*. John Wiley and Sons, Ltd., Chichester, West Sussex.
[7] Loftin L K (1980). *Subsonic aircraft: evolution and the matching of size to performance*. National Aeronautics and Space Administration, Hampton, Virginia.
[8] Howe D (2000). *Aircraft conceptual design synthesis*. Professional Engineering Publishing Limited, London.
[9] Jenkinson L R, Simpkin P, Rhodes D (1999). *Civil jet aircraft design*. Arnold, London.
[10] Sforza P M (2014). *Commercial airplane design principles*. Butterworth-Heinemann, Oxford.
[11] Samuelsson S, Grönstedt T (2021). Performance analysis of turbo-electric propulsion system with fuselage boundary layer ingestion. Aerospace Science and Technology, 109: 106412-106421. https://doi.org/10.1016/j.ast.2020.106412.
[12] Vepa R (2023). *Flight dynamics, simulation, and control*. CRC Press, Boca Raton.
[13] International Civil Aviation Organization (2023). ICAO carbon emissions calculator Methodology. https://www.icao.int/environmental- protection/Carbonoffset/Pages/default.aspx.
[14] European Union Aviation Safety Agency (2023). Easy access rules for large aeroplanes (CS-25). https://www.easa.europa.eu/document-library/easy-access-rules/easy-access-rules-large-aeroplanes-cs-25.
[15] Marksel M, Prapotnik Brdnik A (2022). Maximum take-off mass estimation of a 19-seat fuel cell aircraft consuming liquid hydrogen. Sustainability, 14: 8392-8406. https://doi.org/10.3390/su14148392.
[16] Verstraete D, Hendrick P, Pilidis P, Ramsden K (2010). Hydrogen fuel tanks for subsonic transport aircraft. International Journal of Hydrogen Energy, 35: 11085-11098. https://doi.org/10.1016/j.ijhydene.2010.06.060.
[17] FlyZero (2022), *Fuel cells roadmap report*. Aerospace Technology Institute, Cranfield.
[18] Kadyk T, Winnefeld C, Hanke-Rauschenbach R, Krewer U (2018). Analysis and design of fuel cell systems for aviation. Energies, 11: 375-389. https://doi.org/10.3390/en11020375.
[19] Lee M, Li L K B, Song W (2019). Analysis of direct operating cost of wide-body passenger aircraft: a parametric study based on Hong Kong. Chinese Journal of Aeronautics, 32: 1222-1243. https://doi.org/10.1016/j.cja.2019.03.011.
[20] Scott M (2023). *Analysing the costs of hydrogen aircraft*. Steer, London.

Experimental and numerical analysis of mechanics and fracture characteristics of engine compressor

Guifeng Yu[1], Junfei Guo[2], Jinlong Wang[1*], Zeyu Shi[1], Fengming Du[1] and Yuxing Yang[1]

[1]Marine Engineering College, Dalian Maritime University, Dalian, P.R. China
[2]Zhengzhou Branch of China Nuclear Power Engineering Co Ltd, Zhengzhou, P.R. China

[*]Corresponding author: wjl19890806@dlmu.edu.cn

Abstract: Investigating the fatigue failure behavior of compressor fan blades and their materials in aero-engines is a crucial research area. This paper focuses on the compressor fan blades and their constituent materials of aero-engines. Firstly, fluid-structure interaction (FSI) numerical simulations under various rotational speeds are conducted to determine the stress distribution and identify the maximum stress. Secondly, based on the FSI results, fatigue tests on fan blade materials under different stress amplitudes are performed to explore the relationship between load and life, so as to analyze the fracture characteristics under various stress amplitudes. This research has significant implications for ensuring the stable operation of aero-engine fan blades.

1. Introduction

Compressor fans are essential components in aero-engines, involving interdisciplinary aspects in their design, manufacturing, and maintenance. These blades are characterized by high added value, cost, and processing difficulty [1-2]. During service, they undertake the critical task of compressing inlet air in multiple stages and delivering it to the turbine combustion chamber. Notably, compressor rotor blades operate at high rotational speeds (exceeding 10,000 rpm) [3], enduring fatigue loads caused by substantial centrifugal and aerodynamic forces. Consequently, fatigue cracks are prone to develop, particularly in the region near the blade root, which is a primary location for fatigue failure. Therefore, exploring the stress conditions and analyzing fracture characteristics during compressor fan blade operation are crucial for assessing fatigue performance and ensuring reliable operation.

Previous studies have addressed fatigue failure and life prediction for compressor fan blades and related materials. Bednarz and Misiolek [4-5] investigated the resonant fatigue behavior of compressor blades with surface treatments, utilizing finite element methods (FEM) for modal analysis and fatigue analysis based on Manson-Coffin-Basquin and Ramberg-Osgood models. They determined the effects of different residual stress values on fatigue life and compared the results with literature values. Ball et al. [6] compared local stress predictions for semi-circular and V-notched geometries using nonlinear FEM, revealing limitations of Neuber's rule and the equivalent strain energy density approach. They proposed a generalized notch response curve (GNRC) based on nonlinear FEM results to calculate notch strain responses under any loading conditions. Moradi et al. [7] studied crack growth rates in the 13th-stage blade of a T56 compressor. They performed FEM analysis considering centrifugal and aerodynamic forces to obtain blade stress fields, simulated initial semi-elliptical cracks at maximum stress locations, and calculated stress intensity factors (SIFs) along crack fronts using ANSYS, predicting crack propagation and cycles to failure. Lou et al. [8] introduced an approximate, universal

Journal of Physics: Conference Series **2879** (2024) 012026 doi:10.1088/1742-6596/2879/1/012026

geometric parameter method based on traditional linear elastic fracture mechanics to estimate SIFs at smooth surfaces and V-notch roots. This method, validated by analytical solutions and FEM results, is concise and suitable for engineering applications but has limited accuracy for large opening angles and deep notches.

In this paper, the aero-engine compressor fan blade is taken as the research object, and the FSI numerical simulation is carried out at different speeds to analyze the fluid flow and determine the stress / strain distribution. Subsequently, the fatigue test of TC4 titanium alloy fan blade material under different stress conditions was carried out. The fracture was observed by scanning electron microscopy to determine the source and characteristics of fatigue failure, which provided a theoretical basis for the fatigue performance evaluation and daily maintenance of fan blades.

2. The numerical model

2.1. Finite element modeling

A three-dimensional model of the aero-engine compressor fan is created using ANSYS software, with simplifications to retain only the main shaft and blade structures, as shown in Figure 1(a). The fan blade geometry is meshed using tetrahedral elements with a grid density of 30 mm, resulting in a total of 188,755 elements, as depicted in Figure 1(b) and (c). The material used in this study is titanium alloy TC4, with fundamental parameters including tensile strength, density, and elastic modulus.

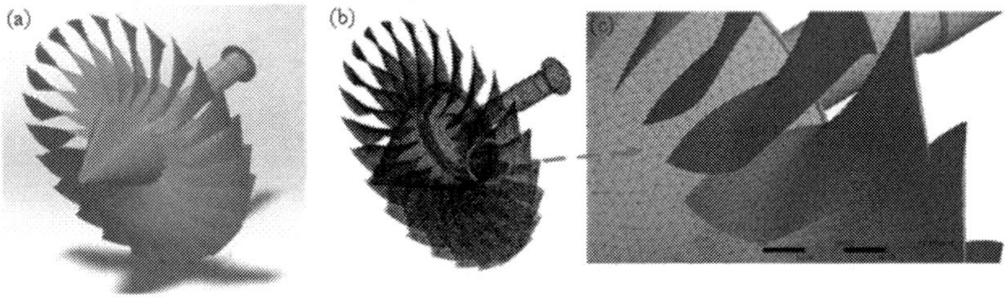

Figure 1. Aero-engine compressor fan ((a) 3D model; (b) model meshing; (c) local mesh).

2.2. Fluid domain modeling

The fluid domain is modeled as a standard cylinder with a base area of 13.704 m² and a length of 4 m, based on the blade dimensions. Boolean operations are performed on the fluid domain and blade models to obtain the blade cavity within the fluid domain, which is further meshed, as shown in Figure 2.

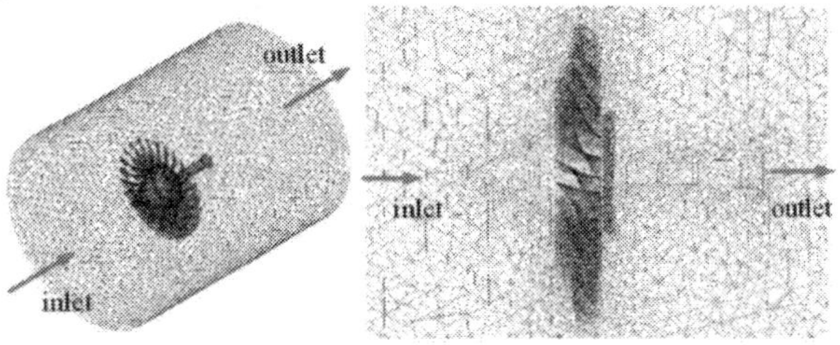

Figure 2. Fluid domain model and meshing.

The inlet pressure is set to standard atmospheric pressure, and the inlet air velocity is 75 m/s. The turbulence model employed is k-epsilon (2 eqn), a widely used two-equation model in engineering fields, including aerospace, automotive engineering, and fluid mechanics.

3. Numerical simulation results

3.1. Aerodynamic simulation results

Figure 3 presents the results of fluid dynamics simulations. As can be observed from the simulation outcomes, particularly the fluid behavior surrounding the blades, different colors of lines represent the air velocity and flow direction. The intensity of line clustering and color depth reflect the velocity variations of the fluid around the blades. Brighter colors or denser lines indicate faster flow velocities in those regions. By comparison, it is evident that the fastest air velocity occurs shortly after the initial contact with the fan blades, approximately 139.5 m/s. Furthermore, the simulation results reveal that regions where airflow undergoes significant changes may lie along critical flow paths, such as at the air intake and on the blade surfaces. To further analyze, a plane is established based on the XY axis, intersecting the entire fan blade, as shown in Figure 3(a). Surrounding the fan blade is a rotating vortex structure, induced by the blade's rotation. The vortex colors gradually transition from red to blue, signifying changes in the direction of gas velocity. Red represents regions of higher velocity and turbulent flow, whereas blue indicates slower, smoother flow, even slower than the initial velocity of air entering the system.

As depicted in Figure 3(b), a pressure contour map of the gas generated on the blade surface is presented. The pressure distribution exhibits a distinct pattern: higher pressure values are observed on the blade surface and its vicinity, reflecting the compression and acceleration of air as it passes through the blades. Conversely, lower pressure values are found on both sides of the blades, likely due to the blade's curved shape causing airflow diversion and vortex formation, thereby reducing pressure in these areas.

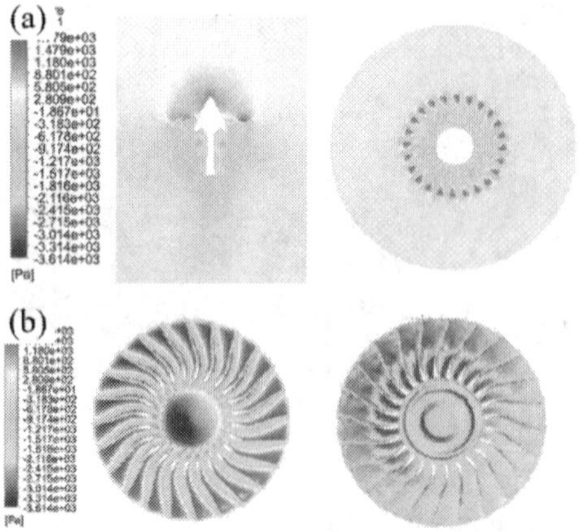

Figure 3. Aerodynamic simulation results ((a) streamline distribution; (b) xy cross-sectional velocity distribution; (c) fluid pressure distribution).

3.2. The stress states under different rotating speeds

The aforementioned aerodynamic simulation results were imported into the Static Structural module, with the gravitational acceleration set to 9.8 m/s² in the -y direction. Based on actual operating conditions, rotational speeds of 5400 rpm and 15000 rpm were configured. By integrating the aerodynamic simulation results, the stress, strain, and overall deformation of the compressor blades under different rotational speeds were obtained. Figures 4(a) and (b) reveal that, given a constant air velocity, the blade's rotation around the hub leads to significant deformation at the blade tip, while the deformation is minimal at the blade root due to its rigid connection with the hub. Deformation gradually increases from the blade root to the tip. Figures 4(c) and (d) show that during rotation, the maximum equivalent elastic strain occurs near the blade root and gradually decreases towards the tip, contrary to the overall deformation pattern. At 5400 rpm, the maximum equivalent elastic strain is 0.10701 mm, with a minimum of 0.033656e-007 mm; while at 15000 rpm, the maximum strain increases to 0.85169 mm, and the minimum to 2.6736×10-6 mm. Figures 4(e) and (f) depict the stress distributions under different rotational speeds, with maximum equivalent stresses also occurring near the blade root, similar to the maximum strain locations. At 5400 rpm, the maximum equivalent stress is approximately 602 MPa, with a minimum of 0.0177 MPa; whereas at 15000 rpm, the maximum stress increases to approximately 618 MPa, and the minimum to 0.1414 MPa.

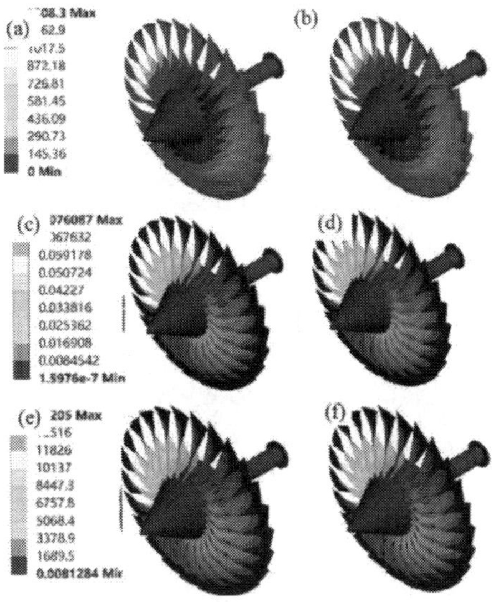

Figure 4. Deformation, strain, and stress distributions under different rotational speeds ((a) deformation at 5400 rpm; (b) deformation at 15000 rpm; (c) strain at 5400 rpm; (d) strain at 15000 rpm; (e) stress at 5400 rpm; (f) stress at 15000 rpm).

4. Fracture characters

Fatigue tests were conducted using TC4 titanium alloy, the material of the compressor blades, on a Shimadzu FTS-2000 ultrasonic fatigue testing machine, where the stress ratio is r=-1, and the test is carried out at room temperature. The specimen shape is shown in Figure 5(a). Based on numerical simulation results, test stress amplitudes were set at 585 MPa, 615 MPa, and 630 MPa. The test results are presented in Figure 5(b), it can be seen that with the decrease of stress amplitude, the increase of fatigue life becomes larger. When the stress amplitude decreases from 630 MPa to 615 MPa, the fatigue life increases by 30 %, and when the stress amplitude decreases from 615 MPa to 585 MPa, the fatigue life increases by nearly 100 %.

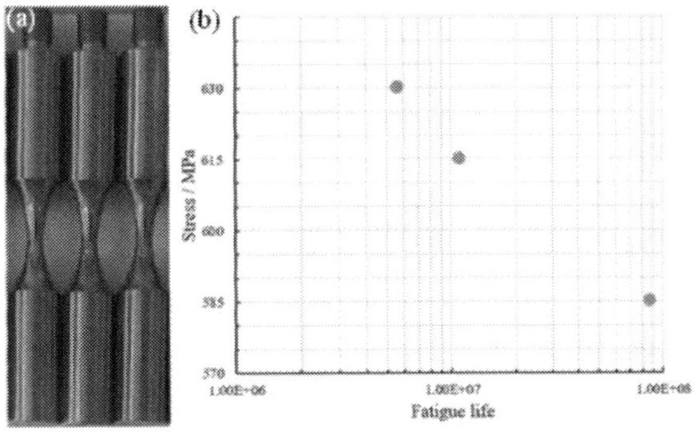

Figure 5. Test specimen and test results ((a) the specimen used in the test; (b) the test results, including the stress amplitude and fatigue life).

The fracture surfaces of the specimens were observed using Scanning Electron Microscopy (SEM), as shown in Figure 6. It can be observed that the fatigue failures of titanium alloy TC4 under different stress amplitudes all originated from the specimen surface and propagated inward. Distinct fatigue crack initiation sites can be discerned on the specimen surfaces, as depicted in Figure 6(a), (c), and (e). In the crack propagation regions, fatigue striations, including terrace-like and river-like patterns, are visible, as illustrated in Figure 6(b), (d), and (f).

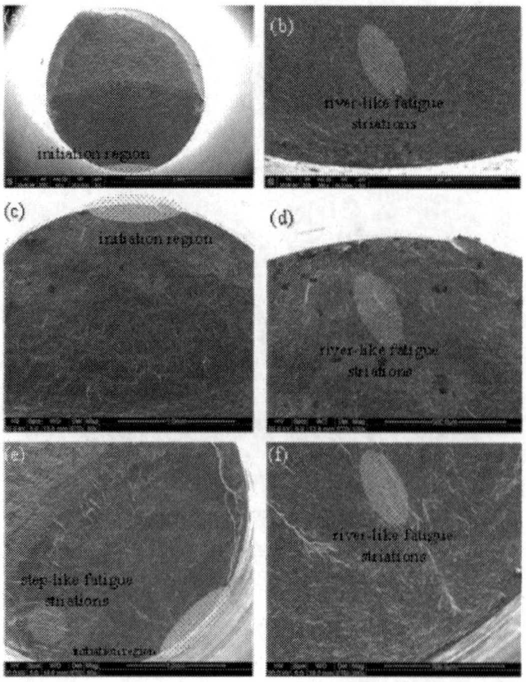

Figure 6. Fracture characteristics under different stress amplitudes ((a) and (b), stress amplitude of 585 mpa; (c) and (d), stress amplitude of 615 mpa; (e) and (f) stress amplitude of 630 mpa).

5. Conclusions

A few conclusions are drawn from this research.

(1) At different rotational speeds, both the maximum stress concentration and the maximum strain locations in TC4 titanium alloy compressor fan blades occur in the blade root region, gradually decreasing from the root towards the blade tip.

(2) The fatigue failure of TC4 titanium alloy, the material used for compressor fan blades, primarily originates from the surface of the specimen.

(3) Distinct step-like and river-like fatigue striations can be observed on the fracture surface of TC4 titanium alloy.

In future studies, more fatigue tests with different stress amplitude and stress ratios will be carried out to explore the *S-N* curve of TC4 alloy, underscoring the necessity of the fatigue failure investigation under different test environments.

Acknowledgment

The authors gratefully acknowledge the financial supported by the Fundamental Research Funds for the Central Universities [3132024212].

References

[1] Kim H, Toledo K K, Jeong Y, et al. A simple approach for evaluating the stress concentration factor of a corroded surface using the fast Fourier transform [J]. Engineering Failure Analysis, 2020, 115: 104612.

[2] Chen J, Kao J, Huang Y, et al. On the stress concentration factor of circular/elliptic hole and rigid inclusion under the remote anti-plane shear by using degenerate kernels [J]. Archive of Applied Mechanics, 2021, pp. 1133-1155 vol. 91.

[3] Zhao F. Discussion on improvement and application of shot peening process and equipment for a certain aeroengine blade tenon [J]. China Plant Engineering. 2024, pp. 15-18 vol.7.
[4] Bednarz A, Misiolek W Z. Assessment of the impact of shot-peening on the fatigue life of a compressor blade subjected to resonance vibrations [J]. Materials, 2020, 5726 vol.13(24).
[5] Bednarz A, Misiolek W Z. Numerical and experimental assessment of the effect of residual stresses on the fatigue strength of an aircraft blade [J]. Materials, 2021, 5279 vol.14(18).
[6] Ball D L. Estimation of elastic-plastic strain response at two-dimensional notches [J]. Fatigue & Fracture of Engineering Materials & Structures, 2020, pp.1895-1916 vol.43(9).
[7] Hamidreza M, Rafiullah R, Muslim A. Crack growth simulation in 13th row of compressor blades under foreign object damage and resonant vibration condition [J]. Journal of Vibroengineering, 2021, pp.57-66 vol.23(1).
[8] Lou B Q, Barltrop N, Zhang J. Geometry parameter method to estimate stress intensity in V-notched plate and a small crack ahead from V-notch apex [J]. Thin-Walled Structures, 2020, 106801 vol.154.

Journal of Physics: Conference Series **2879** (2024) 012027 doi:10.1088/1742-6596/2879/1/012027

Application of PCA-LSTM algorithm in predicting component faults in CNC machining centers

Shuwei Wang

College of Mechanical and Electrical Engineering, Shanxi Datong University, Datong, Shanxi, 037000, China

dtdxsw16@163.com

Abstract: In order to further explore the relationship between component failures and CNC machining center failures, a method of predicting component failures in advance was proposed to reduce the likelihood of machining center failures and improve CNC machining efficiency. In this paper, the principal component analysis method is used to optimize the Long short-term memory neural network to build a network model to study and predict the fault status of NC machining center components in work. The use of Principal Component Analysis (PCA) to extract the main features of machine tool faults can improve the speed of iterative training networks during the machining process and reduce the input dimension and prediction difficulty of LSTM. The extracted data is used as input for the LSTM network. The fault type is output. The comparison of diagnostic results between PCA-LSTM, LSTM, and BP neural networks shows that PCA-LSTM is a high-precision prediction model. The root mean square difference is reduced to 1.572. The predicted error results can effectively reflect the changes that occur in the machining center components with the occurrence of faults. It can be seen that it is feasible to use PCA-LSTM as a tool for predicting the faults of CNC machining center components.

1. Introduction

Traditional research on CNC machining center faults rarely considers the relationship between machining center faults and component faults simultaneously, nor does it consider whether faults are transmitted. Relevant literature also does not involve further exploration [1]. However, in the process of production and processing, faults can spread in several steps. Even if the components are not directly connected, the transmission of faults may be caused by the combination of functions or structures. Therefore, in order to eliminate faults in CNC machining centers, it is necessary to determine the relationship between various components and understand the actual situation of the components in a timely manner. The impact on the machining of CNC machining centers is a very important factor [2].

With the rapid development of modern sensing technology, information about the working status of CNC machining center components can be obtained through sensors and corresponding information collection equipment. The operating status of the components can be monitored, which can achieve signal extraction of CNC machining centers [3]. Based on signal characteristics and the working status of CNC machining centers, a reference can be provided for predicting component faults in CNC machining centers. How to accurately predict the possibility of malfunctions in CNC machining center components is currently a focus that needs to be studied.

In this article, Principal Component Analysis (PCA) is used to reduce the dimensionality of signal features, reduce the prediction difficulty of LSTM, eliminate low correlation features, reduce computational complexity, improve the speed of iterative training networks during the machining process, and provide input for building predictive component fault models. It is combined with long short-term memory (LSTM) to build a network model tool. We create a component fault prediction model and conduct experimental analysis on component faults to obtain the minimum error.

2. The fault analysis of CNC machining center components

2.1 The fault analysis of CNC machining center components

In this article, the VMC850B machining center used is the research object. Based on the functional structure and working principle of this series of machining centers, it is divided into 12 subsystems [4], namely workbench (A), spindle (S), feed (J), tool magazine (T), electrical (D), hydraulic (Y), pneumatic (Q), numerical control (NC), chip removal (P), cooling (L), lubrication (R), and protection (F).

Based on the working principle, functional structure, and fault data collected on site of the machining center, the frequency table of the fault causes of the machining center is compiled and analyzed, as shown in Table 1. Table 1 indicates that component damage, component damage, oil leakage, etc. are the main causes of equipment failures [5]. It is worth noting that part quality, assembly, and quality adjustment are the main factors affecting the reliability of CNC machining centers.

Table 1. Frequency table of fault causes.

The code	Fault cause	Frequency
34	Component damage	0.312 9
20	Oil leakage	0.123 5
33	Component damage	0.076 3
17	Insufficient pressure	0.052 6
14	Loose	0.034 2
04	Burn out	0.264 5
19	Block	0.264 5
31	Misoperation	0.264 5
32	Parameter error	0.264 5
03	Stuck	0.172 8
23	Temperature too high	0.172 8
42	Lack of coolant	0.172 8
01	Fracture	0.006 3
09	Dislocation	0.006 3
15	Fall off	0.006 3

2.2 Correlation analysis of component failures in CNC machining centers

Based on the analysis of faults in CNC machining centers, the types and causes of faults in each component of the CNC machining center are determined, and then the fault rate of the CNC machining center components is calculated [6]. All types of components have different fault modes. When the same fault occurs, each component may have the same or different fault types. There may also be some associations and differences between each fault type.

Based on the fault analysis conducted by the CNC machining center, there was an error when performing component correlation analysis. For example, when problems arise, such as interference with tools, inadequate replacement of the robotic arm, and inability to complete manual tool replacement, although a fault is displayed in the tool magazine, the fault is caused by an abnormality in the pneumatic system [7]. This type of fault is called a related fault, as shown in Figure 1 for CNC machining center components.

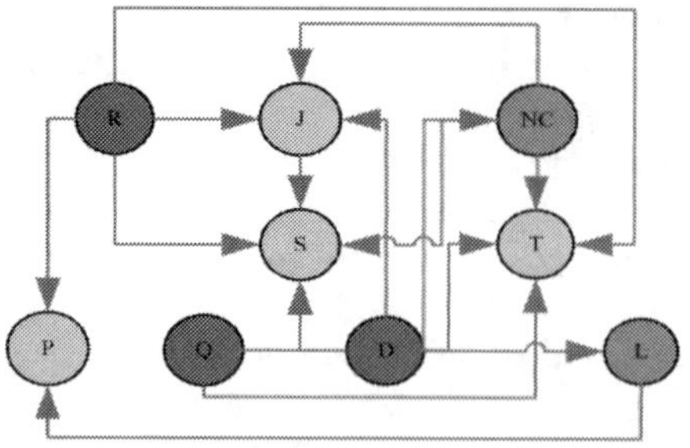

Figure 1. CNC machining center component-related faults.

Due to the complexity of CNC machining centers, the information and data obtained during the machining process are very complex, including some interference and unclear data ownership, which makes signal processing of CNC machining center components more difficult [8]. Therefore, in order to accurately reflect the operating status and performance development trend of CNC machining center components without affecting production and processing efficiency, this article applies advanced sensing technology to monitor the data of the operating status of CNC machining center components and process the data in multiple steps.

Among various CNC machining center failures, mechanical failures account for the largest proportion, accounting for more than half of the total number of failures. At the same time, with the improvement of self-diagnosis functions in CNC systems and electrical systems, the failure rate of CNC systems has become lower and lower [9]. Therefore, this article mainly focuses on monitoring and predicting the status of CNC machining center components when they fail. The spindle system is the main element of this article's research.

3. The model research methods

3.1 Principal component analysis data dimension reduction

With the rapid development of sensing technology and computer technology, obtaining state information becomes more complex, and more information is extracted in the process of predicting the lifespan of CNC machining center components under different working conditions. Therefore, in order to improve the efficiency of predicting the lifespan of machining center components, it is necessary to reduce the dimensionality of high-dimensional signals while maintaining the main signal characteristics of each component of the machining center [10]. The principal component analysis method has been widely applied in research to reduce the dimensionality of signal features.

In this way, a set of related functional features is transformed into a new set of nonlinear features, which are called the main components. This new feature set can effectively reflect the maximum changes in information function signals and reduce the loss of information function signals [11]. Therefore, this article relies on the PCA algorithm to perform dimensionality reduction processing on the signal characteristics of CNC machining center components. The basic working principle is as follows.

There are n samples, each containing i variables, forming $n \times i$-th order matrix:

$$X = \begin{bmatrix} x_{11} & \cdots & x_{1i} \\ \vdots & \ddots & \vdots \\ x_{n1} & \cdots & x_{ni} \end{bmatrix} \tag{1}$$

Using the i vectors $X_1, X_2 \cdots X_i$ of the matrix $X_{n\times i}$ as a linear combination where $k < i$, the following equation is obtained:

$$\begin{cases} F_1 = a_{11}X_1 + a_{12}X_2 + \cdots + a_{1i}X_i \\ F_2 = a_{21}X_1 + a_{22}X_2 + \cdots + a_{2i}X_i \\ \vdots \\ F_k = a_{k1}X_1 + a_{k2}X_2 + \cdots + a_{ki}X_i \end{cases} \tag{2}$$

3.2 Introduction to long short-term memory neural network

A long short-term memory network is a kind of time-recursive neural network that is specially used to solve long-term and commonly dependent problems [12].

The difference between LSTM and recurrent neural network (RNN) is that LSTM has a memory module and memory control function, which can solve the gradient disappearance and explosion problems that cannot be solved in the traditional RNN model training process. This applies to solving long-term dependency problems, and the ability to remember long segments of information is a fundamental skill of LSTM [13]. The internal structure of LSTM is shown in Figure 2.

The LSTM network consists of three neural network layers, namely:

Forgotten Gate: The current moment discards some information after obtaining it from the previous moment. The calculation formula is as follows:

$$f_t = \sigma\left(W_f \times [h_{t-1}, x_t] + b_f\right) \tag{3}$$

$$C_t = i_t \times \hat{C} + f_t \times C_{t-1} \tag{4}$$

Input gate: We confirm the input information and update the current time status. The calculation formula is as follows:

$$i_t = \sigma\left(W_i \times [h_{t-1}, x_t] + b_i\right) \tag{5}$$

$$\hat{C}_t = \tanh\left(W_c \times [h_{t-1}, x_t] + b_i\right) \tag{6}$$

Output gate: We determine the current time state through the Activation function σ, and determine the final output. The calculation formula is as follows:

$$o_t = \sigma\left(W_o \times [h_{t-1}, x_t] + b_o\right) \tag{7}$$

$$h_t = o_t \times \tanh\left(C_t\right) \tag{8}$$

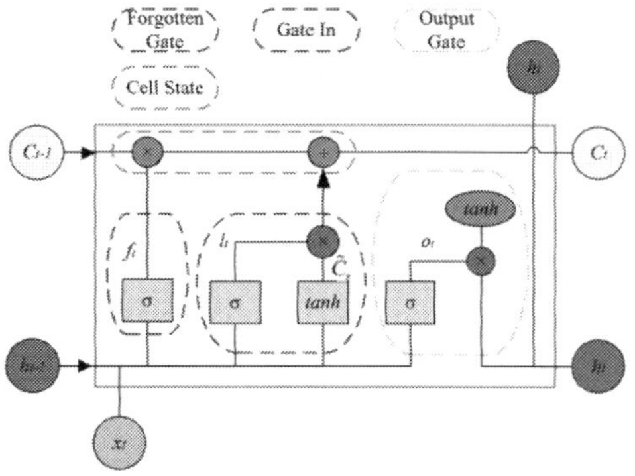

Figure 2. Internal structure of LSTM.

4. The PCA-LSTM prediction model establishment

The modeling process of the PCA-LSTM CNC machining center component life prediction model is as follows:

(1) We collect data and conduct preliminary processing. We process the collected data, including filling in missing values, testing outliers, etc.

(2) Data standardization processing. Data units and dimensions are unified [14] to eliminate the impact on model performance.

(3) We calculate co-variance. We calculate the correlation coefficient of standardized data, solve the matrix eigenvalues and eigenvectors, and arrange them in descending order.

(4) We calculate the cumulative contribution percentage of the main components. If the cumulative contribution percentage of the first K main components is greater than or equal to 80%, it can be used as input data for the model.

(5) We build an LSTM model and import the dimensionality reduction processed data into the LSTM model for training in two sets: a training set and a testing set.

(6) We predict the results, export the predicted results of component fault error fluctuations, and evaluate the performance of the model based on indicators. The PCA-LSTM prediction model is shown in Figure 3.

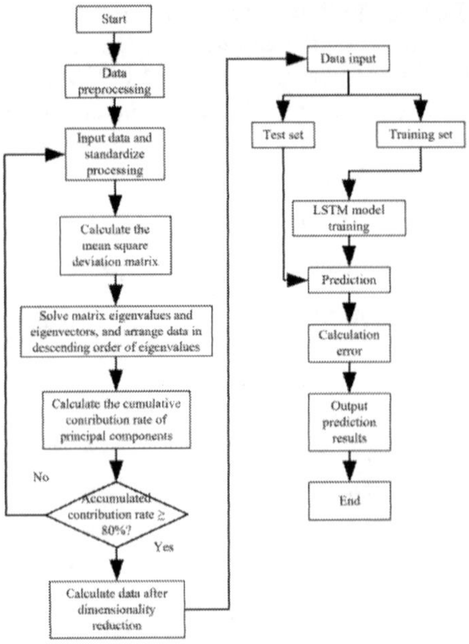

Figure 3. Flow chart of PCA-LSTM prediction model.

5. Experimental testing process and result analysis

5.1 Data extraction and processing

During the operation and processing of CNC machining centers, component failures usually occur in the form of vibration [15]. Therefore, this article chooses to monitor the vibration status of components during the part-cutting process as the basis [16]. Based on the prediction steps of the above model, we analyze the fault data selected from the existing VMC850B CNC machining center experimental platform from 2018 to 2023, organize 200 sets of sample data, and select them. 70% of the data is used as the training set, and 30% of the data is used as the testing set.

For outliers, we first remove the outliers and take the mean. For missing values, we select the previous round of detection values to fill in. Then we standardize the data.

5.2 PCA-LSTM model component fault prediction process

In order to better demonstrate the predictive performance of the PCA-LSTM model studied in this paper, the error fluctuation state between BP neural network and a single LSTM model was compared. The specific training parameters of the three models are shown in Table 2. The comparison of prediction performance is shown in Figure 4.

Table 2. Training parameters for each model.

The model	PCA-LSTM	LSTM	BP
Number of hidden Layer nodes	150	150	10
Initial learning rate	0.001	0.001	0.05
Optimizer	Adam	Adam	SGD
Number of iterations	200	200	200

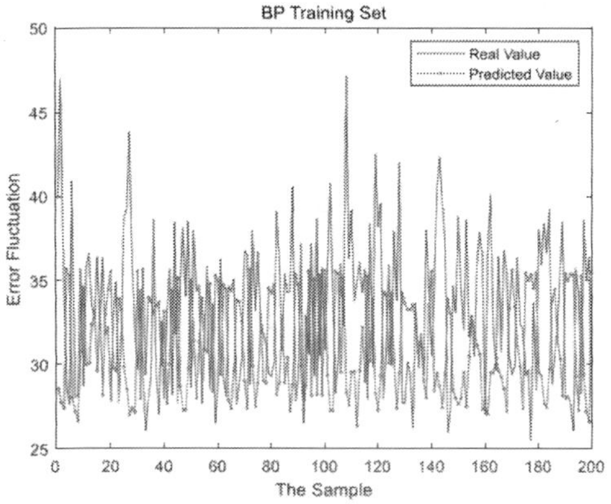

(a) Error fluctuation graph of BP neural network

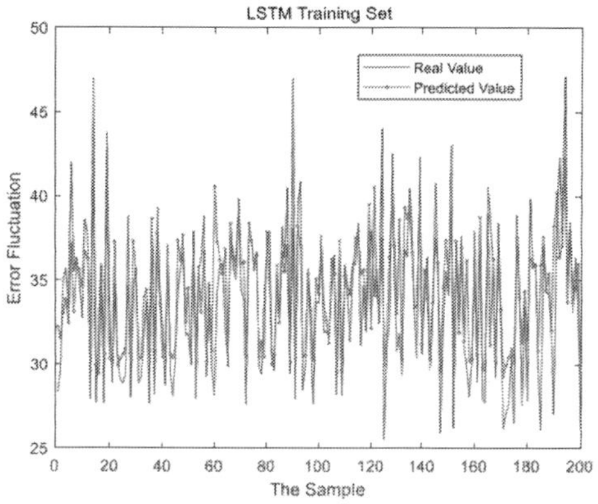

(b) LSTM model error fluctuation chart

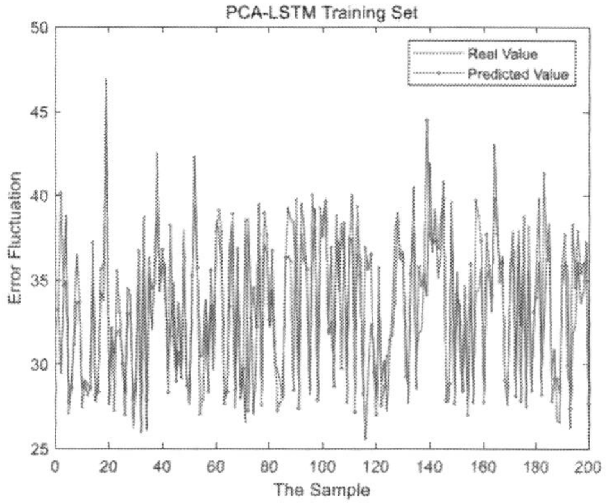

(c) PCA-LSTM model error fluctuation chart

Figure 4. Error fluctuation chart of three prediction models.

From Figure 4(a) and 4 (b), it can be seen that when the BP neural network and LSTM model are trained, not only does the overall trend of expected results show poor performance, but there is also a significant deviation between the expected and actual values.

As shown in Figure 4(c), the PCA-LSTM model performs better in training. Not only is the overall trend basically consistent, but the deviation between the actual value and the expected value is also very small. Overall, the comparison shows that the model has better performance and higher prediction accuracy.

5.3 Error analysis

The error fluctuation diagrams of the three models can intuitively understand the actual and expected states of error fluctuations. In addition, the mean squared error and prediction error of the three models are also compared. The PCA-LSTM model is superior to the data, as shown in Table 3.

Table 3. Comparison of mean square error and prediction error of three models.

Network model	Run time /s	Root mean square difference	Maximum relative error%	Mean absolute error%	Average relative error%
BP	51	4.309	44.282	3.316 6	9.897
LSTM	25	2.209	18.017	1.785 1	3.906
PCA-LSTM	13	1.572	10.161	1.311 6	2.816

Comparing the three models, the main reason for the error is the different types of machine tool faults, which have a high degree of randomness. Predictive models are prone to overfitting or underfitting, which can easily lead to significant biases and imbalances in experimental data. To achieve more optimized results, it is necessary to reduce type complexity by summarizing fault-type data. Further research is still needed.

From this, it can be seen that the PCA-LSTM model not only has a generally consistent overall trend but also has a small deviation between actual values and expected values. Overall, the comparison indicates that the model has better performance and higher prediction accuracy.

6. Conclusion

The following conclusions can be drawn from the experimental results:

By analyzing the types of faults in CNC machining center components, a PCA-LSTM prediction model is constructed. Parameters such as hidden layers and optimal training times are reset to achieve more accurate fault prediction for CNC machining center components.

Compared with other traditional prediction methods, the PCA-LSTM prediction model improves uptime and reduces errors.

The PCA-LSTM model proposed in this article can achieve the prediction of component failures in CNC machining centers.

Acknowledgment

Fund Project: Shanxi Province Higher Education Science and Technology Innovation Plan Project (2024L321)

References

[1] SUN Junmin, Zhang Wentao, Liu Chao, et al. Prediction of Urban Gas Daily Load Based on PCA-LSTM [J]. Gas Technology. 2023(02): 12-19.
[2] LI Yan, ZHANG Jiupeng, CHEN Zi-Xuan, et al. Evaluation of Asphalt Pavement Performance Based on PCA-PSO-SVM [J]. et al. Journal of Jilin University (Engineering and Technology Edition). 2023. 53(6): 1729-1735.
[3] LV Wenwen, WEI Xiuling. Prediction of Residual Life of Lithium-ion Batteries Based on PCA-PSO-ELM [J]. Digital Technology &Application. 2023. 41(3): 74-77.
[4] WANG Keyang, ZHANG Yao, LI Kewen, et al. On-line Detection System of CNC Tool Damage Based on Attention-LSTM [J]. Computer and Modernization. 2022(05): 68-74.
[5] MOU Liming. Remaining Useful Life Prediction of Critical Components of CNC Machine Tool Based on Weibull Regression Model [D]. Jilin University. 2022. 5.
[6] XU Min, WANG Ping. Research on Fault Diagnosis of Rotating Machinery Based on Deep LSTM Residual Network [J]. Machine Tools & Hydraulics, 2023.51 (4): 184-190.
[7] TAN Feng, LI Chengnan, XIAO Hong, et al. A Thermal Error Prediction Method for CNC Machine Tool Based on LSTM Recurrent Neural Network [J]. Chinese Journal of Scientific Instrument. 2020. 41(09): 79-87.
[8] ZHU Yongchao, ZHU Caichao, SONG Chaosheng. PCA-PSO/GS-SVM Combination Method in Wind Power Gearbox Application Research in Fault Prediction [J]. Acta Energize Solaris Sinica. 2021. 42(03): 35-42.

[9] LI Ming, YUAN Yi-Ping, JIA Yi-Daer, et al. Study on Prediction of Wind Turbine Output Power Considering PCA-LSTM [J]. Machinery Design & Manufacture. 2022(9): 145-148.
[10] Zhang Xin, Zhang Tao, Li Yumei, et al. Research on Evaluation Model of Stick-slip Vibration Intensity Based on PCA-LSTM [J]. China Petroleum Machinery. 2023. 51(02):18-25.
[11] LIU Ying, PENG Xinxia, WANG Tong, et al. Non-invasive Load Monitoring Method Based on PCA-LSTM Algorithm [J]. Electrical Measurement & Instrumentation. 2023. 60(03): 53-58.
[12] DAI Jianfei, YANG Peng, ZHU Liyi, et al. A PCA-LSTM Neural Network-integrated Method for Phreatic Line Prediction [J]. China Safety Science Journal. 2020. 30(3): 94-101.
[13] LU Wenxing, RUI Haidong, LIANG Chang Yong, et al. A Method Based on GA-CNN-LSTM for Daily Tourist Flow Prediction at Scenic Spots [J]. Theory and Applications of Information Theoretic Machine Learning. 2020. 22(3), 261(1-18).
[14] GAO Zhongshi, SU Yang, LIU Yudong. Study on Intrusion Detection Based on PCA-LSTM [J]. Computer Science. 2019. 46(S2): 473-476+492.
[15] WANG Yining, XIE Da, WANG Xitian, et al. Prediction of Interaction Between Grid and Wind Farms Based on PCA-LSTM Model [J]. Proceedings of the CSEE. 2019. 39(14): 4070-4081.
[16] YU Da, ZHANG Wei, WANG Hui. Abnormal Voiceprint Diagnosis Method of Oil-immersed Transformer Based on LSTM Neural Network [J]. Smart Power. 2023, 51(02): 45-52.

Failure mechanism analysis of the crack located at the neck of stainless steel flange

Meng Yuan[1,2,3,*], Yuan Yao[1], Wenfang Zhu[4], Fang Zhang[1,2], Fengping Zhong[1,2,3,*], Liuyi Huang[1,2,3], Chao Peng[1] and Korzhyk Volodymyr[3]

[1]National Quality Inspection and Testing Center of Special Metal Structural Materials, Zhejiang Academy of Special Equipment Science, Hangzhou 310020, China
[2]Key Laboratory of Special Equipment Safety Testing Technology of Zhejiang Province, Hangzhou 310020, China
[3]Zhejiang-Ukraine Joint Laboratory for Materials and Welding Testing Technology, Hangzhou 310020, China
[4]Songyang Institute of Measurement and Testing, Lishui 323400, China

[*]Corresponding author's E-mail: yuanmeng@zjtj.org (Meng Yuan); zhongfp@zjtj.org (Fengping Zhong)

Abstract: During new pipeline installation supervision inspections in a large petrochemical project, several dozen flanges, made of S30403, failed due to lateral cracks in the neck areas. A detailed analysis of the failure modes and mechanisms of the flanges was carried out. The chemical composition, hardness, metallographic, microstructure structure and EDS analysis in crack zones were tested by spark discharge optical emission spectrometer, Brinell hardness tester, metalloscope, and scanning electron microscope, respectively. It was proved that in the metallographic structures, excessive wire- and network-like ferrites and a large amount of carbides were observed. The crack originated from the outer surface of the neck of the flange and extended to the center of the flanges along the wire- and network-like ferrites. Meanwhile, a large amount of impurities were found inside the cracks as well, which were confirmed as chromium oxides. It was indicated that the flange may have cracked during the forging process, and then oxidation occurred at a high temperature. Some micro-cracks were formed at the aggregating areas of non-metallic inclusions during the forging of the flange, and masses of ferrites provided the path for rapid crack growth. Therefore, the failure of the flanges should be caused by improper heat treatment processes and excessive non-metallic inclusions in the materials.

1. Introduction

Forged flanges were widely applied in oil and gas transmission, chemical equipment, and other fields, attracting the attention of many engineering scholars. Meanwhile, various engineering failure problems have also been encountered [1-5]. Recently, we received cases of stainless steel flange cracking as well. During the inspection of the stainless steel flange used by a large petrochemical company, it was found that there was a crack in the neck of the flange. To analyze the failure modes and mechanisms of the flange and determine the cause of the crack, the company commissioned us to conduct a failure analysis on the cracked flange. The failed samples are shown in Figure 1 and Figure 2. It was also said that more than 60 flanges experienced similar problems. According to the information provided by the client, the manufacturing standard for these cracked stainless steel flanges was NB/T 47010-2017, the model was PN20-DN150-SCH40S, and the manufacturing material was S30403. In order to identify the true cause of the failure of this batch of stainless steel flanges, we conducted systematic physical and chemical testing and failure mechanism exploration on typical failure samples. This research must be expected to provide engineering and technical support for the processing, manufacturing, installation, and use of similar materials and industrial products.

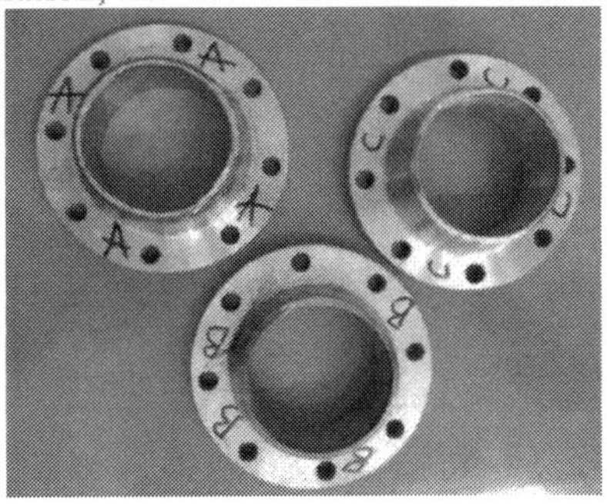

Figure 1. Samples of the failed stainless steel flanges.

2. Materials, methods and results

2.1 Visual inspection

Typically, cracked flange A was selected as the analysis object. A macroscopic inspection was conducted on the cracked flange A. It was found that there was a circumferential crack almost parallel to the butt weld on the outer surface of the flange pipe neck. The length of the circumferential crack was about 285 mm. This crack was about 30 mm away from the flange butt weld, as shown in Figures 2(a)-(c). We took analysis specimens from the neck area of flange A for chemical composition spectrum analysis, hardness testing, metallographic testing, scanning electron microscopy, and spectral analysis. The sampling location and testing items can be seen in Figure 2(d) and Table 1.

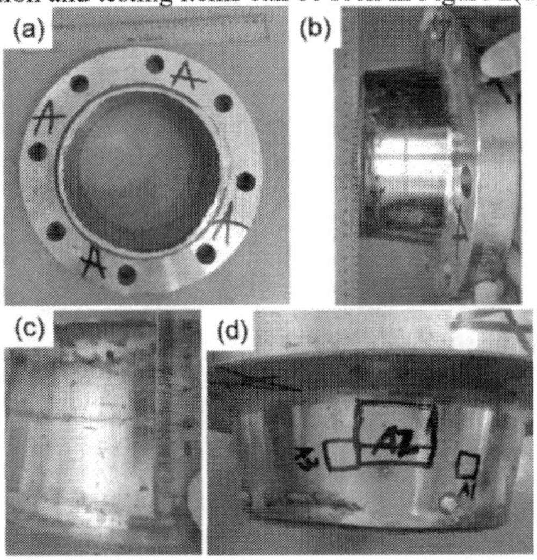

Figure 2. The macro photos of the failed stainless steel flange. (a) The vertical view, (b) The side view, (c) Failure location, and (d) Schematic diagram of the location for taking test specimens.

Table 1. Sampling locations and testing items.

Specimen No.	Sampling location	Testing item
A1	Cracking area of flange neck	Metallographic examination, Scanning electron microscopy and energy spectrum at the crack location
A2	Uncracked area of flange neck	Analysis of chemical composition
A3	Uncracked area of flange neck	Hardness testing

Table 2. Chemical composition of the failure flange (Mass fraction, %).

Element	C	Si	Mn	P	S	Cr	Ni
A2	0.029	0.52	1.55	0.031	0.0067	18.13	8.02
Standard value	≤0.030	≤1.00	≤2.00	≤0.035	≤0.015	18.00~20.00	8.00~12.00

2.2 Chemical analysis

Based on the standard GB/T 11170-2008, the chemical composition analysis was conducted on the A2 sample by spark discharge optical emission spectrometer, and the results are shown in Table 2. It indicated that the compositions of the flange meet the chemical composition requirements for S30403 stainless steel in standard NB/T 47010-2017. The carbon content was close to the upper limit required by the standard.

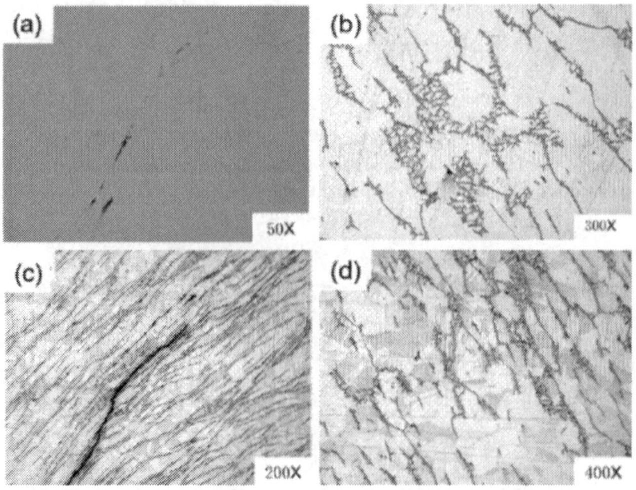

Figure 3. The metallographic images of the failed flange. (a) Non-metallic inclusions, (b) α phase content, (c) and (d) Metallographic morphology 200× and 400×, respectively.

2.3 Hardness test

The Brinell hardness testing was implemented on the A3 sample on the basis of the standard GB/T 231.1-2018, and the results of three randomly measured points were 172 HBW, 172 HBW, and 174 HBW, respectively. The manufacturing standard requirement for flanges was 128~187 HBW, and the test results satisfied the manufacturing requirements.

2.4 Metallography

2.4.1 Non-metallic inclusions, the area content of the α phase and metallographic morphology. Based on the standard GB/T 10561-2005/ISO 4967:1998(E), the A1 specimen was tested for non-metallic inclusions in Figure 3(a). There were coarse silicate inclusions on the outer surface and core of the flange, and the worst field of view was located in the outer surface area. The inclusion level in part A of the flange reached the C3e level.

In line with the standard GB/T13305-2008, after electrolytic corrosion with 20 wt% NaOH solution, the area content of the α phase was determined, and the testing results were shown in Figure 3(b). The results showed that the area content of the α phase was about 15%, which was level 3.0.

Figures 3(c)-(d) are the metallographic structures of the flange sample, which were observed after corrosion with a mixed solution of ferric chloride and hydrochloric acid, on the basis of the standard GB/T 13298-2015. It demonstrated that the metallographic structure of the flange was composed of "austenite + ferrite+ carbide", with the ferrite distributed in a long linear or network shape on the austenitic matrix, and the carbide distributed in a dispersed manner.

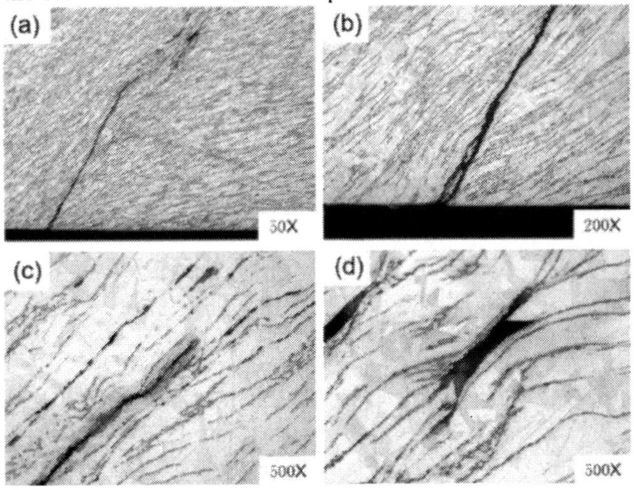

Figure 4. The metallographic images of the crack area at the neck of the flange (A1) at different magnifications.

2.4.2 Characteristics of flange cracks and microstructure of nearby areas. The crack on the flange was located at its neck and propagated in a circumferential direction, as shown in Figure 2(c). The crack initiation point was located on the outer surface of the flange, and a larger volume of inclusions can be observed near the crack tip (Figure 4(a)). In Figure 4(b) and (c), the crack propagation direction and the linear or network distribution of ferrite can be seen, and there were dark gray oxides at the crack tip. The pores formed by the detachment of inclusions near the crack were observed in Figure 4(d).

2.5 SEM and EDS analysis

We took a specimen from the flange crack area. After corrosion and cleaning, we observed the crack characteristics with the scanning electron microscope (SEM). The micro zone composition analysis was carried out using the energy dispersive spectrum (EDS). The results are exhibited in Figure 5. It indicated that the microstructure of the flange was consistent with the metallographic testing, appearing as pits or pores due to corrosion removal of ferrite, carbides, inclusions and so on (Figures 5(a) and (b)). By conducting EDS analysis at different locations (Figure 5(d)), it was found that the weight percentages of chromium and nickel elements in the matrix and pits were similar to the detection results in Table 2 (Point 2 and Point 3). However, the oxygen and chromium elements at the crack were abnormal, which indicated this location had been severely oxidized (Point 1).

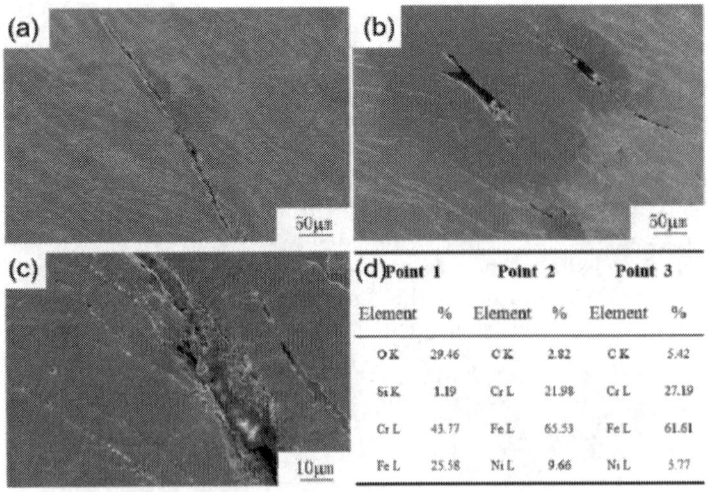

(d)

Point 1		Point 2		Point 3	
Element	%	Element	%	Element	%
O K	29.46	C K	2.82	C K	5.42
Si K	1.19	Cr L	21.98	Cr L	27.19
Cr L	43.77	Fe L	65.53	Fe L	61.61
Fe L	25.58	Ni L	9.66	Ni L	5.77

Figure 5. The SEM images and EDS analysis of the crack area at the neck of the flange.

3. Discussion

According to the above detection data, it can be found that the chemical composition and hardness of the flange meet the technical requirements for S30403 stainless steel forgings in the manufacturing standard.

The metallographic structure of the flange was composed of "austenite+ferrite+carbide", with ferrite distributed in a long linear or network shape on the austenitic matrix, and carbide distributed in a dispersed manner. In the metallographic structure of flanges, the area content of the α phase was about 15%. Under equilibrium state, austenitic stainless steel should exhibit a complete austenitic structure at room temperature, but in actual production, component segregation or heat treatment can cause a certain amount of α phase to exist in the austenite, which can have adverse effects on the material's hot workability and corrosion resistance. The presence of a network of ferrite and a large number of carbides indicated that the insulation time of the flange during solid solution heat treatment was insufficient or the heating temperature was too low, so that the ferrite and dispersed carbides didn't have a sufficient driving force to dissolve into the austenite phase and formed to a single-phase structure [6-7].

The coarse non-metallic inclusions were found in the microstructure of the flange sample, and in some areas, the non-metallic inclusion level reached the C3e level. The worst-case field of view was located in the outer surface area of the flange, and the tip of the inclusion was visible to be connected to the ferrite wire mesh. When there was an accumulation of inclusions in the material (especially excessive inclusions), the continuity of the material would be disrupted, causing stress concentration in a certain area around it, making that area a weak point in the material. There were many inclusions gathering on the outer surface of the pipe neck, according to the metallographic analysis. However, because of the significant differences in deformation rate, elastic coefficient and thermal expansion coefficient between the inclusions and austenitic matrix, it would easily generate microcracks at the interface between inclusions and austenite during forging and heat treatment [8-10].

The crack originated from the outer surface of the flange neck. Then, it extended along the network ferrite direction to the inside of the flange pipe diameter. There were a large number of impurities inside the crack, and EDS results proved that the main product was the oxidation of Cr. However, in a room-temperature environment, the flange oxidation process should not be so fast [11]. Therefore, it was inferred that the crack may have appeared during the forging process, leading to high-temperature oxidation.

4. Conclusions

In light of the above inspections and analysis, we believe that the crack in the flange neck should be caused by improper heat treatment process and the excessive non-metallic inclusions in the stainless steel. During the flange forging process, microcracks formed at the aggregation of non-metallic inclusions. Inappropriate solid solution heat treatment resulted in a large amount of wire mesh ferrite remaining inside the material, which provided a path for rapid crack propagation.

Acknowledgment

We thank the Zhejiang Provincial Natural Science Foundation of China (No. LTGC23E010001) and the Youth Science and Technology Project of Zhejiang Provincial Administration for Market Regulation (No. QN2023427).

References

[1] Jo, H.; Kang, M.; Park, G.-W.; Kim, B.-J.; Choi, C.Y.; Park, H.S.; Shin, S.; Lee, W.; Ahn, Y.-S.; and Jeon, J.B. Effects of Cooling Rate during Quenching and Tempering Conditions on Microstructures and Mechanical Properties of Carbon Steel Flange. Materials 2020, 13, 4186.

[2] Sun, W., Chen, L., Zhang, T. et al. Perform optimization and microstructure analysis on hot precision forging process of a half axle flange. Int. J. Adv. Manuf. Technol. 2018, 95, 2157–2167.

[3] S.S.M. Tavares, J.M. Pardal, B.B. Almeida, M.T. Mendes, J.L.F. Freire, and A.C. Vidal, Failure of superduplex stainless steel flange due to inadequate microstructure and fabrication process, Engineering Failure Analysis 2018, 84, 1–10.

[4] Li, Z., Lai, J., Ren, Z., Shi, Y., and Kong, X. Failure mechanical behaviors and prevention methods of shaft lining in China. Engineering Failure Analysis 2023, 143, 106904.

[5] Alias J, Alang N A, Ahmad A H, and Razak, N. A. Failure Analysis of a Carbon Steel Pipe-Flange Component. Journal of Failure Analysis and Prevention 2023, 23(2): 490-496.

[6] Peng, X. C., Guo, H. J., Zhang, X. F., Luo, Y. W., Sun, Y., Guo, J., & Zheng, X. D. Strengthening and control of second-phase particle precipitation in ferritic/austenitic/martensitic heat-resistant alloys: a review. Journal of Iron and Steel Research International 2024, 1-21.

[7] Zhang, Y., Shen, G., Gu, J., & Zhang, J. Synergistic effect of well/over-tempered hierarchical martensitic structure and carbides (M23C6 and MX) on austenite growth in 9Cr steel. Journal of Materials Research and Technology, 2023, 26, 7381-7394.

[8] Suito H, Ohta H, and Morioka S. Refinement of solidification microstructure and austenite grain by fine inclusion particles. ISIJ international 2006, 46(6): 840-846.

[9] Zhou, Y., Koyama, M., Hojo, T., Ajito, S., and Akiyama, E. Transformation-induced microcracks and their arrest with different deformation temperatures in a medium Mn steel. Metallurgical and Materials Transactions A 2023, 54(1), 153-166.

[10] Hu, L., Zhang, L. Q., Hu, F., Zhang, G. H., & Zheng, K. Fracture mechanism of inclusions in thick plate center under tensile loading. Journal of Iron and Steel Research International 2024, 31(5), 1196-1209.

[11] Borgioli, F. The corrosion behavior in different environments of austenitic stainless steels subjected to thermochemical surface treatments at low temperatures: an overview. Metals 2023, 13(4), 776.

Journal of Physics: Conference Series **2879** (2024) 012029 doi:10.1088/1742-6596/2879/1/012029

Optimization study of air-cooled heat dissipation structure based on waterborne electric airplane

Xinyue Fan[2], Jicheng Liao[1] and Shuli Wang[*,1,2]

[1]Shenyang Aerospace University, Shenyang, Liaoning, 110136, China
[2]Liaoning General Aviation Academy, Shenyang, Liaoning, 110136, China

[*]934800282@qq.com

Abstract. Taking the water electric aircraft as the research object, to solve the problem of high temperature of the motor, controller and radiator in the powerhouse when the aircraft is travelling under the high temperature condition in summer. Firstly, the three-dimensional model of the power module of the electric aircraft is established and simulation experiments are carried out, and then according to the simulation results, the power module is optimized and improved, and after simulation experiments, it is known that the highest temperature of the surface of the electric motor is 432.83K, the highest temperature of the surface of the controller is 370.82K, and the highest temperature of the surface of the radiator is 368.67K, and the final experimental results show that the optimization meets the cooling requirements of the aircraft flight. The final experimental results show that the optimisation meets the heat dissipation requirements of the aircraft flight.

1. Introduction

Electric aircraft have attracted people's attention due to their advantages such as zero emissions, low noise, low cost, and simple structure[1]. During the flight, the aircraft generates a large amount of heat. If it cannot be effectively dissipated, it will affect the efficiency and performance of equipment in the factory, thereby affecting the power, performance, and safety of electric aircraft. Therefore, the heat dissipation design of electric aircraft is very important.

Common heat dissipation methods include air cooling and water cooling. The air-cooled cooling structure is simple, reliable, and does not require the addition of other equipment. Water cooling and heat dissipation require other equipment such as water pumps and pipes. In order to improve the safety and endurance of electric aircraft, and to meet the cooling needs of the aircraft, air-cooled cooling is chosen.

This paper optimizes the air-cooled heat dissipation system of the powerhouse according to the modeling and CFD simulation of the powerhouse of the electric aircraft, and the vector distribution diagram and temperature distribution diagram of the powerhouse are used as references to optimize the air-cooled heat dissipation system of the powerhouse and the optimization meets the heat dissipation requirements of the aircraft, which shows that this research has certain reference value and significance for the heat dissipation design and research of the electric aircraft.

2. Modeling

2.1. Geometric modeling

The paper mainly analyses the thermal performance of the electric aircraft in flight, so only the powerhouse is analysed, including the three parts of the cabin: the motor, the controller and the radiator, which is simplified to save computational resources[2], and the simplified results are shown in Figure 1.

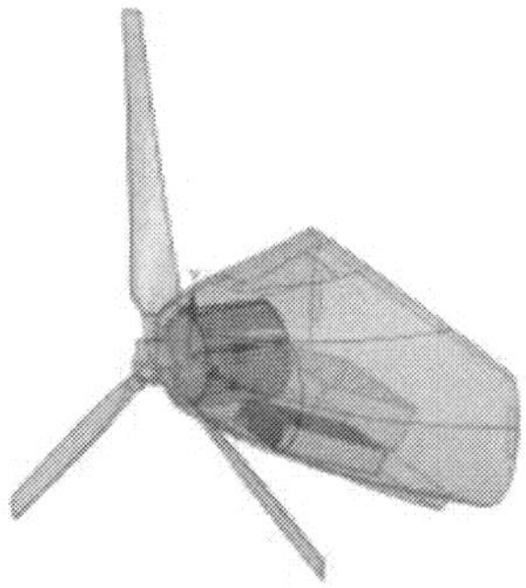

Figure 1. Simplified model drawing of the powerhouse.

2.2. Mathematical model

Because of the relatively low air velocity in the powerhouse, the powerhouse be treated as an incompressible turbulent flow, in accordance with the basic laws of fluid dynamics. The specific equations are as follows[3].

Law of Conservation of Mass:

The mass gain within a fluid element over a given time period equals the net mass influx into that element during the same time period, given by the following equation:

$$\frac{\partial \rho}{\partial t}+\frac{\partial(\rho u)}{\partial x}+\frac{\partial(\rho v)}{\partial y}+\frac{\partial(\rho w)}{\partial z}=0 \tag{1}$$

Equation (1) the density of the fluid is ρ , t is the time, u , v , w are the components of the fluid velocity on x , y , z . When the density of the fluid does not change, the fluid can be when as an incompressible state.

Law of Conservation of Momentum:

The external resultant force on a fluid element is equal to its time rate of change of momentum, which leads to the equation for the conservation of momentum in the x , y and z directions.

The equations for conservation of momentum in the x , y and z directions are as follows:

$$\frac{\partial(\rho u)}{\partial t}+div(\rho u u)=-\frac{\partial p}{\partial x}+\frac{\partial \tau_{xx}}{\partial x}+\frac{\partial \tau_{yx}}{\partial y}+\frac{\partial \tau_{zx}}{\partial z}+Fx \tag{2}$$

$$\frac{\partial(\rho v)}{\partial t}+div(\rho v u)=-\frac{\partial p}{\partial y}+\frac{\partial \tau_{xy}}{\partial x}+\frac{\partial \tau_{yy}}{\partial y}+\frac{\partial \tau_{zy}}{\partial z}+Fy \tag{3}$$

$$\frac{\partial(\rho w)}{\partial t}+div(\rho w u)=-\frac{\partial p}{\partial z}+\frac{\partial \tau_{xz}}{\partial x}+\frac{\partial \tau_{yz}}{\partial y}+\frac{\partial \tau_{zz}}{\partial z}+Fz \tag{4}$$

In Equation(2), (3) and (4), ρ represents the pressure exerted on the microelement, τ_{xx} , τ_{yx} , τ_{zx} denote the components of viscous stress present on the surface of the microelement, while Fx , Fy and Fz signify the body forces acting upon the microelement.

The Law of Conservation of Energy:

The total work done by the outside world on the micrometric body plus the increase in heat is equal to the increase in energy within the micrometric body, and the governing equations are as follows:

$$\frac{\partial(\rho T)}{\partial T}+div(\rho \vec{v} T)=div(\frac{k}{C_\rho}gradT)+S_T \tag{5}$$

In Equation (5) k is the heat transfer coefficient of the fluid, T denotes the temperature, C_ρ is the specific heat capacity of the fluid, and S_T is the residual energy.

There are three types of heat transfer: thermal conduction, thermal convection and thermal radiation. In the study of thermal dissipation in the power cabin of an electric airplane, the convection of heat and radiation of heat have the greatest influence on the cabin temperature, and this paper mainly relies on macroscopic turbulence for heat transfer. Among them, the heat convection equation is shown as follows:

$$q_x = Ah_x(T_w - T_f)_x \tag{6}$$

In Equation (6), q_x is the heat transfer rate of heat convection, A is the heat transfer area of heat convection, h_x is the heat transfer coefficient of heat convection, T_w T is the surface temperature of the object, and T_f stands for fluid temperature[4].

3. CFD experiment and analysis

3.1. Perimental condition

This experiment is for rated operating conditions with the following Table 1:

Table 1. Experimental working condition data.

parameters	numerical value
Environmental temperature /K	300
Environmental pressure /Pa	101325
Motor speed /(r/min)	2450
Motor Losses /kW	4.8
Controller Loss /kW	2

According to relevant information, during normal flight of an aircraft, one-third of the heat generated by the electric motor is dissipated through its own casing, and two-thirds of the heat is dissipated through the heat dissipation holes. 40% of the heat generated by the controller is discharged by the controller itself, and 60% of the heat is released by the radiator[5]. After calculation, the heat flux of each heating wall surface is shown in Table 2.

Table 2. Heat flux parameters.

Wall	Loss of heat/w	Heat dissipation area/m^2	heat flux/W/m^2
Controller	800	0.23083	3465.75
Motor	1600	0.45186	3540.92
Radiator	1200	1.0387	1155.29

3.2. Simulation results

As can be seen in vector Figure 2 and Streamline Figure 3 of the powerhouse:

Problem 1: The blown-in airflow passes over the upper outer surface of the motor before being directed to the inner wall of the powerhouse.

Problem 2: The airflow is directed above on the front of the controller after passing through the surface of the motor, creating a backflow in the nacelle.

Problem 3: The majority of the inflowing air flow is directed upward after passing through the lower surface of the motor, while a small portion flows to the left and right surfaces of the controller and is blocked by the radiator to create a vortex in front of the radiator.

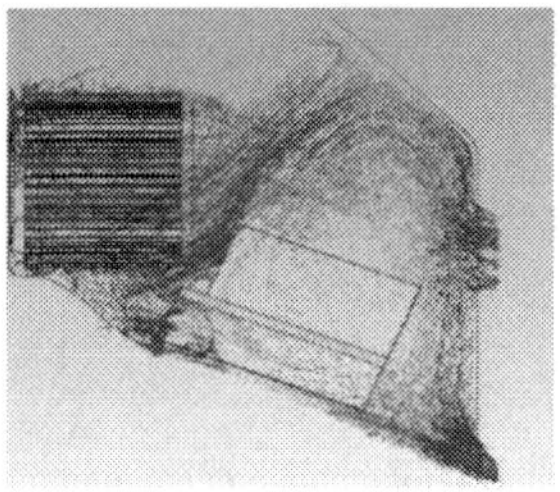

Figure 2. Vector diagram of the powerhouse.

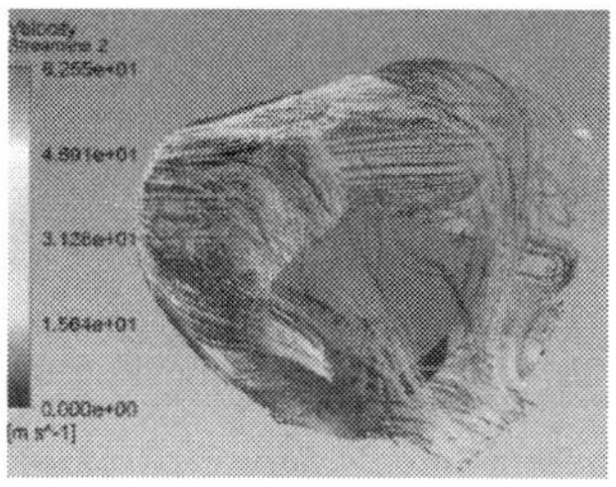

Figure 3. Streamline diagram of powerhouse.

From the temperature distribution diagram in Figure 4, it is seen that the local temperature on the upper surface of the motor reaches 606.85K, the temperature on the surface of the controller reaches 428.313K, and the temperature on the two sides of the radiator reaches 413.17K. It can be seen that the temperature of the parts in the powerhouse is too high to meet the flight requirements of an electric aircraft and therefore needs to be optimised and improved.

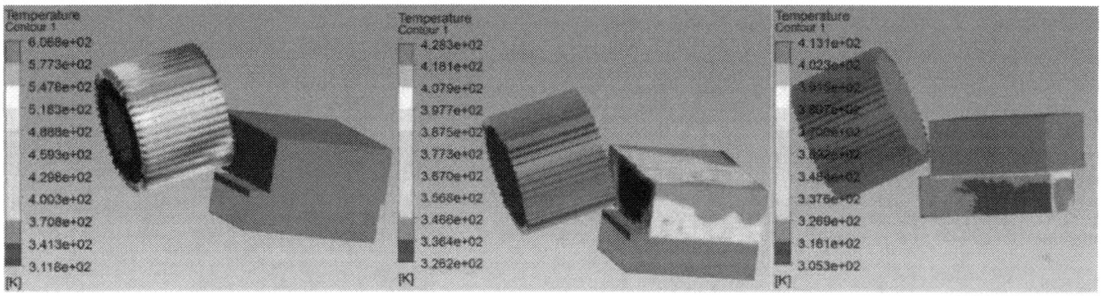

Figure 4. Temperature distribution.

3.3. Optimization program

The following optimizations are made based on the above three problems[6]:

1. add upper inlet 1, inlet 2, and inlet 3 above and on both sides of the powerhouse.The three entrances are all arc-shaped long holes, The area of inlet 1 is 6788.78mm^2 with a circumference of 437.51mm, the area of inlet 2 is 4834.49mm^2 with a circumference of 524.67mm, and the area of inlet 1 is 6788.78mm2 with a circumference of 437.51mm.the model is shown in Figure 5.

2. In order to solve the problems of airflow blockage, backflow, and dispersion in the powerhouse, three deflectors have been added to the powerhouse.The yellow guide plate is a rectangular shape of 180mm × 50mm, placed directly in front of the radiator and parallel to the motor axis. The red guide plate is a curved guide plate that is level with the yellow guide plate and slightly curved at the intersection of the motor and the controller. The front end of the blue guide plate wraps around the upper end of the motor, about 20mm away from the motor, and the rear end is parallel to the upper surface of the controller, about 30mm away from the controller. The specific installation is shown in Figure 6.

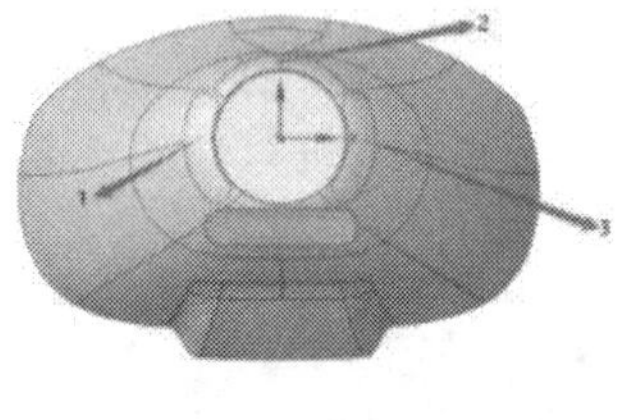

Figure 5. Air intake modification diagram.

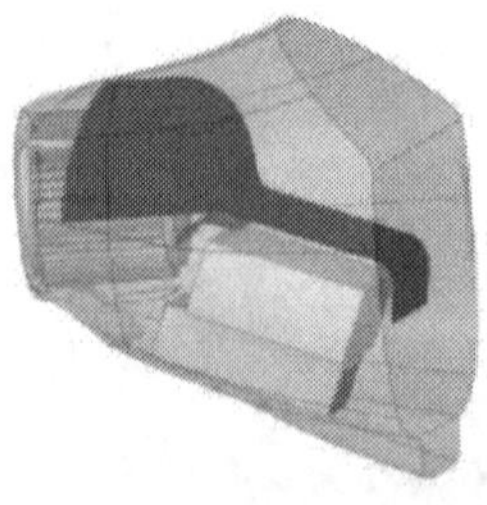

Figure 6. Installation diagram of the deflector plate.

3.4. Optimization results

The optimized vector diagrams of the interior of the powerhouse are shown in Figure 7 and Streamline diagram in Figure 8. The figure shows that the airflow from the upper and side inlets, directed by the guide plate, can closely adhere to the motor surface, effectively removing a substantial amount of heat. The airflow from the lower inlet, influenced by the guide plate, partly combines with the airflow from the upper inlet in front of the controller, then flows out from the controller's upper surface, carrying away heat. The remaining part flows towards the radiator and the sides of the controller. The backflow of air in the surface compartment has been eliminated, and the air blockage in front of the radiator has been

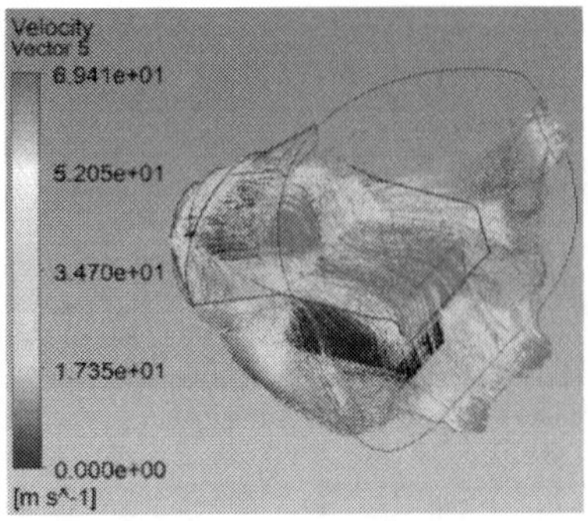

Figure 7. Vector diagram of the powerhouse.

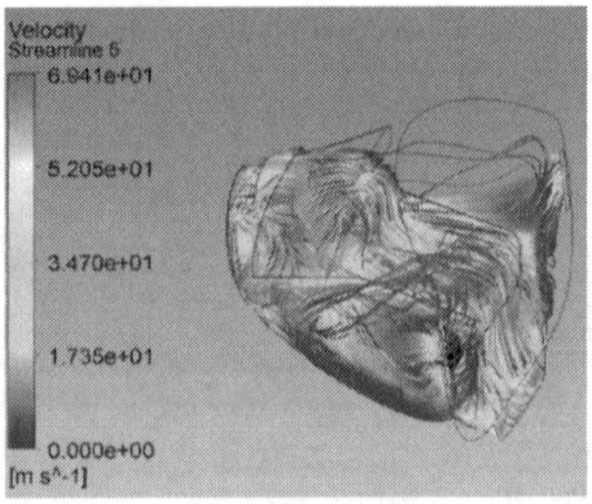

Figure 8. Streamline diagram of powerhouse.

From the temperature distribution diagram in Figure 9, the temperature distribution on the surface of the motor is uniform and decreases significantly after the optimisation, and the temperatures of the controller and radiator are also greatly improved. The maximum temperature of the motor decreased from 606.85K to 432.83 K. The maximum temperature of the controller decreased from 428.313K to 370.82K. The maximum temperature of the heat sink decreased from 413.17Kto 368.67K.

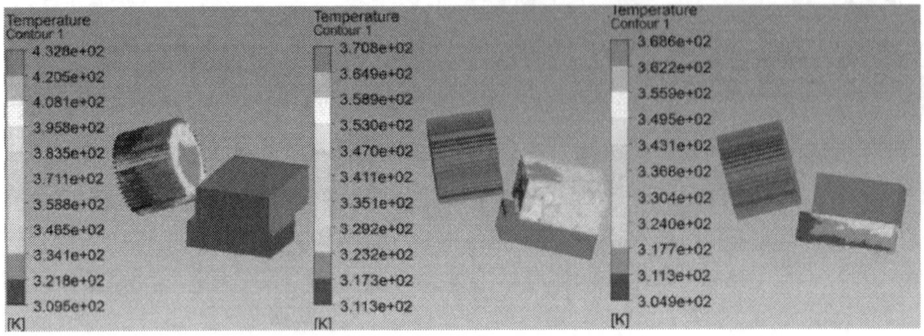

Figure 9. Temperature distribution.

4. Conclusion

The temperature inside the powerhouse is high before optimisation, through CFD simulation and analysis, the optimisation scheme is proposed for the powerhouse heat dissipation system, and finally verified by simulation experiments, the aerodynamic performance inside the powerhouse is improved to a certain extent, and it can reduce the temperatures of the components, which can effectively avoid the occurrence of thermal damage, and improve the heat dissipation efficiency and heat dissipation performance of the powerhouse, which proves that this scheme is effective and feasible.

References

[1] Fengtian Y, Zhenwei F and Song X ,et al. Technological innovation and practice of electric aircraft in China[J]. Journal of Aviation,2021,42(03):7-12.
[2] Jingchun T, Youwei T and Tianyang Z,et al. Analysis and structural optimisation of engine compartment heat dissipation performance of forklift truck[J]. Journal of Hefei University of Technology (Natural Science Edition),2023,46(05):577-582+598.
[3] Liu T.Research on water flow force of single-column type members of cross-sea bridge sorghum[D]. Southwest Jiaotong University,2018.
[4] Guiping J. Research on simulation and analysis of automotive cabin thermal management [D]. Hunan University,2017.
[5] Zhenkun T.Thermal simulation and air cooling design of electric aircraft engine compartment [D]. Shenyang University of Aeronautics and Astronautics, 2018
[6] Lin T, Chenguang L and Libin T. Thermal performance analysis and structural optimisation of motor and controller cooling runners[J]. Journal of Chongqing University of Technology (Natural Science),2023,37(02):104-112.

Journal of Physics: Conference Series **2879** (2024) 012030 doi:10.1088/1742-6596/2879/1/012030

Influence of meshing stiffness variation on the vibration characteristics of main reduction gears

Yunping Pan[1,*], Zhongyin Feng[1] and Cheng Zhang[1]

[1]School of Mechanical and Electrical Engineering, Wuhan University of Technology, Wuhan, Hubei, 430000, China

[*]Corresponding author's e-mail: panyunping@whut.edu.cn

Abstract: The meshing stiffness of the main and passive gears of the main reducer of the rear axle of a commercial vehicle is taken as the object of this study, aiming to optimize the influence of the meshing stiffness of the main and passive gears on the vibration and noise, enhance the meshing performance of the gear teeth, reduce the vibration and noise of the main reducer, and improve the vibration performance of the main reducer. Through UG 3D modeling software and ADAMS dynamics simulation, the dynamics characteristics of hypoid gears under different gear tooth meshing stiffness coefficients are analyzed and simulated. It is concluded that under the meshing stiffness coefficients of 0.8 to 0.9 times, the translational vibration acceleration and angular acceleration of the passive gears are smaller, the translational vibration and torsional vibration are smoother, and it can improve the performance of the meshing vibration of the gear system of the main gear reducer.

1. Introduction

The master reducer is an important part of the automobile rear axle, and its working performance has an important influence on the vibration noise of the chassis, while the hypoid gear, as the most important component in the master reducer, has a direct influence on the comfort of the rear axle drive system in its vibration characteristics at high speeds and under heavy loads. Under the global trend of energy saving and emission reduction, the dynamic characteristics of hypoid gears are a research direction of interest to researchers at home and abroad. Hypoid gears are widely used as automobile drive axle gears due to their large reduction ratio, small size, smooth transmission, low noise, and the ability to change the height of the automobile body [1].

The time-varying meshing stiffness can be understood in a general sense as the ability of the gear teeth to resist deformation [2]. The change in time-varying meshing stiffness results in a dynamic response of the meshing force of the gear system, and the dynamic meshing force feeds back into the gear system itself, resulting in a stiffness excitation that causes vibration and noise to appear in the gear system. Time-varying stiffness excitation is one of the main forms of excitation during gear meshing, and it is also the main factor that affects the stability of the gear transmission system and causes self-excited vibration of the parameter [3]. The meshing stiffness calculation is the basis for gear dynamics analysis [4]. The dynamic load factor, tooth shape correction, and deviation coefficient in the gear strength calculation all involve the meshing stiffness. In the gearing process, meshing stiffness excitation often has an important impact on the performance and life of the gear system [5]. The meshing stiffness of the main reduction gear is one of the important parameters in the dynamics of the drive axle system that determines the gear dynamic loads, the amount of load, the main reduction vibration, and noise [6]. Scholars study the relationship between vibration and stiffness [7]; analyze the system response results and predict the risk of vibration and noise by controlling the main reducer gear meshing stiffness and so on [8].

In this paper, through the establishment of the dynamic model of the master reducer, combined with the simulation software ADAMS comparative analysis of the meshing characteristics under different meshing stiffness coefficients, we derive the range of meshing stiffness coefficients with better vibration characteristics of the main and the driven gears, providing references to the improvement of the vibration characteristics of the rear axle.

2. Three-dimensional modelling of master reducer

Active gear and passive gear are meshed by a variety of factors, the movement is complex, the distribution of the load on the gear teeth is not uniform, and the dynamics of hypoid gears are also subject to the gear's contact deformation, manufacturing errors, assembly errors, which is a very complex state. Gear systems are mainly subject to internal and external excitations. Internal excitations include mesh stiffness excitation, system error excitation, and mesh impact excitation. Manufacturing errors and assembly clearances lead to system error excitation and meshing shock excitation. In this paper, the effects of other excitations on the gear system are not considered, and there will be deviations between the obtained vibration results and the actual values, but the general trend will not change. To facilitate the study of the virtual prototype model of the main gear reducer, we make the following assumptions:

(1) All models are considered as rigid bodies;

(2) We can remove the gearbox shell's large parts simplified into five bearing housing rings instead, with the bearing housing ring fixed on the ground, used to fix the bearings and the main and passive gears, differentials, and so on;

(3) The effect of manufacturing errors and assembly clearances is negligible;

(4) The half-shaft gears and planetary gears in the differential are treated as concentrated loads acting on the differential case, i.e. the differential gears are not considered;

(5) The influence of the rollers and cage of the tapered roller bearings is disregarded, and the tapered roller bearings are replaced by the equivalent of the tapered ring structure matching the inner and outer rings of the bearings [9].

As for the use of UG to establish the three-dimensional model of the main gearbox, some of the parts are simplified, as well as some of the accessories of the main gearbox, such as bushings and adjusting shims. We use the interference function on the assembly of key positions for interference checking, so as not to affect the dynamics of the simulation [10], as shown in Figure 1.

Figure 1. 3D model of master reducer.

3. Dynamics simulation of master reducer

The built model is imported into ADAMS via the Parasolid format supported by UG and ADAMS/View.

3.1. Definition of material properties

We set the material properties of each component of the master reducer respectively, and the relevant parameters of the master reducer components are shown in Table 1.

Table 1. Parameters related to the master reducer components.

Form	Gears	Bear	Differential case
Densities ρ (kg/mm^3)	7.8×10^{-6}	7.8×10^{-6}	7.8×10^{-6}
Modulus of elasticity E (N/mm^2)	2.07×10^5	2.08×10^5	2.07×10^5
Poisson's ratio μ	0.25	0.30	0.26

3.2. Addition of constraints and loads to the model

A rotating sub is applied to the main and passive gears respectively, and the main gear is used as the input. Gear meshing transmission occurs between the gears, the inner and outer rings of the bearing, and the conical ring, enabling frictional movement between the active and passive gears, which adds solid-solid collision forces between the contact vice. In the simulation, the contact vice is added between the inner and outer rings of the bearing and the rollers of the conical ring to simulate their actual movement. The five bearing housing rings are fixed with the earth, the outer rings of the five bearings are fixed with the bearing housing rings, and the inner rings of the bearings are fixed with the corresponding assembly positions of the active gear and differential case respectively. The passive gear and differential assembly are fastened with screws and added with fixing vice [11]. To simulate the actual operation of the main gearbox, there is a certain time course for the combination between the engine and the gearbox, and the STEP drive function is used, adding a drive speed of 1440 r/min to the main gear. Since the ADAMS rotational speed unit is deg/sec, the actual input rotational speed is 8640 (°)/s. Considering the existence of second-order fluctuation in the engine output rotational speed, the STEP drive function expression is STEP (time, 0, 0d, 0.2, (8640d+432 d*sin (2*8640 d*time))). Subsequently, a load torque of 686 N•m is added to the passive gear, and the established virtual prototype model is shown in Figure 2.

Figure 2. Complete virtual prototype model.

3.3. Selection of simulation parameters

A collision force needs to be added between the active and passive gears. ADAMS supports two methods for calculating this contact force: one is called the restitution (compensation) method, and the other is based on the Hertz theory of impact (impact) function model. The impact model uses stiffness and damping coefficients to calculate the collision force. Both of these contact force calculation models come from the penalty function of normal contact constraints. Compared with the two, the compensation method cannot accurately calculate the penalty and compensation coefficients. Therefore, in this paper, the impact function method is chosen to calculate the contact force of the gear pair [12].

$$F=\begin{cases}0 & ,x>x_0\\ K(x_0-x)^e-C\dfrac{dx}{dt}step(x,x_0-d,1,x_0,0) & ,x\le x_0\end{cases} \tag{1}$$

where x is the amount of change in distance during the collision, step is the step function, x_0 is the distance before the collision, e is the characteristic index of the force, C is the maximum damping coefficient, d is the collision penetration, and K is the stiffness coefficient.

The collision stiffness coefficient depends mainly on the material and structural shape of the two gears [13]. The accuracy of the contact stiffness is more important for the convergence of the calculation and can be replaced by the Hertz contact stiffness K, which is calculated as [14]

$$K = \frac{4}{3}\rho^{\frac{1}{2}}E \tag{2}$$

where ρ is the integrated radius of curvature and E is the integrated modulus of elasticity.

$$\frac{1}{\rho} = \frac{1}{\rho_1} + \frac{1}{\rho_2} \tag{3}$$

$$\frac{1}{E} = \frac{(1-{\mu_1}^2)}{E_1} + \frac{(1-{\mu_2}^2)}{E_2} \tag{4}$$

The equivalent radius of the point of contact between the two gears can be replaced by the radius of the indexing circle [15]. The equivalent radii of the active and passive gears are $\rho_1 = 54.53$ mm and $\rho_2 = 192.7$ mm, respectively; μ_1 and μ_2 are the Poisson's ratios of the two objects; E_1 and E_2 are the modulus of elasticity of the two objects, which can be obtained by substituting into the formula. Similarly, other bearing stiffness coefficients can be obtained.

3.4. Dynamic simulation of standard time-varying meshing stiffnesses

During the simulation, the STEP drive function is used so that the active gear reaches its maximum speed in 0.2 seconds, after which it fluctuates steadily. The total simulation time is 1 s, and the number of steps is 5000. In the simulation, the driving speed is directly applied to the active gear shaft. Since the active gear and the passive gear are the largest source of vibration of the main gearbox, the vibration magnitude can be analyzed by counting the vibration condition of the active gear and the passive gear. The time-domain plot of the active and driven gear speed curves is shown in Figure 3; the FFT spectrum of the passive gear speed is shown in Figure 4; the integrated curve of the translational vibration acceleration of the passive gear is shown in Figure 5; and the integrated curve of the rotational angular acceleration of the passive gear is shown in Figure 6.

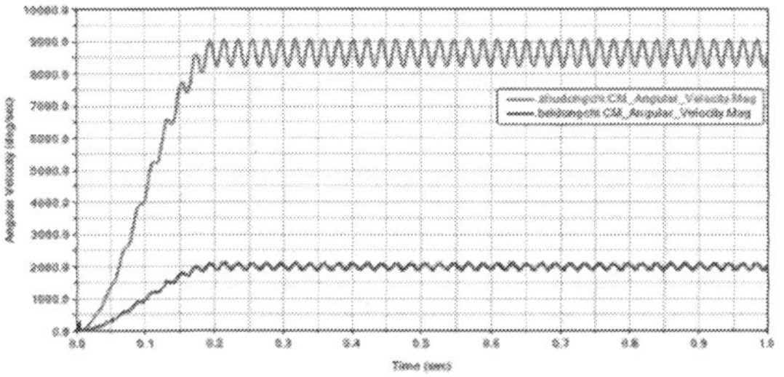

Figure 3. Speed curve of active and passive gears.

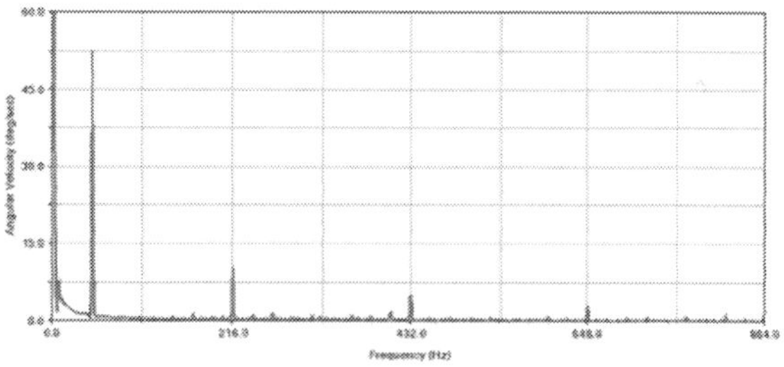

Figure 4. FFT spectrum of passive gear speed.

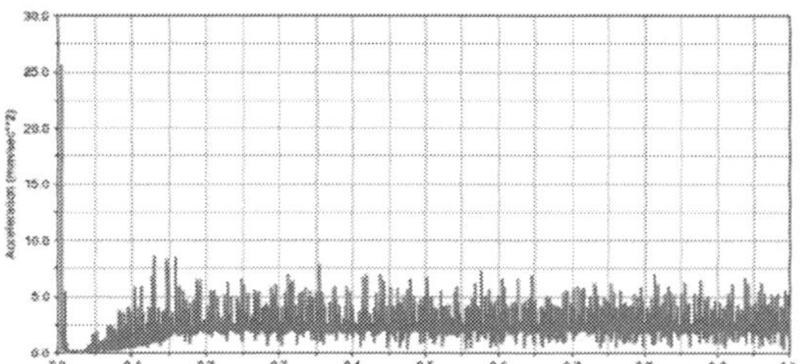

Figure 5. The integrated curve of passive gear translational vibration acceleration.

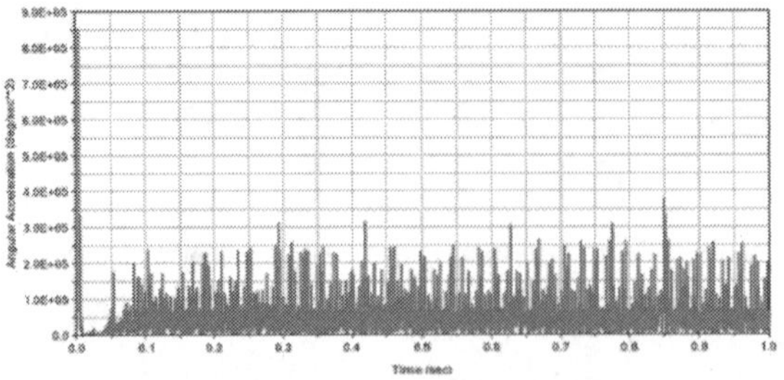

Figure 6. The integrated curve of the angular acceleration of the passive gear.

According to the transmission ratio, the actual speed of the passive gear is 8640×9/39 (°)/s =1993.84 (°)/s, and from Figure 3, the angular velocity of the passive gear is near 2000 (°)/s, which can verify that the speed meets the transmission ratio requirements. There is a shock at the beginning of the movement of the main and passive gears. From 0 to 0.2 s, main and passive gears are in the same trend, rising to the target value. After 0.2 s, because the main and passive gears have begun to rotate steadily, the angular velocity is also stabilized, and the main gear reducer works smoothly. The passive gear stabilizes with periodic small fluctuations. Because of the shock vibration of the main reducer during normal operation, the figure is consistent with the actual movement of the main reducer at work.

From Figure 3 to Figure 6, it can be seen that the amplitude is relatively smooth in the acceleration stage, indicating that its meshing frequency is staggered with the intrinsic frequency of the hypoid gear. After entering the stabilized transmission, its rotational speed fluctuation starts to become larger. In the FFT plot of the passive gear's rotational speed, there are several peaks of frequencies: the 1st order frequency 47.998 Hz, the 2nd order frequency 216.03 Hz, the 3rd order frequency 432.08 Hz, and the 4th order frequency 648.32 Hz. The fluctuation amplitude is the largest when it is at the 1st order frequency 47.998, which is because when performing the dynamics simulation, the drive is applied as a rotational speed containing the 2nd order fluctuation, resulting in the fluctuation of the main. The angular velocity of rotation of the driven gear along its axis also has this fluctuation. Neglecting the effect of this rotational speed fluctuation, 216.03 Hz is the meshing frequency of the main and passive gears, and the frequencies of the following orders are all multiples of the meshing frequency of the main and passive gears. Angular acceleration reflects the magnitude of vibration in the circumferential direction of the gears, which is due to significant shocks during engine acceleration, as well as shock vibrations that occur during gear meshing. The angular acceleration of the passive gear shows a periodic change after 0.2 s, which is consistent with the motion law when the main and passive gear mesh.

By comparing the passive gear translational vibration acceleration and speed acceleration curves, it is found that the vibration trend of the two is consistent, the amplitude of vibration acceleration is small in the acceleration stage of the passive gear, and the fluctuation becomes larger in the later stage of the smooth drive.

It can be seen that during the acceleration stage, the meshing frequency avoids the gear system's intrinsic frequency. However, after entering the smooth meshing stage, the vibration tends to increase when the frequency of 47.998 Hz or the multiples of the driving frequency of 24 Hz is close to the gear system's intrinsic frequency.

3.5. Vibration characterization of gear systems with different simultaneous variable mesh stiffnesses

The meshing stiffness values at standard time were multiplied by 0.8, 0.9, and 1.1 to simulate and analyze the master reducer. The translational vibration acceleration curves of the passive gear with different meshing stiffness are shown in Figure 7 to Figure 10, the angular velocity of the passive gear with different meshing stiffness and its FFT curves are shown in Figure 11 to Figure 14, and the angular acceleration curves of the passive gear with different meshing stiffness are shown in Figure 15 to Figure 18.

 IOP Publishing
doi:10.1088/1742-6596/2879/1/012030

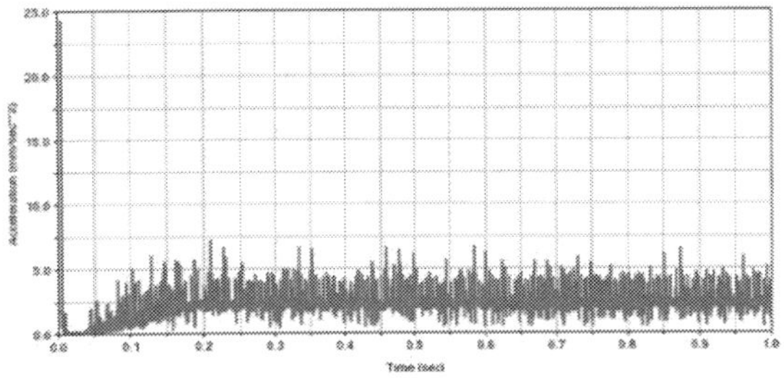

Figure 7. 0.8 times meshing stiffness.

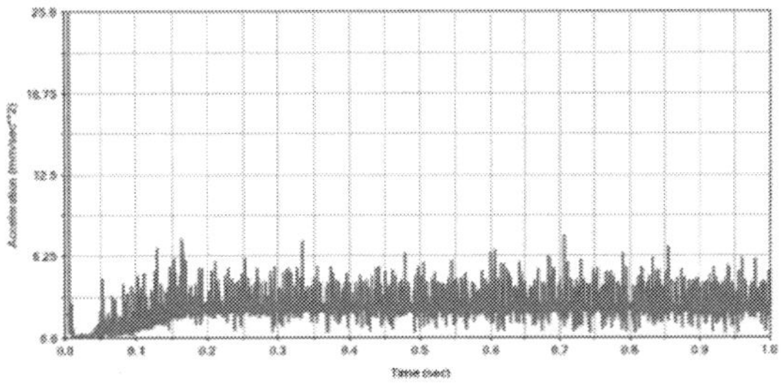

Figure 8. 0.9 times meshing stiffness.

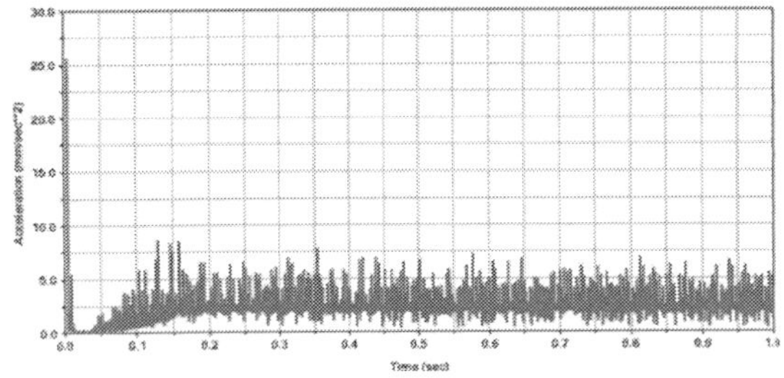

Figure 9. Standard times meshing stiffness.

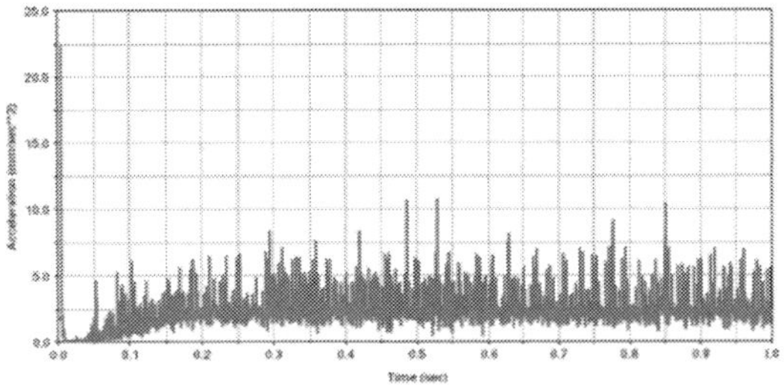

Figure 10. 1.1 times meshing stiffness.

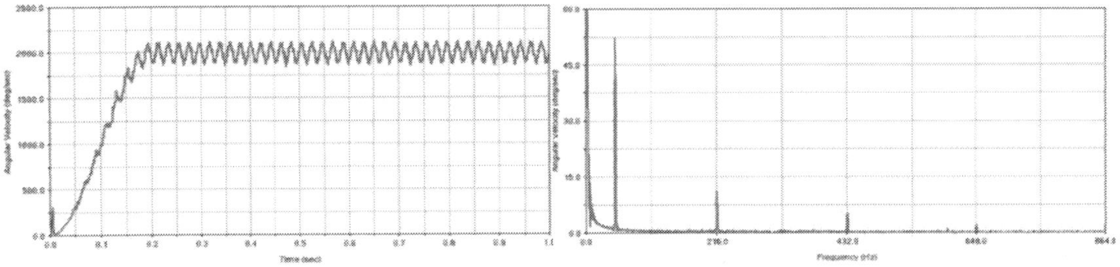

Figure 11. Angular velocity of 0.8 times meshing stiffness passive gear and its FFT curve.

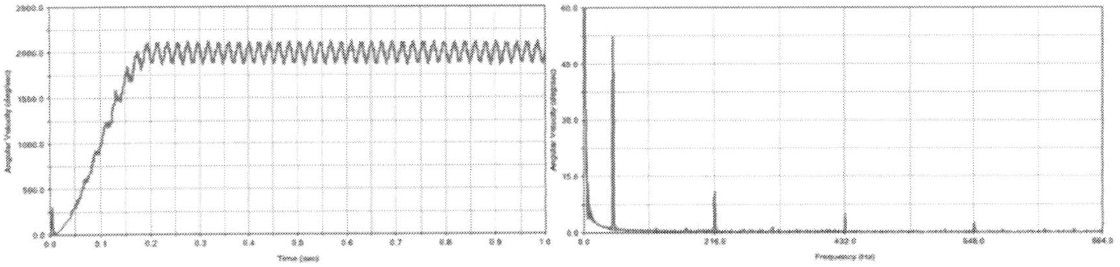

Figure 12. Angular velocity of 0.9 times meshing stiffness passive gear and its FFT curve.

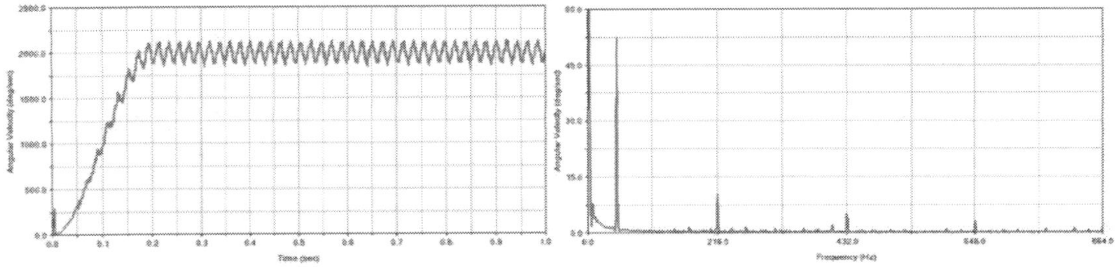

Figure 13. Angular velocity of standard times meshing stiffness passive gear and its FFT curve.

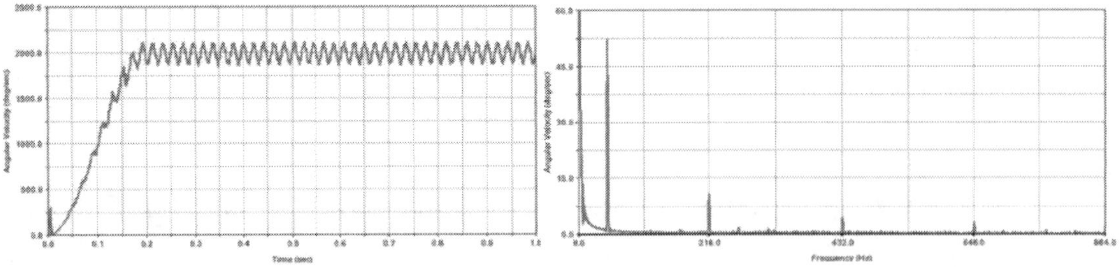

Figure 14. Angular velocity of 1.1 times meshing stiffness passive gear and its FFT curve.

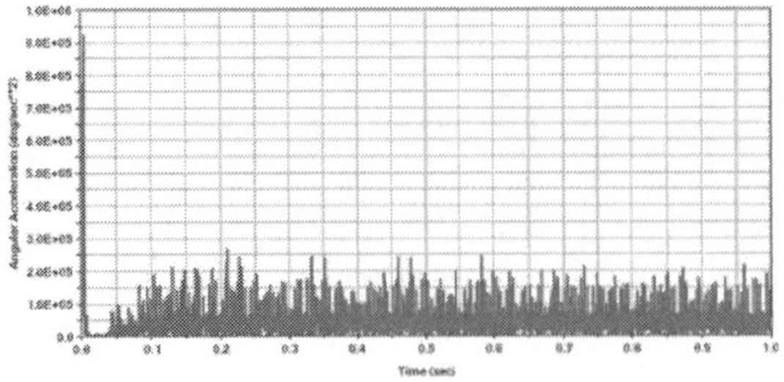

Figure 15. 0.8 times meshing stiffness.

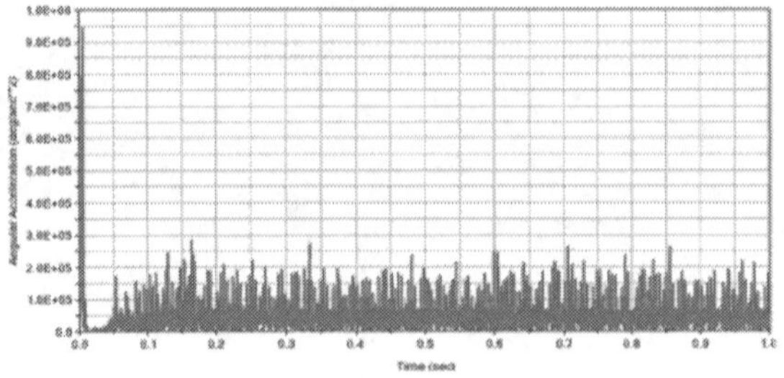

Figure 16. 0.9 times meshing stiffness.

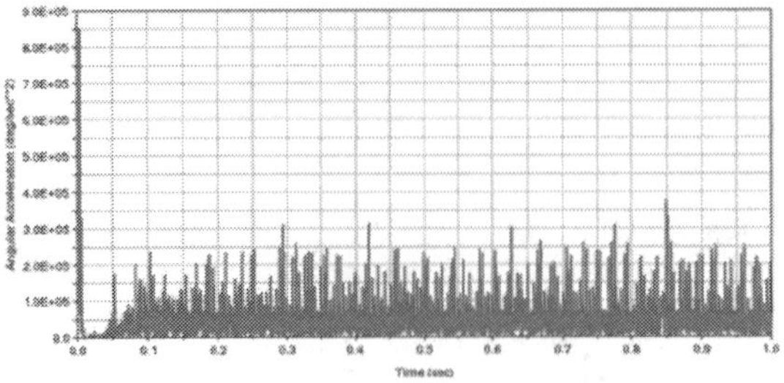

Figure 17. Standard times meshing stiffness.

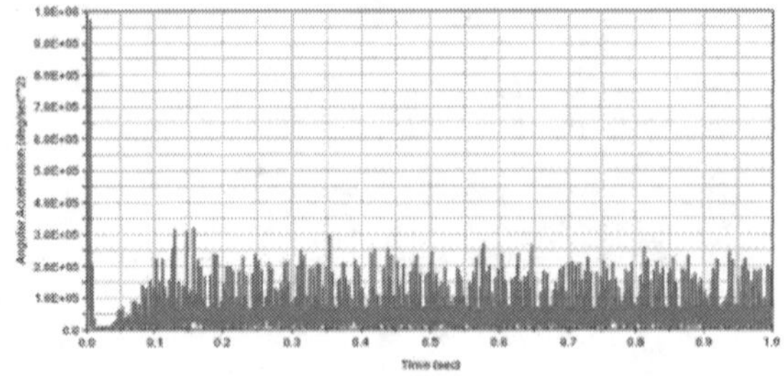

Figure 18. 1.1 times meshing stiffness.

From the comparison of Figure 7 to Figure 10, it can be seen that under the standard meshing stiffness, there is a surge in the individual amplitude of vibration of the gears, and the variation of vibration amplitude for the 0.8 standard meshing stiffness is small. The average vibration amplitude of the four groups is 2.1274 mm/s^2, 2.1576 mm/s^2, 2.2094 mm/s^2, and 2.2335 mm/s^2, respectively. The amplitude of vibration acceleration decreases with the reduction of meshing stiffness, and it can reduce the fluctuation in the transmission process when it is 0.8 to 0.9 times, which is conducive to the smooth transmission of the passive gear and the reduction of the vibration noise of the gear transmission system.

From the comparison of Figure 11 to Figure 14, it can be seen that when the meshing stiffness is 0.8 to 0.9 times, the hypoid gears can better maintain the stability of the angular velocity and reduce the system vibration caused by the angular velocity fluctuation in the transmission process.

As can be seen from the comparison of Figure 15 to Figure 18, by removing the 0.2 s of the acceleration phase and counting the mean angular acceleration for 0.2 s to 2 s, the four groups are 4.4378×10^4 (°)/s^2, 4.5986×10^4 (°)/s^2, 4.8777×10^4 (°)/s^2, and 4.8861×10^4 (°)/s^2 respectively. Under 0.8 to 0.9 times meshing stiffness, the angular acceleration amplitude of the passive gear decreases significantly, which can reduce the torsional vibration of the passive gear. The vibration of the transmission system is relatively smooth, the vibration fluctuation is relatively small, and the noise situation is also correspondingly small.

In summary, the comparative analysis shows that when the meshing stiffness is taken as 0.8 to 0.9 times, it can improve the performance of wheel tooth meshing, reduce the vibration and noise, and improve the vibration of the master reducer. When optimizing the design of the gear system, the meshing stiffness of the gears can be appropriately reduced by improving the machining accuracy, gear tooth trimming, and other methods to make the gear transmission process smoother and improve the meshing condition of the gear system.

4. Conclusion

Through the dynamic simulation analysis, it can be seen that when the meshing stiffness of the hypoid gear of the master reducer is taken as 0.8 to 0.9 times, the angular acceleration of the passive gear as well as the translational vibration acceleration are smaller, and the smoothness is optimal, which effectively reduces the vibration and noise of the gear transmission system of the master reducer, and improves the NVH performance of the car. It is of reference significance to optimize the design of gears, improve the transmission efficiency, and improve the performance of the gear transmission system.

References

[1] Dang Y G, Deng X Z, Wang B and Gao Z S. (2014) A new method for cold swing rolling of hypoid gears and its numerical simulation. J. China Mechanical Engineering (09), 1250-1255.
[2] Jiang P. (2023) Calculation of time-varying meshing stiffness and dynamic characterization of gears with point-contact curve configuration. D. Chongqing Jiaotong University.
[3] Chen S Y, Tan R L and Gou X D. (2021) Research on the calculation method of meshing stiffness of straight-toothed bevel gears. J. Mechanical Drives (09), 62-67.
[4] Li J Q. (2022). Calculation of topological revision stiffness and dynamic simulation analysis of cylindrical gears. D. Henan University of Science and Technology.
[5] Tan J W and Liu L. (2017) Mesh stiffness analysis of ring-faced involute gears. J. Mechanical Manufacturing and Automation (06), 5-9+28.
[6] Chen Z M, Hou L, Wu Y, Wei Y Q, Zhao F, Ma Y Z and Xia L F. (2023) Vibration Characterization of Drive Axle Hypoid Gear System. J. Engineering Science and Technology, 55(04): 224-238.
[7] Jiang J K, Liu Z, Peng X L. (2020) Optimized design of Ease-Off trimmed hypoid gears for vibration damping. J. Journal of South China University of Technology (Natural Science Edition), 48(5): 134-141, 148.
[8] Jin Z B, Zhao X S, Bai X B, Zhang P and Yu B R. (2022) Research on analysis and optimization method of whistling noise of passenger car drive axle. J. Transmission Technology, 36(3): 26-31.
[9] Zou L, Lie Y J, Jia B, Qin A and Xu J L. (2020) Coupled vibration study of driveshaft-rear axle system based on Adams. J. Mechanical Transmission (08), 65-70.
[10] Zhang B W. (2010) Vibration test and dynamics analysis of micro-vehicle main reducer gear. D. Wuhan University of Technology, Wuhan.
[11] Zeng H, Zhang W G and Li Y. (2011) Dynamic Simulation of Curved Bevel Gear Transmission System Based on ADAMS. J. Mechanical Design and Manufacturing (03), 71-72.
[12] Guo H Z, Tan C J and Chen J F. (2013) Simulation of planetary wheel system dynamics based on ADAMS. J. Mechanical Drives (05), 86-89.
[13] Li J Y, Gou Z J and Li Y. (2005) Dynamic simulation of gear force during gear meshing based on ADAMS. J. MACHINE (03), 15-16.
[14] Zhou S L, Zhao X L, Zhao K L, Ren F and Shi Z. (2012) Optimization of planetary carrier structure of travel reducer of bulldozer. J. China Journal of Construction Machinery (03), 297-301.
[15] Ren F, Wang D X, Shi G Q, Liang D, Wang Q, Wang N and Yi Y Y. (2022) Dynamic simulation and fault characterization of double-ring herringbone planetary gear system based on ADAMS. J. Journal of Henan University of Science and Technology (Natural Science Edition) (01), 114-121.

Journal of Physics: Conference Series **2879** (2024) 012031 doi:10.1088/1742-6596/2879/1/012031

Experimental studies on the exhaust emissions of a n-butanol/diesel dual-fuel HCCI engine under different intake temperature

Jiantong Song*, Chunhong Zhu, Yuanfang Shi and Yiwei Fu

Beijing Polytechnic, Beijing, 100176, China

*Corresponding author's e-mail: tjsjt@126.com

Abstract: To improve the exhaust emissions of n-butanol/diesel blends fueled HCCI engines, the influence of intake temperature (T_{in}) on the HC, CO, and NO_x emissions was studied by bench tests on λ of 2.5, engine speed of 1000 r/min under the conditions with the blending ratio of B50 and B90. Experimental results show that the HC and CO decline as T_{in} rises. At the same T_{in}, the CO and HC of B90 are lower than B50. When the intake temperature is lower than 150°C, the NO_x emissions are almost zero.

1. Introduction

Homogeneous Charge Compression Ignition integrates the strengths of gasoline engines and diesel engines and is commonly referred to as HCCI [1]. The combustion mode has no throttling loss during operation, and the mixture in each point of the cylinder ignites simultaneously [2]. The combustion heat release is extremely fast, approaching the ideal constant volume combustion, and the combustion temperature is low [3]. Therefore, compared with conventional engines, the heat efficiency of an HCCI engine is higher, while the NO_x emissions are extremely low. Unlike traditional gasoline and diesel engines, the chemical reaction kinetics of the combustible mixture of HCCI engines control the combustion [4]. Therefore, indirect measures such as controlling excess air coefficient (λ), intake charge temperature, engine speed, etc., can only be taken to adjust the combustion of the HCCI engine.

The combustion process also can be improved by changing the fuel. Ethanol, methanol, n-butanol, and other alcohol fuels have become ideal fuels for many scholars to conduct experimental research on HCCI engines because of their high octane number, good detonation resistance, and easy availability [5]. Alcohol-blended fuels are also an effective way to achieve controlled combustion of HCCI. Compared with common methanol and ethanol, the properties of long-chain butanol are closer to diesel, and the cetane number and energy density are higher. The compatibility of butanol with diesel is better, which is very beneficial for co-combustion with diesel [6]. It is considered one of the most promising biofuels. Butanol can be compatible with existing fuel transportation facilities and can be produced through microbial fermentation using renewable crops.

In order to optimize the exhaust emissions of n-butanol/diesel dual-fuel HCCI engines, the impact of intake charge temperature (T_{in}) on CO, HC and NO_x was analyzed.

2. Test equipment and experimental procedure

2.1. Test equipment

The experimental engine used for the test was modified by a 4-stroke, 2-cylinder, aspirated naturally, water-cooled, CT2100Q diesel engine with direct injection. The displacement is 824.7 cm^3. The bore and stroke are 100 and 105 mm, respectively. The compression ratio is 17. The maximum torque is 110 Nm@2000 r/min. The rated power is 25 kW@2600 r/min. The exhaust valve opens at 47°CA before the bottom dead center. The exhaust valve closes at 17 °CA after the top dead center. The intake valve opens at 17°CA before the top dead center. The intake valve closes at 43°CA after the bottom dead center. The testing system and the HCCI engine test bench are shown in Figure 1 and Figure 2. The first cylinder maintained the diesel combustion mode, while the second cylinder was adjusted to the HCCI combustion mode. The inlet air temperature control device installed on the intake system of HCCI was used to control the T_{in} of the second cylinder. The PFI fuel injection system could provide a homogeneous charge of corresponding concentration. The intake air per cycle was fixed, using the PFI fuel injection system to maintain a certain injection could control the λ. The electric eddy current dynamometer was used to monitor and measure parameters such as engine speed, power, water temperature, and intake air temperature. An AVL DIGAS 4000 exhaust analyzer could measure the exhaust emissions, such as CO, HC, and NO_x.

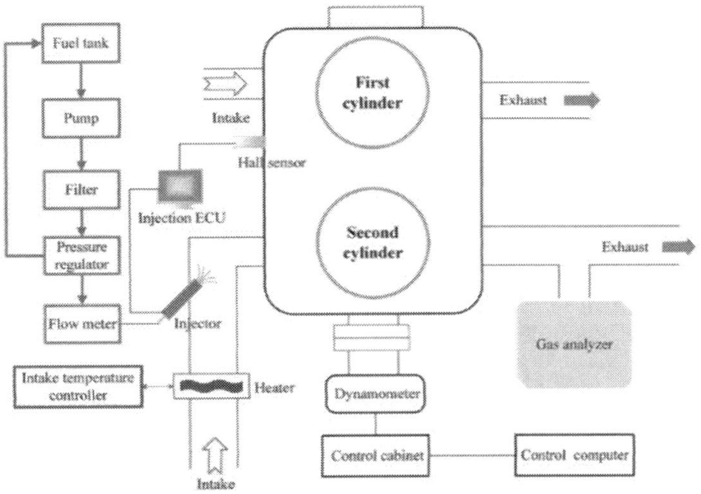

Figure 1. The schematic of the test system based on the HCCI engine.

Figure 2. The HCCI engine test bench.

2.2. Experimental procedures

During the test, diesel was first used to warm up the 1# cylinder. After the cooling water temperature, oil temperature, intake charge temperature, and intake pressure reach the required values for the test and set the speed of the engine to 1000r/min and the λ to 2.5. Because the n-butanol's latent heat of vaporization is relatively high, with a boiling point of around 117°C, the pre-mixing temperature of the intake port is too low to allow good evaporation and mixing of the fuel, resulting in poor ignition and combustion of the mixture when the T_{in} is below 120°C. Therefore, in order to better vaporize and atomize the mixed fuel and form a well-homogenized mixture, during the test, the intake charge temperature was controlled at 120, 130, 140, 150, and 160°C. After all the set points had stabilized, the diesel supply was cut off to the first cylinder. The second cylinder was simultaneously supplied with diesel and n-butanol mixture, allowing the engine to switch to HCCI operation. After the engine ran smoothly in HCCI mode, various emissions data of the engine under this mixture ratio were recorded. Two mixture ratios of B50 and B90 were selected for this test.

2.3. Data processing

Bxx represents the energy proportion of n-butanol in the dual-fuel, which is given by:

$$Bxx = \frac{m_{n-butanol} \times Hu_{n-butanol}}{m_{n-butanol} \times Hu_{n-butanol} + m_{diesel} \times Hu_{diesel}} \times 100\% \quad (1)$$

where m is fuel flow rate and the Hu is the low calorific value of fuel.

3. Results and analysis

3.1. Impact of T_{in} on CO emission

The relationship between intake temperature and CO emission of a n-butanol/diesel blends HCCI engine is presented in Figure 3.

When the λ and the speed of the engine remain constant, the CO emission decreases with the increasing intake charge temperature. This is mainly because with the increasing T_{in}, the maximum temperature of combustion in the engine cylinder increases, the quenching effect gradually decreases, and the combustion rate continues to accelerate, which is very beneficial for CO oxidation to CO_2. Compared with B50, the CO emission of B90 is greater affected by T_{in}. At the same T_{in}, compared with B50, the CO emission of B90 was reduced significantly. This is mainly because with increasing blending ratio, the n-butanol in the fuel increases. Compared with diesel, the volatility of n-butanol is better, and the mixture of n-butanol and air is more uniform. The n-butanol is also an oxygen-containing fuel, which can provide more oxygen during combustion, improve combustion, and reduce CO emissions.

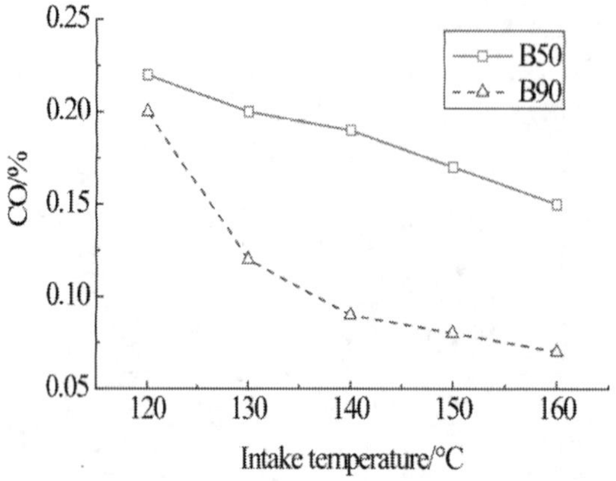

Figure 3. The relationship between intake temperature and CO emission.

3.2. Impact of T_{in} on HC emission

The relationship between T_{in} and HC emission of an n-butanol/diesel blends HCCI engine is presented in Figure 4.

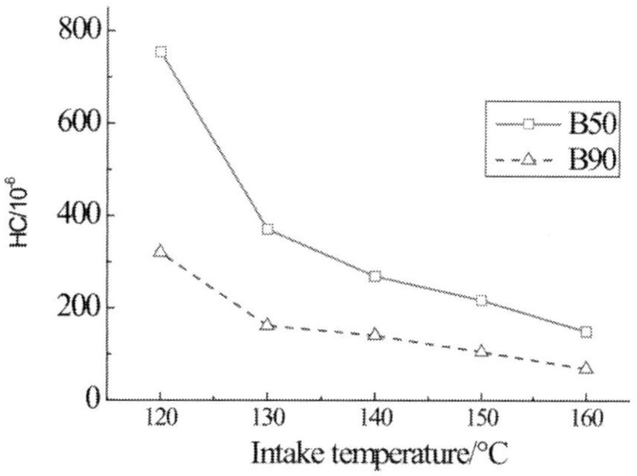

Figure 4. The relationship between intake temperature and HC emission.

When the engine speed and λ are unchanged, the HC emission trend and CO change trend of the n-butanol/diesel mixture are coincident with the increasing intake charge temperature. The gap effect and chilling effect in the combustion process of HCCI engines are the main sources of HC generation. When the T_{in} increases, HC emission decreases. This is mainly because when T_{in} increases, the activity of mixture molecules increases, the thermal peak of combustion in the engine cylinder increases, the combustion efficiency of the mixed fuel also increases, and the residual unburned hydrocarbon in the slit is fully burned, thus reducing incomplete combustion and misfire. The exhaust temperature increases with increasing combustion temperature in the cylinder, which is also conducive to the further oxidation of some unburned HC in the exhaust pipe. At the same intake charge temperature, HC emission of B90 is significantly lower than that of B50, mainly because butanol is more volatile, the mixture is more uniform, and butanol is an oxygen-containing fuel, which improves combustion.

3.3. Impact of T_{in} on NOx

The relationship between intake temperature and NO_x emissions of a n-butanol/diesel blends HCCI engine is presented in Figure 5. The NO_x emissions generated by engine combustion are easy to generate in large quantities under continuous high temperature and oxygen enrichment conditions. The combustion of n-butanol HCCI is fast and has a short combustion duration. Besides, HCCI is a low-temperature combustion with a low average temperature within the cylinder, which is not conducive to NOx generation. Therefore, under the condition that the intake temperature is lower than 150°C, the NO_x emissions of n-butanol/diesel blends are almost zero. However, if the intake air temperature is 160°C, due to knocking combustion, NO_x emissions of B90 rapidly rise to 136×10^{-6}.

4. Conclusions

As the intake temperature increases, the CO emissions of n-butanol/diesel blend fuel decrease. Compared with B50, the CO emission of B90 is greatly affected by the intake temperature. Compared with B50, the CO emission of B90 is significantly reduced at the same intake temperature.

The HC emission trend and CO change trend of the n-butanol/diesel mixture basically coincide with the increasing intake charge temperature. When intake air temperature increases, HC emission decreases. At the same intake charge temperature, the HC emission of B90 is significantly lower than that of B50.

If the intake temperature is lower than 150°C, NO_x emissions of n-butanol/diesel blends are almost zero. When the T_{in} is 160°C, the NO_x emissions of B90 rapidly rise to 136×10^{-6}.

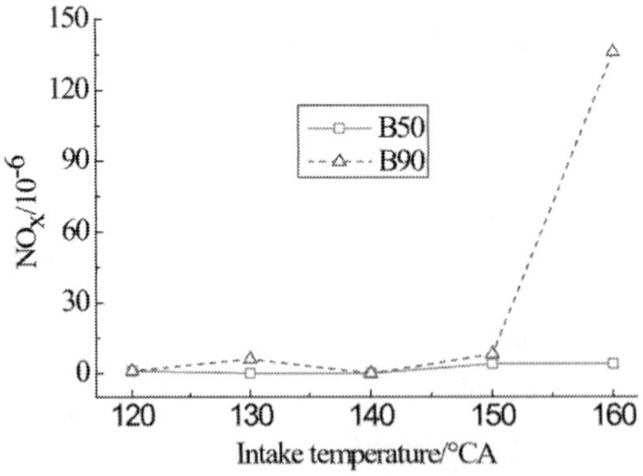

Figure 5. The relationship between intake temperature and NOx emissions.

Acknowledgments
This work is supported by the Beijing Vocational College Teacher Quality Improvement Plan and Key Science and Technology Projects of Beijing Polytechnic (2023X029-KXZ).

References

[1] Bai Y, Wang Y, Hao L (2022). Experimental study on the effects of injection timing and n-butanol energy ratio on combustion and emissions of n-butanol/diesel dfdi engine. Fuel (Lond), 324: 124654.

[2] Zhang Y, Gao S, Zhang Z, Li W, Yuan T, Tan D, et al. (2023). A comprehensive review on combustion, performance and emission aspects of higher alcohols and its additive effect on the diesel engine. Fuel (Lond), 335: 127011.

[3] Tipanluisa L, Thakkar K, Fonseca N, López J (2022). Investigation of diesel/n-butanol blends as drop-in fuel for heavy-duty diesel engines: combustion, performance, and emissions. Energy Convers Manag, 255: 115334.

[4] Kale AV, Krishnasamy A (2024). Experimental study on combustion, performance, and emission characteristics of a homogeneous charge compression ignition engine fuelled with multiple biofuel-gasoline blends. Energy (Oxf), 288: 129621.

[5] Siva Prasad K, Srinivasa Rao S, Raju VRK (2023). Numerical investigation for enhancing hcci combustion characteristics in di-ci engine fueled with butanol/diesel blends. Mağallaẗ al-abḥath al-handasiyyaẗ.

[6] Nguyen VN, Nayak SK, Le HS, Kowalski J, Deepanraj B, Duong XQ, et al (2024). Performance and emission characteristics of diesel engines running on gaseous fuels in dual-fuel mode. Int J Hydrogen Energy, 49: 868-909.

Numerical study on transonic shock wave buffeting of airfoil

Zicheng Hu, Liangchong Zhu, Xin Mo and Keming Shen*

East China University of Technology, Nanchang, Jiangxi, 330013, China

*Corresponding author's E-mail: shen_keming@ecut.edu.cn

Abstract. Modern transportation aircraft typically cruise at transonic speeds. However, under a specific combination of incoming flow Mach numbers and angles of attack, periodic self-excited oscillations of shock waves can lead to unsteady aerodynamic loads, causing sustained vibrations in the aircraft, which is known as transonic shock wave buffeting. Buffeting loads significantly impact the fatigue life of the aircraft's structural materials and its handling performance. In severe cases, it causes structural damage or even leads to flight safety incidents, so we should pay more attention to transonic shock wave buffeting when focusing on aeronautical engineering. This paper employs computational fluid dynamics technology for calculations to solve the Navier-Stokes equations, using the SST k-ω turbulence model to numerically simulate the steady flow field, and further employing the detached eddy simulation (DES) method to study and analyze the buffeting initiation mechanism and flow field characteristics of the NACA0012 airfoil's transonic unsteady flow. By applying the DES method, the results demonstrate an accurate prediction of the buffeting onset boundary of the airfoil compared to wind tunnel experimental data. The research findings on the initiation mechanism and load characteristics of transonic buffeting provide a theoretical basis for optimizing the aerodynamic performance of aircraft during transonic flight and offer guidance for engineering issues such as buffeting load mitigation.

1. Introduction

During the transonic flight, the aircraft may experience large-scale shock wave oscillations along the rigid airfoil model caused by the shock-boundary layer interaction. This phenomenon, known as a transonic buffet, is a result of the interaction between the aforementioned components [1], which was first found by Hilton and Fowler [2] in wind tunnel experiments in 1952. During shock buffet events, the shock wave location on the wing surface experiences oscillations. At the same time, periodic changes also happen to the thickness of the shock wave boundary layer together with the position of flow separation. Shock wave oscillation and unsteady flow lead to pressure pulsations on the wing surface, causing unsteady aerodynamic loads, which can reduce the fatigue life of the aircraft, affecting the aircraft's structure and handling quality. In severe cases, it can lead to structural damage to the aircraft, restrict the flight envelope, and threaten flight safety [3]. Therefore, during the aircraft design process, the buffet is considered an important factor and is taken into account.

Nowadays, computer technology and computational fluid dynamics (CFD) have obtained great development, which results in hot research on the topic of numerical simulation in aeronautical engineering accounting for the study of transonic shock wave buffeting on wings. Many new methods and research ideas have been applied to the study of shock buffet. In the process of analyzing buffeting load characteristics, numerical simulation methods have significant advantages over traditional wind tunnel test methods, as they can provide more detailed flow field information. However, when simulating transonic shock wave buffeting, we need to deal with strong intermittent shock wave motion and large-scale airflow changes, such as separation and reattachment, which pose higher requirements on accuracy and numerical methods. This presents us with a dual challenge: meeting the accuracy requirements of flow field simulation while ensuring the feasibility of numerical calculations, i.e., whether there are sufficient computational resources and reasonable calculation time. Therefore, in current numerical simulation studies of transonic shock wave buffeting, the main methods used are URANS and detached eddy simulation (DES), and efforts are focused on studying the applicability of these two numerical methods and their key parameters. Deck [4] introduced the DES simulation method to predict the buffet boundary of supercritical wings, and the results were consistent with experimental measurements. Xiong and Liu [5] used URANS and DES methods to simulate transonic buffeting on the swept wings of supercritical airfoils, with DES successfully predicting the unsteady flow generated by buffeting. Both transonic buffeting wind tunnel experiments and CFD simulations have shown that the instability of transonic airflow around airfoils or wings is the root cause of buffeting. Therefore, the study of its mechanism has always attracted lots of attention in the research on transonic buffeting of wings [6]. Considering the research progress and engineering applications of transonic buffeting, this work is specifically organized to study the problem of transonic shock wave buffeting on airfoils through numerical simulation methods, which have important theoretical significance for aircraft design, buffeting mechanism analysis, and aircraft performance improvement.

2. Numerical method

2.1. Control equations

This research employs the Navier-Stokes equations to govern the dynamics of viscous separated flow. The corresponding integral conservation form is articulated as follows:

$$\frac{\partial}{\partial t}\int_{\Omega} W d\Omega + \oint_{\partial\Omega} (F_c - F_\upsilon) dS = 0 \tag{1}$$

where W represents the conservation variables; F_c and F_υ represent the convective fluxes and viscous fluxes, respectively.

2.2. Turbulence model

2.2.1. Steady flow field simulation. In CFD, the widely used turbulence models include the one-equation Spalart-Allmaras (S-A) model [7] and the SST k-ω model proposed by Menter [8]. According to the comparison results of different turbulence models on numerical simulation provided by Yao [9], the latter can accurately capture the location of the shock waves. Consequently, this paper uses the SST k-ω model for steady flow field calculations. The governing equations are:

$$\frac{\partial(\rho \mathrm{k})}{dt} + \frac{\partial}{\partial x_j}\left(\rho \mathrm{u}_j \mathrm{k}\right) = \mathrm{P_k} - \rho\beta^* \omega \mathrm{k} + \frac{\partial}{\partial x_j}\left[\left(\mu + \sigma_{\mathrm{k1}}\mu_{\mathrm{t}}\right)\frac{\partial \mathrm{k}}{\partial x_{\mathrm{j}}}\right]$$

$$\frac{\partial(\rho\omega)}{dt} + \frac{\partial}{\partial x_j}\left(\rho \mathrm{u}_j \omega\right) = P_\omega - \beta\rho\omega^2 + \left[\left(\mu + \sigma_\omega \mu_t\right)\frac{\partial \omega}{\partial x}\right] + 2\rho(1 - F_1)\sigma_{\omega 2}\frac{1}{\omega}\frac{\partial k}{\partial x_j}\frac{\partial \omega}{\partial x_j} \tag{2}$$

The turbulent viscosity coefficient is given by:

$$\mu_t = \min\left(\frac{\rho k}{\omega}, \frac{\alpha_1 \rho k}{\Omega F_2}\right), \alpha_1 = 0.31 \tag{3}$$

where $\Omega = \sqrt{2(\frac{1}{2}(\frac{\partial u}{\partial y} - \frac{\partial v}{\partial x}))^2}, P_k = \mu_t \Omega^2, P_\omega = \gamma\rho\Omega^2$, we yield:

$\sigma_\omega = 0.5, \sigma_{\omega 2} = 0.856, \sigma_{k1} = 0.85, \sigma_{k2} = 1.0, \beta_1 = 0.075, \beta_2 = 0.828, \beta^* = C_\mu = 0.09$.

2.2.2. Unsteady flow field simulation. The SST k-ω model is now extensively used for addressing high Reynolds number flow challenges. However, it presents certain limitations for dealing with turbulent large separation flows. Large Eddy Simulation (LES) [10] method is considered an effective numerical simulation method for turbulence, but it demands high grid density and precision, resulting in considerable computational time. Consequently, it has not been widely used in engineering. To address this issue, Johansen et al. [11] proposed the DES method, and Menter and Kuntzm [12] subsequently introduced the SST-DDES model. Hu et al. [13] described the use of the SST-DDES method to simulate the flow along a low-speed triangular wing at big angles of attack, with results showing that SST-DDES is effective in handling complex turbulence. This paper employs the DES method for simulating unsteady flow fields, defining the DES scale as:

$$\tilde{l} = \min\left(l_{k-\omega}, C_{DES}\Delta\right) \tag{4}$$

In Equation (4):

$$l_{k-\omega} = \frac{k^{\frac{1}{2}}}{\beta^* \omega} \qquad \Delta = \max\left(\Delta_x, \Delta_y, \Delta_z\right) \tag{5}$$

$$C_{DES} = (1 - f_1)C_{DES}^{k-\varepsilon} + f_1 C_{DES}^{k-\omega} \tag{6}$$

The grid scale Δ represents the maximum grid step length in three spatial directions, C_{DES} is an adaptive parameter, and f_1 is provided by the turbulence model, $C_{DES}^{k-\varepsilon} = 0.61, C_{DES}^{k-\omega} = 0.78$.

2.3. Validation of numerical simulation capability

2.3.1. Airfoil physical model. Before conducting the simulation on the airfoil buffeting, it is necessary to validate the numerical method under specific incoming flow conditions. The NACA0012 airfoil is a classic case model widely used to validate wind tunnel test techniques, and thus the wind tunnel experimental data for this model are relatively ideal, including the test results of the buffeting onset boundary. McDevitt and Unhook [14] measured the buffeting boundary of this airfoil in wind tunnel tests, with the experimental Mach number spanning the interval of 0.7 up to 0.82. The experimental evidence suggests that the increase in the incoming Mach number results in a reduction in the buffeting onset angle of attack. This paper also selects the NACA0012 airfoil as the research model. Figure 1 provides the geometric configuration of the NACA0012 airfoil, with a maximum thickness equal to 12% of the chord length and located at 30% of it.

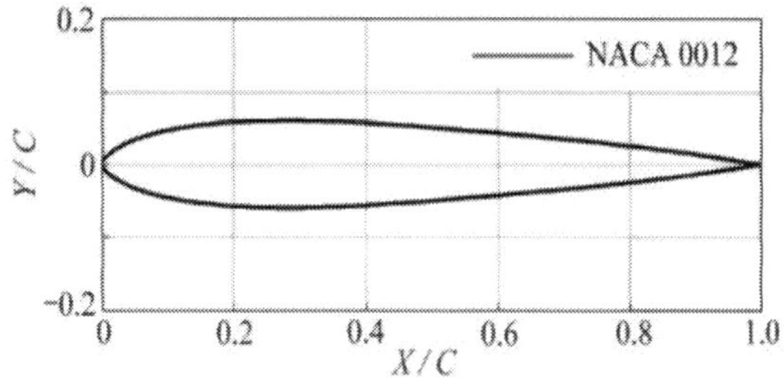

Figure 1. NACA0012 airfoil geometry profile.

2.3.2. Convergence study of computational meshes. To assess the reliability of the numerical simulation approach in transonic flow, this paper conducts a mesh convergence study using three sets of C-type meshes with different densities. The pressure far-field boundary extends to a distance fifty times the airfoil chord. The mesh is carefully refined near critical airfoil features, including the wall surface, leading, and trailing edges, ensuring that the first layer of wall grids corresponds to $y^+<1$, and the minimum quality of the three sets of grids is greater than 0.8. Table 1 illustrates the configuration of the three sets of meshes.

Table 1. NACA0012 airfoil flow field mesh distribution.

Mesh	Grid nodes	Grid elements
A	24810	25122
B	31890	32282
C	68190	68502

According to the wind tunnel test data provided in [14], steady transonic flow field calculations were performed on three sets of grids under the incoming flow conditions of Mach number 0.726, angle of attack 3.91°, and Reynolds number 9.3×10^6. The computed results were compared with experimental data, as depicted in Figure 2. By comparing the distribution data obtained from the three data sets of grids, Figure 2(b) shows that the pressure distributions of the three grids are generally close to the wind tunnel experimental data, except for a large difference in the region downstream of the shock wave. Grids B and C are closer to the wind tunnel experimental data there. Therefore, grid C is selected as the computational flow field for the transonic buffet simulation, which is then revealed in Figure 3.

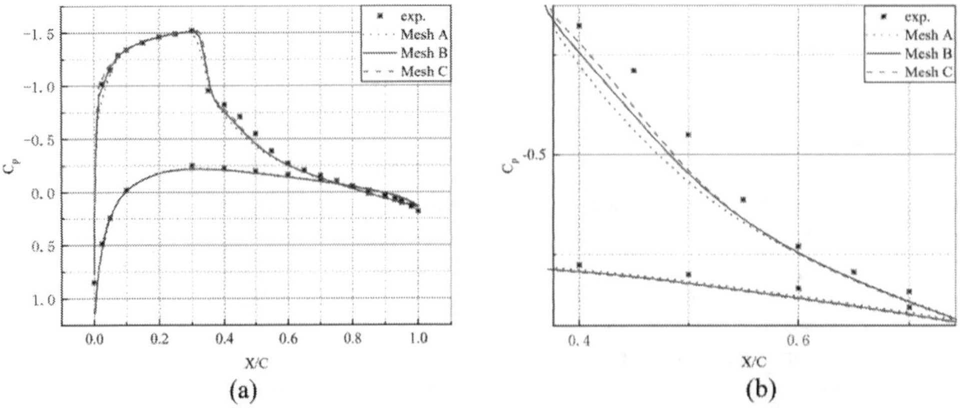

Figure 2. Comparison of average pressure coefficients on airfoil surfaces.

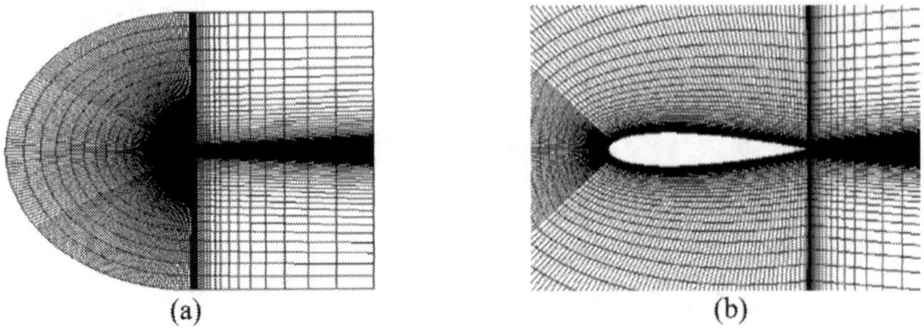

Figure 3. NACA0012 airfoil calculated flow field.

3. NACA0012 airfoil transonic shock wave buffeting steady flow field numerical simulation research

3.1. Analysis of the variation of airfoil aerodynamic characteristics with angle of attack

For M_∞=0.8 and Reynolds number Re=1.8×10^7, the SST k-ω model has been used, and the spatial flux calculation employed the second-order precision Roe-FDS scheme. Through numerical calculations, the aerodynamic characteristic curves of the NACA0012 airfoil were obtained, as shown in Figure 4. The lift coefficient curve in Figure 4(a) reveals that for small angles of attack (α<2°), the Cl of the airfoil exhibits an almost linear correlation with the α. When its angle reaches 2°, the lift coefficient curve begins to deviate significantly, the suppression leads to an accelerated increase of the Cp. Figure 4(b) shows that the drag coefficient curve meets a turning point at an attack angle of 0.5°. Subsequently, the continuous increase of the attack angle makes the Cd vary almost linearly with the change in attack angle. The polar curve shown in Figure 4(d) reveals an interesting trend. At low α (0°~1.5°), the Cl/Cd increases significantly as the angle of attack gradually increases, indicating that in this range the airfoil can generate more lift with minimal drag, resulting in highly efficient performance. The curve peaks at an angle of about 1.5°, with the Cl/Cd of 9.7. This optimum angle represents the best compromise between lift generation and drag minimization for this airfoil, but beyond this critical α, the Cl/Cd begins to decline. This is mainly due to two factors. Firstly, as shown in Figure 4(a), the rate of increase of the Cl slows down as the α exceeds 1.5°. Secondly, as shown in Figure 4(b), the Cd begins to increase significantly with increasing α. This increase in drag is attributed to the development of a larger wake and the formation of a separation bubble on the upper side of the airfoil. In summary, the polar curve highlights the delicate balance between lift generation and drag minimization for the airfoil. The optimum α for maximum efficiency is between 0° and 1.5°, where it achieves its highest Cl/Cd. Beyond this angle, the performance of the airfoil deteriorates due to a slower increase in lift and a rapid increase in drag. Therefore, the Cl/Cd curve shows a turning point at an angle of attack of 1.5°. On the other hand, with the increase of the α, the Cl/Cd of the airfoil gradually decreases.

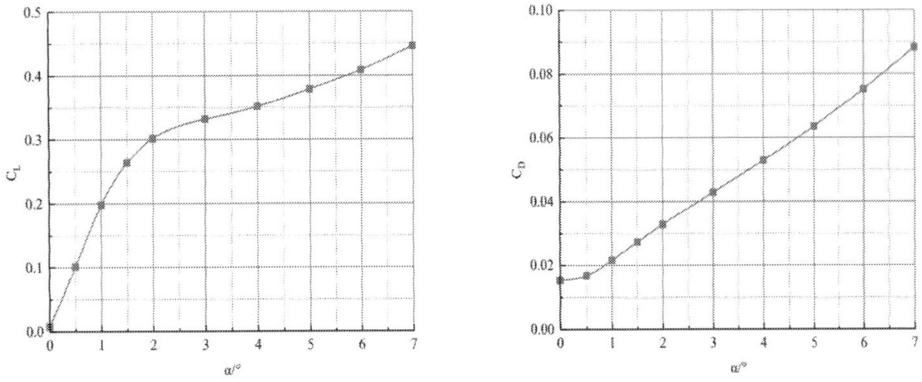

(a) (b)

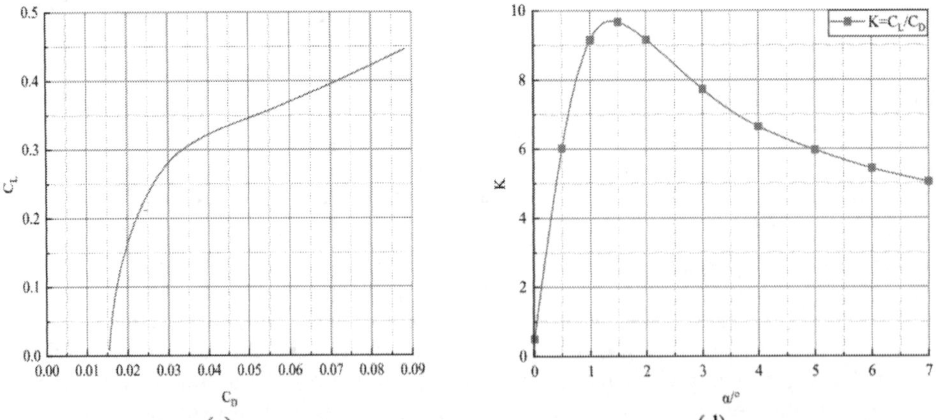

(c) (d)

Figure 4. NACA0012 airfoil aerodynamic characteristics curve.

3.2. Flow field structure variations with the angle of attack

The M_∞ contour plots for the NACA0012 airfoil, as shown in Figure 5, demonstrate the development of shock waves at various angles of attack within the transonic flow regime. For angles of attack less than 2°, shock waves are observed on both sides of the airfoil. As the α increases, the behavior of these shock waves on the two surfaces diverges. The shock wave on the top surface shifts downwards and its M_∞ gradually increases, resulting in a stronger shock. In contrast, the shock wave on the lower surface moves upstream with a decreasing M_∞, resulting in a weaker shock. As previously mentioned, when the attacking angle reaches 2°, the increase in Cl of the airfoil begins to slow down. At this stage, the shock wave on the lower surface vanishes, and a separation bubble forms downstream of the shock wave on the upper surface. As the α of the incoming flow increases further to 7°, the shock wave on the upper surface gradually shifts upstream. The shock wave's interaction with the boundary layer causes the boundary layer downstream to thicken slowly, enlarging the separation zone and leading to a reduction in the airfoil's stability.

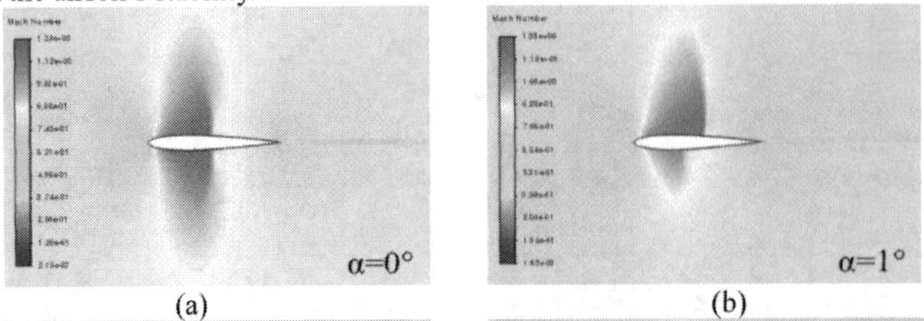

(a) (b)

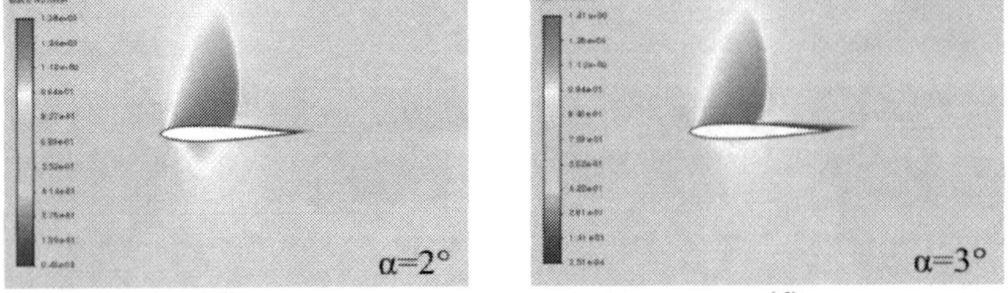

(c) (d)

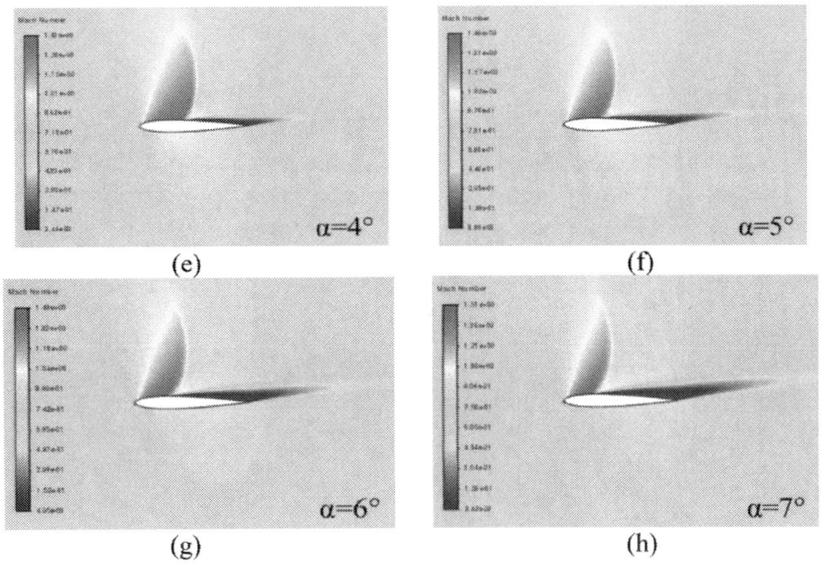

Figure 5. Mach number flow field of the NACA0012 airfoil at different incoming flow angles of attack.

4. NACA0012 airfoil transonic shock oscillation unsteady flow field numerical simulation study

4.1. Prediction of transonic buffet boundary

Under the incoming flow conditions with M_∞=0.725 and Re=9.3×10^6, as given in [14], this paper predicts the onset of the buffet by performing unsteady flow field calculations using the DES method. The Cl curves at various α were consistently obtained through numerical simulations, as displayed in Figure 6. Analysis of the figure reveals that when α reaches 3.9°, the Cl curve of the airfoil exhibits continuous oscillations, indicating the onset of buffet. With the angle continuously increasing to 4.3°, the Cl converges to a constant value. Hence, the onset and offset buffet angles obtained through numerical simulations are 3.9° and 4.3°, respectively. According to the buffet boundary results measured in wind tunnel tests in [14], the onset buffet angle at this M_∞ is 4.0°. The results show that the onset buffet angle determined by the DES method for the airfoil has a relative error of 2.5% compared to wind tunnel test results. By performing the Fourier transform (FFT) on the periodic oscillatory part of the lift coefficient, it is found that the buffet frequency reaches its maximum value of 50.2 Hz at the attacking angle of 4.0°. Figure 7 presents a comparison of the buffet boundary between the numerical and experimental results, showing that with the increase of incoming flow M_∞, the buffet onset angle gradually decreases. The DES method in this work shows a good agreement with wind tunnel results and accurately predicts the buffet onset angle.

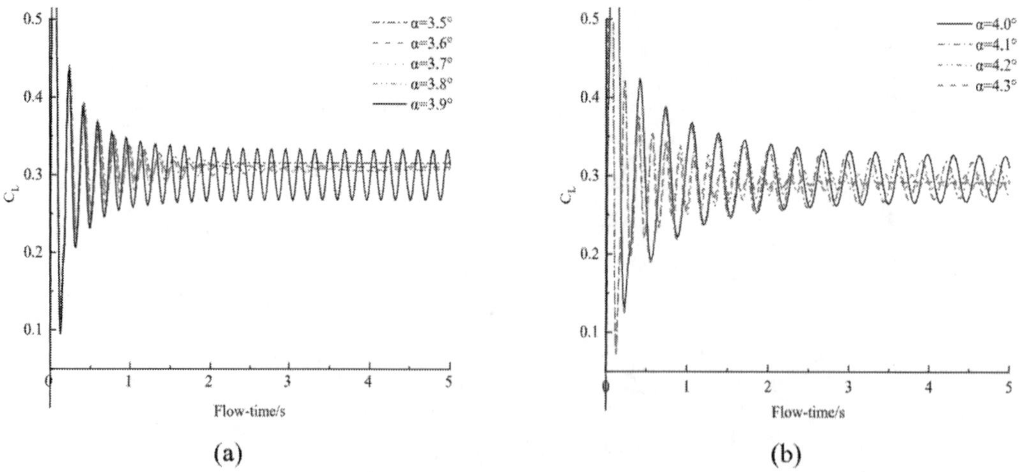

Figure 6. Lift coefficients of NACA0012 airfoils at different angles of attack.

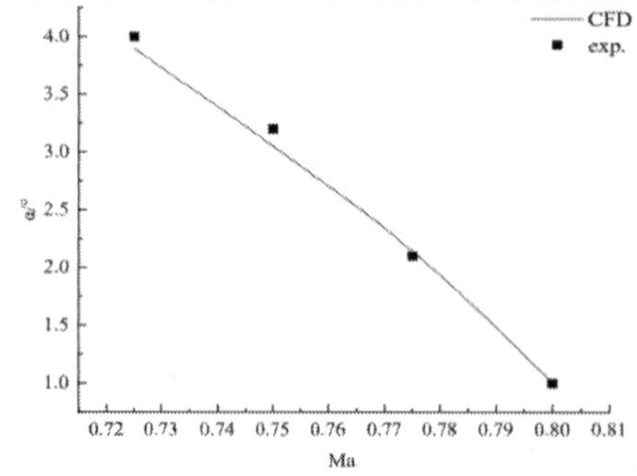

Figure 7. Comparison of shock wave buffet boundary.

4.2.Flow field structure with angle of attack after the onset of buffet

To further investigate the mechanism of changes in the wing buffet state after the aircraft becomes unstable, this paper studied the variation trend of flow field characteristic parameters such as the position and intensity of shock wave, and the intensity of the separated vortex behind the shock wave of this airfoil within a buffet interval, from before the onset of buffet to the end of the buffet state. Figures 8 and 9 show the trend of intensity and position of a shock wave with the increase of the angle of attack after the onset of the buffet. In the figures, it is easily observed that before the occurrence of the buffet, as its angle becomes larger, the intensity of the shock wave gradually increases, and the position of the shock wave starts to move downstream. When α reaches the buffet onset angle α=3.9°, the shock wave begins to gradually move upstream. Concurrently, the escalating intensity of the shock wave fosters a growing interplay between itself and the boundary layer. As the shock wave's intensity and the magnitude of the subsequently separated vortex reach a threshold, further intensification of the shock wave triggers an amplified propensity for the boundary layer to detach behind it. This detachment, in turn, reinforces the strength of the separated vortex in the region adjacent to the shock wave. The enhancement of the separated vortex strength will further increase the shock wave intensity. Therefore, a dynamic reinforcing relationship is established between the shock wave and the separated vortex behind. As the intensity of both increases, this "resonance" effect will be correspondingly enhanced. Therefore, with the gradual increase of the angle of attack, the corresponding interaction between them will show a continuous strengthening trend, as shown in Figure 10.

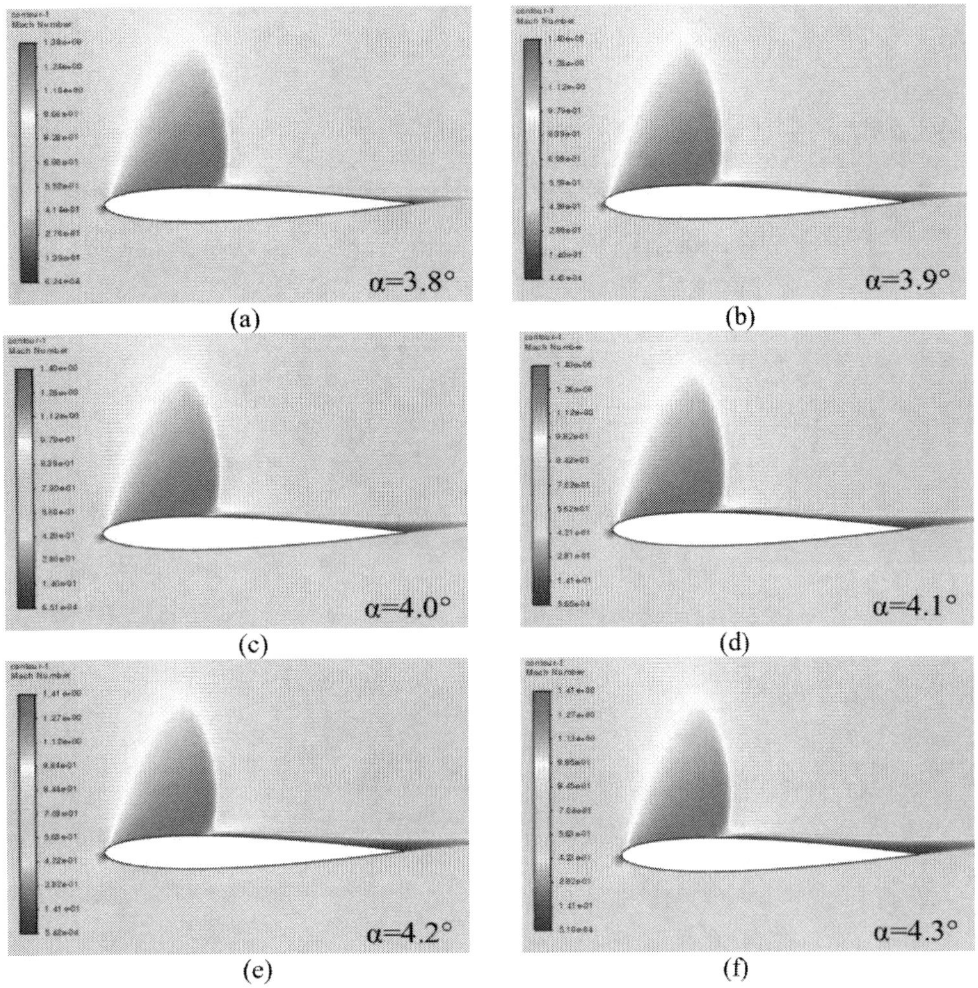

Figure 8. Trends in the variation of excitation position with different angles of attack after the onset of buffet vibration.

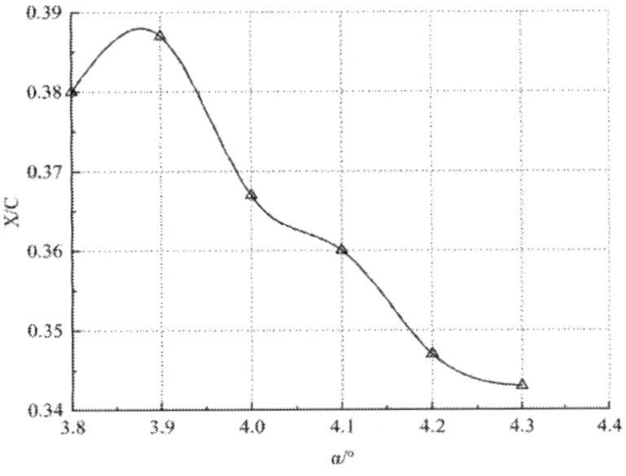

Figure 9. The trend of shock position with different angles of attack.

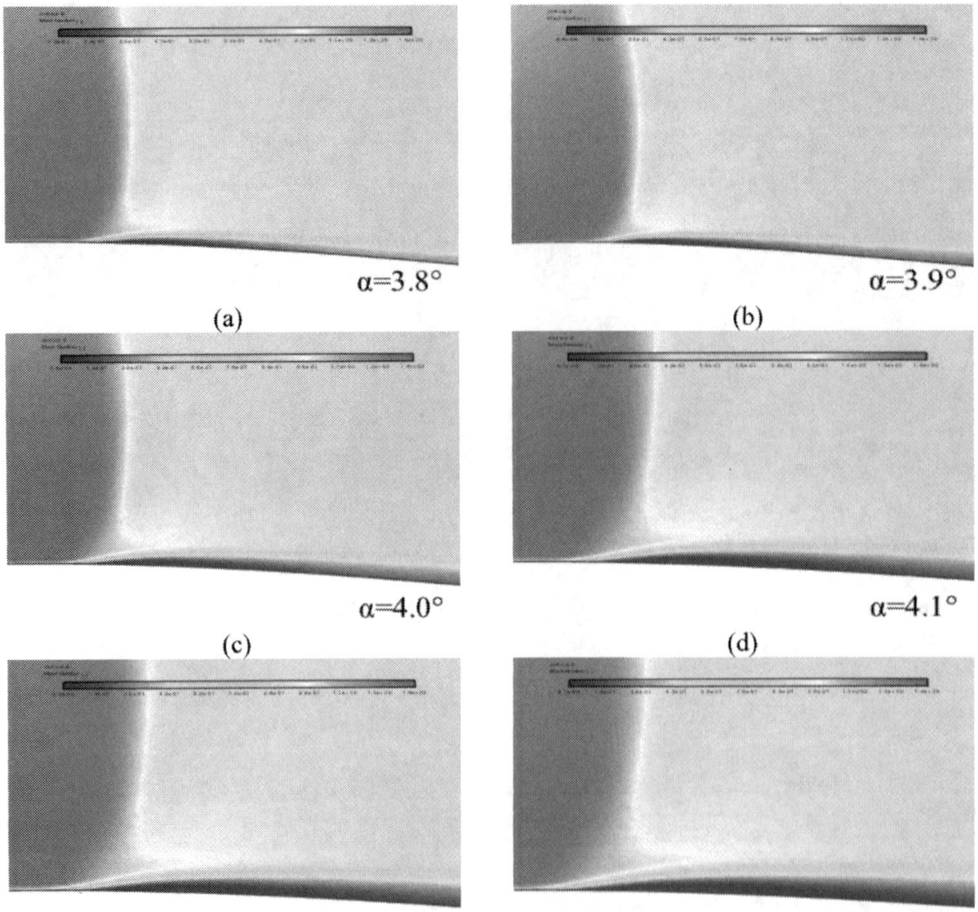

(a) (b) (c) (d) (e) (f)

Figure 10. The trend of separating vortex intensity with different angles of attack after buffet occurs.

4.3. Flow field structure after the onset of buffeting with varying incoming Mach numbers

As the M_∞ of the incoming flow escalates during transonic buffeting, the intensity of the shock wave steadily increases, as shown in Figure 11. At the same time, the shock wave migrates upstream. The increased shock wave intensity catalyzes a robust interaction with the boundary layer, resulting in an increase in airflow velocity in the downstream region of the boundary layer. This increased velocity leads to a relatively lower pressure in the vicinity, as shown in Figure 12. In addition, the low-pressure zone expanding behind the shock wave, combined with the increasing M_∞, promotes an increased tendency for airflow separation in this region. As a result, the intensity of the separation vortex and the thickness of the boundary layer behind it continue to escalate as the M_∞ of the incoming flow increases, as shown in Figure 13.

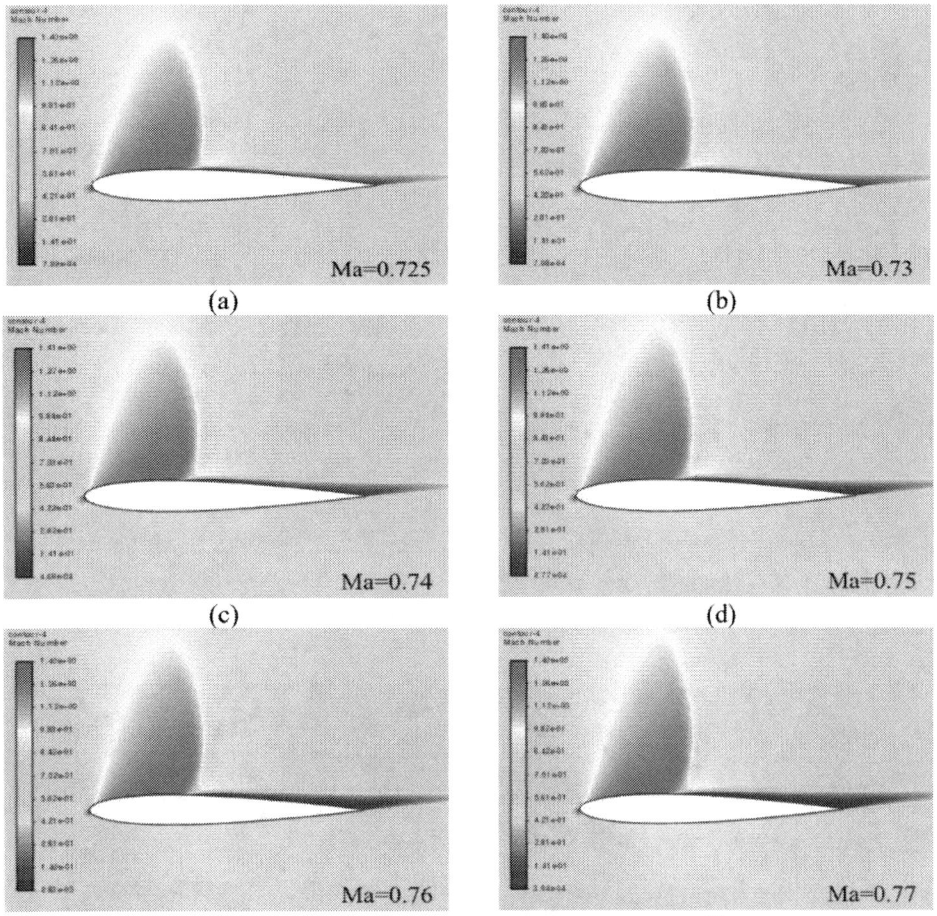

(a) (b) (c) (d) (e) (f)

Figure 11. The trend of shock wave intensity with the incoming flow Mach number.

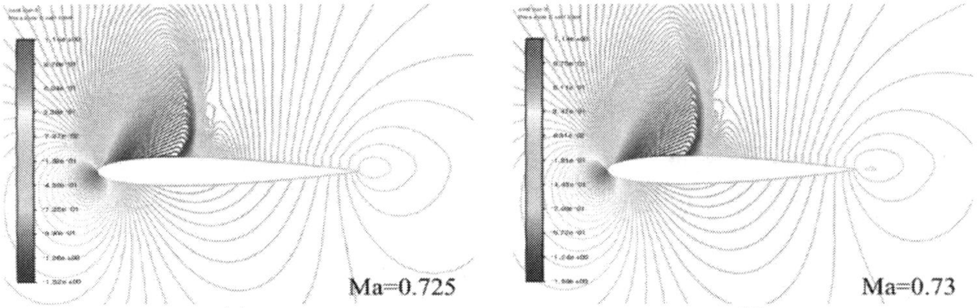

(a) (b)

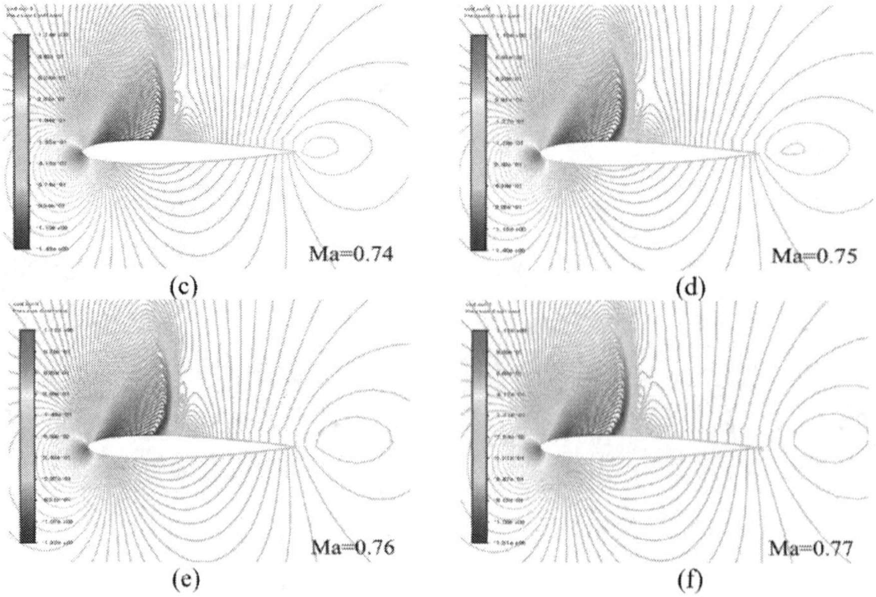

Figure 12. The trend of pressure distribution at the shock wave and the boundary layer behind the shock wave with the incoming flow Mach number.

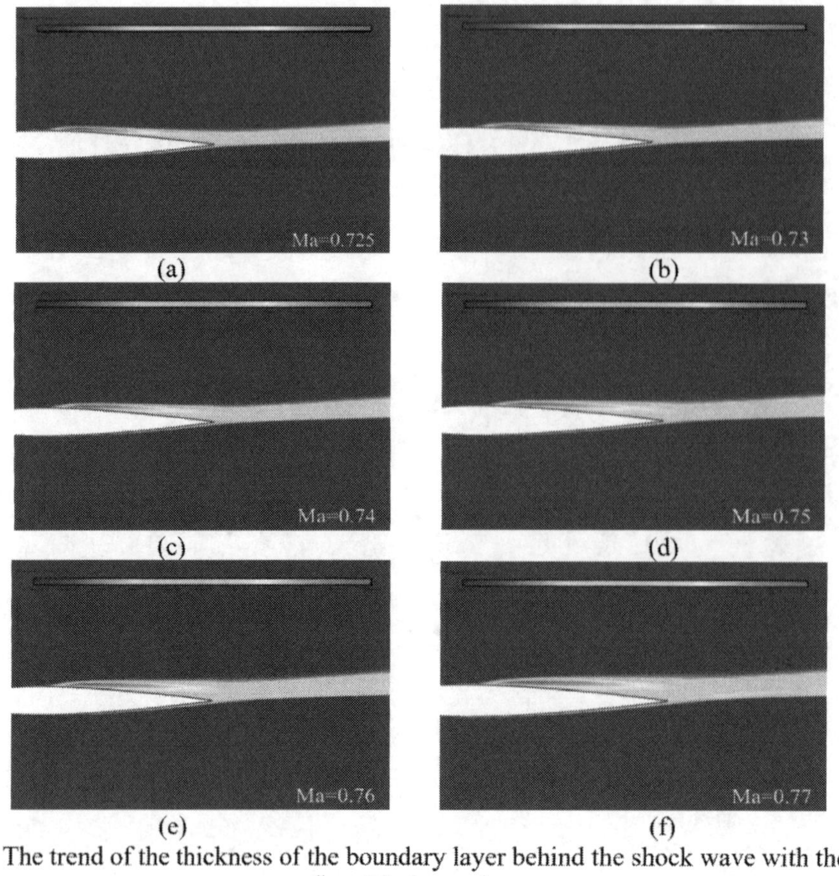

Figure 13. The trend of the thickness of the boundary layer behind the shock wave with the incoming flow Mach number.

4.4. Transonic shock wave oscillation flow field

Figure 14 shows the M_∞ contour plot over a period of shock wave oscillation, enabling a clear observation of the trend of shock wave intensity and position varying over time. As the shock wave moves downstream (Figures 14(a)~(c)), its intensity shows a gradual increase. At Δt/T=0.2460, the shock wave and separation vortex reach their maximum intensity. As the shock wave continues to move downstream, its intensity starts to decrease gradually at Δt/T=0.3721 (Figures 14(d)~(f)), followed by a slight upstream movement of the shock wave. At Δt/T=0.6910, the shock wave and separation vortex intensity reaches its minimum, after which the shock wave intensity gradually increases again with a smaller magnitude, and continues to move downstream slightly.

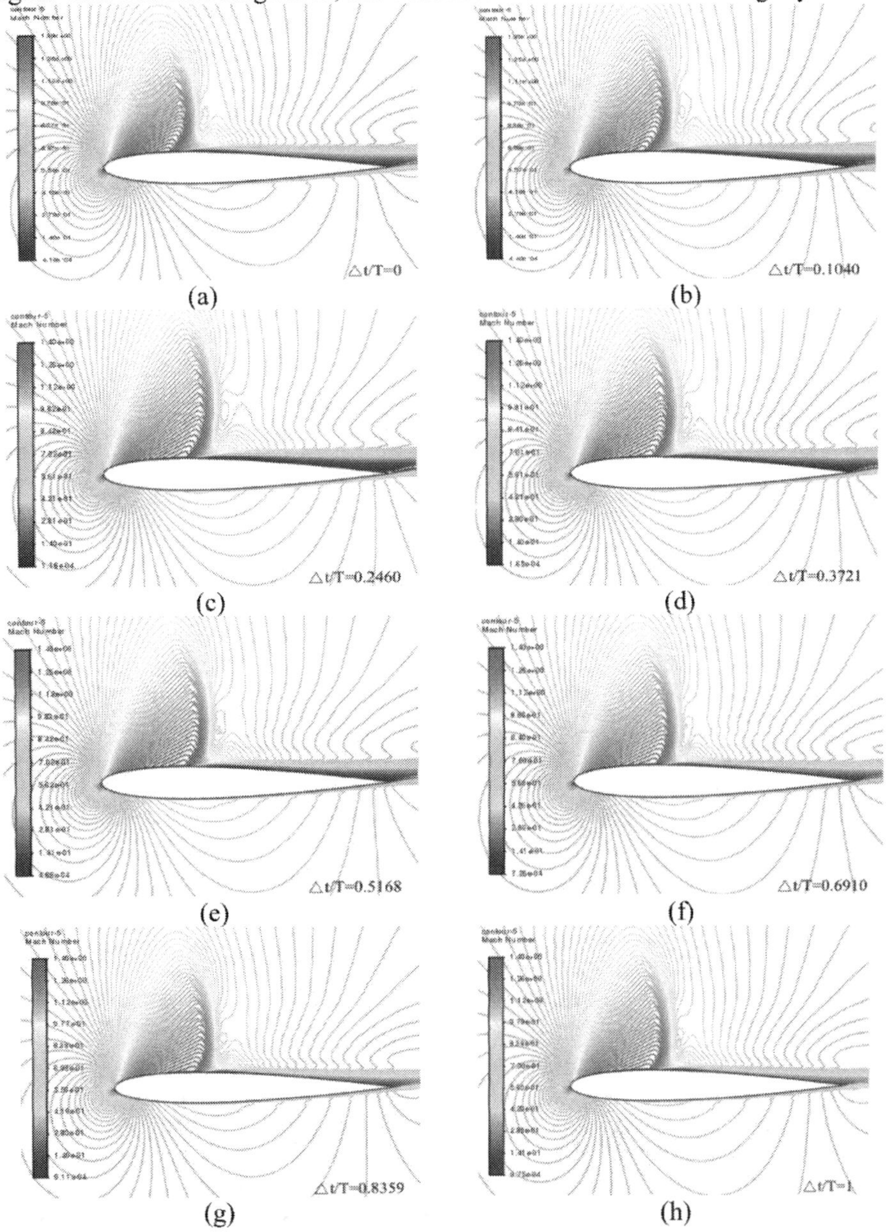

Figure 14. The law of change of shock wave within one shock wave oscillation cycle.

5. Conclusion

This paper employs a validated numerical method to obtain the aerodynamic characteristics of the NACA0012 airfoil under steady states. Additionally, this paper investigates the onset boundary and instability mechanism of transonic shock wave buffeting for this airfoil. The following conclusions are drawn:

(1) Results for the NACA0012 airfoil demonstrate a good agreement between the simulated and wind tunnel experimental data, with accurate prediction of the shock wave. The buffet onset boundary obtained through unsteady flow field calculations using the DES method aligns well with wind tunnel results, providing a reliable prediction of the airfoil's buffet.

(2) Analysis of the airfoil's flow field characteristics reveals a significant increase in the interaction between the shock wave and the separated vortex as the angle of attack or incoming Mach number increases. This enhanced interaction leads to instability in the flow, causing periodic fluctuations in the airfoil's circulation and ultimately resulting in buffeting.

In summary, this research provides valuable insights into the aerodynamic performance of the NACA0012 airfoil under steady-state and the onset characteristics and instability mechanism of transonic shock wave buffeting. These findings have important theoretical implications and offer a reference for the safe flight envelope. By adjusting flight speed and angle of attack, pilots can manage the aircraft's attitude in unstable flight conditions. Future research can build upon this study to explore further advancements in mitigating buffeting loads on this airfoil type.

References

[1] Fulvio Sartor S. Timme. Reynolds-Averaged Navier-Stokes Simulations of Shock Buffet on Half Wing-Body Configuration, 2015.
[2] Hilton W F, Fowler R G. Photographs of Shock Wave Movement[M]. HM Stationery Office, 1952.
[3] Nicholas F. Giannelis, Gareth A. Vio and Oleg Levinski, A Review of Recent Developments in The Understanding of Transonic Shock Buffet, PROGRESS IN AEROSPACE SCIENCES, 2017.
[4] Sébastien Deck. Numerical Simulation of Transonic Buffet Over A Supercritical Airfoil, AIAA JOURNAL, 2005.
[5] Juntao Xiong and Feng Liu. Numerical Simulation of Transonic Buffet on Swept Wing of Supercritical Airfoils, AIAA 2013-3205. 43rd Fluid Dynamics Conference. June 2013.
[6] Liu Jinli. Study on Mechanism of Transonic Shock Oscillation and Alleviation of Buffet Load on the Airfoil [D]. Northwestern Polytechnical University, 2018. (in Chinese)
[7] Spalart P R, Allmaras S R. A One-Equation Turbulence Model for Aerodynamic Flows[C]. 30th Aerospace Sciences Meeting & Exhibit, January 1992, Reno NV, AIAA 92-0439.
[8] Menter F R. Two-equation Eddy-viscosity Turbulence Models for Engineering Applications[J]. AIAA Journal, 1994, 32(8): 1598-1605.
[9] Yao Meng. Numerical Study of Airfoil Transonic Buffet Characteristics[D]. Nanjing University of Aeronautics and Astronautics, 2018. (in Chinese)
[10] Sandip Ghosal. Mathematical and Physical Constraints on Large-Eddy Simulation of Turbulence[J]. AIAA Journal, 1999.
[11] Johansen J, Sørensen N, Michelsen J and Schreck S. Detached-eddy Simulation of Flow Around The NREL Phase VI Blade, Wing Energy, 2002.
[12] Menter F R, Kuntzm. Development and application of a zonal DES turbulence model for CFX-5 [R]. CFX-VAL17/0703,2003.
[13] Hu Ou, Zhao ning and Shen Zhiwei. Simulation of Large Separated Flows with SST-DDES Model[J].Journal of Nanjing University of Aeronautics and Astronautics, 2017, 49(02): 206-211. DOI:10.16356/j.1005-2615.2017.02.009. (in Chinese)
[14] McDevitt J B, Unhook A F. Static and dynamic pressure measurements on a NACA 0012 airfoil in the high Reynolds number facility [R]. USA: National Aeronautics and Space Administration, 1985: NASA TP-2485.

Journal of Physics: Conference Series **2879** (2024) 012033 doi:10.1088/1742-6596/2879/1/012033

Overall design and structural finite element analysis of deep-sea mining test device

Jiancheng Liu[1], Jiwei Liu[2], Lei Li[1], Liangwan Rong[*3] and Yong Ma[2,4]

[1]China Merchants Marine Equipment Research Institute Co., Ltd., Shenzhen, Guangdong, 518067, China
[2]School of Ocean Engineering and Technology, Sun Yat-sen University, Zhuhai, Guangdong, 519082, China
[3]College of Civil Engineering and Transportation, South China University of Technology, Guangzhou, Guangdong 510640, China
[4]Southern Ocean Science and Engineering Guangdong Provincial Laboratory (Zhuhai), Zhuhai, Guangdong 519080, China

[*]Corresponding author's e-mail: ctlwrong@scut.edu.cn

Abstract: In order to support the experimental research and optimization of the deep-sea mining acquisition device, the overall design of a test device capable of carrying multiple acquisition heads was carried out. The design is divided into three parts: the bearing system of the device, the acquisition system, and the data collection system. It also includes the layout scheme of instruments and equipment such as flowmeter and pressure sensor. The experimental platform of simulation acquisition is constructed. The strength and stiffness of the main structure of each system are analyzed and optimized. The analysis results show that the strength and stiffness of the test device can meet the test requirements. The structural arrangement of the device meets the requirements of the test function, which provides a reference for the application design of similar devices.

1. Introduction

The challenges faced by the mining of land mining areas are becoming increasingly severe. As the mining volume of new mining areas gradually decreases, the reserves of mined mining areas also find it difficult to meet future needs, which leads to many problems, such as rising mining costs, declining resource quality, and increasing safety risks. Deep-sea mining is, therefore, seen as a key means of addressing these issues. As shown in Figure 1, the deep ocean is rich in mineral resources, which are widely distributed around the world, providing a reliable guarantee for future mineral demand and becoming an important way for human beings to obtain natural resources [1-2].

Since the beginning of the 21st century, global research investment in deep-sea mining technology has continued to grow. Global Sea Mineral Resources (GSR) in Belgium successfully developed the Patania I deep-sea mining test device in 2017 to study the ability of mining vehicles to travel on the seabed. In the same year, GSR joined forces with the MiningImpact2 (MI2) team to launch the ProCat 2 program, complete the development of an upgraded version of the mining test device Patania II in 2021, and conduct technical tests in the deep Pacific Ocean [3-4]. In the same year, the Dutch Delft University of Technology also carried out experiments on the deep-sea mining test device it developed and studied the influence of factors such as travel speed and jet speed on acquisition efficiency [5]. Zhang et al. [6] carried out the research and development of a deep-sea acquisition test device in 2022. By changing the diameter and angle of the acquisition head, the influence of structural parameters on the acquisition efficiency was studied [6-7].

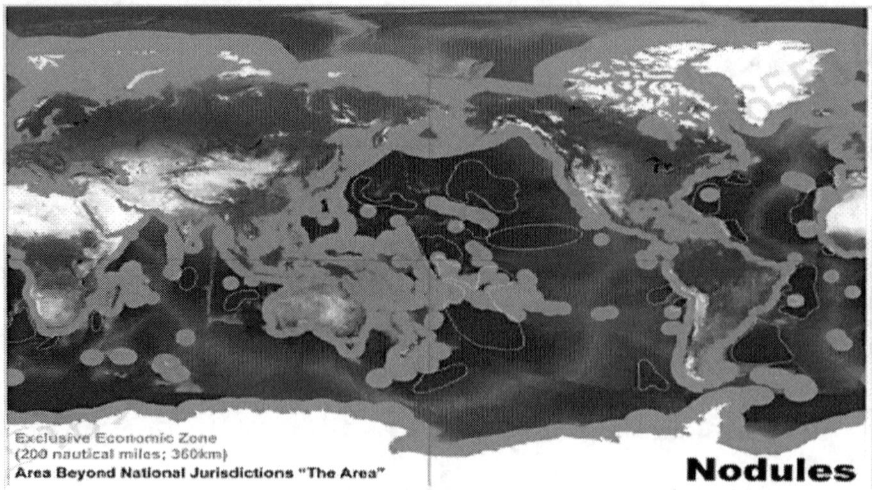

Figure 1. Global distribution of polymetallic nodule mineral resources [3].

The development of these test devices has undoubtedly promoted the progress of deep-sea mining technology research, but there is still room for further optimization in terms of universality. Therefore, it is urgent to develop a deep-sea hydraulic ore collection device suitable for multiple collection heads to support the collection experimental research better. The test device designed in this paper is composed of multiple systems. The actual deep-sea mining operation is simulated by carrying different types of acquisition heads and changing the test conditions. The crane carries the acquisition head to move along the direction of the flume at the set speed of the working condition, collects the ore particles laid in the test flume into the acquisition head, and falls into the collection bin along the transportation pipeline.

Compared with similar devices, the device has more comprehensive simulation conditions. It can be connected to a variety of acquisition heads through the ore transportation pipeline and change the driving speed, the height of the acquisition head from the bottom, the nozzle flow, the ore laying abundance, the test water depth, and other variables, so as to simulate the actual deep-sea operation. The working conditions of different acquisition heads in different acquisition areas. At the same time, the side wall of the designed flume is made of transparent toughened glass material, which can make the experimenter observe the test situation more intuitively and carry out more accurate and efficient data analysis combined with the data collected by various sensors arranged in the design scheme. Using this device to carry out simulation tests can not only deepen the exploration of related fields but also provide a solid equipment foundation for the development of deep-sea mining technology in the future [8-10].

2. The overall structure design of deep-sea mining acquisition test device

In order to support the research of deep-sea mining, it is necessary to design a universal analog acquisition device. Therefore, this paper studies the structural design of the acquisition test device.

First of all, according to the design requirements, the acquisition device should have the following functions:

(1) The traveling simulation can be controlled to ensure that the device is running at a specified speed.

(2) The water level of the flume can be adjusted to ensure the balance of the water level of the two flumes.

(3) Pallet height can be adjusted to adapt to different test conditions and needs.

(4) The flow rate of the jet pump can be adjusted to match different working conditions.

In addition, the acquisition device should also have data acquisition and performance parameter monitoring functions in the work and can ensure the smooth operation and operation safety of the equipment.

Therefore, according to the functional requirements, a deep-sea mining simulation acquisition device with strong adaptability, good stability, and a multi-type hydraulic mining device prototype is designed to realize the simulation and monitoring of the acquisition process. The acquisition test device is composed of three parts: bearing system, traveling system, and data acquisition system. The three-dimensional structure diagram of the device is shown in Figure 2.

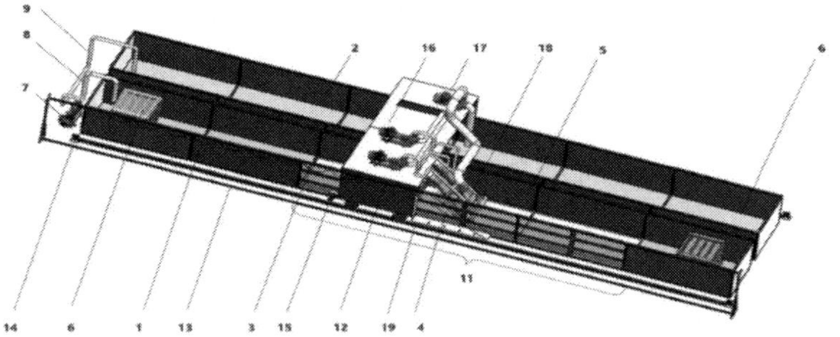

1. Test tank; 2. Water storage tank; 3. Steel support structure; 4. Tempered glass; 5. The material tray; 6. Rectifier plate; 7. Suction pump; 8. Test the water tank pipeline; 9. Water storage tank pipeline; 10. Connecting the pipeline; 11. Experimental area; 12. Driving; 13. Slide rail; 14. Stop the block; 15. On-board motor; 16. Nozzle jet pump; 17. Annular jet pump; 18. Collecting head; 19. Observation platform

Figure 2. Structure diagram of deep-sea mining acquisition device.

2.1. Overall design of the bearing system

2.1.1. The main scale and structure form of the system. In the deep-sea mining acquisition test, the cooperative operation between the systems is the key to ensuring the smooth progress of the test. As the bearing foundation of the test device, the load-bearing system is composed of two parts: the test flume and the test flume, which are connected to each other.

The overall design of the bearing system is divided into four parts: test tank, water storage tank, bottom tray, and suction pump. The overall structure is shown in Figure 3 and Figure 4:

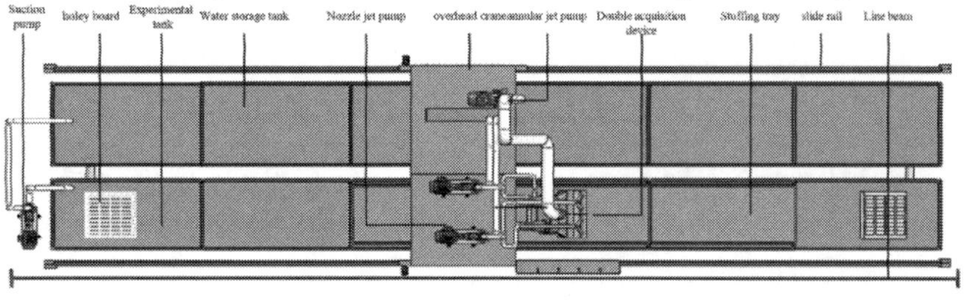

Figure 3. Overall equipment top view.

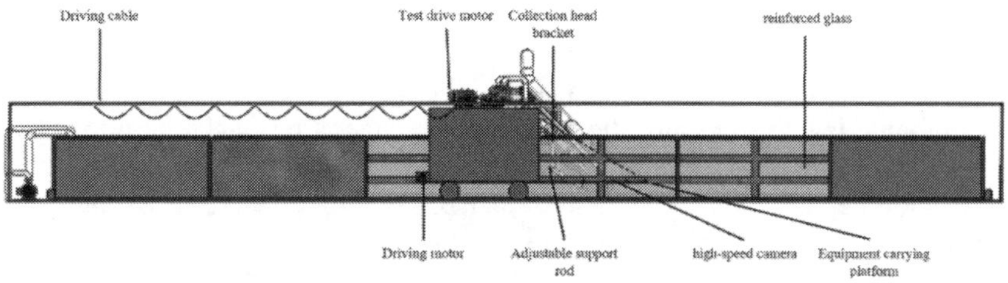

Figure 4. Overall equipment side view.

2.1.2. Design of bearing system. The test flume is mainly used for the collection and observation of nodules. The combination design of steel skeleton and tempered glass is adopted on the observation side. While ensuring structural strength and stability, researchers can use high-speed cameras and other precision instruments to monitor the operation of the hydraulic mining device in real time through tempered glass so as to realize the direct observation of the collection process.

The water storage tank is mainly used to accommodate the water discharged from the test tank during the test and as a supplement to the water body of the test tank under the new working conditions. The overall design parameters of the sink are shown in Table 1.

Table 1. Flume design parameters.

Name	Experimental tank	Water storage tank
Length/m	30	30
Width/m	2.5	3
Height/m	2.0	2
Wall thickness/mm	10	10
Operating flux level/m	1.5	1.5
Water-bearing capacity/m^3	150	180
Connecting pipe diameter/mm	30	

As shown in Table 2, the filling tray is mainly used to adjust the relative position between the collecting head and the nodule particles. During the test, the distance between the nozzle and the filling tray can be changed by changing the number of pads under the tray so as to achieve the purpose of changing the test conditions to meet the test requirements of different working conditions.

The design of the suction pump and the connecting pipe is mainly used to solve the problem of water level balance between the test flume and the water storage flume. One end of the pump pipe is connected to the test tank. The other end is extended into the water storage tank. Before the start of the test, the pump discharges the water in the water storage tank into the test tank; during the test, according to the water absorption and drainage of the driving pump system, the pump works to maintain the water level balance; after the test, the pump discharges the water in the test tank back to the water storage tank to prepare for the next test. This function of the suction pump is very important to maintain the stability of the test environment in the test tank.

Table 2. Design parameters of tray and suction pump.

Name	Parameter
Pallet size/m	15×2.4×0.05
Stand-off height/mm	20
Pallet height adjustment range/mm	80-220
Suction pump selection	IS 200-150-250B
Pump flow/ m^3/h	300

2.2. The overall design of the acquisition traveling system

2.2.1. Driving structure form. As shown in Figure 5, the acquisition travel system is composed of a driving, motor, frequency conversion box, brake device, and slide rail. The motor drives the driving, the frequency conversion box adjusts the travel speed, and the brake ensures the rapid stop of the travel. The auxiliary equipment in the system, such as the jet pump and collecting hopper, is connected to the vehicle in a reasonable layout to ensure the overall stable operation of the system. The equipment is equipped with a platform in the observation area, which provides sufficient installation space for the data acquisition equipment. The height angle is adjustable so as to adapt to different test conditions.

At the same time, in order to improve the durability and maintenance convenience of the system, all structures are made of corrosion-resistant, high-strength structural steel materials to ensure that good performance can be maintained during long-term use. At the same time, sufficient maintenance space and channels are reserved in the design to facilitate the daily inspection and maintenance of the equipment. The design parameters are shown in Table 3.

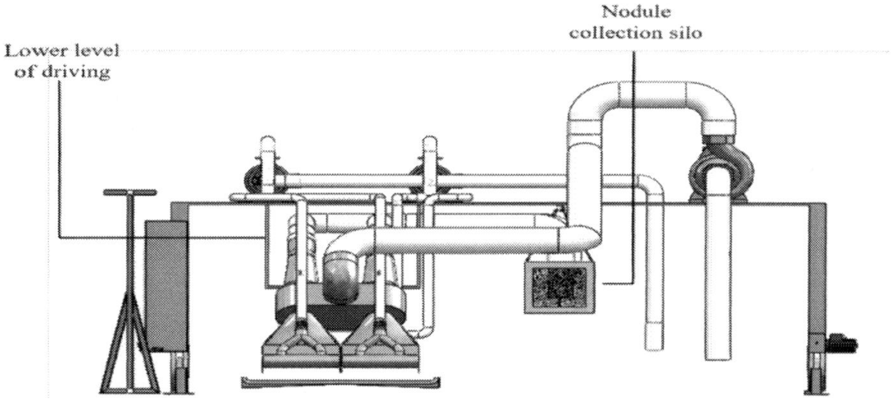

Figure 5. Schematic diagram of vehicle carrying structure.

Table 3. Traffic design parameters.

Name	Parameter
Length/m	7
Width/m	3.5
Height/m	2.5
Upper platform thickness/mm	30
Thickness of lower platform/mm	30

2.2.2. Acquisition head structure design. The acquisition head is an indispensable key component in deep-sea mining experiments. Its main function is to efficiently extract ore particles from the seabed and transport them to the acquisition device's pipeline system. In order to achieve this goal, this design adopts a double-connected acquisition head structure, as shown in Figure 6 and Figure 7. During the collection process, the water flow enters through two nozzle jet inlets and an annular jet inlet. These inlets are connected to the corresponding pumps, respectively. The precise guidance and control of the water flow are realized through the pipeline system. The mixed flow is then transported through the outlet to the nodule collection silo in the water storage tank. The network structure inside the silo is designed to intercept the nodule particles and discharge the excess water effectively.

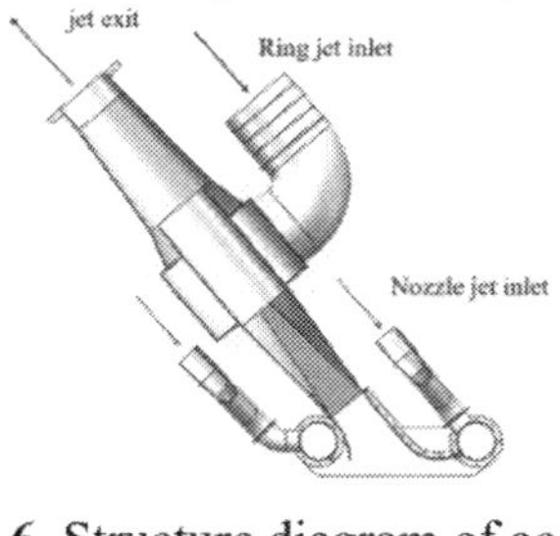

Figure 6. Structure diagram of acquisition head.

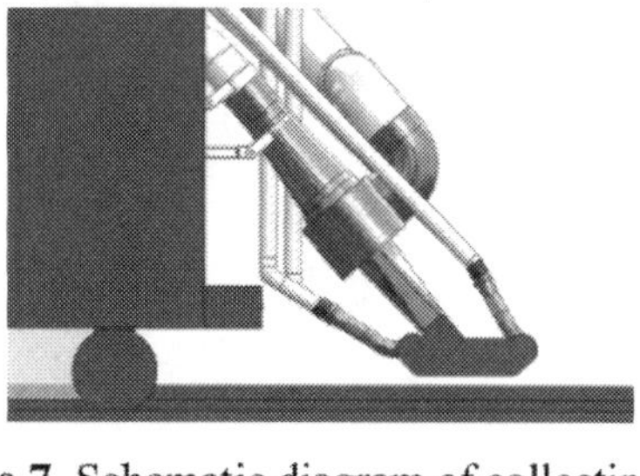

Figure 7. Schematic diagram of collecting head and driving fixed bracket.

2.2.3. Injection pump. In order to ensure the efficient and continuous operation of the dual acquisition head during the operation, a stable water supply must be provided. Therefore, two nozzle jet pumps and an annular jet pump are installed on the crane in this design. The fixed installation of these pumps can ensure the continuity and stability of the water flow. Furthermore, in order to meet the specific requirements of the flow during the test, the pump is equipped with a flow adjustment function. The use of an electric valve or frequency conversion control box as a means of regulation not only makes the operation more flexible and convenient but also ensures the accuracy of regulation. They can be adjusted in time according to the specific needs of the test and flexibly respond to various changes so as to ensure the fluency and efficiency of the collection process and greatly improve the work efficiency and accuracy of the test.

2.2.4. Nodule collection silo. As shown in Figure 8, in the final stage of the collection process, the ore particles are transported to the silo. The silo is composed of wire mesh, which is closely combined with the driving system to ensure the synchronous movement of the silo with the vehicle during the test. It is designed to be detachable, with a funnel-shaped movable plate door at the bottom to facilitate the smooth discharge of nodule particles. The mixed flow output pipe of the collection head extends to the inside of the silo. While ensuring the effective collection of particles, the excess water and sediment can be discharged through the sieve to maintain the silo clean. After the completion of the test, it is only necessary to open the movable plate door at the bottom and dump the nodule particles into the collection barrel below so as to prepare for the next round of tests.

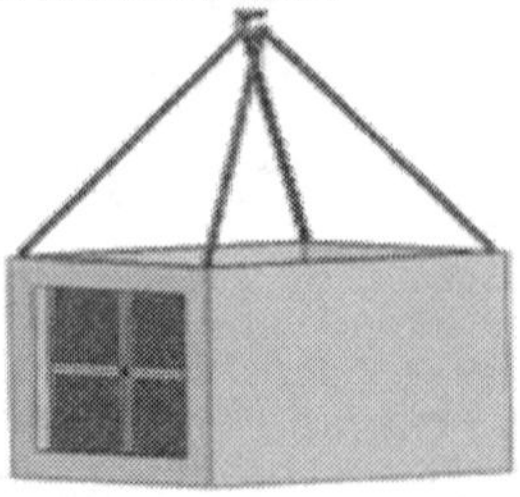

Figure 8. Schematic diagram of nodule collection silo.

2.3. Overall design of data acquisition system

In the deep-sea mining simulation test, in order to better monitor and control the test process, a variety of sensors and equipment are used in the design scheme.

2.3.1. Flowmeter. The flow meter is responsible for monitoring the water flow dynamics in the pipeline system to ensure that the water flow is accurately measured and regulated. As shown in Figure 9, two flowmeters are located in the bifurcated pipe between the outlet of the small flow centrifugal pump and the jet inlet of the acquisition head and are responsible for monitoring the annular jet flow; a flowmeter is arranged between the outlet of the large flow centrifugal pump and the inlet of the annular jet of the collection head. The annular jet is also monitored. The remaining flowmeter is installed in the middle of the pipeline at the output end of the mixed flow of the acquisition head to ensure effective monitoring of the mixed flow of ore-bearing particles. Such a layout can ensure the accuracy of the entire system flow monitoring.

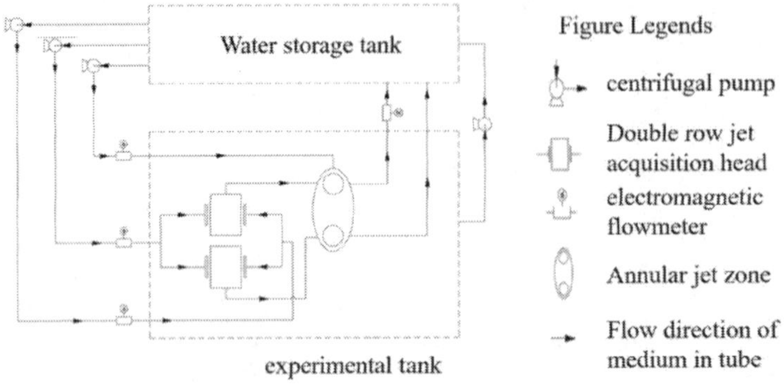

Figure 9. Flowmeter layout scheme.

2.3.2. Pressure transducer. The pressure sensor is used to monitor the pressure change in the tray and the collection head pipe in real-time in the test so as to analyze the specific influence of pressure fluctuation on the test results. As shown in Figure 10 and Figure 11, 13 pressure sensors are fixed on the tray in the form of an array, of which five are laterally distributed and eight are longitudinally arranged, all located in the collection area near the tempered glass observation area. In addition, the arrangement of the sensor inside the acquisition head is selected on the side wall surface rather than the bottom to avoid the direct impact of the sediment, ensure measurement accuracy, and prevent equipment damage. Because the heavier particles in the mixed flow tend to gather under the pipe, the sensors arranged on the side wall can capture the flow field data more accurately and reduce the error, thus providing more reliable pressure monitoring results.

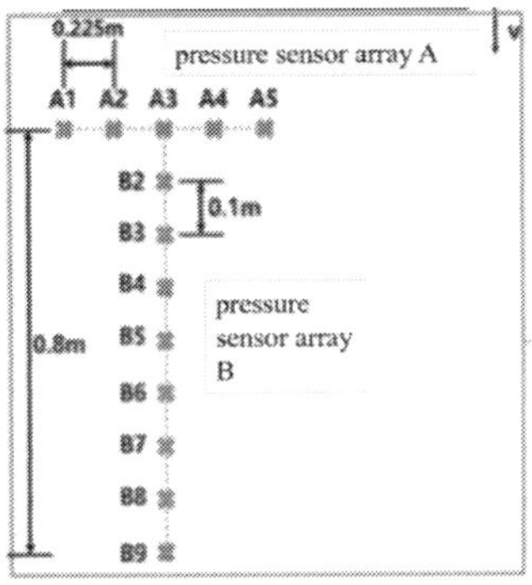

Figure 10. Layout scheme of the tray pressure sensor.

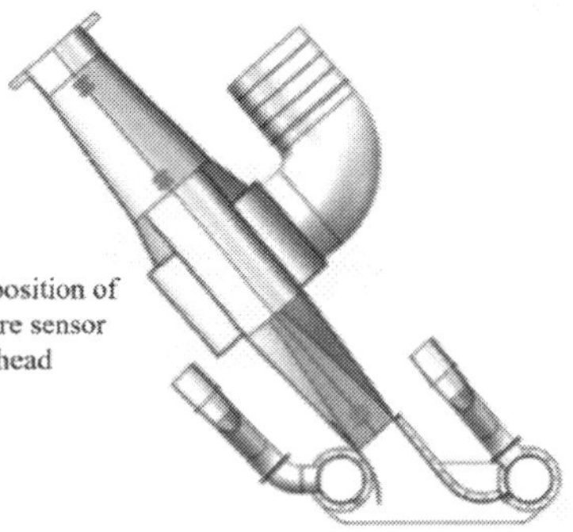

Figure 11. Arrangement scheme of the internal pressure sensor of the acquisition head.

2.3.3. Motion detection system. In order to capture the motion of nodules and the lifting process of ore particles fully, a motion monitoring system is arranged in the scheme, which mainly includes two parts: a high-speed camera and an underwater camera. The high-speed camera is mounted on the vehicle's monitoring equipment platform, which is used to record the dynamic behavior and particle lifting of the nodules in real-time. At the same time, the underwater camera equipped with IP68 waterproof grade focuses on recording the rising trajectory and distribution of nodule particles, providing a three-dimensional monitoring perspective to ensure the accuracy of the test results. The cooperative operation of these two camera devices ensures the complete recording of key data and the accuracy of subsequent analysis during the test.

3. Finite element analysis and optimization of the key structure of the acquisition device

To ensure that the test device can support the test, it is necessary to check and optimize the key parts of the design to verify the reliability and safety of the device. In this paper, the finite element analysis of the main structure of the design is carried out using the static analysis module in ANSYS Workbench software. The deformation and stress are checked to ensure the rationality of the design.

3.1. ANSYS Workbench optimization method

As shown in Figure 12, finite element analysis is the main optimization method for ANSYS Workbench. This process includes key steps such as model establishment, meshing, solution, and verification optimization.

Through the above steps, we can use ANSYS Workbench to conduct a comprehensive analysis and optimization of the structure, thereby improving its performance, saving materials, or reducing manufacturing costs.

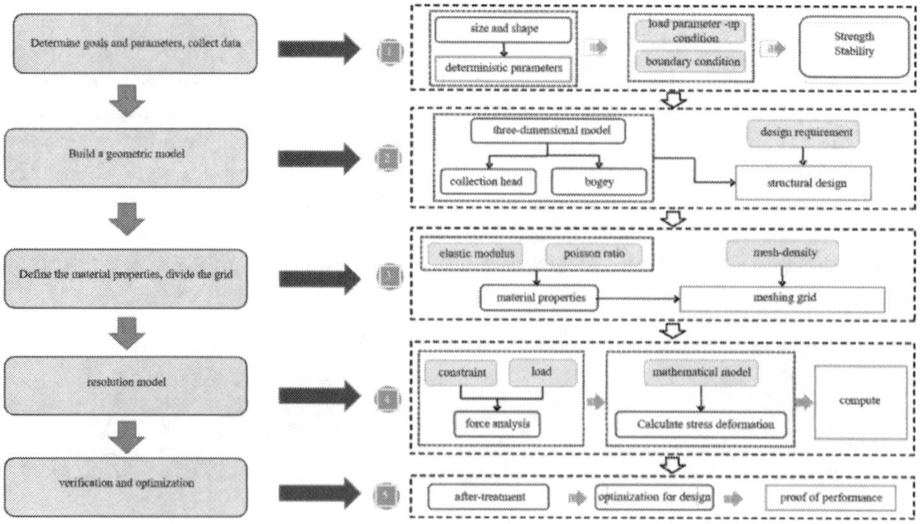

Figure 12. ANSYS Workbench numerical optimization process.

3.2. Structure analysis and optimization of the device

3.2.1. Definition of material properties. As the main bearing mechanism of the test device, the strength, stiffness check, and modal analysis of the tank and crane are important components of structural design. In order to improve the speed of finite element simulation calculation, it is necessary to simplify the three-dimensional model before importing the model into Workbench. The simplified contents are as follows:

(1) We Remove the small parts without force, ignoring the chamfer, rounded corners, holes, and other process characteristics.

(2) The glass side plate and the steel side wall are combined as a whole for analysis.

(3) We remove small gaps such as welds and fill in positions with smaller radians or angles.

When selecting the material of the device, because Q235 carbon steel has high strength and good plasticity, toughness, and weldability, its high yield strength and tensile strength make it able to withstand large mechanical loads and ensure the stability of the device. At the same time, its good plasticity can ensure the safety of steel structures under the action of unfavorable factors such as overload, impact, welding, and temperature stress. Therefore, the device is made of Q235 carbon steel. The simplified water tank model was imported into Workbench. The material was Q235, a high-strength carbon steel. Its density is 7850 kg.m^3. Yield strength is 235 Mpa, tensile strength is 375 Mpa, elastic modulus is 200 Gpa, hardness is 30 HRC, and Poisson's ratio is 0.3. Its material parameters are shown in Table 4:

Table 4. Material parameters.

Name	Elastic modulus/GPa	Poisson ratio	Density/(Kg/m^3)	Yield strength/Mpa	Tensile strength/Mpa	Hardness/HRC
Q235	200	0.3	7850	235Mpa	375Mpa	30

3.2.2. Selection and optimization of flume grid unit. In the static strength analysis of structures, the selection of grid elements is very important. At present, Solid186 and Solid187 are two main types of mesh units. Among them, Solid186 is a hexahedron unit, while Solid187 is a tetrahedral unit. At the same mesh size, the number of mesh elements and nodes of the tetrahedron is more than that of the hexahedron. This means that the use of tetrahedral meshes provides greater flexibility in simulating complex geometries. In addition, due to the more gentle transition characteristics of the tetrahedral element, it can better adapt to various complex geometric shapes and boundary conditions. Therefore, based on the above considerations, the mesh unit of this simulation is selected as Solid187, as shown in Figure 13.

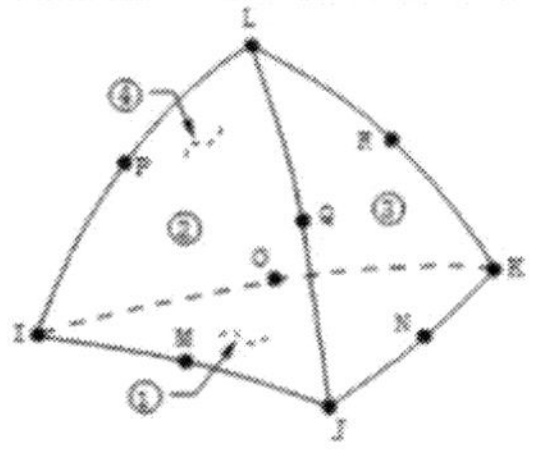

Figure 13. Solid187 tetrahedral mesh.

The mesh size is an important factor affecting the finite element calculation. In order to reduce the sensitivity of the mesh and improve the operation speed under the premise of accurate results, it is necessary to verify the independence of the flume mesh and determine a mesh with the most suitable size. The grid independence verification of the flume uses stress and deformation as the objective function. The specific function is:

$$\alpha = \frac{|S_m - S_n|}{S_m} \times 100\% \tag{1}$$

$$\beta = \frac{|R_m - R_n|}{R_m} \times 100\% \tag{2}$$

In the formula, the ultimate stress change rate; the ultimate deformation rate of change; the ultimate stress (Mpa) when the number of elements is m and n; is the ultimate deformation (mm) when the number of elements is m and n.

In order to determine the appropriate mesh size of the flume, the finite element models with mesh sizes of 80 mm, 60 mm, 40 mm, 20 mm, and 10 mm were established in turn to solve the stress and deformation. At the same time, according to the grid independence verification Formulas (1) and (2), the limit change rate of different grid sizes is calculated. The results are shown in Table 5.

Table 5. Quality table of flume grid.

Grid size (mm)	Number field of the element	Hop count	α(×100%)	β(×100%)
1000	4203	8672	-	-
100	46753	89276	18.226	80.96
80	63953	115762	0.068	1.024
60	127049	221310	1.166	1.563
40	155695	261682	2.586	1.053

The finite element takes the nodes of the grid element as the basic calculation unit, so the number of nodes can reflect the speed of ANSYS finite element calculation. The calculation accuracy can be obtained by the limit change rate. The smaller the sum is, the smaller the influence of the mesh size on the solution result is, and the higher the solution accuracy is. From Table 5, it can be seen that with the decrease of the grid, the number of nodes in the sink is increasing; that is, the finite element calculation speed of the sink is decreasing, and the limit change rate of the grid is decreasing, and the finite element calculation accuracy is increasing. When the mesh size decreases from 80 mm to 60 mm, the number of sink nodes increases by 91.2 %, while the limit change rate only decreases by 1.166 % and 1.563 %. At this time, the finite element calculation speed drops sharply, but the calculation accuracy is not greatly improved. Therefore, a better finite element calculation model can be obtained by setting the mesh size to 80 mm. At this time, the number of elements is 63953. The number of nodes is 115762. The mesh model is shown in Figures 14 and 15.

Figure 14. Grid division of water storage tank.

Figure 15. Test flume grid division.

3.2.3. Structure analysis of water tank body. In the installation process of the test device, in order to ensure the accuracy and reliability of the test results, it is necessary to install and constrain the flume accurately. Because the bottom of the water storage tank and the test tank are fixed, it can be directly regarded as a fixed constraint during installation. The side wall of the flume is the key part that bears the load. In order to ensure the structural safety of the side wall of the flume, it needs to be checked emphatically. The scheme uses uniform load simulation to simulate and analyze the side wall of the tank to evaluate its carrying capacity and stability. Through this simulation analysis method, the bearing capacity of the side wall of the flume can be evaluated more accurately. The safety and stability of the test device can be ensured. Therefore, in the process of structural simulation, uniform loads of 117.6kN and 98kN are applied to the side walls of the water storage tank and the test tank, respectively. After the above settings are completed, the solution is solved. The results are shown in Figure 16 and Figure 17.

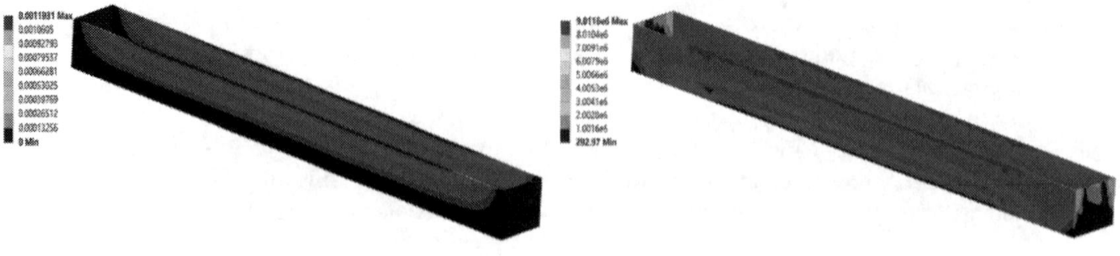

Figure 16. Water storage tank cloud map.

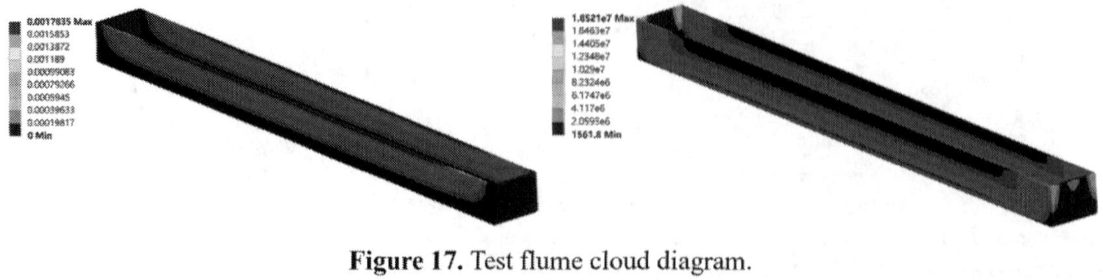

Figure 17. Test flume cloud diagram.

It can be seen from Figure 16 that the overall deformation amplitude of the water storage tank is relatively uniform. The maximum deformation occurs at the midpoint of the side wall, with a maximum deformation of 1.1931 mm. The deformation of the remaining positions is small. The overall deformation fully meets the requirements. At the same time, it can be seen from the stress distribution map that the overall stress distribution is also reasonable. The maximum stress appears at the junction of the side wall and the bottom surface of the sink. The maximum stress value is 9.0116 MPa, which is far less than the yield limit of the material of 235 MPa.

It can be seen from Figure 17 that, similar to the water storage tank, the maximum deformation also appears at the midpoint of the side wall, with a value of 1.7835 mm, and the maximum stress appears at the junction of the side wall and the bottom surface of the tank, with a value of 18.521 MPa. It also meets the safety requirements of the design. Due to the conservative side wall thickness, the overall structure of the sink is highly reliable and safe.

3.2.4. Selection and optimization of driving grid unit. The driving machine is an important part of the acquisition test device. It needs to be arranged above the nozzle jet pump, annular jet pump, motor, frequency conversion box, and other equipment. It is responsible for carrying the acquisition head for uniform motion during the test. Therefore, the structural characteristics of the crane will directly affect the collection efficiency. In order to ensure the reliability of the whole machine's performance, a finite element analysis of the crane is necessary.

Firstly, according to the principle of flume model simplification, the driving model is simplified and imported into Workbench. The same material and flume are set as Q235. The finite element analysis model with grid sizes of 140mm, 120mm, 100mm, 80mm, 60mm, 40mm, and 20mm is established in turn. The grid independence is verified. The verification results are shown in Table 6.

Table 6. Traffic grid quality table.

Grid size (mm)	The number field of the element	Hop count	α(×100%)	β(×100%)
140	31517	63056	-	-
120	32554	64218	0.543	11.37
100	33386	64169	0.095	5.605
80	56424	105649	0.994	5.73
60	105451	190026	0.308	6.99
40	275381	474371	0.326	5.72
20	2538743	1592879	0.229	18.65

From Table 6, it can be obtained that when the grid size is reduced from 40mm to 20mm, the number of driving nodes increases by 235.79 %, while the sum of limit change rates only decreases by 0.229 % and 18.65 %. At this time, the number of nodes to be calculated by the finite element method increases sharply, but the calculation accuracy is not greatly improved. Therefore, the grid size is set to 40 mm, which can ensure the speed and accuracy of the finite element calculation of the vehicle. At this time, the number of elements is 275391, and the number of nodes is 474371. The grid model is shown in Figure 18.

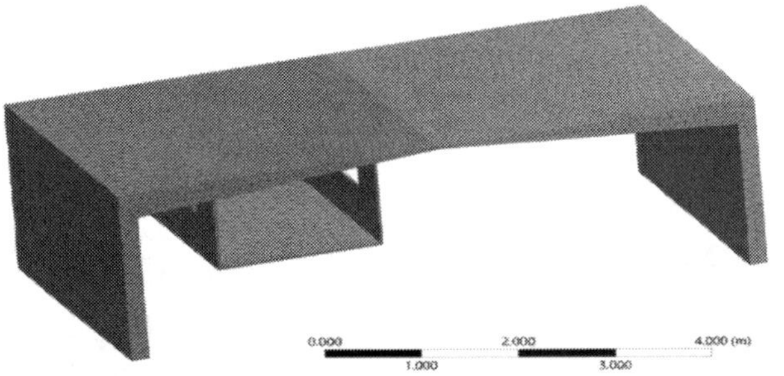

Figure 18. The meshing of traffic structure.

3.2.5. Optimization analysis of driving structure. When the vehicle is in actual operation, the support force is provided by the slide rail below. The linear motion is carried out along the moving direction. There is no force in other directions except the vertical direction. Therefore, a fixed constraint is applied at the bottom. A fixed load of 23128 N and 29400 N perpendicular to the horizontal plane of the platform is applied in the upper and lower layers, respectively, to simulate the real situation of the vehicle operation.

After completing the above settings, the driving model is solved. The results are shown in Figure 19 and Figure 20.

It can be seen from Figure 19 and Figure 20 that the overall deformation of the vehicle is concentrated at the bearing platform. The maximum deformation occurs at the edge of the lower bearing platform. The maximum deformation is 26.971 mm. The deformation of the remaining positions is small. The maximum deformation of the upper platform appears in the center of the platform. The deformation is

Journal of Physics: Conference Series **2879** (2024) 012033 doi:10.1088/1742-6596/2879/1/012033

25.7 mm. At the same time, it can be seen from the stress distribution map that the upper platform bears a large load. The maximum stress appears at the welding point of the upper and lower platforms. The maximum stress value is 143.97 MPa. The maximum stress of the upper platform appears at the rear of the vehicle. The maximum value is 59.178 MPa. Although the stress on the vehicle is less than the yield limit of the material of 235 MPa, its deformation is too large, and the stability of the bearing cannot be guaranteed, so it still needs to be optimized.

Aiming at the problem of excessive deformation of the bearing platform, the optimization method adopted is to increase the triangular reinforcement on the front and rear surfaces of the vehicle and increase the contact area at the connection of the upper and lower platforms to solve the excessive deformation of the upper and lower platforms. The results of the finite element analysis again after the optimization are shown in Figure 21 and Figure 22.

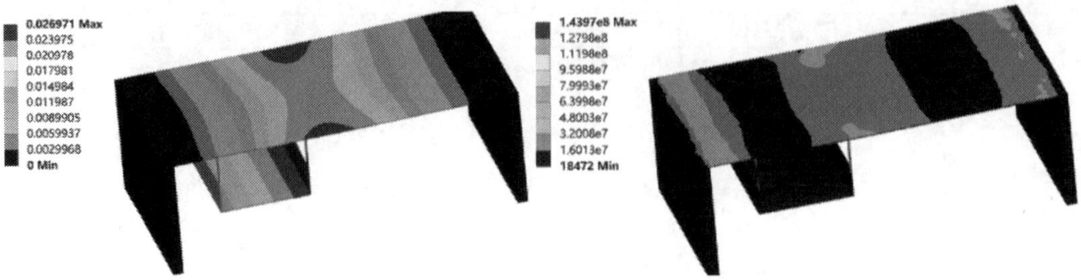

Figure 19. Driving deformation cloud map.

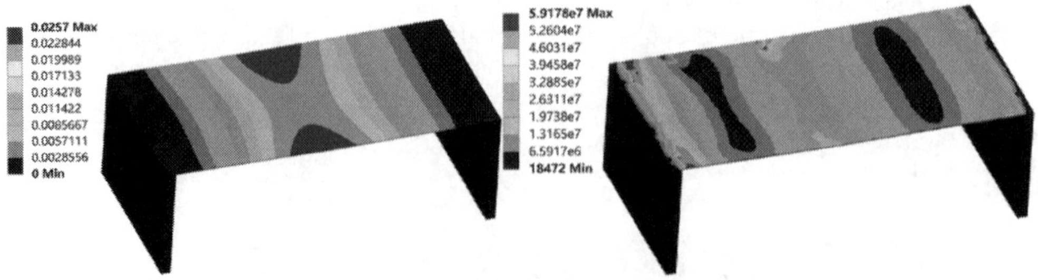

Figure 20. Driving stress distribution cloud map.

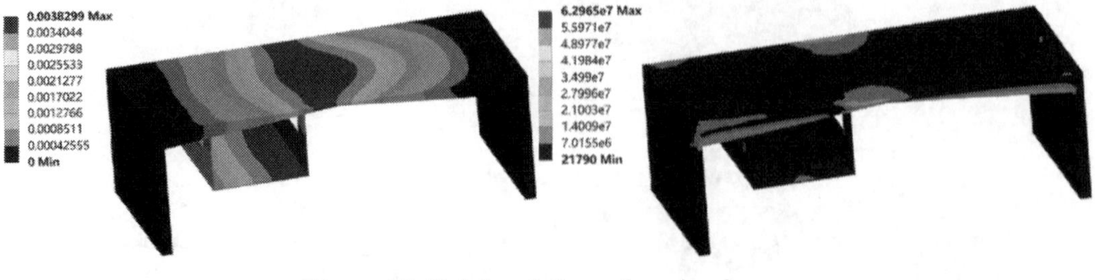

Figure 21. Driving deformation cloud map.

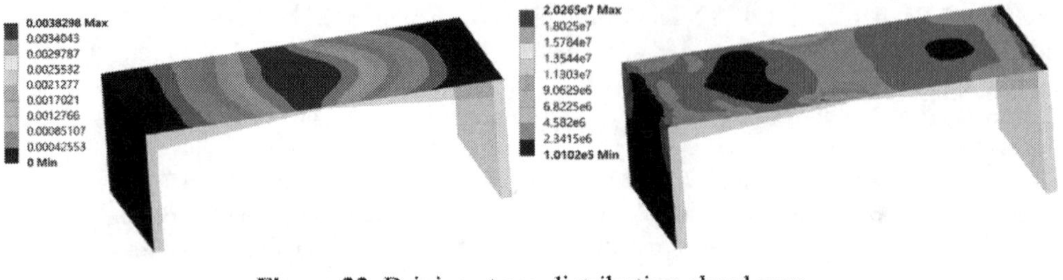

Figure 22. Driving stress distribution cloud map.

Figures 21 and 22 clearly show that the vehicle's overall deformation range has been significantly reduced after optimization. The maximum deformation is 3.8299 mm at the edge of the lower bearing platform. In the upper platform, the maximum deformation is 3.8298 mm in the center of the platform.

In terms of stress distribution, we found that the maximum stress appeared at the stiffener of the upper platform. Its value was 62.965 MPa. Further analysis of the two-layer platform shows that the maximum stress of the upper platform appears at the rear of the vehicle, with a value of 20.265 MPa. It is worth noting that this stress value is much smaller than the yield limit of the material of 235 MPa.

After optimization, the total deformation is reduced by 85.8 %. The maximum stress value is reduced by 56.3 % compared to before optimization. These significant data changes fully show that the optimized driving has higher reliability and safety and can better ensure the smooth progress of the test process.

3.2.6. Test verification and analysis. In order to verify the effectiveness and feasibility of the design scheme of the deep-sea mining test device, we developed the prototype based on the optimized three-dimensional model and carried out the performance test of the device under different working conditions to ensure the stability and reliability of the device.

As shown in Figure 23, according to the different settings of the test conditions, the ore particles with the specified abundance are laid in the test tank of the test device. The water is added to the specified height. At the same time, the crane is connected to the acquisition head. The test verification is carried out after the debugging is completed. The real seabed environment is simulated by changing the driving speed, nozzle flow, ore abundance, water depth, and other factors to determine the influence of different variables on the acquisition efficiency and then verify the feasibility of the design of the test device. Some representative test results are shown in Table 7.

Figure 23. Deep-sea mining test device prototype.

Table 7. Working condition setting and test results.

Driving speed (m/s)	Front and rear nozzle flow (m^3/h)	Annular jet flow (m^3/h)	Ore laying abundance (kg/m^3)	Water depth of flume (m)	Collecting efficiency
0.3	200	1700	15	1.5	93.47%
0.3	200	1700	15	1.65	89.24%
0.3	200	1700	30	1.5	93.03%
0.3	200	1300	15	1.5	80.07%
0.4	200	1300	15	1.5	81.129%
0.5	200	1300	15	1.5	83.421%

The results show that the acquisition test device can maintain relatively stable and efficient acquisition efficiency under different test conditions. In the test of more than 60 groups of different working conditions, each structure of the device works stably. The test verification is successfully completed, which further verifies the feasibility of the device design scheme.

4. Conclusion

In this paper, the overall scheme design and main structural strength and stiffness analysis of the deep-sea mining test device are given. Through the finite element method, the water tank and the driving structure are calculated, and the deformation and stress distribution are obtained. According to the calculation results, the following conclusions are obtained:

1) The design scheme of the test device composed of the main structures, such as the sink and the vehicle, has good rationality and practicability and can better meet the test functional requirements of the acquisition simulation test.

2) In the middle of the upper layer and the inner side of the lower layer, the stress at the connection between the side of the flume and the bottom surface is relatively high. Local reinforcement is needed in the design.

3) After optimizing the driving structure based on the simulation results, the prototype is developed and tested according to the design scheme. The test results show that the design meets the use requirements and has high practical application value.

In summary, before the actual deep-sea mining operation is carried out, this test device can be used to simulate the real deep-sea environment. At the same time, it can also be equipped with a variety of acquisition heads for performance testing so as to avoid the problems that may occur in deep-sea operations in advance and provide a reference for real deep-sea mining operations. In the more extreme deep-sea environment exploration in the future, high efficiency and safety will become the future direction of the improvement of the test device. Through the improvement of the existing structure and the design of more auxiliary modules, the collection work will be safer and more efficient to cope with the more extreme and complex working environment.

Acknowledgments

Guangdong Natural Capital Partners (GDNCP)[2024] No. 44 :In 2024, the six major marine industries' special projects of provincial-level promotion of high-quality economic development (marine economic development) include key technology research and equipment development of a 6000-meter deep-sea polymetallic nodule mining vehicle.

References

[1] Yu M, Deng X.G, Yao H.Q, et al. (2018) World survey and research progress of submarine polymetallic nodules [J]. Geology of China, 45 (01): 29-38.
[2] Chen Y.C. (1999) Prospective evaluation of mineral resources in major metallogenic belts in China [M]. Beijing: Geological Publishing House.
[3] GSR (2018), https://www.deme-group.com/gsr.
[4] GSR (2023), https://deme-gsr.com/news/first-report-of-the-gsr-patania-ii-expedition-published-technical-achievements-and-scientific-learnings/.
[5] Alhaddad S, Mehta D, Helmons R. (2023) Mining of deep-seabed nodules using a coandă-effect-based collector [J]. Results in Engineering,17: 100852.
[6] Zhang D.K, Liu M.L, Xia J. X. (2023) Experimental study on sediment disturbance during deep-sea polymetallic nodule collection [J]. Mining and Metallurgy Engineering, 43 (03): 20-23.
[7] Guan L, Zhang D.K, Xia Y.C, etc. (2021) Comparative analysis of collection methods and structural parameters of deep-sea polymetallic nodules [J]. Journal of Marine Technology, 40 (05): 62-70.
[8] Kang Y.J, Liu S.J. (2021) The development history and prospect of deep-sea polymetallic nodule mining technology [J]. Chinese Journal of Nonferrous Metals, 31 (10): 2848-2860.
[9] Zhao Y.Y, Zeng X.G, Lang S.Y. Status and Prospect of Deep-sea Mining System [J]. (2016) Ship Materials and Market, (06): 39-41.
[10] Yang J.M, Liu L, Lv H. N, et al. (2020) Development status and prospect of deep-sea mineral resources development equipment in China [J]. China Engineering Science, 22 (06): 1-9.

Journal of Physics: Conference Series **2879** (2024) 012034 doi:10.1088/1742-6596/2879/1/012034

Failure analysis on cracking and leakage of 304 stainless steel flange in petrochemical butyl device

Lei Cheng[1], Chenyang Du[1*], Xiaowei Li[1], Baolin Liu[1] and Guoju Cui[2]

[1]China Special Equipment Inspection and Research Institute, Beijing 100029, China.
[2]SINOPEC Shengli Oilfield Company Petrochemical Plant, Dongying 257001, China.

*chenyangdu@126.com

Abstract: During the pressure test process in a petrochemical butyl device in China, it was found that the 304 stainless steel flange of the ethylene pipeline appeared to be cracking and leakage. This article analyzed the reasons for failure using macroscopic inspection, chemical composition analysis, metallographic structure analysis, scanning electron microscope, energy dispersive spectroscopy, mechanical properties test, and intergranular corrosion test. The results show that the carbon content of the flange base material exceeds the standard, many carbides precipitate at the grain boundary, the material itself is seriously sensitized, and the intergranular corrosion resistance is reduced. At the same time, under corrosive media such as sulfur and chlorine and high stress, the flange displays stress corrosion cracking, which leads to cracking and leakage failure. This article reports the reasons for the failure of the flange, and some reasonable suggestions are put forward to avoid similar leakage accidents.

1 Introduction

As a kind of synthetic rubber, butyl rubber possesses good wear resistance and high damping [1], which is mainly used to manufacture automotive tires, sealing elements, vulcanized capsules, shock-absorbing support fields, etc [2, 3]. Due to excellent corrosion resistance, mechanical properties, machinability, and weldability, 304 austenitic stainless steel is generally adopted as a flange material in synthetic rubber synthesis devices [4]. Although 304 austenitic stainless steel has many advantages, it is prone to stress corrosion in the industrial environment containing sulfur and chlorine media, which may result in cracking and leakage failure of petrochemical equipment. Therefore, performing an analysis to clarify the reasons for failure is an important route to take preventive measures to avoid the recurrence of failure and improve quality reliability.

There is an actual instance of cracking and leakage failure of 304 stainless steel flange in a petrochemical butyl device. During the pressure test process in a petrochemical synthetic rubber factory in China, it was found that the flange of the ethylene pipeline appeared to be cracking and leakage, which caused the pressure test to be interrupted. The elementary information on the failure flange is illustrated below. This flange was used in 2005 and was made of 304 austenitic stainless steel. Its specification was DN50, operation pressure was 1.8MPa, and operation temperature was -35 °C. The working medium was ethylene. The wall thickness grade and connecting pipe value were sch40s and 7.1 mm.

This paper sampled materials in intact and cracked areas of 304 stainless steel to perform related physical and chemical performance testing. With the tests of chemical composition, metallographic analysis, scanning electron microscopy (SEM) with energy dispersive spectrum (EDS) analysis, mechanical properties test, and intergranular corrosion test, the reasons for cracking and leakage failure of 304 stainless steel flange in petrochemical butyl device is determined and the relevant protection suggestions are further put forward.

2 Physical and chemical experiments characterization of flange failure

2.1 Macroscopic examination

Firstly, the failure flange was observed macroscopically, and the macroscopic failure characteristics of the failure flange were mastered, which laid a good foundation for the subsequent experimental analysis. Figure 1 and Figure 2 are the macroscopic of flange crack morphology. Macro-inspection indicates no obvious corrosion on the inner and outer surfaces. The cracks are distributed along the circumference of the flange neck in a zigzag and intermittent way, almost parallel to the weld. The cracks on the outer surface of the flange neck are extended in a full circle along the circumferential direction, and some of them have penetrated. Besides, it is displayed that the crack gap on the outer surface is larger than on the inner surface. Considering that the crack position is not only a structural discontinuity but also a weld heat-affected zone with high-stress characteristics, it can be judged from the foregoing phenomena that the cracks originate from the outer surface and extend to the inner surface.

Figure 1. External crack morphology of flange.

Figure 2. Morphology of internal cracks in flange.

2.2 Material chemical composition analysis

Next, the chemical composition of the material is analyzed to check whether the flange material meets the relevant design and manufacturing requirements. The specimens far away from the cracked location are sampled to verify whether the chemical composition of the flange is reasonable. Table 1 shows the results of the chemical composition elements analysis. Compared with the measured value in the Chinese NB/T 47010-2017 standard, the results indicate that the C content is 0.16%, which is larger than the upper limit of the standard value, while the Cr content is 16.66%, which is lower than the lower limit of the standard value. Due to Cr, Ni is an essential corrosion-resistant element of stainless steel. Cr can be formed on the surface of the stainless steel chromium-rich oxide film, which effectively protects the stainless steel from the medium of further Oxidation and corrosion and improves the corrosion resistance of stainless steel [5, 6]. Consequently, this flange possesses the chemical composition features of high-carbon and low-Cr, which may lead to decreased intergranular corrosion resistance and increased sensitization possibility.

Table 1. Chemical composition analysis of flange material (wt.%).

Element	C	Cr	Ni	Mn	S	P
Standard value (NB/T47010-2017)	≤0.08	18.00~20.00	8.00~11.00	≤2.00	≤0.03	≤0.045
Measured value	0.16	16.66	9.03	0.75	0.01	0.02

2.3 *Metallographic analysis*

Metallographic analysis was carried out to observe the microstructure characteristics of the crack. Figure 3 displays the microscopic metallographic morphology of flange cracks. It can be seen the microstructure is austenite, and there are twins and granular carbides distributed in the grain, which leads to a significant electrochemical property difference between the grain and the grain boundary. Besides, the cracks display dendritic morphology with obvious trunk and branches and spread along the grain boundary with sharp tips, which is a typical feature of stress corrosion cracking.

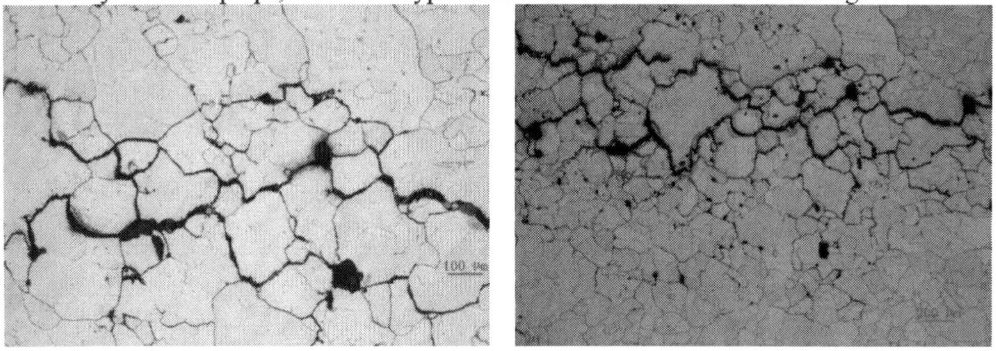

Figure 3. Microscopic morphology of cracks.

2.4 *SEM and EDS analysis*

SEM and EDS were used to analyze the crack position of the flange, and the micro-morphology characteristics of the crack position were observed. The corresponding corrosion element information was found. As shown in Figure 4, the fracture surface of the flange was further observed with SEM. The results illustrate that the fracture surfaces have a polyhedral feature and crystalline grains are angular and smooth to show a typical rock sugar shape. The results reveal that the flange cracking is an obvious intergranular fracture. Table 2 exhibits the energy spectrum analysis results of the corrosion products on the fracture surface. Compared with the chemical composition result in Table 1, the contents of chlorine and sulfur increase significantly, respectively, at 3.02% and 0.49%. This may be attributed to the fact that there are many corrosive elements in the industrial environment, and the increase in sulfur and chlorine content is related to the exposed contact between the flange and the atmospheric environment. The increase of chloride and sulfur contents will destroy the passive film formed on the grain boundary, resulting in intergranular corrosion or pitting corrosion and easily causing stress corrosion cracking of the flange.

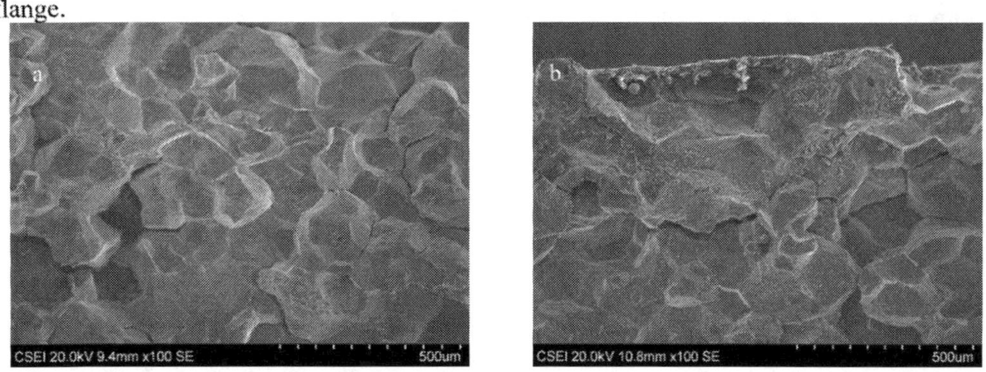

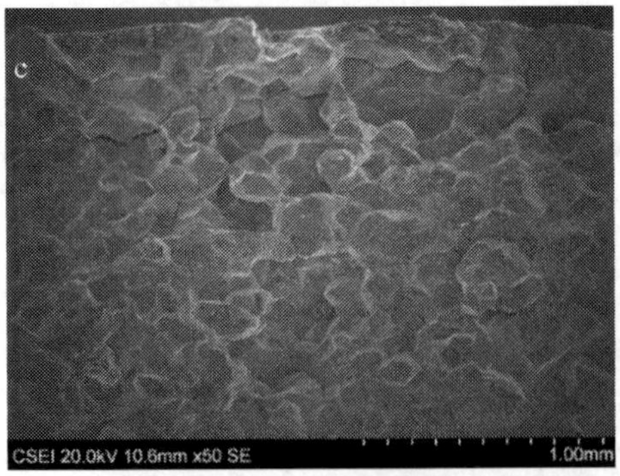

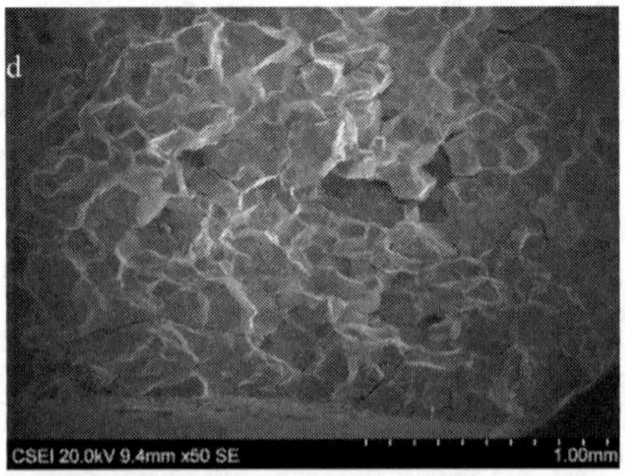

Figure 4. Fracture morphology of flange.

Table 2. Content of elements in energy spectrum analysis of corrosion products

Element	Mass/%	Atom/%
C	34.44	50.46
O	35.86	39.44
S	0.49	0.27
Cl	3.02	1.50
Cr	3.66	1.24
Fe	22.22	7.00

2.5 Mechanical properties and intergranular corrosion tests

The mechanical properties of the failed flange were tested, and the mechanical properties of the current materials were mastered. The samples from the flange cracks were taken for impact and tensile tests, respectively. Tables 3 and 4 are the test results corresponding to the aforementioned tests, respectively. It can be seen that the mechanical indexes such as impact absorption energy, yield strength, tensile strength, and area shrinkage all meet the requirements of China NB/T 47010-2017 standard.

Table 3. Impact test results.

Sample No.	Sample size/mm	Test temperature/°C	Impact absorbing energy/J
1	9.98×9.99×55	23	217
2	9.98×9.99×55	23	228

Table 4. Tensile test results.

Sample No.	Lu/mm	Yield strength Rp0.2/MPa	Tensile strength Rm/MPa	Reduction of area A/%
1	38.47	355	663	54.0
2	37.49	367	664	50.0

Samples were taken at the crack and the base of the flange, and intergranular corrosion tests were carried out to determine whether the stainless steel material of the flange was sensitized. As indicated in Figure 5, it can be seen that there are a lot of carbides precipitated at the grain boundary accompanied by pitting corrosion, and corrosion grooves surround some grains. According to GB/T 4334-2020, it can be judged as three types of groove structure, and the material itself is seriously sensitized. The possible reasons for sensitization may be that the flange has not been treated by a solid solution process before leaving the manufacturing plant.

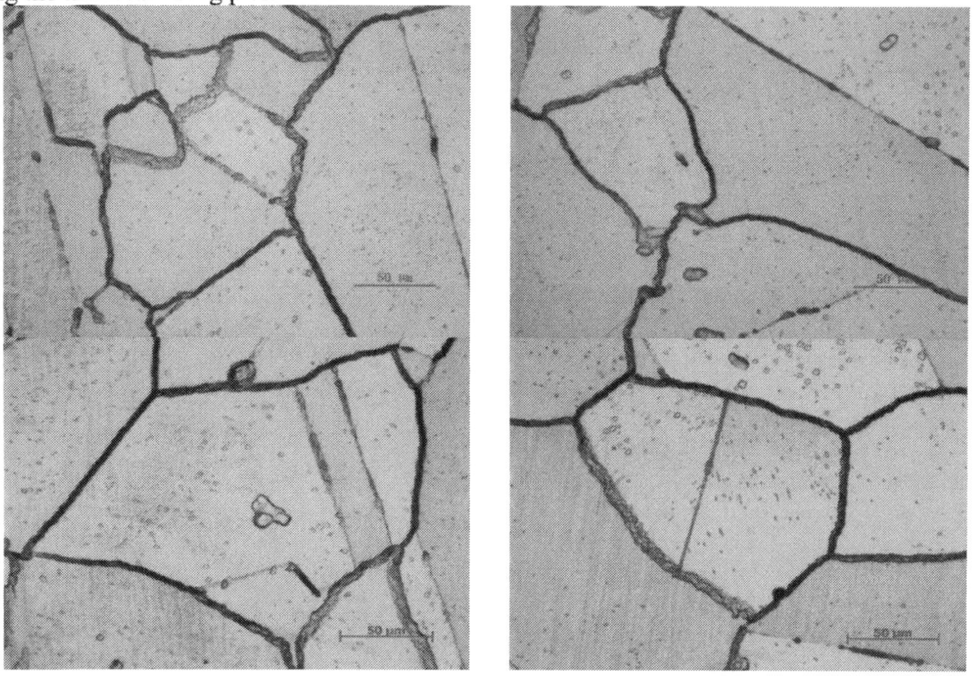

Figure 5. Morphology of intergranular corrosion.

3 Analysis and discussion

Based on the above physical and chemical inspection results, the flange cracking failure is analyzed as follows.

3.1 Sensitive materials

According to material chemical composition analysis, the chemical composition of the element is beyond the limits of related standards. The carbon content is higher than the maximum value, and the chromium content is lower than the minimum value specified in the standard. At the same time, there are a lot of carbides precipitated on the grain boundary of the flange microstructure, which is related to the high carbon content of the flange and the hot working process system. Excessive carbon element provides a material basis for generating grain boundary precipitates.

The higher the carbon content, the higher the sensitization sensitivity, the greater the tendency of intergranular carbide precipitation, and the more prone to intergranular corrosion [7, 8]. It is well known that if the carbon content in stainless steel is lower than 0.08%, the amount of carbide precipitation is less; if the carbon content is higher than 0.08%, the amount of carbide precipitation will increase rapidly [9]. Besides, the solubility of carbon in austenite at room temperature is very small, at only about 0.02%; thus, supersaturated carbon is dissolved in austenite [10]. After solution treatment, when the temperature stays in the range of 450-850℃ for some time, the supersaturated carbon continuously diffuses to the austenite grain boundary. It combines with chromium to form $Cr_{23}C_6$ compounds. Because the diffusion speed of chromium in grains is slower than that along grain boundaries, the internal chromium can't diffuse to grain boundaries. The chromium needed for chromium carbide formed between grains mainly comes from the vicinity of grain boundaries, greatly reducing the chromium content near grain boundaries. When the chromium content at grain boundaries is less than 12%, it forms a chromium-poor region and causes a sensitizing phenomenon [11].

The above process is further validated by the results of the intergranular corrosion test, which shows that the flange material itself has been seriously sensitized. The occurrence of sensitization leads to a significant difference in electrochemical performance between the chromium-poor region and the crystal grain itself, which further forms an activated-passive battery with great potential difference between the chromium-poor region (anode) and the substrate (cathode). The chromium-poor area at the grain boundary is rapidly corroded with corrosive media such as sulfur and chlorine combined in the atmospheric environment. Consequently, the intergranular bonding force is significantly weakened, and the grain boundary's mechanical strength and intergranular corrosion resistance are significantly reduced [12-14].

3.2 Specific corrosive medium
Energy spectrum analysis shows that the corrosion products on the fracture surface contain corrosive elements such as chlorine, sulfur, oxygen and matrix elements such as iron and chromium. The appearance of corrosive elements is related to the direct contact between the flange and the atmospheric environment, the erosion of rain, and the existence of this element in the atmospheric environment of the factory.

3.3 Welding residual stresses
In the process of flange welding, due to the physical properties of austenitic stainless steel itself, the thermal conductivity is small, and the linear expansion coefficient is large. Significant welding residual stresses will be generated in the heat-affected zone of the welding process [15, 16], which provides conditions for intergranular stress cracking.

From the above analysis, it can be seen that the flange neck cracking meets the three conditions of stress corrosion cracking. Under normal circumstances, stress corrosion cracking of austenitic stainless steel caused by chloride ions is intergranular [17, 18]. When there is a precipitated phase of carbides such as $Cr_{23}C_6$ on the grain boundary of austenitic stainless steel, the electrode potential of the precipitate is different from that of solid solution austenitic grain, which is easy to form many micro-batteries. With the precipitation of chromium, the austenite grain boundary has a lower potential. It forms the anode of the micro-battery, so the grain boundary becomes the preferential channel for corrosion. At the same time, the crack opens under stress, accelerating the crack propagation, and finally, the intergranular stress corrosion cracking morphology is produced.

4 Conclusions and recommendations
This article reports the cracking and leakage failure analysis of 304 stainless steel flange in a petrochemical butyl device. The chemical composition of the flange material is unqualified. The content of C is higher than the upper limit, while the content of Cr is lower than the lower limit of the standard values. Serious sensitization occurs, displaying that a large number of carbon and chromium compounds are precipitated on the grain boundary, which reduces the grain boundary strength and corrosion resistance, and there is an obvious tendency for intergranular corrosion. Under the combined action of corrosive media such as sulfur and chlorine in an atmospheric environment and high welding residual stress, intergranular corrosion occurs in the welding heat-affected zone, which results in flange fracture failure with typical stress corrosion cracking. The problems found in this failure analysis should also attract the attention of enterprise managers. The reinspection of incoming materials is the first checkpoint to control the quality. If we don't pay attention to it, it will lead to many subsequent problems.

Based on the above analysis, preventive measures and effective suggestions are provided to avoid recurrence failure for other flanges. Firstly, newly purchased 304 stainless steel flanges should be treated by solution treatment after forging; thus, the carbides dissolve in austenite after heating to 1100-1150°C, and then they are rapidly cooled to obtain a single-phase austenite structure. The ultra-low carbon 304L covered electrode is used for welding, such as A002, etc, which can avoid the precipitation of weld carbide during welding. Secondly, the quality of materials can be controlled by strengthening the supervision of incoming flange materials through related tests, such as chemical composition analysis, metallographic examination, and intergranular corrosion test. Thirdly, the interlayer temperature of weldments should be kept low, preferably not exceeding 150°C. During field welding, the residence time at 450-850°C can be reduced by shortening the heating time, reducing the weld cross-section, reducing the linear energy, and cooling rapidly to avoid the material sensitization caused by welding.

References

[1] Shen, M., Xia, L., Feng, Q., Zhang, J., Li, J., Guo, S. Damping characteristics of a multilayered constrained beam using viscoelastic butyl rubber layer with wide temperature range. Mater. Express, 2021, 11 (3),372-380.
[2] Shiva, M, Kamkar Dallakeh, M, Ahmadi, M, Lakhi, M. Effects of silicon carbide as a heat conductive filler in butyl rubber for bladder tire curing applications. Marer Today Commun, 2021, 29,102773.
[3] Lee, S. R., Bae, K. M., Baek, J. J., Kang, M. C., Lee, T. I. Adhesion enhancement between aluminum and butyl rubber by (3-mercaptopropyl) trimethoxy silane for vibration damping plate. J. Adhesion Sci. Technol,2021,35 (10),1114-1124.
[4] Sharma R. K., Mohanty S., Gupta V. Advances in butyl rubber synthesis via cationic polymerization: An Overview [J]. Polymer International. 2021.DOI:10.1002/pi.6180.
[5] Sun, J., Tang, H., Wang, C., Han, Z. & Li, S. (2022). Effects of alloying elements and microstructure on stainless steel corrosion: A review. steel research international, 93(5),2100450.
[6] Sun, M., Du, C., Liu, Z., Liu, C., Li, X. & Wu, Y. (2021). Fundamental understanding on the effect of Cr on corrosion resistance of weathering steel in simulated tropical marine atmosphere. Corrosion science, 186, 109427.
[7] Shit, G., Mariappan, K. & Ningshen, S. (2023). Improvement of sensitization and intergranular corrosion of AISI type 304L stainless steel through thermo-mechanical treatment. Corrosion Science, 213, 110975.
[8] Bai, S., Zhang, S., Ma, J., Li, J., & Mou, L. (2024). Understanding the effect of decreasing C contents and increasing solid-solution time on intergranular corrosion resistance of 304 austenitic stainless steel. Journal of Materials Research and Technology, 30, 4750-4761.
[9] Hwang E. R., Kang, S. G. Intergranular corrosion of stainless steel in molten carbonate salt [J]. Journal of Materials Science Letters,1997,16 (16): 1387- 1388.
[10] Bi H. J. Influence of heat treatment to microstructures and localized corrosion in several stainless steels [D]. Lanzhou: Lanzhou University of Technology. 2011.
[11] Marikkannan, S. K., Gnanasekaran, S., Rakkiyannan, J., Veluchamy, M., Subramanian, P. & Sujatha, V. I. (2023, July). Heat treatment effect on sensitization behavior of 304 grade stainless steel. In AIP Conference Proceedings (Vol. 2788, No. 1). AIP Publishing.
[12] Fujii, T., Kawarabayashi, K. & Shimamura, Y. (2024). Intergranular corrosion in sensitized austenitic stainless steel subjected to tensile loading and unloading in the elastic-plastic region. Corrosion Science, 230, 111896.
[13] Vukkum, V. B., Delvecchio, E., Christudasjustus, J., Storck, S. & Gupta, R. K. (2023). Intergranular Corrosion of Feedstock Modified—Additively Manufactured Stainless Steel After Sensitization. Corrosion, 79(PNNL-SA-186235).
[14] Zatkalíková, V., Uhríčik, M., Markovičová, L., Pastierovičová, L. & Kuchariková, L. (2023). The Effect of Sensitization on the Susceptibility of AISI 316L Biomaterial to Pitting Corrosion. Materials, 16(16), 5714.
[15] Das Banik, S., Kumar, S., Singh, P. K. & Bhattacharya, S. (2023). Influence of weld repair on the residual stresses induced in austenitic stainless steel weld joints. Production Engineering, 17(1), 81-94.

Journal of Physics: Conference Series **2879** (2024) 012034 doi:10.1088/1742-6596/2879/1/012034

[16] Djeloud, H., Moussaoui, M., Kouider, R., Al-Kassir, A. & Carrasco-Amador, J. P. (2023). Study of the heat exchange and relaxation conditions of residual stresses due to welding of austenitic stainless steel. Energies, 16(7), 3176.
[17] Rhodes, P. R. (1969). Mechanism of chloride stress corrosion cracking of austenitic stainless steels. Corrosion, 25(11), 462-472.
[18] Wells, D. B., Stewart, J., Davidson, R., Scott, P. M., & Williams, D. E. (1992). The mechanism of intergranular stress corrosion cracking of sensitised austenitic stainless steel in dilute thiosulphate solution. Corrosion Science, 33(1), 39-71.

Simplified modeling study of the forward fuselage structure based on the overall structural rapid design software for civil aircraft

Feng Li[1] and Lijun Pan[1*]

[1]Shanghai Aircraft Design and Research Institute, Shanghai, 201210, China

[*]Corresponding author's e-mail: panlijun@comac.cc

Abstract: In order to solve the problems of poor weight estimation accuracy and difficult layout design trade-offs in the concept demonstration and overall program definition stage of civil aircraft, Aircraft Layout, a civil aircraft overall program structure rapid design software, is used to establish a numerical calculation-based layout program rapid design simulation platform. Based on the forward fuselage structure, this paper gives the results of rapid modeling and finite element simulation and analysis of the forward fuselage isotropic section of a certain type of aircraft, which provides a reference for the optimization design of the structural components of the forward fuselage section of a certain type of aircraft in the conceptual scheme design stage.

1. Introduction

The fuselage structure of an airplane, whether it is a military or civil aircraft, is a typical thin-walled structure. The fuselage structures are all composed of longitudinal elements (long trusses, truss beams) and transverse elements perpendicular to the fuselage (spacer frame, skin)[1]. The main function of the skin is to form the aerodynamic shape and keep the surface smooth. It bears and transmits the shear force and torque in the fuselage. As the longitudinal member of the fuselage, the long truss is mainly used in the honed fuselage to bear the axial force generated by the bending of the fuselage, and the long truss has a supporting effect on the skin. Spacer frames are divided into two categories: ordinary frames and reinforced frames. Ordinary frames, typically designed in the form of rings, serve to uphold the cross-sectional shape of the fuselage, providing support for both the long truss and the skin. Reinforced frames, on the other hand, encompass the functionality of ordinary frames while additionally bearing loads within the frame plane. These loads, originating from various sources and components attached to the fuselage via joints, are transmitted by the reinforced frames and distributed into shear flow before being transferred to the shell [2-9].

1.1. Software function introduction

Aircraft Layout software is located in the design stage of aircraft fuselage structure. Compared with traditional design methods, it takes the overall aerodynamic shape as the basis through the comprehensive integration of modeling and computational analysis tools of aircraft structural design disciplines, designing, modeling, and multidisciplinary analysis and optimization in an integrated environment to realize the integration of structural layout design between various disciplines and specialties, and to improve the quality and efficiency of the aircraft structural design. The biggest advantage is that it allows for rapid iteration in a short period of time. The main tool modules of the software are shown in Figure 1.

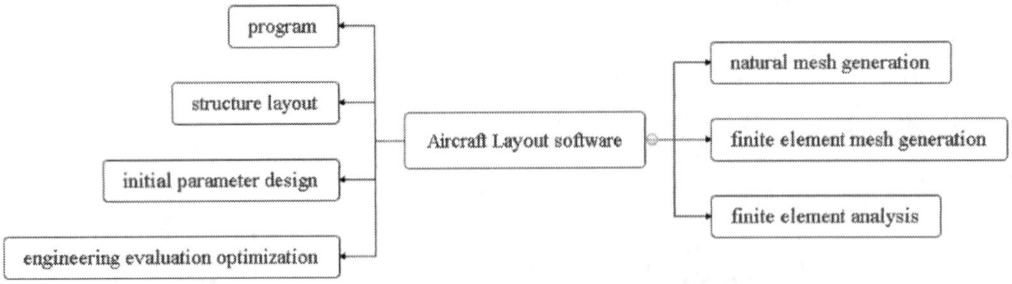

Figure 1. Schematic diagram of the functions realized by the software.

2. Structural rapid modeling

In this paper, the front fuselage of a certain aircraft is taken as the object, 11 reinforcing frames are selected, and in order to simplify the research content, the variable-diameter cylinder section connected with the nose is straightened to an equal-diameter cylinder section. According to the main geometric model of a certain aircraft, the theoretical shape of the front fuselage and its booth surface in the main skeleton of the whole aircraft structure are modeled, and the fuselage reinforcing frames, long trusses, skin wall panels, cabin floor beams, slide rails, cargo floor beams, and columns are constructed to complete the fuselage tube section structural modeling, as shown in Figure 2.

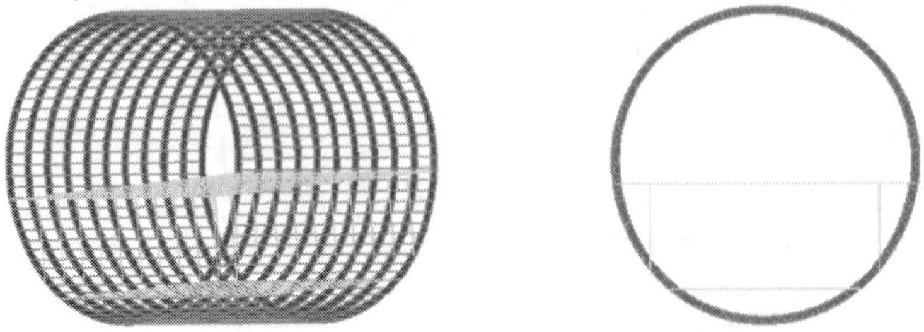

(a) (b)

Figure 2. Front fuselage section skeleton model of a certain type of aircraft.

Reinforcement frames are defined using station faces. The components in this paper use a two-point type to arrange the fuselage long truss. Wall plate is used for local or whole skin structure arrangement on the airplane fuselage. The fuselage components use the station type to arrange the passenger cabin floor beams, slide rails, columns, and cargo floor beams.

2.1. Definition of metal materials

According to the overall target requirements, the structure of the airplane airliner, including the fuselage, adopts composite materials on a large scale. The long modeling cycle of composite materials and the use of composite materials in the model can impede the subsequent optimization process. Consequently, for the front fuselage of the aircraft, a natural mesh is exclusively employed for the aluminum alloy material, as illustrated in Table 1.

Table 1. Material properties.

Material name	7075 aluminum	2024 aluminum	Material name	7075 aluminum	2024 aluminum
E/MPa	71347	73177	σ0.2/MPa	399	440
μ	0.33	0.33	σb/MPa	468	481
G/MPa	26822.2	27510.2	σ-b/MPa	715	743
ρ/kg/m^3	2795.7	2795.7	τ/MPa	282	282

2.2. Material unit definitions

The material library itself does not record the unit system of the data. However, it aims to ensure the correctness of upstream and downstream data connections. When editing material parameters in the material library, you need to strictly follow the unified unit system, as shown in Table 2. The corresponding unit system in the following table is used for the corresponding quantities in the paper.

Table 2. Unit system for material parameters.

Number	Dimension	Unit system
1	Modulus/Stress	MPa
2	Density	Kg/m^3
3	Length	mm

3. Natural mesh modeling

Natural meshing is to intersect the wireframe model according to the model geometry itself, and divide the feature surfaces and lines naturally. The natural mesh model connects the wireframe model and the finite element model, inherits the material and geometric properties of the wireframe model, facilitates the automatic mesh division of the finite element mesh, and describes the material and geometric properties of the structure in more detail than the wireframe model.

3.1. Natural meshing

Based on the wireframe model, a "narrow" natural mesh is generated, as shown in Figure 3.

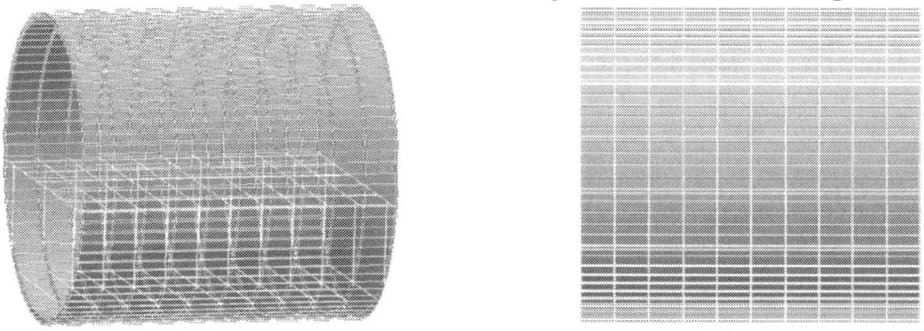

(a) (b)

Figure 3. A natural mesh model of the forward fuselage section of the airplane.

3.2. Attribute inheritance

The wireframe model is cut into a number of sub-elements through natural meshing, and the attributes of the sub-elements should be inherited from the original wireframe elements. Attribute inheritance is the automatic identification and matching of these elements to automatically generate the attribute data of the natural mesh 1D and 2D cells.

4. Finite element mesh

The finite element mesh is generated by firstly importing the structural mesh model into Hypermesh, automatically dividing the finite element mesh through the secondary development of CAE software, and then reading the mesh divided by Hypermesh and the element attributes in the structural mesh model by Patran, to generate the Patran finite element model which contains the finite element mesh and cell attributes.

4.1. Finite element meshing

The finite element meshing is performed using Hypermesh. The working method is to export the parts to build the finite element model. The structural mesh model will be exported first and then imported into the finite element software to perform the mesh delineation. The finite element mesh model is constructed based on the structural model of the front fuselage and other straight sections. Since the frame spacing of the reinforcing frame is 610mm, the grid size of 122mm is selected for the finite element meshing so that the adjacent reinforcing frames have the whole number of cells, as shown in Figure 4.

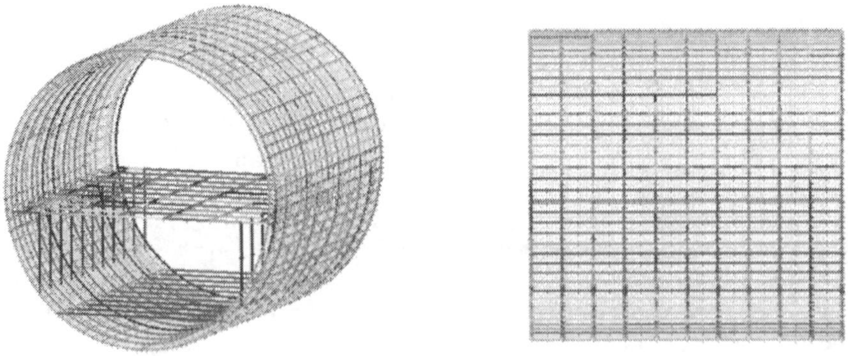

(a) (b)

Figure 4. A finite element mesh model of the forward fuselage section of the airplane.

4.2. Definition of cell properties
After the definition is completed, the finite element model file (.bdf) and the group file that can be read by Patran are output, and then the mesh and attributes divided by Hypermesh are read by Patran to generate the Patran finite element model that contains the finite element mesh and cell attributes.

4.3. Model checking
Using the Modelcheck function in the Hypermesh panel, the minimum length of 1d, 2d cells, the maximum and minimum interior angles of triangle and quadrilateral cells are checked. The minimum finite element mesh size is 9.04mm, the minimum angle is 28.98°, the total number of nodes is 12205, and the total number of cells is 20318.

5. Finite element analysis
The finite element analysis module contains functions such as finite element model assembly, boundary condition definition, load application, working condition definition, calculation and analysis, and result viewing. In the front fuselage skeleton model, the skin wall plate and web plate are two-dimensional parts, and the edge strip, long truss, and short beam are one-dimensional parts, so all the nodes in the natural mesh model are located on the outer surface of the skin, i.e., the theoretical shape surface. The finite element mesh model constructed based on the natural mesh model also contains only two-dimensional or one-dimensional mesh cells[10].

5.1. Boundary loads
The displacement of the nodes at the ends of the barrel segments in three directions (X, Y, Z) is limited in the finite element model of the forward fuselage. A uniform pressure of 0.13 MPa was applied to the fuselage skin, as shown in Table 3.

Table 3. Load condition.

Number	Condition name	Load name	P(MPa)	Safety factor
1	loadcase1	press	0.13	1

5.2. Static analysis

The finite element analysis process was carried out iteratively with the optimization process for engineering evaluation and the parameters optimized included the thickness of the material, cross sectional area. After two rounds of static strength optimization of the FEA results the factor of safety of the component is updated according to the recommended values without any significant change. The results are shown in Table 4.

Table 4. Static analysis results.

Number of iterations	Model weight/t	Maximum displacement/mm	Maximum stress/MPa
1	1.34206	338	178
2	0.96331	644	175

As can be seen from Table 4, for the finite element model after the first engineering evaluation, the total weight is 1.34206t, the maximum displacement is 338mm, and the maximum stress is 178MPa. After the second iteration optimization according to the default suggested values of the components, the total weight of the model is 0.96331t, the maximum displacement is 664mm, the maximum stress is 175MPa. Through the optimization of the thickness of the material, the weight is reduced while the model is able to withstand greater displacement and stress in terms of static strength. However, combined with the engineering practice, according to the optimization of the system proposed value, after many rounds of iteration, the weight of the structure will converge to a relatively fixed or almost no longer fluctuate in the value. At this time, it is considered the optimal solution of the theoretical model.

The final engineering evaluation of the optimized displacement cloud is shown in Figure 5, and the stress cloud is shown in Figure 6.

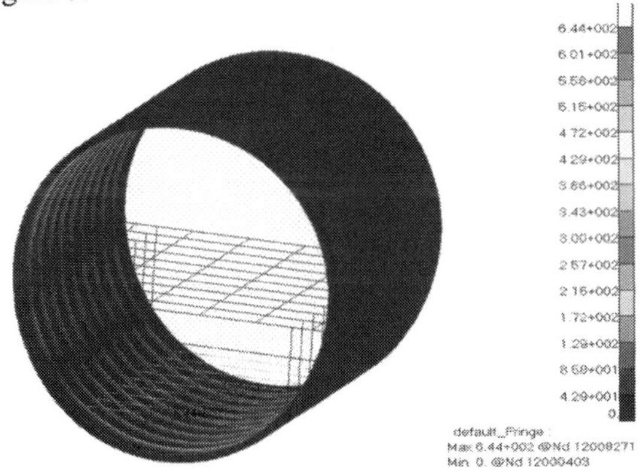

Figure 5. Front fuselage displacement cloud.

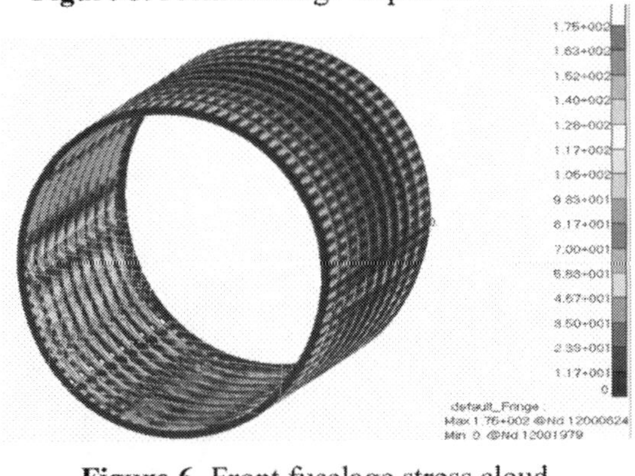

Figure 6. Front fuselage stress cloud.

Among them, the deformation of the cabin floor is the maximum of 644mm; since the one-dimensional components do not bear compression, the skin between the column and the floor beam is the maximum stress area, and the maximum stress value is 175MPa.

6. Conclusion

Since the model described in this paper is only a simplified model for engineering practice, this paper mainly proposes a method of modeling the forward fuselage structure based on the layout scheme rapid design simulation platform of numerical computation. With the forward fuselage structure as the basis, it gives the results of a certain type of aircraft forward fuselage isotropic section rapid modeling and finite element simulation and analysis. After obtaining the stress cloud map of the structure, the stress characteristics of the components are analyzed to provide a reference for structural optimization.

References

[1] Junhao G (2020) Modeling and Finite Element Analysis of a Typical Front Fuselage Structure. Manufacture & Upgrading Today, 09: 53–55.

[2] Zhijin W and Weixing Y (2007) Structural Design of Aircraft. National Defense Industry Press, Beijing.

[3] Zhengbao D, Liguo Y, Yucan W, Xiaoming H and Guichu D (2021) Modeling and study of stiffness during machining of aerospace aluminum alloy bulkhead plates. Journal of Binzhou College, 02: 11–16.

[4] Junhao G (2020) Modeling and Finite Element Analysis of a Typical Front Fuselage Structure. Manufacture & Upgrading Today, 09: 53–55.

[5] Junhao G (2020) Modeling and Finite Element Analysis of Typical Wing Structures.Engineering & Test,03: 27–29.

[6] Ke S, Zhaozhong Y, Yin H and Ming Z (2021) Finite element analysis and optimization design of front rack flip spreader.Wind Energy,01: 100–103.

[7] Zhenyu L, Ziyu Z and Qiang W(2023) A finite element parametric modeling method applied to the composite fuselage of civil aircraft. Machine Design and Manufacturing Engineering,03: 67–72.

[8] Da Z, Guanyi L, Yunpeng Z, Ying W, Xuefei W and Peng D (2023) Finite Element Analysis of Strength of Aircraft Composite Plate Spring Landing Gear and Fuselage Joints.Manufacturing and Upgrading Today,05: 123–126.

[9] Heng L, Guowei T and Xiailing Z(2023) Research and strength analysis of hoop bending test of composite fuselage integral frame.Progress in aeronautical engineering,001: 104–113.

[10] Feng L and Ming Z (2020) Random acoustic fatigue life analysis of typical fuselage structures. Machine Design and Manufacturing Engineering, 07:49–54.

Cyclic creep behavior and life prediction of DZ125 alloy under different stress ratios

Haowen Chen[1], Xuming Niu[1*], Zhigang Sun[1], Yingdong Song[2] and Lu Zhang[1]

[1]College of Energy and Power Engineering, Nanjing University of Aeronautics and Astronautics, Nanjing, Jiangsu, 210016, China
[2]Harbin Engineering University, Harbin, Heilongjiang, 150009, China

[*]Corresponding author's e-mail: xumingniu@nuaa.edu.cn

Abstract: In order to study the creep behavior and mechanism of DZ125 alloy under cyclic creep under different stress ratios, cyclic creep experiments are performed under a temperature of 850°C and peak stress of 500 MPa, with stress ratios of 0.1, 0.3, 0.5, 0.7, 0.9, and 1, and the mechanical behavior and microstructure of creep samples under different stress ratios are compared and analyzed. The accuracy of several life prediction models is compared. The results show that cyclic hardening or cyclic softening will occur under different stress ratios due to grain boundary slip and dislocation plugging. According to the results of fracture analysis, as the stress ratio rises, the proportion of creep damage increases, and the fracture mode shifts from transgranular fracture to intergranular fracture. Several life prediction models are selected, and the results show that the FS frequency separation model and M-G model have better life prediction accuracy.

1. Introduction

DZ125 is a directionally solidified [1-4] nickel-based cast superalloy known for its outstanding comprehensive properties at medium to high temperatures, as well as its outstanding thermal fatigue resistance. It is a crucial material for the production of turbine blades used in aero engines. Creep represents the primary mode of damage in these blades, and its detrimental effects are significantly exacerbated when combined with fatigue. Therefore, investigating the creep behavior and underlying mechanisms of DZ125 alloy under cyclic creep conditions is of paramount importance.

Research has shown that under specific load conditions and holding times, cyclic creep can exhibit a longer test life compared to static creep, a phenomenon commonly referred to as the cyclic hardening effect. Kim et al. [5] conducted a series of cyclic creep experiments on Gr.91 steel at 600°C and discovered that, at different stress ranges, when the maximum load was held for 10 minutes with a stress ratio of 0.1, the creep rate was lower and the fracture life was extended in comparison to static creep tests. Similarly, Ying et al. [6] performed static and cyclic creep tests on ultra-pure 409 ferritic stainless steel, revealing that cyclic creep specimens demonstrated extended fracture lifespans and reduced minimum creep rates compared to pure creep. Furthermore, they found that shorter peak load holding times corresponded to longer fracture lifetimes and reduced minimum creep rates. The deceleration observed in cyclic creep was generally attributed to anelastic strain. However, alongside this prolonged life effect, there could also be a phenomenon known as cyclic softening, where creep life is shortened. Aktaa et al. [7] investigated this cyclic softening in low-activation martensitic steel EUROFER97 by selecting two loading stress amplitude values. Their findings indicated that cyclic creep could accelerate creep rates, with increased stress leading to a more pronounced rise in creep rates. Kim et al. [5] carried out a large number of cyclic creep experiments on Gr.91 steel. Their findings revealed that, in contrast to static creep, cyclic creep resulted in a shorter creep life and a faster creep rate. The behavior of static creep was mainly influenced by factors like stress ratio and maximum stress, whereas temperature did not seem to play a significant role. Hu et al [8]. focused on DZ125 alloy as the subject of the research. They found that introducing a load holding time at maximum stress significantly decreased the material's final fracture life. Moreover, when the peak load holding time increased, the proportion of creep damage rose, and the specimen's durability gradually approached that of pure creep conditions. Yang [9] et al. conducted a large number of DZ125 cyclic creep tests, taking into account factors such as temperature, strain ratio, and holding time. The experimental findings indicated that, under identical conditions, the fracture life was shorter when the strain ratio was negative. This can be attributed to the fact that the stretching process tends to induce cyclic softening more readily than the compression process, resulting in a longer lifespan.

Currently, most studies on cyclic creep focus on peak load holding time as a variable to investigate the creep behavior of materials, with relatively few examining cyclic creep under varying stress ratios. DZ125 is an alloy recognized for its exceptional overall performance at medium to high temperatures, as well as its remarkable resistance to fatigue, resulting in its widespread application in the manufacture of aero-engine turbine blades [10-12]. The hot-end components of these engines operate under prolonged exposure to high temperature and high-pressure conditions. The service environment presents a complex array of loads. In addition to the holding time, the aircraft faces a variety of stress loads when performing different flight tasks. Therefore, it is essential to explore the cyclic creep characteristics of the DZ125 superalloy across various stress ratios. In this paper, we conducted cyclic creep tests at a stress of 500 MPa with stress ratios of 0.1, 0.3, 0.5, 0.7, 0.9, and 1. The aim was to compare the mechanical behavior and microstructure of the creep samples across these different stress ratios. Additionally, several life prediction models were evaluated to assess their accuracy in predicting material life.

2. Experimental procedures

The test material is directionally solidified superalloy DZ125. The main chemical compositions are shown in Table 1. The configuration of the specimen is demonstrated in Figure 1.

Table 1. Main chemical composition of DZ125 alloy (wt./%).

Element	C	Cr	Al	Co	W	Mo	Ti	Ni
Content	0.07~0.12	8.4~9.4	5.2~6.2	9.5~10.5	6.5~7.5	1.5~2.5	0.7~0.12	Bal.

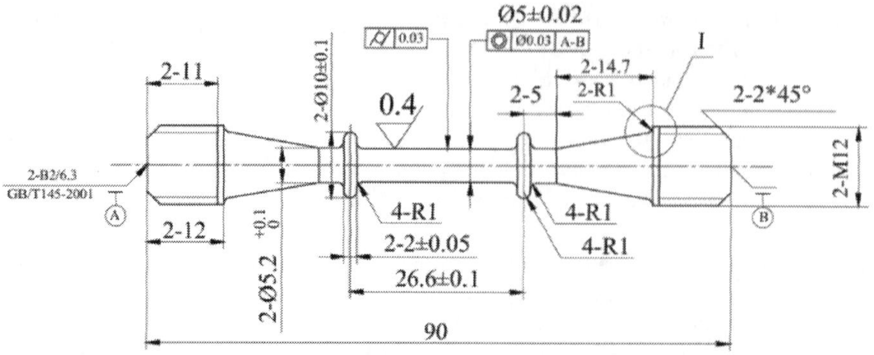

Figure 1. The configuration of the specimen (Unit: mm).

In accordance with the typical operating conditions of aero-engine high-pressure turbine blades, the test temperature was set at 850°C, utilizing an electronic high-temperature creep endurance testing machine, as illustrated in Figure 2. To achieve consistent heating of the sample, three thermocouples were employed to measure the temperatures at the top, center, and bottom sections. During the test, temperature fluctuations were maintained within ±3°C. After reaching the target temperature, the test commenced following a 60-minute stabilization period. The test adopted a stress control mode, sustaining a steady maximum stress of 500 MPa, with stress ratios set at 0.1, 0.3, 0.5, 0.7, and 0.9. The loading rate was established at 200 MPa/s, and the test waveform was configured as a trapezoidal wave, as shown in Figure 3. Following the test, scanning electron microscopy (SEM) was used to observe the fracture morphology of the sample, thereby revealing the damage mechanisms associated with cyclic creep under different stress ratios.

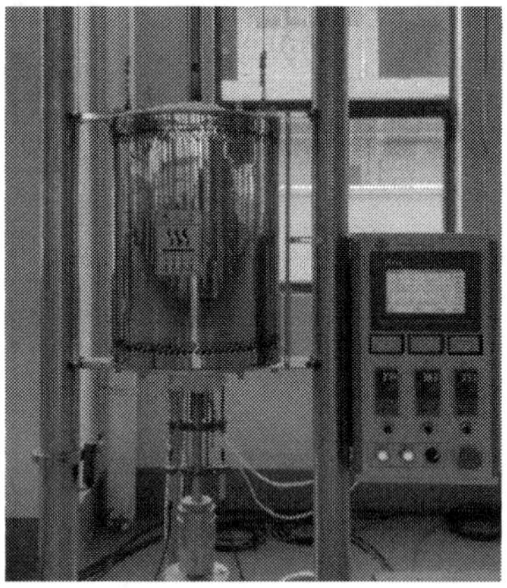

Figure 2. High-temperature electronic creep testing machine.

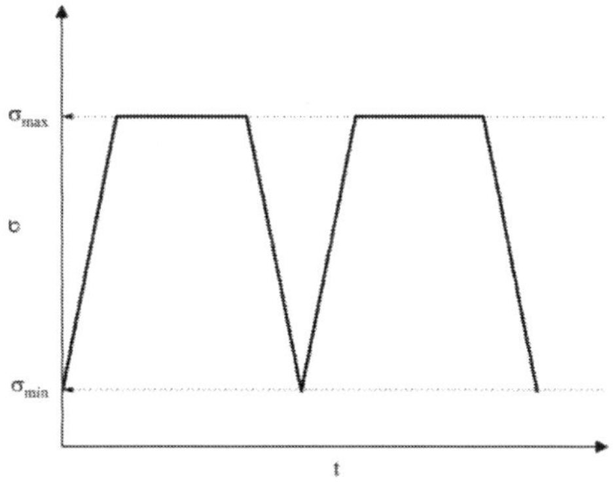

Figure 3. Load wave.

3. Experimental results and analysis

3.1. Impact of stress ratio on fracture life

Test data under different stress ratios are presented in Table 2 and Figure 4. It is evident that similar to pure creep, there is a significant correlation between the minimum creep rate and the rupture life. Generally, as the stress ratio increases, the lowest creep rate decreases while the creep fracture life increases. Notably, when the stress ratio is 0.1, 0.3, 0.5, cyclic softening occurs, resulting in a cyclic creep life that is lower than that observed under pure creep conditions, with a corresponding smaller maximum strain range. As the stress ratio continues to increase, the cyclic creep life gradually approaches that of pure creep. At stress ratios of 0.7 and 0.9, cyclic hardening is observed, where the cyclic creep life surpasses that of pure creep, and the strain range is also greater. This indicates that the maximum strain range is closely related to the creep life. The common explanation for the reduction in creep life due to cyclic loading is that it periodically induces the formation of cavities within the material, leading to accelerated failure. Conversely, the increase in creep life associated with cyclic loading can be attributed to an anelastic mechanism. During loading, the material stores anelastic strain, which is partially recovered during unloading. This storage of anelastic strain must occur before any non-recoverable creep takes place, thereby delaying the onset of subsequent non-recoverable creep strain.

Table 2. Cyclic creep test data of DZ125 in different stress ratios.

Stress Ratio	Minimum Creep Rate (h^{-1})	Maximum Strain Range (%)	Fracture Life (h)	Cycle
0.1	0.099123184	9.248222222	29.25861306	718
0.3	0.078315867	11.47244444	35.03973806	983
0.5	0.073260644	12.58777778	36.40736361	1201
0.7	0.063198566	13.92133333	41.99083389	1693
0.9	0.060847569	15.11111111	44.40779528	2294
1	0.07851862	13.40577778	38.43167056	—

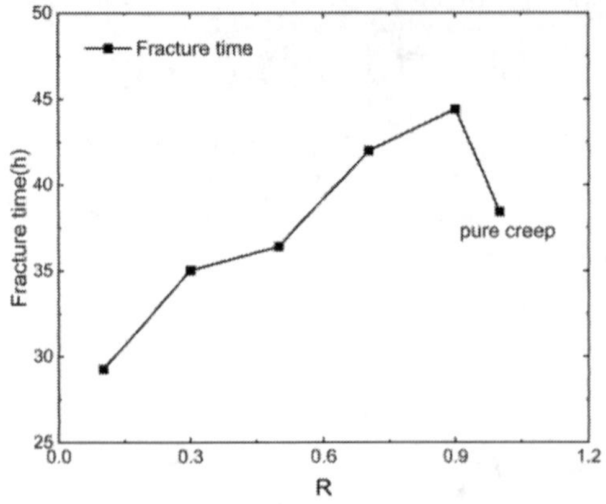

Figure 4. Relationship between fracture time and stress ratio.

Table 3 shows the cyclic creep behavior of FGH96 alloy under different holding times [13]. The relationship between the fracture time and the holding time is shown in Figure 5. It can be observed that as the holding time increases, the number of fracture cycles for the FGH96 alloy decreases. Specifically, when the holding time is less than 10 seconds, the fatigue creep life declines exponentially. Beyond this point, as the holding time continues to increase, the cyclic creep life decreases at a slower rate and eventually stabilizes. This trend indicates that with longer holding times, the proportion of damage incurred during each cycle of creep increases, thereby accelerating material failure. Once the holding time reaches a certain threshold, the proportion of damage from a single cycle approaches a fixed value, resulting in a gentler reduction rate of cyclic creep life. At present, the research on cyclic creep behavior mainly focuses on the influence of holding time on cyclic creep behavior. This paper focuses on the effect of different stress ratios on the cyclic creep behavior of superalloys.

Table 3. Fatigue-creep life of FGH96 under different holding time [13].

No.	Holding Time	Cycle Number
1	1	83906
2	5	19571
3	10	7781
4	15	4165
5	20	2592
6	25	2048

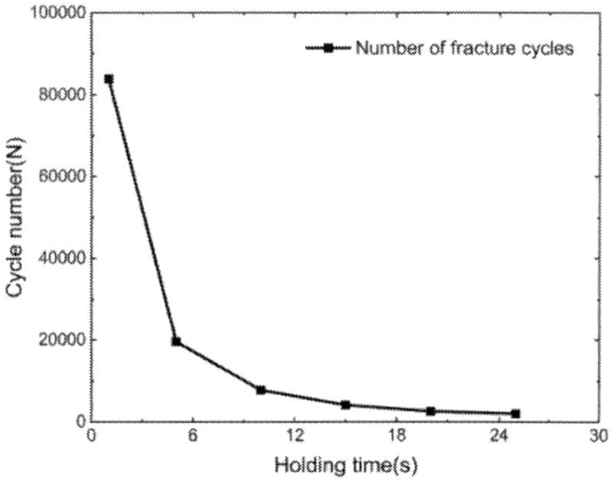

Figure 5. Relationship between fracture cycles and holding time of FGH96.

3.2. Influence of stress ratio on cyclic strain response

The value of the maximum strain in a single cycle is defined as the envelope strain. In the cyclic creep test, it can well reflect the relationship between inelastic strain and cycle number. While the envelope strain rate refers to the rate of variation in envelope strain. Figure 6 illustrates the envelope strain response curves of DZ125 alloy at stress ratios of 0.1, 0.3, 0.5, 0.7, and 0.9. It is evident that the creep curves under cyclic creep conditions exhibit characteristics similar to those of static creep curves [14-15]. The failure process consists of three stages. At the beginning of the initial stage, the material experiences rapid plastic deformation, with a relatively high deformation rate. As time progresses, the deformation rate gradually decreases, leading to the steady-state creep stage, where the material's deformation rate stabilizes. Eventually, after prolonged exposure, the material enters the accelerated creep stage, characterized by the formation of micro-cracks and voids. During this stage, the deformation rate increases, culminating in the material's fracture failure. Figure 7 indicates that the envelope strain rate exhibits minimal variation with variations in the stress ratio. This indicates that as the stress ratio increases, the overall process of cyclic creep is neither accelerated nor slowed down. However, the maximum strain range increases, which contributes to an extended final cyclic creep fracture life.

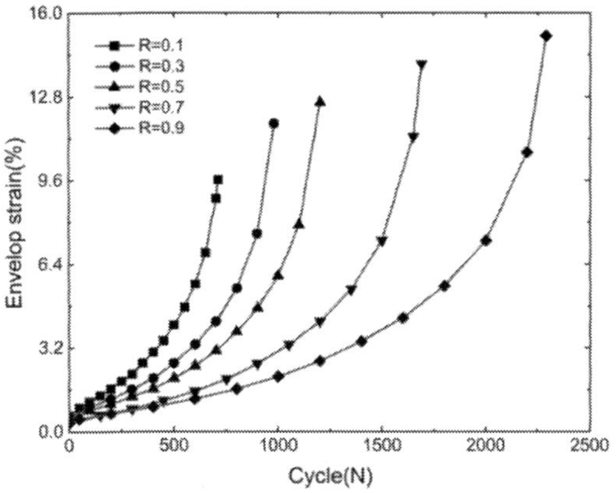

Figure 6. The variation of envelope strain with cycle life.

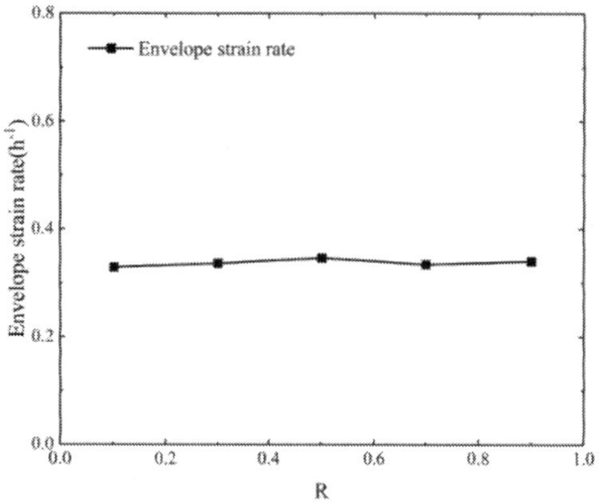

Figure 7. The variation of envelope strain rate with stress ratio.

3.3. Research of strain rate

The stress-strain curve under cyclic loading is of great significance in examining the low cycle fatigue creep properties, which illustrate how materials deform in cyclic creep experiments. The second cycle in the case of R = 0.1, 1/3 of the total number of cycles, 2/3, and the cycle before fracture are selected, as depicted in Figure 8. It is evident that the increase of creep strain throughout the process is nonlinear, with the creep rate generally accelerating over time. The creep rates recorded every 50 cycles are illustrated in Figure 9. This data aligns with the findings in the stress-strain curve, showing that the creep rate initially increases slowly during the steady-state stage, exhibiting slight fluctuations throughout this period. As the material approaches fracture, it transitions into the accelerated creep stage, where the creep rate sharply rises until failure occurs. The phenomena of cyclic hardening and cyclic softening can be attributed to the interplay of creep-fatigue, which induces slip and dislocation pinning at the grain boundaries of superalloy materials. Grain boundary slip tends to accelerate the creep rate, while dislocation pinning acts to slow it down. These two microscopic mechanisms coexist and compete with one another, leading to observable fluctuations in the creep rate on the creep displacement curve during the steady creep stage. This behavior is consistent with the observations presented in Figure 9.

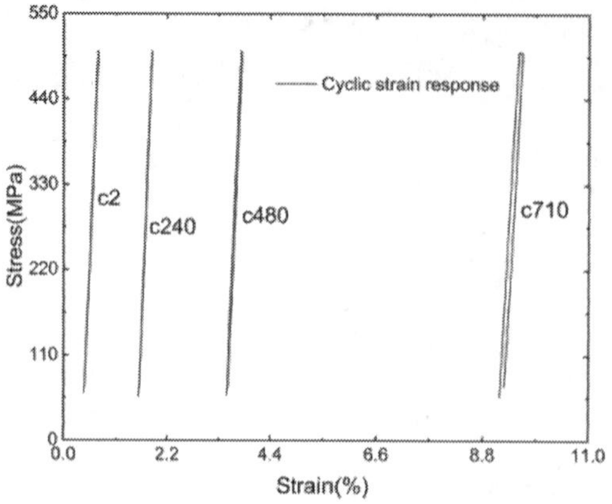

Figure 8. Cyclic strain response under different cycle numbers when R = 0.1.

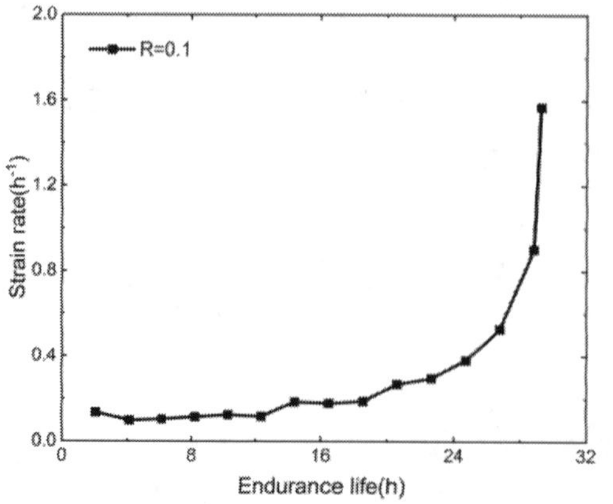

Figure 9. Relation between strain rate and endurance life when R = 0.1.

Figure 10 illustrates the creep rate curves over time for stress ratios R = 0.3, 0.5, 0.7, and 0.9. Analysis of the previous test data indicates that the cyclic softening effect is most pronounced when the stress ratio R = 0.1. The strain rate-time curve for R = 0.1 shows significant fluctuations during the cyclic creep test, suggesting that interactions between grain boundary slip and dislocation pinning are occurring, with grain boundary slip being the dominant mechanism. This interaction accelerates the failure and fracture of the sample. In contrast, for stress ratios R = 0.3, 0.5, 0.7, and 0.9, the fracture life is similar to that observed in pure creep, indicating that the effects of cyclic softening and cyclic hardening are not pronounced in these cases. The creep rate-time curves show no significant fluctuations

doi:10.1088/1742-6596/2879/1/012036

during this period. Notably, for R = 0.3 and 0.5, cyclic softening is observed, while for R = 0.7 and 0.9, cyclic hardening occurs. When cyclic hardening is present, the creep rate increases relatively gently during the steady-state creep stage, suggesting that dislocation pinning becomes the dominant mechanism. From this analysis, it can be concluded that the slip and accumulation of dislocations at the grain boundary are responsible for the observed cyclic softening and cyclic hardening phenomena. The differing behaviors under various stress ratios are attributed to the varying proportions of influence exerted by grain boundary slip and dislocation accumulation.

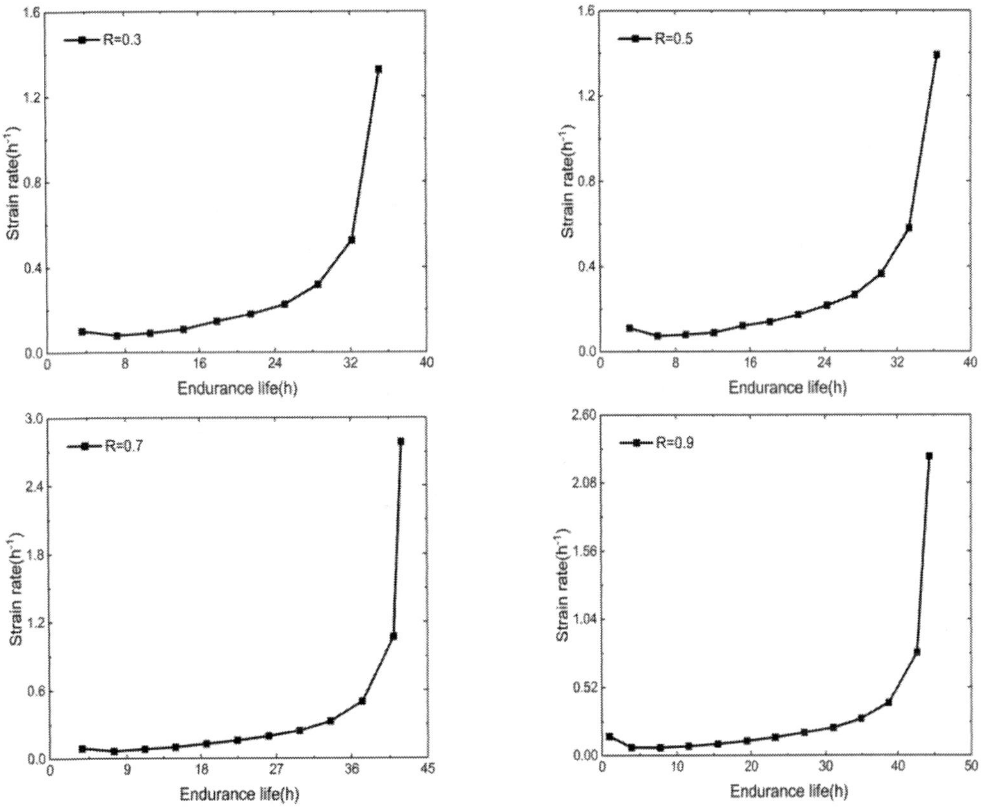

Figure 10. Relation between strain rate and endurance life when R = 0.3, R = 0.5, R = 0.7, R = 0.9.

3.4. Port microanalysis

The macro cross-sections of the samples at stress ratios R = 0.3, 0.5, 0.7, and 0.9 are presented in Figure 11. The fractures exhibit an undulating profile with distinct step characteristics, the morphology of dendritic fracture characteristics can be observed in each fracture, and a noticeable shear lip zone is evident at the edges of the fractures. This indicates that the DZ125 alloy retains high fracture toughness even at elevated temperatures. As the stress ratio increases from low to high, the fracture surface gradually becomes smoother, with fewer steps formed during the fatigue crack propagation process, resulting in a reduced proportion of fatigue damage. The first three sections display clear characteristics of fatigue fracture morphology, with distinct fatigue source areas, extension zones, and transient fault regions, predominantly exhibiting transgranular fracture. However, at R = 0.9, the fatigue features diminish, and there are no distinct characteristics observed in the fatigue source area or the crack propagation region. At this stage, fatigue damage constitutes a minor portion of the overall failure fracture process. The fracture surface is relatively flat, with the majority exhibiting intergranular fracture characteristics, along with the presence of a relatively small crack. This observation suggests that when the stress ratio is elevated, the fracture mode shifts from transgranular fracture to intergranular fracture, and the fatigue damage characteristics of the fracture gradually decrease, which is more manifested as creep damage-dominated failure.

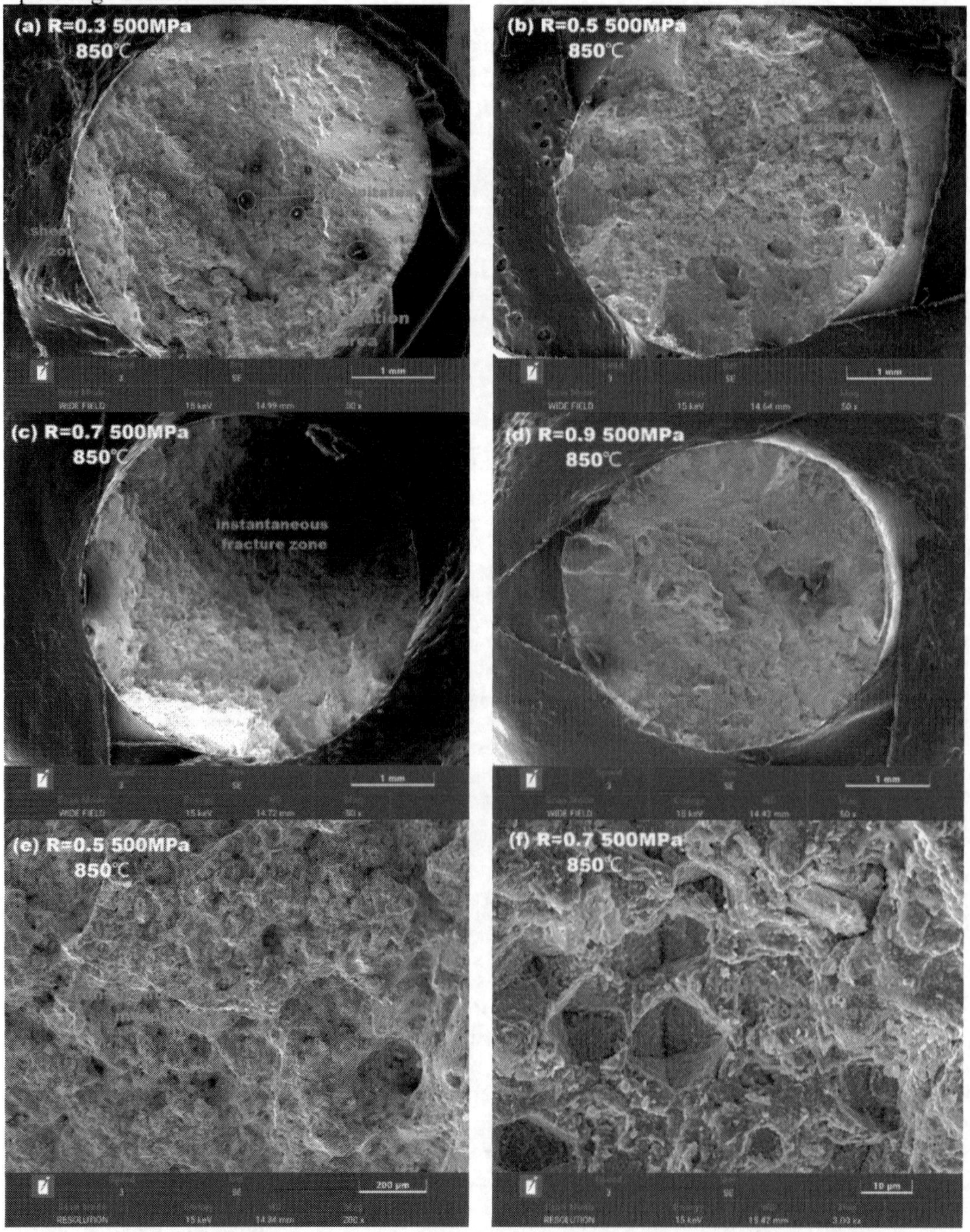

Figure 11. Fatigue and creep characteristics observed on various fracture surfaces. (a) The microscopic port morphology when R = 0.3 (b) The microscopic port morphology when R = 0.5 (c) The microscopic port morphology when R = 0.7 (d) The microscopic port morphology when R = 0.9 (e) Transgranular fracture morphology (f) Intergranular fracture morphology.

The macroscopic fracture images for R = 1, representing pure creep, and R = 0.1 are presented in Figure 12, based on observations from SEM. In contrast to pure creep, the fatigue characteristics observed at R = 0.1 are pronounced, evidenced by distinct regions: the fatigue source region, fatigue expansion region, and transient fault region. The crack initiates from a specific direction characterized by a stepped morphology, aligning with a particular crystallographic surface, before transitioning into a relatively flat crack growth area. This flat surface displays prominent river patterns, indicative of the crack propagation process. Additionally, features of interdendritic fracture, formed by the intersections of primary and secondary dendrite arms, are clearly visible. Small dimples with carbide precipitation are observed, resulting from the nucleation of carbide particles. As the main fatigue crack expands, cracks initiate at the grain boundaries on the sample surface and continue to grow under shear stress, ultimately interacting with the transient fault zone and the primary fatigue crack to form a significant crack, as illustrated in the accompanying figure. The transient fault zone is characterized by numerous dimples and holes of varying depths and sizes. In the case of pure creep, the fracture surface exhibits more holes, presenting a rough and uneven macro-section with a small shear lip area located at the fracture edge, which is distributed discontinuously. A large, shallow equiaxial dimple is noted in the fracture source area, primarily formed by the growth of nucleated cavities. The substantial deformation during the crack growth stage provides the driving force for the rapid enlargement of the dimple. Furthermore, as oxygen infiltrates along with crack expansion during cyclic loading, a significant number of oxide particles gradually form and adhere to the surface of the dimples.

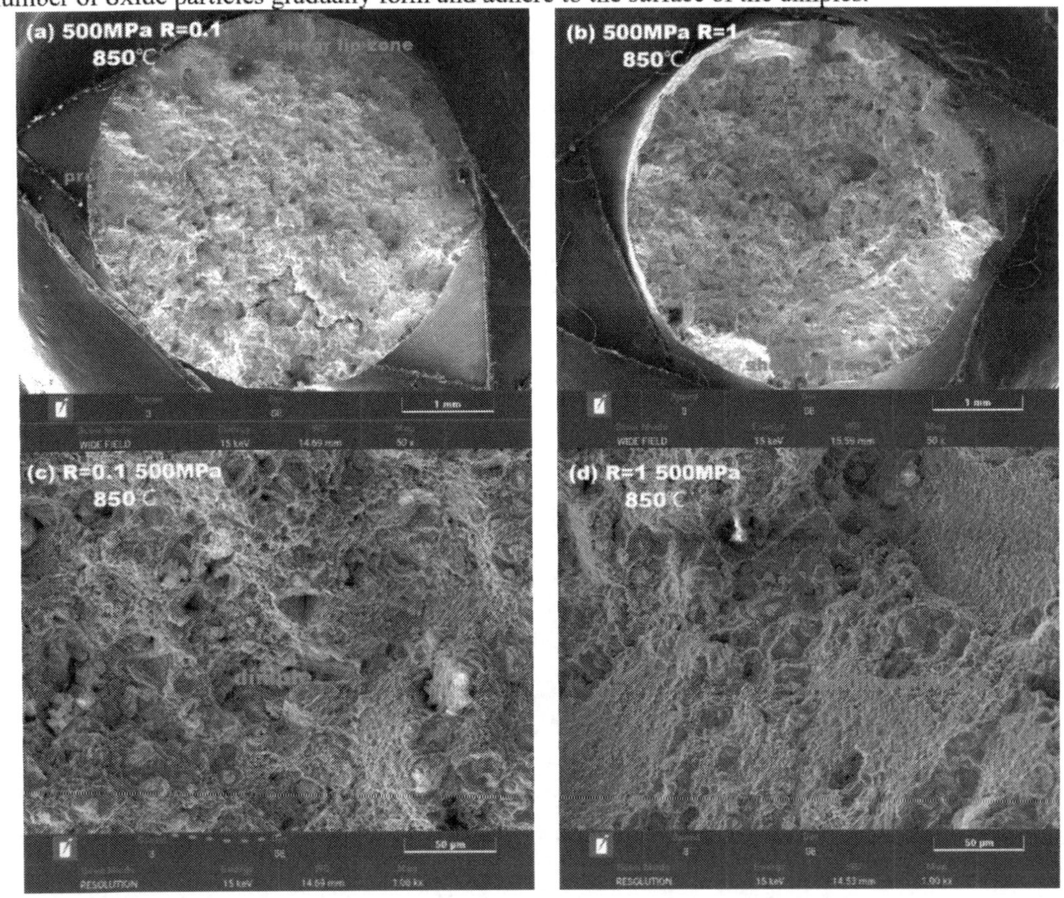

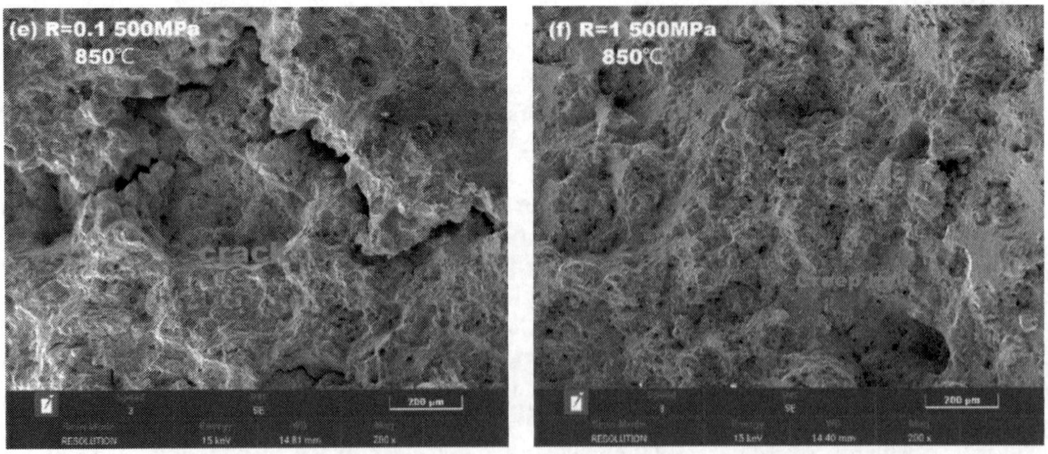

Figure 12. Fatigue and creep characteristics observed on various fracture surfaces. (a) The microscopic port morphology when R = 0.1 (b) The microscopic port morphology when R = 1 (c) Toughness nests morphology (d) Interconnecting dimples (e) Large crack (f) Creep cavity.

4. Damage evaluation and life modeling

Several commonly used engineering models were employed to forecast the cyclic creep life of DZ125 alloy in various stress ratios, including the Miner linear model, FS frequency separation model, Walker model, and Monkman-Grant model. The predicted lifespans from these models were compared, and the expressions for each model are as follows:

(1) Miner linear model

Miner's [16] theory believes that under cyclic load, fatigue damage and load cycle number of structures always have the same linear relationship, and fatigue damage generated by stress cycles at different times is independent and unrelated to each other. When damage accumulates to the critical value, fatigue damage occurs, as shown in Equation (1):

$$D_c + D_f = \sum (t_i / t_f) + \sum (N_i / N_f) = 1 \tag{1}$$

In this paper, cyclic creep under different stress ratios is considered. Therefore, fatigue damage can be calculated after equivalent stress is calculated by the Goodman iso life curve, as shown in Equation (2):

$$\frac{\sigma_a}{\sigma_{N(R=-1)}} + \frac{\sigma_m}{\sigma_u} = 1 \tag{2}$$

where σ_a is the stress amplitude, σ_m is the average stress, σ_u is the ultimate strength, and $\sigma_{N(R=-1)}$ is the equivalent stress. The stress values under each stress ratio are as Table 4 shown:

Table 4. The Stress Equivalent Value under Each Stress Ratio.

Stress Peaks	Stress Valley Value	Stress Amplitudes	Mean of Stress	Equivalent Stress
500	50	225	275	325
500	150	175	325	275
500	250	125	375	215
500	350	75	425	143
500	450	25	475	53

(2) Frequency separation model

A frequency correction method was proposed by Coffin [17], who believed that the main damage in low-cycle fatigue was caused by plastic strain. On this basis, Ecker proposed the following Equation (3):

$$v^K t_f = v^K \frac{N_f}{v} = N_f V^{K-1} = f(\Delta\varepsilon_p) = C \tag{3}$$

where t_f represents the fracture time, K is the temperature-dependent material constant, v represents the frequency, and $\Delta\varepsilon_p$ represents the plastic strain range.

Substituting the above equation into the Manson-Coffin equation gives an expression that takes into account frequency correction as Equation (4):

$$V^{K-1} N_f (\Delta\varepsilon_p)^a = C \tag{4}$$

Based on the equation mentioned above, taking into account the impact of load holding duration on lifespan at high temperatures, another frequency separation method is proposed by Coffin as Equation (5):

$$N_f = C\Delta\varepsilon_{in}^{-a} v_f^{m} (v_c / v_f)^K \tag{5}$$

where v_f and v_c are the frequency of tensile and compressive load holding, and $\Delta\varepsilon_{in}$ is the inelastic strain.

(3) Walker strain model

Walker's [18] equivalent strain is defined as Equation (6):

$$\varepsilon_\omega = \frac{\sigma_{\max}}{E} \cdot \left(\frac{\Delta\varepsilon \cdot E}{\sigma_{\max}}\right)^m \tag{6}$$

where ε_ω represents the strain parameter, m represents the equivalent strain parameter, and $\Delta\varepsilon$ represents the strain range.

Walker's model is a phenomenological model based on the development of strain parameters defined by Walker. There are two common models, one is Wang Weiguo's power function Walker strain life prediction model [19], and the other is Jaske's inverse hyperbolic tangent function prediction model [20], as shown in Equation (7) and Equation (8):

$$\varepsilon_\omega = u \cdot N_f^{-v} \tag{7}$$

$$\lg N_f = A_0 + A_1 \arctan h \left\{ \frac{\lg(\varepsilon_e \varepsilon_u / \varepsilon_w^2)}{\lg(\varepsilon_u / \varepsilon_w)} \right\} \tag{8}$$

In the equation, u, v, A_0 and A_1 are model parameters; $\varepsilon_u = \varepsilon_w |_{N=10} + 0.0025$ is the strain at 1/4 of the cycle; $\varepsilon_e = \varepsilon_w |_{N=10^8} - 0.0005$ is the ultimate fatigue strain. In this paper, Wang Weiguo's power function Walker strain life prediction model is selected for calculation.

(4) Monkman-Grant model

Monkman and Grant put forward the hypothesis that the minimum creep strain rate has a direct hyperbolic relationship with the total rupture life [21]. Additionally, it is viewed depending on both stress and temperature, as shown in Equation (9) and Equation (10):

$$M = \dot{\varepsilon}_m t_f \tag{9}$$

$$M / t_f = \dot{\varepsilon}_m = A\sigma^n \exp(-Q/(RT)) \tag{10}$$

where M is the material parameter; A and n represent material parameters that are influenced by stress and temperature; Q represents the creep activation energy.

The calculated and test values obtained by using the above life prediction models are shown in Table 5:

Table 5. Model predictions and experimental values.

R	Miner	FS	Walker	M-G	Test
0.1	28475	792	916	550	718
0.3	28491	1030	816	1049	983
0.5	28498	1320	957	1259	1201
0.7	28499	1740	1347	1887	1693
0.9	28500	2590	3936	2094	2294

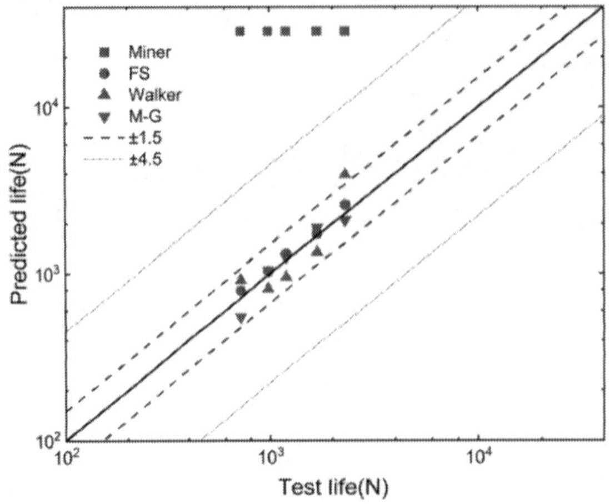

Figure 13. Life prediction results.

As shown in Figure 13, the prediction results indicate that the Miner linear model performs poorly, with data points falling outside the ±4.5 times dispersion band. This inadequacy arises because the model only accounts for linear accumulation under varying damage levels while neglecting the coupling effects of different types of damage. In subsequent studies, a nonlinear damage model can be developed and compared with the linear model. Furthermore, to provide a clearer representation of the error dispersion and the accuracy of each life model's predictions, *Error* —defined as the difference between the logarithmic predicted lifespan and the logarithmic tested lifespan[22], will be utilized for analysis, as shown in Equation (11)

$$Error = \lg(N_p) - \lg(N_t) \tag{11}$$

where N_p is the predicted value for each life model; N_f is the test life value.

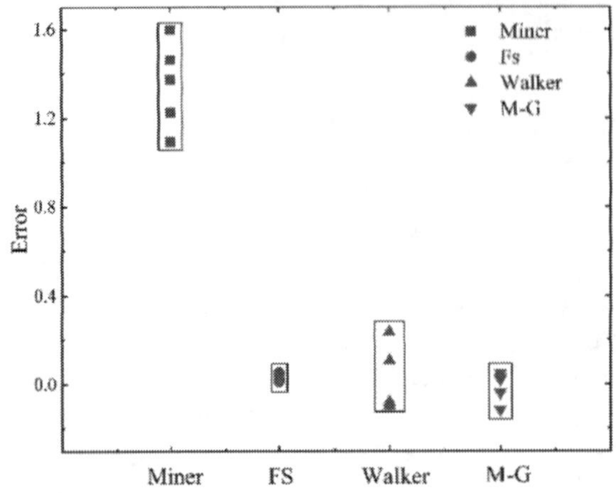

Figure 14. Life prediction error of different models of DZ125.

Figure 14 illustrates the dispersion of the logarithmic lifespan prediction errors for each model. From the figure, it is evident that the Miner model exhibits a larger dispersion and mean error in its lifespan predictions. In contrast, both the Walker model and the M-G model show smaller mean errors, although they exhibit relatively high dispersion. The FS model stands out by demonstrating both a low mean error and minimal dispersion, with its logarithmic error confined within ±0.2. Through comparison and validation, it is clear that among these four commonly used models, the FS model achieves the smallest dispersion in the cyclic creep lifespan prediction error for the DZ125 alloy, resulting in higher prediction accuracy.

5. Conclusions

In order to study the cyclic creep behavior and life prediction of DZ125 alloy under different stress ratios, cyclic creep experiments were carried out at 850°C and stress ratios of 0.1,0.3,0.5,0.7 and 0.9. Based on the results and discussions from these tests, several conclusions can be reached:

1. The envelope strain response of DZ125 alloy under cyclic creep loading resembles its static creep behavior.
2. Compared with static creep, cyclic hardening or cyclic softening occurs under fatigue creep load, and different stress ratios may be different.
3. Grain boundary slip can accelerate the creep rate, and a dislocation plug can slow down the creep rate. These two microscopic mechanisms coexist and compete with each other, leading to cyclic hardening and cyclic softening.
4. As the stress ratio increases, the fracture fatigue characteristics of specimens gradually decrease, creep characteristics gradually increase, and the fracture type gradually shifts from transgranular fracture to intergranular fracture.
5. For the prediction of cyclic creep life under different stress ratios, after transforming the stress by the Goodman curve, the prediction effect of the Miner linear model is poor, and the prediction result of the FS model is the most reasonable.

References

[1] Chen C, Sun J, Diao A, Yang Y, Li J and Zhou Y (2022). On the dendrite deformation and evolution mechanism of Ni-based superalloy during directional solidification. Journal of Alloys and Compounds., 891: 161949. [CrossRef]

[2] Wang H, Zhang X, Meng J, Yang J, Yang Y, Zhou Y and Sun X (2021). A new model of competitive grain growth dominated by the solute field of the Nickel-based superalloys during directional solidification. Journal of Alloys and Compounds., 873: 159794. [CrossRef]

[3] Sun Q Y, Ren Y and Liu D R (2019). Numerical investigations of freckles in directionally solidified nickel-based superalloy casting with abrupt contraction in cross-section. Results in Physics., 12: 1547-1558. [CrossRef]

[4] Wang H, Long H, Sun M, Yang G, Wei H, Mao S and Han X (2023). Effect of directional solidification methods on solid solution window in Ni-based single-crystal superalloy. Journal of Materials Research and Technology., 24: 8307-8319. [CrossRef]

[5] Kim W G, Park J Y, Ekaputra I M W, Kim S J and Jang J (2017) Influence of hold time and stress ratio on cyclic creep properties under controlled tension loading cycles of grade 91 steel. Nuclear Engineering and Technology., 49(3): 581-591. [CrossRef]

[6] Ying C H, Chen L J, Liu T L and Guo L Q (2016) Effect of hold time on cyclic creep behavior of 409 ferritic stainless steel. Journal of Shenyang University of Technology., 38(6): 634-639. [CrossRef]

[7] Aktaa J, Walter M, Angella G and Cippo E P (2019) Creep-fatigue design rules for cyclic softening steels. International Journal of Fatigue., 118: 98-103. [CrossRef]

[8] Hu X, Zhang Q, Jiang Y, Rao G, Miao G, He W, and Nie X (2020) The effect of cyclic loading on the creep-fatigue life and creep strength of a DS superalloy: Damage mechanism and life modeling. International Journal of Fatigue., 134: 105452. [CrossRef]

[9] Huang J, Yang X, Shi D, Yu H, Dong C, and Hu, X. (2014) Systematic methodology for high-temperature LCF life prediction of smooth and notched Ni-based superalloy with and without dwells. Computational materials science., 89: 65-74. [CrossRef]
[10] Li Z, Xu H, Shi D, Huang J, Xu G and Yang X (2022). Combined tensile and bending fatigue behavior and failure mechanism of a blade-like specimen at elevated temperature. International Journal of Fatigue., 164: 107163. [CrossRef]
[11] Zhao Z, Qian N, Ding W, Wang Y, and Fu Y (2020). Profile grinding of DZ125 nickel-based superalloy: Grinding heat, temperature field, and surface quality. Journal of Manufacturing Processes., 57: 10-22. [CrossRef]
[12] Zhang J, Zhao Z, Kong Y, Zhang Z, and Zhong Q (2019). Crack initiation and propagation mechanisms during thermal fatigue in directionally solidified superalloy DZ125. International Journal of Fatigue., 119: 355-366. [CrossRef]
[13] Xiao Y, Qin H, Xu K, and Wang Y (2020). Experimental Study on Fatigue-Creep of P/M FGH96 Superalloy with Different Holding Time. Journal of Northwestern Polytechnical University., 38(4): 873-880. [Cross Ref]
[14] Liang Z, Xiao S, Li Q, Li X, Chi D, Zheng Y and Chen Y (2023). Creep behavior and related phase precipitation of a creep-resistant Y2O3-bearing high Nb containing TiAl alloy. Materials Characterization., 198: 112767. [CrossRef]
[15] Wu S, Song H Y, Peng H Z, Hodgson P D, Wang H, Wu X H and Huang A J (2022). A microstructure-based creep model for additively manufactured nickel-based superalloys. Acta Materialia., 224: 117528. [CrossRef]
[16] Miner M A (1945) Cumulative damage in fatigue. J Appl Mech.,12: 159-164.[CrossRef]
[17] Coffin Jr L F (1974) Fatigue at high temperature-prediction and interpretation. Proceedings of the Institution of Mechanical Engineers., 188: 109-127. [CrossRef]
[18] Xiao G, Jiang J, Wang Y, Liu Y, Zhang Y, and Huang M (2021) High-temperature creep behavior of thermoformed nickel-based superalloy parts. Materials Science and Engineering: A., 814: 141216. [CrossRef]
[19] Wang W G, Gu Y X, Qing H, Teng G and Wen W D (2006) Investigation of disk LCF life prediction and its experimental verification. Journal of Aerospace Power., 21(5): 862-866. [CrossRef]
[20] Jaske C E, Feddersen C E, Davies K B and Rice R C (1973) Analysis of fatigue, fatique-crack propagation, and fracture data (No. NASA-CR-132332). [CrossRef]
[21] Sundararajan G (1989) The monkman-grant relationship. Materials Science and Engineering: A., 112: 205-214. [CrossRef]
[22] Bai S, Huang H Z, Li Y F, Yu A and Deng Z (2021). A modified damage accumulation model for life prediction of aero-engine materials under combined high and low cycle fatigue loading. Fatigue & Fracture of Engineering Materials & Structures., 44(11): 3121-3134.[CrossRef]

Buckling analysis of nanoscale beams based on nonlocal Timoshenko beam theory

Gen Li*

China Aircraft Flight Test Technology Institute, Chinese Flight Test Establishment, Xi'an 710089, China

*Corresponding author's e-mail: 1262164815@qq.com

Abstract: Eigen-buckling problems of nanoscale beams continue to be of great research interest, and the nonlocal theory is widely used. However, the existing research generally adopted some simplified assumptions of nonlocal effects. This article studies the buckling behaviors of the nonlocal Timoshenko beam, where the nonlocal effects are considered both on the governing equations and boundary conditions. The variational principle is adopted to obtain the nonlocal governing equations and boundary conditions. The buckling solutions for nanoscale beams are obtained analytically. Numerical comparisons validate the correctness of the present results. Parameter study shows that the buckling characteristic of nonlocal beams is different from that of classical beams, and the nonlocal effect is dependent on boundary conditions and geometry size of nanoscale beams.

1. Introduction

As the most basic nanostructures, nanoscale beams have been studied extensively because of their superior mechanical properties. To fulfill their wider application, researchers have done some nanoscale experiments [1] and molecular dynamics simulation work [2] to study the buckling characteristics of nanoscale beams, showing obvious small-scale effects.

However, both nanoscale experiments and MD simulations are difficult and expensive, thus continuum mechanics theories are widely used in the buckling analysis of nanoscale beams, including Eringen's nonlocal elasticity theory[3-5]. In 2003, Peddieson et al. [6] first used the nonlocal theory to demonstrate the small-scale effect on the bending characteristics of microscale beams. Since then, researchers have used various nonlocal elasticity models to solve the buckling problems of nanoscale beams, including the nonlocal differential, nonlocal integral, and local/nonlocal mixture models.

Among these models, the differential model is the simplest and the most widely used, and some relevant studies are reviewed below. Wang et al. [7] conducted the buckling analysis of a nonlocal Timoshenko beam and presented the analytical buckling solutions for beams with various boundary conditions (BCs). However, this work neglected the nonlocal effect on the shear constitutive relation. Reddy [8] combined the nonlocal theory with beam theories to propose several nonlocal beam theories, including the nonlocal Timoshenko beam theory. Then Reddy and Pang [9] achieved the bending, vibration, and buckling solutions for nanoscale beams. In the buckling analysis of nonlocal Timoshenko beams, the higher-order term in governing equations is omitted. Hosseini-Ara [10] employed the modified nonlocal elasticity theory in the buckling analysis of micro Timoshenko beams, where the higher-order terms in nonlocal constitutive relations are neglected. Zhang et al. [11] employed the method of weighted residuals to analyze the buckling behaviors of simply supported nonlocal

Timoshenko beams. Ansari et al. [12] used the generalized differential quadrature (GDQ) method to study the axial buckling and post-buckling problem of the nonlocal Timoshenko beam.

Besides, Xu et al. [13] employed the nonlocal integral model in the buckling analysis of nonlocal Timoshenko beam. Tang and Qing [14] have employed nonlocal strain gradient theory to study the vibration and buckling problems of functionally graded Timoshenko beam, where the nonlocal effects are considered in the integral model. Furthermore, Ren et al. [15-16] employed a local/nonlocal mixture model to study the bending, buckling, and vibration behaviors of functionally graded Timoshenko nanobeam.

From the above review, one can see that there are usually some simplifications in the buckling analysis of nonlocal differential Timoshenko beam models. Wang et al. [7] and Hosseini-Ara [10] used simplified nonlocal constitutive relations; Reddy and Pang [9] neglected the higher-order term of governing equations; Zhang et al. [11] and Reddy [8] only presented the buckling solutions for simply-supported nonlocal beams.

Thus this paper aims to conduct a comprehensive study on the buckling behaviors of nonlocal Timoshenko beams. Firstly, both the nonlocal bending moment and nonlocal shear force constitutive relations for the Timoshenko beam are adopted. Then the variational principle is employed to derive the nonlocal governing equations and BCs. Finally, the analytical solutions for beams with any homogenous BCs are obtained, which satisfy the nonlocal governing equations and BCs simultaneously. The numerical results are achieved and used to conduct a parameter study to analyze the nonlocal effects on the critical buckling loads of nanobeams.

2. Theoretical model

For a Timoshenko beam, the displacements are:

$$u_x = z\phi(x), u_y = 0, u_z = w(x) \tag{1}$$

where w is the deflection, and ϕ is the rotation, as in Figure 1.

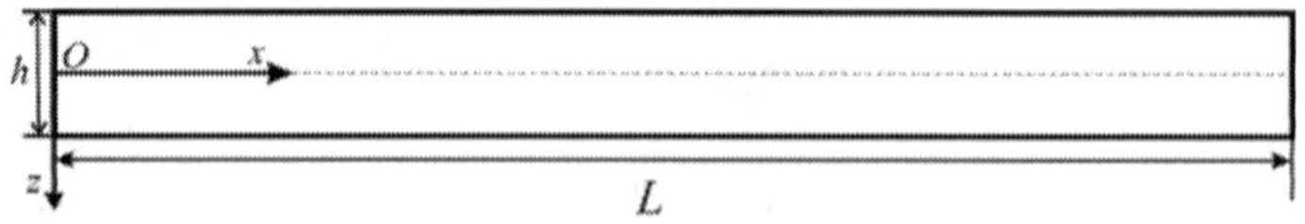

Figure 1. A nanoscale beam.

The non-zero strains are:

$$\varepsilon_{xx} = z\frac{\mathrm{d}\phi}{\mathrm{d}x}, \gamma_{xz} = \frac{\mathrm{d}w}{\mathrm{d}x} + \phi \tag{2}$$

where ε_{xx} and γ_{xz} are normal strain and shear strain, respectively. The classical stress field is:

$$\sigma_{xx}^{\mathrm{L}} = E\varepsilon_{xx}, \sigma_{xz}^{\mathrm{L}} = G\gamma_{xz} \tag{3}$$

where E is Young's modulus, $\mathrm{G} = E/2(1+\nu)$ is the shear modulus, and ν is Poisson's ratio.

From the virtual displacement principle, we have the governing equations of Timoshenko beams under compressive axial load P as:

$$\begin{aligned} &\frac{\mathrm{d}M}{\mathrm{d}x} - Q = 0 \\ &\frac{\mathrm{d}Q}{\mathrm{d}x} - P\frac{\mathrm{d}^2 w}{\mathrm{d}x^2} = 0 \end{aligned} \tag{4}$$

where resultant forces M and Q are:

$$M = \int_A z\sigma_{xx}\mathrm{d}A, Q = K_s\int_A \sigma_{xz}\mathrm{d}A \tag{5}$$

where k_s is the shear correction factor and A is the cross-section area. Note that Equation (4) is applicable for both classical and nonlocal beams.

The nonlocal constitutive relation is:

$$\left(1-k^2\nabla^2\right)\sigma^{\mathrm{NL}}=\sigma^{\mathrm{L}} \tag{6}$$

where $k = e_0 a$ is the nonlocal parameter, e_0 is a constant, a is the characteristic length, ∇^2 is the Laplacian operator, and σ^{NL} and σ^{L} are the nonlocal stress and classical stress, respectively.

Substituting Equations (2) and (3) into Equation (6) generates:

$$\sigma_{xx}^{\mathrm{NL}}-k^2\frac{\mathrm{d}^2\sigma_{xx}^{\mathrm{NL}}}{\mathrm{d}x^2}=E\varepsilon_{xx},\sigma_{xz}^{\mathrm{NL}}-k^2\frac{\mathrm{d}^2\sigma_{xz}^{\mathrm{NL}}}{\mathrm{d}x^2}=G\gamma_{xz} \tag{7}$$

Substituting Equation (7) into Equation (5) generates the constitutive relations of nonlocal Timoshenko beam as:

$$\begin{aligned}M^{\mathrm{NL}}-k^2\frac{\mathrm{d}^2M^{\mathrm{NL}}}{\mathrm{d}x^2}&=EI\frac{\mathrm{d}\phi}{\mathrm{d}x}\\ Q^{\mathrm{NL}}-k^2\frac{\mathrm{d}^2Q^{\mathrm{NL}}}{\mathrm{d}x^2}&=GAK_s\left(\frac{\mathrm{d}w}{\mathrm{d}x}+\phi\right)\end{aligned} \tag{8}$$

where $(A,I)=\int_A (1,z^2)\mathrm{d}A$.

From Equations (4) and (8), we can obtain:

$$\begin{aligned}M^{\mathrm{NL}}&=EI\frac{\mathrm{d}\phi}{\mathrm{d}x}+k^2P\frac{\mathrm{d}^2w}{\mathrm{d}x^2}\\ Q^{\mathrm{NL}}&=GAK_s\left(\frac{\mathrm{d}w}{\mathrm{d}x}+\phi\right)+k^2P\frac{\mathrm{d}^3w}{\mathrm{d}x^3}\end{aligned} \tag{9}$$

Thus, the energy functional of the nonlocal Timoshenko beam is:

$$\Pi=\frac{1}{2}\int_0^L M^{\mathrm{NL}}\frac{\mathrm{d}\phi}{\mathrm{d}x}\mathrm{d}x+\frac{1}{2}\int_0^L Q^{\mathrm{NL}}\left(\frac{\mathrm{d}w}{\mathrm{d}x}+\phi\right)\mathrm{d}x-\frac{1}{2}\int_0^L P\left(\frac{\mathrm{d}w}{\mathrm{d}x}\right)^2\mathrm{d}x \tag{10}$$

Substituting Equation (9) into Equation (10) and conducting variation calculation $\delta\Pi = 0$, one can obtain the governing equations as:

$$\begin{aligned}&\mathrm{EI}\frac{\mathrm{d}^2\phi}{\mathrm{d}x^2}-GAK_s\left(\frac{\mathrm{d}w}{\mathrm{d}x}+\phi\right)=0\\ &\mathrm{GAK}_s\left(\frac{\mathrm{d}^2w}{\mathrm{d}x^2}+\frac{\mathrm{d}\phi}{\mathrm{d}x}\right)+k^2P\frac{\mathrm{d}^4w}{\mathrm{d}x^4}-P\frac{\mathrm{d}^2w}{\mathrm{d}x^2}=0\end{aligned} \tag{11}$$

Besides, the corresponding boundary terms at $x = 0$ and $x = L$ are:

$$\begin{aligned}&\left(2\mathrm{EI}\frac{\mathrm{d}\phi}{\mathrm{d}x}+k^2P\frac{\mathrm{d}^2w}{\mathrm{d}x^2}\right)\delta\phi=0\\ &\left[2GAK_s\left(\frac{\mathrm{d}w}{\mathrm{d}x}+\phi\right)+2k^2P\frac{\mathrm{d}^3w}{\mathrm{d}x^3}-2P\frac{\mathrm{d}w}{\mathrm{d}x}\right]\delta w=0\\ &k^2P\frac{\mathrm{d}^2w}{\mathrm{d}x^2}\delta\frac{\mathrm{d}w}{\mathrm{d}x}=0\\ &k^2P\left(\frac{\mathrm{d}w}{\mathrm{d}x}+\phi\right)\delta\frac{\mathrm{d}^2w}{\mathrm{d}x^2}=0\end{aligned} \tag{12}$$

3. Analytical buckling solutions

By introducing the following non-dimensional parameters and variables, we can obtain:

$$X=\frac{x}{L},W=\frac{w}{L},\Omega=\frac{EI}{GAK_sL^2} \tag{13}$$

Equation (11) changes into:

$$\Omega\frac{\mathrm{d}^2\phi}{\mathrm{d}X^2}-\frac{\mathrm{d}W}{\mathrm{d}X}-\phi=0$$
$$EI\left(\frac{\mathrm{d}^2W}{\mathrm{d}X^2}+\frac{\mathrm{d}\phi}{\mathrm{d}X}\right)+k^2P\Omega\frac{\mathrm{d}^4W}{\mathrm{d}X^4}-P\Omega L^2\frac{\mathrm{d}^2W}{\mathrm{d}X^2}=0 \quad (14)$$

By eliminating ϕ or W from Equation (14), one can obtain:

$$A_1\frac{\mathrm{d}^6W}{\mathrm{d}X^6}+A_2\frac{\mathrm{d}^4W}{\mathrm{d}X^4}+A_3\frac{\mathrm{d}^2W}{\mathrm{d}X^2}=0$$
$$A_1\frac{\mathrm{d}^5\phi}{\mathrm{d}X^5}+A_2\frac{\mathrm{d}^3\phi}{\mathrm{d}X^3}+A_3\frac{\mathrm{d}\phi}{\mathrm{d}X}=0 \quad (15)$$

$$A_1=k^2P\Omega, A_2=EI-P\Omega L^2-k^2P, A_3=PL^2 \quad (16)$$

The general solutions for Equation (15) can be written as:

$$W(X)=D_1\sin(\beta X)+D_2\cos(\beta X)+D_3\sin(\gamma X)+D_4\cos(\gamma X)+D_5X+D_6$$
$$\phi(X)=Q_1D_2\sin(\beta X)-Q_1D_1\cos(\beta X)+Q_2D_4\sin(\gamma X)-Q_2D_3\cos(\gamma X)-D_5 \quad (17)$$

$$\beta^2=\frac{A_2+\sqrt{A_2^2-4A_1A_3}}{2A_1},\gamma^2=\frac{A_2-\sqrt{A_2^2-4A_1A_3}}{2A_1} \quad (18)$$

With Equation (14), the coefficients are determined as:

$$Q_1=\frac{\beta}{1+\beta^2\Omega},Q_2=\frac{\gamma}{1+\gamma^2\Omega} \quad (19)$$

To determine the mode coefficients D_1-D_6, only three sets of BCs at both ends are needed. Four sets of BCs are derived from the energy function, as shown in Equation (12). Thus, the nonlocal BCs need to be derived, which must meet the following limiting conditions:

1) Equation (12) must be satisfied;
2) The nonlocal BCs can degenerate into the BCs for the classical Timoshenko beam, as shown in Table 1.

Table 1. BCs of classical Timoshenko beam.

BCs	Deflection w	Rotation ϕ	Shear force $Q-P\frac{\mathrm{d}w}{\mathrm{d}x}$	Bending moment M
Simple support (S)	0			0
Clamp (C)	0	0		
Free (F)			0	0

From Equation (12) and Table 1, we can obtain the following BCs, as shown in Table 2. One can easily know that the expression in Table 2 meets both limitation conditions. By substituting the general solutions Equation (17) into BCs in Table 2, we can obtain a matrix equation for unknown constants D_1-D_6, from which a transcendental equation of P is obtained. Thus, we can obtain the critical buckling load P_{cr} and mode functions by solving the characteristic equation.

Table 2. BCs of nonlocal Timoshenko beam.

BCs	Deflection w	Rotation ϕ	Slope $\frac{\mathrm{d}w}{\mathrm{d}x}$	Shear force $Q^{NL}-P\frac{\mathrm{d}w}{\mathrm{d}x}$	Resultant moment $\frac{\mathrm{d}^2w}{\mathrm{d}x^2}$	Resultant moment $\frac{\mathrm{d}\phi}{\mathrm{d}x}$
S	0				0	0
C	0	0	0			
F				0	0	0

4. Results and discussion

We consider a nanobeam with Diameter $d = 1$ nm and the following mechanical parameters [7]:

$$E = 1\text{TPa}, \nu = 0.19, K_s = 0.9 \tag{20}$$

The boundary conditions are expressed as following: C-S beam means that edges of $x = 0$ and L are clamped (C) and simply-supported (S), respectively.

Table 3 lists the critical buckling loads P_{cr} for this nanobeam, verifying the correctness of the present results. Wang et al. [7] neglected the nonlocal shear force, while the nonlocal forces are considered fully in this paper. Thus, the present results are smaller because nonlocal shear force can reduce P_{cr} further.

Table 3. Critical buckling loads P_{cr}(nN) of nonlocal Timoshenko beam with $L/d = 10$.

		S-S	C-C	C-S	C-F
$k = 0.5$ nm	Present	4.6522	16.558	9.1010	1.1989
	Wang et al. [7]	4.6540	16.649	9.1179	1.1989
$k = 1$ nm	Present	4.3388	13.043	7.9578	1.1772
	Wang et al. [7]	4.3450	13.273	8.0055	1.1773
$k = 1.5$ nm	Present	3.9007	9.6342	6.5728	1.1428
	Wang et al. [7]	3.9121	9.9200	6.6520	1.1431

Then some benchmark results are presented and discussed in detail. Table 4 lists the critical buckling loads for nanobeams with several aspect ratios L/d and BCs. The following conclusions can be obtained from the sensitivity analysis:

1) P_{cr} decreases when k increases for all cases;
2) With the same aspect ratio and the same increment of k, the reduction of P_{cr} is different for beams with different BCs. The relation is C-C>C-S>S-S>C-F, meaning that P_{cr} of beams with stiffer BCs are more sensitive to the nonlocal effect;
3) With the same BC and the same increment of k, the reduction of P_{cr} is larger for beams with smaller aspect ratios, meaning that P_{cr} for beams with shorter lengths are more sensitive to the nonlocal effect.

Table 4. Critical buckling loads P_{cr}(nN) for nanobeams.

L/d	$k = 0$ nm	$k = 0.5$ nm	$k = 1$ nm	$k = 1.5$ nm	$k = 2$ nm
			S-S beam		
10	4.7670	4.6522	4.3388	3.9007	3.4177
12	3.3267	3.2707	3.1133	2.8822	2.6109
14	2.4514	2.4209	2.3339	2.2019	2.0404
16	1.8805	1.8625	1.8107	1.7304	1.6292
18	1.4878	1.4766	1.4438	1.3924	1.3262
20	1.2063	1.1989	1.1772	1.1428	1.0979
			C-C beam		
10	18.192	16.558	13.043	9.6342	7.0535
12	12.874	12.048	10.104	7.9625	6.1405
14	9.5687	9.1099	7.9645	6.5846	5.2992
16	7.3817	7.1077	6.3955	5.4802	4.5655
18	5.8631	5.6898	5.2263	4.6015	3.9419
20	4.7670	4.6522	4.3388	3.9007	3.4177
			C-S beam		
10	9.5605	9.1010	7.9578	6.5728	5.2883
12	6.7118	6.4846	5.8865	5.1019	4.2990
14	4.9638	4.8393	4.5003	4.0297	3.5153
16	3.8168	3.7431	3.5377	3.2416	2.9016

18	3.0247	2.9783	2.8476	2.6527	2.4212
20	2.4553	2.4247	2.3373	2.2051	2.0428
			C-F beam		
10	1.2063	1.1989	1.1772	1.1428	1.0979
12	0.8387	0.8352	0.8246	0.8076	0.7849
14	0.6167	0.6147	0.6090	0.5997	0.5871
16	0.4724	0.4712	0.4679	0.4623	0.4548
18	0.3734	0.3726	0.3705	0.3671	0.3623
20	0.3025	0.3020	0.3006	0.2984	0.2952

A critical buckling load ratio ξ is defined as:

$$\xi = \frac{P_{cr}(k \neq 0)}{P_{cr}(k = 0)} \tag{21}$$

Figure 2 shows the variations of ξ with k for nanobeam with several BCs, and the specific values are listed in the form of labels. We can see that under different BCs, ξ is different with the same k, and the relation is C-C<C-S<S-S<C-F. Besides, with the same increment of k, the relation of reduction of ξ is C-C>C-S>S-S>C-F. The reason is that the nonlocal effect is greater for beams with stiffer BCs, which is consistence with the second sensitivity analysis result of Table 4.

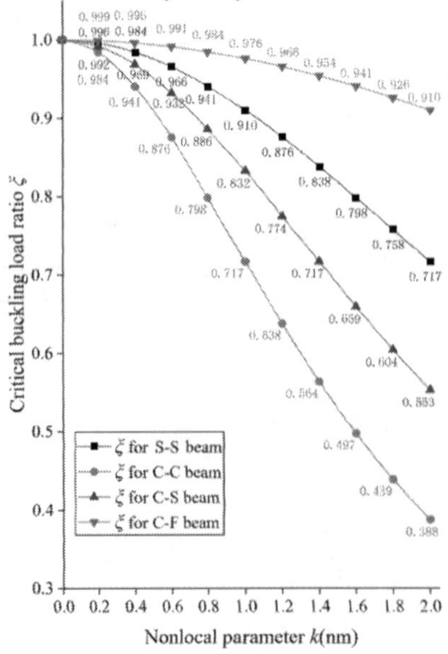

Figure 2. Critical buckling load ratios ξ under various BCs with $L/d = 10$.

We present the variations of critical buckling loads P_{cr} with various aspect ratios in Figure 3. We can see that when L/d increases, critical buckling load decreases under all BCs. This phenomenon is similar to the local results because a longer beam is subject to instability than a shorter one.

The variations of ξ with L/d are depicted in Figure 4. It is known that as L/d increases, the nonlocal effect is reduced. Besides, with the same BC and the same increment of $k = 1$ nm, the reduction of ξ is larger for beams with smaller aspect ratios. The reason is that the nonlocal effect is stronger for shorter beams, as explained in the sensitivity analysis of Table 4.

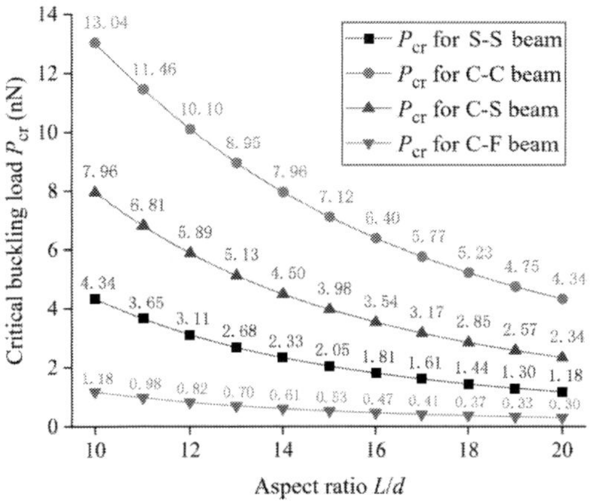

Figure 3. Critical buckling loads P_{cr}(nN) with various aspect ratios L/d when $k = 1$ nm.

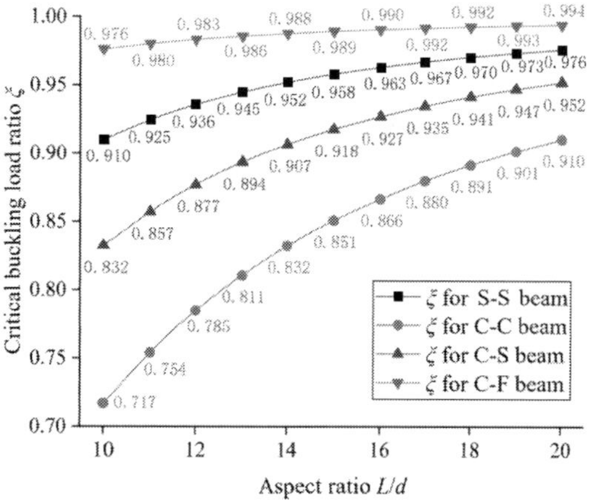

Figure 4. Critical buckling load ratios ξ with various aspect ratios L/d when $k = 1$ nm.

5. Conclusions

The buckling problem of nonlocal Timoshenko beams is studied in this paper. We employ the variational principle to derive the governing equations and the nonlocal BCs, and the buckling solutions are obtained analytically. The effects of nonlocal parameters, BCs, and aspect ratios on critical buckling loads are discussed. Numerical results show that the nonlocal effects lower the critical buckling load, and the nonlocal effects are greater for nanoscale beams with stiffer BCs and shorter lengths.

This work demonstrates that the buckling behavior of nanoscale beams is significantly size-dependent. In nanoscale beam design and engineering, the nanoscale effect needs to be considered from the very beginning, and the design experience of classic beam structures can be used with caution. Specifically, the experience of classical beams is more referable for nanobeams with lower stiffness, such as longer beams with weaker BCs, which can reduce the design cost. The present numerical results can be also used to verify the buckling results of other numerical and analytical methods.

References

[1] Yeau-Ren Jeng, Ping-Chi Tsai, and Te-Hua Fang. (2007). Experimental and numerical investigation into buckling instability of carbon nanotube probes under nanoindentation. *Appl. Phys. Lett.* **90** 161, 913.
[2] Guoxin Cao and Xi Chen (2006). Buckling of single-walled carbon nanotubes upon bending: Molecular dynamics simulations and finite element method. *Phys. Rev. B.* **73** 155, 435.
[3] A. Cemal Eringen (1972). Nonlocal polar elastic continua. *Int. J. Eng. Sci.* **10** 1-16.
[4] A. Cemal Eringen (1983). On differential equations of nonlocal elasticity and solutions of screw dislocation and surface waves. *Journal of Applied Physics* **54** 4, 703-10.
[5] A. Cemal Eringen and D. G. B. Edelen (1972). On nonlocal elasticity. *Int. J. Eng. Sci.* **10** 233-48.
[6] John Peddieson, George R., Buchanan, and Richard P. Mcnitt (2003). Application of nonlocal continuum models to nanotechnology. *Int. J. Eng. Sci.* **41** 305-12.
[7] C. M. Wang, Y. Y. Zhang, Sai Sudha Ramesh, and S Kitipornchai (2006). Buckling analysis of micro- and nano-rods/tubes based on nonlocal Timoshenko beam theory. *J. Phys. D: Appl. Phys.* **39** 3, 904-9.
[8] J. N. Reddy (2007). Nonlocal theories for bending, buckling and vibration of beams. *Int. J. Eng. Sci.* **45** 288-307.
[9] J. N. Reddy and S. D. Pang (2008). Nonlocal continuum theories of beams for the analysis of carbon nanotubes. *J. Appl. Phys.* **103** 023511.
[10] Reza Hosseini-Ara (2018). Nano-scale effects on nonlocal boundary conditions for exact buckling analysis of nano-beams with different end conditions. *J. Braz. Soc. Mech. Sci.* **40** 144.

[11] Z. Zhang, N. Challamel, and C. M. Wang (2013). Eringen's small length scale coefficient for buckling of nonlocal Timoshenko beam based on microstructured beam model. *J. Appl. Phys.* **114** 114, 902.

[12] R. Ansari, M. Faghih Shojaei, V. Mohammadi, R. Gholami, and H. Rouhi (2015). Buckling and postbuckling of single-walled carbon nanotubes based on a nonlocal Timoshenko beam model. *ZAMM-Z. Angew. Math. Mech.* **95** 939-51.

[13] Chi Xu, Yang Li, Mingyue Lu, and Zhendong Dai (2022). Stress-driven nonlocal Timoshenko beam model for buckling analysis of carbon nanotubes constrained by surface van der Waals interactions. *Microsyst. Technol.* **28** 1, 115-27.

[14] Yuan Tang and Hai Qing (2021). Elastic buckling and free vibration analysis of functionally graded Timoshenko beam with nonlocal strain gradient integral model. *Appl. Math. Model.* **96** 657-77.

[15] Yanming Ren and Hai Qing (2022). Bending and buckling analysis of functionally graded Timoshenko nanobeam using Two-Phase Local/Nonlocal piezoelectric integral model. *Compos. Struct.* **300** 116, 129.

[16] Pei Zhang, Peter Schiavone, and Hai Qing (2022). Stress-driven local/nonlocal mixture model for buckling and free vibration of FG sandwich Timoshenko beams resting on a nonlocal elastic foundation. *Compos. Struct.* **289** 115, 473.

Exploration of the damage evolution of HTPB propellant adhesive interface based on trilinear damage model

Pengchen Jiang[1], Chuntao Li[2,3], Fangyi Xu[4], Xiangyu Song[2,3] and Wei Li[2,3,*]

[1]Nanjing University of Aeronautics and Astronautics, Nanjing 210016, China
[2]National Key Laboratory of Aerospace Chemical Power, Xiangyang 441003, China
[3]Hubei Institute of Aerospace Chemotechnology, Xiangyang 441003, China
[4]China Airborne Missile Academy, Luoyang 471000, China

[*]Corresponding author's e-mail: gracelee@mail ustc.edu.cn

Abstract: In this paper, the HTPB propellant was used as the research object. A uniaxial tensile test was carried out at four rates. A mesoscopic model of propellant was established using the molecular dynamics method. The trilinear cohesion model was used to describe the damage behavior of the particle/matrix interface, which revealed the damage evolution of HTPB propellant under different strain rates. The results show that the interfacial damage of HTPB propellant is diffused at a high strain rate, the damage degree is lower, and the number is higher at a lower strain rate. The method adopted in this paper can reasonably describe the damage evolution phenomenon of the adhesive interface of HTPB propellant.

1. Introduction

The composite propellant has the advantages of stable performance, convenient storage, and transportation, and it is widely used in aviation and aerospace fields as the energy source of solid rocket engines [1]. In terms of composition and structure, composite propellant is a kind of particle-reinforced composite material with oxidant, metal, and other particles as the reinforcing phase, as well as polymers such as adhesives in the matrix. The use of composite propellant is seriously restricted under the different mechanical responses with the varying load and loading history conditions by its enhanced particles, polymer matrix, and adhesive interface properties directly affecting the mechanical properties of the propellant [2].

Early research on the mechanical behavior of propellants was mainly based on continuum mechanics, which regarded propellants as homogeneous materials. It also described the mechanical response of propellants by establishing phenomenological constitutive equations. It cannot be ignored to the influence of the mesostructure of propellants, and methods such as Eshelby equivalent inclusion theory [3], self-consistency theory, Mori-Tanaka method, and differential method are gradually applied to the prediction of propellant equivalent stiffness with the deepening of research [4]. It has become a research hotspot with the continuous development of computational mechanics methods, the study of propellant fine damage, and macroscopic mechanical response based on the mesostructure model of propellant [5]. Peng et al. [6] established a cylindrical axisymmetric viscoelastic finite element model of composite propellant, which was to discuss the influence of particle modulus and particle fraction on the mechanical properties of propellant. For the first time, Knott et al. [7] applied a molecular dynamics approach to a mesoscopic model of a composite solid propellant. Marur [8] used the method of uniform periodic distribution of cell elements to calculate and analyze the effective elastic properties of particle-reinforced composites and the stress state of spherical inclusions under uniaxial tensile load. Jackson et al. [9] used a molecular dynamics approach to study the maximum volume fraction of solid propellant particles filled by taking into account the multi-level distribution of particles. Zhang et al. [10] and Wu et al. [11] used the bilinear cohesive bonding interface model to calculate the stiffness and strength of the propellant and analyzed the influence of interface damage on elongation. On this basis, Zhang and Ala [12], Cui et al. [13] and Lo et al. [14] introduced kinds of nonlinear cohesive models to explore the influence of interface damage on macroscopic mechanical response.

In this study, the HTPB composite propellant was taken as the research object to obtain the stress-strain response characteristics under different conditions. A mesoscopic model of propellant was established using the molecular dynamics method. The trilinear cohesion model was used to describe the damage behavior of the particle/matrix interface, which is to numerically analyze the damage evolution of the composite propellant under different strain rates. It is to gain the relationship between propellant interface damage and tensile strain rate, which was to reveal the evolution law of composite propellant damage under different strain rates.

2. Propellant tensile test

In this paper, the tensile rate of the propellant specimen is controlled by a high-speed tensile testing machine to achieve uniaxial tensile of the propellant at different rates. The test specimen used in this test is a standard specimen of HTPB propellant dumbbell at the atmospheric pressure conditions, which are 100 mm/min, 1000mm/min, 2100mm/min, and 4200mm/min from low to high. The propellant specimen and test device are shown in Figure 1.

Figure 1. Propellant specimen and tensile test device.

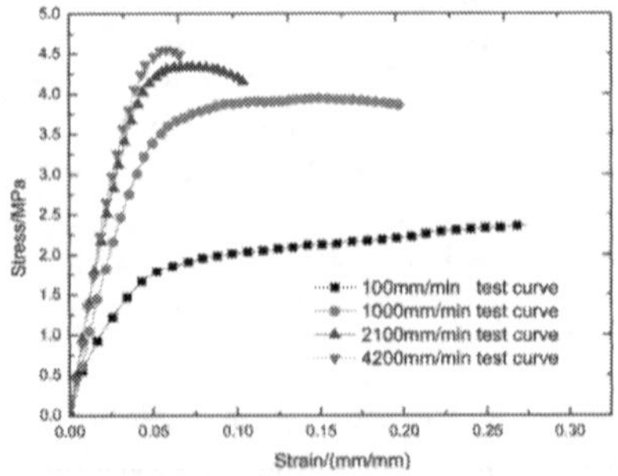

Figure 2. Results of propellant constant speed tensile test.

Figure 2 shows the results of the propellant tensile test at constant velocity, from which it can be seen that the tensile rate of HTPB propellant increases from 100 mm/min to 4200 mm/min at atmospheric pressure, the maximum elongation decreases significantly with the increase of strain rate, and the tensile strength increases with the increase of strain rate. When the tensile rate exceeds 1000mm/min, the damage propagation stage of the stress-strain curve gradually becomes less obvious. The stress begins to decrease immediately after the end of the dehumidification damage section. The curve bends rapidly until the propellant breaks. It can be seen that the HTPB propellant exhibits different mechanical responses at different levels of tensile rate, indicating that there is an obvious "rate correlation" in the damage to its mesostructure. The damage develops slowly at low-speed tensile and spreads rapidly at high-speed tensile. The propellant specimen breaks rapidly.

3. Mesoscopic damage model and result analysis

3.1. Propellant mesostructure model

In this paper, the method proposed by Knot et al. [7] was used to establish a periodic mesoscopic model that meets the target fill rate and gradation relationship according to the molecular dynamics method. The particle size relationship of the propellant selected and the resulting mesostructure are shown in Figure 3 and Figure 4. The model filling rate is 65.7%.

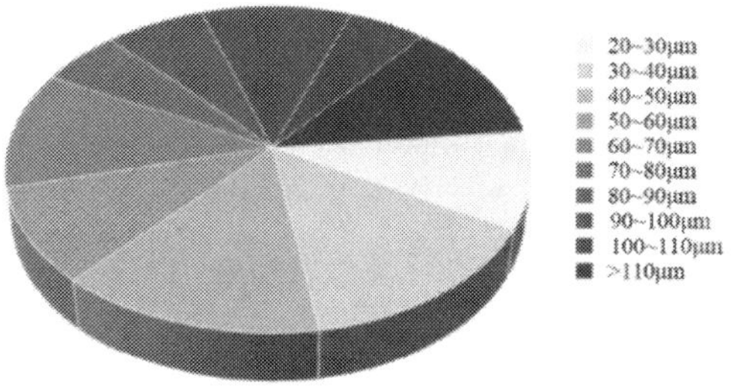

Figure 3. Propellant gradation relationship.

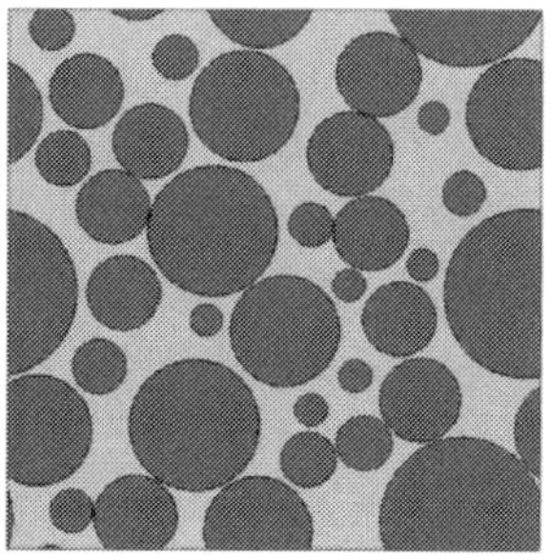

Figure 4. Mesoscopic structure of propellant.

3.2. Interfacial damage model

It is difficult to accurately describe the phenomenon of rapid detachment of the propellant stress-strain curve from the linear elasticity in the high-speed tensile state due to the linear reduction of the interfacial stiffness of the commonly used bilinear model. In this paper, the damage at the interface between the solid particles and the matrix is simulated in a trilinear form, as shown in Figure 5.

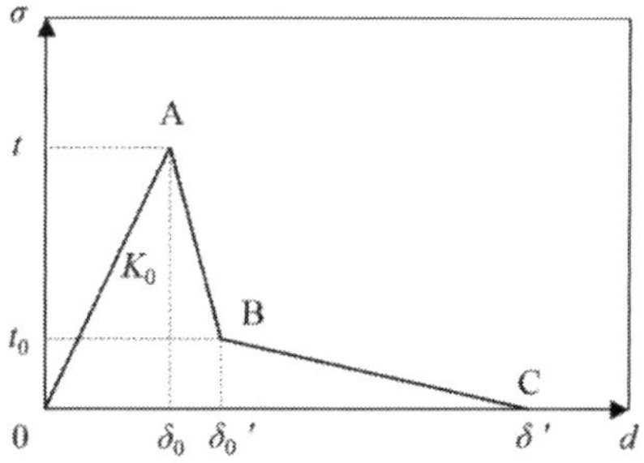

Figure 5. Three-linear damage model.

There are five damage parameters in the model. When the interfacial displacement is less than δ_0, the interfacial stiffness matrix is a constant value of K_0. Increasing K_0 will enhance the initial modulus of the curve, which will be delayed when the stress and strain of the curve reach the dehumidification damage point. The propellant strength will increase significantly when increasing the t. When the interface displacement is greater than δ_0, the interface stiffness matrix begins to decay. Increasing t_0 and $\delta_0{}'$ is closer to the bilinear model. The stress-strain curve shows slower dehumidification damage and higher failure stress. When the displacement is equal to $\delta_0{}'$, the decay rate of the interface stiffness decreases until the displacement reaches $\delta'$and the interface is completely broken. Increasing the δ' will make the transition section of the stress-strain curve smoother and the stress slightly larger. The cohesive force model in this form can achieve a rapid reduction in the tensile curve stiffness of the propellant after the damage point is reached. Then, the bending rate slows down, which makes it easier to reflect the complex mechanical behavior of some propellants during tension.

The interfacial mechanical response in the model can be described as:

$$t = \begin{pmatrix} t_n \\ t_s \end{pmatrix} = \begin{pmatrix} K_{nn} & K_{ns} \\ K_{sn} & K_{ss} \end{pmatrix} \begin{pmatrix} \delta_n \\ \delta_s \end{pmatrix} \tag{1}$$

where t represents the nominal stress matrix of the interface, which contains two contents, representing the normal and tangential nominal stresses, respectively. The K-matrix is represented as the interfacial stiffness matrix.

The damage initiation criterion is the maximum nominal stress criterion, which is:

$$max\left\{\frac{\langle t_n \rangle}{t_{no}}, \frac{t_s}{t_{so}}\right\} = 1 \tag{2}$$

Among them, t_n is the normal stress and t_s is the tangential stress; t_{n0} and t_{s0} are the corresponding critical stresses respectively; $\langle\ \rangle$ is the Macaulay symbol.

$$t_n' = \begin{cases}(1-D)t_n, & t_n \geq 0 \\ t_n, & t_n < 0\end{cases} \tag{3}$$

$$t_s' = (1-D)t_s \tag{4}$$

where t_n and t_s are the stress components calculated without damage; D is the damage factor.

According to Figure 5, the damage factors can be deduced as:

$$D = \begin{cases} 0, & \delta \leq \delta_0 \\ 1 - \dfrac{t\delta_0(\delta_0'-\delta)+t_1\delta_0(\delta-\delta_0)}{t\delta(\delta_0'-\delta_0)}, & \delta_0 \leq \delta \leq \delta_0{}' \\ 1 - \dfrac{t_1\delta_0(\delta'-\delta)}{t\delta(\delta'-\delta_0')}, & \delta_0{}' \leq \delta \leq \delta' \\ 1, & \delta \geq \delta' \end{cases} \tag{5}$$

3.3. Results & analysis

In this paper, because the modulus of AP particles is much larger than that of the matrix modulus, AP particles are regarded as completely elastic materials, E=32450MPa, and ν=0.1433. The matrix is modeled as a linear viscoelastic material. Its relaxation modulus can be expressed as a Prony series:

$$E(t) = E_\infty + \sum_{i=1}^{n} E_i \cdot e^{-\frac{t}{\tau_i}} \tag{6}$$

The parameters of the Prony series can be obtained by fitting the stress relaxation test data of the propellant matrix. As shown in Table 1, the initial modulus of the matrix are E_∞=6.6294MPa, ν=0.499.

Table 1. Relaxation modulus fitting data.

i	1	2	3	4	5
E_i	51.50331	12.41362	7.49219	5.59053	4.410796
τ_i	0.0003	0.0056	2.7285	0.124	60.0262

The propellant tensile curve was used to invert the damage parameters and after several rounds of iterations, which was to obtain the interface damage parameters of the composite solid propellant finally, thereinto, t=0.2MPa, K=5000, MPa/mm, $\delta_0{}'$=0.0037mm, t_0=0.07MPa, and δ'=0.25mm. The fitting results of the tensile curve of HTPB propellant are shown in Figure 6.

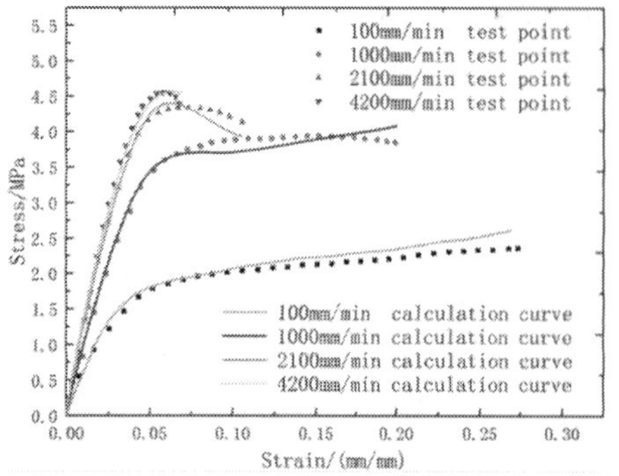

Figure 6. Comparison of the stress-strain calculation line and the test curve of the propellant different tensile rates.

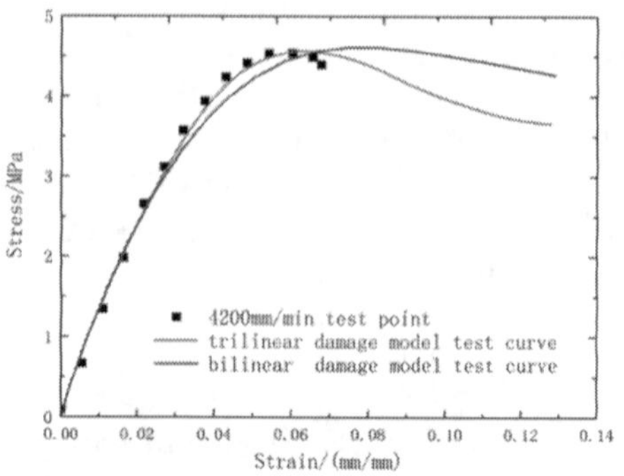

Figure 7. Comparison of the trilinear cohesion model and the bilinear cohesion model of the 4200mm/min tensile rates.

The linear segment of the test and the prediction curve fit well. The prediction error of the area enclosed by the curve and the x-axis is less than 7.58%. The prediction error of the initial modulus is 12.6% under different stretching rates. As the loading progresses, the location of the dehumidification point is accurately predicted by the simulation curves at different stretching rates. The strain error is controlled within 8.6%. The prediction error for propellant strength is not more than 11.2%. It can be seen that the molecular dynamics method combined with the trilinear cohesion model can reasonably describe the tensile behavior of the propellant at different rates. The calculation curve is in good agreement with the test curve.

Figure 7 is the comparison of the trilinear cohesion model and the bilinear cohesion model of the 4200mm/min tensile rates. The simulation curve of the trilinear model is closer to the test curve. The stiffness prediction is more accurate compared with the bilinear cohesion model at the tensile rate of 4200mm/min. It was flatter for the bilinear cohesion model simulation curves, while both models accurately predicted the strength of the propellant. The strain at the dehumidification point is predicted to be low. The trilinear cohesion model is more reasonable for predicting the rapid bending and decline of the test curve after the dehumidification point at a high strain rate.

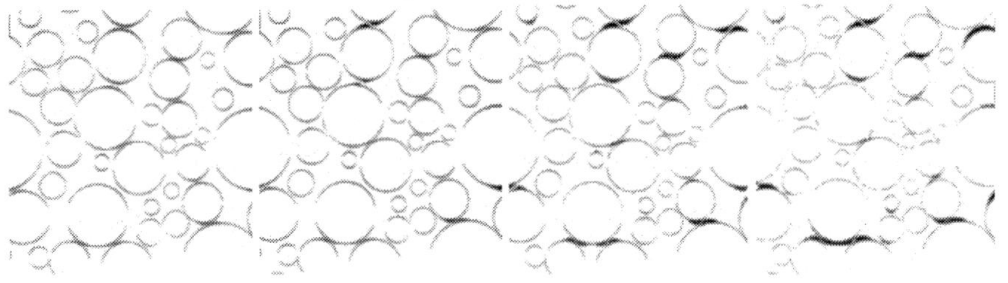

Figure 8. Dehumidification point damage area.

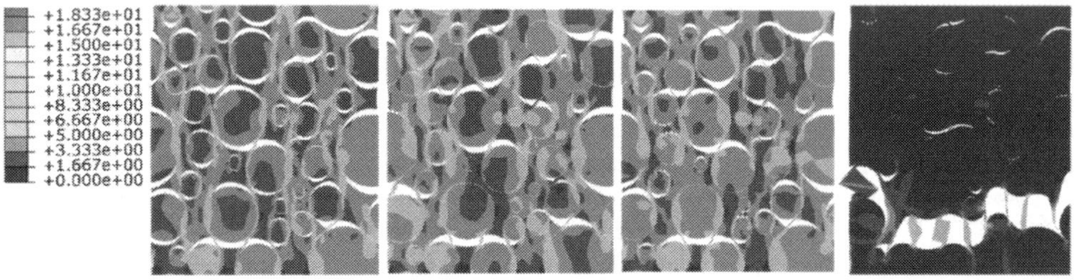

Figure 9. Mises stress of 15% strain propellant.

Figure 8 and Figure 9 are the interface damage at the propellant dehumidification point and the Mises stress diagram at 15% strain, respectively. It can be seen that with the decrease of the tensile rate, the number of interfaces where damage occurs decreases, but the degree of damage increases significantly. The propellant load is mainly borne by the matrix at high strain rates at the same strain rate, but the stress of propellant AP particles is greater at low strain rates. According to the calculation results, there are two main factors affecting the macroscopic mechanical properties of propellants due to particle dehumidification damage: the number of interfaces that cause damage and the degree of evolution after interface damage. With the increase of the strain rate, the number of interfaces that produce damage at the dehumidification point increases, but the damage degree is significantly lower, and the dehumidification damage of the propellant is inhibited.

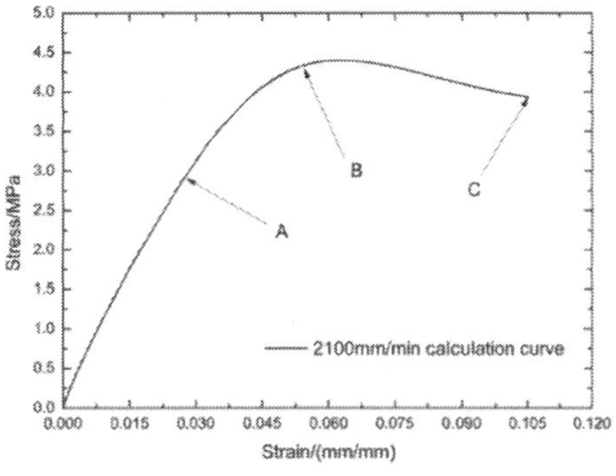

Figure 10. Diagram of the damage evolution at different locations at the tensile rate of 2100 mm/min.

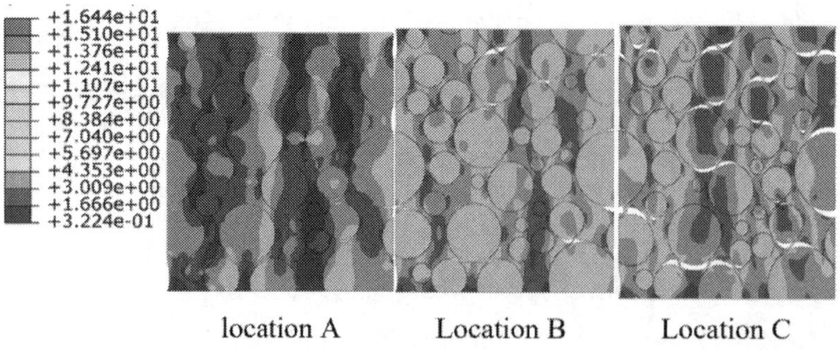

Figure 11. The damage evolution of the propellant at the tensile rate of 2100 mm/min.

The damage evolution of the propellant at the tensile rate of 2100 mm/min is shown in Figure 10 and Figure 11. In the initial loading stage (location A), there is a certain range of cracking damage at the particle interface and the matrix, but the damage degree is low. The curve is still linear as a whole. As the loading progresses to the dehumidification point (location B), the damage degree of the matrix gradually increases. The overall stiffness gradually decreases. The curve is obviously deflected. After the dehumidification damage point (location C), the degree of damage is further increased with the large local deformation of the matrix. If you continue to stretch, the damage will continue to aggregate and break.

When the strain rate is high, the degree of interface dehumidification is low, and the external load is mainly borne by the polymer matrix. With the increase of the tensile rate, the modulus and strength of the viscoelastic matrix increase, which explains the reason why the stress-strain curve of the propellant rises approximately linearly with a high modulus. When the stress in the matrix reaches a certain critical value, cracks appear in the matrix and continue to expand, and the stress-strain curve bends rapidly until the test piece breaks. The results indicate that the high strain rate delays the evolution of particle dehumidification damage, and the corresponding matrix deformation is increasing. When the strain rate is high enough, the matrix reaches its deformation limit, which directly leads to propellant breakage.

It is different from the meso-damage form of the propellant at low strain rates. The dehumidification of the particle/matrix interface is the main component of the meso-damage. As the propellant continues to stretch, the stress at the interface reaches the initial stress of damage, the particle/matrix interface begins to damage and continues to evolve, and the pores of the interface continue to increase. This process corresponds to a curve with a low strain rate. The curve bends smoothly from the obvious dehumidification point without obvious abrupt changes.

4. Conclusion

(1) In this paper, it is proposed that the trilinear cohesion model can reasonably be described as the interfacial damage behavior of propellants at different rates. The calculation curve is in good agreement with the test curve.

(2) The interfacial damage of HTPB propellants has obvious rate-dependent properties; it appears as a diffuse, low-damage interface at high-speed rate stretching. At low strain rates, there are fewer high-damage interfaces.

(3) The HTPB propellant damage is inhibited at high tensile and is more loaded by the viscoelastic matrix, while it is more borne by the particles at low tensile.

(4) The degree and amount of damage at the interface directly affect the macroscopic mechanical properties of the propellant. When the strength of the interface and the separation distance are small, the number of the damage will increase significantly. The dehumidification damage will appear earlier, the propellant stiffness will be reduced earlier, and the tensile strength will decrease. The elongation of the propellant is related to the degree of local damage. The greater the degree of damage within the tolerance of the matrix is, the greater the elongation at break is.

(5) The research in this paper is of great significance for predicting the mechanical properties of propellant formulations in the early stage. Its mechanical properties can be preliminarily judged through the formulation, which can reduce the number of iterations and greatly improve the efficiency of propellant development.

Acknowledgment
This work was financially supported by a Study on STACPL220221B05 of the National Key Laboratory of Aerospace Chemical Power Foundation.

References

[1] Bozic V and Jankovski B. Some civilian applications of solid propellants. J. Chemical Rocket Propulsion, 2017: 935–962.
[2] Chengfeng Wu, Xuejin Fu, Shaoqing Hu, et al. Research Progress on Meso-damage Evolution of Composite Propellants. J. Chinese Journal of Explosives & Propellants, 2023, 46(12): 1043-1058.
[3] Junfang Yu, Xuhai Ni, and Zhiqiang Zheng. Equivalent properties prediction of the composite ceramics with an arbitrary-shaped inclusion. J. Strength of Materials, 2021, 53(2): 342-352.
[4] Jianliang Gong, Peijin Liu and Qiang Li. Nonlinear Constitutive Relation of HTPB Propellant Based on the First Law of Thermodynamics. J. Chinese Journal of Energetic Materials, 2013, 21(3): 325-329.
[5] Jiaxiang Wang, Hongfu Qiang, and Zhejun Wang. Research progress on mesomechanics of composite solid propellants. J. Journal of Solid Rocket Technology, 2020, 43(6): 788-798.
[6] Wei Peng, Jianping Zhou and Weichang Zhao. A finite-element analysis of microscopic stress distribution and microdamage growth in composite solid propellant. J. Journal of Solid Rocket Technology, 2002, 25(1): 12-15.
[7] Knott GM, Jackson TL and Buckmaster J. The random packing of heterogeneous propellants. J. AIAA, 2001, 39(04): 678-686.
[8] Marur PR. Estimation of effective elastic properties and interface stress concentrations in particulate composites by unit cell methods. J. Acta Materialia, 2004, 52(05): 1263-1270.
[9] Jackson T, Stafford DS. Random Packs and Their Use in Grain-Scale Modeling, with Applications to Energetic Materials. J. International Journal for Multiscale Computational Engineering, 2010, 8(05): 473-487.
[10] Liang Zhang, Zhibin Shen, Kuahai Yu. Faliure Machanism Analysis of Solid Propellant under Confining Pressure. J. Journal of National University of Defense Technology, 2022, 44(6): 6-13.
[11] Wu Ding, Jinsheng Xu, Changsheng Zhou, et al. Type I Failure Temperature-dependent Properties of HTPB Propellant/Liner Interface. Chin. J. Energetic Mater. 2022, 30, 146–154.
[12] Wenlong Zhang, Tabiei Ala. Improvement of an exponential cohesive zone model for fatigue analysis. J. Journal of Failure Analysis and Prevention, 2018, 18: 607-618.
[13] Huiru Cui, Zhibin Shen and Haiyang Li. A novel time dependent cohesive zone model for the debonding interface between solid propellant and insulation. J. Meccanica, 2018, 53(14): 3527-3544.
[14] Girado LO, Spring DW, Paulino GH, et al. An efficient mixed-mode rate-dependent cohesive fracture model using sigmoidal functions. J. Engineering Fracture Mechanics, 2018, 192: 307-327.

Wear analysis of solid state forging for high-strength ball head hanging ring

Meihua Bian*[1], Dengjie Zhu[2], Deqin Zhang[3], Yongli Liao[2] and Xingsen Zhang[1]

[1]Electric Power Research Institute of Guangxi Power Grid Co., Ltd., Guangxi Key Laboratory of Intelligent Control and Maintenance of Power Equipment, Nanning, China
[2]Electric Power Research Institute, CSG, Guangzhou, China
[3]Guangxi Power Grid Company Limited Beihai Power Supply Bureau, Beihai, China

*Corresponding author: 27250060@qq.com

Abstract: In the process of die service, mold wear failure is the most common failure form. This paper analyzes the mold wear during the solid-state forging process of a certain high-strength aluminum alloy ball head hanging ring. Based on Archard's wear theory, finite element analysis software was used to determine the influence of main process parameters on mold wear through orthogonal tests. Among these parameters, the forging temperature had the greatest impact on mold wear, while the forging speed and mold temperature had lesser impacts. The optimal process parameters for the solid-state forging process were determined to be a die forging temperature of 540°C, a mold temperature of 150°C, and a die forging speed of 300 mm/s.

1. Introduction

Since the late 1980s, the construction of large-scale transboundary power grids such as "west-east transmission" has been in full swing and achieved good results in China [1, 2]. The operating mileage of transmission lines in China has been increasing continuously. With the increase in the trans-distance of transmission lines in China, it is inevitable that transmission lines will pass through some heavy ice areas. The power system puts forward higher requirements for the mechanical properties, reliability, and durability of the power fittings [3]. Therefore, the use of a variety of new materials to connect the fittings came into being. The application of high-strength materials in fittings also puts forward new requirements for machining molds.

In the process of metal processing, the mold has three failure forms: wear, plastic deformation, and fracture [4]. Among them, mold failure caused by plastic deformation and fracture can be reduced by reasonable mold design. However, mold wear is inevitable, and wear is caused by the contact between the mold and the workpiece. Therefore, mold failure caused by excessive wear is difficult to control [5]. Therefore, this paper addresses the actual needs of transmission lines by developing a high-strength aluminum alloy material. Through orthogonal experiments, it analyzes the influence of various process parameters on mold wear during the solid-state forging process of the newly researched high-strength aluminum alloy ball head hanging rings. The aim is to determine suitable solid-state forging process parameters to reduce mold wear, achieve a match between new materials and processes, and ensure the strength performance of high-strength aluminum alloy ball head hanging rings while improving production economics.

2. Archard model and modified model

Archard wear theory [6] describes the relationship between wear depth and material hardness, sliding speed, and pressure. Its expression is shown in Equation (1).

$$W=\int K\frac{pv}{H}dt \tag{1}$$

In the equation, W is the wear depth, P is the positive pressure on the mold surface, v is the sliding speed, K is the wear coefficient, and H is the mold hardness.

Archard believes that the wear coefficient K and hardness H are constants, but in fact, both indexes are closely related to temperature change. This phenomenon is more obvious in the process of hot working. After the temperature rises, the mold will soften, resulting in a decrease in mold hardness and the roughness of the contact surface between the mold and the workpiece, thus increasing the wear of the mold [7]. In view of this phenomenon, Jahanmir et al. [8] modified the Archard model by taking H13 die steel as the object, taking temperature as the influence factor of wear coefficient and mold hardness into consideration, and proposed a wear calculation formula considering the change of wear coefficient and hardness with temperature. Lee et al. [9] fitted $K(T)$ and $H(T)$ through a high-temperature wear experiment and a high-temperature and high-hardness wear experiment and obtained $K(T)$ and $H(T)$ expressions.

The pressure, temperature, and speed in the solid state forging process of hardware are different with the position and time. So Equation (1) can be changed to:

$$\Delta W_{ij}=k_{ij}(T)\frac{\sigma_{ij}u_{ij}}{H_{ij}(T)} \tag{2}$$

In the equation, W_{ij} is the wear depth at the i position of the die at the j time; Both σ_{ij} and u_{ij} can be obtained by forging a finite element process.

After modification, the wear depth of single die forging at the i position point of the mold can be expressed by Equation (3):

$$W_i=\sum_{j=1}^{j=n}k_{ij}(T)\frac{\sigma_{ij}u_{ij}}{H_{ij}(T)} \tag{3}$$

The modified Archard model is not in the existing wear model of finite element analysis software. The modified formula can be introduced into finite element analysis software through secondary development.

3. Orthogonal test analysis of solid state forging

Orthogonal testing is an experimental method often used for multi-factors and multi-levels, and it has the characteristics of high efficiency [10]. Die forging temperature, mold temperature, and die forging speed are three key process parameters. These three key process parameters have an impact on mold wear in the process of solid-state forging. In this experiment, the orthogonal table of $L16(4^3)$ is used. The die forging temperature, mold temperature, and die forging speed are taken as three factors, each of which corresponds to the values of four levels. A total of 16 groups of tests were conducted.

According to the field production experience, the range of orthogonal test parameters is determined: the die forging temperature range is 480-540°C, the mold temperature range is 100-250°C, and the die forging speed range is 300-600mm/s. The die forging temperature, mold temperature, and die forging speed are taken as three factors. The three factors are divided into four levels. The die forging temperatures are 480, 500, 520, and 540°C. The mold temperatures are 100, 150, 200, and 250°C. The die forging speeds are 300, 400, 500, and 600 mm/s. The orthogonal test parameter setting table is shown in Table 1.

Journal of Physics: Conference Series **2879** (2024) 012039 doi:10.1088/1742-6596/2879/1/012039

Table 1. Orthogonal test parameter setting table.

Factor	Level			
	1	2	3	4
Die forging temperature (°C)	480	500	520	540
Mold temperature (°C)	100	150	200	250
Die forging speed (mm/s)	300	400	500	600

Simulation model parameters are set as follows:
(1) Die material: H13 steel.
(2) Workpiece material: Particle-reinforced 6082 aluminum alloy.
(3) Mold hardness: 50HRC.
(4) Workpiece hardness: 15HRC.
(5) Friction coefficient: shear friction, coefficient of 0.3.
(6) Heat exchange coefficient: 11.
(7) Simulation type: Lagrangian Incremental.
(8) Object type: Workpiece is plastic; bottom die and top die are rigid.

Finite element analysis software was used to conduct simulation analysis of solid-state forging. The finite element model of solid-state forging is shown in Figure 1. Among them, the billet is a round rod type. The size is 50mm in diameter and 80mm in length. The main part of the upper and lower die is a cuboid with a length of 270mm, a width of 185 mm, and a height of 140 mm. The cavity part is set according to the outer contour of the QP-7 ball head hanging ring; a convex table and a concave table are made, respectively, on the parting surface of the top die and the bottom die.

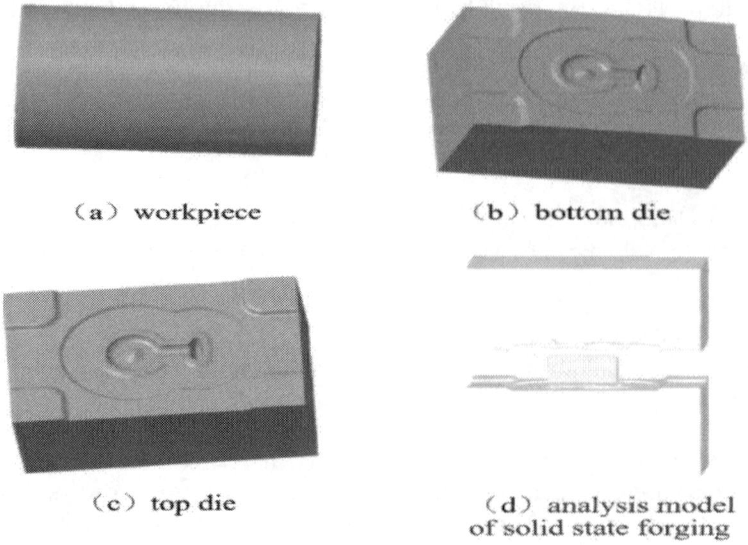

Figure 1. Solid state forging finite element model diagram.

The steps of the mold wear test with finite element analysis software are as follows:
(1) Creo established the geometric model of the top die, bottom die, and workpiece. The geometric position relationship was determined. The model was exported in STL format.
(2) We import the STL model into the finite element analysis software.
(3) We define the workpiece temperature, bottom die temperature, and top die temperature. We define the forging speed and direction of the top die. We define the hardness of the workpiece and die.
(4) We define workpiece, top die, and bottom die object type.
(5) We divide the workpiece and die mesh.
(6) We define heat transfer and friction types and enter correlation coefficients.

(7) We generate the database and perform calculations.

(8) The amount of mold wear depth is derived in the post-processing interface.

Then, one test is over. We change the die forging temperature, mold temperature, and die forging speed for the next test until the end of all 16 tests.

The 16 groups of orthogonal test results obtained by finite element analysis software simulation were sorted out. The orthogonal test data are shown in Table 2.

Table 2. Orthogonal test data table.

Experiment number	Die forging temperature (°C)	Mold temperature (°C)	Die forging speed (mm/s)	Mold wear depth (mm)
1	480	100	300	0.021376
2	480	150	400	0.020650
3	480	200	500	0.021107
4	480	250	600	0.021564
5	500	100	400	0.021094
6	500	150	500	0.020170
7	500	200	600	0.020281
8	500	250	300	0.020149
9	520	100	500	0.020599
10	520	150	600	0.020264
11	520	200	300	0.019484
12	520	250	400	0.020027
13	540	100	600	0.019433
14	540	150	300	0.018729
15	540	200	400	0.019143
16	540	250	500	0.018808

After fitting the test results, the fitting curve was obtained, as shown in Figure 2. The fitting curve equation of this orthogonal test is $y = 5.1935\times10^{-8}x^5 - 2.213\times10^{-6}x^4 + 3.439\times10^{-5}x^3 - 2.418\times10^{-5}x^2 + 6.071\times10^{-4}x + 0.021$. When using the fitting formula, the sum of squared residuals is 1.805×10^{-5}, and the R^2 value is 0.76626.

According to the test results, the wear amount of the mold in Group 14 is the smallest, which is 0.01873mm. Compared with the test result in Group 4, the wear amount of the mold is reduced by 11.04%. Therefore, the optimum solid state forging process parameters of this test are die forging temperature 540°C, mold temperature 150°C, and die forging speed 300 mm/s.

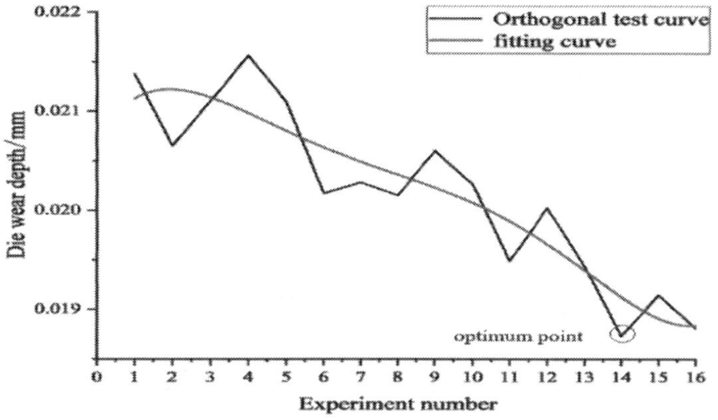

Figure 2. Orthogonal test curve and fitting curve.

The orthogonal test results were visually analyzed. The analysis results are shown in Table 3. The results show that the die forging temperature has a greater influence on mold wear during the solid-state forging process, far greater than the mold temperature and die forging speed, which has little influence on mold wear.

Table 3. Orthogonal test range analysis data table.

Factor	Level	Die forging temperature	Mold temperature	Die forging speed
K	1	0.084699	0.082504	0.079741
	2	0.081697	0.079814	0.080917
	3	0.080376	0.080018	0.080686
	4	0.076116	0.080551	0.081544
Kavg	1	0.021175	0.020626	0.019935
	2	0.020424	0.019954	0.020229
	3	0.020094	0.020138	0.020171
	4	0.019029	0.020138	0.020386
R		0.002146	0.000672	0.000451

4. Conclusions

In this paper, the influence of various process parameters on the wear of the die in the solid-state forging process is studied by orthogonal test. The optimum process parameters of high-strength ball head hanging ring solid die forging and the influence level of parameters on wear were studied, which provided guidance for the production enterprises to determine the production process parameters and saved the manpower and material resources of the production enterprises. The following conclusions are obtained:

(1) The process parameters of solid-state forging with minimum mold wear are obtained by orthogonal test: die forging temperature 540°C, mold temperature 150°C, and die forging speed 300 mm/s. Under this parameter combination, the mold wear is the smallest, only 0.01873mm. The wear is reduced by 11.04% compared with the result of Orthogonal Test 4.

(2) The range analysis method is used to analyze the obtained test data. It is found that die forging temperature has the greatest influence on mold wear during solid-state forging, while mold temperature and die forging speed have little influence on mold wear.

Acknowledgements

The authors would like to acknowledge the financial support provided for this research work by fund project: Science and Technology Project of Guangxi Power Grid Co., Ltd. (GXKJXM20220064).

References

[1] Zhong L, " "12th Five-Year Plan" China's power grid scale ranks first in the world". Guangxi Electricity Power, (10)4-5(2015).

[2] Zhou X Q, "Development history, planning and implementation of China's "West-east power transmission". Power System Technology, 27(5), 1-5+36. (2003).

[3] Zou G L, Cong H X, and Lü Q G, "Development and application of overhead transmission line hardware technology in China". Electric Power Engineering Technology, 31(06), 82-84. (2012).

[4] Li B J, Wang Z H, Tan L, Yang J T, and Qin Z H. "Finite element analysis of influence of hot forging process parameters on die wear", Die & Mould Industry, 40(9), 6-11. (2014).

[5] Kang J H, Park I W, Jae J S, and Kang S S. "A Study on Die Wear Model considering Thermal Softening(I): Construction of Wear Model". Journal of Materials Processing Technology, 96(1/3), 53-58. (1998).

[6] Wang Q W, Li Z P, and Li J. "Analysis of hard seal wear life based on Archard theory". Fluid Machinery, 049(011), 86-91. (2021).

[7] Lin G Y, Feng D, Zheng X Y and Yang W, Sun L P. "Analysis of the influence of extrusion times on die wear based on Archard theory". Journal of Central South University (Science and Technology), 40(5), 1245-1251. (2009).
[8] Jahanmir S and Suh N P. "The delamination theory of wear and the wear of a composite surface". Wear, 32(1), 33-49. (1975).
[9] Lee R S and Jou J L. "Application of numerical simulation for wear analysis of warm forging die". Journal of Materials Processing Tech, 140(1-3), 43-48. (2003).
[10] Cai L G, Liu H D, Cheng Q, Liu Z F, Yang C B, and Qi Y. " Mold Wear During Die Forging Based on Variance Analysis and Prediction of Die Life ". Journal of Beijing University of Technology, 46(1), 1-9 (2020).

Study on the bending mechanical properties of composite tube and metal joint bonding structure

Liang Yan[1], He Xu[2,*], Biao Li[1], Dong Liu[1] and Yongwen Lin[3]

[1]Beijing Institute of Spacecraft System Engineering, Beijing 100094, China
[2]Beijing spacecraft, Beijing 100094, China
[3]Institute of Telecommunication and Navigation Satellite, Beijing 100094, China

[*]Corresponding author's e-mail: 103354928@qq.com

Abstract: The study of the mechanical properties of composite adhesively bonded joints is a very important issue, which directly affects the safety and reliability of the structure. This article studies the bending performance of carbon fiber composite tubes and metal joint bonding structures, as well as obtains the structural bearing capacity, influencing factors, and structural failure modes. Through mechanical analysis and static bending load tests, it is shown that the failure model of composite tube adhesively bonded metal joints under bending load is manifested as the first layer peeling or interlayer failure of the composite material in the bonding area between the tube and the joint, leading to local crushing at the end of the tube. There is no gap between the tube and the joint, which has little effect on the bending bearing capacity. The thickness gradient treatment at the end of the joint does not help improve the bending bearing capacity. The adhesive layer is at a 90 ° angle, which makes it easy for the first layer to peel off; it should be avoided. Other angle layers have little impact on load-bearing capacity. The longer the bonding length is, the stronger the bending capacity is. The adhesive J133 or 420 has a similar load-bearing. Wrapping a carbon fiber unidirectional layer around the ends of tubes and joints is helpful in improving their bending load-bearing capacity.

1. Preface

Adhesive bonding is the most widely used form of connection in composite structures [1-3]. It has the characteristics of uniform stress distribution, good fatigue and corrosion resistance, a lightweight, simple bonding process that can shorten the production cycle, and suitability for connecting components of different materials, thicknesses, ultra-thin specifications, and complex configurations. It has become an indispensable connection technology in aerospace technology. Practice has shown [4-5] that adhesive bonding is an ideal connection method for composite material structures.

Composite structures often fail at the joints. Due to their much more complex damage and failure mechanisms compared to traditional metal structures, the mechanical performance analysis of composite bonding structures and their reinforcement structures is still incomplete.

As shown in Figure 1, the truss structure adopts the form of composite material tube fittings bonded with metal joints. In the complex force transmission path of the structure, bending loads are often accompanied. This paper studies the bending performance of the composite material tube fittings metal joint connection and obtains the structural bearing capacity, influencing factors, and structural failure modes.

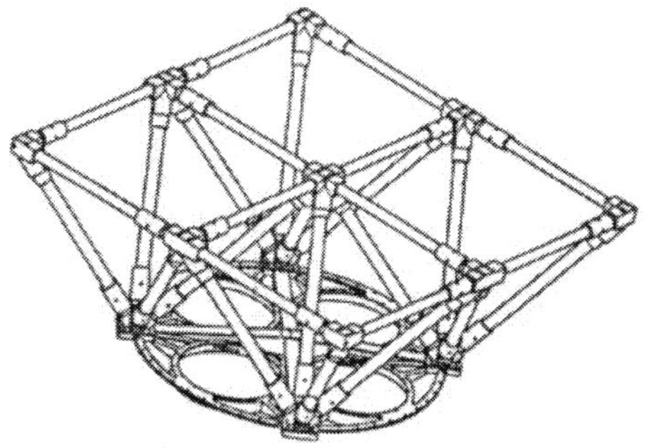

Figure 1. Schematic diagram of truss structure.

There are many studies on the tensile performance of composite tube and metal joint adhesive structures, but few on their bending performance. This article uses mechanical analysis and experimental data to obtain the failure mode of carbon fiber composite tube and metal joint adhesive structures under bending load and the influence of factors such as bonding length and layer angle of composite tube on bending performance.

2. Test piece design

As shown in Figure 2, the test piece of composite tube with an adhesively bonded metal joints structure adopts an M40J carbon fiber/epoxy resin winding composite material tube with an outer diameter of φ 15mm and a wall thickness of 1.5mm.

As shown in Figure 3, the metal joint is made of titanium alloy material. The wall thickness of the joint bonding part is 1.5mm.

The overall size and length of the test piece are 650mm.

We consider several influencing factors on the bending load of the specimen:

- The gap between the tube and the joint, mainly considering whether there will be compression deformation between the stepped end face of the joint and the composite tube, which will affect the bending load of the structure.
- The stiffness of the metal joint end in composite tube fittings mainly considers whether the metal joint end of the specimen will crush the composite tube due to excessive stiffness under bending load.
- The effect of the layer angle of the composite tube.
- The effect of different bonding lengths.
- The effect of adhesives.
- The effect of reinforcing the unidirectional layer wrapped around the end region on improving the load-bearing capacity.

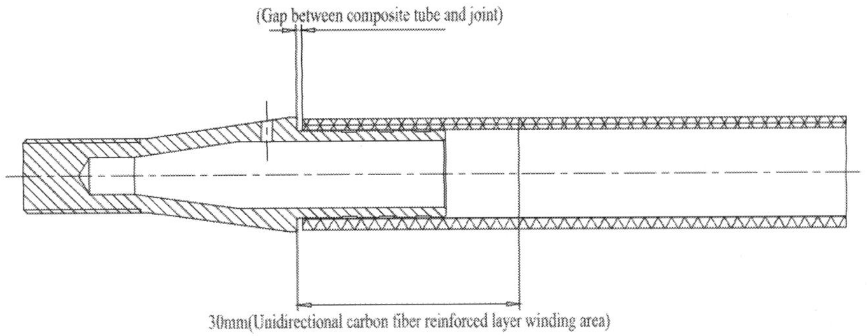

Figure 2. Sectional view of composite tube and metal joint.

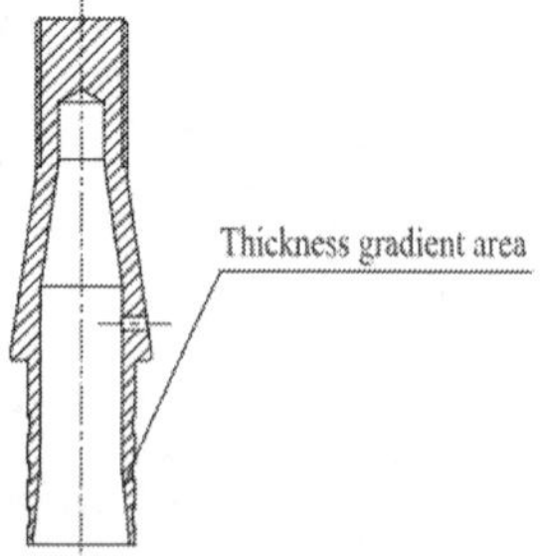

Figure 3. Schematic diagram of thickness gradient at the end of the joint.

Based on the analysis of influencing factors, the test piece status has been formulated, as shown in the table 1 below.

Table 1. Test piece status.

Test piece number	Test piece status
Comparative Test Piece 1	The gap between the tube and the joint is 0 or 1mm
Comparative Test Piece 2	The thickness gradient is applied at the end of the joint or not
Comparative Test Piece 3	The layer design is $[\pm30°]_5$, $[\pm45°]_5$, $[\pm60°]_5$, $90°/[\pm45°]_4/90°$, respectively
Comparative Test Piece 4	The bonding length is 20mm, 40mm, and 60mm, respectively
Comparative Test Piece 5	The adhesive uses J133, 420, respectively
Comparative Test Piece 6	The gap between the tube and the joint is 1mm, and two unidirectional layers of carbon fiber are wrapped around the 30mm range of the end or not

3. Structural mechanics analysis

Carbon fiber composite tube fittings use shell elements, and joints use solid elements.

The boundary condition is that the metal joint is fixedly supported at one end of the fixture. We apply forced displacement at the end of the tube fittings and observe the stress state of the structure.

The stress cloud map is shown in Figure 4, it can be seen that the maximum stress occurs at the root of the composite tube and the corresponding area at the joint end of the composite tube.

This analysis aims to identify the weak links of structural components under bending loads through finite element analysis. The stress values are for reference only.

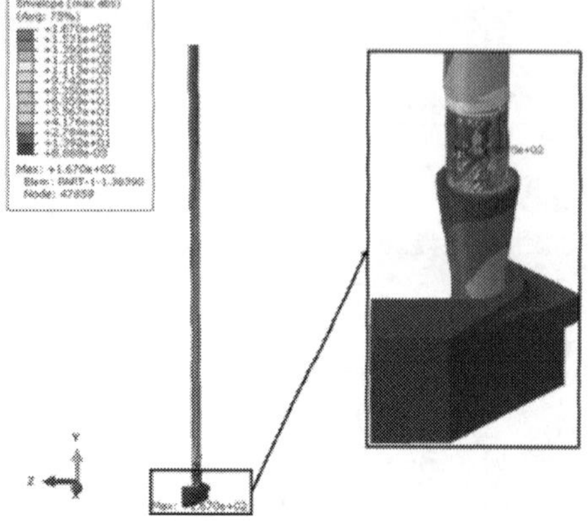

Figure 4. Stress cloud map of composite tube bonded metal joint under bending load.

4. Structural testing verification

We obtain the bending bearing capacity of the test piece. We conduct full-scale bending strength and deformation tests on tube fittings. Testing failure load, maximum displacement during failure, and studying failure mode can provide a more accurate and comprehensive evaluation of the overall performance of the composite tube.

The bending test involves fixing one end of the tube fitting and applying a load P to the other end until strength failure occurs. Loading and strain/displacement measurement point status is shown in Figure 5.

We record the failure load and maximum displacement of the tube end during the test.

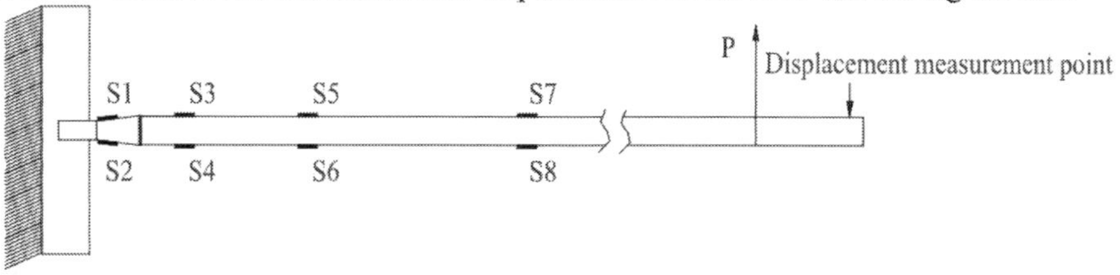

Figure 5. Loading and strain/displacement measurement point status of bending test piece.

The bending load loading method is as follows: continuous loading, recording strain, and displacement data until failure, and recording the failure mode and failure load. Clamping state of the test piece is shown in Figure 6.

Figure 6. Clamping state of the test piece.

4.1. Test result

The destructive load and the displacement during destruction in the static bending test are shown in Table 2~7.

Table 2. Static test data for bending load capacity (Comparative test piece 1).

Test piece status	Destructive load/N (Average value)	Destructive load/N •m (Average value)	Displacement during destruction/mm (Average value)
The gap between the tube and the joint is 0	13.5	78.7	68.5
The gap between the tube and the joint is 1mm (Basic type)	13.0	75.8	67.0

From the table above, it can be concluded that there is no gap between the tube and the joint, which has little effect on the bending bearing capacity.

Table 3. Static test data for bending load capacity (Comparative test piece 2).

Test piece status	Destructive load/N (Average value)	Destructive load/N •m (Average value)	Displacement during destruction/mm (Average value)
The thickness gradient is not applied at the end of the joint (Basic type)	13.0	75.8	67.0
The thickness gradient is applied at the end of the joint	12.0	70.0	61.2

From the table above, it can be concluded that the thickness gradient treatment at the end of the joint does not help improve the bending bearing capacity.

Table 4. Static test data for bending load capacity (Comparative test piece 3).

Test piece status	Destructive load/N (Average value)	Destructive load/N •m (Average value)	Displacement during destruction/mm (Average value)
The layers are $[\pm 30°]_5$(Basic type)	13.0	75.8	67.0
The layers are $[\pm 45°]_5$	13.5	78.71	67.6
The layers are $[\pm 60°]_5$	12.8	74.62	63.3
The layers are $90°/[\pm 45°]_4/90°$	11.2	65.30	59.5

From the table above, it can be concluded that the adhesive layer is at a 90° angle, which makes it easy for the first layer to peel off; it should be avoided. Other angle layers have little impact on load-bearing capacity.

Table 5. Static test data for bending load capacity (Comparative test piece 4).

Test piece status	Destructive load/N (Average value)	Destructive load/N •m (Average value)	Displacement during destruction/mm (Average value)
The bonding length is 20mm (Basic type)	13.0	75.8	67.0
The bonding length is 40mm	15.5	90.36	75.5
The bonding length is 60mm	17.0	99.11	83.0

From the table above, it can be concluded that the longer the bonding length is, the stronger the bending capacity is.

Table 6. Static test data for bending load capacity (Comparative test piece 5).

Test piece status	Destructive load/N (Average value)	Destructive load/N •m (Average value)	Displacement during destruction/mm (Average value)
The adhesive is J133 (Basic type)	13.0	75.8	67.0
The adhesive is 420	13.2	76.96	67.8

From the table above, it can be concluded that the adhesive J133 or 420 has similar load-bearing properties.

Table 7. Static test data for bending load capacity (Comparative test piece 6)

Test piece status	Destructive load/N (Average value)	Destructive load/N •m (Average value)	Displacement during destruction/mm (Average value)
No carbon fiber wrapped (Basic type)	13.0	75.8	67.0
Two unidirectional layers of carbon fiber are wrapped around the 30mm range of the end	14.5	84.5	71.0

From the table above, it can be concluded that wrapping two layers of unidirectional cloth within a 30mm range of the tube fittings and joint ends can help improve the bending bearing capacity.

4.2. Destruction mode

The typical failure mode of the test specimen is shown in Figure 7. According to the experimental results, the specimen's failure mode is a mixed failure, including interface failure, fiber tearing failure, and cohesive failure. Fiber tearing failure is manifested as the first layer peeling or interlayer failure of the composite material in the junction area between the tube and the joint, which then leads to local crushing at the end of the tube.

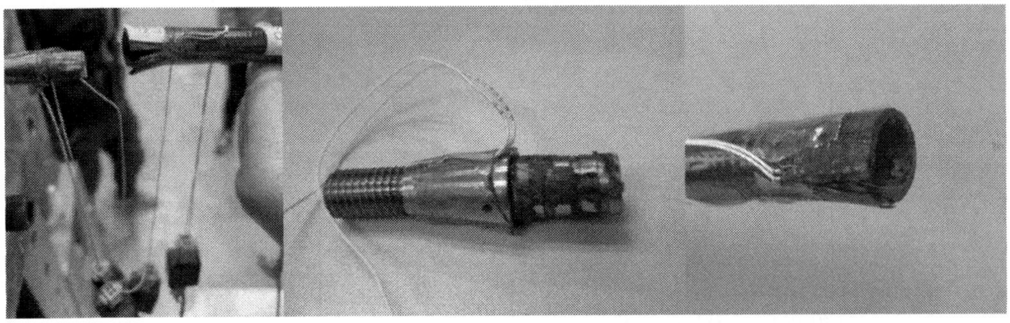

Figure 7. Failure mode of bending load static test.

The load strain curve curve of the static bending test is shown in Figure 8.

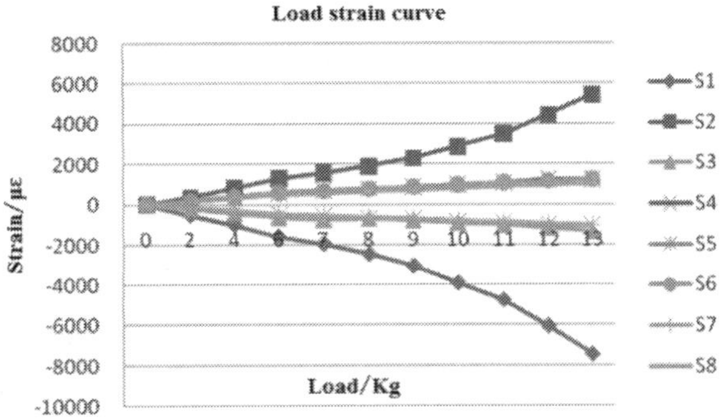

Figure 8. Strain load curve of bending load static test (Basic type).

Journal of Physics: Conference Series **2879** (2024) 012040 doi:10.1088/1742-6596/2879/1/012040

5. Conclusion

Based on the above analysis and experiments, the following conclusions can be drawn:

- The failure model of composite tube fittings bonded to metal joints under bending load is characterized by the peeling of the first layer of composite material or interlayer damage in the bonding area between the tube fittings and the joint. This leads to local crushing at the end of the tube fittings.
- There is no gap between the tube and the joint, which has little effect on the bending bearing capacity.
- The thickness gradient treatment at the end of the joint does not help improve the bending bearing capacity.
- The adhesive layer is at a 90° angle, which makes it easy for the first layer to peel off; it should be avoided. Other angle layers have little impact on load-bearing capacity.
- The longer the bonding length is, the stronger the bending capacity is.
- The adhesive J133 or 420 has a similar load-bearing.
- Wrapping two layers of unidirectional cloth within a 30mm range of the tube fittings and joint ends can help improve the bending bearing capacity.

References

[1] Tuo Hongliang, Lu Zhixian, Ma Xiaoping, Guo Hongyu. Study on delamination damage evolution of composite L-shaped adhesive joint based on cohesive behavior [J]. Northwestern polytechnical university, 2021, 39(2): 309-316

[2] Mao Zhengang, Hou Yuliang, Li Cheng.Tie Ying, Sun Liang Effect of lap length and stacking sequence on strength and damage behaviors of adhesively bonded CFRP composite laminates [J]. Acta Materiae Compositae Sinica, 2020, 37(1): 121-131

[3] Jinlu, Zhang Feiyang, Tian Dake Thermal-structure analysis of modular space deployable antenna support truss structures [J] China Space Science and Technology 2022, 42 (5): 78-90

[4] Zou Tianchun, Li Longhui, Fu Ji, Li Ye, Ju Yuezhang Effect of fiber layup methods on tensile properties of CFRP-aluminum alloy single-lap adhesive joint, Journal of Aerospace Power, 2022 37(9): 1905-1914

[5] Zou Tianchun, Ju Yuezhang, Fu Ji, Liu Zhihao, Li Ye Fatigue life and failure mode of CFRP-aluminum alloy single-lap adhesive joints, Acta Materiae Compositae Sinica, 2022 39(12): 6078-6087

MATMA-2024 IOP Publishing
Journal of Physics: Conference Series **2879** (2024) 012041 doi:10.1088/1742-6596/2879/1/012041

Study on the suppression of secondary electron emission characteristics based on TiN

Guanghui Miao[1,2,*], Penghui Duan[1], Xiongfei Li[1] and Wanzhao Cui[1]

[1]China Academy of Space Technology (Xi'an), Shaanxi, 710100, China
[2]College of Aeronautics Nanjing University of Aeronautics and Astronautics, Nanjing, 210016, China

[*]Corresponding author's e-mail: miao_guanghui@163.com

Abstract: With the continuous development of satellite communications, the problem of high-frequency high-power multipactor caused by the miniaturization of microwave passive components in the aerospace field has become more and more serious. In this paper, for the problem of multipactor suppression of load passive devices, we carried out research on the preparation of TiN thin films based on PEALD and put forward a new method for suppressing SEE with good suppression effect, high consistency, low surface resistance, and other aspects of the unity of the multipactor effect. The use of 20nm TiN coating can make the silver-plated surface secondary electron emission coefficient down to 1.59. The RF surface resistance change rate is less than 3%. In the field of communication, satellite and navigation and other fields have a wide range of application prospects.

1. Introduction

Multifactor phenomenon refers to the creation of secondary electrons on a solid surface due to the interaction of free electrons with a radiofrequency field in a vacuum environment, leading to the multiplication of the breakdown effect. In recent years, the development of satellite communications tends to be higher capacity and greater flexibility, thus requiring more and more miniaturization and integration of microwave passive components in the field of load [1]. As a result, the problem of multifactor in high-frequency and high-power microwave components has become more and more serious. It is urgent to improve the multifactor threshold and find new techniques to suppress secondary electron emission characteristics [2].

At present, there are two main trends in the research of multifactor suppression technology for material surfaces. One is surface roughening treatment, i.e., the formation of deeper hole arrays on the surface of the component, creating "traps" to inhibit the emission of secondary electrons, thus suppressing the multifactor effect. However, the introduction of the "trap" structure will lead to the deterioration of the surface properties, especially the degradation of the electrical properties of high-frequency devices and the increase in insertion loss, affecting the normal operation of the device. The other is the surface coating treatment, that is, the surface of the substrate directly on the surface of the low SEY material thin film coating, which can inhibit the micro-discharge effect at the same time, effectively avoiding the deterioration of the surface electrical properties of the components. Researchers conducted a large number of experiments to screen out the commonly used low SEY coating materials, which are titanium nitride (TiN), graphene, organic small molecule materials, carbon, and Alodine [3-5]. TiN is an ideal plating material to inhibit secondary electron multiplication as the surface of microwave passive devices due to its low secondary electron emission coefficient, low loss, low gas release rate, dielectric, not easy to oxidize, harmless, etc. It can maintain high stability under high-frequency fields and high temperatures. Atomic layer deposition (ALD) of TiN film has excellent three-dimensional conformality, large area uniformity and precise sub-single layer thickness control, etc. It can be directly used for component-level nano-size coating, and the engineering application can realize high [6].

In this paper, to address the above problems, we first carry out the preparation of TiN films based on plasma-enhanced atomic layer deposition (PEALD); subsequently, we carry out the study of the secondary electron emission characteristics of the silver-plated surface of aluminum alloys, and screen out the optimal thickness of the surface plating of the components; finally, a large area of TiN was coated on a silver-plated microwave proof part with good uniformity.

2. Preparation of TiN thin films

Atomic layer deposition technology, a special chemical vapor deposition technique, is a method of forming thin films by alternating the pulse of gas precursor through the reaction chamber and the chemical adsorption reaction on the surface of the deposited substrate. The proposed ALD equipment for the experiment is the Elegant II-Y200 plasma-enhanced atomic layer deposition system produced by Nanjing Yuanlei Nano Co. In order to obtain high-purity and high-quality TiN thin films, high-purity (99.9999%) titanium tetrachloride was used as the Ti precursor source, and high-purity (99.9999%) ammonia (NH3) plasma as the N precursor plasma as the N precursor source to prepare TiN films. The carrier gas was high-purity (99.999%) nitrogen. During the reaction process, the titanium precursor source and ammonia plasma alternately enter the reaction chamber. The chemical reaction occurs on the surface to produce TiN films.

It was shown that the carrier gas concentration, cavity temperature, carrier gas flow rate, plasma power and pulse time during the whole reaction process affect the quality of film growth. In order to obtain highly stable, large-area, high-quality TiN film layers on the surface of microwave components, this experiment was carried out to study the effects of argon gas flow, reaction temperature, and plasma power on the growth thickness per cycle (GPC), uniformity of film formation and conductivity.

2.1. Argon flow rate on film quality

A 6-inch single-throw <100> crystal-oriented P-type silicon wafer was selected as the reaction substrate material for the experiment. The temperature of the reaction chamber was set at 250 degrees Celsius at first. The wafer was fed into the chamber after preheating for 600 s at the end of the temperature rise. The carrier gas N_2 pressure was 0.3Mpa, followed by the carrier gas pulsed 22 °C gas precursor titanium tetrachloride for 0.3 seconds; at this time, the TiCl4 pulse pressure was 1.5Torr, followed by the entire cavity for 8 seconds 300sccm nitrogen purge. Subsequently, the Plasma power was set at 2000W, and 150sccm of ammonia was passed into the plasma glow reaction under the action of 60/70/90/100sccm of argon (99.999%) into the reaction chamber, respectively, with the plasma working for 15 seconds. Then 10 seconds of blowing was carried out, and so on for 200 cycles. Finally, ellipsometry was used to characterize the thickness, homogeneity, and conductivity per cycle.

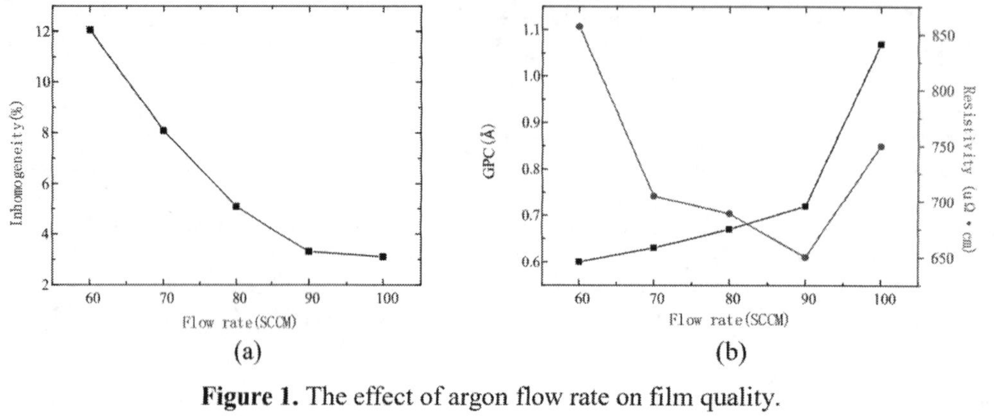

Figure 1. The effect of argon flow rate on film quality.

As shown in Figure 1(a), with the argon flow rate from 60sccm to 90sccm change, the unevenness of TiN grew gradually from 12.06% to 3.31%, argon flow rate and then increased to 100sccm, the unevenness reached 3.11%. This is due to the fact that when the argon flow rate is below 90sccm, the ammonia may not be able to fill the whole cavity and react completely with the substrate. At the same time, it tends to be saturated at 100sccm, which is able to ensure the uniformity of the reaction of a large area very well.

In addition, the relationship between the thickness of the film growth per cycle and the conductivity with the increase of argon flow rate can be seen in Figure 1(b). With the increase of argon flow rate, the thickness of the film growth per cycle increases sequentially. The resistivity generally decreases, but then increases at 100 sccm. This is because the growth rate is too fast, which leads to the loosening of the film's structure and an increase in resistivity.

2.2. Temperature on film quality

In recent experiments, it was observed that the growth rate of the TiN film is unstable and has poor resistivity. To address this issue, a literature review was conducted to find ways to improve the reaction temperature in order to increase the nitriding degree. This, in turn, would stabilize the film composition and increase the thickness of the film deposition, leading to more reliable test results. We selected a 6-inch single throw <100> crystal direction of a P-type silicon wafer as the reaction substrate material. We set the reaction chamber temperature to 250/300/350°C and the Plasma power to 2000W. We used argon (99.999%) at 100 sccm for 300 cycles. Then, ellipsometry was used to carry out the measurement of the thickness, homogeneity, and conductivity of the plasma for each cycle. Thickness, homogeneity and conductivity were characterized.

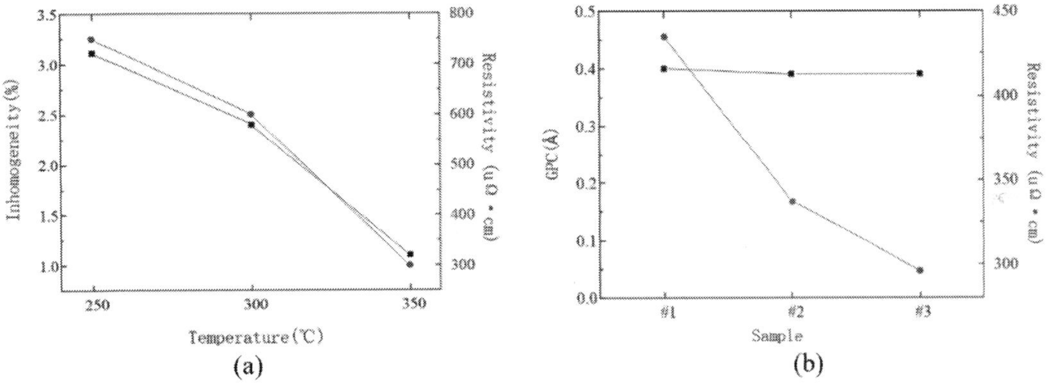

Figure 2. The quality of film growth at different temperatures.

As can be seen from Figure 2(a), the temperature has a great influence on the quality of the film, and as the temperature increases, the film inhomogeneity decreases sequentially. At 350°C, it has reached about 1%. With the increase in temperature, the resistivity also decreases sequentially, and it can reach 300 uΩ·cm at 350°C.

However, as can be seen from Figure 2(b) below, with the increase in temperature, it turns out that the growth rate can be stabilized at 350°C and the growth rate can be stabilized at 0.4Å. It is closed to the relevant literature data. Different batches of TiN growth, growth rate, and inhomogeneity have been basically stable. Still, its resistivity is not stable, suspected to be the lower activity of nitrogen ions or the higher oxygen content in the cavity leads to the higher oxygen content in the film of TiN. Subsequent XPS analyses were performed for samples at different temperatures. The results proved that the ratio of surface O molecules in the TiN films of different samples was not stable at 350°C and Plasma 2000W. The TiO_2 generated by O and Ti atoms affected the quality of TiN films to a certain extent. Especially for the same sample, the ratio of Ti atoms and N atoms at different positions is also between 1.1 and 1.5.

2.3. Plamsa power on film quality

In order to control the oxygen content in the cavity, the study of the effect of Plamsa power on the quality of the film was carried out. For the experiment, 6-inch single-throw <100> crystal-oriented P-type silicon wafers were selected as the reaction substrate material. The temperature of the reaction chamber was set at 350°C. The Plasma power was set to 2000W/2500W/3000W. 150sccm of ammonia was passed into the plasma glow reaction under the action of 100sccm of argon (99.999%) into the reaction cavity. After 300 cycles, the thickness, homogeneity, and conductivity were characterized by ellipsometry for each cycle.

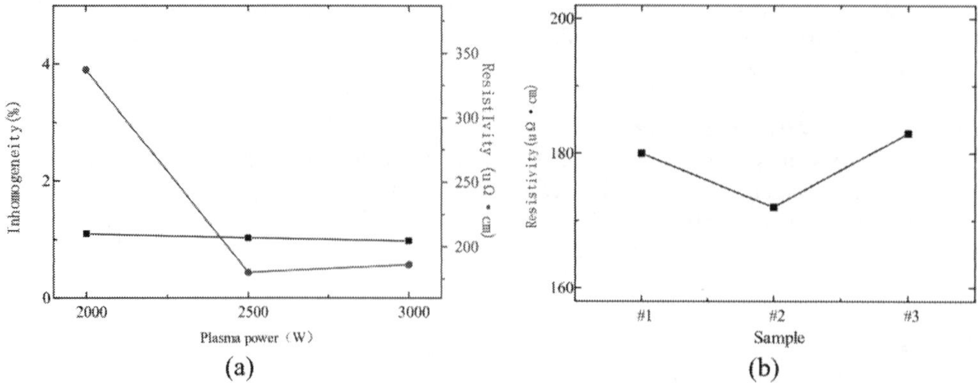

Figure 3. Effect of plasma power on film quality.

As can be seen in Figure 3(a), with the increase of power, the inhomogeneity stabilizes at about 1%. The conductivity stabilizes close to 180uΩ·cm. It turns out that 2500W of Plasma power can already saturate the conductivity. Subsequently, the process stability test was carried out, using 2500W power for a total of three batches of deposition. The results are shown in Figure 3(b). Compared with the conductive silicon 1000uΩ·cm, different batches of atomic layer deposition of TiN film can be stabilized at 180uΩ·cm, which can reach the international level. Finally, XPS was used for compositional analysis. The three batches of samples prepared under Plasma2500W conditions showed that the surface C, O, Ti and N atoms were stabilized. The ratio of Ti and N tended to be 1~1.1. It is worth noting that in the subsequent sample preparation process, more preheating time can be taken to minimize the oxygen content of the reaction chamber. At the same time, the NH3 pulse time can be increased to make the sample surface fully react to improve the ratio of Ti and N.

Finalized process: temperature 350°C, Plasma power 2500W, ammonia action time 17s, and argon carrier gas flow 100sccm.

3. Secondary electron emission properties of TiN films

Based on the optimal process conditions, a 12cm x 12cm x 1cm specimen with a surface roughness of less than 0.8um, consisting of an aluminum alloy surface coated with 7um of silver, was utilized to prepare samples for atomic layer deposition of TiN coating. The aim was to investigate the inhibitory effect of TiN films with varying nanometer thicknesses on the characteristics of secondary electron emission. The samples were ultrasonically cleaned with acetone and anhydrous ethanol before coating the thin films. After drying, the samples with different thicknesses of 5nm, 10nm, 15nm, 20nm, and 30nm were prepared, respectively.

For the measurement of the secondary electron emission yield of TiN nanofilms on silver surfaces, the tests were carried out using a self-developed secondary electron emission characterization test platform with a system vacuum below 10^{-8} Torr and incident energy of 20-3000eV [7]. The bias current method is used for testing. Firstly, +500V bias voltage is added to the sample. The incident electrons bombard the surface of the sample, and the measured current value can be approximated as the current of the incident electrons; subsequently, a -20V bias voltage is added to the sample. At this time, the current value measured through the ammeter should be the difference between the current of the incident electrons and that of the true secondary electrons. The following Formula (1) can be used to calculate the secondary electron emission coefficient of the measured sample. δ is the secondary electron emission coefficient; I_R is the difference between the value of the incident current on the sample surface and the current value of the secondary electrons; I_0 is the value of the incident current on the sample surface.

$$\delta = 1 - \frac{I_R}{I_0} \tag{1}$$

Firstly, the same sample with 10nm and 20nm silver-based TiN-plated samples were used for the SEY uniformity test at different locations, respectively. The inhomogeneity of the secondary electron emission coefficient at different locations of the same sample after TiN coating is less than 5%. The samples have good uniformity and stability.

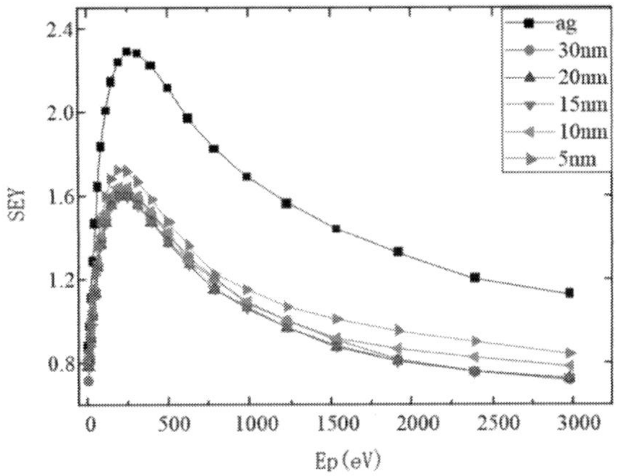

Figure 4. SEY of TiN films with different nanometer thicknesses.

Subsequently, the secondary electron emission coefficient of TiN films with different nanometer thicknesses was tested. The experimental results are shown in Figure 4. Comparing the SEY curves of silver plating on the metal surfaces of commonly used microwave components, the SEY decreased significantly after plating TiN. With the increase of the thickness of TiN films, the secondary electron emission coefficient gradually decreased and tended to be saturated at 15nm; the secondary electron emission coefficient stabilized at about 1.6. The atomic layer deposition of TiN film compared with the original silver surface showed that the secondary electron emission coefficient decreased from 2.3 to 1.59, a decrease of 31%.

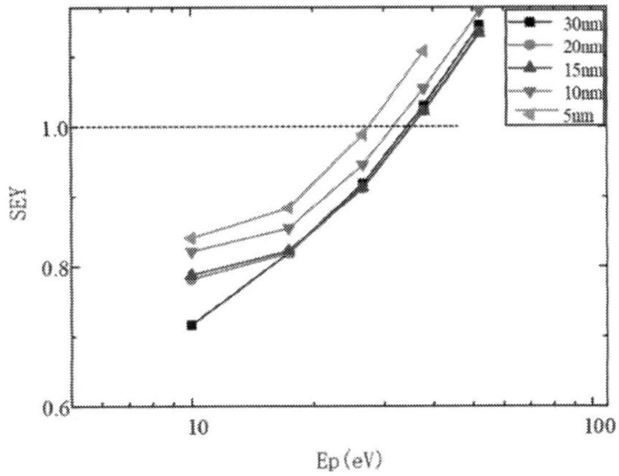

Figure 5. E1 of TiN films with different nanometer thicknesses.

Meanwhile, we pay more attention to the effect of the E1 point than to the effect of the maximum value of the secondary electron emission coefficient on the multifactor threshold. Figure 5 shows graphs of low-energy secondary electron coefficients at different thicknesses. As can be seen from the figure, as the film thickness increases from 0 to 30 nm, E1 gradually increases from 18.7 eV to 35.3 eV.

When the film thickness is 15nm, E1 is 34.6eV, which tends to be saturated. The SEY is 1.6, which, side by side, confirms that the optimal thickness of the film inhibition is 15nm. However, in consideration of guaranteeing the reliability of the aerospace engineering experiments, the subsequent microwave components surface large-area atomic layer deposition is proposed to be used in 20nm TiN film.

Then, for the problem of multifactor between parallel plates of high-power microwave components, the multifactor threshold is generally inversely proportional to the square of the incident energy at the E1 point, and the ALD-coated TiN thin film will significantly improve the multifactor threshold of microwave passive components. After processing two sets of Ku-band impedance converters, we conducted a test to see how applying a 7um layer of silver plating followed by a 20nm ALD coating of TiN thin film to one of the sets would impact the suppression of multiple factors. As a result of this treatment, the multifactor threshold increased from the original 2000W to 5000W. The experiment proved that the atomic layer deposition of TiN on the surface of microwave components inhibits multifactor problems, but it also solved the engineering application of large-area film uniformity and bypassed the problem.

4. Conclusion

In view of the new requirements for high-power microwave components in subsequent space engineering missions, there is an urgent need for continuous and in-depth research on secondary electron emission suppression technology for microwave materials for aerospace applications. Combined with the diversity of suppression objects, a new method for suppression of the secondary electron emission characteristics, which is harmonized in many aspects such as good suppression effect, high consistency, and low surface resistance, has been provided by taking advantage of cross-discipline and new material properties in materials science. The use of 20nm TiN coating can make the silver-plated surface secondary electron emission coefficient down to 1.59. The RF surface resistance change rate is less than 3%. The process method has good scientific research value and engineering applicability, provides important technical reserve and support for the continuous improvement of high-power microwave performance, and ensures that there are technical options and technologies available for the subsequent space engineering missions to suppress the multifactor effect of high-power microwave components.

References

[1] Zhou Y. et al. “An overview of flexible payload technologies for communication satellites.” Space Electronics Technology 20. 3(2023): 29-38.

[2] Kazemi F, Mostajeran M, Romanov G. The influence of secondary electron yield uncertainty on the single-sided multipacting in dielectrics [J]. The European Physical Journal Plus, 2023, 138. DOI: 10.1140/epjp/s13360-023-04124-9.

[3] D. Wang, Y. N. He, M. Ye, et al. Secondary electron emission characteristics of nanostructured silver surfaces. Journal of Applied Physics, 122, 153302(2017).

[4] J. M. Sattler, R. A. Coutu, R. Lake, et al. Modeling micro-porous surfaces for secondary electron emission control to suppress multipactor. Journal of Applied Physics, 122, 055304(2017).

[5] Yuting Zhang, Jing Yang, Zhaolun Yang, Guanghui Miao, Na Zhang, & Wanzhao Cui. (2024). Secondary electron emission characterization of molybdenum. Rare Metal Materials and Engineering (002), 053.

[6] Nguyen V H, Akbari M, Sekkat A, et al. Atmospheric atomic layer deposition of SnO2 thin films with tin (II) acetylacetonate and water [J]. Dalton Transactions, 2022, 51. DOI:10.1039/D2DT01427K.

[7] He, Y., et al. “High-performance multifunctional dielectric secondary electron emission characterization platform.” Intense Laser and Particle Beam 32. 3(2020): 5.

Anti-icing performance of SLIPSs injected with different lubricants

Bo Li[1], Jie Bai[1], Zengxin Pu[1], Bo Liu[2], Xujiang Hua[3], Tao Zhu[3], Xu Dai[4], Huiying Xiang[3] and Yuan Yuan[3,*]

[1]Electric Power Research Institute of Guizhou Power Grid Co., Ltd., Guiyang 550002, China
[2]Guiyang Power Supply Bureau of Guizhou Power Grid Co., Ltd., Guiyang 550005, China
[3]College of Materials Science and Engineering, Chongqing University, Chongqing 400044, China
[4]State Key Laboratory of Power Transmission Equipment & System Security and New Technology, Chongqing University, Chongqing 400044, China

[*]Corresponding author's e-mail: yuany@cqu.edu.cn

Abstract: Transmission line icing is seriously threatening the power grid to operate safely. Slippery Liquid-infused Porous Surface (SLIPS) is a latent application to solve transmission line ice cover. The matrix structure and lubricant are important components of SLIPS. However, a variety of lubricants have been used to prepare SLIPS, and there is a lack of certain theoretical guidance for selecting lubricants to prepare anti-icing SLIPS. In this paper, the effects of three lubricants on the wettability, anti-icing performance, and anti-icing durability of SLIPS were investigated. It was found that dimethyl silicone oil was more favorable for preparing anti-icing SLIPS. The SLIPS prepared from silicone oil had a complete frosting time of about 10 min longer than that of mineral oil and safflower oil and had the lowest ice adhesion strength of 4.5 ± 0.8 kPa. It also has better anti-icing durability; even after 30 repetitions of icing/deicing, ice adhesion strength is still below 20 kPa.

1. Introduction

Transmission line icing is seriously threatening the power grid to operate safely, affecting people's normal use of electricity and bringing huge economic losses [1]. Slippery Liquid-infused Porous Surface (SLIPS) absorbs and stores slippery liquid through porous structures, forming a smooth surface with low contact angle hysteresis (CAH) of less than 2.5°. There have been reports that SLIPS can retard ice crystal nucleation and the lubricant can hinder ice-substrate contact, leading to a decrease in the adhesion strength between ice and matrix [2]. Therefore, SLIPS is of great research value in the application of anti-icing and a potential technical means to solve transmission conductor icing.

The matrix structure and lubricant of SLIPS are the main factors affecting their anti-icing properties. Many researchers have investigated durable SLIPSs with low deicing force. By modifying the matrix structure, it is possible to reduce the deicing force of SLIPS, increase lubricant storage, and reduce the loss of lubricant. For example, Subramanyam et al. have prepared SLIPS with a microcolumn structure on a silicon surface using laser etching [3]. It was found that the deicing force of SLIPS gradually decreased with the increase of microcolumn density in the range of $0\text{-}50\times10^{8}/m^2$. Ice adhesion strength was significantly affected by microcolumn density. In addition, there are many structures have been studied such as hierarchical micro/nanostructures [4], network structures [5], and dendritic pore structures [6]. Besides the matrix structure, the lubricant is also a crucial factor affecting the anti-icing ability of SLIPS. Commonly used lubricants are dimethyl silicone oil (SO) [7-8], perfluoropolyether oil (PFPE) [9], and liquid paraffin [10]. Sasidharanpillai and Lee [7] found that lubricants significantly affect the durability of SLIPS. The selection of lubricants is of critical importance in the preparation of anti-icing SLIPS, but there is a lack of clear theoretical guidance.

In this paper, an aluminum matrix with a porous structure was prepared by using an anodic oxidation method and SLIPSs were prepared by injecting different kinds of lubricants (dimethyl silicon oil, safflower oil, and mineral oil). The microstructure of anodized aluminum oxide (AAO) was characterized through a scanning electron microscope. The wettability of SLIPSs was characterized by a contact angle test. The frosting test was used to investigate the anti-frost performance of SLIPS in a low-temperature environment. The ice push experiment was used to measure the deicing force of SLIPS, which in turn was used to calculate the ice adhesion strength. Repeated icing and deicing tests were used to characterize the long-term anti-icing ability of SLIPSs.

2. Materials and Methods

2.1. Materials

Shanghai Zhihengda Aluminum was the supplier of pure aluminum sheets. Shanghai Macklin Biochemical Technology Co., Ltd. was the supplier of dimethyl silicone oil (SO) and safflower oil (SFO). N-octadecyl trimethoxy silane (OTS) and phosphoric acid (H_3PO_4) were supplied by Sinopharm Chemical Reagent Co., Ltd., and Mineral oil (MO) was supplied by Changsha Zhixuan Chemical Products Co., Ltd. Yonghua Chemical Co., Ltd. was the supplier of sodium hydroxide (NaOH), which was used to remove the oxide film on Al.

2.2. Fabrication of SLIPS

The aluminum was cut into 2 cm × 3 cm pieces using a slicer. The aluminum sheets were then ultrasonically cleaned in distilled water and ethanol for 5 min, respectively. After the cleaning was completed, the surface oxide film on the Al sheet was removed by putting the Al sheet in NaOH solution and immersing it for 5 minutes. The treated aluminum sheet was oxidized in oxalic acid solution (0.3 M/L) for 15 min with a stainless steel plate as a cathode and an aluminum sheet as an anode to obtain anodized aluminum oxide.

O-AAO was obtained by placing AAO into an ethanol solution of OTS at a concentration of 3 wt.% for 2 h. The O-AAO was dried at 100°C for 3 h. Subsequently, O-AAO was put into different lubricating oils and vacuum immersed for 12 h, thus preparing SLIPSs. According to the type of lubricant, they were named S-SLIPS, F-SLIPS, and M-SLIPS.

2.3. Characterizations

A camera (Canon 200d) was utilized to photograph the macroscopic morphology of the product surface. The surface morphology and sectional morphology of the samples were observed by microscope (SEM, Zeiss Auriga). A contact angle measuring instrument (OCA25, Dataphysics) was used to measure the CA and CAH of the samples. The motion of water droplets on SLIPSs was characterized by a high frame rate camera.

The frost protection performance of the samples was measured by the frosting platform, and the frosting on the sample surface was photographed every 10 min. The ambient temperature was 11.5 ± 0.5°C, the humidity was 60 ± 3%, and the temperature of the cooling platform was -10°C. The deicing force of the samples was measured utilizing the horizontal shear method and the ice adhesion strength. The anti-icing durability of the samples was characterized by repeated icing and deicing tests. The adhesion force between ice and SLIPS of the samples was recorded at five-cycle intervals over 30 icing/deicing times.

3. Results and discussions

3.1. Morphology of the matrix

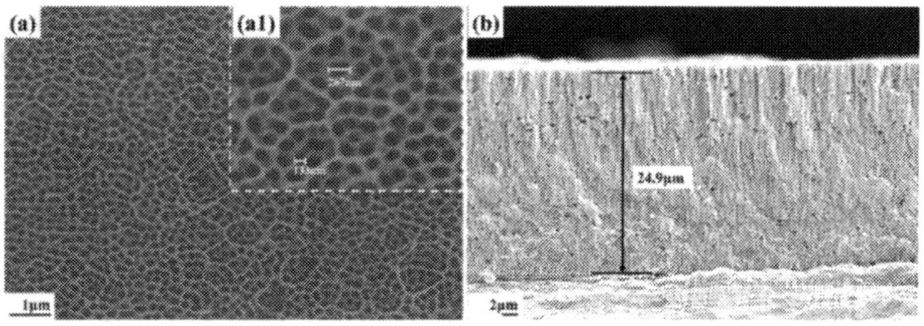

(a) Surface morphology (10, 000×); (a1) Surface morphology (20, 000×); (b) Sectional morphology.

Figure 1. SEM pictures of the surface morphology and sectional morphology of matrixes.

Porous structure is a necessary component for the preparation of SLIPS. In this paper, we used an aluminum sheet as a matrix and constructed a porous structure successfully by anodic oxidation method. Figure 1 shows the surface morphology and cross-section morphology of the substrate. As illustrated in Figure 1 (a), there are plenty of rounded pores existing on the samples, which are more uniformly distributed (the porosity of AAO was 43.7%) and have a pore size between 130 nm and 270 nm. As shown in Figure 1 (b), the surface of AAO is relatively flat, and the thickness of the porous structure is about 24.9 μm. The homogeneous porous structure provides the basis for the stockpile of the lubricant and the preparation of a smooth surface.

3.2. Macroscopic morphology

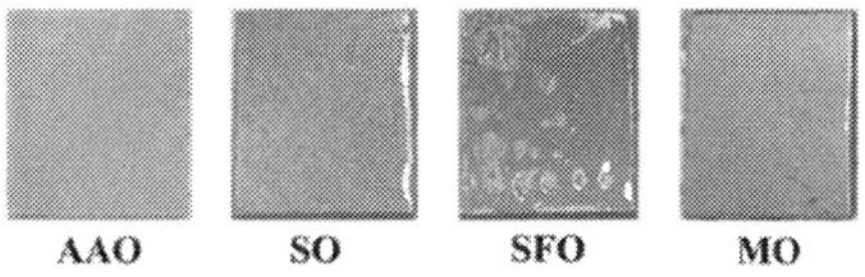

Figure 2. Macroscopic morphology of SLIPSs after injection of different lubricants.

The distribution of the lubricant on the substrate surface is a significant determinant of the anti-icing efficacy of SLIPS. Figure 2 shows that SO and MO can form a continuous and uniform oil layer on AAO, while SFO has poor uniformity on the surface of AAO. There may be two possible reasons for the non-uniformity of the SFO oil film. One is that SFO becomes viscous when placed in the air, and the poor fluidity causes the oil film to be difficult to spread. The other reason could be that SFO is chemically less compatible with the modifier OTS, and the chemical interaction between the lubricant and the modifier affects the spread of the lubricant.

3.3. Wettability & slid ability

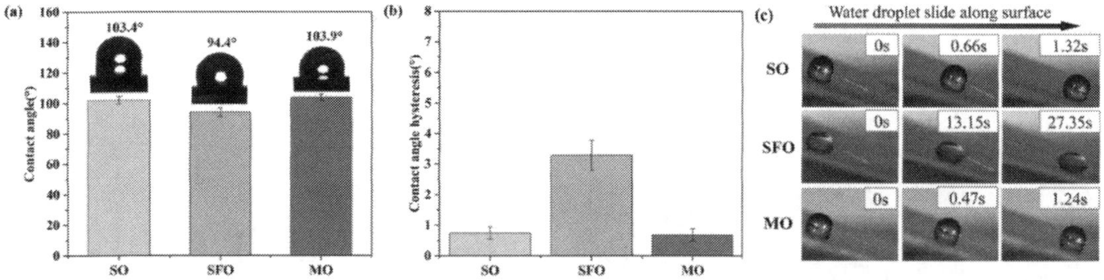

(a) CA; (b) CAH; (c) Water droplet sliding image

Figure 3. The CA, CAH, and water droplet sliding image of different SLIPSs.

CA and CAH are important parameters for characterizing the wettability of SLIPS. The CA and CAH for different SLIPS are shown in Figure 3. The results show that F-SLIPS has the smallest CA (94.4° ± 2.7°), while S-SLIPS and M-SLIPS have close CAs of 103° ± 2.5° and 103° ± 2.1°, respectively. Due to the proximity of the matrix structure, the wettability of the three SLIPSs was mainly determined by the property of the lubricants, which may be the result of the fact that SFO is more hydrophilic than SO and MO leading to the smallest CA of F-SLIPS. In addition, as illustrated in Figure 3 (b), the CAH of F-SLIPS is much larger than that of S-SLIPS and M-SLIPS, which suggests that the water droplets are more challenging to slide on SFO than on SO and MO. As shown in Figure 3 (c), when the droplet slid the same distance, the droplet needs 1.24 s and 1.32 s on S-SLIPS and M-SLIPS, respectively, which is much shorter than F-SLIPS (27.35 s). Combined with the result of the CAH (Figure 3 (b)), this result indicated that the adhesion force of SO and MO to water is close, significantly lower than SFO.

3.4. *Anti-frosting performance*

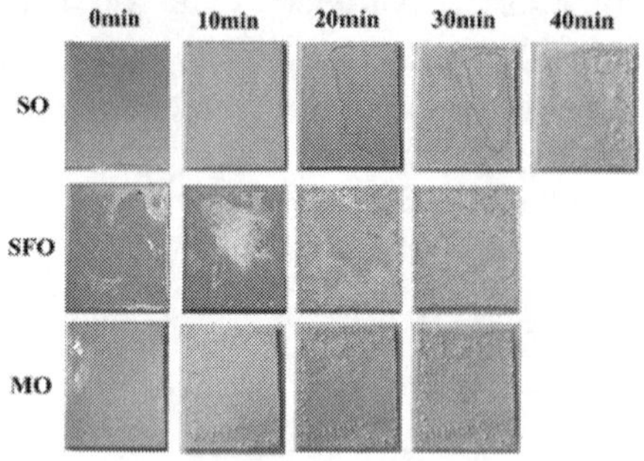

Figure 4. Frosting images of different SLIPSs (The red line box shows the unfrosted area).

The frosting is one of the most important factors in the degradation of SLIPS anti-icing performance, and the formation of frost crystals causes ice to accumulate more easily on the surface of SLIPS. Frosting images of SLIPSs prepared with different lubricants over time are shown in Figure 4. In this case, S-SLIPS was completely frosted at about 40 minutes. With the increase of time, the frost crystals began to grow from the periphery of S-SLIPS, and then gradually expanded with the oil film shrinking until it completely spread over the surface of the sample. However, both F-SLIPS and M-SLIPS were completely frosted in about 30 min. Seo et al. [2] have reported that the spread of the frost crystal is related to the mobility of water on SLIPS, which was affected by viscosity. The higher the viscosity is, the poorer the mobility of water is [11]. Though the mobility of water on SO and MO is close (Figure 3 (c)) at room temperature, the viscosity of SO is less affected by temperature and the viscosity of MO increased at -10°C. Thus, the mobility of water on M-SLIPS was decreased. The frosted crystal was easier to spread, resulting in the anti-frosting performance of S-SLIPS being greater than M-SLIPS. This suggests that the anti-frosting performance of SO is better than that of SFO and MO.

3.5. *Ice adhesion strength*

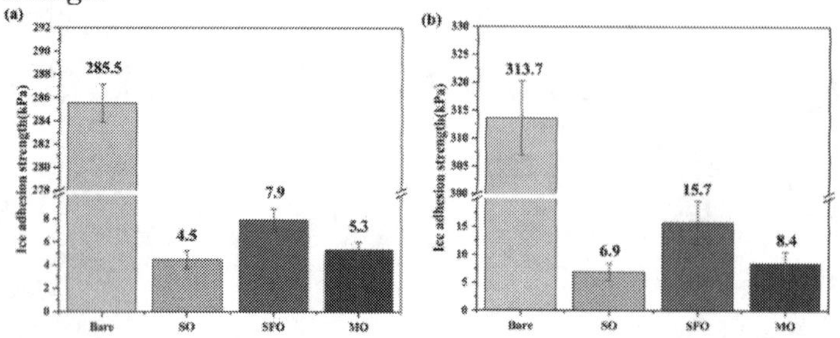

Figure 5. The adhesion strength between SLIPSs and ice in different icing times: (a) 20 min; (b) 40 min.

Ice adhesion strength is a significant criterion for assessing the anti-icing ability of SLIPS. To investigate the effect of three lubricants on the anti-icing ability of SLIPS, the ice adhesion strength of SLIPSs was measured. Figure 5 shows the ice adhesion strength of SLIPS prepared with different lubricants. The results show that all SLIPSs are easy to deicing due to the low adhesion between ice and samples which is below 20 kPa. When the freezing time was 20 minutes, F-SLIPS had the maximal ice adhesion strength of 7.9 ± 1.0 kPa (Figure 5 (a)). Due to the SFO not spreading completely on the samples (Figure 1), the ice adhesion force to the matrix is greater, along with the possibility of ice to grow into the pores, resulting in a significant enhancement in adhesion strength between ice and F-SLIPS. So, defective oil films are not suitable for the preparation of anti-icing SLIPS. M-SLIPS was 5.3 ± 0.7 kPa, and S-SLIPS was the smallest at 4.5 ± 0.8 kPa, which showed excellent anti-icing properties. For M-SLIPS and S-SLIPS, both SO and MO formed a continuous uniform oil film on the surface of the substrate. Thus, ice nucleated on the surface of the lubricant without contacting the matrix, and the deicing occurred at the ice-oil interface, and both the adhesion between the lubricant and ice and the low-temperature fluidity of the lubricant affect the adhesion between ice and SLIPS. Temperature-induced variation in the viscosity of MO may be one of the reasons that M-SLIPS are more difficult to deice than S-SLIPS. Increased icing time of all SLIPS is more difficult to deice (Figure 5 (b)), so this phenomenon is probably related to denser icicles due to increased icing time.

3.6. *Anti-icing durability*

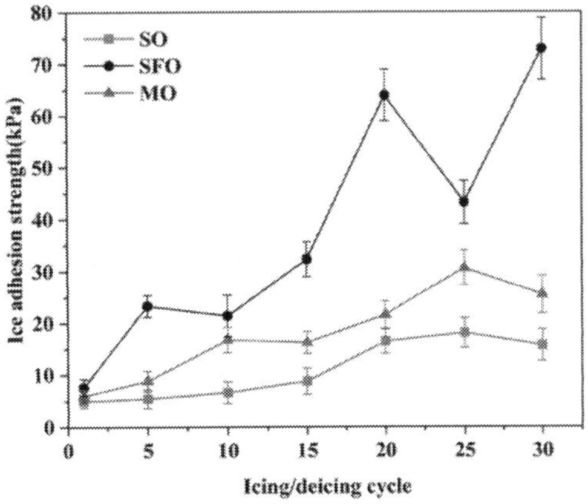

Figure 6. The adhesion variation between ice and SLIPSs during 30 repetitions of icing/deicing.

To investigate the long-term anti-icing performance of SLIPSs prepared from three lubricants, the SLIPSs were tested for 30 icing/deicing cycles. From Figure 6, the adhesion between ice and all SLIPSs exhibited a gradual increase with the number of icing/deicing cycles. This phenomenon is due to the depletion of lubricants during the deicing process, the gradual contact of ice with the matrix, and the gradual change of the deicing interface from an ice-oil interface to an ice-solid interface. In addition, the lubricant in the hole is replenished to the matrix surface by capillary forces. At the early stages of the repetitions of icing/deicing, the ice adhesion strength of F-SLIPS increased rapidly and dramatically more than that of M-SLIPS and S-SLIPS, indicating that the depletion of SFO was greater than that of MO and SO. It may be that the adhesion of SFO to ice is large and greater than the intermolecular forces of SFO, and a large amount of SFO was taken away by ice. On the other hand, the adhesion of MO and SO to ice is smaller than the intermolecular forces of MO and SO, and only a small amount of lubricant was taken away by ice. Generally, 20 kPa is considered the maximum ice adhesion strength for natural deicing. Among them, the ice adhesion strength of F-SLIPS exhibited a notable increase, rising from 7.6 kPa to 23.3 kPa (>20 kPa) after five cycles and reaching 72.9 kPa after 30 cycles, which indicated that the anti-icing durability of F-SLIPS was poor. M-SLIPS failed after 20 cycles, at which point the ice adhesion strength was 21.5 kPa. S-SLIPS showed the least increase, which increased from 5.0 kPa to

15.7 kPa after 30 icing/deicing cycles, without exceeding 20 kPa, for the best anti-icing durability. When the depletion rate is close to the repair rate, the adhesion between ice and SLIPS changes less. The S-SLIPS had the smallest ice adhesion strength enhancement after 30 cycles, suggesting that the depletion rate of SO and the repair rate of SO were closest to each other. In conclusion, SO is more favorable to preparing SLIPS with great anti-icing properties.

4. Conclusion

In this paper, porous structures were constructed on Al sheets used in the anodic oxidation method, and SLIPSs were successfully prepared by injecting three lubricants. Due to the same structure of the matrix, the properties of SLIPSs are mainly determined by the properties of the lubricants. Among the three lubricants, SO is more suitable for the preparation of anti-icing SLIPS. SO and MO were able to form a continuous and uniform oil film on the AAO with approximate CA and CAH. However, the SO-prepared SLIPS have even better anti-icing performance than MO-prepared. S-SLIPS was completely frosted after about 40 min. Not only the initial ice adhesion strength is 4.5 ± 0.8 kPa, but also lower than 20 kPa even after 30 icing/deicing cycles, which provides wonderful long-term anti-icing performance. To prepare excellent anti-icing SLIPS, in addition to the basic requirements for the preparation of SLIPS, a lubricant with a small effect of temperature on viscosity should be selected. Meanwhile, lubricants with close simultaneous depletion and repair rates facilitate the preparation of SLIPS with more stable anti-icing performance.

Acknowledgment

This work was supported by the National Natural Science Foundation of China (Project Nos. U23B20133) and the Electric Power Research Institute of Guizhou Power Grid Co., Ltd., China (Contract No. 066600202203010lHX00001).

References

[1] Wu X., Y. Liao, L. Yao, Y. Zha, R. Li, Y. Wang, D. Zhao, S. Mo, and W. Huang (2022). A non-percolative rGO/XLPE composite with a high electrothermal performance at high voltage and effective de/anti-icing for transmission lines. Composites Science and Technology. 230: p. 109, 772.

[2] Seo D., S. Oh, B. Moon, H. Kim, J. Kim, C. Lee, and Y. Nam (2019). Influence of lubricant-mediated droplet coalescence on frosting delay on lubricant-impregnated surfaces. International Journal of Heat and Mass Transfer, 2019. 128: p. 217-228.

[3] Subramanyam S. B., K. Rykaczewski, and K. K. Varanasi (2013). Ice adhesion on lubricant-impregnated textured surfaces. Langmuir. 29. (44): p. 13, 414-13, 418.

[4] Chung J., H. Cho, H. Yong, D. Heo, Y. S. Rim, and S. Lee (2020). Versatile surface for solid-solid/liquid-solid triboelectric nanogenerator based on fluorocarbon liquid-infused surfaces. Science and technology of advanced materials. 21 (1): p. 139-146.

[5] Liu C., Y. Li, C. Lu, Y. Liu, S. Feng, and Y. Liu (2020). Robust slippery liquid-infused porous network surfaces for enhanced anti-icing/deicing performance. ACS applied materials & interfaces. 12 (22): p. 25, 471-25, 477.

[6] Xiang H., Y. Yuan, T. Zhu, X. Dai, C. Zhang, Y. Gai, and R. Liao (2020). A novel durable anti-icing slippery surface with a dendritic porous structure. Materials Today Physics. 35: p. 101, 137.

[7] Sasidharanpillai A., Y. Lee, and S. Lee (2022). Design of stable liquid infused surfaces: Influence of oil viscosity on stability. Colloids and Surfaces A: Physicochemical and Engineering Aspects. 646: p. 128, 923.

[8] Wei C., G. Zhang, Q. Zhang, X. Zhan, and F. Chen (2016). Silicone oil-infused slippery surfaces based on sol-gel process-induced nanocomposite coatings: a facile approach to the highly stable bioinspired surface for biofouling resistance. ACS applied materials & interfaces. 8 (50): p. 34, 810-34, 819.

[9] Liu Y., S. Gao, J. Liu, and Q. Zhang (2023). Biomimetic slippery liquid-infused porous surfaces fabricated by porous fluorinated polyurethane films for anti-icing properties. Progress in Organic Coatings. 179: p. 107, 524.
[10] Meng X., Z. Wang, L. Wang, L. Heng, and L. Jiang (2018). A stable solid slippery surface with thermally assisted self-healing ability. Journal of Materials Chemistry A. 6 (34): p. 16, 355-16, 360.
[11] Xiang H., Y. Yuan, C. Zhang, T. Zhu, X. Dai, G. Liu, and L. Song (2022). Effect of Lubricant Viscosity on Wetting Behaviors and Durability of Anti-icing Slippery Liquid-Infused Porous Surfaces. In 2022 International Conference on New Energy, Energy Storage, and Power Engineering, NESP 2022, July 15, 2022 – July 17, 2022. Kunming, China: Institute of Physics.

Journal of Physics: Conference Series **2879** (2024) 012043 doi:10.1088/1742-6596/2879/1/012043

Investigation of torsional and fatigue resistance properties based on triply periodic minimal surfaces (TPMS)

Jijun Jiang[1], Zhanlong Cao[1,*], Jing Wu[1], Hongwei Lu[1] and Jiahao Zhang[1]

[1]College of Mechanical Engineering, Hubei University of Automotive Technology, Shiyan, Hubei, 442002, China

[*]Corresponding author's e-mail: 1017170082@qq.com

Abstract: Triply periodic minimal surfaces (TPMS) are algebraic surfaces found in nature, defined by implicit functions, and exhibit exceptional properties in various fields such as mechanics, computer graphics, and tissue engineering. This study focuses on four common minimal surface structures: Gyroid, Diamond, I-WP, and Schwarz P structures, and presents two methods for converting these surfaces into TPMS cells. These four surface structures are converted into TPMS cells, which are then organized into TPMS cylindrical units for finite element analysis. Through a comparison of the torsional mechanical properties of different minimal surfaces using the finite element method, the stress-strain curves of TPMS were generated. The findings suggest that the Schwarz P cylindrical structure remains within the elastic deformation stage under the same load, with a stress-load curve exhibiting a smaller slope, indicating superior torsional resistance. Additionally, a fatigue strength analysis of various minimal surface structures revealed that, under equivalent load conditions, the fatigue life of the Schwarz P cylindrical structure tends towards infinity.

1. Introduction

With the advancement of science and technology, significant progress has been achieved in the fields of materials science and manufacturing. In industries, the design, fabrication, and utilization of various complex structures have become common, most of which are solid entities. However, nature presents numerous intricate porous structures like bones, honeycombs, corals, sponges, and cork. The distinctive characteristics of these natural porous structures have piqued the interest of researchers globally, leading to the exploration of various biomimetic porous structures. In recent years, TPMS (triply periodic minimal surfaces) porous structures have shown exceptional performance in research and have been found to have numerous practical applications. TPMS represent a unique class of surfaces that display periodicity in three directions and do not intersect themselves.

In addressing the challenges of modeling and structural design of minimal surfaces. Zhang et al. [1] focused on designing porous structures for tissue engineering bone scaffolds. They proposed a bone scaffold design approach that integrates the three-periodic minimal surface theory, moving cube algorithm, template method, and isoparametric element method, as well as a method based on solid model voxelization and three-periodic minimal surface. Yoo [2] introduced a three-dimensional porous human tissue scaffold design method that leverages the distance field algorithm, Boolean algorithm, and TPMS. Yu et al. [3] presented a modeling technique for complex contour TPMS lattice structures based on TPMS and voxel distance field calculations. This method enables the creation of intricate TPMS lattice structures by computing the distance field of boundary points and performing Boolean operations with the TPMS distance field. While the methods discussed offer a relatively comprehensive overview of modeling and structural design for minimal surfaces, there is a notable lack of strategies for converting surfaces into solid units.

Research on the mechanical properties of minimally curved surface structures has been on the rise in recent years. Lin et al. [4] developed a design approach for gradient porous bone scaffolds with porosity varying radially and axially, based on the Network-type Gyroid porous structure. They examined the design method's impact on mechanical properties and permeability through finite element simulations, finding that the permeability of the radially gradient scaffold outperformed both homogeneous and axially gradient scaffolds. Abueidda et al. [5-6] conducted finite element calculations and experimental studies on the mechanical properties of three-period minimal surface structures. Their analysis revealed that the Gyroid structure demonstrated superior mechanical properties compared to other TPMS structures. Additionally, the Neovius and IWP structures displayed similar mechanical response performance, with higher stiffness and strength than the Primitive structure. Afshar et al. [7] investigated the tensile and bending responses of P-type and D-type TPMS structures. Their findings indicated that the P-type structure excelled in tensile properties, whereas the D-type structure performed remarkably well in bending properties. Gao et al. [8] effectively leveraged the superior mechanical properties of the Schwarz P structure to redesign the diagonal braces of specific aircraft landing gear types. They analyzed the parameters associated with varying outer wall thicknesses of the structure. Simulation calculations demonstrated that optimizing the minimum surface structure significantly alleviated stress concentration and reduced the overall weight, thereby offering an innovative approach to lightweight structural design.

The three-period minimal surface demonstrates significant potential for engineering applications. For instance, the p-type exhibits excellent tensile and compressive resistance while remaining lightweight. Torsion, a common working condition encountered in engineering problems, warrants attention. Although much of the scholarly focus has been on the compressive performance of minimally curved surface structures, research on their torsional performance and torsional fatigue performance remains scarce. The study of minimum surface torsion and fatigue analysis offers valuable insights into stability and deformation analysis, which can be directly utilized in the design and analysis of components like bone tissue, Aerospace components, helical structures, and other elements exposed to torsional forces. Additionally, these minimal surface structures are crucial in mechanical analysis.

This article explores four common minimal surface structures: Gyroid, Diamond, I-WP, and Schwarz P surfaces. It introduces two methods for converting TPMS surfaces into TPMS units, transforming minimal surface structures into minimal surface units. The TPMS solid unit is further transformed into a TPMS columnar structure through an array. The study utilizes the finite element method to compare and summarize the anti-torsion mechanical properties of various minimally curved columnar structures, presenting the stress-strain curve of the TPMS columnar structure to determine its torsion resistance. Finally, fatigue strength analysis is conducted on different minimal surface structures. This research is significant as it introduces a novel approach to designing diverse TPMS columnar structures and predicting their torsional mechanical properties and fatigue strength.

2. Minimal surface mathematical model

A minimal surface is defined as a surface with an average curvature of 0 throughout, expressed as $H = \frac{1}{2}(k_1 + k_2)$, where k_1 and k_2 represent the principal curvatures. The three-periodic minimal surface is a specific type of minimal surface that exhibits periodicity in three vertical directions within space. This surface can be precisely described mathematically through either the Weierstrass formula or by approximating it using implicit periodic surfaces.

For the Weierstrass formula, it is defined as follows:

$$\begin{cases} x = Re\int_{\omega_0}^{\omega} e^{i\theta}(1-\tau^2)R(\tau)d\tau \\ y = Re\int_{\omega_0}^{\omega} e^{i\theta}(1-\tau^2)R(\tau)d\tau \\ z = Re\int_{\omega_0}^{\omega} e^{i\theta}(1-\tau^2)R(\tau)d\tau \end{cases} \tag{1}$$

The formula represents Re as the real part of the contour integral under Cartesian coordinates at any point. Here, ω is a complex variable, i is an imaginary number, θ is the angle, and $R(\tau)$ is the function of surface change.

For approximation using implicit periodic surfaces, the definition is as follows [9]:

$$\varphi(r) = \sum_{k=0}^{K} A_k \cos\left[\frac{2\pi h_k \cdot r}{\lambda_k} + P_k\right] = C \tag{2}$$

In the formula, r represents the position vector in Euclidean space; h_k denotes the k-th grid vector in reciprocal space; A_k stands for the amplitude factor; λ_k signifies the wavelength of the period; P_k represents the phase; and C is a constant.

Implicit periodic surfaces are commonly utilized for approximation due to their simpler calculation process compared to the Weierstrass formula and their ability to represent a wider range of minimal surfaces. Various types of three-period minimal surfaces, such as Diamond [10], Schwarz P [11], Gyroid [12], and I-WP [13], exist. The formulas for approximating minimal surfaces with implicit periodic surfaces can be found in Table 1, while the typical shapes of minimal surfaces are illustrated in Figure 1.

(a) Diamond (b) Gyroid (c) I-WP (d) Schwarz P

Figure 1. Four common surface shapes of minimal surfaces.

Table 1. Approximate TPMS implicit function formula.

TPMS category	Trigonometric function implicit function expression
Gyroid	$\varphi(x,y,z) = \cos(x)\sin(y) + \cos(y)\sin(z) + \cos(z)\sin(x) = C$
Diamond	$\varphi(x,y,z) = \cos(x)\cos(y)\cos(z) - \sin(x)\sin(y)\sin(z) = C$
SchwarzP	$\varphi(x,y,z) = \cos(x) + \cos(y) + \cos(z) = C$
I-WP	$\varphi(x,y,z) = 2[\cos(x)\cos(y) + \cos(y)\cos(z) + \cos(z)\cos(x)] - [\cos(2x)\cos(2y) + \cos(2y)\cos(2z) + \cos(2z)\cos(2x)] = C$

3. Minimal surface solid modeling

3.1 Solid element modeling

Due to the insufficient minimum surface thickness, elements without thickness are not suitable for mechanical analysis and finite element calculations, preventing torsional fatigue performance analysis. To address this issue, two methods can be employed: thickening the minimum surface to create a solid body for analysis, or adding a certain thickness to the surface during finite element calculations. This study opted for the former approach. The two techniques for thickening a minimum surface into a solid are the offset thickening method and the section closure thickening method. The offset thickening methods include two-way thickening and ordinary offset thickening. On the other hand, the section closure thickening method involves utilizing the outer surface and vertices of a minimum surface to generate solid elements, followed by using the isocaps function to produce outer caps and the patch function to fill the volume. In contrast, the offset thickening method entails offsetting minimal surfaces to create new surfaces and then filling the gaps between them to form solid elements. For instance, a

Gyroid surface can be transformed into a sheet-like Gyroid porous structure through offsetting and thickening, or into a network-type Gyroid porous structure by sealing and thickening specific parts of the surface.

3.2 TPMS columnar solid modeling

The TPMS columnar entity is generated by combining the TPMS unit with a columnar body. To ensure the reliability of finite element analysis results and provide meaningful comparisons, four common types of minimal surfaces are modeled with identical dimensions. The modeling dimensions include a base width and length of 3 mm each, and a column height of 15 mm.

4. Torsional performance analysis

4.1 Material properties

The TPMS columnar body in this study is made of 40Cr, a steel commonly used in the machinery manufacturing industry. Following quenching and tempering, it exhibits favorable comprehensive mechanical properties, low-temperature impact toughness, and low-notch sensitivity. The detailed material properties of 40Cr can be found in Table 2.

Table 2. 40Cr material properties.

Density (kg/m^3)	Elastic Modulus (MPa)	Poisson's ratio	Yield Strength (MPa)
7850	211	0.29	290

4.2 Mechanical properties simulation settings

This study utilized Hypermesh software to conduct tetrahedral meshing on a minimal surface tetrahedral mesh with a mesh size set at 0.2 mm. Upon completion of the meshing process, the finite element mesh model in inp format was exported. The segmented finite element mesh was then imported into ANSYS Workbench software for static simulation of torsion resistance. To assess the anti-torsion mechanical properties of four common types of minimal surfaces, identical boundary conditions were applied to all four types, including fixed constraints and a consistent torsional load. The loading schematic diagram of the TPMS columnar entity can be observed in Figure 2.

Figure 2. TPMS columnar load diagram.

4.3 Twist analysis

Torsion-resistant finite element analysis was conducted on various minimal curved surfaces to determine the stress distribution of four types of TPMS columns, as illustrated in Figure 3. By examining the stress distribution and comparing the maximum stresses outlined in Table 3, it is evident that the Schwarz P-type columnar body exhibits the lowest maximum stress value. Therefore, the torsion resistance performance ratio of the Schwarz P-type columnar body surpasses that of the other three types of TPMS columns.

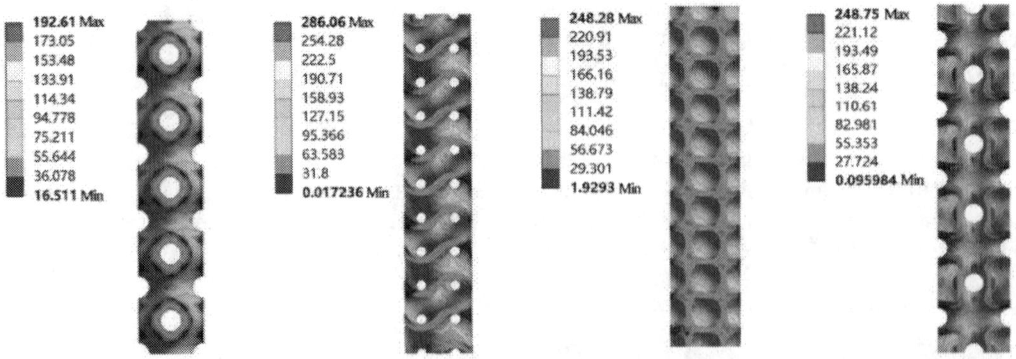

(a) Schwarz P (b) Gyroid (c) Diamond (d) I-WP

Figure 3. Stress distribution cloud diagrams of 4 types of TPMS columns.

Table 3. Stress values of 4 types of TPMS columns.

TPMS category [a]	Schwarz P	Gyroid	Diamond	I-WP
Maximum stress (MPa)[b]	192.61	286.06	248.28	248.75

[a] Four types of minimal surfaces.
[b] Maximum stress after applying load.

The stress-strain curves of various types of TPMS columns obtained from finite element numerical simulations were analyzed in Figure 4. Prior to reaching the elastic limit, the stress-strain curves of the four types of TPMS columns exhibited consistent trends. However, it was evident that the Gyroid, Diamond, and I-WP type columns entered the yield stage, whereas the Schwarz P-type column maintained a linear trend, indicating elastic deformation without reaching the yield stage. Notably, the Gyroid column showed a significant increase in stress post-yield stage, suggesting the onset of plastic deformation. Comparing the torsion resistance, it can be inferred that the Schwarz P-type columns exhibited superior performance, followed by Diamond and I-WP columns with similar resistance, while Gyroid-type columns showed the least torsion resistance.

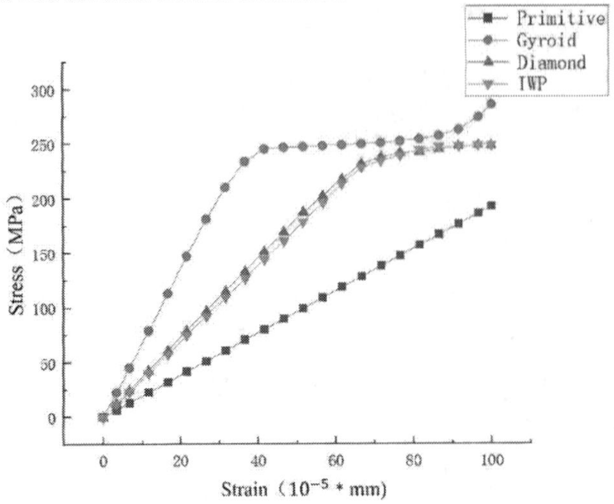

Figure 4. TPMS columnar stress-strain curve.

5. Torsional fatigue performance analysis

5.1 S-N curve of material

Fatigue can be categorized into low-cycle fatigue and high-cycle fatigue based on the number of failure cycles. Low-cycle fatigue typically involves less than 1×10^4cycles, while high-cycle fatigue involves more than 3×10^6cycles, typically ranging between 10^9~10^{10}cycles [14]. In the case of a columnar body, the applied load is symmetrical with a stress ratio of -1. Figure 5 illustrates the S-N stress fatigue curve of the material 40Cr utilized in the columnar body.

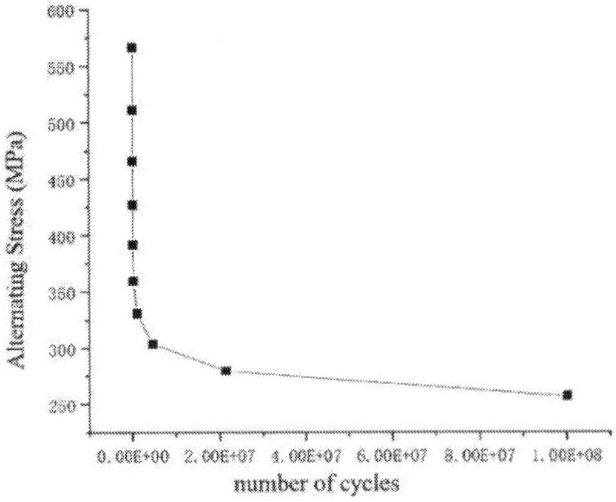

Figure 5. S-N stress fatigue curve of 40Cr.

5.2 TPMS column fatigue analysis

The fatigue test load conditions and the material S-N curve were established in the previous chapter. Fatigue analysis and calculations were conducted using the fatigue tools of ANSYS Workbench on four types of TPMS columnar structures, resulting in the determination of fatigue lives. The coefficients can be found in Figure 6. Analysis of the fatigue life cloud diagram reveals that the Schwarz P-type columnar body exhibits an infinite fatigue life under the same torsional load, indicating superior torsional fatigue life compared to the other structures.

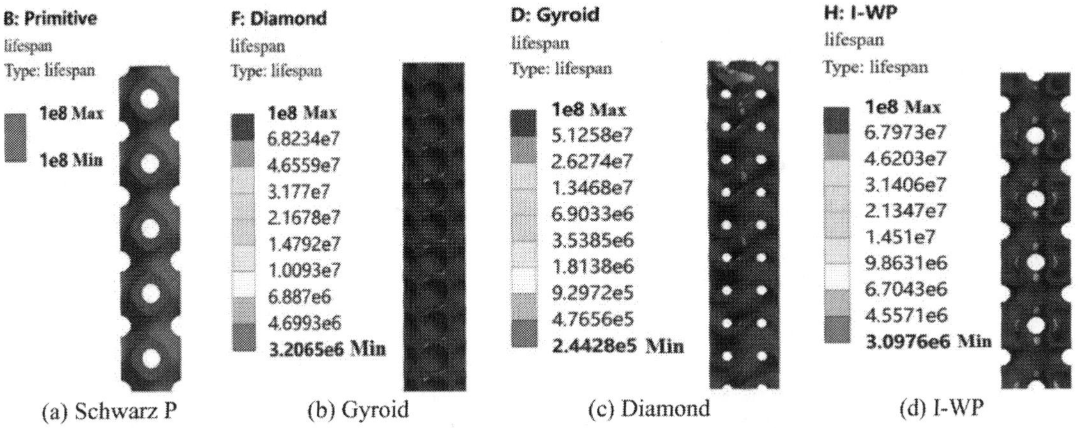

Figure 6. Four types of TPMS columnar fatigue analysis cloud charts.

6. Discussion

This paper investigates the mechanical properties of four common TPMS structures by converting their surface geometries into TPMS units and conducting finite element analyses of their performance under torsion and fatigue. The torsional analysis reveals significant differences among the TPMS structures. The stress-strain cloud diagram indicates that stress concentration occurs in all four structures during the compression process. The areas of stress concentration align closely with the regions exhibiting larger strain, suggesting that damage initially occurs in localized areas with elevated stress. Notably, the Schwarz P structure markedly mitigates the stress concentration phenomenon, distributing stress more uniformly across the outer wall and thereby reducing the risk of fracture in these critical areas. During the linear phase of the curve, no damage to the lattice structure is observed, as all materials experience elastic deformation. The cylindrical shape demonstrates a higher resistance to deformation under load and exhibits excellent torsional resistance. Additionally, the analysis indicates that, under identical loading conditions, the Schwarz P structure remains within the elastic deformation stage, signifying its capability to absorb more energy before reaching the yield point, which is a crucial attribute for applications demanding high torsional resistance. In terms of fatigue strength, Schwarz P structures display superior performance owing to their unique geometric properties, which facilitate a more even distribution of stress and diminish the likelihood of fatigue failure. Consequently, Schwarz P structures are ideally suited for components subjected to repeated stresses, such as artificial bones and components in the automotive and aerospace industries.

7. Conclusions

This article examines four common minimally curved surface structures and evaluates their torsion and fatigue properties. The Schwarz P type demonstrates strong torsion resistance and fatigue resistance, which are precisely the mechanical properties required for the construction of artificial bones, as well as automotive and aerospace components. The study's findings indicate that:

(1) The TPMS minimal surfaces can be converted into TPMS solid units using offset thickening and section sealing thickening methods. Arraying solid units can form TPMS columnar units.

(2) Torsion-resistant mechanical properties of different minimally curved columnar structures are compared and summarized using finite element analysis. Stress-strain and stress-load curves of TPMS columnar units are analyzed, revealing that Schwarz P-type columnar bodies exhibit superior torsion resistance under similar load conditions.

(3) Fatigue analysis of four TPMS columnar structures shows that the fatigue life of Schwarz P-type tends towards infinity under the same torsional load, indicating the best torsional fatigue life.

Acknowledgments

This research was supported by the National Natural Science Foundation of China Project (No.52375508), the Hubei Provincial Department of Education Key Project (No. D20211803), and the Hubei University of Automotive Technology Doctoral Fund Project (No. BK202001).

References

[1] Zhang Z Y, Zhao K and Li Y S (2020), et al. Design of porous bone scaffold structures based on voxel and three period minimal surfaces. Computer Integrated Manufacturing System, 26 (03): 697-706.

[2] Yoo D J (2021). Porous scaffold design using the distance field and triply periodic minimal surface models. Biomaterials, 32(31): 7741-7754.

[3] Yu X W, Li L H and Zhang T (2022), et al. Modeling of Complex Contour Lattice Structures Based on Three Period Minimal Surfaces and Voxel Distance Fields. Journal of Chongqing University, 45 (06): 75-84.

[4] Lin X, Yang B, Zhu K P and Zhang R T (2023). Gradient structure design and mechanical performance analysis of porous bone scaffolds based on Gyroid three period minimal surfaces. Journal of Wuhan University of Science and Technology, 46 (02): 109-117.

[5] Abueidda D W, Bakir M and Al-Rub R K A (2017), et al. Mechanical properties of 3D printed polymeric cellular materials with triply periodic minimal surface architectures. Materials & Design, 122: 255-267.

[6] Abueidda D W, Elhebeary M and Shiang C S A (2019), et al. Mechanical properties of 3D printed polymeric Gyroid cellular structures: Experimental and finite element study. Materials & Design, 165: 107597.
[7] Afshar M, Anaraki A P and Montazerian H (2016), et al. Additive manufacturing and mechanical characterization of graded porosity scaffolds designed based on triply periodic minimal surface architectures. Journal of the mechanical behavior of biomedical materials, 62: 481-494.
[8] Gao H, Wang Q and Liu X C (2022). Structural design of landing gear diagonal brace based on three-period minimal curved surface. Nanchang Hangkong University, Nanchang, China.
[9] Yu X W (2021). Modeling and Design of Lattice Structures Based on Three Period Minimal Surfaces. Chongqing University, Chongqing, China.
[10] Gandy P J F, Cvijović D and Mackay A L (1999), et al. Exact computation of the triply periodic D (diamond') minimal surface. Chemical Physics Letters, 314(5-6): 543-551.
[11] Gandy P J F and Klinowski J (2000). Exact computation of the triply periodic Schwarz P minimal surface. Chemical Physics Letters, 322(6): 579-586.
[12] Gandy P J F and Klinowski J (2000). Exact computation of the triply periodic G (Gyroid') minimal surface. Chemical Physics Letters, 321(5-6): 363-371.
[13] Cvijović D and Klinowski J (1994). The computation of the triply periodic I-WP minimal surface. Chemical physics letters, 226(1-2): 93-99.
[14] Chen Y C, Zhou R P and Huang G B (2022), et al. Research on Fatigue Analysis of Ship Propulsion Shafting in Ice Regions. China Shipbuilding, 63 (04): 93-106.

Journal of Physics: Conference Series **2879** (2024) 012044 doi:10.1088/1742-6596/2879/1/012044

Establishment of ductile fracture locus of 2219 aluminum alloy at cryogenic temperature

Qiwei Hu[1], Tongliang Du[2], Fangxing Wu[1], Zhichao Zhang[2] and Xiaobo Fan[1*]

[1]School of Mechanical Engineering, Dalian University of Technology, Dalian 116000, China

[2]Shanghai Aerospace Equipment Manufacturer Co., Ltd., Shanghai, China

*xbfan@dlut.edu.cn

Abstract. Cryogenic forming has been used to manufacture intricate curved surface aluminum components. The special geometry and mold constraints cause the stress state to continuously evolve. Coupling with the cryogenic temperature makes the deformation process more complicated. Especially for the final fracture, which greatly determines the forming quality of the components. Thus, the cryogenic fracture behavior of three typical stress states was obtained by conducting tensile experiments. Fracture strains and stress paths were obtained using a hybrid numerical-experimental method. Finally, the cryogenic fracture locus was obtained based on the MMC ductile fracture criterion. The results indicate that the AA2219-O exhibits superior ductile fracture properties at cryogenic temperature. The fracture strain of AA2219-O was increased by 52.6% at the uniaxial tensile stress state, and by up to 80.0% at under plane strain stress state. This research has a guiding significance for cryogenic forming.

1. Introduction

The 2219 aluminum alloy is the primary frame material for rockets, planes, and other aerospace devices. However, forming complex structures at room temperature (RT) is very difficult due to low elongation limitations. Researchers have recently discovered that aluminum alloys at cryogenic temperatures (CT) generally exhibit a double enhancement effect (DEE) [1], in which both plasticity and hardening increase simultaneously. Several cryogenic forming processes have been developed, including deep drawing [2] and dielectric pressure forming [3]. The forming limit of the sheet is significantly increased due to DEE.

The formability of sheet metal can be evaluated using experimentally determined forming limits (FLC). However, it should be noted that the experimentally tested FLCs only cover uniaxial to biaxial tensions, and no investigation has been conducted on shear or compressive stress states. Therefore, its applicability for forming complex parts is limited. The use of theoretical models such as the Swift and M-K models can be employed to predict the formability of sheet metal [4]. However, the varying internal damage cannot be accurately represented using these models. To solve these problems, researchers have developed the Modified Mohr's Coulomb (MMC) ductile fracture model by describing the change in stress state through stress triaxiality and Lode parameters. This model demonstrates the capacity to forecast fracture within a wide spectrum of stress triaxiality.

Researchers have paid considerable attention to predicting the formability of CT. Yuan et al. predicted forming limit curves of AA2024-O at different temperatures using the M-K model. Their findings indicate that the formability of metal materials improves as the forming temperature decreases [5]. Wang et al. successfully predicted the tube of AA6061 deformation limit from RT to -196°C using the

Journal of Physics: Conference Series **2879** (2024) 012044 doi:10.1088/1742-6596/2879/1/012044

dynamically changing hardening capacity modified M-K model [3]. The prediction of damage and failure in AA2219-W can be accurately achieved by utilizing low-temperature FLD and FLSD, ranging from uniaxial to biaxial tensile strains. In addition, theoretical analysis and rigid die bulging tests were used to determine the low-temperature flow stress curves [6]. However, there are currently no appropriate models or theories for predicting the forming and damage of sheets under complicated stress states at CT.

In this paper, the cryogenic fracture behavior of the aluminum alloy 2219-O is investigated through cryogenic fracture experiments at different stress states. Firstly, the cryogenic Swift model is corrected using the inverse method to obtain the cryogenic stress-strain relationship and enable reliable large-deformation finite element (FE) simulation. Secondly, the fracture strains of uniaxial tensile, plane strain, and pure shear stress states were obtained using a hybrid numerical-experiment method. Finally, this study establishes a cryogenic fracture locus that covers a broad state range of stresses based on the three-parameters MMC ductile fracture criterion. This study guides further applications of cryogenic forming.

2. MMC ductile fracture criterion

To quantitatively evaluate the ductility of AA2219-O, the three parameters of MMC were selected. Equation (1) elucidates the manner in which the occurrence of damage in plastic and metal materials is contingent upon the prevailing stress states. The model was an extension of the initial Mohr-Coulomb model, which was originally designed for geotechnical materials, owing to its ability to capture the impact of both the stress triaxiality and the lode angle parameter on the deformation behavior. The fracture strain equivalent is determined by the stress triaxiality and the Lode angle parameter and decreases exponentially as the stress triaxiality increases. C_1, C_2, and C_3 could be obtained through subsequent fracture experiments.

$$\bar{\varepsilon}_f = \left\{ \frac{K}{C_2}\left[C_3 + \frac{\sqrt{3}}{2-\sqrt{3}}(1-C_3)\left(sec\left(\frac{\pi}{6}\bar{\theta}\right)-1\right)\right] \times \left[\sqrt{\frac{1+C_1^2}{3}}cos\left(\frac{\pi}{6}\bar{\theta}\right)+C_1\left(\eta+\frac{1}{3}sin\left(\frac{\pi}{6}\bar{\theta}\right)\right)\right] \right\}^{-\frac{1}{n}} \tag{1}$$

The parameters K and n were obtained using the Swift hardening model. The model calibration parameters are C_1, C_2, and C_3. The Lode angle parameter is represented by $\bar{\theta}$, the stress triaxiality by η and fracture strain $\bar{\varepsilon}_f$ were considered.

3. Experiments

3.1. Materials

Selecting a 2 mm AA2219-O aluminum alloy sheet, uniaxial tensile experiments were performed at RT and -196°C. The Swift hardening model was employed to align the true stress-strain curve with the observed data. Stress values are overestimated due to the oversaturation of the swift model. Based on the FE simulation, simulations of pure shear specimens are aligned with the test load-displacement curves. Swift's K and n are corrected by the inverse method, as shown in Figure 1.

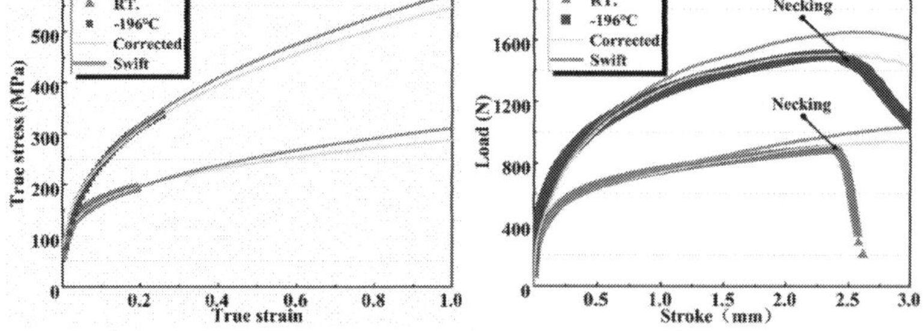

Figure 1. True stress-strain curves and load-stroke curves before and after correction (a) true stress-strain curve (b) load-stroke curve.

Figure 2 shows the flowchart of the inverse method. The true stress-strain curve of the specimen was obtained through uniaxial tensile testing before necking occurred. The resulting true stress-strain data was then extrapolated. The force-displacement data under large deformation were obtained through pure shear experimentation. The extrapolated true stress-strain data was used to perform finite element simulation of the pure shear experiment, resulting in simulated pure shear force-displacement data. The shear force-displacement data obtained from the experiment in the village is compared with the pure shear force-displacement data obtained from the simulation. If they match, the extrapolated stress-strain curve is considered reliable. If they do not match, the extrapolated curves are corrected until a reliable stress-strain curve is obtained.

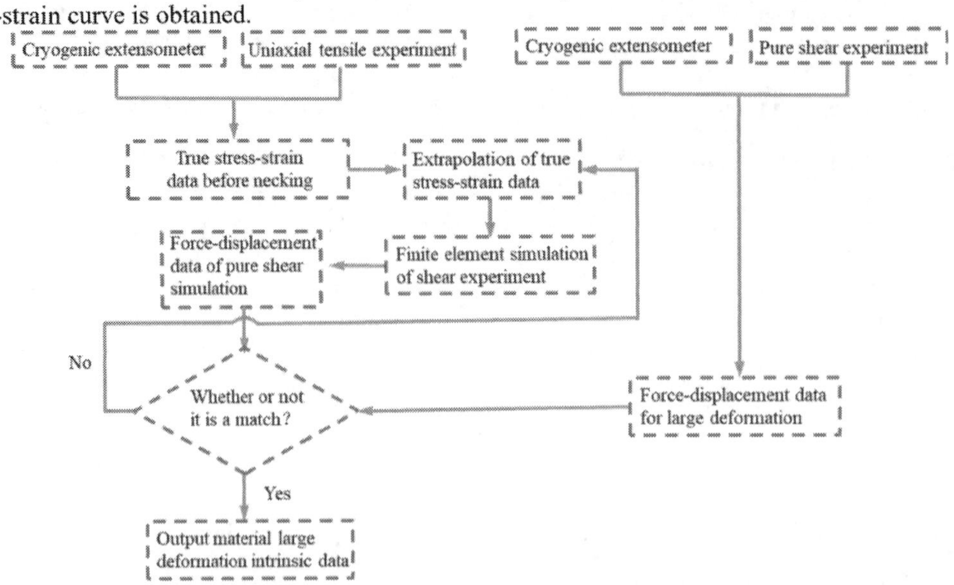

Figure 2. The flowchart of the inverse method.

3.2. Ductile fracture experiment

To calibrate the MMC ductile fracture criterion, the tensile dog-bone specimen (uniaxial tensile), the notched specimen (plane strain), and the in-plane shear specimen (pure shear) were designed. Three types of specimens were tested using a cryogenic universal tensile testing machine to complete RT and -196°C ductile fracture experiments. A cryogenic extensometer was used to record load-stroke data at 25 mm intervals during the experiment. The tensile rates for the three types of specimens were 4.2, 0.5, and 0.6 mm/min for uniaxial tension, plane strain, and pure shear, respectively.

4. Establishment of cryogenic fracture locus

The ductile fracture criterion requires the determination of several fracture parameters. These parameters are extracted from specific specimens and include fracture strain, stress triaxiality, and the Lode angle parameter. The parameters are then optimized using the least-squares method to establish the ductile fracture locus of the AA2219-O at RT and -196°C.

A FE simulation is performed on the load-stroke curves obtained from testing specimens of different shapes. The fracture strain is obtained from the numerical simulation results, specifically, the core fracture strain derived from the hybrid method and representative of the fracture location at the time of fracture inception. Figures. 3, 4, and 5 are the load-stroke and equivalent strain histories of different specimens which was obtained from the experiments and the FE results at different temperatures. The red curve represents the load-stroke curve obtained from the experiment, while the black square scatter represents the load-stroke curve extracted from FE. The data are in good agreement with each other, indicating that the established FE model accurately describes the deformation behavior of the specimen. Additionally, stress triaxiality and Lode angle parameters are obtained against the equivalent strain. To calibrate the ductile fracture criterion, it is common practice to normalize these two values using the fracture strain. The equations for normalizing stress triaxiality and the Lode angle parameter are listed in Equations (2) and (3), respectively. Table 1 shows the fracture strains and the averaged stress triaxiality and Lode angle parameters obtained at different temperatures under the three stress states, which can be substituted into Equation (1) to derive the three unknown parameters C_1, C_2, and C_3. Table 2 shows the MMC ductile fracture criterion parameters for AA2219-O at both RT and -196°C.

$$\eta_{av} = \frac{1}{\varepsilon_f}\int_0^{\varepsilon_f} \eta \mathrm{d}\varepsilon \tag{2}$$

$$\bar{\theta}_{av} = \frac{1}{\varepsilon_f}\int_0^{\varepsilon_f} \bar{\theta} \mathrm{d}\varepsilon \tag{3}$$

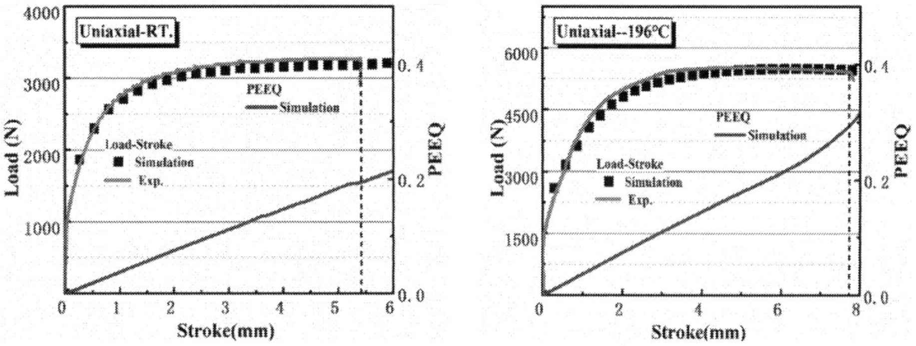

Figure 3. Equivalent plastic strain and load-stroke curves of dog-bone specimens comparison of FE results with experimental results at RT and -196°C.

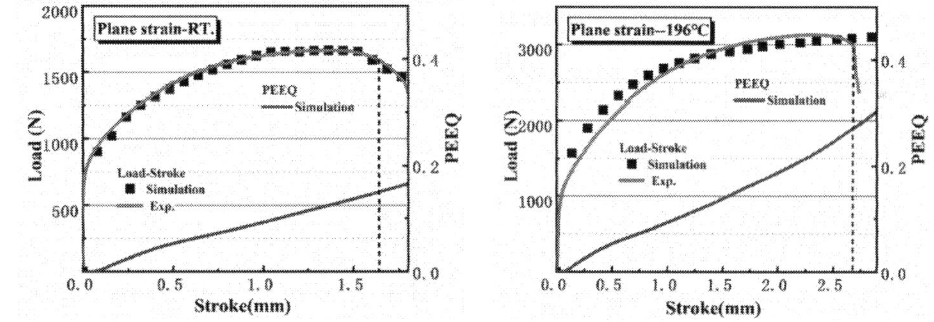

Figure 4. Equivalent plastic strain and load-stroke curves of notched specimens comparison of FE results with experimental results at RT and -196°C.

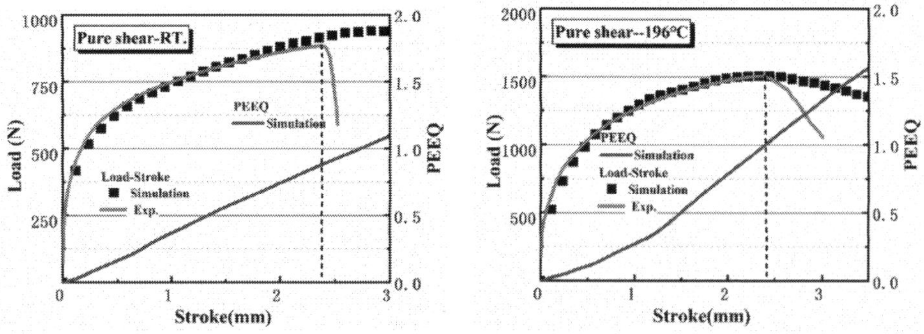

Figure 5. Equivalent plastic strain and load-stroke curves of in-plane shear specimens comparison of FE results with experimental results at RT and -196°C.

The experiment and the FE simulations described above show the correspondences of the fracture strains under two different temperatures. It is obvious that the enhancement effects of different stress states are significantly different at CT. The fracture strain in the uniaxial tensile stress state is 0.29 at -196°C, which is an increase of 56% compared to RT. The fracture strain in the plane strain state is 0.27 at -196°C, which is an increase of 80% compared to RT. The fracture strain in the pure shear stress state is 1.00 at -196°C, which is an increase of 14.9% compared to RT.

Table 1. Fracture parameters of three specimens at RT and -196°C.

Stress state	Temperature	$\bar{\varepsilon}_f$	η_{av}	$\bar{\theta}_{av}$
Uniaxial	RT.	0.19	0.3328	0.9946
	-196°C	0.29	0.3330	0.9976
Plane strain	RT.	0.15	0.5292	0.2409
	-196°C	0.27	0.5236	0.2748
Pure shear	RT.	0.87	0.0761	0.2191
	-196°C	1.00	0.0873	0.2495

Table 2. MMC criterion parameters of AA2219 -O at RT and -196°C.

Temperature	K	n	C_1	C_2	C_3
RT.	296	0.26	1.6181	337.8707	0.9244
-196°C	550	0.33	1.4951	601.4071	0.8831

The fracture locus of AA2219-O at RT and -196°C were obtained by taking three data points and solving the fracture criterion coefficients using the least squares method. Figure. 6 is the AA2219-O ductile fracture locus. The plotted fracture locus shows that the AA2219-O has better ductile fracture properties at -196°C than at RT. Based on cryogenic fracture trajectories, splitting defects in complex stress states can be predicted and analyzed.

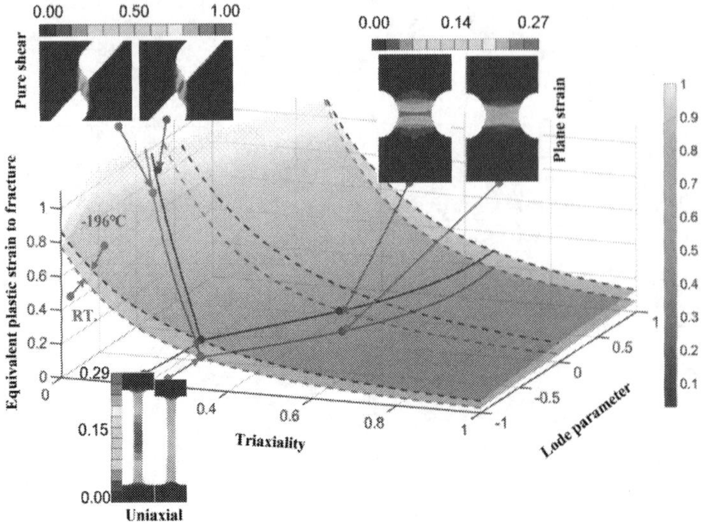

Figure 6. Ductile fracture locus of AA2219-O state at RT and -196°C.

5. Conclusions

In this study, the cryogenic ductile fracture locus of 2219-O was established. Conclusions can be drawn as follows:

(1) Wide range stress-strain data were obtained at cryogenic temperatures. The fracture experiments and numerical simulations were conducted under various stress states. The deformation simulation requirements under different stress states were met.

(2) The enhancing effects of different stress states on fracture strain at cryogenic temperatures were studied. The fracture strains are obtained by the hybrid numerical experiment method. The fracture strains of specimens under different stress states are increased at cryogenic temperature. The fracture strains under uniaxial tensile and plane strain stress states were increased by 56% and 80%, respectively. It illustrates the potential of cryogenic forming.

(3) The cryogenic ductile fracture locus was established based on the MMC3 ductile fracture criterion. The ductile fracture performance of AA2219-O significantly increased at CT. This highlights the superiority of cryogenic forming and provides guidance for subsequent numerical simulations and process tests in complex stress states.

Acknowledgments

This work was financially supported by the National Key Research and Development Program of China (2019YFA0708804). National Natural Science Foundation of China [Grant No. 52375311]; and Shanghai Rising-Star Program.

References

[1] Innovation for forming aluminum alloy thin shells at ultra-low temperature by the dual enhancement effect - IOPscience[EB/OL]. [2024-03-12]. https://iopscience.iop.org/article/10.1088/2631-7990/ac6b62.

[2] FAN X, CHEN X, YUAN S. Novel forming process for aluminum alloy thin shells at ultra-low temperature gradient[J/OL]. International Journal of Machine Tools and Manufacture, 2023, 185: 103992. DOI:10.1016/j.ijmachtools.2022.103992.

[3] WANG X, FAN X, CHEN X, et al. Forming limit of 6061 aluminum alloy tube at cryogenic temperatures[J/OL]. Journal of Materials Processing Technology, 2022, 306: 117649. DOI:10.1016/j.jmatprotec.2022.117649.

[4] SIGNORELLI J W, BERTINETTI M A, ROATTA A. A review of recent investigations using the Marciniak-Kuczynski technique in conjunction with crystal plasticity models[J/OL]. Journal of Materials Processing Technology, 2021, 287: 116517. DOI:10.1016/j.jmatprotec.2019.116517.

[5] YUAN S, CHENG W, LIU W. Cryogenic formability of a solution-treated aluminum alloy sheet at low temperatures[J/OL]. Journal of Materials Processing Technology, 2021, 298: 117295. DOI:10.1016/j.jmatprotec.2021.117295.

[6] LIU W, CHENG W, YUAN S. Analyses on formability and flow stress of an Al-Cu-Mn alloy sheet under biaxial stress at cryogenic temperatures[J/OL]. International Journal of Mechanical Sciences, 2021, 195: 106266. DOI:10.1016/j.ijmecsci.2021.106266.

Journal of Physics: Conference Series **2879** (2024) 012045 doi:10.1088/1742-6596/2879/1/012045

Airport runway FOD detection model based on improved YOLOv10m

Jinlei Wang[1], Ruifeng Meng[1,*], Yi Zhang[1] and Zhi Qiao[2]

[1]School of Aviation, Inner Mongolia University of Technology, Hohhot 010021, China
[2]Inner Mongolia Comprehensive Transportation Science Research Institute Co., Ltd., Hohhot 010021, China

[*]Corresponding author's e-mail: mrfnmgcn@imut.edu.cn

Abstract: Foreign Object Debris (FOD) poses a significant safety risk to aircraft operations, making efficient and accurate detection methods crucial. Existing detection techniques often struggle to meet the demands of busy airports due to limitations in real-time performance and detection accuracy. To address these challenges, we proposed a lightweight FOD detection model based on YOLOv10. By incorporating PConv and EMA attention mechanisms, our model enhances detection accuracy, achieving a 0.4% increase in mAP@50 and a 0.7% improvement in mAP@50-95 compared to the original YOLOv10m. Additionally, our model reduces the number of parameters by 19.8% and decreases computational complexity by 20.9%, thanks to the integration of Feature Pyramid Shared Convolution (FPSConv) and a lightweight shared detection head (Detect-LSCD). These improvements make our model more suitable for real-time FOD detection on airport runways, offering a more efficient and accurate solution than the original YOLOv10m.

1. Introduction

FOD refers to foreign substances left in airport operational areas that do not belong to the runway surface, such as batteries, plastic parts, and metal plates. This debris can pose serious threats to aircraft operations, making FOD detection critical for ensuring flight safety. However, FOD detection has long been a significant challenge in aircraft safety assurance. Currently, many airports still rely on manual patrols for FOD inspection, typically closing the runway during aircraft takeoff and landing, with staff forming a circle to visually inspect the runway. This manual inspection method is not only time-consuming and labor-intensive, but also costly, and it imposes significant restrictions on airports with high flight frequencies, making it difficult to accurately detect FOD between frequent flights. Traditional object detection methods, such as radar-based technologies, although somewhat effective, cannot detect debris immediately.

With the advancement of object detection technology, deep learning algorithms have been widely applied in this field. Two-stage algorithms, such as Faster R-CNN, have been preliminarily applied in debris detection due to their high accuracy. These two-stage algorithms first generate candidate regions and then classify and localize these regions. Despite their high detection accuracy, their complex processing pipeline results in slower detection speeds, making them unsuitable for applications that require real-time performance, especially in airport runway FOD detection where quick response is critical.

To address the limitations of two-stage algorithms, we propose an improved YOLOv10m [1] single-stage algorithm. This algorithm not only enhances detection accuracy but also reduces the number of parameters, lowers computational complexity, and increases detection speed. The single-stage approach provides a more efficient and precise solution for real-time and accurate FOD detection on airport runways. This paper makes the following contributions:

• We propose a FeaturePyramidSharedConv (FPSConv) with a shared convolutional layer. Compared with the original SPPF, it not only retains the key feature information but also gives the model stronger expressive power while maintaining computational efficiency.

• We design a lightweight shared detection head (Detect-LSCD) that balances minimal accuracy drop and lightweight model.

• We introduce a novel Cf-Faster-EMA for YOLOv10m, which leverages PConv and EMA attention mechanisms to improve the detection accuracy of the model while reducing the number of parameters and computational complexity.

2. Improvement of the YOLOv10m algorithm

2.1 EMA lightweight bottleneck

In deep learning networks, achieving higher computational speed is crucial for building faster and more resilient models. A recent development in this area is the lightweight neural network, FasterNet [2], which introduces a new concept called partial convolution. For spatial feature extraction, this method only applies standard convolution to part of the input channels, while other channels remain unchanged. This unused information is used by subsequent convolutional layers. Therefore, we can combine the C2f structure of YOLOv10m with it to create a C2f-Faster model. This model not only improves the computational efficiency but also reduces the computational complexity of the original C2f structure.

Incorporating components from FasterNet significantly reduces the model's parameter count but does have an impact on accuracy. Therefore, it is necessary to integrate a module that enhances accuracy without increasing the parameter count, where the attention module comes in.

In many computer vision tasks, attention mechanisms are often used to enhance their feature representation. However, cross-channel relationship modeling with dimensionality reduction may inadvertently have a negative impact on deep visual representation. Therefore, the efficient multi-scale attention (EMA) module represents the latest progress in cross-space learning to solve this problem. This module preserves the integrity of channel types without dimensionality reduction and effectively combines multi-scale spatial information, which can greatly improve the feature representation ability of the model. By combining the partial convolution method of FasterNet with the multi-scale attention mechanism, we introduce an EMA Faster Block to replace the Bottleneck structure in C2f of YOLOv10m. This new design improves detection accuracy while also reducing computational load. Figure 1 illustrates the specific details of this new structure.

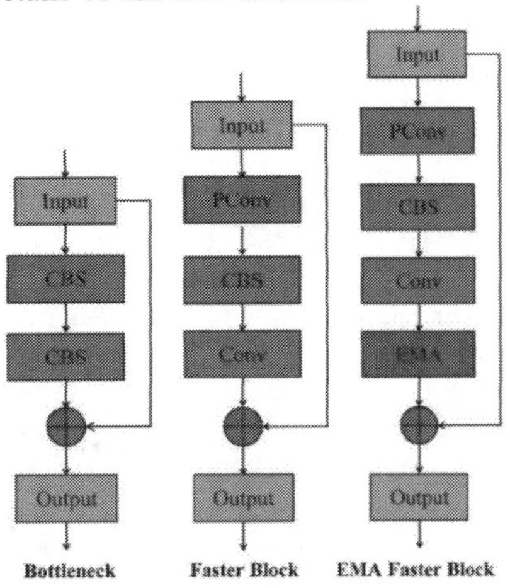

Figure 1. Comparison of bottleneck, faster block, and EMA faster block structures.

In this study, we introduced the C2f-Faster-EMA structure in place of the traditional C2f design, illustrated in Figure 2. This updated configuration also substitutes the previous Bottleneck structure. The C2f-Faster-EMA now incorporates three EMA Faster Blocks, as demonstrated in the provided example.

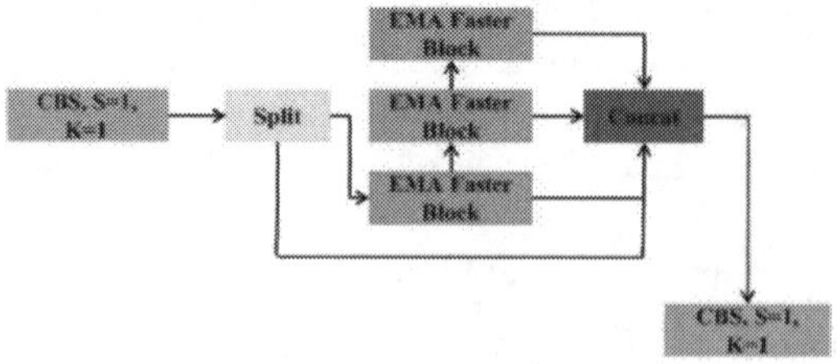

Figure 2. C2f-Faster-EMA module diagram.

2.2 FPSConv

Multi-scale feature extraction is crucial in computer vision tasks [3]. By incorporating varying dilation rates in convolutional layers, the module can extract features at different scales, which is highly beneficial for capturing objects of varying sizes and contextual information within images. Specifically, convolutional layers with low dilation rates are adept at capturing fine local details, such as edges and textures, while those with high dilation rates excel in capturing global context, including the relative positions of objects and background information. This strategy of multi-scale feature extraction allows the model to achieve a deeper understanding of image content, resulting in enhanced performance across a range of vision tasks.

As shown in Figure 3, to enhance model efficiency and reduce redundancy, parameter-sharing techniques have been widely adopted. By employing shared convolutional layers (ShareConv), the model no longer requires separate convolutional layers to be trained for each dilation rate. This sharing mechanism greatly reduces the number of trainable parameters, leading to decreased storage and computational demands, and significantly enhancing computational efficiency. Additionally, by utilizing two 1×1 convolutional layers, the module can efficiently adjust the number of channels and perform feature fusion in the process. This approach not only preserves critical feature information but also empowers the model with greater expressive capability while maintaining computational efficiency. In comparison, the convolution operations in FeaturePyramidSharedConv (FPSConv) can capture finer-grained features, whereas traditional pooling operations like SPPF may result in the loss of some details. Consequently, convolution operations exhibit greater flexibility and expressiveness in feature extraction, allowing for the better capture of complex patterns and subtle distinctions within images.

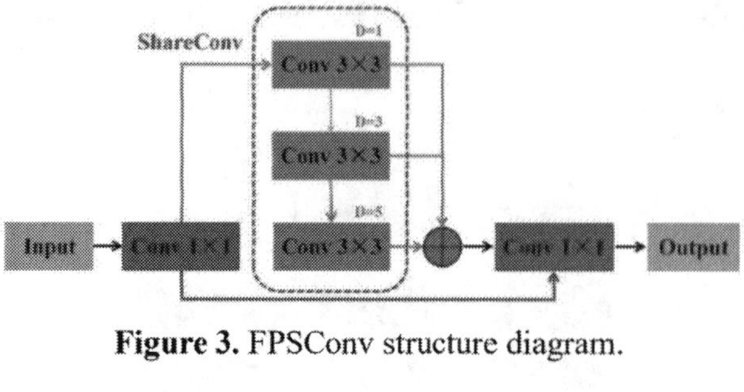

Figure 3. FPSConv structure diagram.

2.3 Detect-LSCD

For the head structure, we developed a dynamic detection head known as Detect-LSCD (Lightweight Shared Convolutional Detection Head), which effectively balances minimal reduction in detection accuracy with a lightweight model, as depicted in Figure 4. To enhance airport security, we incorporated the shared convolution technique Conv-GN (Group Normalization) [4] which has been shown to improve the detection head's positioning and classification performance. GN [5] is an advanced deep-learning normalization method that can replace BN (Batch Normalization), offering significant benefits in terms of faster training and convergence. Unlike BN, which relies on batch size for normalization, GN overcomes this limitation. Convolutions typically extract unstructured features or vectors, where features learned from the same image should ideally share a common distribution, enabling features with similar distributions to be grouped. This means that within each layer, multiple convolutional kernels learn features that are not entirely independent, with some sharing distributions that help accelerate training and convergence. By incorporating shared convolutions in Detect-LSCD, the model significantly reduces parameters, computational complexity, and overall size, making it ideal for deployment on resource-constrained devices. Moreover, to address variations in target sizes detected by each detection head, a scale layer is employed to adjust the scale of features accordingly.

Overall, the Pose-LSCD design ensures the detection head maintains high accuracy while minimizing parameters and computational complexity. The finalized structure of the improved YOLOv10m is illustrated in Figure 5.

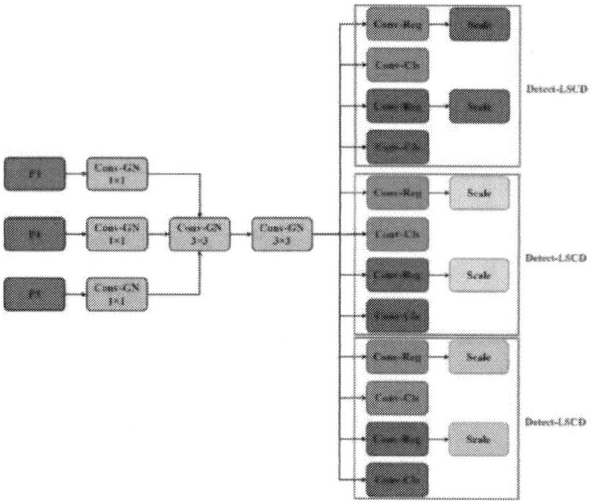

Figure 4. Diagram of the detect-LSCD structure.

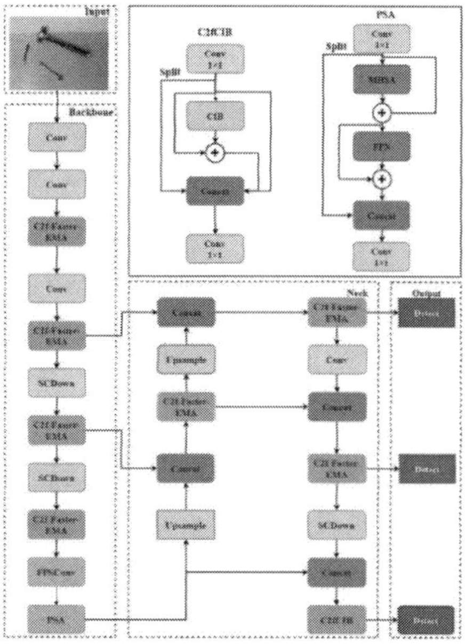

Figure 5. Improved YOLOv10m structure diagram.

3. Result and analysis

3.1 Experimental environment and metric

To validate the proposed method's effectiveness, we conducted experiments on a system running Ubuntu 20.04 utilizing Python 3.8. The hardware setup included an Intel Core i7-13700K CPU, an NVIDIA GeForce RTX 4080s GPU, and 32 GB of RAM. The training was performed with a learning rate of 0.01 and SGD as the optimizer. The model was trained for 200 epochs with a batch size of 128. This configuration was carefully selected to balance computational efficiency with the desired performance outcomes.

We used different key metrics for algorithm evaluation. mAP (mean average precision) is employed to assess the algorithm's accuracy. The term parameters denote the total number of trainable parameters within the network model. GFLOPS (giga floating-point operations per second) measures the model's computational complexity, indicating the number of billions of floating-point operations executed each second. Additionally, detection speed is evaluated by measuring the number of frames per second (FPS) the model can process, which reflects its real-time performance. The model size refers to the storage space required to save the trained model, which is crucial for deployment, especially in resource-constrained environments.

3.2 Dataset

The selection of object categories for FOD-A[6] was guided by FAA documents and relevant research to ensure the dataset's representativeness and practical applicability. The dataset, comprising 33, 793 images, was divided into training, validation, and test sets in an 8:1:1 ratio, with all images having a resolution of 300×300 to balance image quality and computational efficiency. This rigorous partitioning ensures the model gains sufficient diversity and generalization capability for accurate and reliable detection in real-world applications. To enhance robustness across different environments, we implemented an online data augmentation strategy, simulating complex weather conditions such as fog, rain, snow, shadows, and sun glare. Additionally, we not only incorporated potential adverse weather scenarios but also applied various data augmentation techniques to the images, such as random cropping, center cropping, and random rotation. This process allows the model to learn and recognize features under various conditions, significantly improving its adaptability in real-world scenarios. Figures 6 and 7 respectively show the distribution of FOD labels and the different augmentation methods used.

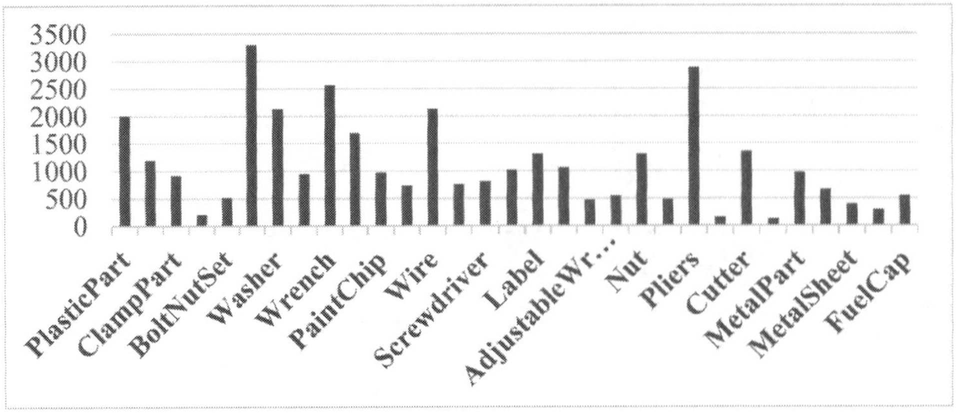

Figure 6. The distribution of images among the 31 classes in the dataset.

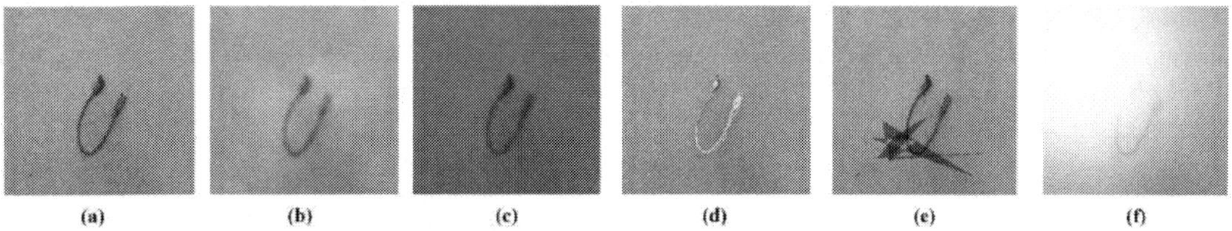

Figure 7. Online dataset enhancement methods for various severe weather conditions. (a) Original; (b) RandomFog; (c) RandomRain; (d) RandomSnow; (e) Randomshadow; (f) RandomSunFlare.

3.3 Comparison of C2f-Faster with different attention mechanisms

To demonstrate the effectiveness of our C2f-Faster-EMA structure, we conducted experiments by using C2f-Faster with various attention mechanisms. As demonstrated in the experimental results in Table 1, while the accuracy of the improved YOLOv10m with C2f-Faster-EMA is slightly lower than the original version, there is a significant reduction in model size, parameter count, and computational complexity.

Table 1. The comparative experiment of C2f-Faster and different attention mechanisms.

Attention	mAP@50-95	Module size (M)	Para (M)	GFLOPs
None	**0.906**	33.5	15.33	59.7
+C2f-Faster-SE [7]	0.885	26.5	11.75	38.9
+C2f-Faster-GC [8]	0.849	26.6	11.75	39.0
+C2f-Faster-TripletAttention [9]	0.881	26.5	**11.74**	39.0
+C2f-Faster-ECA [10]	0.865	26.5	11.99	39.0
+C2f-Faster-GAM [11]	0.872	35.0	13.49	34.3
+C2f-Faster-SGE [12]	0.849	26.6	11.79	39.0
+C2f-Faster-EFA [13]	0.877	26.7	11.80	39.2
+C2f-Faster-MCAttn [14]	0.884	26.6	11.79	39.1
+C2f-Faster-EMA	0.898	**26.5**	11.78	39.3

3.4 Ablation experiment

As shown in Table 2, from the experimental results, when we introduce the C2f-Faster-EMA module alone, although the accuracy is slightly reduced, the number of model parameters and computational complexity have a clear downward trend; although the FPSConv module improves the detection accuracy, the cost is a slight increase in computational complexity. The Detect-LSCD module demonstrated the best performance in terms of accuracy improvement but required more computational resources. Ultimately, when we combined the C2f-Faster-EMA, FPSConv, and Detect-LSCD modules, the model achieved near-optimal detection accuracy while maintaining the lowest computational complexity and parameter count. This demonstrates that the synergistic effect of these modules successfully balances lightweight design with efficient detection performance.

Table 2. The ablation experiment of the module.

C2f-Faster-EMA	FPSConv	Detect-LSCD	mAP@50	mAP@50-95	Module size (M)	Para (M)	GFLOPs
			0.991	0.906	33.5	15.33	59.7
√			0.992	0.898	26.5	11.78	39.3
	√		0.993	0.915	41.7	19.44	64.1
		√	0.994	0.923	37.3	18.46	72.0
√	√		0.991	0.895	28.0	12.52	39.3
	√	√	0.993	0.907	38.4	19.22	72.3
√	√	√	0.993	**0.912**	**25.1**	12.29	47.2

3.5 Performance comparison of different models

As shown in Table 3, we compared several algorithms, including Faster-RCNN [15], CenterNet, RT-DETR-L, YOLOv9m, and YOLOv10m, in terms of detection accuracy and computational resource requirements. Although Faster-RCNN delivers high accuracy, its considerable computational complexity renders it impractical for real-time applications. CenterNet [16] and RT-DETR-L [17] reduce the demand for computational resources but at the expense of accuracy. The YOLO series algorithms, particularly YOLOv9m [18] and YOLOv10m, strike a relatively balanced performance, significantly reducing computational complexity while maintaining high accuracy. Ultimately, our model surpasses all other algorithms with the highest accuracy, coupled with the lowest parameter count and computational complexity, demonstrating exceptional detection performance and efficiency.

Table 3. The performance table of different models.

Algorithms	mAP@50	mAP@50-95	Module size (M)	Para (M)	GFLOPs
Faster-RCNN	0.986	0.870	108.2	136.70	940.85
CenterNet	0.963	0.843	124.8	32.70	70.2
RT-DETR-L	0.981	0.868	59.1	56.40	105.5
YOLOv3m	0.990	0.887	30.9	47.91	121.8
YOLOv5m	0.985	0.899	40.8	25.06	64.1
YOLOv6m	0.988	0.877	72.6	52.00	161.3
YOLOv8m	0.990	0.898	49.7	25.89	78.9
YOLOv9m	0.991	0.910	38.8	20.36	78.3
YOLOv10m	0.991	0.906	33.5	15.33	59.7
Our	**0.995**	**0.912**	**25.1**	**12.29**	**47.2**

4. Conclusion

In this study, we presented an enhanced version of YOLOv10m, tailored for real-time Foreign Object Debris (FOD) detection on airport runways. Our model integrates PConv and EMA attention mechanisms, resulting in improved detection accuracy with a mAP@50 of 0.995, a 0.4% increase, and a mAP@50-95 of 0.912, marking a 0.7% improvement over the original YOLOv10m. These advancements were achieved while significantly reducing the model's complexity; the parameter count was decreased by 19.8%, from 15.33 million to 12.29 million, and GFLOPs were reduced by 20.9%, from 59.7 to 47.2. Additionally, the incorporation of Feature Pyramid Shared Convolution (FPSConv) and a lightweight shared detection head (Detect-LSCD) ensures that the model remains lightweight without compromising precision. These enhancements offer a more efficient and accurate solution compared to the original YOLOv10m, making our model particularly well-suited for the rigorous demands of busy airports. While our tests were conducted on public datasets, these datasets have been designed to closely simulate the complexity of real airport environments, including various weather conditions and lighting variations. We employed an online data augmentation strategy to incorporate potential scenarios and applied multiple data augmentation techniques to further approximate the complexities of real-world applications. Therefore, we believe that the model's performance on public datasets is a reliable indicator of its effectiveness and accuracy in real environments. However, we acknowledge the importance of further testing and feedback collection in actual airport settings. We plan to carry out more field validations and optimizations in future work.

Acknowledgments

This research was funded by the Inner Mongolia Autonomous Region Science and Technology Plan Project "Research on Key Technologies for Intelligent Analysis and Decision-making of Big Data in 5G Smart Highway Operations" (2022YFSJO040) and the Autonomous Region's Directly Affiliated Universities Basic Scientific Research Business Expenses Project "Research on Intelligent Analysis Technology of Road Conditions Big Data of G6 Expressway Huhhot-Baotou Section" (JY20220170).

References

[1] A. Wang et al. "YOLOv10: Real-Time End-to-End Object Detection", May 23, 2024, *arXiv*: arXiv: 2405.14458. Doi: 10.48550/arXiv.2405.14458.
[2] W. Wang et al. "InternImage: Exploring Large-Scale Vision Foundation Models with Deformable Convolutions", Apr. 17, 2023, *arXiv*: arXiv: 2211.05778. Doi: 10.48550/arXiv.2211.05778.
[3] W. Lu, S. B. Chen, J. Tang, C. H. Q. Ding, and B. Luo. "A Robust Feature Downsampling Module for Remote-Sensing Visual Tasks", *IEEE Transactions on Geoscience and Remote Sensing*, Vol. 61, pp. 1-12, 2023. Doi: 10.1109/TGRS.2023.3282048.
[4] Z. Tian, C. Shen, H. Chen, and T. He. "FCOS: Fully Convolutional One-Stage Object Detection", in *2019 IEEE/CVF International Conference on Computer Vision (ICCV)*, Seoul, Korea (South): IEEE, Oct. 2019, pp. 9, 626-9, 635. Doi: 10.1109/ICCV.2019.00972.
[5] Y. Wu and K. He. "Group Normalization", presented at the Proceedings of the European Conference on Computer Vision (ECCV), 2018, pp. 3-19. Accessed: Mar. 15, 2024. [Online]. Available: https://openaccess.thecvf.com/content_ECCV_2018/html/Yuxin_Wu_Group_Normalization_ECCV_2018_paper.html.
[6] T. Munyer, P. C. Huang, C. Huang, and X. Zhong. "FOD-A: A Dataset for Foreign Object Debris in Airports", Jan. 26, 2022, *arXiv*: arXiv: 2110.03072. Accessed: Jan. 26, 2024. [Online]. Available: http://arxiv.org/abs/2110.03072.
[7] J. Hu, L. Shen, S. Albanie, G. Sun, and E. Wu. "Squeeze-and-Excitation Networks", May 16, 2019, *arXiv*: arXiv: 1709.01507. Doi: 10.48550/arXiv.1709.01507.
[8] Y. Cao, J. Xu, S. Lin, F. Wei, and H. Hu. "GCNet: Non-local Networks Meet Squeeze-Excitation Networks and Beyond", Apr. 25, 2019, *arXiv*: arXiv: 1904.11492. Doi: 10.48550/arXiv.1904.11492.
[9] D. Misra, T. Nalamada, A. U. Arasanipalai, and Q. Hou. "Rotate to Attend: Convolutional Triplet Attention Module", Nov. 05, 2020, *arXiv*: arXiv: 2010.03045. Accessed: Dec. 09, 2023. [Online]. Available: http://arxiv.org/abs/2010.03045.
[10] Q. Wang, B. Wu, P. Zhu, P. Li, W. Zuo, and Q. Hu. "ECA-Net: Efficient Channel Attention for Deep Convolutional Neural Networks", Apr. 07, 2020, *arXiv*: arXiv: 1910.03151. Doi: 10.48550/arXiv.1910.03151.
[11] Y. Liu, Z. Shao, and N. Hoffmann. "Global Attention Mechanism: Retain Information to Enhance Channel-Spatial Interactions", Dec. 10, 2021, *arXiv*: arXiv: 2112.05561. Doi: 10.48550/arXiv.2112.05561.
[12] X. Li, X. Hu, and J. Yang. "Spatial Group-wise Enhance: Improving Semantic Feature Learning in Convolutional Networks", May 25, 2019, *arXiv*: arXiv: 1905.09646. Doi: 10.48550/arXiv.1905.09646.
[13] Q. Zhou, Q. Wang, Y. Bao, L. Kong, X. Jin, and W. Ou. "LAEDNet: a lightweight attention encoder-decoder network for ultrasound medical image segmentation", *Computers and Electrical Engineering*, Vol. 99, p. 107, 777, 2022.
[14] W. Dai et al. "Exploiting Scale-Variant Attention for Segmenting Small Medical Objects", Aug. 05, 2024, *arXiv*: arXiv: 2407.07720. Doi: 10.48550/arXiv.2407.07720.
[15] S. Ren, K. He, R. Girshick, and J. Sun. "Faster R-CNN: Towards Real-Time Object Detection with Region Proposal Networks", in *Advances in Neural Information Processing Systems*, Curran Associates, Inc., 2015. Accessed: Dec. 03, 2023. [Online]. Available: https://proceedings.neurips.cc/paper_files/paper/2015/hash/14bfa6bb14875e45bba028a21ed38046-Abstract.html.
[16] K. Duan, S. Bai, L. Xie, H. Qi, Q. Huang, and Q. Tian. "CenterNet: Keypoint Triplets for Object Detection", Apr. 18, 2019, *arXiv*: arXiv: 1904.08189. Doi: 10.48550/arXiv.1904.08189.
[17] W. Lv et al. "DETRs Beat YOLOs on Real-time Object Detection", arXiv.org. Accessed: Jan. 12, 2024. [Online]. Available: https://arxiv.org/abs/2304.08069v2.
[18] C. Y. Wang, I. H. Yeh, and H. Y. M. Liao. "YOLOv9: Learning What You Want to Learn Using Programmable Gradient Information", Feb. 28, 2024, *arXiv*: arXiv: 2402.13616. Doi: 10.48550/arXiv.2402.13616.

Space approaching assessment of SVOM satellite to other space objects

Dong Li[1,2], Hu Jiang[1,2,*], Xiaofeng Zhang[1,2], Ruifeng Su[1,2] and Junwang He[1,2]

[1]Innovation Academy for Microsatellites, Chinese Academy of Sciences, Shanghai 201304, China
[2]Shanghai Engineering Center for Microsatellites, Shanghai 201304, China

[*]Corresponding author's e-mail: jiangh@microsate.com

Abstract: The SVOM satellite is one of the joint space projects between China and France. It was successfully launched into the preferred orbit in June 2024. Its goal is to monitor the Gamma Ray Burst in space. The mission consists of 4 main instruments: ECLAIRs, MXT, GRM, and VT. The ECLAIRs telescope is used to detect and position gamma bursts in the X-ray band and low-energy gamma rays. Micro-channel X-ray Telescope (MXT telescope) will make the observation of gamma bursts in the soft X-ray range. GRM (Gamma Ray Burst Monitor) is used to measure the spectrum of high-energy bursts. VT telescope (Visible Telescope) operates in the visible range to detect and observe the visible emission produced immediately after a gamma burst. Based on recent orbit elements for the SVOM satellite, which was available via TT&C, SOCRATES Plus software is used to carry out space collision assessment for the SVOM satellite. SVOM and the accompanying satellite shared the same rocket before they were sent into their own preferred orbits. During the simulations, the mean main orbit elements are derived from the GNSS PV data sets, which can help to know the orbit health of SVOM. On June 27, 2024, the distance between SVOM and the accompanying satellite varies between 509 km and 620 km, and the distance between SVOM and the rocket residual varies between 441 km and 5263 km. On July 2, 2024, the distance between SVOM and the accompanying satellite varies between 1067 km and 1180 km, and the distance between SVOM and the rocket residual varies between 357 Km and 5107 Km. In addition, space approaching the SVOM satellite and other objects is simulated for five-day intervals; it turns out that there are three space objects approaching the SVOM satellite with about 4 km ~ 5 Km separation.

1. Introduction

On June 22, 2024, Space Variable Objects Monitor (SVOM) was launched into low Earth orbit by a CZ-2C rocket in the Xichang satellite launch center [1]. Its weight is about 930 kg; its lifespan is at least three years. Its scientific goal is to detect, position, and study Gamma-ray bursts and other high-energy cosmic phenomena. SVOM includes platform and payload; the payload is composed of four apparatuses [2], which include the Gamma-ray monitor (GRM), Visual Telescope (VT) [3], the soft γray imager (ECLAIR), and Micro-channel X-ray Telescope (MXT) [4]. Attitude Determine Accuracy is ±40 arcsec (3σ) in the Xs component, ±20 arcsec (3σ) in the Ys component, and ±20 arcsec (3σ) in the Zs component. Relative Performance Error (RPE) is 12 arcsec /100s in the Xs component, 0.8 arcsec/100s in the Ys component, and 0.8 arcsec in the Zs component [5]. Such high precision attitude performance provides a stable platform for payloads. During the first six months, the SVOM team will carry out in-orbit testing of all SVOM satellites; as one of the testing tasks, the space safety assessment of SVOM will be important. Herein, orbit derived from telemetry is considered and analyzed; the relative distance between SVOM and rocket residual and the relative distance between SVOM and accompanying satellite are simulated, respectively. In addition to accompanying satellite and rocket residuals, simulations of the approach between SVOM and other space objects are also carried out. The authors of this research have the privilege of having access to the most precise orbit elements for SVOM, the accompanying satellite, and the rocket residual. In addition, SOCRATES Plus software has embedded comprehensive and rich data sets of orbit elements for a great number of in-orbit space objects. These advantages will lead to helpful research results, which will benefit the safe operation and in-orbit testing of the SVOM satellite. SVOM space engineering needs high-quality reliability, including space safety, in order for scientists to carry out efficient scientific observations by means of the payloads on the SVOM platform. Within one month after SVOM was deployed in space, scientists have successfully detected 3events of a Gamma-ray burst, which might lead to possible new discoveries. The work of this paper is one of the prerequisites for a successful SVOM mission. The work of this paper has a more practical engineering meaning than a scientific meaning. However, the reliable engineering of SVOM will support greater scientific discoveries for scientists of the SVOM mission.

2. SVOM orbit element feature

At a certain epoch of 2024/7/2 7:12:35 UTC (denoted by T0), GNSS-based position and velocity are presented as followings: X=-3645.895km, Y=5982.625Km, Z=238.132km, Xdot=-5.707558Km/s, Ydot=-3.331566Km/s, Zdot=-3.636772Km/s. The mean orbit element for SVOM is derived from GNSS-based position and velocity. The main conversion process includes the following steps: First, six instantaneous orbit elements can be derived from the given position and velocity of the satellite according to basic two-body theory; second, six mean orbit elements can be derived by subtracting the short-term variations from the six instantaneous orbit elements.

The horizontal axes for Figure 1, Figure 2, and Figure 3 are the time that is accumulated with respect to the epoch of 2024/7/2 7:12:35 UTC. From Figures 1, 2, 3, we can see the semi-major axis varies between 7012.6Km and 7012.95Km during the day considered, and the eccentricity varies between 0.00047 and 0.00058, and the inclination varies between 28.875°and 28.883°. The orbit elements satisfy the requirements of SVOM engineering.

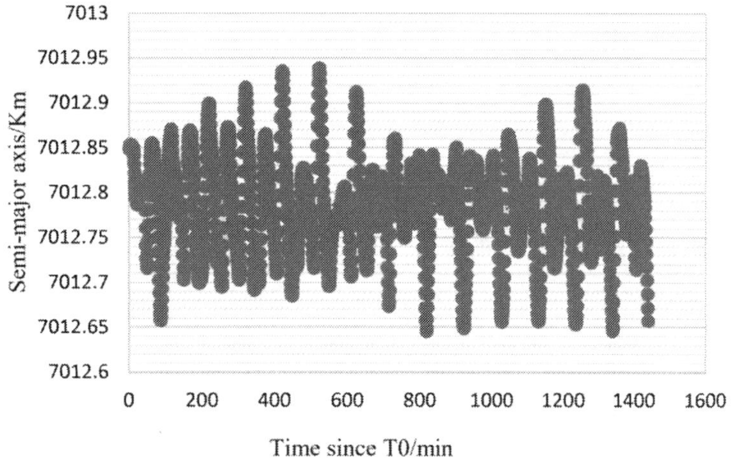

Figure 1. Semi-major axis variation with time.

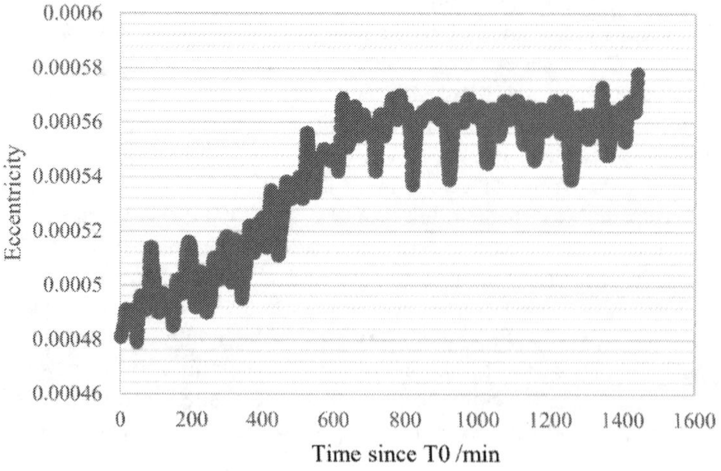

Figure 2. Eccentricity variation with time.

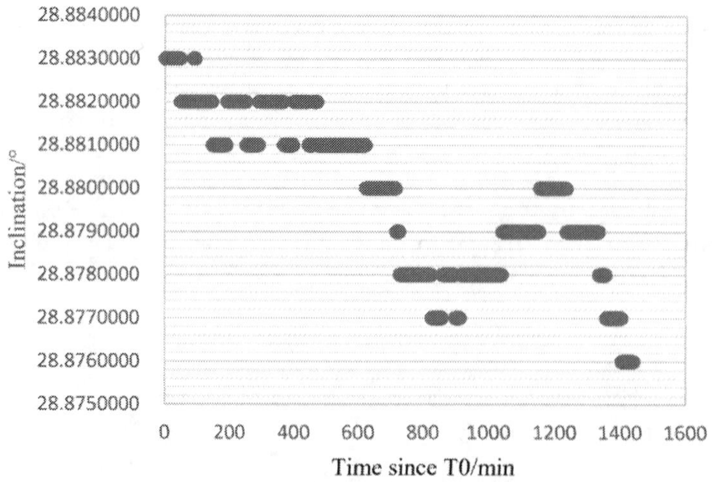

Figure 3. Inclination variation with time.

3. Relative distance simulation

During the simulation, the positions for SVOM, accompanying satellite, and rocket residual are expressed in J2000 coordinates (X, Y, Z), which are converted from instantaneous orbit elements of SVOM, accompanying satellite, and rocket residual. The step size is set to be one minute. Given coordinates (X1, Y1, Z1) for SVOM and coordinates (X2, Y2, Z2) for the accompanying satellite, the relative distance between SVOM and the accompanying satellite can be derived by sqrt ((x1-x2)^2+(y1-y2)^2+(z1-z2)^2). Assuming that (X3, Y3, Z3) is the coordinates for the rocket residual, the relative distance between SVOM and the rocket residual can be derived by sqrt ((x1-x3)^2+(y1-y3)^2+(z1-z3)^2).

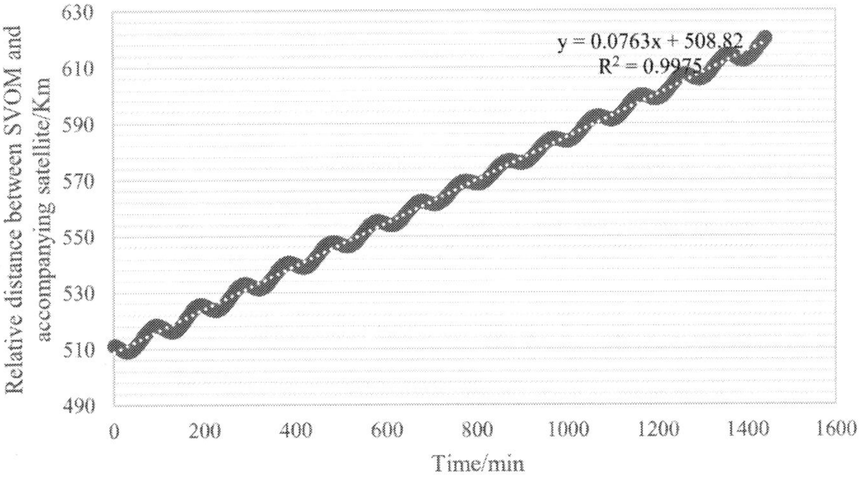

Figure 4. Relative distance between SVOM and accompanying satellite On Jun. 27.

From Figure 4, the relative distance between SVOM and the accompanying satellite on June 27 varies between 509 km and 620 km. There is an obvious increasing trend quasi-linearly. The trend line is fitted as y=0.0763x+508.82 based on the measurements.

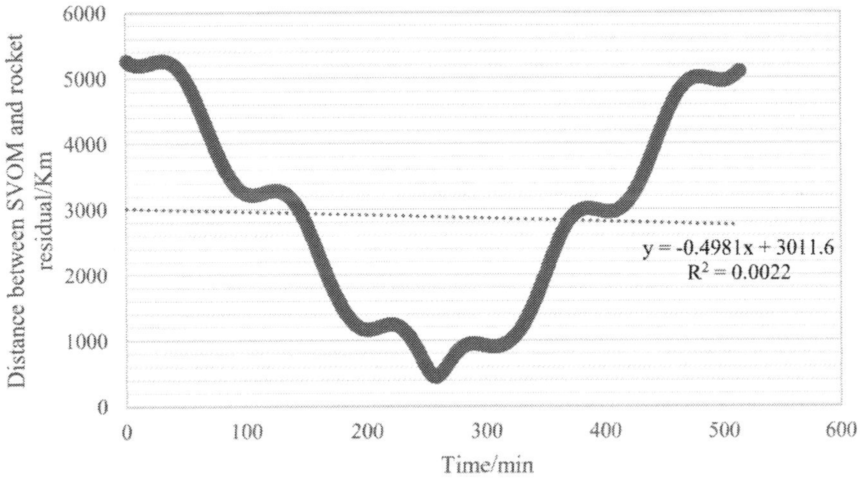

Figure 5. Relative distance between SVOM and rocket residual On Jun. 27.

From Figure 5, the relative distance between SVOM and rocket residual on Jun. 27 varies between 441 km and 5263 km. There is an obvious "V" shape variation, which shows a one-day period. In order to re-confirm the relative distance variation trend, we made another simulation with another day's orbit element data set and space objects' data set on July 2, 2024.

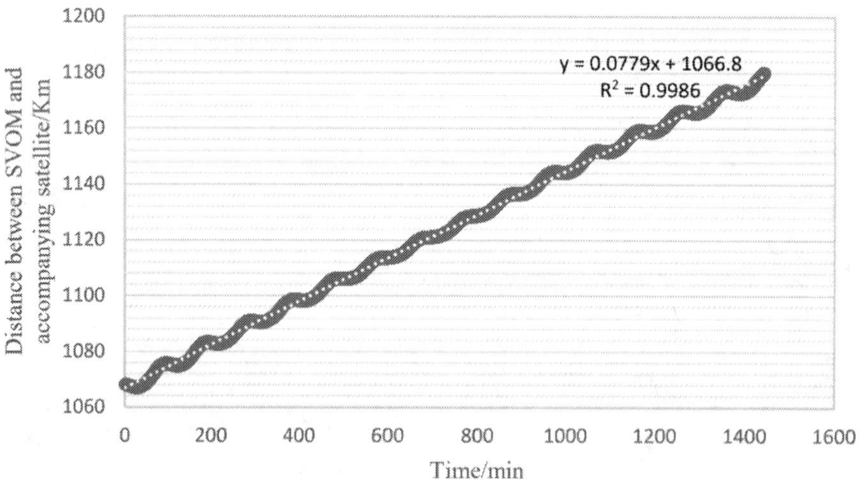

Figure 6. Relative distance between SVOM and accompanying satellite on July 2.

From Figure 6, the relative distance between SVOM and accompanying satellite on July 2 varies between 1067 km and 1180 km. There is an obvious increasing trend quasi-linearly.

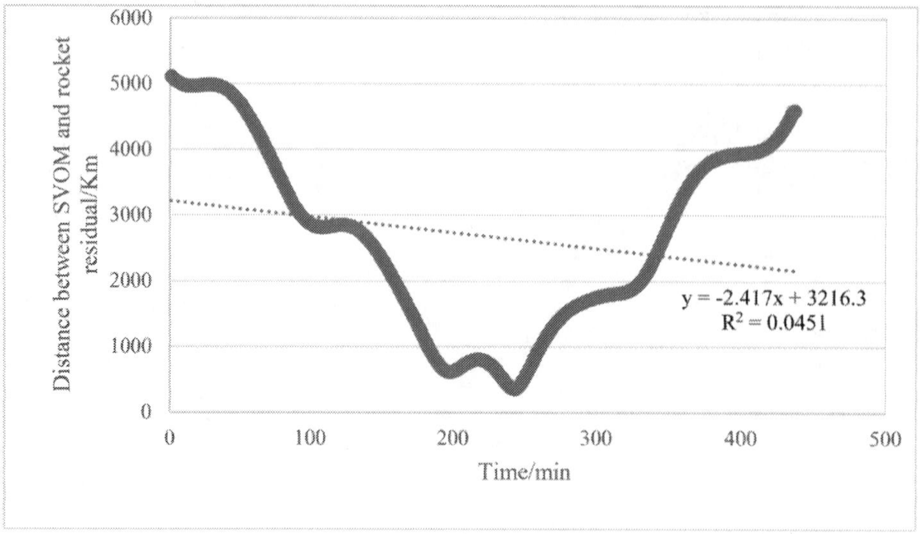

Figure 7. Relative distance between SVOM and rocket residual On July 2.

From Figure 7, the relative distance between SVOM and rocket residual on July 2 varies between 357 Km and 5107 Km. There is an obvious "V" shape variation, which shows a one-day period.

In addition to the rocket residual and accompanying satellite, we have carried out further simulations of the approach between SVOM and another space object. Orbit elements dataset of space objects published on the website (https://celestrak.org/) are considered [6]. Also, online software for predictions of space objects approaching is used.

Orbit elements in the form of TLE for SVOM are listed as follows:

1 60089U 24116B 24211.84729825 00002642 00000+0 32825-3 0 9998
2 60089 28.9951 131.5473 0005645 240.6981 119.3047 14.79724329 5539

Orbit elements in the form of TLE for object 43641 are listed as follows:

1 43641U 18076A 24211.81114327 00000938 00000+0 12436-3 0 9995

2 43641 97.8893 37.4190 0001513 89.9361 270.2026 14.82161092314285
Orbit elements in the form of TLE for object 58496 are listed as follows:
1 58496U 23187A 24211.83184512 00004173 00000+0 55769-3 0 9994
2 58496 97.9856 287.4860 0012039 216.2338 143.8060 14.80015975 35272
Orbit elements in the form of TLE for object 58498 are listed as follows:
1 58498U 23187C 24211.83104364 00003644 00000+0 48883-3 0 9994
2 58498 97.9862 287.5012 0011408 213.1743 146.8755 14.79935822 35269
Based on the above orbit elements, the simulation results are presented in Table 1.

Table 1. Space approaching between SVOM satellite and other objects.

NORAD Catalog Number	Object Name [Ops Status]	Time of closest approach/UTC	Minimum range/km	Maximum probability
43641	SAOCOM 1A [+]	2024-07-06 20: 42: 41.261	3.998	1.892E-07
58496	OBJECT A [+]	2024-07-07 07: 35: 03.808	4.314	2.003E-07
58498	OBJECT C [+]	2024-07-08 09: 31: 30.132	4.997	1.495E-07

Table 1 shows that there are three objects approaching the SVOM satellite within 5 km on July 6, July 7, and July 8. Such events are dangerous warming of space collision, although the probability is only at a level of 10^{-7}. SVOM satellite has the capability of orbit maneuvering, so it can help itself out of mishaps.

4. Conclusion

Based on the above simulations, we can come to the following conclusions. 1) The distance between SVOM and the accompanying satellite becomes quasi-linearly greater with time. 2) The distance between SVOM and the rocket residual has shown that there is an obvious "V" shape variation within a one-day period. 3) Space approaching the SVOM satellite and other objects is simulated for five-day intervals; it turns out that there are three space objects approaching the SVOM satellite with about 4 ~ 5 Km separation, which indicates a possible future collision between SVOM and other objects. The more precise prediction of possible space collision is, the more reliable countermeasures the SVOM task team can plot. The simulations support and benefit the efficient in-orbit testing of SVOM and also provide a reliable warning of possible space collision crises. It is suggested that weekly reports of space safety be made to prevent SVOM from colliding with space objects.

Acknowledgments

The part of simulations for this research is carried out by the online software SOCRATES Plus; Thank Kelso, T. S. for his website sources [6].

References

[1] Huaxia, China launches new astronomical satellite developed in cooperation with France, 2024, https://english.news.cn/20240622/c8a8e5bfe1194c1bbe925f2e63aaa552/c.h tml.
[2] Wang J, Qiu Y L, Wei J Y, 2020, A pilot study of catching high-z GRBs and exploring circumburst environment in the forthcoming SVOM era, Res. Astron. Astrophys. 20(8): 124(6pp). https://doi.org/10.1088/1674-4527/20/8/124.
[3] Pan Y, 2022, Research on radiometric calibration of the SVOM Visible Telescope PASP 134: 037001(4pp). https://doi.org/10.1088/1538-3873/ac5849.
[4] Shaymaa Hussein, Florent Robinet, Martin Boutelier, et al., 2023, Analysis Methods to Localize and Characterize X-Ray Sources with the Microchannel X-Ray Telescope on Board the SVOM Satellite, ApJ, 944: 170(12pp). https://doi.org/10.3847/1538-4357/acb539.

Journal of Physics: Conference Series **2879** (2024) 012046 doi:10.1088/1742-6596/2879/1/012046

[5] Liu S, Qin G J, Li D, et al., 2021, Attitude Control System Design and Simulation for the SVOM Mission, J. Phys.: Conf. Ser. 1861: 012091(7pp). https://doi.org/10.1088/1742-6596/1861/1/012091.
[6] Kelso T S, 2024, NORAD GP Element Sets, https://celestrak.org/NORAD/elements/.

Journal of Physics: Conference Series **2879** (2024) 012047 doi:10.1088/1742-6596/2879/1/012047

Parameter optimization control of multi-rotor unmanned aerial vehicles based on adaptive differential evolution

Yinglong Hu[1,2,*]

[1]Zhengzhou Electromechanical Engineering Research Institute, Zhengzhou, Henan, 450015, China
[2]North University of China, Taiyuan, Shanxi, 030051, China

[*]Corresponding author's e-mail: tailiula@163.com

Abstract: Aiming at the problems of instability and underactuation encountered in the control process of traditional multi-rotor unmanned aerial vehicles, an adaptive differential evolution algorithm is designed to optimize the parameters of the sliding mode controller. The advantages of the fast convergence of the sliding mode controller and the global search of the evolutionary algorithm are combined for parameter optimization. The simulation results show that the sliding mode controller, after parameter optimization, has good stability.

1. Introduction

Multi-rotor aircraft are widely used in various scenarios due to their advantages, such as low noise, small size, high safety, and strong mobility, such as power inspection in figure 1, logistics transportation in figure 2, agricultural plant protection, and post-disaster search and rescue, etc. It has brought great convenience to people's lives. However, the research process of multi-rotor aircraft is restricted by its own physical structure, and there are also many control difficulties, such as underactuation, high coupling, nonlinearity, multivariable and other factors, making the research of multi-rotor controllers need to be continuously deepened.[1]

The differential evolution algorithm was proposed by Storn and Price. It has three key control parameters, namely population size, scaling factor, and crossover rate. After selecting the appropriate control parameters, the population is continuously iterated to accumulate the optimal value to achieve the optimal selection of parameters. The traditional differential evolution algorithm has high computational costs and different crossover and mutation strategies. The adaptive differential evolution algorithm used in this article can adaptively select crossover and mutation strategies through continuous iterative learning, thereby significantly reducing the computational cost and ensuring the optimal selection of parameters.[2]

Figure 1. Electric power inspection and patrol.

Figure 2. Transportation of takeout.

2. Adaptive differential evolution algorithm

The differential evolution algorithm initializes the population ***NP*** and generates a matrix of a certain population size according to the number of required parameters ***D***, as shown in equation (1)

$$X_{i,G}=\left\{x_{i,G}^{1},\cdots,x_{i,G}^{D}\right\},i=1,\cdots,NP \tag{1}$$

Meanwhile, the matrix elements are randomly generated within the prescribed minimum and maximum boundaries, and through multiple iterations ***G***, the entire search space is covered as much as possible, as given by equation (2).

$$\begin{cases}X_{\max}=\left\{x_{\max}^{1},\cdots,x_{\max}^{D}\right\}\\ X_{\min}=\left\{x_{\min}^{1},\cdots,x_{\min}^{D}\right\}\end{cases} \tag{2}$$

First, the initial population is generated in equation (3):

$$x_{i,0}^{j}=(x_{\max}^{j}-x_{\min}^{j})\cdot\mathrm{rand}(0,1)+x_{\min}^{j},j=1,2,\cdots,D \tag{3}$$

The random variable from 0 to 1 is given by rand(0,1).

After the population initialization, the mutation strategy $V_{i,G}$ is used to operate on $X_{i,G}$ to generate the corresponding mutation matrix, as follows

$$V_{j,G}=X_{r_1^j,G}+\mathrm{F}\cdot\left(X_{r_2^j,G}-X_{r_3^j,G}\right)+\mathrm{F}\cdot\left(X_{r_4^j,G}-X_{r_5^j,G}\right) \tag{4}$$

In equation (4), r_1^j, r_2^j, r_3^j, r_4^j, and r_5^j is the randomly defined index number during the calculation process, and ***j*** is the index of a certain row of the matrix. **F** is the mutation scaling factor.

After the population undergoes mutation, next, the matrix generated after the mutation $V_{i,G}$ undergo crossover processing with the population matrix $X_{i,G}$ to obtain a new matrix[3]

$$u_{i,G}^{j}=\begin{cases}v_{i,G}^{j},\mathrm{if}(\mathrm{rand}_j[0,1)\leq\mathrm{CR})\mathrm{or}(j=j_{rand})\\ x_{i,G}^{j},otherwise\end{cases} j=1,2,\cdots,D \tag{5}$$

In equation (5), ***CR*** is the crossover probability factor, which affects the diversity of the population. Increasing its value can increase the probability of new individuals in the population, and reducing its value is conducive to stable search.

The matrix generated after the population undergoes mutation and crossover can be used for the selection of parameters. The fitness function, as a performance indicator for parameter selection, also plays an important role. This article selects the equation (6) as the performance indicator[4]

$$f(x_i)=\int t\left|e_j(t)\right|dt \tag{6}$$

3. Mathematical model of multi-rotor unmanned aerial vehicle

3.1. State equation

The following state-space form is selected:

$$X = [\phi \quad \dot{\phi} \quad \theta \quad \dot{\theta} \quad \psi \quad \dot{\psi} \quad x \quad \dot{x} \quad y \quad \dot{y} \quad z \quad \dot{z}] \tag{7}$$

The state-space form rewriting of the multi-rotor mathematical model is carried out, and the state equation (8) is obtained as follows (equation (9) is the attitude state equation):

$$\begin{cases} \dot{x}_1 = x_2 \\ \dot{x}_2 = \dfrac{1}{J_x}\left\{x_4 x_6 \left(J_y - J_z\right) - F_{dx} x_2^{\ 2} - J_r \varpi x_4 + d\tau_x\right\} \\ \dot{x}_3 = x_4 \\ \dot{x}_4 = \dfrac{1}{J_y}\left\{x_2 x_6 \left(J_z - J_x\right) - F_{dy} x_4^{\ 2} + J_r \varpi x_2 + d\tau_y\right\} \\ \dot{x}_5 = x_6 \\ \dot{x}_6 = \dfrac{1}{J_z}\left\{x_2 x_4 \left(J_x - J_y\right) - F_{dz} x_6^{\ 2} + c_m \tau_z\right\} \end{cases} \tag{8}$$

Equation (9) is the position state equation,

$$\begin{cases} \dot{x}_7 = x_8 \\ \dot{x}_8 = \dfrac{1}{m}\{(\cos\phi\sin\theta\cos\psi + \sin\phi\sin\psi)f - M_{dx} x_8\} \\ \dot{x}_9 = x_{10} \\ \dot{x}_{10} = \dfrac{1}{m}\{(\cos\phi\sin\theta\sin\psi - \sin\phi\cos\psi)f - M_{dy} x_{10}\} \\ \dot{x}_{11} = x_{12} \\ \dot{x}_{12} = \dfrac{1}{m}\{(\cos\phi\cos\theta)f - M_{dz} x_{12}\} - g \end{cases} . \tag{9}$$

The mass of the multi-rotor unmanned aerial vehicle is given by m, and the moment of inertia of the airframe is given by $J \triangleq \left[J_x \quad J_y \quad J_z\right]$. The drag coefficients of the three positions are given by F_{dx} 、 F_{dy} and F_{dz}. Damping torque coefficients are given by M_{dx}, M_{dy}, and M_{dz}. The shaft diameter of the unmanned aerial vehicle is given by d. Motor constant is given by J_r. Gravitational acceleration is given by g. The difference in rotational speeds of the four motors is given by ϖ. The lift force generated by the four motors is given by f. [5]

3.2. Control equation

The selected sliding surface is in equation (10):

$$s_1 = z_2 \tag{10}$$

The roll channel is taken as an example to define the error function and the virtual control law in equation (11):

$$\begin{cases} z_1 = x_1 - x_{d1} \\ z_2 = x_2 - \dot{x}_{1_d} - \alpha_1 z_1 \end{cases} (\alpha_1 > 0) \tag{11}$$

We design a backstepping sliding mode controller and the following reaching law in equation (12)[6],

$$\dot{s} = -\varepsilon\,\mathrm{sgn}(s) - ks, (\varepsilon > 0, k > 0) \tag{12}$$

Combining equations (9) and (11), the position control equation of the roll channel is derived in equation (13):

$$\tau_x = \frac{1}{b_1}\left\{-\varepsilon_1 \operatorname{sign}(s_1) - k_1 s_1 - a_1 x_4 x_6 - a_2 x_2^2 - a_3 \varpi x_4 + \ddot{\phi}_d + \alpha_1(\dot{\phi}_d - x_2)\right\} \tag{13}$$

Similarly, the control equations of the position and attitude of the multi-rotor are derived in equation (14):

$$\begin{cases} \tau_x = \frac{1}{b_1}\left\{-\varepsilon_1 \operatorname{sign}(s_1) - k_1 s_1 - a_1 x_4 x_6 - a_2 x_2^2 - a_3 \varpi x_4 + \ddot{\phi}_d + \alpha_1(\dot{\phi}_d - x_2)\right\} \\ \tau_3 = \frac{1}{b_2}\left\{-\varepsilon_2 \operatorname{sign}(s_2) - k_2 s_2 - a_4 x_2 x_6 - a_5 x_4^2 - a_6 \varpi x_2 + \ddot{\theta}_d + \alpha_3(\dot{\theta}_d - x_4)\right\} \\ \tau_4 = \frac{1}{b_3}\left\{-\varepsilon_3 \operatorname{sign}(s_3) - k_3 s_3 - a_7 x_2 x_4 - a_8 x_6^2 + \ddot{\psi}_d + \alpha_5(\dot{\psi}_d - x_6)\right\} \\ f_x = \frac{m}{U_1}\left\{-\varepsilon_4 \operatorname{sign}(s_x) - k_4 s_x - a_9 x_8 + \ddot{x}_d + \alpha_6(\dot{x}_d - x_8)\right\} \quad / f_z \neq 0 \\ f_y = \frac{m}{U_1}\left\{-\varepsilon_5 \operatorname{sign}(s_5) - k_5 s_5 - a_{10} x_{10} + \ddot{y}_d + \alpha_9(\dot{y}_d - y_{10})\right\} \quad / f_z \neq 0 \\ f_z = \frac{m}{\cos\phi\cos\theta}\left\{-\varepsilon_6 \operatorname{sign}(s_6) - k_6 s_6 - a_{11} x_{12} + \ddot{z}_d + \alpha_{11}(\dot{z}_d - x_{12}) + g\right\} \end{cases} \tag{14}$$

The parameter definitions in equation (14) can be found in equation (15).

$$\begin{cases} a_1 = \left(\frac{J_y - J_z}{J_x}\right), a_2 = \frac{-M_{dx}}{J_x}, a_3 = \frac{-J_r}{J_x}, a_5 = \frac{-M_{dy}}{J_y} \\ a_4 = \left(\frac{J_z - J_x}{J_y}\right), a_6 = \frac{J_r}{J_y}, a_7 = \left(\frac{J_x - J_y}{J_z}\right), a_8 = \frac{-M_{dz}}{J_z} \\ a_9 = \frac{-F_{dx}}{m}, a_{10} = \frac{-F_{dy}}{m}, a_{11} = \frac{-F_{dz}}{m} \\ b_1 = \frac{d}{J_x}, b_2 = \frac{d}{J_y}, b_3 = \frac{1}{J_z} \end{cases} \tag{15}$$

In equation (15), ε, k and α are the parameters that differential evolution needs to optimize iteratively. The parameter values of the multi-rotor UAV mathematical model used in this paper are shown in table 1.

Table 1. Model parameters of multi-rotor unmanned aerial vehicle.

Constant parameter	Value	Constant parameter	Value
m	0.720	J_r	1.42×10^{-5}
J_x	3.62×10^{-3}	g	9.8
J_y	3.62×10^{-3}	d	0.125
J_z	7.043×10^{-3}	g	9.8
F_{dx}	5.567×10^{-4}	M_{dx}	5.567×10^{-4}
F_{dy}	5.567×10^{-4}	M_{dy}	5.567×10^{-4}
F_{dz}	6.354×10^{-4}	M_{dz}	6.354×10^{-4}

4. Simulation results

Building the simulation model based on equations (8), (9), and (14), the design of the adaptive differential evolution algorithm for parameter optimization selection is in figure 3.

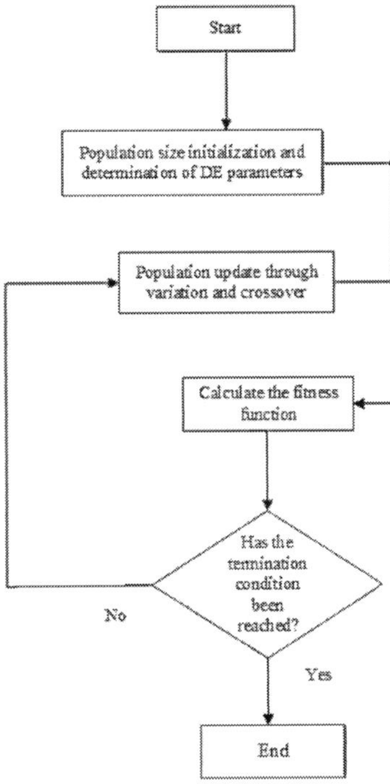

Figure 3. Schematic diagram of model parameters regulated by the differential algorithm[7-8].

We define ε, k and α in the controller as iterative parameters and build parameters ***D*=3, *NP*=15.** The number of iterations is 100, the scaling factor is 0.7, and the crossover factor is 0.5. Keeping the above data unchanged, we conduct parameter optimization 100 times for the traditional differential evolution algorithm DE and the adaptive differential evolution algorithm SaDE. It can be seen that the optimal value found by the adaptive differential evolution algorithm is faster in figure 4, and the required number of iterations is less. This dramatically reduces the computational cost of parameter optimization.

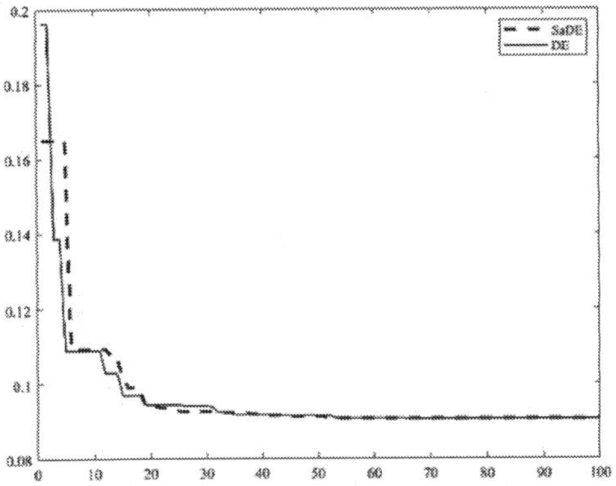

Figure 4. Comparison of the iterative convergence graphs of DE and SaDE.

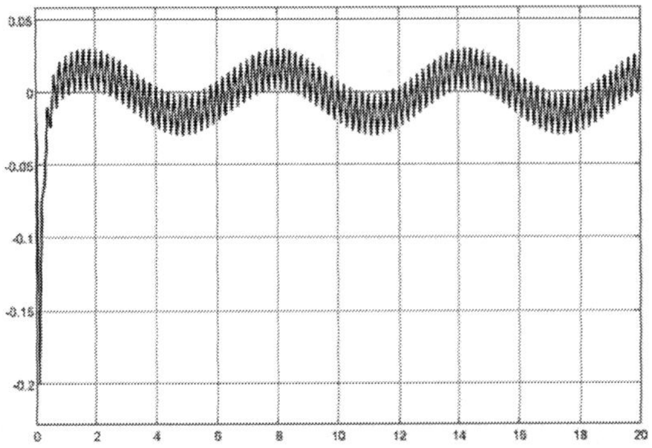

Figure 5. Real-time graph of position error x changing with time.

Journal of Physics: Conference Series **2879** (2024) 012047 doi:10.1088/1742-6596/2879/1/012047

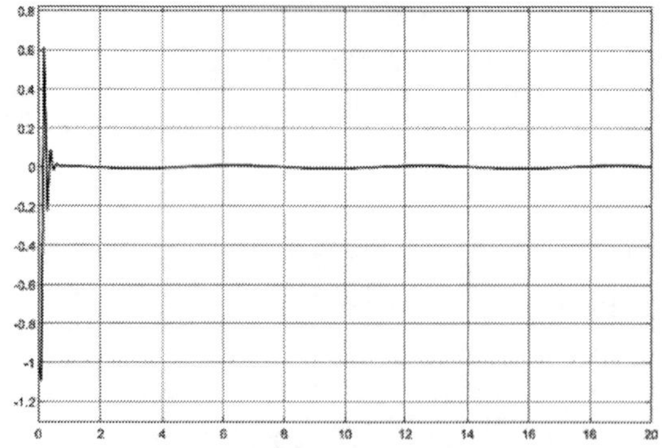

Figure 6. Real-time graph of position error ***y*** changing with time.

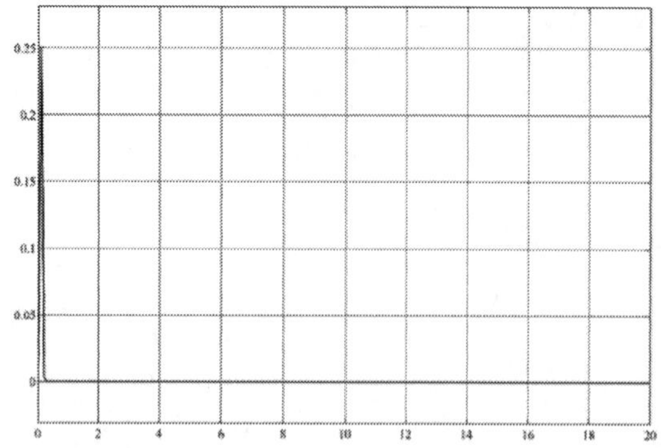

Figure 7. Cumulative integral of position *z* error.

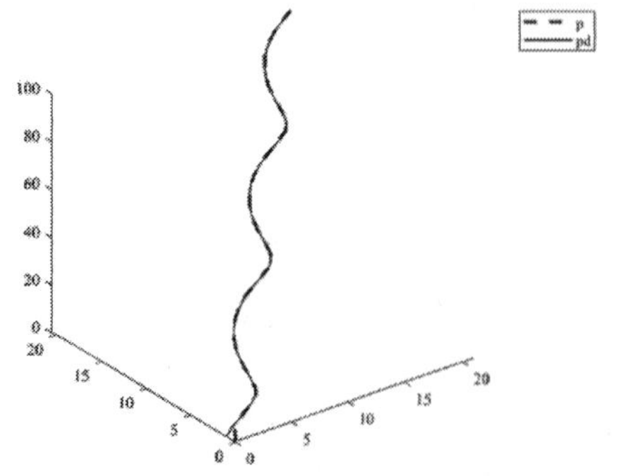

Figure 8. The curve comparison chart of ***y*** and ***y*d.**

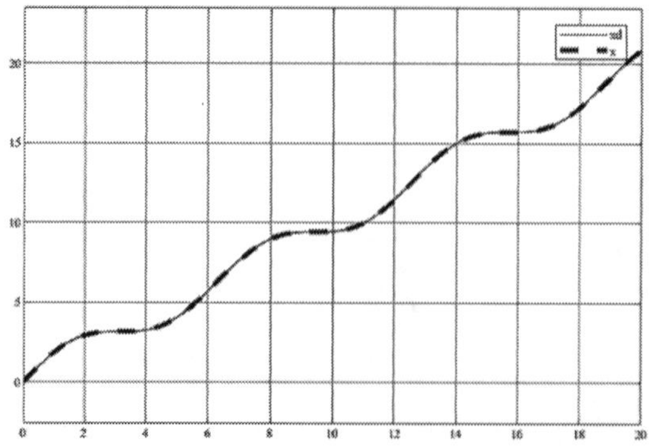

Figure 9. The curve comparison chart of ***x*** and ***x*d.**

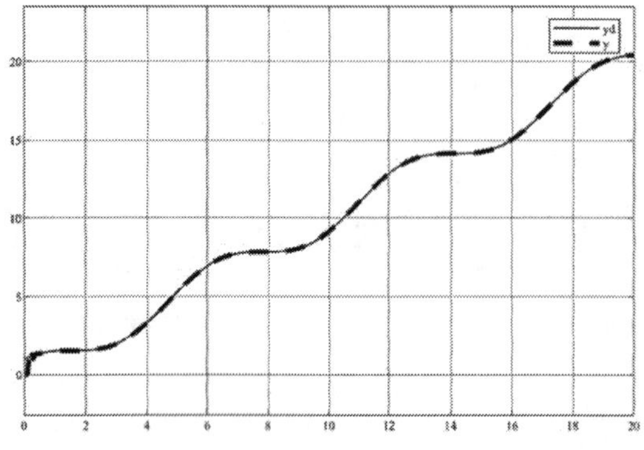

Figure 10. The curve comparison chart of ***y*** and ***y*d.**

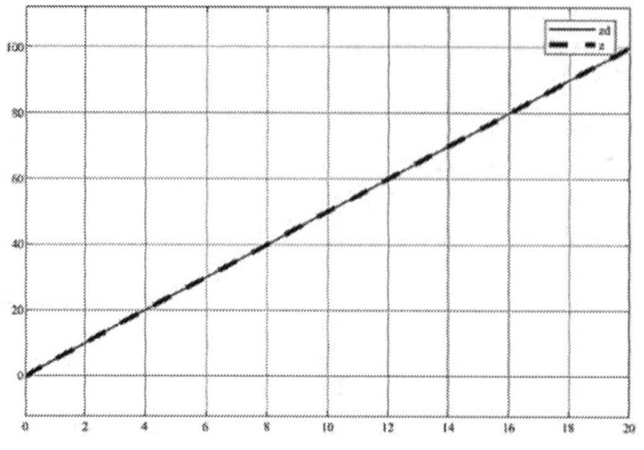

Figure 11. The curve comparison chart of *z* and *zd.*

The optimized parameters in table 2 are substituted into the three channels of the sliding mode controller to obtain schematic diagrams of three position errors changing with time, as shown in figures 5, 6, and 7. It can be seen that the state value reaches the expected value in a very short time and remains stable. Figure 8 is a graph of the changes in the actual position and the expected position, which can more intuitively compare the real-time state of the drone and the expected trajectory. Figures 9, 10, and 11 more intuitively show the convergence state of the actual value of each position channel, which is convenient for readers to observe. Compared with the traditional differential evolution algorithm, the algorithm in this paper not only reduces the computational cost of the algorithm, but also ensures the stability of the controller, and realizes the convergence to the expected value in a short time.

Table 2. The optimal values of the three position parameters after differential optimization.

	Parameter ε	Parameter k	Parameter α
Position x	1.9966	10.8559	4.4267
Position y	3.9077	6.4343	3.4716
Position z	0.0625	6.5258	17.9455

5. Conclusion

This paper designs an adaptive differential evolution algorithm. After multiple iterations, the optimized parameters of the sliding mode controller are obtained. The convergence of the parameters is verified by simulation, which proves the effectiveness of the designed algorithm. It can be applied to the parameter tuning of common models, significantly reducing the computational cost while ensuring the optimality of the parameters. It is an advanced and effective parameter optimization method.

References

[1] Quan Q. (2018). Design and Control of Multi-Rotor Aircrafts. Publishing House of Electronics Industry, Beijing.
[2] Manlika R, Xu H, L. (2022). Parameter Identification of Stochastic Delay Differential Equations Using Differential Evolution. In: 2022 37th International Technical Conference on Circuits/Systems, Computers and Communications (ITC-CSCC). Phuket. pp. 1-3.
[3] Fu-S H. (2020). Ridesharing based on a Discrete Self-adaptive Differential Evolution Algorithm. In: 2020 11th IEEE Annual Information Technology, Electronics and Mobile Communication Conference (IEMCON). Vancouver. pp. 696-700.
[4] Luis B P T, Javier A R H.(2021). Support vector machine for the implementation of the fitness function of genetic algorithms. In: 2021 16th Iberian Conference on Information Systems and Technologies (CISTI). Chaves. pp. 1-6.
[5] Zhou S, Huang X G. (2023) Research on Modeling and Control of Multi-Rotor Unmanned Aerial Vehicle System. In: 2023 China Automation Congress (CAC). Chongqing. pp. 7431-7436.
[6] Dinesh P; Binoy K R. (2021). Robustness Test in Conventional and Integral Sliding Mode Control. In: 2021 International Conference on Communication, Control and Information Sciences (ICCISc). Idukki. pp. 1-5.
[7] Zhong X Y, Ren J Q. (2022). Optimal design of quadrotor PID parameters based on differential evolution algorithm. Computer Simulation. J. 47-52.
[8] Ge Y Y, Chen D B, Zhou F. (2023). Multi-group and multi-strategy differential evolution algorithm for large-scale multi-objective optimization. Control and Decision. J. 429-439.

Journal of Physics: Conference Series **2879** (2024) 012048 doi:10.1088/1742-6596/2879/1/012048

Research on intelligent control system of dynamic magnetic tubular linear compressor

Hui Liu[1,2], Qisheng Ren[1,*], Xin Wang[1] and Xianyang Liu[1]

[1]School of Mechanical Engineering, Liaoning Technical University, Fuxin 123000, China

[2]Ordos Institute of Liaoning Technical University, Ordos 017000, China

[*]Corresponding author's e-mail: 1335623882@qq.com

Abstract: Addressing the cooling power control issue of linear compressors, a PID controller based on a cubic spline improved grey wolf optimization (tri-GWO) algorithm was designed. The mechanical and electromagnetic characteristics of the linear compressor were used to construct the cooling power transfer function of the linear compressor. The effects of different control algorithms on the cooling power of the linear compressor were studied in Simulink, and the response speed, settling time, and robustness of different algorithms were compared based on the simulation results. Further analysis was conducted on the impact of different algorithms in PID control. The results show that in terms of stability and robustness, the tri-GWO-PID controller has an overshoot close to 0 and quickly stabilizes at the target cooling power; in terms of tracking, the tri-GWO-PID controller exhibits the best tracking performance with the smallest cooling power deviation. The research results provide a reference for the study of cooling power control of linear compressors.

1. Introduction

The linear compressor is a key component for cooling operations in many household appliances, such as air conditioners and refrigerators. It has the ability to adjust cooling power according to voltage changes, thus requiring reliable control methods to ensure stable operation. In practice, PID (Proportional-Integral-Derivative) or intelligent PID control systems are usually employed to ensure precise control and regulation of cooling power.

With the development of linear compressors, extensive research on their control systems has been conducted by scholars both domestically and internationally. Tang et al.[1][2][3] analyzed the energy transfer process of linear compressors, proposed principles for high-efficiency control, and conducted experimental verification. Sheng et al.[4] designed a robust adaptive active controller for compressor surge, which improved the effectiveness of the control law and tracking performance, though surge can damage the compressor. Chun et al. [5] studied the efficiency and power factor under changes in mechanical and electrical resonance frequencies. By controlling the mean value of the stroke current product to be zero, they adjusted the inverter frequency to the mechanical resonance frequency, thus improving the efficiency of the linear compressor. Xie et al.[6] designed an integral separation control algorithm that can reduce system overshoot, improve the dynamic and static performance of the system, and enhance the position control performance of the linear motor. However, the separation of the integral makes the terminal rise significantly slower. Zou et al.[7] researched data monitoring and variable operating condition performance optimization techniques in the control system of linear compressors for refrigerators and developed control strategies suitable for the refrigeration operation of refrigerators, ensuring the efficient operation of the linear compressor.

In summary, current research on the control of linear compressors is mainly based on their inherent physical characteristics, with few designs of control systems from the perspective of control. This paper designs a cubic spline improved grey wolf optimization algorithm (tri-GWO) to make the research more representative. Four algorithms, including genetic algorithms, were selected for comparative analysis to study the impact of intelligent algorithm optimization of PID parameters on control performance. These efforts aim to provide new perspectives and methods for the design of control systems for linear compressors.

2. Construction of the state equation of the control system

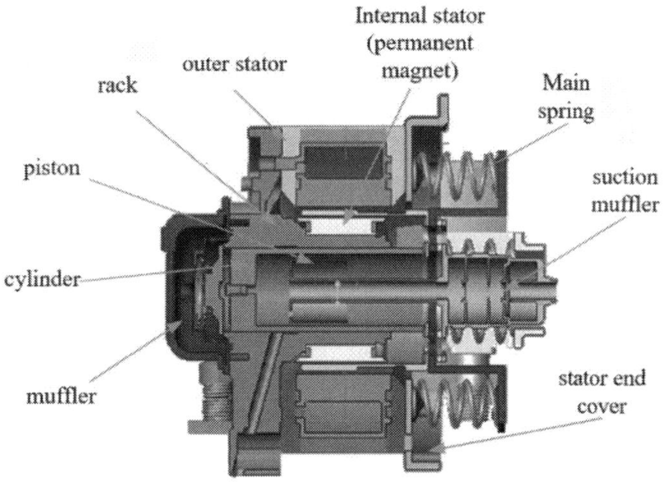

Figure 1. Structure diagram of a linear compressor.

A linear compressor is a special type of compressor that has a relatively simple structure and stable operation. Its main components include the outer stator, inner stator, permanent magnet, piston, and spring [8]. The outer stator uses wound electromagnetic coils to generate a varying magnetic field through alternating current. The inner stator, which consists of permanent magnets, is acted upon by the electromagnetic force generated by the outer stator, pushing the piston to reciprocate within the chamber [9]. The piston works in coordination with the main spring to complete the intake and exhaust processes, thus achieving the compression and circulation of the refrigerant. The structure is shown in Figure 1.

A moving magnet tubular linear compressor can be divided into mechanical and electromagnetic systems. The mechanical system can be simplified into a spring-damping system[10], as shown in the simplified model in Figure 2.

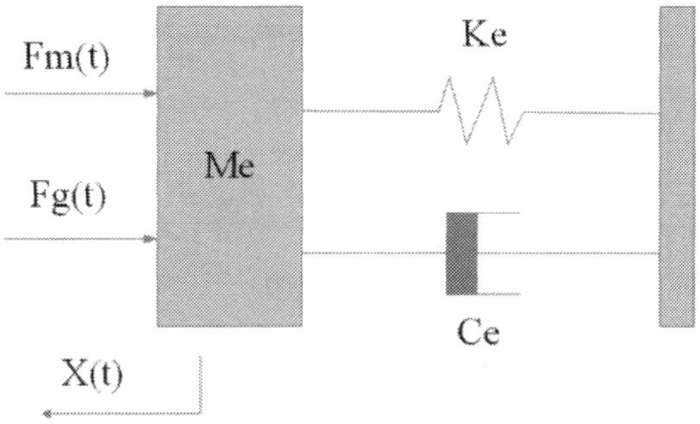

Figure 2. Simplified model of the mechanical system.

Journal of Physics: Conference Series **2879** (2024) 012048 doi:10.1088/1742-6596/2879/1/012048

Based on Figure 2, the dynamic equation of the mechanical system of the moving magnet tubular linear compressor is constructed:

$$M_e\ddot{x}(t)+C_e\dot{x}(t)+K_ex(t)=F_m(t)+F_g(t)M_e\ddot{x}(t)+C_e\dot{x}(t)+K_ex(t)=F_m(t)+F_g(t) \quad (1)$$

where Me is the equivalent mass of the mover, including the permanent magnet, piston, inner stator, etc., in kg; Ke is the equivalent stiffness of the main spring, in N/m; Ce is the equivalent damping during piston motion, in Ns/m; Fm is the electromagnetic force, in N; Fg is the gas force, in N.

$$F_m = NBLi \quad (2)$$

where B is the magnetic field strength inside the linear compressor, in T; L is the effective length of the coil under the magnetic field force, in m; N is the number of coil turns.

$$F_g(t)=A[p(t)-p_0] \quad (3)$$

where A is the cross-sectional area of the piston, in m^2; p0 is the back pressure, in Pa; p(t) is the instantaneous pressure inside the chamber.

Applying the Laplace transform to Equations (1)–(3) gives:

$$\frac{X(s)}{G(s)}=\frac{1}{M_e\bullet s^2+C_e\bullet s+K_e} \quad (4)$$

The equivalent circuit of the electromagnetic system of the linear compressor is shown in Figure 3 [11][12].

$$U(t)=R_0i(t)+L_0\frac{di(t)}{dt}+K_0\frac{dx(t)}{dt} \quad (5)$$

where U(t) is the driving voltage, in V; i(t) is the excitation current, in A; R0 is the system equivalent resistance, in Ω; L0 is the system equivalent inductance, in H; K0 is the electromagnetic force coefficient.

Applying the Laplace transform to equation (5) gives:

$$\frac{X(s)}{G(s)}=\frac{1}{R_0\bullet s+L_0} \quad (6)$$

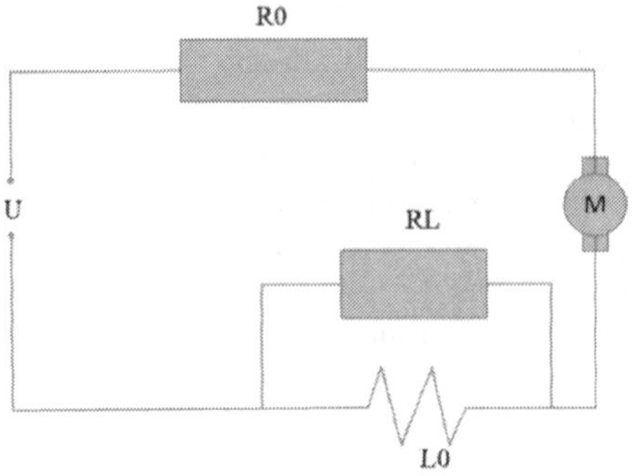

Figure 3. Equivalent circuit diagram.

The cooling capacity of the linear compressor per unit time is:

$$Q_c=\rho\pi r^2 fx\Delta h \quad (7)$$

where ρ is the density of the refrigerant R134a; r is the radius of the linear compressor cylinder; h is the enthalpy difference.

Applying the Laplace transform to equation (7) gives:

$$\frac{X(s)}{G(s)}=\rho_{suc}\pi r^2 f(h_1-h_4)X(s) \quad (8)$$

3. Linear compressor control algorithm

A traditional PID controller primarily consists of three components: proportional, differential, and integral. Its transfer function can be expressed as:

$$G_{pid} = K_p + K_i / s + K_d s \tag{9}$$

In traditional PID control algorithms, the system performance is controlled by *Kp*, *Ki*, and *Kd*. However, in classical PID algorithms, the tuning of parameters relies heavily on manual adjustment and experience, leading to overall lower control efficiency.

In recent decades, with the advancement of artificial intelligence, the concept of intelligent PID has been proposed. Intelligent PID control systems have gradually become a research and application hotspot in industrial control and automation fields. Currently, intelligent PID control systems are still in the development stage. An intelligent PID controller is an evolution of the traditional PID controller, integrating artificial intelligence technology with traditional PID control algorithms to improve the performance of control systems. This type of controller utilizes optimization algorithms such as genetic algorithms to search for the optimal parameter combination, adjusting parameters based on changes in the system's dynamic characteristics to achieve better control effects.

The basic control process of the algorithm is shown in Figure 4.

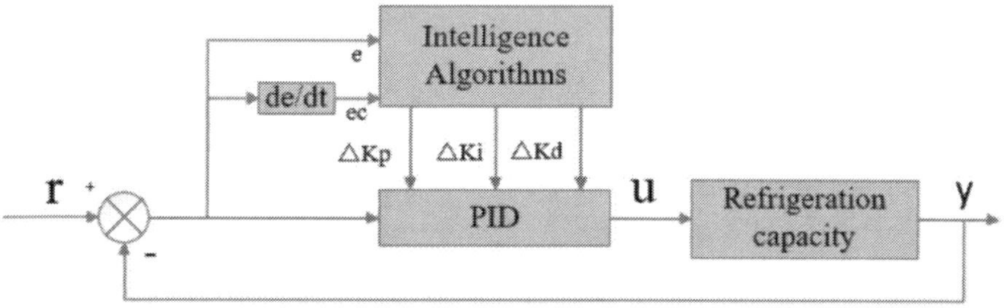

Figure 4. Algorithm flowchart.

3.1. Evaluation function

The fitness function is a crucial performance indicator for evaluating intelligent algorithms, directly affecting the control performance of the intelligent algorithm. For the linear compressor control system, response speed and cumulative error impact the choice of the fitness function. Therefore, the fitness function in this system adopts ITAE (Integral of Time-weighted Absolute Error).

$$ITAE = \int_0^{\infty} t \,|\, e(t) \,|\, dt \tag{10}$$

3.2. Grey wolf algorithm

Inspired by the hunting strategies and hierarchical roles of grey wolves in nature, the grey wolf algorithm classifies the grey wolf population into four levels: α wolves, β wolves, δ wolves, and ω wolves. Among them, α wolves, β wolves, and δ wolves search and guide the remaining ω wolves to simulate grey wolf behaviors such as tracking, surrounding, and attacking in the solution space to locate the optimal solution.

Surrounding strategy:

$$\vec{D} = |\vec{C} \cdot \vec{X}_P(t) - \vec{X}(t)| \tag{11}$$

$$\vec{X}(t+1) = \vec{X}_P(t) - \vec{A} \cdot \vec{D} \tag{12}$$

where $\vec{D}$ is the distance between the individual gray wolf and its prey; $\vec{X}_P$ is the location of the prey; $\vec{X}$ is the location of the grey wolf

Coefficient vector:

$$\vec{A} = a(2\vec{r}_1 - 1) \tag{13}$$

$$\vec{C} = 2\vec{r_2} \tag{14}$$

where $\vec{r_1}, \vec{r_2}$ are [1,1] random vectors.

Grey Wolf Algorithm Convergence Factor:

$$a = 2 - 2(t / T) \tag{15}$$

Hunting strategy:

$$\begin{cases} \vec{D}_\alpha = |\vec{C} \cdot \vec{X}_\alpha(t) - \vec{X}(t)| \\ \vec{D}_\beta = |\vec{C} \cdot \vec{X}_\beta(t) - \vec{X}(t)| \\ \vec{D}_\delta = |\vec{C} \cdot \vec{X}_\delta(t) - \vec{X}(t)| \end{cases} \tag{16}$$

$$\begin{cases} \vec{X}_1 = \vec{X}_\alpha - \vec{A}_1 \cdot D_\alpha \\ \vec{X}_2 = \vec{X}_\beta - \vec{A}_2 \cdot D_\beta \\ \vec{X}_3 = \vec{X}_\delta - \vec{A}_3 \cdot D_\delta \end{cases} \tag{17}$$

where $\vec{X}_\alpha, \vec{X}_\beta, \vec{X}_\delta$ represent the position of the leading wolf.

Grassroots wolf to head wolf position update step size:

$$\vec{X}(t+1) = (\vec{X}_1 + \vec{X}_2 + \vec{X}_3) / 3 \tag{18}$$

3.3. Improvement strategy of cubic spline grey wolf algorithm

(1) Improve the initial population

In the traditional grey wolf algorithm, the initial individuals are generated using a random population generation method, which results in an uncertain convergence speed. Based on maintaining the randomness of the initial generation, this paper starts from the practical situation of PID control parameters. Generally, PID parameters exist around the median value, so a normally distributed initial vector generation method is created. Using the generated function matrix to initialize the population makes the initial population closer to the median value, accelerating the convergence speed. The normal distribution equation is as follows:

$$Z_{k+1} = \begin{cases} Z_k \times 0.9, Z_k \in (0, 0.5] \\ (1 - Z_k) / 0.8, Z_k \in (0.5, 1] \end{cases} \tag{19}$$

Compared with the initial population generation method of random sampling, the parameter domain of the PID controller is obtained through the empirical method, and the parameter values of this domain structure tend to be biased towards the median. Compared with the random sampling of existing algorithms, normal sampling is more in line with the PID control strategy and can accelerate convergence speed.

(2) Convergence Factor Adjustment

In the grey wolf algorithm, the convergence factor A is linearly decreasing, which is not conducive to finding the global optimal solution and is prone to getting stuck in local optimal solutions. The convergence factor changes linearly, and the iteration step size gradually decreases. If it falls into a local optimal solution, it will iteratively converge near the local optimal solution and cannot obtain the global optimal solution. Therefore, this paper proposes the following cubic spline convergence strategy:

$$A = 2 - (2t^3 / T^3) \tag{20}$$

where t is the current number of iterations, and T is the maximum number of iterations.

Compared with the linear convergence factor of existing algorithms, the convergence factor adjustment strategy for cubic spline values has a smaller change in iteration step size in the early stage, which can avoid local optima. In the later stage of calculation, the convergence factor decreases rapidly, and the iteration step size decreases quickly, which is beneficial for accurately solving the optimal solution.

(3) Algorithm process and steps

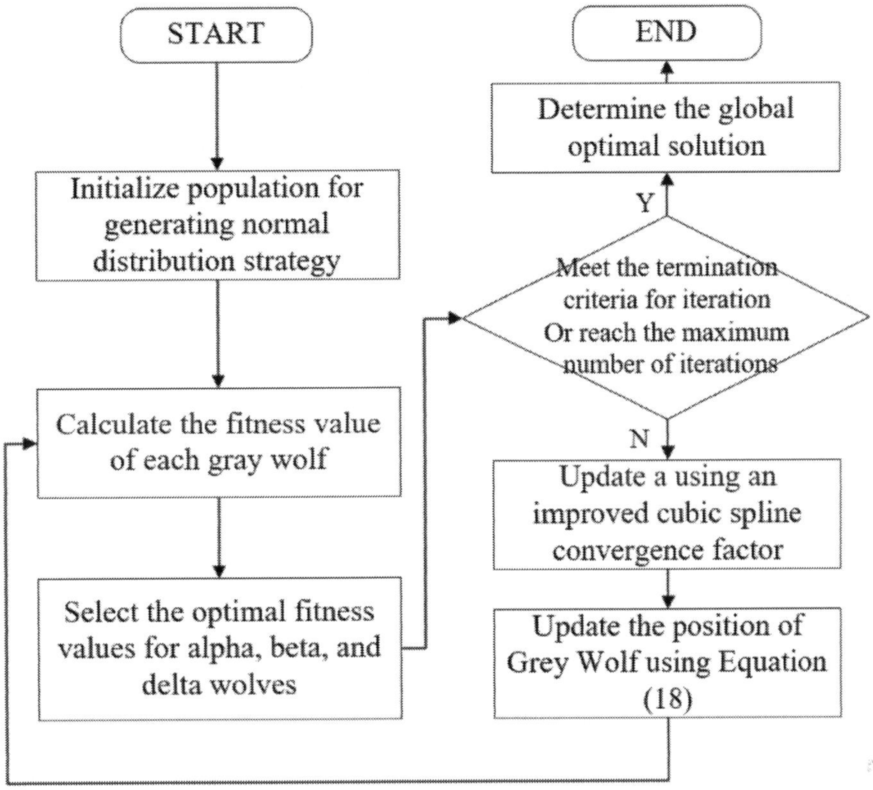

Figure 5. Algorithm flow chart.

a. Set the size of the grey wolf population, search dimensions, and iteration times;

b. Generate an initialized population using equation (19);

c. Calculate the individual fitness values of grey wolves and record the alpha, beta, and delta wolves with the best fitness values;

d. Determine whether the iteration termination condition is met. If so, output the result and end the calculation; if not, proceed to the next step;

e. Adjust the convergence factor a using equation (20);

f. Update the position of the grey wolf using equation (18) and return to step c.

4. Simulation experiments and results analysis

To verify the control effect of the designed tri-GWO-PID algorithm, a linear compressor model and a tri-GWO-PID control system model were built in the Simulink module of Matlab based on the mathematical model and parameters of the linear compressor, including input, output, and parameter calculation modules. The simulation model of the linear compressor is shown in Figure 5.

The key parameters of the linear compressor model in the simulation are set as follows: equivalent mass of the mover *Me*, equivalent damping of the piston motion *Ce*, equivalent stiffness of the main spring *Ke*, magnetic field intensity *B*, number of coil turns *N*, effective length of the coil under magnetic force *L*, effective piston stroke, equivalent resistance of the system R_0, equivalent inductance of the system L_0, piston area AA, exhaust pressure *Pe*, suction pressure *Ps*, gas density at intake temperature and pressure ρ, piston radius *r*, and specific heat difference of the refrigerant $\Delta h = 180.8kj/kg$.

To verify the control effect of the designed tri-GWO-PID algorithm, a linear compressor model and a tri-GWO-PID control system model were built in the Simulink module of Matlab based on the mathematical model and parameters of the linear compressor, including input, output, and parameter calculation modules. The simulation model of the linear compressor is shown in Figure 6.

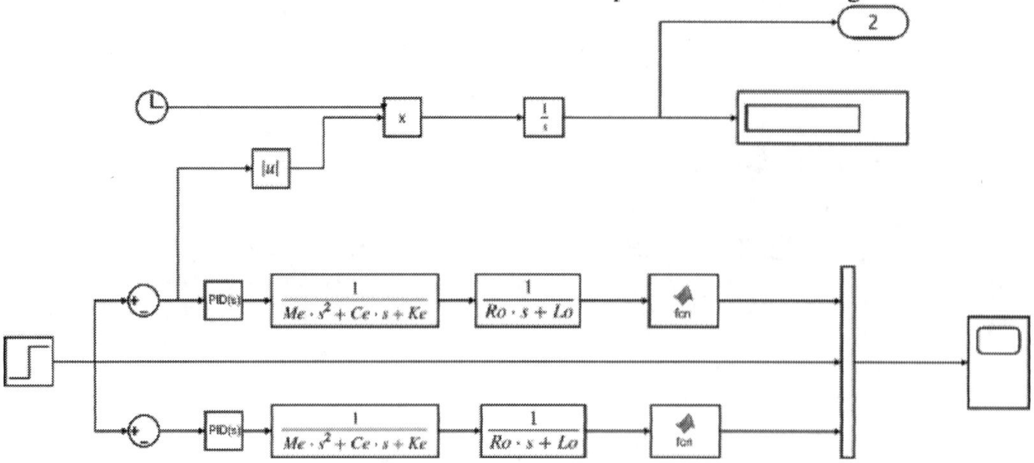

Figure 6. Simulation model diagram.

We set key parameters for the linear compressor model in simulation: equivalent mass of the moving element $M_e = 0.4kg$, equivalent damping in piston motion $C_e = 0.02N \cdot s / m$, equivalent stiffness of the main spring $K_e = 12000N / m$, induced magnetic field strength $B = 0.9T$, turns $N = 200$, effective length of coil subjected to magnetic field force $L = 0.2m$, effective stroke of piston $d = 8 \times 10^{-3} m$, equivalent resistance of the system $R_0 = 16.6\Omega$, equivalent inductance of the system $L_0 = 0.064H$, piston area $A = 7.64 \times 10^{-4} m^2$, exhaust pressure $P_d = 3.2 \times 10^5 Pa$, suction pressure $P_s = 1.01 \times 10^5 Pa$, gas density at intake temperature and pressure $\rho = 1181.1kg / m^3$, piston radius $r = 18mm$, and specific heat tolerance of refrigerant $\Delta h = 180.8kj / kg$.

(1) To further verify the effectiveness of the iterative PID controller parameters proposed in this paper, genetic algorithm (GA) [13], rat swarm optimization (RSO) [14], optical microscope algorithm (OMA) [15], and grey wolf optimization algorithm improved by dynamic weight and probability perturbation strategy (IGWO) [16-17] were used to iterate the PID controller parameters. The third-order spline improved grey wolf optimization PID (tri GWO-PID) controller designed in this paper was compared and analyzed for the step response curve of the controlled linear compressor (Figure 7).

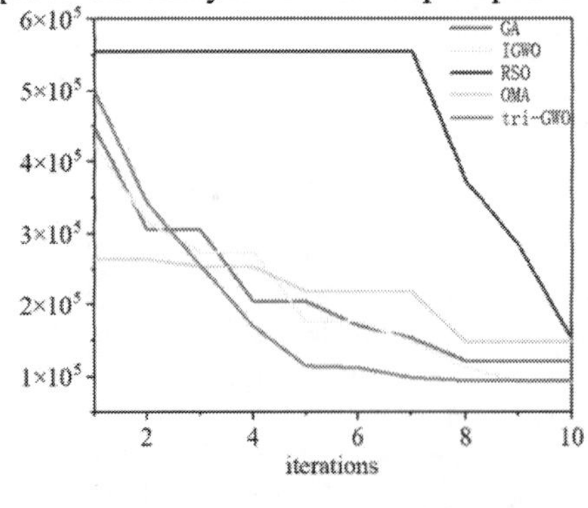

Figure 7. Convergence curve of fitness value.

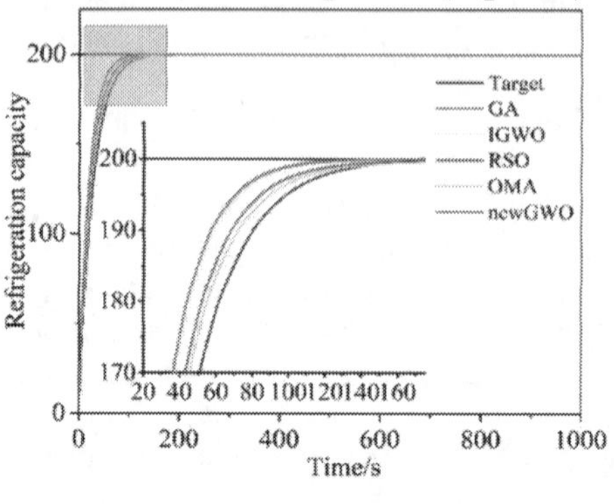

Figure 8. Step response curve of the system.

Under the ITAE criterion, with the population size, number of iterations, and boundary values for Kp, Ki, and Kd kept consistent, the convergence curves of the optimal individual fitness values obtained by the algorithm for iterating PID controller parameters are shown in Figure 8, and the system step response curves are shown in Figure 7. As can be seen from Figures 8 and 9, the RSO algorithm reached the maximum number of iterations after ten iterations, the new GWO algorithm converged after nine iterations, and the GA, OMA, and tri-GWO algorithms converged after eight iterations.

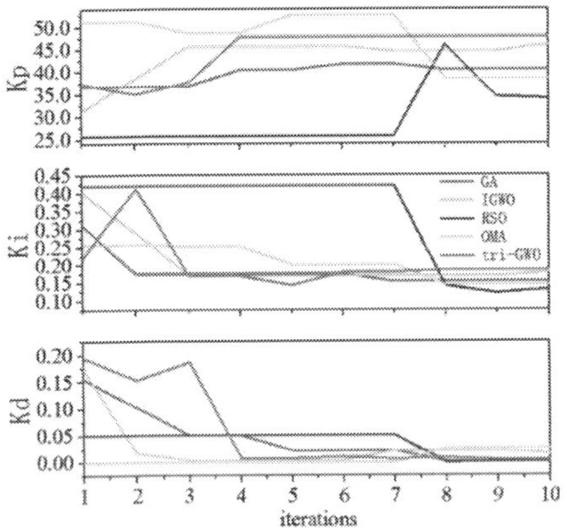

Figure 9. Parameter iteration plot.

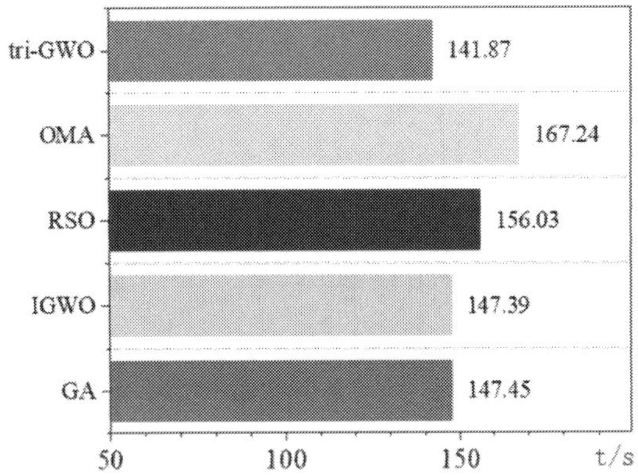

Figure 10. Steady-state time.

Based on the results of Figures 7 and 10, we can conclude that the tri-GWO algorithm has a faster response speed and performs best in terms of settling time and overshoot. The method of optimizing PID controller parameters using the tri-GWO algorithm outperformed other algorithms in terms of settling time, overshoot, and steady-state error. Regarding the steady-state error, the tri-GWO algorithm exhibited a 0.019% error (see Figure 11). Therefore, according to the ITAE performance criterion, the PID controller designed using the tri-GWO algorithm demonstrates superior parameter optimization performance.

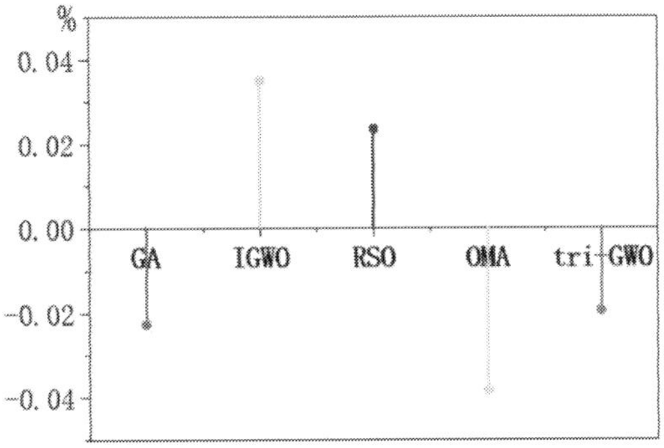

Figure 11. Steady-state error.

(2) To further validate the superiority of the algorithm presented in this paper, simulations were conducted under both sinusoidal load and sudden cooling capacity changes. The resulting step response curves are shown in Figures 8 and 9. Analyzing the results in Figures 11 and 12 from the perspective of performance indicators, it can be observed that, in terms of both tracking and deviation, the tri-GWO algorithm performed the best, while the RSO algorithm performed the worst, with other algorithms falling between the tri-GWO and RSO algorithms. Additionally, in response to sudden changes in cooling capacity, the tri-GWO algorithm demonstrated the best regression performance.

doi:10.1088/1742-6596/2879/1/012048

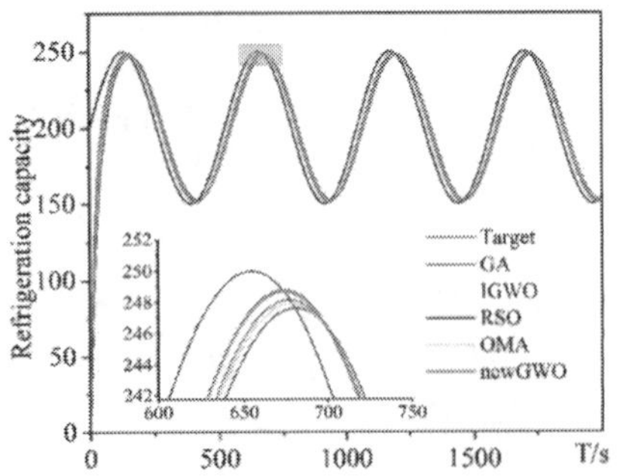

Figure 12. Sinusoidal load step response curve.

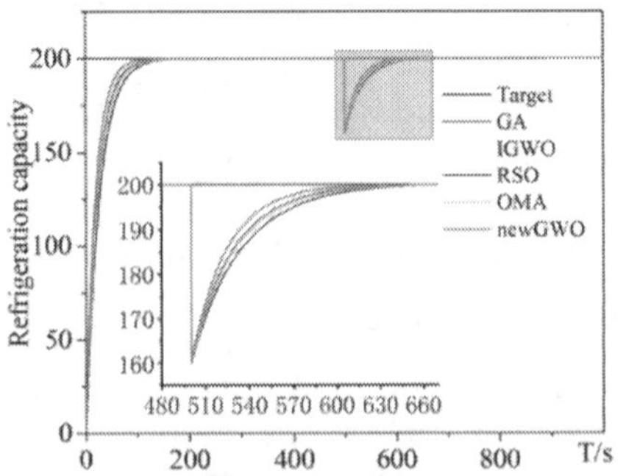

Figure 13. Step response curve of sudden load change change.

As shown in Figure 12, the tri-PID controller achieved the highest control accuracy, with a delay time of 18.56 seconds, a peak cooling capacity of 248.79 W, and an error of only 1.21 W. Figure 13 indicates that the introduction of a sudden load causes a sudden change in cooling capacity; however, in terms of correcting the cooling capacity, the tri-PID controller quickly returned to the target cooling capacity, rapidly increasing the cooling capacity within 93.88 seconds. The second-best regression in cooling capacity was achieved by the IGWO algorithm, which took 111.87 seconds.

From the above simulations and comparative analysis, the conclusion can be drawn that an initial population with a normal distribution better aligns with the parameter selection direction for PID control, and the convergence factor of the cubic spline difference is more conducive to escaping local optima and finding the global optimal solution. Therefore, the tri-GWO optimization algorithm has a certain advantage in PID parameter optimization and tuning.

5. Experimental validation

To verify the control effect of the tri-GWO-PID controller proposed in this paper on the refrigeration system of a linear compressor, a test platform for a linear compressor was built independently, as shown in Figure 13. The platform includes a host computer, NF power supply, mass flow meter, constant temperature water tank, and linear compressor. During the experiment, the linear compressor is connected to components such as the expansion valve and condenser. A PID optimization model was established on the host computer, with the power of the constant temperature water tank used as a feedback signal, transmitted to the host computer. The host computer monitors the feedback signal in real-time, compares it with the set value, and uses intelligent optimization algorithms to optimize the PID parameters. The output signal is then sent to the NF power supply to control the voltage and frequency of the linear motor. The optimized parameters are applied to the linear compressor to achieve precise control of the refrigeration capacity. This test platform can verify the application effect of the tri-GWO-PID controller in the actual linear compressor refrigeration system, providing reliable theoretical and practical support for system performance optimization.

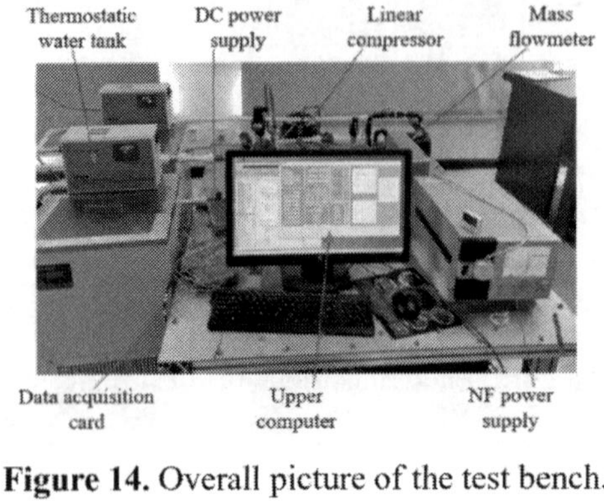

Figure 14. Overall picture of the test bench.

The main performance parameters of the selected moving-magnet linear compressor are shown in Table 1.

Table 1. Main performance parameters.

Parameter	Value	Unit
Intake Pressure	0.1	MPa
Exhaust Pressure	0.3-0.6	MPa
Displacement	10	cc
Rated Power	180	w
Operating Frequency	Variable Frequency	
COP	1.95	

(1) On the test platform, the algorithm described in this paper is used to control the refrigeration capacity of the linear compressor. Under sinusoidal input conditions, the measured refrigeration capacity error curve of the linear compressor and the expected value are shown in Figure 15, and Figure 16 shows the tracking deviation time of the linear compressor.

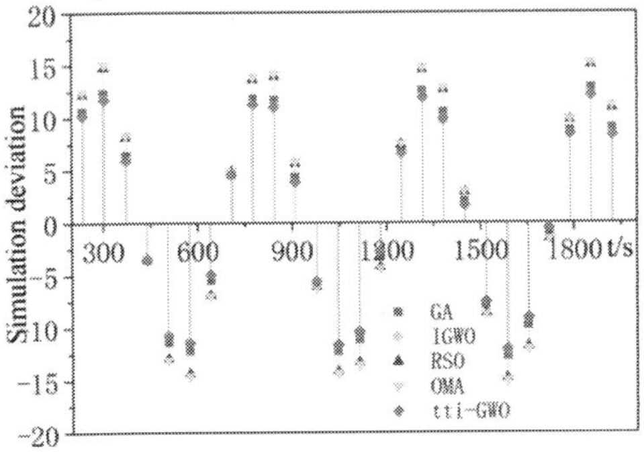

Figure 15. Error of expectation value.

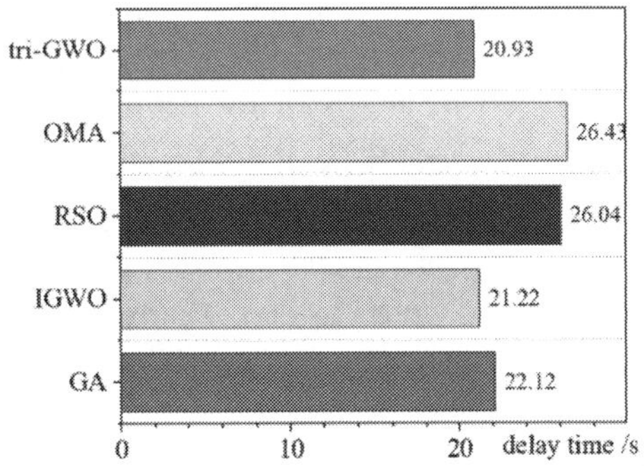

Figure 16. Tracking time deviation.

Through the analysis of Figures 14 and 15, it can be concluded that in the case of variable demand, the tri GWO algorithm for linear compressors has the smallest delay of 20.93s and the best tracking performance. In terms of overall error, using the tri GWO algorithm control program, the overall tracking error is within 6%.

(2) To ensure generality and reliability, the target cooling capacity was changed, and simulations and experimental error analyses were conducted using the tri-PID controller. The cooling capacities were set to 50 W, 100 W, 150 W, and 200 W for verification and analysis.

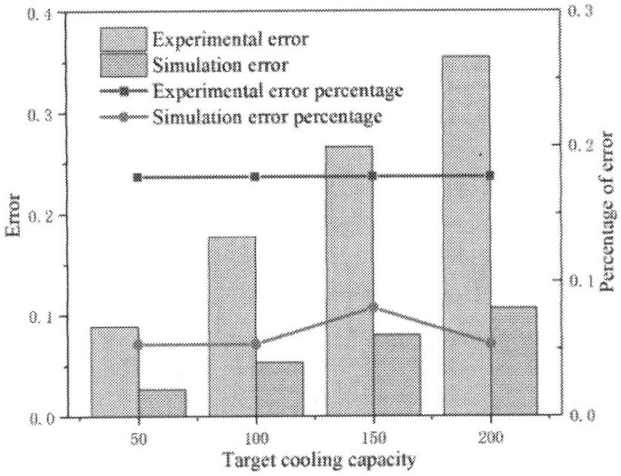

Figure 17. Cooling capacity error.

As shown in Figure 17, the experimental error is greater than the simulation error, and the error increases as the target cooling capacity increases. However, the error percentage remains relatively stable, with the simulation error percentage showing slight fluctuations, while the experimental error remains very stable, indicating a high level of control accuracy. This experiment demonstrates that the tri-PID controller exhibits robustness in controlling linear compressors.

6. Conclusion

This article proposes a cubic spline improved grey wolf algorithm and introduces a linear compressor mathematical model. The research results of this model show that:

(1) Comparing the tri-GWO algorithm with other algorithms, the tri-GWO algorithm demonstrates fast convergence speed, minimal overshoot, and excellent optimization performance. It is applied to the optimization of PID controller parameters for the refrigeration capacity of a linear compressor.

(2) In terms of refrigeration capacity control, five algorithms were compared using the ITAE index. The simulation results show that the tri-GWO algorithm used in this paper exhibits better optimization performance. To validate its superiority, simulations were conducted under sinusoidal and sudden load conditions, confirming the advantages of the tri-GWO algorithm.

(3) The improved grey wolf algorithm (tri-GWO) PID controller was applied to the control of the refrigeration capacity system of a linear compressor. Experimental validation showed that the maximum relative error is less than 6%, proving the accuracy of the control algorithm proposed in this paper.

The cubic spline-improved grey wolf algorithm proposed in this paper, utilizing a normal distribution algorithm suitable for PID controllers, was applied in the design of the tri-PID controller and tested through simulations on a linear compressor. The results have a certain reference value and provide a new approach to industrial control. Additionally, the position update strategy used in this algorithm may be refined in the future to find even better computational methods, offering opportunities for further improvement.

References

[1] Mingsheng Tang, Huiming Zou, Xuan Li, Changqing Tian. The analysis of energy conservation employed in the control of linear compressor[A]. International Conference on Compressors and their Systems,2019. [CrossRef]

[2] Liang K. A review of linear compressors for refrigeration[J].International Journal of Refrigeration, 2017, 84. [CrossRef]

[3] Hanying Jiang, Zhaohua Li, Kun Liang. Performance of a Linear Refrigeration Compressor with Small Clearance Volume[J]. International Journal of Refrigeration (2019), 2019.10.013. [CrossRef]

[4] Hanlin Sheng, Qian Chen, Jiacheng Li, Zefan Li, Zhe Wang, Tianhong Zhang. Robust adaptive backstepping active control of compressor surge based on wavelet neural network[J]. Aerospace Science and Technology,2020. [CrossRef]

[5] Xiaokuan You, Limin, et al. Study on the stroke amplitude of the linear compressor[J]. Applied thermal engineering: Design, processes, equipment, economics, 2018, 129:1488-1495. [CrossRef]

[6] XIE Y B, GUI L M, XU Z X, et al. Research on the integral separation PID control of a linear compressor[C]//Proceedings of 2007 Annual Meeting of China Refrigeration Society. Hangzhou: China Refrigeration Society, 2007: 836-838 (in Chinese) [CrossRef]

[7] ZOU H M, TANG M S, XU H B, et al. Control system of linear compressor for refrigerator[J]. CIESC Journal, 2014, 65(S2): 175-180 (in Chinese) [CrossRef]

[8] Liang X Y, Mao J, Wang X. Research on variable-speed sliding mode control method of linear compressor[J]. Mechanical Science and Technology, 2021, 40(11): 6. [CrossRef]

[9] Mao J, Wang N, Chen H Y. Dynamic characteristics of linear compressor with dynamic magnetic[J]. Journal of Mechanical Design and Research, 2022, 38(02): 205-209+214. [CrossRef]

[10] Fan S T. Performance analysis and multi-objective optimization design of armature components for linear compressor with dynamic magnetic[J]. Liaoning Technical University, 2021. [CrossRef]
[11] Chen H Y, Chen C, Li W, et al. Analysis and optimization of electromagnetic system for linear compressor with dynamic magnetic[J]. Journal of Mechanical Design and Research, 2022, 38(04): 192-197. [CrossRef]
[12] Xu W, Li X, Zhu J, et al. 3D Modelling and Testing of a Stator-Magnet Transverse-Flux Linear Oscillatory Machine for Direct Compressor Drive[J]. IEEE Transactions on Industrial Electronics, 2020, PP (99):1-1. [CrossRef]
[13] Xiong A, Wang Q. Optimal controller design of demolding machine rotation system based on genetic algorithm[J]. Machinery, 2024, 51(01): 67-73. [CrossRef]
[14] Xue J, Shen B. Dung beetle optimizer: a new meta-heuristic algorithm for global optimization[J]. The Journal of Supercomputing, 2022:1-32. [CrossRef]
[15] M.-Y. Cheng and M.N. Sholeh, Optical microscope algorithm: A new metaheuristic inspired by microscope magnification for solving engineering optimization problems[J]. Knowledge-Based Systems (2023).2023.110939002E[CrossRef]
[16] Li Y, Li W G, Zhao Y T, et al. Grey wolf algorithm based on Levy flight and random walk strategy[J]. Computer Science, 2020, 47(08): 291-296. [CrossRef]
[17] Xie S P, Wu B S, Zhao X T, et al. Structural damage identification based on improved grey wolf optimization algorithm[J/OL]. Journal of Computational Mechanics, 1-8 [2024-04-11]. [CrossRef]

Journal of Physics: Conference Series **2879** (2024) 012049 doi:10.1088/1742-6596/2879/1/012049

Research on load prediction of carrier-based aircraft landing test based on measured data

Feng'an Xie[1*] and Whenchao Xu[1]

[1]China Flight Test Establishment, Xi'an, Shaanxi, 710089, China

[*]Email: 1946141397@qq.com

Abstract: As one of the most important parts in the ground (carrier) landing process of carrier-based aircraft, undercarriage is directly related to the safety of the aircraft, the crew, and even the landing platform with its strength and stiffness. The landing test to evaluate the strength and stiffness of the undercarriage of the carrier-based aircraft has strict regulations on the sinking velocity of the aircraft during landing. Still, in consideration of safety, it is difficult for the aircraft to reach the sinking velocity specified in the test under normal conditions. In this thesis, based on a large number of measured flight data, the artificial neural network was used to reasonably predict the undercarriage landing load of carrier-based aircraft and improve the landing test points, thus providing a new idea for the research on undercarriage load strength.

1. Introduction

1.1. The purpose of carrier-based aircraft landing test and the requirement of sinking velocity

The undercarriage is one of the most important parts of the ground landing/carrier landing stage of carrier-based aircraft, which plays a role in buffer power-absorbing, sliding, and braking deceleration during the aircraft landing. In the process of aircraft landing, the undercarriage almost absorbs all the vertical and lateral energy during the carrier-based aircraft landing period and part of the energy of the course so that the aircraft can safely land on the airfield or carrier. This explains that the strength and stiffness performance of the undercarriage is directly related to the safety of the aircraft and the crew. To judge whether the strength and stiffness of the undercarriage meet the requirements, it is necessary to analyze the load under impact. The sinking velocity of an aircraft is the most critical core parameter that affects the load on the undercarriage under impact. In the carrier-based aircraft landing test, the aircraft must land with the specified landing weight, sinking velocity, and attitude. It is stipulated in the "Flight Tests for Strength and Stiffness Specifications of Carrier-based Aircraft" (GJB2759-1996) [1] that in the undercarriage structural verification test, the sinking velocity at the center of gravity of the aircraft shall meet the requirements in Table 1.

Table 1. Aircraft structure verification test.

Test	Sinking velocity m/s	Description
a	$0.0885V_A+1.86$	Only once for the specified conditions, or three times in accordance with other conditions, but the sinking velocity shall not be less than 0.8 times the specified value.
b		
c		

The symbols in the table are defined as follows [2]

V_A refers to the aircraft's approach velocity.

a. It refers to touch-and-go landing, aircraft-carrier blocked landing.

b. It refers to touch-and-go landing and airfield-blocked landings at trimmed and semi-trimmed airfields.

c. It refers to carrier landing practice on a trimmed airfield runway.

However, due to safety considerations, three tests are usually performed based on the 80% specified value of the additional conditions in the instructions to avoid risks and ensure flight safety in actual test flights.

1.2. Research objectives

The actual three times of flight tests reaching the 80% sinking velocity specified value can satisfy the requirements of the "National Military Standards of the People's Republic of China", but it can be considered that the purpose of verifying whether the design indicators of the strength and stiffness of the undercarriage meet the requirements has been achieved. However, it will be more convincing to obtain the load obtained by the undercarriage at 100% sinking velocity limit value in Table 1.

Big data analysis is increasingly applied to engineering problems with the development of computational methods. In the landing test of carrier-based aircraft in this thesis, the artificial neural network will be trained using the measured flight data during the flights. The trained network will be used for load prediction when the sinking velocity reaches the 100% limit value in Table 1. If the predicted results are within a certain error range, then the predicted results of the neural network are credible, the test can be perfected, and the test points can be completed, which can more convincingly verify whether the strength and stiffness design of the undercarriage meets the performance requirements [3].

To sum up, this research aims to screen the real flight data measured by a certain model of carrier-based aircraft test and find out the influence parameters related to the impact load suffered by the undercarriage during landing. Part of the data was used for the training of artificial neural networks, and the other part was used for the verification of the training effect of the network model. If the training results achieve the expected results, the model can be used to predict the load at the specified sinking velocity so that the landing test of carrier-based aircraft can be improved. The flow diagram of the research idea in this thesis is shown in Figure 1.

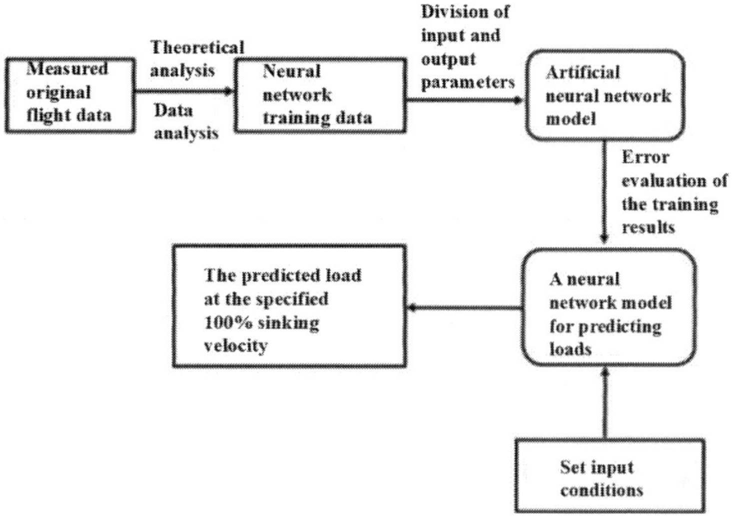

Figure 1. Research idea flow chart.

2. Analysis of factors related to landing loads of carrier-based aircraft

According to the equation $e=\frac{1}{2}mv^2$, the primary factors influencing the impact of landing gear on carrier-based aircraft during landing are the descent speed v at the center of gravity and the weight m of the aircraft.

The sinking velocity at the center of gravity of an aircraft is the most direct and important factor affecting the landing load. The seeking velocity measured at the center of gravity is affected by many factors.

An aircraft flying in space can be decomposed into three-way motion at the GPS antenna position and rotation around a fixed point at the GPS antenna position. The absolute velocity of any point in a rigid body is equal to the algebraic sum of the three-dimensional velocity caused by the point's rotation around the GPS antenna position and the three-dimensional velocity at the GPS antenna position. The relevant coordinate system and angle definition are shown in Figure 2 [4].

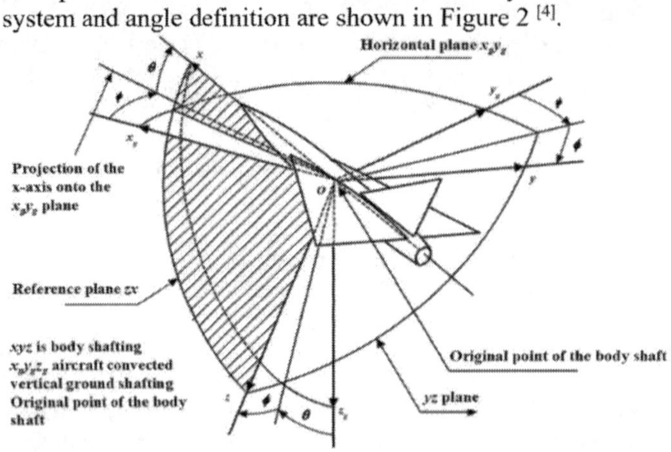

Figure 2. Diagram of correlative coordinate system of space flight aircraft.

ϕ, θ, and ψ represent the roll angle, pitch angle, and yaw angle of the aircraft, respectively.

Under the body convected coordinate system OGPSXYZ with the GPS antenna position as the origin, the velocity formula of each point in the fixed point moving rigid body or the Euler formula of velocity was used to calculate the three-way velocity of the center of gravity of the body under the axis system relative to the GPS antenna position caused by the rotation of the rigid body:

$$\vec{v}_b = \vec{\omega}\times\vec{r} \ or \begin{cases} v_x = \omega_y z - \omega_z y \\ v_y = \omega_z x - \omega_x z \\ v_z = \omega_x y - \omega_y x \end{cases} \tag{1}$$

where $\vec{v}_b$ is $\{v_x \quad v_y \quad v_z\}^{-1}$.

$\vec{r}$ is the coordinate vector for the position of the center of gravity of the aircraft under $O_{GPS}XYZ$ $\{x, \quad y, \quad z\}^{-1}$.

The coordinate rotation matrix T_b^i was used to convert $\vec{v}_b$ into a vector composed of the three-way velocity under $OX_gY_gZ_g$ of the aircraft convected vertical ground shafting:

$$\vec{v} = T_b^i \vec{v}_b \tag{2}$$

T_b^i is the conversion matrix of body shafting to convected earth shafting, as shown in Equation (3).

$$T_b^i = \begin{bmatrix} \cos\theta\cos\psi & \sin\theta\sin\phi\cos\psi - \sin\psi\cos\phi & \sin\theta\cos\phi\cos\psi + \sin\psi\sin\phi \\ \cos\theta\sin\psi & \sin\psi\sin\theta\sin\phi + \cos\psi\cos\phi & \sin\psi\sin\theta\cos\phi - \cos\psi\sin\phi \\ -\sin\theta & \sin\phi\cos\theta & \cos\phi\cos\theta \end{bmatrix} \tag{3}$$

Finally, on the basis of the three-way ENU velocity measured by the GPS antenna, the upward velocity of the center of gravity was obtained by superimposing. The three-way velocity under the convected vertical ground shafting is caused by the rigid body rotation.

$$v_v = v_{T\text{-GPS}} + v_z' \tag{4}$$

Therefore, the relatively accurate sinking velocity at the center of gravity of the aircraft is the sinking velocity measured by the antenna after a certain matrix change. The relevant influencing factors include roll angle ϕ, pitch angle θ, yaw angle ψ, and angular velocity of roll, pitch, and yaw, which reflect the landing attitude of the aircraft.

To sum up, the load on the undercarriage when the aircraft is landed is related to the sinking velocity at the center of gravity of the aircraft, the attitude angle of the aircraft and the attitude angular velocity.

Another factor that impacts the landing load of carrier-based aircraft is the weight of the aircraft. When the sinking speed and attitude are consistent, a heavier aircraft will exert a greater impact on the landing gear, resulting in a higher landing load.

In conclusion, the load on the landing gear during aircraft grounding is influenced by factors such as sinking speed at the center of gravity, aircraft weight, attitude angle, and attitude angular velocity.

3. Application of MATLAB neural network model in load prediction

3.1. Application idea of neural network model

a) Firstly, a model suitable for this research in the MATLAB neural network module was selected. The sinking velocity V_y, roll angle ϕ, pitch angle θ, yaw angle ψ, roll angular rate ϑ, pitch angular rate ζ, pitch angular rate A and yaw angular rate A at the time of landing in the measured flight data of a certain model of aircraft were taken as the inputs of the training network model. The vertical load F_z on the undercarriage was used as the output of the training network model to train the model.

b) Secondly, other data of different sorties of the same aircraft type were selected to predict the output results using the trained model.

c) Thirdly, the vertical load predicted by the model was verified by the measured vertical load in real flight data. If the error between the two was within a certain range, the training effect of the model was considered to be ideal.

d) Finally, the trained network model was used to predict the undercarriage vertical load when the sinking velocity limit value was 100%. The predicted result was obtained.

The input and output of the neural network model were calculated as follows according to the above analysis:

Table 2. Inputs and outputs of network model training.

Input	V_y	ϕ	θ	ψ	φ	ϑ	ζ	m
Output					F_z			

3.2. Classification of MATLAB neural network model

For the problem studied in this paper, the network model has multiple input parameters. The relationship between the input parameters and the output parameters is non-linear, so the appropriate neural network can be selected for prediction according to the characteristics of this data. MATLAB neural network toolbox provides a variety of network models to choose from. Different models are suitable for different engineering problems, such as single-layer perceptron, BP neural network, radial basis function network, etc. [5-7].

3.3. Flight data processing

In the process of training neural networks, if the absolute value difference between the values of each input parameter is large, the weight of each parameter may be affected, resulting in an unsatisfactory training effect or slow convergence velocity. A better solution is to normalize the processing of the data.

One of the common methods of data normalization processing is to divide data by a constant so that it is all contained within the interval [0, 1] or [-1, 1]. It can be written as [8]:

$$x' = \frac{x - \min A}{\max A - \min A} \tag{5}$$

where x is the original value of the data and the value after the normalization of x, min$A$ is the minimum value, and max$A$ is the maximum value of the data in this group.

The selected measured flight data were preprocessed according to the above method. A certain model of aircraft studied in this thesis had a total of 37 landing test data, among which three aircraft reached the specified value of 80% sinking velocity. The relevant parameters of landing moment in these measured data were screened out and normalized according to Equation (5) and divided into an input data matrix and an output data matrix according to Table 2.

3.4. Training and testing of RBF NN model

Several different neural network models are introduced in Section 3.2. The data processed in Section 3.3 is imported into the network according to the division in Table 2 in Section 3.1. The network is trained. The training results are shown in Table 3:

Table 3. Training results of different types of network models.

Network type	Train		Test	
	Training error	Regression	Training error	Regression
Single-layer perceptron	0.0322	0.5904	0.1277	-0.6381
BP neural network	0.0049	0.9773	0.0192	0.8792
Radial basis function network	0.0020	0.9737	0.0019	0.9678

As can be seen from Table 3, the single-layer perceptron is not suitable for the research in this paper. The error and regression of BP network in the test process are not as good as that of RBF network, so the RBF network is chosen as the network model for training in this paper.

The RBF NN model structure is shown in the figure 3 below:

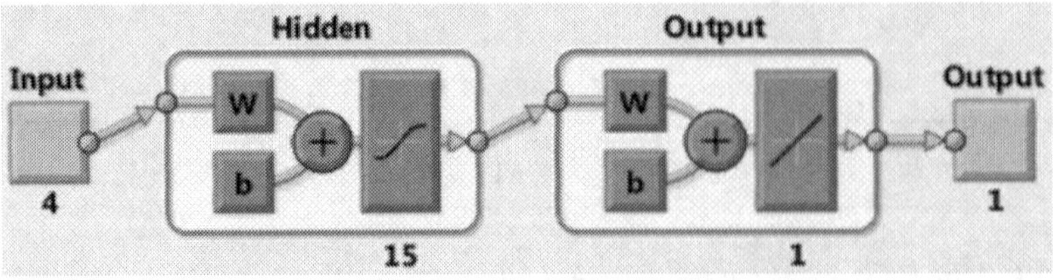

Figure 3. RBF network model structure.

In the figure, w is the weight of the neuron, and b is the bias.

An RBF NN was established in the MATLAB neural network toolbox. The number of hidden layers was set to 15. The training samples accounted for 70%, the test samples accounted for 30%, and the hidden layer algorithm function was selected as a Bayesian regularization algorithm.

The data processed in Section 3.3 was imported into the network according to the division in Table 2 in Section 3.1. The network was trained to obtain the training results as shown in Figure 4 and Figure 5:

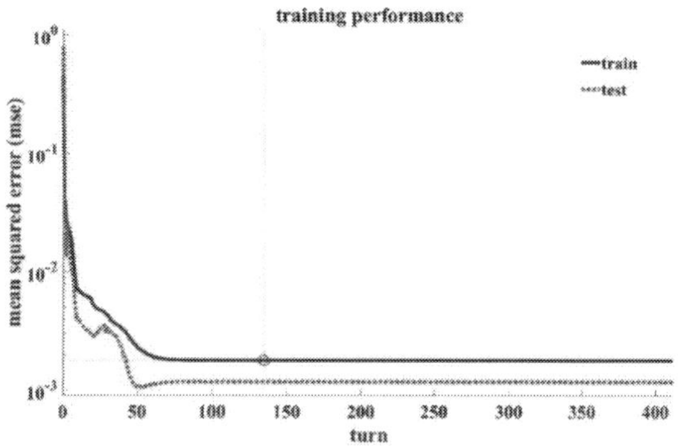

Figure 4. Network training and testing performance.

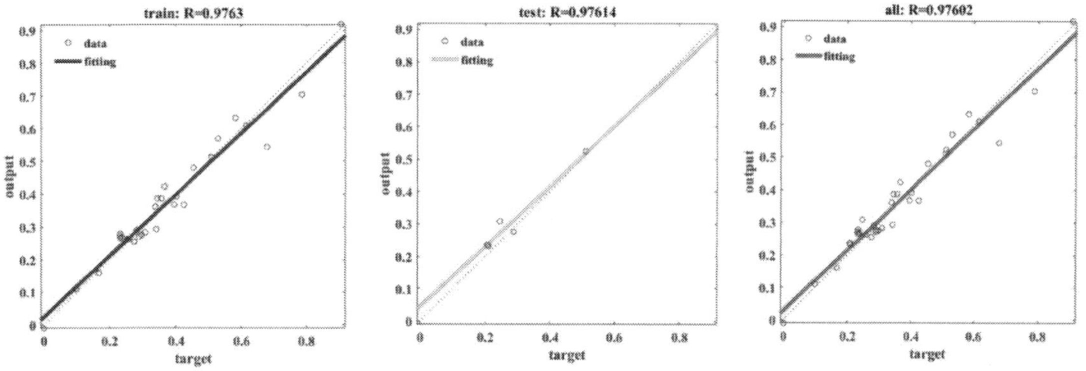

Figure 5. Training and test regression.

Figure 4 shows the mean square error of neural network model training and test. It can be seen from Figure 5 that the network conducted more than 200 rounds of training and testing. The final results showed that the mean square error was in the order of magnitudes of 10^{-2}. Considering the physical meaning of the object parameters studied in this thesis, the error of the order of magnitudes met the research needs. Figure 5 shows the fitting effect of measured values and predicted output of training samples, test samples, and all samples. It can be seen from Figure 4 that most points of the overall samples are distributed near the middle line. The fitting effect is relatively ideal.

The authenticity of the network prediction was verified after the completion of the network training. The measured data of three landing test aircraft that did not participate in network training were selected. Error analysis and calculation were carried out between the Fz value predicted by the network and the Fz value in the measured data to verify the network model. The measured value in the table was normalized according to Section 3.3. The obtained results are shown in Figure 6 and Table 4:

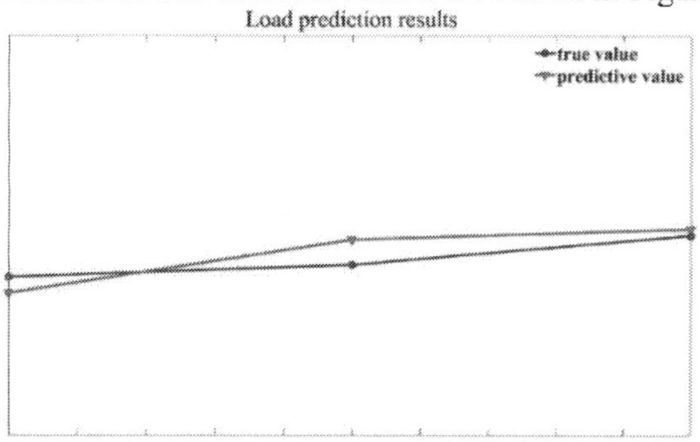

Figure 6. Comparison of measured load and predicted load.

Table 4. Neural network verification results.

Aircraft	Measured value	Predicted value	Relative error	Average relative error	Maximum relative error
1	0.8963	0.8564	0.044503		
2	0.9267	0.9892	-0.067408	0.042676	0.067408
3	1	1.0161	-0.016118		

It can be seen from Figure 6 that the measured load value during model verification is almost the same as that predicted by the neural network. It can be seen from Table 4 that the error between the undercarriage load predicted by the neural network and the measured value is within 10%, indicating that the network training effect is ideal and the prediction result is real and credible.

3.5. Load prediction at 100% sinking velocity limit value

For the neural network model that has been trained and verified, the undercarriage load prediction corresponding to the 100% sinking velocity limit value in the National Military Standard could be carried out.

When setting the network input value, the 100% sinking velocity limit value was taken as the network input. Then, the sinking velocity was converted into the normalized value through Equation (5); the remaining parameters were statistically analyzed by referring to the measured values of the parameters of other aircraft participating in the training to obtain the network input values.

After setting the input value of the network model, it was imported into the trained neural network to obtain the predicted value of the network, which is shown in Table 5.

Table 5. Neural network prediction results.

Sinking velocity	Load predicated value
100% limit value	1.1177

The predicted load value with physical significance after inverse normalization could be obtained by putting the predicted result into Equation (5), which can be used as the real value of the vertical load on the undercarriage when the sinking velocity of the aircraft reaches 100% limit value in the actual landing test. This value can be used to improve the test points in the landing test of carrier-based aircraft and to evaluate the strength and stiffness of the undercarriage more comprehensively.

3.6. Analysis of model errors

The main causes of the errors in the trained model in the aforementioned section are analyzed as follows:

a. The sample size of the data involved in the model training is insufficiently abundant, leading to the inadequate generalization ability of the model training outcomes.

b. The selection of the hidden layer algorithm function during model training requires optimization, and the number of hidden layers is not rational.

c. There are also certain factors influencing the landing load that are not encompassed in the training of the model.

For the abovementioned reasons, the following improvement approaches can be adopted in the subsequent research to optimize the training results:

a. We augment the sample size of the training data and incorporate more measured data in the training of the model.

b. We endeavor to employ diverse hidden layer algorithm functions and the number of hidden layers for training, compare the training results, and optimize the network model.

c. We conduct an in-depth analysis of the factors influencing landing loads at the theoretical level, increase the types of input parameters, and optimize the training process.

4. Conclusion

Sinking velocity is a key parameter index. In the landing test, to evaluate the strength and stiffness of the aircraft undercarriage, whether the sinking velocity reaches the limit value stipulated by the standard during landing directly affects the effect of the landing test. However, for the sake of security, it is difficult for test flight researchers and test pilots to control the sinking velocity at 100% of the prescribed limit value when designing the test point of landing test, which is a defect for the strength and stiffness assessment of undercarriage.

In this thesis, based on a large number of measured flight data, the artificial neural network model is applied to the aircraft landing test. The key parameters affecting the subsidence velocity were selected through a reasonable physical model that simplifies the aircraft landing process. The input and output of the model were allocated for training, verification, and application to realize the prediction of the vertical load on the aircraft undercarriage under the 100% sinking velocity limit value. The verification results of the model are in good agreement with the measured flight data, which verifies the reliability of the model. Finally, the prediction results are obtained to improve the test point of carrier-based aircraft landing tests and provide a new idea and method for the strength and stiffness assessment of aircraft undercarriage.

References

[1] Rui Hu, Xiaochuan Liu, Chunyu Bai et al. Progress of the kinetic test research on the takeoff and landing device of shipborne aircraft [J]. Aviation Science and Technology, 2022, 33, (1).
[2] COSTIND. (1996) Carrier-based airplane strength and rigidity specification: Flight Tests. GJB2759-96.
[3] COSTIND. (1996) Carrier-based airplane strength and rigidity specification: Ground Loads. GJB2753-96.
[4] Muqiao Li. Analysis of Artificial Neural Network and Its Application Model [A]. Science and Technology Communication 2019, 08.
[5] Fei Wu, Binbing Liao, Yahong Han. Interpretability of deep learning [A]. Aviation Arms, 2019, 26(1).
[6] Zhenping Fang. Aircraft Flight Dynamics [M]. Beijing: Beijing University of Aeronautics and Astronautics Press, 2005.
[7] Ming Chen et al., MATLAB Neural Network Principle and Case analysis [M]. Beijing: Tsinghua University Press, 2021.
[8] pSomng.(2020)Common Method of Data Standardization . http://blog.csdn.net/weixin_42792088/article/details/103971069.

Journal of Physics: Conference Series **2879** (2024) 012050 doi:10.1088/1742-6596/2879/1/012050

Analysis of multipath effects on Low Earth Orbit navigation satellites

Menglin Tong[1,*], Yansong Meng[1] and Tao Yan[1]

[1]Xi'an Branch, China Academy of Space Technology, Xi'an 710000, Shaanxi, China

[*]Corresponding author's e-mail: 15832520621@163.com

Abstract: With the large-scale emergence of Low-Earth Orbit (LEO) internet constellations, the direct or indirect utilization of LEO communication constellations for navigation and positioning has become a hot topic of current research. However, the multipath effect remains a critical factor influencing positioning accuracy. Compared to traditional Medium-Earth Orbit (MEO) satellites, LEO satellites move swiftly, resulting in faster multipath variations, which renders traditional multipath envelope analysis methods inapplicable. This study proposes a novel analytical model tailored to the multipath effect characteristics of LEO satellites. Through this model, we have observed that the multipath variations of high-frequency signals in LEO rapidly fluctuate near zero, offering the potential to reduce errors through smoothing methods. Furthermore, the study reveals that the impact of the multipath effect is positively correlated with orbital altitude and inversely correlated with carrier frequency. This study provides a new perspective for understanding the multipath effect characteristics of LEO satellites and offers necessary theoretical support and practical guidance for improving the accuracy of navigation and positioning systems based on LEO satellites. Future research can further explore multipath suppression methods and promote the widespread application of LEO satellites in the field of navigation and positioning.

1. Introduction

In the traditional Global Navigation Satellite System (GNSS) field, researchers have conducted extensive and in-depth studies on multipath effects, aiming to mitigate their impacts on navigation signals through technological innovations and method optimizations. These studies encompass various aspects, ranging from receiver algorithm improvements (e.g., employing adaptive filters and multipath mitigation algorithms) to antenna design optimizations (e.g., adopting multi-antenna arrays and special-shaped antennas to enhance directionality and multipath resistance) [1-7]. However, with the continuous advancements in technology and the emergence of new navigation and positioning technologies, particularly the rapid development of large-scale LEO internet constellations, traditional multipath analysis methods, and mitigation strategies face new challenges.

LEO satellites, with their low orbital altitudes, extensive coverage, and fast data transmission speeds, offer not only communication services but also demonstrate immense potential in the field of navigation and positioning, leading to a plethora of research exploring the direct or indirect use of LEO constellations for navigation [8-14]. Nevertheless, compared to MEO satellites, the rapid motion characteristics of LEO satellites result in more drastic dynamic changes in multipath effects, rendering traditional multipath envelope analysis methods based on static or slowly varying environmental assumptions inapplicable. This necessitates a re-examination and construction of multipath effect

Journal of Physics: Conference Series **2879** (2024) 012050 doi:10.1088/1742-6596/2879/1/012050

models tailored to LEO satellite characteristics, to more accurately describe their multipath variation features and provide a theoretical foundation for subsequent error compensation, mitigation, and even elimination strategies. However, existing works have only conducted qualitative analyses on the multipath effects of LEO satellites, pointing out that the rapid motion of LEO satellites leads to rapid variations in multipath, which contributes to the whitening of multipath effects. None of these works have proposed specific models for analyzing the multipath effects of LEO satellites. To fill this gap, this paper will analyze and demonstrate the multipath effects of LEO satellites from the aspects of model establishment, scenario simulation, and data analysis.

2. Multipath error envelope

For GNSS signals in medium and high Earth orbits, the occurrence of multipath signals is particularly frequent and complex due to their long transmission paths and susceptibility to factors such as the atmosphere, ionosphere, and surface reflections. As a result, when analyzing and evaluating such signals, two key indicators, namely the multipath error envelope and the average multipath error, are commonly employed to quantify the deviations caused by multipath signals during the tracking process. The generation of multipath errors primarily stems from the interaction between the direct signal and one or more multipath signals. These multipath signals, having traversed different propagation paths, exhibit distinct delay times, amplitude attenuations, and phase shifts compared to the direct signal. Consequently, the multipath signals processed at the receiver feature different multipath delays, multipath-to-direct ratios, and relative phases relative to the direct signal, making it challenging to accurately lock onto the direct signal and resulting in tracking biases.

The multipath error envelope represents an extreme manifestation of multipath errors under specific conditions. It typically refers to the range of multipath errors that occur when the relative phases are 0° and 180°, respectively, i.e., when the multipath signals are in-phase or out-of-phase with the direct signal. These two-phase points represent the most significant impacts of multipath effects on signal tracking, and thus, the multipath error envelope provides an intuitive reflection of the potential impact of the multipath environment on the reception performance of GNSS signals.

When considering only one direct signal and one multipath signal, the complex representation of the received multipath signal can be expressed as:

$$s(t)=s_{direct}(t)+a_1e^{j\theta_1}s_{direct}(t-\tau_1) \tag{1}$$

where θ_1 represents the phase of the multipath reflected signal, a_1 represents the amplitude of the multipath signal, and τ_1 represents the time delay of the multipath reflected signal. When a multipath signal is present, the correlation function between the local reference signal and the received signal is given as follows:

$$R(\Delta\tau)=R_{direct}(\Delta\tau)+a_1e^{j\theta_1}R_{direct}(\Delta\tau-\tau_1) \tag{2}$$

In the design of GNSS receivers, the correlation function between the local reference signal and the received signal is crucial for determining the signal arrival time and achieving precise tracking. However, in the presence of multipath signals, this correlation function becomes distorted, with the originally smooth S-curve shape becoming irregular and its zero-crossing point shifting. This shift directly leads to tracking errors, impacting positioning accuracy. Taking the example of a BPSK(2) signal, which employs Binary Phase Shift Keying modulation with a data rate that is twice the spreading code rate, and assuming a receiver bandwidth set at 6*1.023 MHz (typical for C/A code signals), and a multipath-to-direct ratio of -10 dB (indicating that the multipath signal is attenuated by a factor of 10 compared to the direct signal), the multipath error envelope is illustrated in Figure 1.

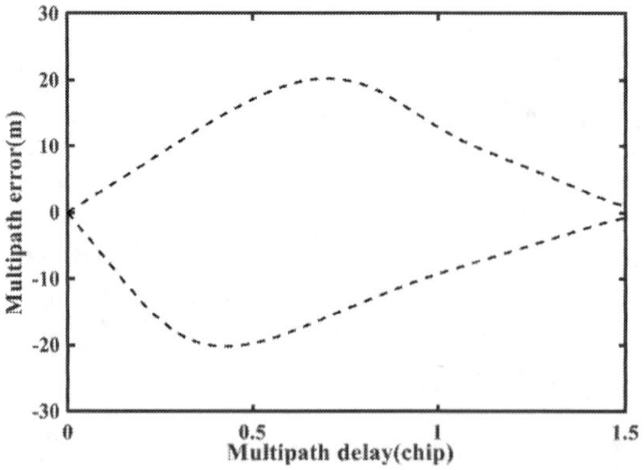

Figure 1. Multipath error envelop.

For medium earth orbit (MEO) navigation satellites, operating in geosynchronous orbits or higher, their relatively low velocities directly result in slow changes in the relative positions between the satellites and ground receiving stations. In such scenarios, when signals encounter obstacles during propagation and generate multipath effects, the multipath delays, multipath-to-direct ratios, and phase differences between these multipath signals and the direct signal vary relatively slowly on the order of seconds. This stability contributes to a tendency for the distortion of the correlation function formed by multipath signals at the receiver to also be stable, making it less prone to being masked by rapidly changing signal characteristics. In contrast, LEO satellites have orbital altitudes significantly lower than those of MEO satellites, leading to rapid changes in the relative positions between the satellites and ground receiving stations. This rapid motion causes significant variations in multipath delays, multipath-to-direct ratios, and relative phases between multipath signals and the direct signal within extremely short periods. Due to the rapid transformation of these parameters, multipath effects at the receiver do not form stable deviation patterns but manifest as dynamic and random interference. When these dynamic multipath signals are averaged over long-time scales, their impacts tend to cancel each other out, effectively reducing their overall impact on navigation accuracy. Especially in the Ka-band, where the higher frequency and shorter wavelength increase the signal's sensitivity to minute path changes, the rapid variation of multipath effects results in even more negligible averaged multipath errors. This paper assesses the multipath error characteristics of low-orbit L/Ka navigation signals by establishing signal models and conducting simulation analyses. Furthermore, it quantifies the effects of multipath errors through detailed analysis.

3. Multipath reception signal and simulation model

To evaluate the multipath error characteristics of low orbit L/Ka navigation signals, the following multipath received signal model is constructed as follows:

$$s(t) = s_{direct}(t) + a_1 e^{j\theta_1(t)} s_{direct}(t - \tau_1(t)) \tag{3}$$

This model is based on traditional multipath signal modeling, with the addition of variables in the non-line-of-sight signal component to analyze the impact of multipath variation rate on the multipath effects. In the above equation, we only consider the impact of a single multipath, as well as the multipath delay and phase difference of the multipath signal relative to the direct signal vary over time, while we assume that the multipath to direct ratio remains unchanged, so we have:

$$\begin{aligned} \tau_1(t) &= \tau_0 + \tau(t) \\ \theta_1(t) &= 2\pi f_{carrier} \tau_1(t) \end{aligned} \tag{4}$$

To reveal the variations in multipath effects, it is necessary to construct correlation functions for analysis. It is assumed that τ_0 represents the initial delay and $f_{carrier}$ represents carrier phase. According to Equation (3), the relevant functions are obtained as follows:

$$\begin{aligned} R(\Delta\tau) &= R(\tau - \tau_1(t)) \\ &= R_{direct}(\tau) + a_1 e^{j\theta_1(t)} R_{direct}(\tau - \tau_1(t)) \\ \Delta\tau &= \tau - \tau_1(t) \end{aligned} \tag{5}$$

To achieve tracking of navigation signals, locally generated early/prompt/late reference signals are used. The code phase discriminator function used here is the commonly employed non-coherent Early Minus Late Power (EMLP) method in the field of navigation. During the implementation process, the relevant outputs are divided into in-phase and quadrature components, satisfying:

$$\begin{aligned} I_E + jQ_E &= \left(\frac{2CT_p}{N_0}\right)^{1/2} R\left(\Delta\tau - \frac{d}{2}\right) + n_{I_E} + jn_{Q_E} \\ I_P + jQ_P &= \left(\frac{2CT_p}{N_0}\right)^{1/2} R(\Delta\tau) + n_{I_P} + jn_{Q_P} \\ I_L + jQ_L &= \left(\frac{2CT_p}{N_0}\right)^{1/2} R\left(\Delta\tau + \frac{d}{2}\right) + n_{I_L} + jn_{Q_L} \end{aligned} \tag{6}$$

where n_{I_E}, n_{Q_E}, n_{I_P}, n_{Q_P}, n_{I_L}, and n_{Q_L} are standard normal distribution N (0, 1) and they are independent of each other. The code tracking multipath error of the non-coherent EMLP loop is independent of the carrier phase error, so we can simplify the code ring authentication function used into:

$$\frac{1}{2}\left[\left(I_E^2 + Q_E^2\right)^{1/2} - \left(I_L^2 + Q_L^2\right)^{1/2}\right]\left[\left(I_E^2 + Q_E^2\right)^{1/2} + \left(I_L^2 + Q_L^2\right)^{1/2}\right]^{-1} \tag{7}$$

Based on Equations (3)-(7), we construct the simulation model as shown in Figure 2.

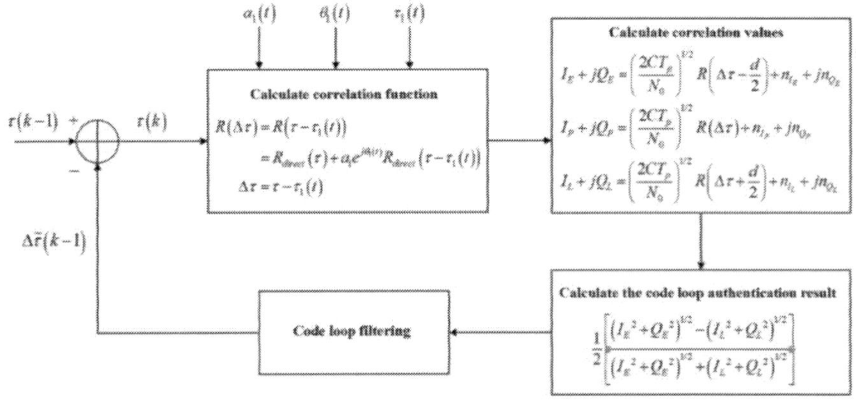

Figure 2. Simulation model.

4. Simulation and analysis of multipath scenario

4.1. Scenario without multipath

In this section, we simulate the scenario without multipath effects, setting the initial delay to 0.3 chips, carrier-to-noise ratio to 45 dB-Hz, code length to 2, 046, code period to 1 ms, and updating the code loop every 1 ms. These parameter values are the most commonly used and have high accuracy under the current navigation system. We calculate the relevant results after tracking with the code loop and plot the curve of the code pseudorange over time within 10 seconds. We calculate the mean and standard deviation of the code delay within the last 5 seconds. The simulation results are shown in

Figure 3. In scenarios without multipath effects, after tracking through the code loop, the local code phase can align with the direct signal.

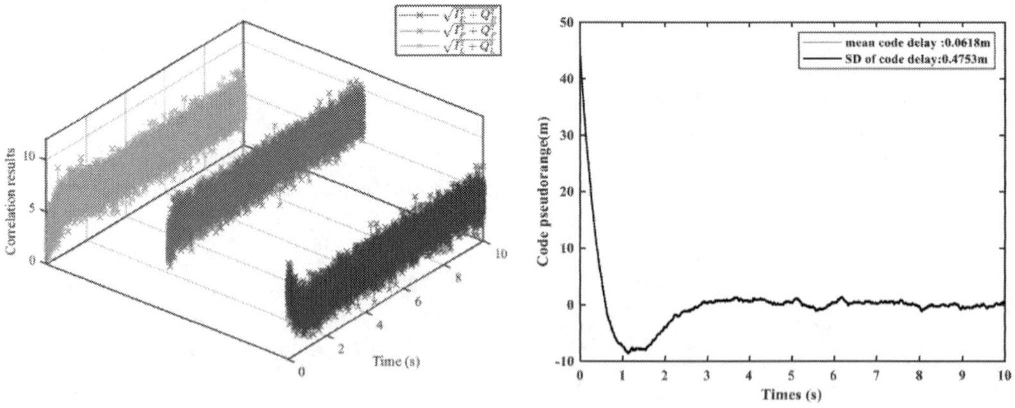

Figure 3. Correlation results and code pseudorange error.

4.2. Scenario with one multipath path

4.2.1. MEO L-frequency band. In this section, we analyzed the simulation of a scenario where there is a multipath in the mid to high orbit. We assume that the delay of the multipath signal relative to the direct signal varies linearly, which simplifies the complexity of the model and reflects to some extent the general trend of multipath signal delay with time or spatial position in the actual environment. If the satellite speed is v m/s, then there is:

$$\tau(t) = \frac{vt}{c} \tag{8}$$

For a typical MEO satellite, we suppose $v = 0.001\text{m/s}$ and τ_0 in Equation (4) is taken as 0.1365 Tc, and the carrier frequency is taken as 1, 575.42 MHz, which is the most commonly used in the L-band. To conduct a parallel comparison, the multipath direct ratio is set to -10 dB, and the other parameter settings are the same as in Section 4.1. The distance and phase differences of the multipath signal relative to the direct signal are shown in Figure 4.

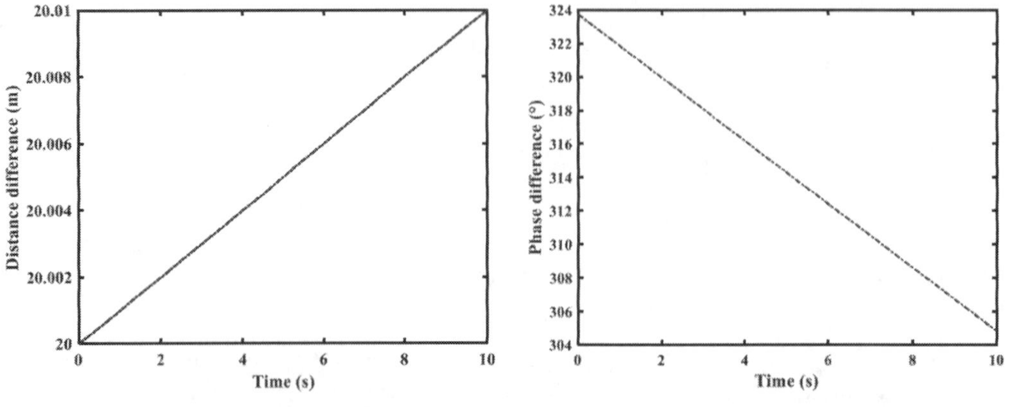

Figure 4. Distance and phase differences between direct and multipath paths.

We calculate the relevant results after tracking with the code loop and plot the curve of the code pseudorange over time within 10 seconds. We calculate the mean and standard deviation of the code delay within the last 5 seconds as shown in Figure 5. There is an estimated deviation of 3.9180 m.

IOP Publishing
doi:10.1088/1742-6596/2879/1/012050

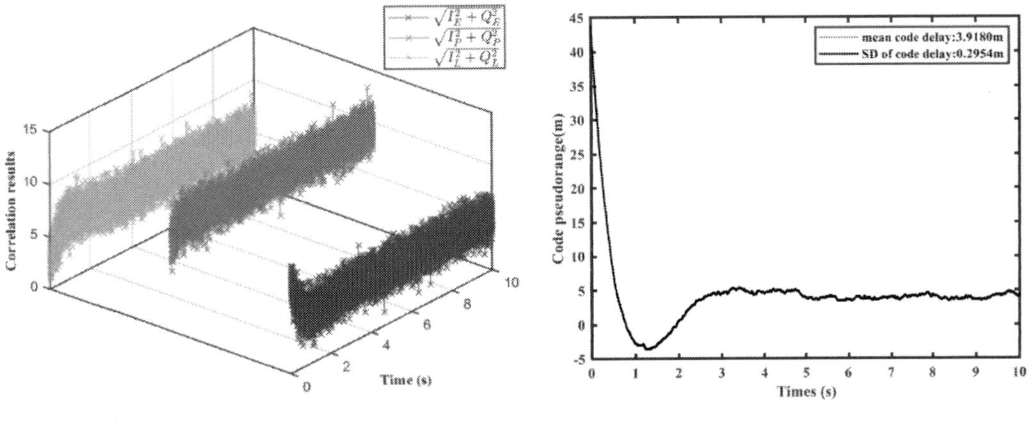

Figure 5. Correlation results and code pseudorange error.

Theoretically, the results obtained at an initial delay of 0.1365 Tc are shown in Figure 6. It can be seen that the simulation results are consistent with the theoretical analysis results. This indicates that for medium to high-orbit L signals, the deviation of multipath signals relative to direct signals is stable and can be approximated as a constant.

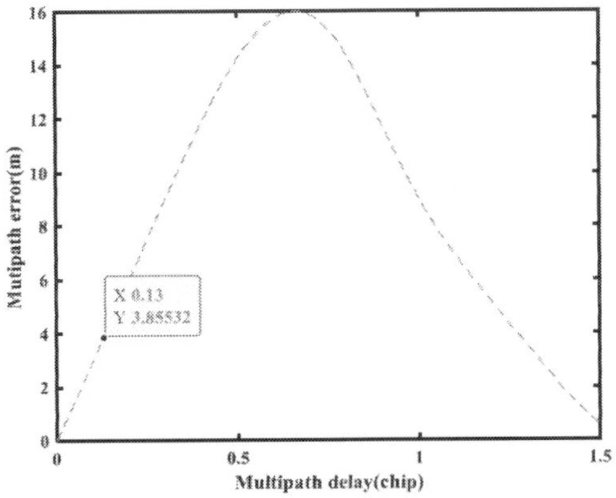

Figure 6. Theoretical multipath error.

4.2.2. LEO L- frequency band. For low Earth orbit satellites, based on the LEO satellite movement simulation results from Satellite Tool Kit, we take $v = 0.03$m/s in Equation (8) and τ_0 in Equation (4) as 0.1365 Tc, and the carrier frequency is also taken as 1, 575.42 MHz, which is the most commonly used in the L-band. The multipath direct ratio is set to -10 dB, and the other parameter settings are the same as in Section 4.1. The distance and phase differences of the multipath signal relative to the direct signal are shown in Figure 7. The phase of the multipath signal changes by nearly two cycles compared to the direct signal.

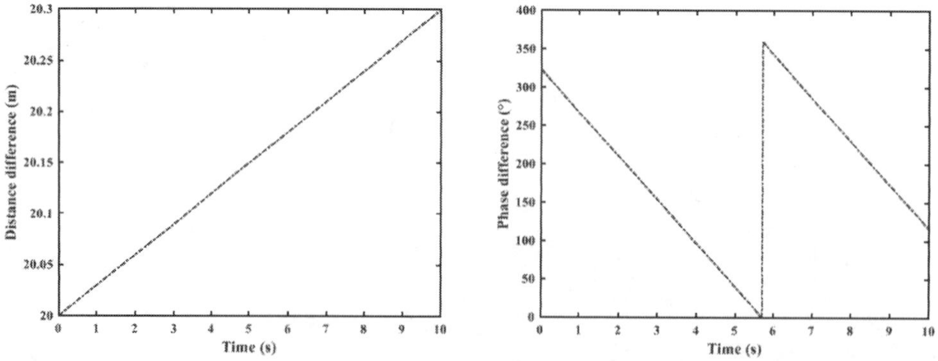

Figure 7. Distance and phase differences between direct and multipath paths.

We calculate the relevant results after tracking with the code loop and plot the curve of the code pseudorange over time within 10 seconds. We calculate the mean and standard deviation of the code delay within the last 5 seconds as shown in Figure 8. At this point, the code delay fluctuates around 0 and varies with the relative phase. The average estimated value within the last 5 seconds is close to 0, theoretically smoothing out the impact of multipath errors on time delay estimation.

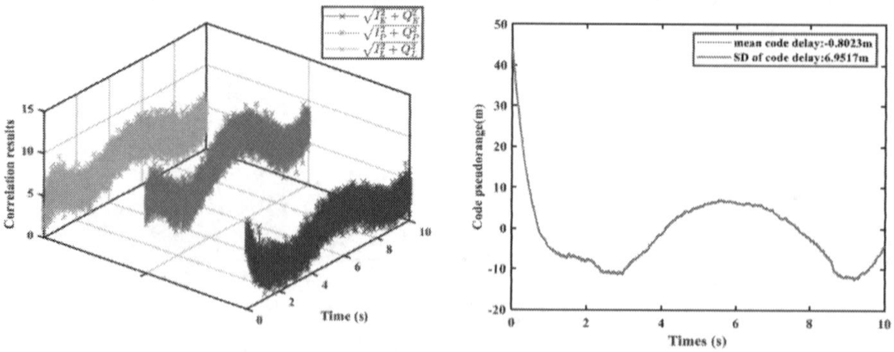

Figure 8. Correlation results and code pseudorange error.

4.2.3. LEO Ka- frequency band. For a typical LEO satellite, we suppose $v = 0.03$m/s in Equation (8) and τ_0 in Equation (4) as 0.1365 Tc, and the carrier frequency is taken as 20 GHz, which is the most commonly used in the L-band. The multipath direct ratio is set to -10 dB, and the other parameter settings are the same as that in Section 4.1. The distance and phase differences of the multipath signal relative to the direct signal are shown in Figure 9.

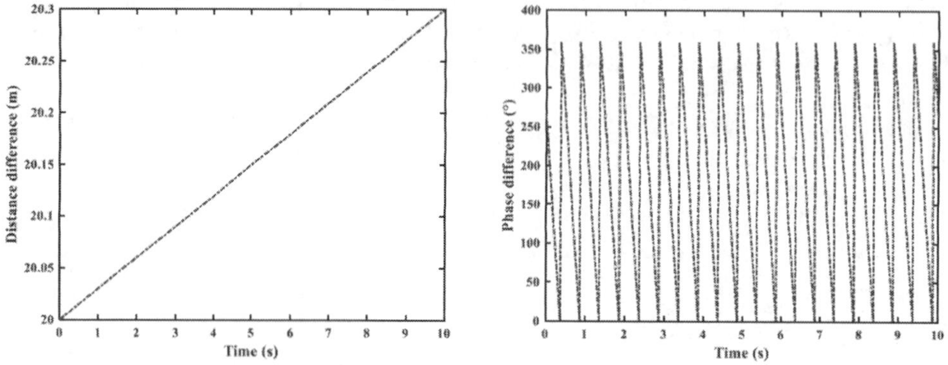

Figure 9. Distance and phase differences between direct and multipath paths.

We calculate the relevant results after tracking with the code loop and plot the curve of the code pseudorange over time within 10 seconds. We calculate the mean and standard deviation of the code delay within the last 5 seconds, as shown in Figure 10. At this point, the code delay fluctuates around 0, varying with the relative phase. The average estimated value in the last 5 seconds is close to 0, with a significantly smaller standard deviation than the L-band signal. The multipath effect is smaller, making it easier to smooth out the impact of multipath errors on time delay estimation.

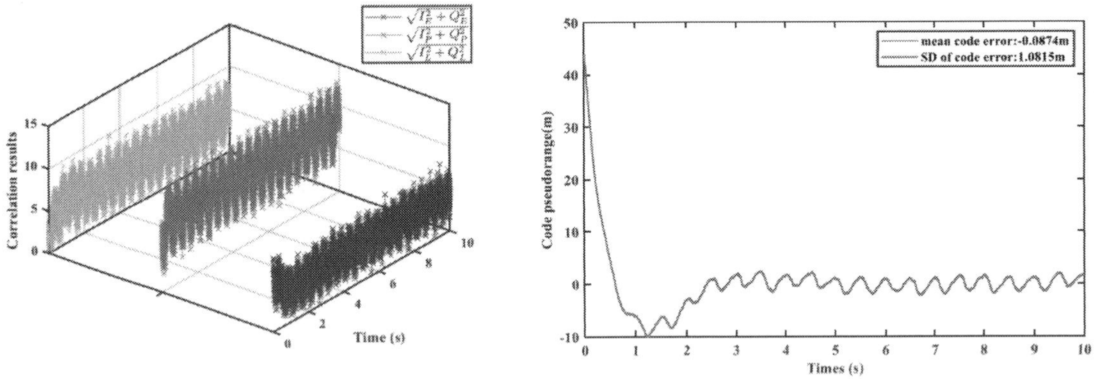

Figure 10. Correlation results and code pseudorange error.

4.3. Simulation result analysis

In scenarios without multipath effects, after tracking through the code loop, the local code phase can align with the direct signal. Based on the above simulation results, we can find that multipath error is closely related to the carrier frequency and orbit altitude (which determines satellite velocity). To further analyze their relationship, we changed the simulation parameters and listed the relationship between multipath error (the mean code delay of the last 5 seconds) and carrier frequency and orbit altitude (represented by satellite speed here) as shown in Table 1. Due to the inverse relationship between satellite line velocity and orbital altitude, it can be assumed that the impact of multipath effects is positively correlated with orbital altitude and inversely correlated with carrier frequency.

Table 1. Change in code pseudorange error (mean value).

Satellite speed	Carrier frequency		
v (m/s)	1, 575.42 MHz	20 GHz	40 GHz
0.001	3.9180 m	1.0282 m	0.6238 m
0.030	-0.8023 m	-0.0874 m	0.0427 m
0.100	0.6576 m	0.0381 m	0.0125 m

For LEO satellites, the currently popular Ka-band offers abundant available bandwidth resources, which can support carriers at higher frequencies. This implies that the influence of multipath effects can be more easily mitigated through smoothing techniques. In the design of LEO constellations, it is also necessary to fully consider the impacts of coverage and signal characteristics. At the same time, this poses new requirements for modulation and demodulation methods. In the past, there was a view that the rapid movement of low earth orbit satellites may contribute to the "whitening" of multipath effects, by reducing their correlation through rapid changes and simplifying processing. However, this study challenges this viewpoint by revealing that the multipath variations of LEO high-frequency signals fluctuate frequently near zero. This discovery provides a new perspective and possibility for eliminating multipath errors through smoothing techniques, and the impact of multipath effects is positively correlated with orbital altitude and inversely correlated with carrier frequency, which is an important factor that needs to be balanced and considered in the design of LEO navigation constellations.

5. Conclusion

This paper presents an innovative framework analyzing LEO satellite multipath characteristics and designs a tailored model that captures their dynamic behavior, deepening understanding of multipath effects. It bridges theoretical-qualitative gaps with empirical-quantitative methods. The insights enhance positioning accuracy, stabilizing systems and LEO navigation design innovation. Future work will focus on developing efficient, precise multipath mitigation algorithms tailored to LEO environments, especially for high-speed dynamics. This optimization promises improved positioning accuracy, which is ideal for precision-critical industries like UAVs and autonomous driving, unlocking the vast potential for navigation advancements.

References

[1] Cho S., Seok H. W., and Kong S. H. MPCNet: GNSS Multipath Error Compensation Network via Multi-task learning, 2023 IEEE Intelligent Vehicles Symposium (IV), Anchorage, AK, USA, 2023, pp. 1-6.

[2] Chen X. C., Yu J. M., Yu W., and Truong T. K. (2020). GPS L1CA/BDS B1I Multipath Channel Measurements and Modeling for Dynamic Land Vehicle in Shanghai Dense Urban Area, in IEEE Transactions on Vehicular Technology, Vol. 69, No. 12, pp. 14, 247-14, 263.

[3] Liu J., Yan Y., Yu H., and Ma B. (2023). Approximate BER Performance of LoRa Modulation with Heavy Multipath Interference, in IEEE Wireless Communications Letters, Vol. 12, No. 5, pp. 853-857.

[4] Ma X., Han C., Jin R., Wang D., et al. Study on GNSS Spoofing Interference Detection Method in Urban Multipath Environment Based on CNN and Clustering Model, 2023 Cross-Strait Radio Science and Wireless Technology Conference (CSRSWTC), Guilin, China, 2023, pp. 1-3.

[5] Matera E. R., Ekambi B., and Chamard J. A Comparative Analysis of GNSS Multipath Error Characterization Methodologies in an Urban Environment, 2024 International Conference on Localization and GNSS (ICL-GNSS), Antwerp, Belgium, 2024, pp. 1-7.

[6] Tan Y., Li F., and Qi Y. C/N0-Based GNSS Multipath Mitigation Methods: Research, Challenges, and Prospects, 2022 IEEE International Workshop on Electromagnetics: Applications and Student Innovation Competition (IWEM), Narashino, Japan, 2022, pp. 101-102.

[7] Yan T., Wang Y., Qu B., et al. Weighted discrimination function-based tracking method for binary offset carrier signals [J]. Chinese Space Science and Technology, 2018, 38 (3): 45-53 (in Chinese).

[8] Reid T. G. R., Chan B., Goel A., et al. Satellite navigation for the age of autonomy [C]// 2020 IEEE/ION Position, Location and Navigation Symposium (PLANS), Portland, OR, USA, April 20-23, 2020, 342-352.

[9] Michel F., Trevisan M., Giordano D., et al. A first look at Starlink performance [C]// Proceedings of the 22nd ACM Internet Measurement Conference. 2022: 130-136.

[10] Lawrence D., Cobb H. S., Gutt G., et al. Navigation from LEO's current capability and future promise [J]. GPS World, 2017: 42-49.

[11] Reid T. G. R., Neish A. M., Walter T., et al. Broadband LEO constellations for navigation [J]. Navigation: Journal of the Institute of Navigation, 2018, 65 (2): 205-220.

[12] Iannucci P. A. and Humphreys T. E. Economical fused LEO GNSS [C]// Proceedings of 2020 IEEE/ION Position Location and Navigation Symposium, 2020.

[13] J. Khalife, M. Neinavaie, and Z. M. Kassas. The First Carrier Phase Tracking and Positioning Results with Starlink LEO Satellite Signals [J]. IEEE Transactions on Aerospace and Electronic Systems, Vol. 58, No. 2, pp. 1, 487-1, 491.

[14] Wu C. L., Lang X. K., Wang T., et al. LE0-based communication and navigation fusion security positioning technology system and performance verification [J]. Space Electronic Technology, 2023, 20 (6): 118-124.

Research on mechanical properties of HTPB propellants

Chang Liu*, **Zhiqiang Chang, Chao Li, Fujun He and Yake Wu**

Qingzhou Hi-Tech Institute, Qingzhou 262500, Shandong Province, China

*Corresponding author's e-mail: h14chang@163.com

Abstract: Slat propellants with different sizes and shapes are designed to investigate the strength criteria of HTPB propellants. According to the calculation and simulation results, with the increase of multiple, the designed size slat specimens can achieve equivalent biaxial tensile stress ratios of 1:2, 1:2.5, 1:3, 1:3.5, and 1:4 under uniaxial tensile conditions. The biaxial mechanical properties of the slab with a stress ratio of 1:2 were tested at low temperatures (25°C, - 30°C, - 50°C, and 0.4 s^{-1}, 4.0 s^{-1}, 14.29 s^{-1}, 42.86 s^{-1}). It was found that as the temperature and the strain rate rise, an increase is observed in the maximum tensile strength and the initial modulus, a decrease is observed in the maximum elongation of the propellant, and the strength envelope is constructed according to the typical mechanical parameters: the strength envelope of HTPB propellant increases as the temperature decreases. The research findings are expected to provide statistical support for the analysis of the structural integrity of HTPB propellants under low-temperature ignition.

1. Introduction

As an engineering component, the solid propellant grain must withstand various types of loads including thermal stress, impact, vibration, acceleration, and ignition pressure during the manufacturing, storage, transportation, ignition, and flight of engineering facilities. Under the action of the above loads, any damage to the solid propellant grain will cause serious deterioration of the SRM performance. Therefore, the propellant selected for SRM design should have such mechanical properties that the deformation and strength of the propellant should be sufficient to resist the internal stress of the grain under various possible temperatures and forces, so as not to cause damage to the structural integrity of the grain and changes in the internal ballistic performance beyond the allowable limits. The deformation and strength of solid propellants under different loads are usually determined by conducting mechanical performance experiments under corresponding loading conditions.

As reported by Pang et al., the initial modulus and the maximum tensile strength of solid propellants are inversely proportional to the temperature, but there is no significant proportional relationship between the maximum elongation. Lai Jianwei and Wang Xiaoying[1] both conducted low-temperature and low-temperature recovery room temperature tensile tests to study the effect of low temperatures on propellants. Through low-temperature tensile tests, Wang Zhejun[2] studied the performance changes of propellants under uniaxial and biaxial tensile tests. Shekhar compared the changes in propellant performance caused by various temperatures and found through experiments that low temperatures promote an increase in the maximum tensile strength of the propellant.

Regarding the experiment mainly relying on changes in the tensile or compression rate of the test piece, Lai Jianwei et al. compared the mechanical properties of propellants under different compression rates by changing their compression rate and found that the magnitude of compression rate can change the strength of propellants. Sun et al. conducted mechanical tests on modified double-base solid propellants in different strain rate ranges and established constitutive equations based on mechanical performance parameters. Sun Chaoxiang conducted research on solid propellants by changing the strain rate and found that high strain rates have a greater impact.

To simulate the true stress state of propellants during SRM operation, domestic and foreign researchers have proposed experimental schemes ranging from one-dimensional to multi-dimensional for research. For uniaxial tensile testing, the JANNAF standard and GJB 770B-2005 standard can be used as references, but there is still a lack of standards that can be referenced for multi-axial mechanical testing. To better study the mechanical properties of propellants under multi-dimensional stress conditions, Zhang Ya conducted biaxial tensile and compression tests using a biaxial testing machine. Compared with his studies on uniaxial tensile and compression, he found that the biaxial tensile strength increased, but the elongation decreased; Wang Zhejun also found that biax will enhance the maximum tensile strength of propellant through conducting Flat noodles test pieces, and found that solid propellant will be constrained in two directions under biaxial conditions, so its elongation will decrease[3-5].

The stability of solid propellants plays a crucial role in SRM, and during the ignition process of SRM, the propellant is subjected to multi-axial forces, with its mechanical properties simultaneously affected by the combined effect of the temperature and the strain rate. Therefore, mechanical tests performed under varied conditions can better reveal the structural integrity of the propellant grains in complex environments.

At present, the low-temperature dynamic tensile test is only carried out under the condition of approximately 2:1, and no design and test has been carried out for other Flat noodles. In this paper, Flat noodles' dimensions are designed for different proportions, dynamic mechanical property tests are carried out for approximately 1:2 Flat noodles test pieces, and strength criterion envelopes at different temperatures are constructed. The test results can provide a statistical foundation for further investigations on the structural integrity of low-temperature ignition transient solid rocket motor grains.

2. Size selection of Flat noodles propellant specimen

In this paper, tests were carried out on HTPB Flat noodles propellant. Because this paper did not analyze the influence of the formula on it, the filling mass fraction of its solid particles (AP/Al) was 88% the same as that of Wang Zhejun and others. According to Wang Zhicun, through finite software analysis, it is shown that because of the specificity of its structure, the Flat noodles are in a plane strain state when it is under tension, and can obtain an approximate biaxial tensile effect under uniaxial tension. The size of Flat noodles used in the test is shown in Figure 1.

When W>>H, the stress and strain of Flat noodles propellant at the central position conform to the following relationship:

$$\sigma_y \neq 0\,,\ \varepsilon_y \neq 0 \tag{1}$$

$$\sigma_x \neq 0,\ \varepsilon_x = 0 \tag{2}$$

$$\sigma_z = 0\,,\ \varepsilon_z \neq 0 \tag{3}$$

$$\sigma_x = \nu\sigma_y \tag{4}$$

According to Equation (4), when L2>>H, due to the solid propellant being an approximately incompressible material with a Poisson's ratio close to 0.5, it is approximately 2:1, which can achieve quasi-biaxial stretching in the y-direction and x-direction with a ratio of 2:1.

Linear viscoelasticity is used to analyze the size of the Flat noodles test piece. Prony parameters are used as shown in Table 1. The grid adopts hexahedral structural units, the unit type is C3D8R, and the number of units is 21, 660, as shown in Figure 2.

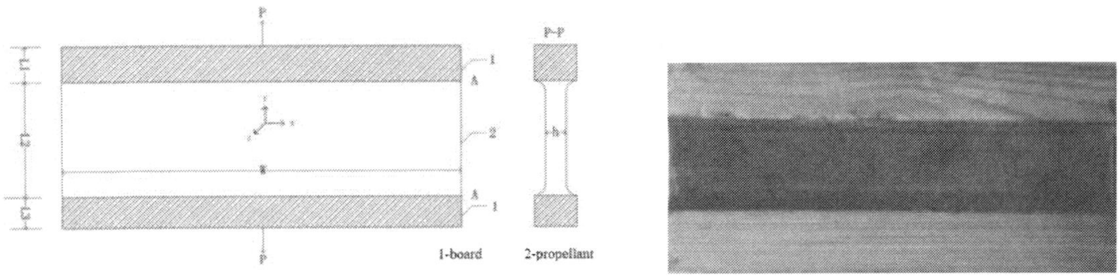

(a) Sketch of slat size (mm) (b) Physical drawing of lath

Figure 1. Experimental use of experimental pieces.

Figure 2. Finite element mesh of propellant.

The top of the test piece is fixed in the x direction ($U1 = UR2 = UR3 = 0$), and the bottom is completely fixed. When its L2 and H are unchanged, we calculate the displacement load of the Flat noodles whose multiples are 0.54, 0.475, 0.42, 0.32, and 0.2 respectively. Calculations reveal that the shear stress in the central region of the specimen is negligible. The bi-directional stress ratio no longer satisfies the proportional relationship in Equation (4). Table 2 shows the parameters of the central lines of the test specimens.

Table 1. Material Prony performance parameters.

E0	E1	E2	E3	E4	E5	E6
3.131 46	16.425 9	3.536 3	3.0756	1.441 6	0.845 8	0.978 0
	τ_1	τ_2	τ_3	τ_4	τ_5	τ_6
	5.920 0E-01	1.148 5	12.0812	55.578 0	217.588 3	1, 236.395 4

Table 2. The value of σ_x and σ_y on the center line of the biaxial strip test piece.

ε /%	$\sigma_x \times 10^{-2} / MPa$					$\sigma_y \times 10^{-2} / MPa$				
	0.54	0.475	0.42	0.32	0.2	0.54	0.475	0.42	0.32	0.2
5.0	4.17	5.04	5.83	7.43	8.73	17.15	17.37	17.58	18.01	18.38
10.0	8.33	10.06	11.72	14.86	17.65	34.29	34.76	35.17	36.03	36.77
15.0	12.49	15.27	17.38	22.25	26.41	51.36	52.10	52.74	54.04	55.13
20.0	16.82	20.36	23.18	29.70	35.18	68.10	69.52	70.38	72.05	73.53
25.0	21.03	25.22	29.30	37.08	44.05	85.58	86.92	87.97	90.05	91.89
30.0	25.62	30.05	35.01	44.59	52.79	102.39	103.56	105.83	108.09	110.29
σ_y / σ_x						4	3.5	3	2.5	2

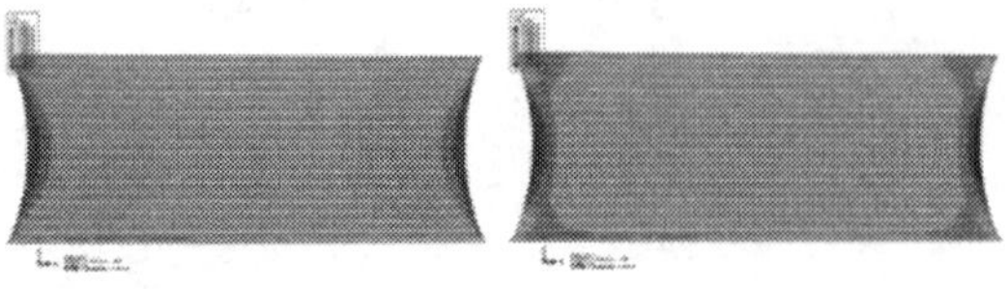

(a) The stress ratio of 1:2

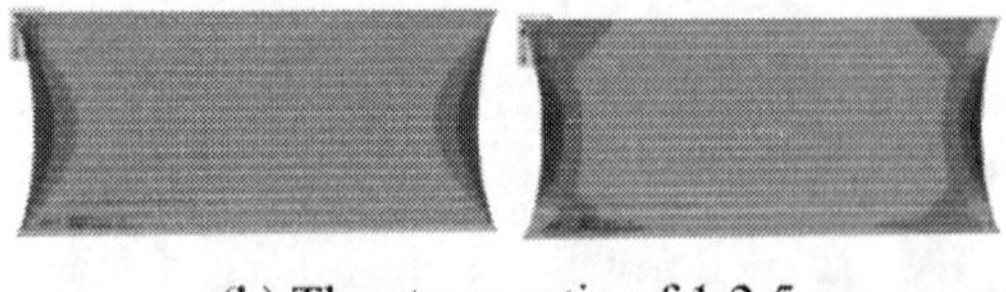

(b) The stress ratio of 1:2.5

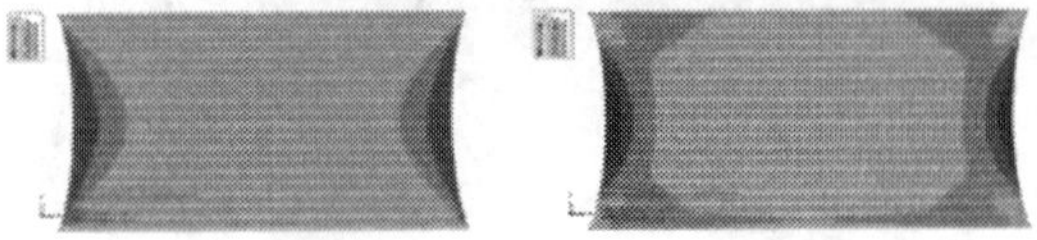

(c) The stress ratio of 1:3

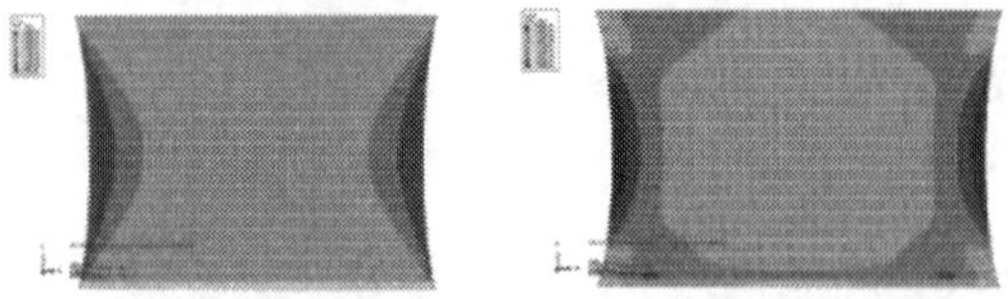

(d) The stress ratio of 1:3.5

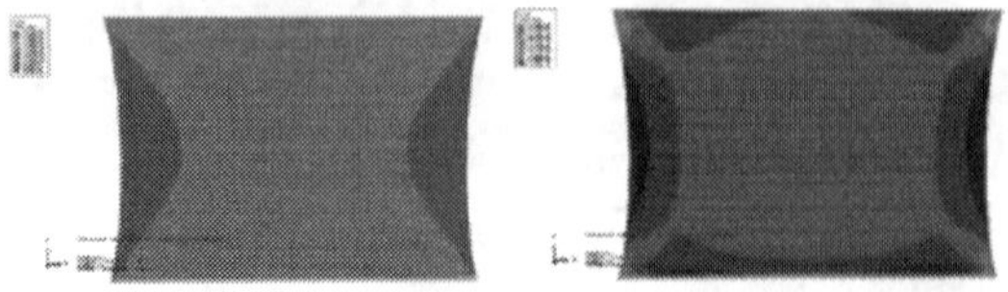

(e) The stress ratio of 1:4

Figure 3. Force nephogram of S11 and S22 in the direct loading direction.

According to the calculation in the table, biaxial tensile results with stress ratios of 1:2, 1:2.5, 1:3, 1:3.5, and 1:4 can be obtained when the values are 0.2, 0.32, 0.42, 0.475, and 0.54. To verify the uniformity of the central region of propellants of different sizes, simulation calculations were conducted by using ABAQUS on different sizes, and the stress changes in different size regions were obtained as shown in Figure 3.

According to the calculation and simulation, with the increase of the multiple, the designed size Flat noodles specimen meets the biaxial tension with equal stress ratios of 1:2, 1:2.5, 1:3, 1:3.5, and 1:4.

3. Experimental implementation and rationality verification

Taking the HTPB propellant with 88% mass fraction filled with solid particles (AP/Al) for SRM of the typical tactical missile as the research object, the Flat noodles test piece was made according to the test method of Wang Zhejun, et al. We refer to GJB770B-2005 "Test Methods for Gunpowder" and conduct tensile tests at 25, -30, and -50°C and strain rates of 0.40, 4.0, 14.29, and 42.86 s^{-1}. Before conducting tensile tests at 30 and -50°C, the propellant was subjected to low-temperature freezing for 24 hours, and stress-strain curves of the HTPB propellant as the temperature and the strain rate change were drafted in Figure 4.

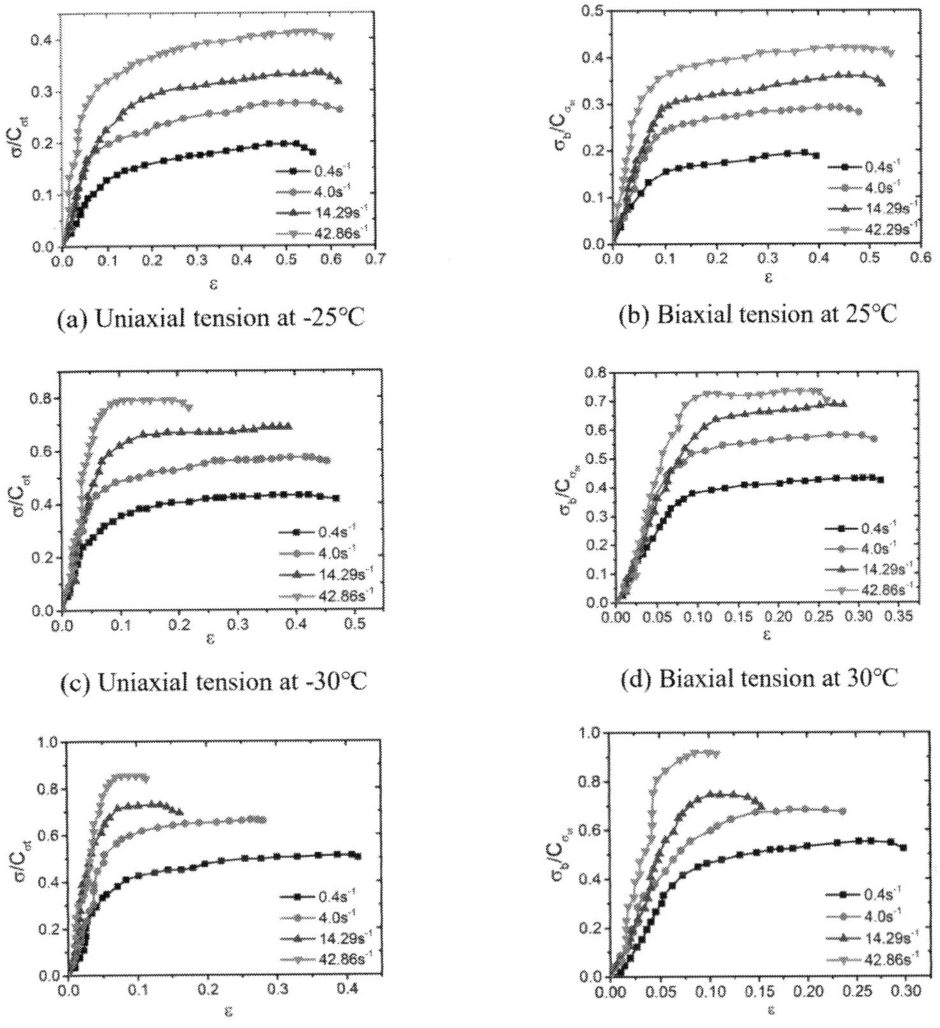

(a) Uniaxial tension at -25°C (b) Biaxial tension at 25°C

(c) Uniaxial tension at -30°C (d) Biaxial tension at 30°C

(e) Uniaxial tension at -50°C (f) Biaxial tension at 50°C

Figure 4. Stress-strain curve of HTPB propellant under uniaxial and quasi-biaxial tension along the tensile direction of the testing machine.

4. Conclusion

According to the calculation and simulation, with the increase of the multiple, the Flat noodles test pieces with the design size meet the biaxial tension with the stress ratio of 1:2, 1:2.5, 1:3, 1:3.5 and 1:4. At present, only the 1:2 Flat noodles test pieces have been processed and verified, and the Flat noodles test pieces with other proportions need to be further verified by the force test.

When the temperature and the strain rate remain unchanged, the maximum tensile strength ratio of quasi-biaxial tension and uniaxial tension is both greater than 1, while the maximum elongation under quasi-biaxial tension is approximately 50% of the value under uniaxial tension. The change in elongation under quasi-biaxial tension can be used as a criterion for the structural integrity analysis of tactical missile SRM grains during low-temperature ignition.

Journal of Physics: Conference Series **2879** (2024) 012051 doi:10.1088/1742-6596/2879/1/012051

References

[1] Wang Xiaoying, Wang Yue, Yin Xinmei, et al. Research on low-temperature mechanical performance of NEPE propellant [J]. Chemical Propellants & Polymeric Materials, 2016, 14 (5): 64-67.
[2] WANG Zhejun. Experimental and Theoretical Research on Mechanical Behaviors of HTPB Propellant at Low Temperatures under Dynamic Loading [D]. Xi'an: Xi'an Hi-Tech Institute, 2016.
[3] LIU Chang, QIANG Hongfu, WANG Zhejun, et al. Strength criterion of aged HTPB propellant at low temperature under dynamic loading [J]. Journal of Propulsion Technology, 2022, 39 (11): 2， 581-2， 587.
[4] SUN Chaoxiang, JU Yutao, ZHENG Jian, et al. Experimental Research on the Compressive Properties of Modified Double Base Propellant at Low and High Strain Rates [J]. Acta Armamentarii, 2023, 34 (6): 698-703.
[5] SUN C., XU J., CHEN X., et al. Strain rate and temperature dependence of the compressive behavior of a composite modified double-base propellant [J]. Mechanics of Materials, 2021, 89: 35-46.

Journal of Physics: Conference Series **2879** (2024) 012052 doi:10.1088/1742-6596/2879/1/012052

Three-dimensional CFD model and aerodynamic analysis of dragonfly's corrugated wing during gliding

Minghao Xu[1], Wei Xia[1,*], Xin Zhang[1] and Shuling Hu[1,*]

[1]State Key Laboratory for Strength and Vibration of Mechanical Structures, School of Aerospace Engineering, Xi'an Jiaotong University, Xi'an 710049, China

*Corresponding author's e-mail: xwei@mail.xjtu.edu.cn(Wei Xia); slhu@mail.xjtu.edu.cn(Shuling Hu)

Abstract: The aerodynamic performance of micro air vehicles (MAVs) in low-speed flight would be improved by mimicking the dragonfly's wing with corrugated airfoil. Research on two-dimensional corrugated airfoils has revealed that the local vortex in the corrugated structure increases the flow speed and the lift-to-drag ratio in low Reynolds numbers. However, studies seldom focus on the effects of three-dimensional corrugated structures on aerodynamics. In this paper, the mechanism of high aerodynamic performance in a dragonfly's wing is studied considering a three-dimensional corrugated structure. The high-fidelity dragonfly forewing model is established through reverse engineering. The computational fluid dynamics (CFD) simulation is performed for the gliding corrugated wing with angles of attack (AOA) of 0°~24° under three Reynolds numbers. The flow characteristics of corrugated wings and the aerodynamic gain from it are analyzed through the simulated streamlines and pressure contours, as well as the comparison between the corrugated wing, flat wing, and flat plate. The results show that the aerodynamic characteristics of corrugated wings are generally superior. Although the introduction of the corrugated structure increases some drag, it brings a higher lift and lift-to-drag ratio than the flat wings.

1. Introduction

Miniaturization of unmanned aerial vehicles turns MAVs' Reynolds number (Re) into 10^2-10^5. At such low Reynolds numbers, traditional wings will experience increased drag, decreased maximum lift, and increased stall speed due to viscous effects and turbulence [1-5]. In the natural world, dragonflies are known as flying fighter jets in the insect kingdom due to their excellent flight performance. Their exceptional gliding ability is very eye-catching. According to records, the highest gliding speed of a dragonfly can reach 10 m/s. The excellent gliding ability of dragonflies cannot be separated from their wings with uneven features, known as corrugated structures, which have been widely studied by scholars in recent years. Vargas et al. [6] found through numerical simulation that the presence of corrugated structures can improve the lift-to-drag ratio of airfoils. Buckholz [7] tested a corrugated wing model in a wind tunnel and found that the corrugated structure can increase the lift of dragonfly wings (Re=1, 500). Kesel [8] established 2D corrugated wing models through wind tunnel experiments and found that the aerodynamic performance of the envelope airfoils was not as good as that of the corrugated airfoils. Dao et al. [9] studied 2D airfoils by changing the position distribution and amplitude of corrugated structures and found that different wrinkles have different effects on aerodynamic performance, especially on lift coefficient. Peng et al. [10] studied the flapping performance of wings by imitating the outer contour of dragonfly wings and found that there is an inter-wing interference effect between front and rear wings which can improve hovering efficiency. In summary, most of the research focuses on two-dimensional airfoils, neglecting the three-dimensional effects. To better understand the specific advantages of corrugated wings, this paper establishes a high-fidelity dragonfly forewing model and conducts numerical simulations and comparisons with the established control model at different Reynolds numbers and angles of attack to analyze the aerodynamic characteristics of corrugated wings at low Reynolds numbers.

2. Model and method

2.1 Simplification and establishment of model

To reduce the difficulty of model establishment and computational costs, this article simplifies the geometric model of dragonfly forewings based on the following:

(1) Neglecting the size of the forewing vein structure and the variation of the wing membrane size along the chordal and spanwise directions, the average size of the wing membrane is taken as 0.04 mm;

(2) Neglecting the influence of dragonfly torso on the aerodynamic characteristics of dragonfly forewings is included.

Firstly, the high-definition image of the dragonfly's forewings was imported by using the reverse modeling function of CATIA. The wing vein structure was fitted with spline curves to obtain the planar contour. Then, the high-precision CT scanning instrument was used to obtain the deflection angles between each wing membrane, forming the corrugated structure. Finally, the wing membrane thickness was given, and the geometric model of the dragonfly's forewings was shown in Figure 1 (a). We establish computational fluid domain as shown in Figure 1 (b), with all surfaces and boundary conditions of the dragonfly forewing geometric model set to "Wall", except for the labeled "Inlet" and "Outlet".

2.2 Validation of method

Polyhedral meshes can adapt to complex geometric shapes and are more effective for modeling irregular objects. So, polyhedral mesh is used to generate the aerodynamic mesh model, and encrypted boundary layers are added around the dragonfly's forewings. The mesh model is shown in Figure 2 (a), and the turbulence model uses a k-ω SST low Reynolds number model because it applies to a wider range of flow situations and produces more accurate results.

To eliminate the influence of grid quantity on the calculation results, the lift coefficients of four sets of dragonfly forewing models with grid quantities of 3.76×10^6, 5.36×10^6, 7.12×10^6, and 9.6×10^6 were calculated at Re=10, 000 and α=4° respectively, and the four different grids have varying densities.

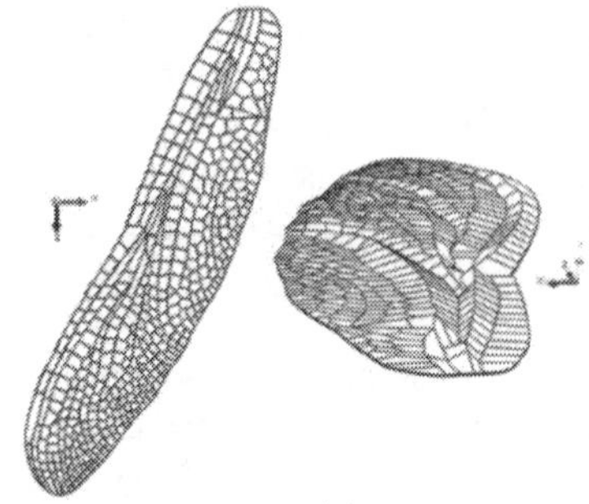

(a)

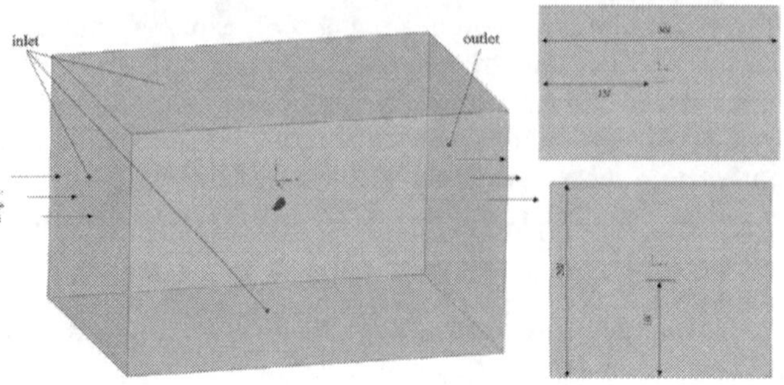

(b)

Figure 1. Geometric model. (a) and computational domain; (b) of dragonfly forewings.

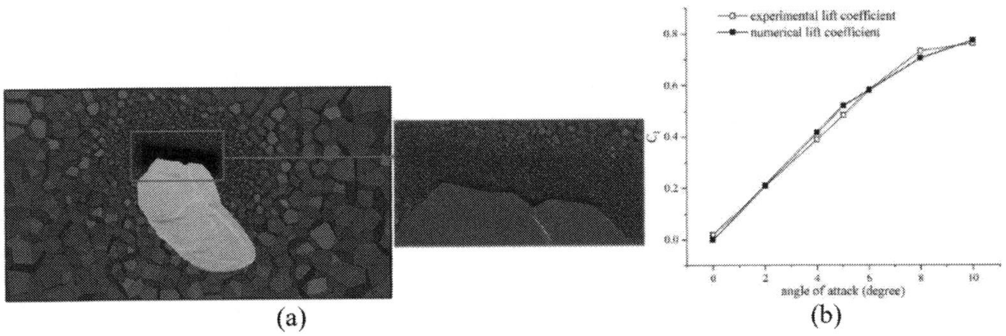

Figure 2. Grids of the corrugated dragonfly forewing (a) and comparison between the numerical and the experimental lift coefficient of Kesel [8] (b).

In the vicinity of the forewings, the difference in lift coefficients between the four models is 6.07%, 0.62%, and 0.09%, respectively. Taking into account the calculation accuracy and cost, the second set of grids, with the number of grids of 5.36×10^6, will be used for subsequent calculations in this paper. The second set of grids was used to calculate the lift coefficient of a flat plate at Re=10, 000 and AOA=0°-10°. The results were compared with Kesal's [8] experimental results. As shown in Figure 2 (b), it can be seen that the calculated value of the lift coefficient is similar to the experimental value, and the trend of change is consistent, thus verifying the reliability of this grid and turbulence model.

3. Analysis of aerodynamic performance

To analyze the aerodynamic performance of corrugated wings, a flat wing model and a flat plate model were set as control models for comparison. The geometric models of the three wings and their geometric are shown in Figure 3. The parameters are as follows: Their thickness (d/mm)=0.04, mean geometric chord (*c*/mm)=6.12, wingspan(*l*/mm)=36.4, and wing area(S/mm2)=222.7.

Firstly, the lift coefficient, drag coefficient, and lift-to-drag ratio of the three models were calculated at Re=1, 500, 4, 500, and 7, 500. The calculation results are shown in Figure 4. From Figure 4, it can be seen that the lift-to-drag ratio of the corrugated wing is superior, only inferior to the flat wing at high Reynolds numbers (4, 500, 7, 500) and low angles of attack (4°). At the same time, although the introduction of the corrugated structure increases drags, with an average increase of 7.19% and a maximum increase of 26.9%, it also brings a significant increase in lift, with an average increase of 13.04% and a maximum increase of 21.47%, making the lift to drag ratio higher than the flat wing. This increase in lift-to-drag ratio is more significant at low Reynolds numbers.

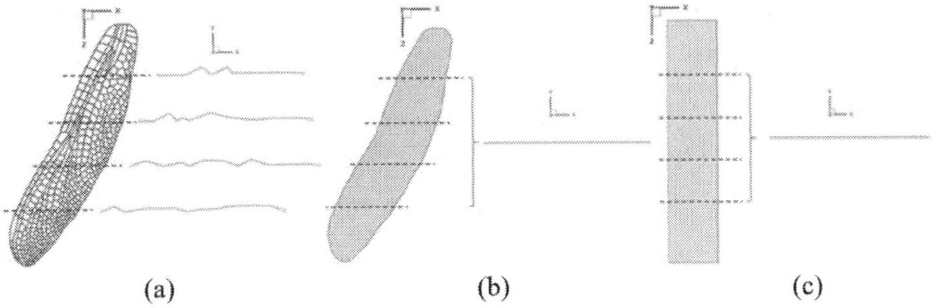

Figure 3. Geometric model and sectional view of the corrugated wing (a), flat wing (b), and flat plate (c).

To investigate the reasons why corrugated wings have more advantages in lift and drag at low Reynolds numbers and small angles of attack, pressure contours were drawn for three models near the wing at different span positions at Re=1, 500 and 4 degrees of attack, as shown in Figure 5.

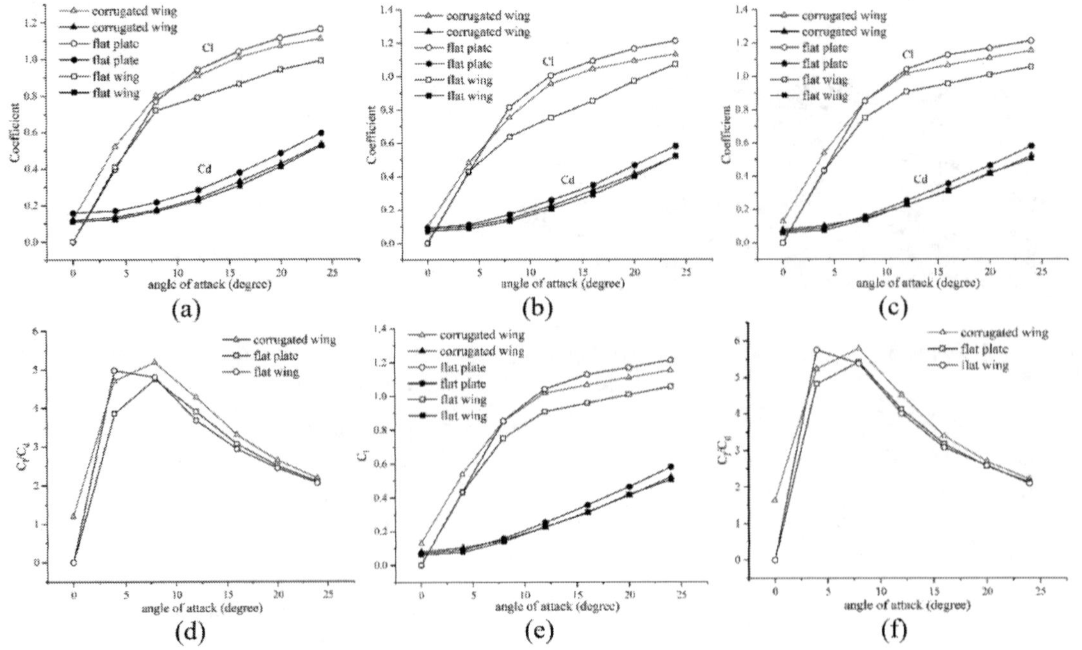

Figure 4. The lift coefficient, drag coefficient, and lift-drag ratio of wings at Re=1, 500 in (a) and (b), 4, 500 in (c) and (d), and 7, 500 in (e) and (f).

Firstly, from the pressure contours, compared to the flat wing and the flat plate, the corrugated wing not only has positive pressure below the leading edge of the wing but also has positive pressure zones in some corrugated areas. At the same time, there is a larger range of high negative pressure zones above the wing surface, which makes the lift coefficient of the corrugated wing greater.

As shown in Figure 6, due to the corrugated structures, vortices exist at these zones of the corrugated wing, which accelerate the airflow above and explain why there is a larger negative pressure zone above the corrugated wing. At the same time, we noticed that vortex at corrugated structures caused the positive pressure at the leading edge of the corrugated wing to shift backward, thereby reducing the pressure drag of the corrugated wing and explaining why the drag coefficient of the corrugated wing is lower.

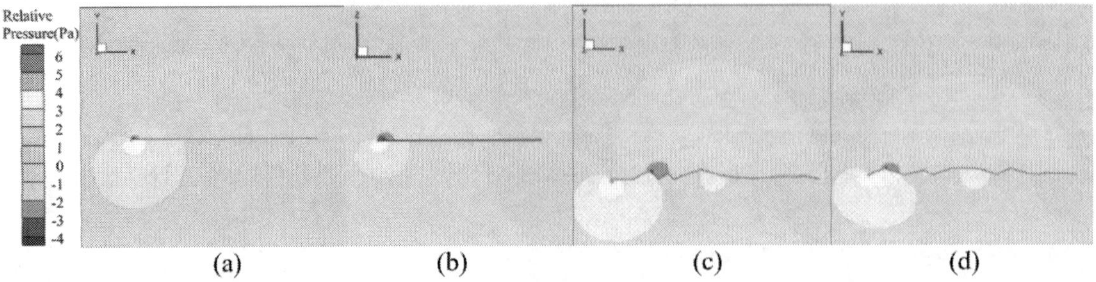

Figure 5. At Re=1, 500 and AOA=4°, the pressure contours of the flat wing at lengths z/l=0.2 (a), the flat wing at z/l= 0.2 (b), corrugated wing at z/l= 0.2 (c), and 0.6 (d).

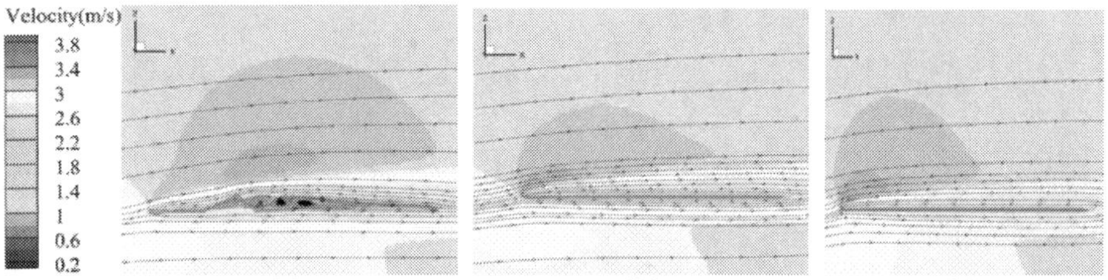

(a) (b) (c)

Figure 6. At Re=1, 500, AOA=4°, and z/l=0.2, the velocity contours overlaid with streamlines of corrugated wings (a), flat wings (b), and flat plates (c).

Finally, using a flat plate as a comparison, a line graph was plotted for the lift, drag coefficient, and lift-to-drag ratio of the corrugated wing and the flat plate at AOA=8° and Re=0-50, 000, as shown in Figure 7. It was found that at a Re=4, 500, which is the highest gliding speed that dragonflies can achieve, the lower the speed is, the higher the lift coefficient is, which is opposite to a flat plate. At the same time, it was found that before the Re=15, 000, the lift-to-drag ratio of the corrugated wing was higher than the flat plate.

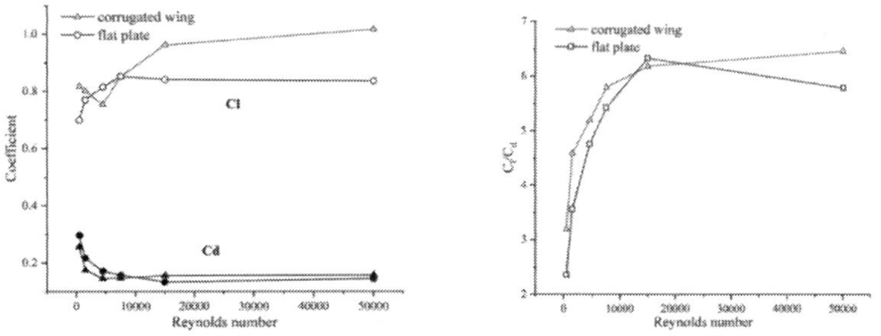

(a) (b)

Figure 7. At AOA=8°, the lift coefficient, drag coefficient (a), and lift-to-drag ratio (b) of corrugated wings and flat plates vary with the Reynolds number.

4. Conclusions

1) At Re=1, 500, 4, 500, and 7, 500, the lift-to-drag ratio of the corrugated wing is superior, only inferior to the flat wing at high Re (4, 500, 7, 500) and low AOA (4°). Although the introduction of a corrugated structure will increase drag, with an average increase of 7.19% and a maximum increase of 26.9%, it also brings a significant increase in lift, with an average increase of 13.04% and a maximum increase of 21.47%.

2) The increase in lift after the introduction of corrugated structure is mainly due to the acceleration of nearby airflow by the vortices between corrugated structure, which creates additional positive pressure zones in some corrugated areas of the corrugated wing, and a larger range of high negative pressure zones above the wing surface. Compared to flat plates, the smaller resistance is due to the vortex at corrugated structures causing the positive pressure at the leading edge of the corrugated wing to shift backward, thereby reducing the pressure drag.

3) Compared to a flat plate, the flight advantage of a corrugated wing lies in the R= 0-15, 000, within which the corrugated wing has a higher lift-to-drag ratio.

Acknowledgment
This work was supported by the National Natural Science Foundation of China (12172275), and the "111" Program (B18040). The calculations were performed by using the HPC Platform at Xi'an Jiaotong University.

Reference

[1] Gad E. MAVs: Can They Be Controlled Better? [J]. Journal of Aircraft, 2001, 38 (3): 419-429.
[2] Zhao M. Research progress on the energy consumption of bionic flapping-wing aerial vehicles [J]. Chinese Journal of Engineering, 2022, 44 (12): 2, 111-2, 123.
[3] Hou D., Tan B., and Zhong Z. Effects of Corrugation and Camber on Aerodynamics of Dragonfly Wing in Forward Flight [J]. Chinese Quarterly of Mechanics, 2024, 45 (01): 19-29.
[4] Xu N., Zhou S. Z., and Mou X. L. Influence of corrugation position on the aerodynamic performance of dragonfly gliding airfoils [J]. Journal of Aerospace Power, 2021, 36 (07): 1, 434-1, 442.
[5] Wang L., Song B., Sun Z., et al. Review on ultra-lightweight flapping-wing nano air vehicles: Artificial muscles, flight control mechanism, and biomimetic wings [J]. Chinese Journal of Aeronautics, 2023, 36 (06): 63-91.
[6] Vargas A., Mittal R., and Dong H. A computational study of the aerodynamic performance of a dragonfly wing section in gliding flight [J]. Bioinspiration & Biomimetics, 2008, 3 (2): 13-0.
[7] Buckholz R. H. The Functional Role of Wing Corrugations in Living Systems [J]. Journal of Fluids Engineering, 1986, 108 (1): 93-97.
[8] Kesel A. B. Aerodynamic characteristics of dragonfly wing sections compared with technical aerofoils [J]. Journal of Experimental Biology, 2000, 203 (Pt 20): 3, 125-35.
[9] Dao T. T., Loan Au T. K., Park S. H., et al. Effect of wing corrugation on the aerodynamic efficiency of two-dimensional flapping wings [J]. Applied Sciences, 2020, 10 (20): 7, 375.
[10] Peng L. S., Zheng M. Z., and Pan T. Y. Effects of tandem-wing interactions on the aerodynamics of hovering dragonflies [J]. Acta Aeronautica et Astronautica Sinica, 2021, 42 (07): 271-280.

Journal of Physics: Conference Series **2879** (2024) 012053 doi:10.1088/1742-6596/2879/1/012053

Research on spacecraft environment control and astronaut health evaluation based on Modelica

Tong Ge[1*]

[1]Majestic International College, Foshan, China

*Corresponding author's e-mail address: tongge0614@outlook.com

Abstract. The environmental control system of a manned spacecraft plays a pivotal role in maintaining the internal environment and temperature control, and it constitutes a fundamental component of the spacecraft. In this study, a Modelica-based model for the environmental control and health assessment of astronauts was examined with the objective of analysing the impact of the internal environment of the spacecraft on human physiological changes. The model employs multi-domain simulation techniques, integrating thermodynamic and hydrodynamic models to describe airflow and temperature changes, while incorporating human physiology models to capture the body's response to these environmental changes. In particular, thermodynamic models are employed to calculate the transfer and distribution of heat within the cabin air, while hydrodynamic models are utilised to simulate airflow and pressure changes, which subsequently influence the breathing patterns and thermoregulatory mechanisms of the astronauts. The human physiology models include simulations of the cardiovascular, respiratory, and thermoregulatory systems. The coupling of these models enables the accurate prediction of the physiological state of astronauts under various environmental conditions. By modifying the environmental control parameters within the model, including air flow rate, temperature set point, and humidity level, the impact of disparate environmental control strategies can be evaluated. This method has the potential to be employed in the prediction of the health status of astronauts during various flight phases.

1. Introduction

In order to enhance the efficacy of mathematical and semi-physical simulation verification throughout the process of spacecraft development and guarantee the accuracy of design and manufacturing, it is imperative to undertake a comprehensive simulation analysis of the spacecraft at the system level. System-level simulation serves to mitigate risk and enhance reliability at all stages of the development process. This is achieved by verifying the vehicle's performance parameters, identifying potential issues, and optimising the design during the conceptualisation phase [1]. Consequently, the role of simulation in spacecraft development is becoming increasingly significant, encompassing all stages from preliminary design to detailed engineering analysis and final validation and testing.

As a pivotal subsystem of environmental control and thermal management in manned spacecraft, the environment control system necessitates a comprehensive understanding of diverse fields, including thermodynamics and fluid mechanics, and represents a pivotal aspect of spacecraft design [2]. In the case of manned spacecraft, the quality of the design of the climate control system has a direct impact on the success or failure of the mission. In the event of an inadequately designed system or a system failure during the mission, the spacecraft may be unable to fulfil its intended purpose, with potentially fatal consequences for the astronauts on board. A failure of the climate control system will have a catastrophic impact on the functionality of the entire spacecraft [3]. It is therefore imperative that the stability and reliability of the system are guaranteed at each stage of the spacecraft design and manufacturing process. By employing sophisticated simulation techniques, engineers can accurately predict and optimise the performance of climate control systems during the design phase, thereby ensuring their effective operation under all potential mission conditions [4].

Once the structure and internal composition of a manned spacecraft have been determined, the setting of the parameters for each component of the environmental control system becomes a crucial issue. If the parameter setting is not reasonable, it may affect the function and performance of the environmental control system, and in serious cases, it may even lead to an accident in a manned spacecraft, endangering the work, life and even life safety of astronauts [5]. The conventional methodology entails the utilisation of ground trials or the emulation of prior experiences when engineering an environmental control system. Nevertheless, ground tests are often lengthy and expensive, and the experience gained from past trials may no longer be applicable when a new model is introduced. There has been a dearth of expedient and efficacious methodologies for the verification of pivotal parameters in the design of environmental control systems for an extended period [6].

The innovation of the model lies in its application of multi-domain simulation techniques, which integrate thermodynamic, fluid dynamic and human physiological models, thereby enabling a more comprehensive analysis. The advantage of this approach over traditional models is that it can more accurately predict the physiological state of astronauts under different environmental conditions. In particular, the model simulates the flow and temperature of air within the cabin by combining thermodynamic and hydrodynamic models.

2. Related work

Marco Molina and Paolo Vercesi employed classical fuzzy control methodologies to ascertain the optimal strategy for regulating temperature and humidity in manned spacecraft [7]. The results of their research demonstrate that the application of fuzzy control theory enables the precise adjustment of environmental conditions within the spacecraft, while simultaneously enhancing the response speed and accuracy of the control system. Conversely, Williams of the National Aeronautics and Space Administration (NASA) of the United States outlined a standard methodology for the design of temperature and humidity control for Node 1 in his work on the validation of the temperature and humidity control subsystem for Node 1 of the International Space Station [8]. The objective of this method is to maintain a stable temperature and humidity level in the complex space environment, thereby ensuring the comfort and safety of the astronauts.

Furthermore, Alptekin et al. researched the rapid removal of carbon dioxide and water, and proposed an effective technique to accelerate the removal process of carbon dioxide and water [9]. This technology is of great importance for the maintenance of air quality and environmental hygiene in manned spacecraft, with the notable reduction of carbon dioxide concentrations and humidity levels, and the assurance of an appropriate environment within the spacecraft. The ongoing exploration of these studies and methods has led to the continuous optimisation and improvement of the environmental control system of manned spacecraft, thereby providing a more reliable guarantee for space missions.

3. Methodologies

3.1. Establishment of environmental parameters model

The initial stage of the process entails the utilisation of the theory of thermodynamics and fluid dynamics in order to construct a model of the transfer of heat and mass within the internal environment of the spacecraft. The primary environmental control parameters are set to air flow, temperature, humidity and pressure. A thermodynamic model is employed to elucidate the process of heat exchange occurring within the cabin environment. In consideration of the modes of heat transfer, which include convection, conduction, and radiation, the energy balance equation can be expressed as Equation (1).

$$\frac{dQ}{dt} = mc_p \frac{dT}{dt} = Q_{convection} + Q_{conduction} + Q_{radiation} \tag{1}$$

where m is the air quality, c_p is the specific heat capacity of the air, T is the air temperature, and $Q_{convection}$, $Q_{conduction}$, and $Q_{radiation}$ represent the heat transfer rates of convection, conduction, and radiation, respectively. Fluid dynamics models are employed to simulate the flow of air and associated changes in pressure. The conservation of momentum for air flow can be described in accordance with the Navier-Stokes equation. The equation is expressed as Equation (2).

$$\rho\left(\frac{\partial u}{\partial t} + u \cdot \nabla u\right) = -\nabla p + \mu \nabla^2 u + f \tag{2}$$

where ρ is the density of the air, u is the velocity field, p is the pressure, μ is the dynamic viscosity of the air, and f is the action of an external force such as gravity. These equations are solved in sequence in order to simulate the complex behaviour of airflow within a spacecraft, including changes in the airflow velocity field and pressure distribution.

3.2. Construction of a model of human physiological response

The human physiological response model is employed to simulate the physiological response of astronauts to alterations in their environment, with a specific emphasis on the cardiovascular system, respiratory system, and thermoregulatory mechanisms. The human physiological response is primarily characterised by the following three sub-models. The Windkessel model is employed to describe the haemodynamic properties, with the objective of simulating the pumping of the heart and the flow of blood in the circulatory system. The equation is expressed as Equation (3).

$$P(t) = P_0 e^{-\frac{1}{RC}} + \frac{Q(t)}{C} \tag{3}$$

where the function $P(t)$ is the arterial pressure, P_0 is the initial pressure, R and C are the capacitance of peripheral resistance and arterial volume, respectively, and $Q(t)$ is the blood flow pumped out by the heart. The gas exchange equation was employed to simulate the dynamic alterations in gas exchange within the alveoli. The respiratory rate and ventilation are influenced by the ambient oxygen concentration and carbon dioxide levels. The corresponding model equation is expressed as Equation (4).

$$V_A = \frac{V_T - V_D}{T_{resp}} \tag{4}$$

where V_A is alveolar ventilation, V_T is tidal volume, V_D is ineffective cavity volume, and T_{resp} is the respiratory cycle. Fanger's Thermal Comfort Model (PMV-PPD) was used to estimate the body's comfort response to temperature changes is expressed as Equation (5).

$$PMV = (0.303e^{-0.036M} + 0.028)\begin{pmatrix}(M - W) - 3.96 \times 10^{-8} f_{cl}(T_{cl}^4 - T_r^4) - f_{cl}h_c(T_{cl} - T_a) \\ -0.0014M(34 - T_a) - 0.0173M(5.87 - P_a)\end{pmatrix} \tag{5}$$

where M is the metabolic rate, W is the external work, f_{cl} is the garment factor, T_{cl} is the garment surface temperature, T_r is the average radiation temperature, h_c is the convective heat transfer coefficient, T_a is the air temperature, and P_a is the partial pressure of the air.

In the human body, blood flow can facilitate the transfer of heat, making it necessary to couple the thermodynamic model with the hydrodynamic model. In order to simulate the circulation of blood, it is necessary to provide the hydrodynamic model with data regarding the flow rate and flow of blood. The thermodynamic model, on the other hand, is responsible for calculating the heat exchange between the blood and the surrounding tissues.

3.3. Multi-domain simulation and coupling

In this subsection, the multi-domain simulation and coupling method was employed to integrate the model of the internal environmental parameters of the spacecraft and the model of the human physiological response of the astronaut, thereby forming a comprehensive simulation platform. Through the co-simulation of multiphysics (such as thermodynamics, fluid dynamics, and human physiology), the method enables a comprehensive analysis of the effects of different environmental conditions on the physiological state of astronauts. The fundamental objective of the multi-domain simulation framework is to facilitate close coupling and dynamic interaction between the models of disparate subsystems. By employing the multiphysics modelling language on the Modelica platform, the following items can be achieved:

- In order to realize the interaction between the environmental model and the physiological model, it is necessary to define a coupling interface to ensure data transmission and synchronous updates between various subsystems. Interface variables include air temperature (T_a), Humidity (RH), pressure (p), oxygen concentration (O_2), carbon dioxide concentration (CO_2), as well as astronauts' body temperature (T_{body}), heart rate (HR), respiratory rate (RR), etc.
- During the simulation, the environmental model and the physiological model are exchanged through a coupled interface. At each simulation time step, the environmental model first calculates new environmental states, and then feeds these state variables into the physiological model, which then calculates new physiological responses, such as heart rate and body temperature, based on the current environmental state. The process is implemented through iterative calculations until the convergence condition is reached.

The systems of equations involved in multi-domain simulations typically include partial differential equations (PDEs) and ordinary differential equations (ODEs). In order to achieve efficient calculation, these equations are discretized by numerical solving methods. The coupling of heat transfer and thermoregulation is expressed as Equation (6), where T_{body} is the body temperature of the astronaut, m_{body} is the mass of the human body, c_{body} is the specific heat capacity of the human body, Q_{met} is the heat generated by metabolism, Q_{resp} is the heat lost through breathing, Q_{skin} is the heat dissipation of the skin, and Q_{env} is the heat exchange with the environment.

$$\frac{dT_{body}}{dt} = \frac{1}{m_{body}c_{body}}(Q_{met} + Q_{resp} + Q_{skin} - Q_{env}) \tag{6}$$

Gas exchange coupled with respiratory rate is expressed as Equation (7), where RR is the respiratory rate, and the function f describes the effect of changes in carbon dioxide (CO_2) and oxygen (O_2) concentrations on respiratory rate.

$$\frac{dRR}{dt} = f(RR, CO_2, O_2) \tag{7}$$

Cardiovascular response coupled to environmental stress is shown in Equation (8), where HR is the heart rate, and the function g describes the regulatory effect of changes in ambient stress (p) on heart rate.

$$\frac{dHR}{dt} = g(HR, p) \tag{8}$$

Initial values are set based on the spacecraft's environmental conditions at the start of the mission and the astronaut's initial state of health, which include the initial air temperature, humidity, oxygen concentration, as well as the astronaut's body temperature and heart rate. At each simulation time step, the environment model first updates the internal environment state. These state updates are achieved by solving thermodynamic and fluid dynamics equations. Updated air temperature is expressed as Equation (9).

$$T_a(t + \Delta t) = T_a(t) + \frac{\Delta t}{m_{air}c_{p,air}}(Q_{in} - Q_{out}) \tag{9}$$

where Δt is the time step, m_{air} is the air quality, $c_{p,air}$ is the specific heat capacity of the air, and Q_{in} and Q_{out} represent the input and output heat, respectively. Using updated environmental state variables, physiological models calculate astronauts' physiological responses, such as updated respiratory rate and heart rate. For example, the equation for updating respiratory rate is Equation (10). T_{resp} is the time constant of the respiratory system and RR_{set} is the set target value of the respiratory rate, which is usually calculated from the current CO_2 and O_2 concentrations.

$$RR(t + \Delta t) = RR(t) + \frac{\Delta t}{T_{resp}}(RR_{set} - RR(t)) \tag{10}$$

Journal of Physics: Conference Series **2879** (2024) 012053 doi:10.1088/1742-6596/2879/1/012053

4. Experiments

4.1. Experimental setups

The specific data used in the experiment are mainly derived from public databases and research literature on the International Space Station and other manned space missions. This data includes the internal environmental parameters of the spacecraft, such as temperature, humidity, air pressure, oxygen concentration, and carbon dioxide concentration, published by the National Aeronautics and Space Administration (NASA) and the European Space Agency (ESA) in the United States. In addition, data on the physiological indicators of astronauts at different mission stages are used, which are usually derived from medical monitoring equipment and post-flight physical examination records, including heart rate, respiratory rate, oxygen saturation and core body temperature.

The experimental setup involved the adjustment of parameters, including airflow rate, temperature (initially set at 22°C), humidity (initially 40%), air pressure (initially 101.3 kPa), oxygen concentration (initially 21%), and carbon dioxide concentration (initially 0.04%), with the objective of simulating a range of cabin environmental conditions. The initial conditions are established in a standard environment that simulates the astronaut's physiological parameters, including body temperature, heart rate, and respiratory rate. During the simulation, the stability and responsiveness of the environmental control system were tested by adjusting these parameters, such as gradually increasing or decreasing the temperature, increasing or decreasing the humidity to simulate human evaporation, air pressure simulating air leakage or replenishment, and oxygen and carbon dioxide concentrations simulating respiratory metabolism changes.

4.2. Experimental analysis

Environmental parameter stability is used to assess the ability of a spacecraft's internal environmental control system to maintain temperature, humidity, air pressure, and oxygen concentrations within safe limits. These include temperature stability coefficients, humidity control accuracy, barometric pressure fluctuation ranges, and oxygen concentration retention. Figure 1 illustrates the stability of temperature, humidity, air pressure, and oxygen concentration in the environmental control system of a manned spacecraft. The change curve of each parameter shows their dynamic change over a certain period of time and their stability, helping to evaluate the performance and adjustment effect of the environmental control system.

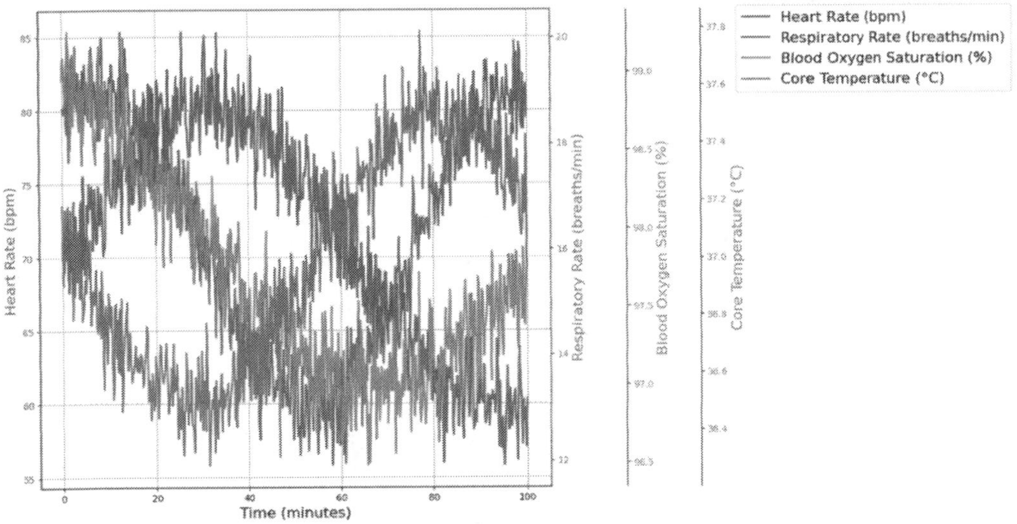

Figure 1. Environmental parameters stability over time.

Journal of Physics: Conference Series **2879** (2024) 012053 doi:10.1088/1742-6596/2879/1/012053

The physiological parameter response index is used to assess whether the astronaut's physiological response is within the safe and expected range, especially the effects of different environmental conditions on heart rate, respiratory rate, blood oxygen saturation and core body temperature. Evaluation criteria include the rate of change of physiological parameters, mean deviation and fluctuation range. Figure 2 is a graph of the experimental results of physiological parameter response indicators, showing the dynamic changes in heart rate, respiratory rate, oxygen saturation and core body temperature over a certain period of time.

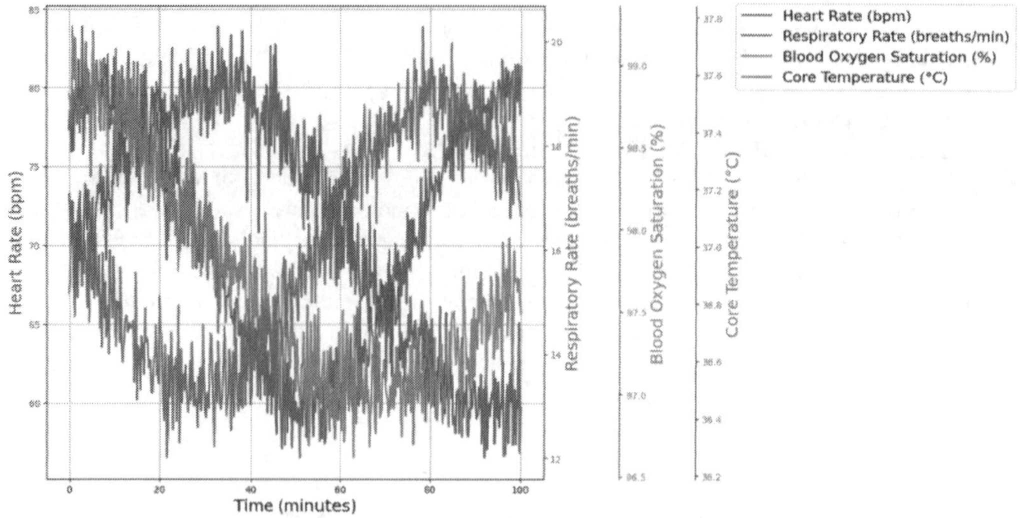

Figure 2. Physiological parameters response over time.

System response time evaluates the speed at which the environmental control system recovers to a steady state in the event of environmental changes or emergencies. This includes the system's response time to changes in temperature and humidity, the speed at which the air pressure is regulated, and the efficiency of the removal of carbon dioxide concentrations. Figure 3 shows the results of response time, showing the response time of each parameter when environmental disturbances at different time points.

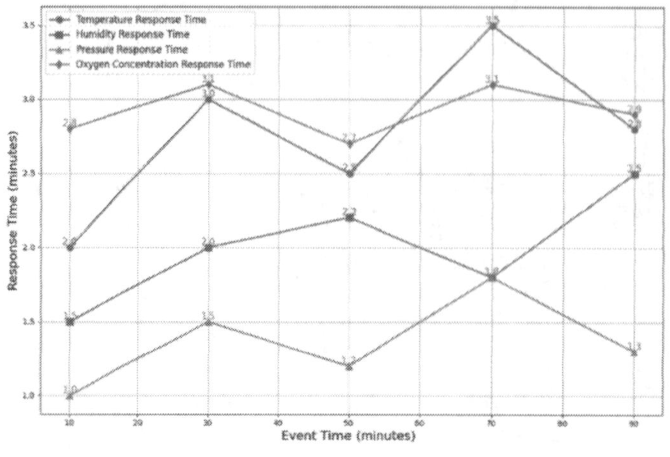

Figure 3. System Response Times to Environmental Disturbances.

The veracity of the model can be validated through its application to disparate flight phases or environmental conditions, through the simulation of varying environmental parameters and astronaut physiological responses, and through a comparison with authentic space mission data. By means of this comparative analysis, any discrepancies in the model can be identified and optimised in order to enhance the accuracy of the model under a variety of flight conditions. Furthermore, this will provide a more reliable basis for the design of an environmental control system for future space missions. Figure 4 illustrates the comparison between the model simulation results and the actual flight data. The graph illustrates the discrepancy between the simulated and actual data over time, with respect to varying environmental parameters, including temperature, humidity, air pressure, and oxygen concentration.

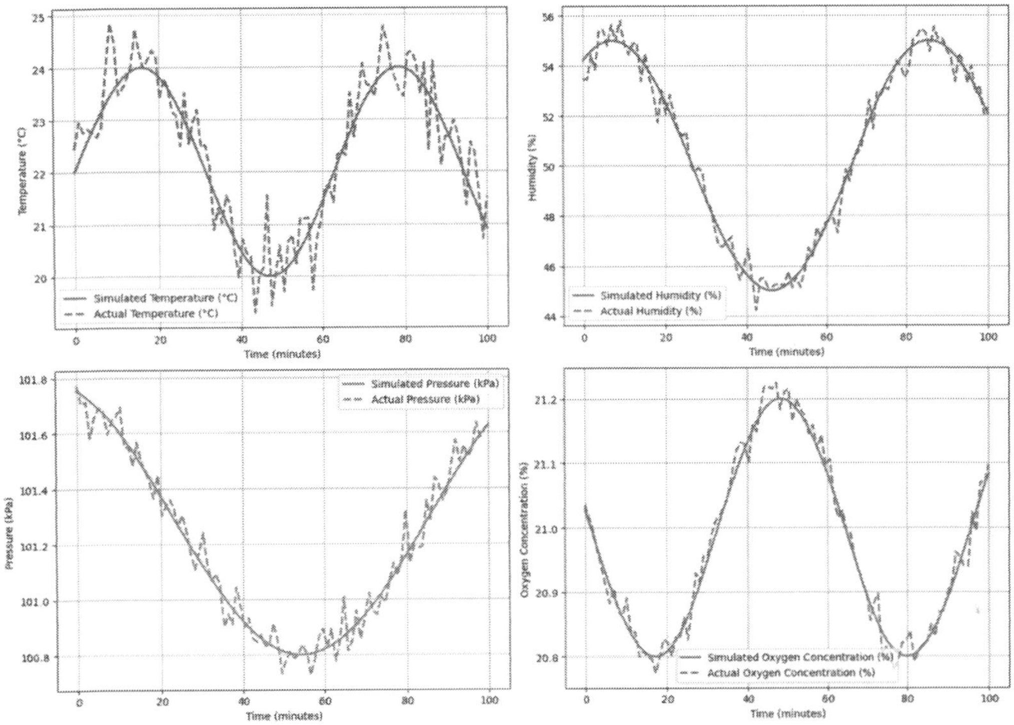

Figure 4. Comparison of model simulation results with actual flight data.

5. Conclusion

In conclusion, it is evident that the spacecraft environmental control system exerts a discernible influence on the stability of pivotal environmental parameters, and that these parameter alterations engender corresponding shifts in the physiological indicators of astronauts. The findings demonstrate that the environmental control system plays a pivotal role in ensuring the well-being of astronauts and optimising the operational environment in space. Nevertheless, future research should seek to further optimise models and simulation techniques in order to cope with more complex changes in the space environment. Additionally, there is a need to explore more refined control strategies and health monitoring methods with a view to further improving the safety and quality of life of astronauts.

References

[1] Shirobokov, Maksim, Sergey Trofimov, and Mikhail Ovchinnikov. "Survey of machine learning techniques in spacecraft control design." Acta Astronautica 186 (2021): 87-97.
[2] Hu, Haitao, et al. "Design and thermal characteristic test of a temperature control system for spacecraft precision instrument." International Journal of Refrigeration 146 (2023): 462-470.

Journal of Physics: Conference Series **2879** (2024) 012053 doi:10.1088/1742-6596/2879/1/012053

[3] Riano-Rios, Camilo, et al. "Aerodynamic and gravity gradient based attitude control for CubeSats in the presence of environmental and spacecraft uncertainties." Acta Astronautica 180 (2021): 439-450.
[4] Riano-Rios, Camilo, Alberto Fedele, and Riccardo Bevilacqua. "Roto-translational control of spacecraft in low earth orbit using environmental forces and torques." Applied Sciences 11.10 (2021): 4606.
[5] Oguri, Kenshiro, and Jay W. McMahon. "Robust spacecraft guidance around small bodies under uncertainty: Stochastic optimal control approach." Journal of Guidance, Control, and Dynamics 44.7 (2021): 1295-1313.
[6] Wei, Caisheng, et al. "An overview of prescribed performance control and its application to spacecraft attitude system." Proceedings of the Institution of Mechanical Engineers, Part I: Journal of Systems and Control Engineering 235.4 (2021): 435-447.
[7] Molina, Marco, and Paolo Vercesi. Fuzzy Logic Algorithm for Manned Modules Temperature and Humidity Control using EcosimPro R. No. 2004-01-2320. SAE Technical Paper, (2004): 2320.
[8] Williams, David E. "International space station temperature and humidity control subsystem verification for node 1." 37th International Conference on Environmental Systems. (2007).
[9] Alptekin, Gokhan, et al. "Rapid cycling CO2 and H2O removal system for EMU." SAE International Conference on Environmental. (2006).

Research on cabin VOC controlling of civil aircraft

Jinqiao Xue[*]

Shanghai Aircraft Design and Research Institute, No. 5188, Jinke Road, Pudong New Area, Shanghai, China

[*]Corresponding author's e-mail: xuejinqiao@comac.cc

Abstract: This article outlines the main sources of VOC in civil aircraft cabins. We compared the current state of control at home and abroad in this field and found there remains a huge blank in relevant domestic standards. To further clarify the baseline of VOC levels in domestic civil aircraft cabins, the VOC contents in the cabin of a domestic aircraft were tested and analyzed by referring to relevant domestic and foreign standards, after which several control suggestions were given. This work is of great significance for establishing VOC control standards in domestic civil aircraft cabins.

1. Introduction

Recently, the comfort of civil aviation transportation has drawn much more attention than ever before [1]. Aircraft design nowadays focuses not only on structure and safety but also on comfort as the people-oriented design concept has long penetrated all aspects of aircraft research and development. With the continuous improvement of people's attention to air quality, creating a good cabin air environment has gradually become one of the factors to enhance the competitiveness of flights. Different from the general indoor environment of a building, the uniqueness of an aircraft cabin is that it exposes passengers and crew to a closed environment with low relative humidity, low air pressure, high population density, and high pollutant concentrations, all of which may hurt the physiological and psychological well-being of passengers and crew.

Existing domestic and international airworthiness regulations are not comprehensive in terms of cabin air quality requirements, especially the lack of a clear limit on volatile organic compounds (VOC). Despite this, advanced European and American countries have been ahead of China in the field of actual measurement research and standard establishment. ASHRAE worked with CSS to test the range of parameters, including thermal environment and contaminants, on eight flights in 1999 [2]. Spicer et al. reported test data from four flights, and analysis of 54 target substances showed that alcohol and acetone were the two most concentrated substances detected in the cabin [3]. Spengler et al. reported the test results of 83 flights of different aircraft types from three airlines and compared them with the pollutant limits in the building interior [4].

However, the domestic passenger cabin VOC measurement and standard development work started late. To fill the gap in the establishment of domestic cabin VOC standards, it is particularly urgent to carry out research on cabin VOC levels for domestic civil aircraft, to build a corresponding database to cope with the increasingly fierce competition in the civil aircraft environmental protection market.

Journal of Physics: Conference Series **2879** (2024) 012054 doi:10.1088/1742-6596/2879/1/012054

2. Current situation of VOC control in the cabin of civil aircraft

2.1. Sources of VOC in civil aircraft cabin

VOC in the cabin of civil aircraft mainly includes aldehydes, benzene, carbon tetrachloride, trichloroethylene, and other gases. These trace harmful gases can induce irreversible effects on the body such as causing the imbalance of the human immune system, damaging the hematopoietic and liver systems, and leading to chest distress, dizziness, nausea and vomiting, loss of appetite, etc. These gases are usually studied as a whole and referred to as VOC by researchers. At present, the main sources of VOC in civil aircraft cabins are as follows:

a) Airborne pollutants in surface airports. The exhaust gas of aircraft and cars is the main source of pollutants in the waiting period for flights. Since the outdoor air can directly enter the cabin without filtration, pollutants such as VOC in the air will be brought into the cabin;

b) Engine bleed air pollutants. The evaporation and combustion products of aircraft fuel and hydraulic oil may also enter the cabin via the environmental control system and cause VOC pollution.

c) Cabin interior material volatiles. Composite components such as cabin inner panel, ceiling, seat, and carpet, as well as adhesives used to connect components and topcoats will emit VOC and cause pollution;

d) Cabin service process volatiles. Regular cabin cleaning is essential for a comfortable flight experience. The residue of seat cover surface cleaning agents and the insecticide sprayed during the entry and exit of long-distance flights can become a short-term strong emission source of gaseous chemical pollutants in the cabin.

e) Volatile items of cabin personnel. High personnel density is one of the typical characteristics of civil aircraft cabins. The irritating odor gases produced by cosmetics and perfumes used by passengers and crew in the cabin are also specific VOC;

f) Other chemical reactions. Although existing aircraft are generally equipped with ozone converters to regulate the ozone concentration, it remains to be determined whether the ozone concentration in the cabin is low enough that the occurrence of ozonation can be ignored. In addition, squalene and ozone can react in the semi-closed environment of the cabin to produce ozone by-products, such as aldehyde VOCs (nonaldehyde and caprinaldehyde) and ultrafine particles, causing air pollution and affecting human health.

2.2. Current situation of VOC control in the cabin of civil aircraft

Cabin air quality controlling of civil aircraft is mainly reflected in the form of airworthiness standards, such as the United States Federal Aviation Regulation FAR-25, the European joint airworthiness Regulation JAR-25, China's transport aircraft airworthiness standard CCAR-25-R3, and other mandatory norms and standards [5]. The limits of the main parameters of cabin air quality standards are listed in Table 1. The existing standards only have clear provisions on the control level of physical environmental factors such as cabin pressure and ventilation volume and inorganic chemical environmental factors such as carbon monoxide, carbon dioxide, and ozone. However, the limit values for VOC, which is widely concerned, remain almost blank.

Table 1. Cabin air quality limits reference.

Parameter	FAR [a], JAR [b], CCAR [c]	ASHRAE [d]	GB 9673 [e]	GB 18883 [f]	EPA, NAAQS [g]	OSHA PEL [i]	ACGIH, TLV [j]
T (°C)	/	25 (ground) 24 (cruise)	24~28 (summer);18~20 (winter)	22~28 (summer); 16~24 (winter)	/	/	/

Parameter	FAR [a], JAR [b], CCAR [c]	ASHRAE [d]	GB 9673 [e]	GB 18883 [f]	EPA, NAAQS [g]	OSHA PEL [i]	ACGIH, TLV [j]
Relative humidity (%)	/	14% (average)	40~60	40~80 (summer); 30~60 (winter)	/	/	/
Wind speed (m/s)	/	0.1~0.35	≦0.5	0.3 (summer); 0.2 (winter)	/	/	/
Fresh air volume	250 g/(min·person)	/	25 m^3/ (h· person)	30 m^3/(h·person)	/	/	/
Carbon dioxide (ppm)	5, 000	1, 000	1, 500 (daily mean)	1, 000 (daily mean)	/	5, 000	5, 000 (8 h)/ 30, 000 (15 min)
Carbon monoxide (ppm)	50	/	8 (1 h)	8 (1 h)	35 (1 h)/ 9 (8 h)	50	25
Ozone (ppm)	0.1 (3 h)/ 0.25 (8 h)	/	/	0.075 (1 h)	0.075 (8 h)	0.1	0.1
Volatile organic compounds (ppm)	/	/	/	0.24	/	/	/
Total colony (cfu/m^3, percussion method)	/	/	2, 500	2, 500	/	/	/

Note: [a]: Federal Aviation regulation of the United States;
[b]: Joint European Civil Aviation Authority flying mark and operation regulation;
[c]: Civil Aviation Regulation of China;
[d]: American Society of Heating, Refrigeration, and Air Conditioning Engineers;
[e]: The hygienic standard of public transport in China;
[f]: China's indoor air quality standard;
[g]: U.S. Environmental Protection Agency's National Air Quality Standard;
[i]: U.S. Occupational Safety and Health Administration's allowable exposure limit;
[j]: U.S. Government Association of Industrial Hygienists threshold.

In addition to mandatory airworthiness regulations, environmental pioneers such as the European Union and the United States have long been far ahead in the research and development of industry standards for civil aircraft cabin air quality control [6]. ASTM D 6399, Standard Guide for Measuring Instruments and Methods for Detecting Air Quality in Aircraft Cabins, issued by the American Society for Testing Materials, provides unified guidance for detecting VOC [7]. Especially, the European Aviation Industry Association issued EN 4618 "Aerospace Series − Aircraft Internal Air Quality Standards, Conditions, and Measurement Methods" based on the common air pollutants in some civil aircraft cabins, which classified aircraft cabin air quality into safety, health, and comfort values and controlled different items in aircraft cabins accordingly. It has important reference significance for the study of VOC in actual civil aircraft cabins, and the relevant standards are listed in Table 2 [8].

Table 2. Aircraft internal air quality standards and criteria in EN 4618.

Items	Safety limits (mg·m^{-3})	Health limits (mg·m^{-3})	Comfort limits (mg·m^{-3})
Formaldehyde	2.47 (15 min) 0.93 (TWA^{a} 8 h)	< 0.1 (30 min)	/
Acetaldehyde	45 (15 min)	1.8 (24 h)	/
Acrolein	0.75 (15 min) 0.125 (TWA 8 h)	0.05 (TWA 30 min)	/
Acetone	3, 630 (STEL^{b} 15 min) 1, 210 (TWA 8 h)	1, 782 (STEL 15 min) 1, 188 (TWA 8 h)	240
Butanone	897.8 (15 min) 1, 795.5 (TWA 8 h)	589.5 (TWA 8 h) 897.8 (STEL 15 min)	/
Benzene	3.2 (TWA 8 h)	12.8 (15 min)	/
Toluene	760 (15 min)	190 (8 h)	153
Dichloromethane	< 3 (24 h)	< 3 (24 h)	/
Carbon dioxide	9, 130	36, 520 (15 min)	3, 650
Carbon monoxide	58.1	29.1 (TWA 1 h) 11.6 (TWA 8 h)	/
Ozone	/	0.5 (peak value, FAR) 0.2 (TWA 3 h, FARs) < 0.12 (TWA 8 h)	/
PM2.5	/	0.1 (TWA 1 h) 0.04 (continuous)	/
PM10	/	0.15 (TWA 24 h)	/

Note: ^{a}: Time-Weighted Average; ^{b}: Short-Term Exposure Limit.

However, at present, domestic standards for cabin air quality of civil aircraft remain a blank slate. The only two referable criteria are GB/T 18883: Indoor Air Quality Standards [9] and GB/T 18204.2: Public Place Health Inspection Methods Part 2: Chemical Contaminants [10], which are primarily geared towards workplace and public space. Among them, the GB/T 18883 standard specifies the determination methods and requirements of physical, chemical, biological, and radioactive indicators inside residential and office buildings, to which some limits can be applied to VOC control in the aircraft cabin (Table 3).

Table 3. Air quality control standards in GB/T 18883.

Standard number	Standard name	Scope of application	Regular items	Limits (mg/m^3)
GB/T 18883	Indoor air quality standards	Indoor air quality in residential and office buildings, etc.	formaldehyde benzene toluene xylene TVOC ^{b}	0.1 (mean 1 h ^{a}) 0.03 (mean 1 h) 0.2 (mean 1 h) 0.2 (mean 1 h) 0.6 (mean 8 h)

Note: ^{a}: The arithmetic mean of any 1-hour pollutant concentration; ^{b}: Total volatile organic compounds, i.e., we sampled by using Tenax TA or equivalent packing adsorption tubes and analyzed on non-polar or weakly polar capillary columns (polarity index less than 10) with retention times between *n*-hexane and *n*-hexadecane volatile organic compounds.

3. Study on VOC level in the cabin of a domestic civil aircraft

3.1. Sampling method

The cabin air of a domestic civil aircraft was selected as the research object. The aircraft under test was placed in a daily environment and kept closed for a certain period. Gas was then collected in multiple locations such as the cockpit and cabin of the aircraft. Sampling conditions are listed in Table 4.

Table 4. Environmental conditions for sampling.

Environmental condition	T (°C)	Humidness (%)	Atmospheric pressure (Kpa)	Sampling requirement
Normal temperature	18.1	54	101.0	Seal for 12 h before sampling
Sample number	Trap velocity (mL/min)	Capture time (min)		Trap volume (L)
Tenax sampling tube	100	30		3
DNPH sampling tube	400	30		12

Gas sampling was performed by using a gas sample collection device consisting of a constant flow gas sampling pump, a primary flow meter, a sampling catheter, and a filled column sampling tube. Besides, VOC was sampled by using Tenax sampling tubes, and aldehydes and ketones were sampled by using DNPH (2, 4-dinitrophenylhydrazine) silica gel sampling tubes.

The sampling procedure was carried out according to the provisions of GB/T 18883. The sampling points are arranged according to the requirements in the standard. At least 5 points are set on the diagonal in the room with an area of more than 100 m^2, and the relative height is about 0.5 m from the ground, which is consistent with the height of the passenger's breathing belt. The actual sampling points include the cockpit, cabin door, Front 1, Front 2, middle, left, and tail, as shown in Figure 1.

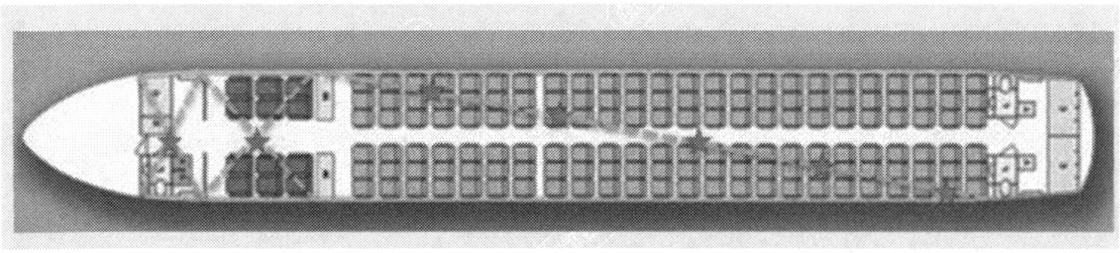

Figure 1. Location diagram of sampling points.

3.2. Test method

The samples test and analysis mainly involved volatile organic compounds (including benzene, toluene, xylene, ethylbenzene, styrene, TVOC, and other VOC substances), aldehydes, and ketones (including formaldehyde, acetaldehyde, acrolein, and other aldehydes and ketones components). Volatile organic compounds were analyzed by using ISO 16000-6, and the analytical instrument was gas chromatography-mass spectrometry (GC-MS). Aldehydes and ketones were analyzed by using ISO 16000-3, and the analytical instrument was a high-performance liquid chromatograph (HPLC). The analysis conditions of volatile organic compounds, aldehydes, and ketones are shown in Table 5.

Table 5. Analysis conditions of volatile organic compounds, aldehydes, and ketones.

Analysis condition	Instrument	Pillar	Parameter	Remark
TDS-GC/MS	Agilent (7890B/5977B)	HP-ULTRA 2 50 m×0.32 mm×0.52 μm	Shunt ratio126:1	TDS: PE (ATD 350)
HPLC	SHIMADZU (LC-20AT)	AkzoNobel kromasil 100-5-C18 4.6 mm×250 mm	Injection volume 20 μL	Mobile phase: acetonitrile/water

3.3. Results and analysis

The VOC test results of a civil aircraft cabin are shown in Table 6. The level of cabin air quality can be determined via comparing the analysis results with the VOC limits stipulated in the current domestic and foreign industry standards.

Table 6. VOC level of a civil aircraft cabin.

Items	GB/T 18883	EN 4618 [a]	Detection results (mg/m³)						
			Cockpit	Cabin door	Front 1	Front 2	Middle	Left	Tail
TVOC (C6-C16)	0.6 (mean 8 h)	/	0.536	0.359	0.724	0.718	0.8	0.844	0.85
Formaldehyde	0.1 (mean 1 h)	<0.1 (30 min)	0.03	0.023	0.034	0.037	0.04	0.04	0.041
Acetaldehyde	/	1.8 (24 h)	0.022	0.021	0.026	0.029	0.028	0.027	0.027
Acrolein	/	0.05 (TWA 30 min)	/	/	/	/	/	/	/
Acetone	/	1, 188 (TWA 8 h) 1, 782 (STEL 15 min)	0.099	0.091	0.13	0.143	0.15	0.141	0.146
Propanal	/	/	/	/	/	/	/	/	/
Crotonal	/	/	/	/	/	/	/	/	/
Isobutanal	/	/	/	/	/	/	/	/	/
Butanone	/	598.5 (TWA 8 h) 897.8 (STEL 15 min)	0.021	0.018	0.029	0.034	0.037	0.034	0.035
Butyraldehyde	/	/	/	/	/	/	/	/	/
Benzaldehyde	/	/	/	/	/	/	/	/	/
Valeraldehyde	/	/	/	/	/	/	/	/	/
p-methylbenzaldehyde	/	/	/	/	/	/	/	/	/
Caproaldehyde	/	/	/	/	/	/	/	/	/
Benzene	0.03 (mean 1 h)	12.8 (15 min)	/	/	/	0.004	0.004	0.004	0.004
Toluene	0.20 (mean 1 h)	190 (8 h)	0.03	0.027	0.048	0.048	0.055	0.055	0.058
Ethylbenzene	/	/	0.008	0.008	0.012	0.012	0.013	0.013	0.014
p-m-xylene	/	/	0.023	0.021	0.034	0.035	0.038	0.038	0.039
o-xylene	/	/	0.01	0.009	0.014	0.014	0.016	0.015	0.016
Xylene	0.20 (mean 1 h)	/	0.033	0.03	0.048	0.049	0.054	0.053	0.055
Styrene	/	/	/	/	/	/	/	/	/
Dichloromethane # [b]	/	<3 (24 h)	/	/	/	/	/	/	/

Note: [a]: The "health value" in EN 4618 is used for reference here; [b]: The test results were a semi-quantitative analysis based on the toluene TIC linear regression curve, and the semi-quantitative analysis results were based on the NIST spectrum library search results attached to the mass spectrum.

As shown in Table 6, TVOC content in the cockpit of a civil aircraft well meets the limit of 0.6 mg/m^3 stipulated in GB/T 18883. However, the TVOC content in the cabin (including Front 1, Front 2, middle, left, and tail of the aircraft) slightly exceeds the standard. The contents of other pollutants in aircraft cabins, such as formaldehyde, acetaldehyde, acetone, acrolein, butanone, benzene, toluene, and dichloromethane, are in accordance with the European EN 4618 and GB/T 18883 limits.

The excess of TVOC indicates the sufficient necessity to strengthen control in the selection of raw materials and components and purification post-treatment in the later stage, which may induce pollutants redundant like alkanes, olefins, and complex VOC mixtures.

In terms of raw material control, environmentally friendly materials with low VOC, low odor, and high purity should be selected as much as possible, or VOC adsorbents and catalysts should be added to raw materials to effectively remove small molecules VOC substances such as alkanes, olefins, phenols, aldehydes, and ketones. Selection of adhesives, paints, coatings, and other materials should avoid solvent-based products as much as possible, and water-based or hot-melt adhesive products should be used. Materials purchased such as rubber and synthetic resin should also prefer better environmental-performance products with low harmful additives residue to improve the air quality in the aircraft cabin.

Regarding component control, it is necessary to focus on the processing technology, which should utilize pure production processing technology and post-treatment methods such as ventilation or high-temperature baking. During the production process of parts, the environment must be ensured to be pollution-free, and the eco-friendly performance of parts should be optimized via enhancement of the storage environment. Direct exchange of air can significantly reduce the concentration of VOC in components and materials by extending the ventilation time. In addition, the application of high-temperature post-treatment may also accelerate the release of the low boiling point or small molecules of VOC substances contained in the parts, resulting in a reduction in VOC release within the use of parts.

As to air purification and post-treatment control of the civil aircraft cabin, ventilation exposure, passive adsorption purification, and active adsorption purification can be adopted correspondingly. When exposed to sunlight, the cabin temperature increases, and VOCs in components and materials accelerate volatilization into the air and are discharged out of the cabin along with the circulation system, thereby reducing VOC levels and improving air quality. Passive adsorption purification means relying on the product's adsorption force to guide the airflow without mechanical external force assistance and finally complete the adsorption of impurities in the air. Typical passive adsorption purification products include activated carbon products and photocatalytic products. Active adsorption purification refers to the use of mechanical equipment to assist and guide the airflow and complete the adsorption (filtration) of impurities in the air, oppositely, and the most typical product is the negative ion air purifier.

4. Conclusions

The current situation of air quality and control standards in the cabin of civil aircraft was summarized, which was used as a guideline for testing and evaluating the VOC concentration in the cabin of a domestic civil aircraft. The conclusions can be summarized as follows:

1) It was found that the content of TVOC in the cockpit met well with the limit requirements stipulated in GB/T 18883, while the content of TVOC in the cabin (including the fuselage and tail) slightly exceeded via sampling and analysis of the cabin air of a domestic civil aircraft.

2) The contents of other single substances in the aircraft cabin, such as formaldehyde, acetaldehyde, acrolein, acetone, butanone, dichloromethane, benzene, and toluene, were all below the limits of European EN 4618 and GB/T 18883.

As a result, the control of raw materials, components, and post-purification treatment in the cabin should be strengthened to fully improve the cabin air quality and comfort of domestic aircraft. This work is of great significance for clarifying the cabin VOC baseline of domestic civil aircraft, establishing the cabin VOC database, as well as further building up and improving the relevant standard system.

Journal of Physics: Conference Series **2879** (2024) 012054 doi:10.1088/1742-6596/2879/1/012054

References

[1] Chen L. Y. and Xu J. H. (2023). Research on comfort design of business class seats of large civil aircraft [J]. Design, 36: 121-123.
[2] CSS. Related air quality and other factors to symptoms reported by passengers and crew on commercial transport category aircraft [R]. ASHRAE Research Project 957-RP, Consolidated Safety Service. Atlanta, GA, American Society of Heating, Refrigerating, and Air-Conditioning Engineers, 1998.
[3] Spicer C. W., Murphy M. J., and Holdren M. W. (2004). Relating Air Quality and Other Factors to Comfort and Health Symptoms Reported by Passengers and Crew on Commercial Transport Aircraft (Part I) (ASHRAE Project 1262-TRP). Atlanta, GA, American Society of Heating, Refrigerating, and Air-Conditioning Engineering.
[4] Spengler J. D., Vallarino J., and Mcneely E. (2012). In-Flight/Onboard Monitoring: ACER's Component for ASHRAE 1262, Part 2. Boston, MA. National Air Transportation Center of Excellence for Research in the Intermodal Transport Environment (RITE).
[5] Qiu B., Bai G. Y., and Li L. L. (2013). Comparison of standard limits and detection methods of air quality in civil aircraft cabins [J]. Journal of Environmental Health, 3: 515-523.
[6] Lordo R. A., Buehler S. S., and James R. R. (2023). Relating air quality and other factors to comfort and health-related symptoms reported by passengers and crew on commercial transport aircraft (ASHRAE RP-1262) [J]. Science and Technology for the Built Environment, 29: 268-279.
[7] American Society for Testing Material. ASTM D 6399-2010 standard guide for selecting instruments and methods for measuring air quality in aircraft cabins [S], 2010.
[8] AECMA.BS EN 4618: 2009 Aerospace series. Aircraft internal air quality standards, criteria, and determination methods [S], 2009.
[9] State Administration for Market Regulation, GB 18883-2022. Indoor air quality standards [S], 2022.
[10] Ministry of Health of the People's Republic of China. GB/T 18204.2-2014 Hygienic inspection methods in public places – Part 2: Chemical contaminants [S], 2014.

Performance evaluation of vibration-damping bucking bar in human-robot collaboration

Yingzhao Zhan[1*]

[1]School Institute of Technology Shanghai, Shanghai, 510000, China

[*]Corresponding author's e-mail: 372017605@qq.com

Abstract: The riveting process in aircraft manufacturing generates large vibrations, and if it is exposed to vibration for a long period of time, riveting workers may suffer from Hand-Arm Vibration Syndrome causing irreversible damage to the robot. In this study, firstly, the vibration mechanical model of human-robot collaboration was established, and then the vibration characteristics of human-robot collaboration under squeezing force were deduced. Finally, an experiment was performed to evaluate the vibration damping performance of vibration-damping bucking bars by evaluating the mean peak squeezing force with different rivet guns and input pneumatic pressure. Compared with the conventional stainless steel bucking bar, the PU-cylinder bucking bar has a lower mean peak squeezing force. The results show that the utilization of the PU-cylinder bucking bar in human-robot collaborative riveting operations can considerably reduce the vibration transmitted from the rivet gun to the robot and the operator.

1. Introduction

As a classical mechanical connection, riveting is widely used in the assembly of aircraft components because of its advantages of simple process, good sealing, and impact resistance. The riveting process will produce a large impact force, and engaging in manual riveting work is likely to lead to workers' arm system vascular, neuropathy, and musculoskeletal disease, suffering from "Hand-Arm Vibration Syndrome" (HAVS). Workers with HAVS suffer from poor hand-foot coordination, tactile insensitivity, and an inability to perform fine manipulation, leading to partial loss of work capacity and reduced quality of life. Hand-arm vibration syndrome is not only a serious health hazard for workers but also an economic loss for their families and society [1].

Scholars in various countries have researched how to suppress vibration to eliminate the negative impacts of vibration shocks. Shargawi et al. found that the use of D3O materials in anti-vibration gloves can reduce the vibration force of impact tools in the aircraft manufacturing environment [2]. Chen et al. used a new vibration damping system with a buffer to solve the problems of overweight, high recoil, and inconvenience of Electromagnet Riveter, and completed the miniaturization of Electromagnet Riveter [3]. Jorgensen and Khan investigated the effect of different rivet guns and bucking bars on riveting vibration and selected a vibration-reducing rivet gun and found that vibration on a tungsten bucking bar was reduced by 34% compared to a steel bucking bar [4].

To reduce the incidence of HAVS in workers as well as to minimize the damage to the robot caused by squeezing force, it is of great value to design vibration-damping bucking bars. In this paper, human-machine collaborative riveting in the field of aerospace assembly is taken as the object of study. The objective of this study is to model the vibration mechanics of human-robot collaboration, derive the vibration characteristics of human-robot collaboration under squeezing force, and evaluate the peak squeezing force of stainless steel bucking bar, cylinder bucking bar, and PU-cylinder bucking bar to identify vibration-damping bucking bar that reduces the vibration exposure of riveting workers and robots.

2. Mechanical model

Due to the relatively high force on the bucking bar during riveting, the robot is easily damaged during the human-robot collaborative riveting (Figure 1), so it is necessary to reduce the maximum squeezing force during the riveting process. In this paper, a vibration-damping bucking bar consisting of a cylinder and a PU damping pad is used. The cylinder is mounted on the robot, and a PU damping pad is placed between the cylinder and the bucking bar to improve vibration damping (Figure 2).

Figure 1. Human-robot collaborative riveting.

Figure 2. A vibration-damping bucking bar.

Human-robot collaborative riveting is considered a collision process. Collision is a complex process, which is not only related to the mass, hardness, elasticity, and other basic properties of the object, but also closely associated with the relative speed, the geometry of the contact surface, and the time of the contact [5]. To analyse the vibration-damping bucking bar of the human-robot collaborative riveting, it is necessary to simplify the system and establish a suitable mechanical model of the system based on the simplification. Two objects colliding multiple times with a certain period and frequency are regarded as a vibration system. To analyse a vibration system, it is generally simplified into a system composed of three elements: Mass, Spring, and Damper.

When studying vibration models, preference is given to the possibility of reducing them to a linear model, since linear vibration theory is well established and the analysis methods are simple and reliable. The robot, the operator, and the vibration damping bucking bar are riveted together with the force along the direction of riveting so they can be linearized to simplify the mechanical model to facilitate the study. Considering the robot, the operator, and the vibration-damping bucking bar as a vibration system, the human-robot collaborative system can therefore be simplified to a mass-spring-damping system as shown in Figure 3.

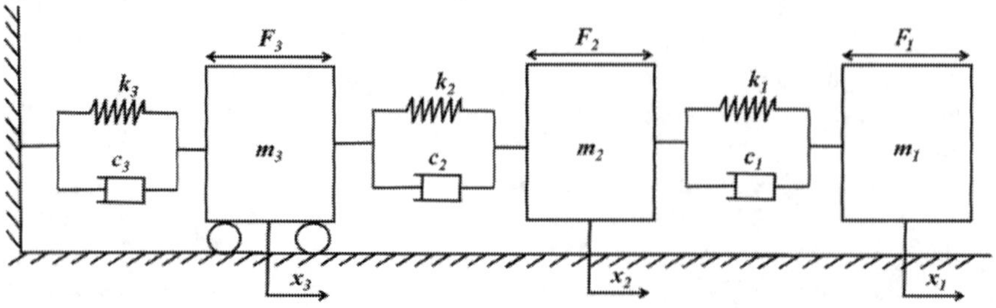

Figure 3. Human-robot collaborative vibration models.

In Figure 3, c_1 and k_1 represent the damping coefficient and spring stiffness of the PU damping pad and bucking bar, c_2 and k_2 represent the damping coefficient and spring stiffness of the cylinder, and c_3 and k_3 represent the damping coefficient and elastic stiffness of the robot and worker arm, respectively. The vibrator m_1 represents the mass of the PU damping pad and bucking bar, m_2 represents the mass of the cylinder, and m_3 represents the mass of the worker arm and robot. x_1, x_2, and x_3 represent the displacement of the PU damping pad and bucking bar, the cylinder, and the robot and worker, respectively; F_1, F_2, and F_3 represent the forces on the PU damping pad and bucking bar, the cylinder, and the robot arm and worker, respectively.

In linear vibration theory, a linear damping assumption is commonly used to analyze vibration systems, where the damping force is assumed to be proportional to the first power of the velocity during the vibration process. In the analysis of a three-degree-of-freedom system, the damping matrix in the system of motion differential equations is typically a 3×3 square matrix. According to the principles of mechanical vibration theory and Newtonian mechanics, the system of motion differential equations can be established as follows:

$$\begin{cases} m_1\ddot{x}_1 + c_1\dot{x}_1 - c_1\dot{x}_2 + k_1x_1 - k_1x_2 = F_1 \\ m_2\ddot{x}_2 - c_1\dot{x}_1 + (c_1+c_2)\dot{x}_2 - c_2\dot{x}_3 - k_1x_1 + (k_1+k_2)x_2 - k_2x_3 = F_2 \\ m_3\ddot{x}_3 - c_2\dot{x}_2 + (c_2+c_3)\dot{x}_3 - k_2x_2 + (k_2+k_3)x_3 = F_3 \end{cases} \tag{1}$$

In Equation (1), the dot (.) above the variables represents the derivative concerning time t.

Based on the data collected from the riveting experiment, it was found that the value of the peak squeezing force does not change significantly during the riveting process, and the frequency of the riveting impact remains constant. Therefore, the excitation force acting on this system can be modeled as a simple harmonic force:

$$\begin{cases} F_1 = p_1 sin(\Omega T + \tau) \\ F_2 = p_2 sin(\Omega T + \tau) \\ F_3 = p_3 sin(\Omega T + \tau) \end{cases} \tag{2}$$

To better analyze its motion the general differential Equation (1) is written in matrix form:

$$M\ddot{X} + C\dot{X} + KX = F \tag{3}$$

In the equation, M, C, and K represent the mass matrix, damping matrix, and stiffness matrix, respectively. X represents the displacement matrix, and F is the force matrix for this system, X= $[x_1, x_2, x_3]^T$, F= $[p_1 sin(\Omega T + \tau), p_2 sin(\Omega T + \tau), p_3 sin(\Omega T + \tau)]^T$.

To eliminate the effect of magnitude difference between different features on the calculation results, it is necessary to dimensionless the motion differential equations and the dimensionless matrix equation can be obtained by the dimensionless treatment of Equation (3) as follows.

$$\begin{bmatrix} 1 & 0 & 0 \\ 0 & m_2 & 0 \\ 0 & 0 & m_3 \end{bmatrix} \begin{Bmatrix} \ddot{x}_1 \\ \ddot{x}_2 \\ \ddot{x}_3 \end{Bmatrix} + 2\Upsilon \begin{bmatrix} 1 & -1 & 0 \\ -1 & 1+k_2 & -k_2 \\ 0 & -k_2 & k_2+k_3 \end{bmatrix} \begin{Bmatrix} \dot{x}_1 \\ \dot{x}_2 \\ \dot{x}_3 \end{Bmatrix} + \begin{bmatrix} 1 & -1 & 0 \\ -1 & 1+k_2 & -k_2 \\ 0 & -k_2 & k_2+k_3 \end{bmatrix} \begin{Bmatrix} x_1 \\ x_2 \\ x_3 \end{Bmatrix} = \begin{Bmatrix} f_{10} \\ f_{20} \\ f_{30} \end{Bmatrix} sin(\omega t + \tau) \tag{4}$$

Let Ψ denote the modal matrix of Equation (4), ω_1, ω_2, and ω_3 represent the natural frequencies of the linear vibration system. Using the modal superposition method, the general solution of Equation (4) can be obtained as follows (for i=1, 2, 3):

$$\begin{cases} x_i = \sum_{j=1}^{3} \psi_{ij} \left[e^{-\eta_j t}(a_{j1}\cos\omega_{dj}t + b_{j1}\cos\omega_{dj}t) + A_j sin(\omega t + \tau) + B_j cos(\omega t + \tau) \right] \\ \dot{x}_i = \sum_{j=1}^{3} \psi_{ij} \left\{ e^{-\eta_j t}\left[(b_{j1}\omega_{dj} - a_{j1}\eta_j)\cos\omega_{dj}t\text{-}(a_{j1}\omega_{dj} + b_{j1}\eta_j)\sin\omega_{dj}t \right] \right. \\ \left. \quad + A_j\omega\cos(\omega t + \tau) - B_j\omega\sin(\omega t + \tau) \right\} \end{cases} \tag{5}$$

In the equation, ψ_{ij} represents the element of the normalized modal matrix Ψ, $\eta_j = \Upsilon\omega_j^2$, $\omega_{dj} = \sqrt{\omega_j^2 - \eta_j^2}$, a_{jk} and b_{jk} are the integration constants (k=1, 2), which are determined by the initial conditions and modal parameters. A_j and B_j are the amplitude constants, which are usually given in the following form [6]:

$$\begin{cases} A_j = \dfrac{\omega_j^2 - \omega^2}{\left(\omega_j^2 - \omega^2\right)^2 + \left(2\eta_j\omega\right)^2}\bar{f}_j \\ B_j = \dfrac{-2\eta_j\omega^2}{\left(\omega_j^2 - \omega^2\right)^2 + \left(2\eta_j\omega\right)^2}\bar{f}_j \end{cases} \tag{6}$$

In the equation, $\bar{f}_j = \sum_{k=1}^{3} \psi_{kj} f_{k0}$, ψ_{ij} represents the element of the normalized modal matrix.

3. Experiment

A riveting experiment platform (Figure 4) was set up to conduct riveting experiments on the designed vibration-damping bucking bars. The platform uses solenoid valves to control the input pneumatic pressure and the riveting preload during riveting, respectively, and a force sensor at 30 kHz to capture the impact force on the robot and the worker during riveting. Three types of vibration-damping bucking bars were subjected to forces during riveting to evaluate the potential ability of vibration-damping bucking bars to reduce impact forces.

The force sensor is a NOS F306 sensor with a measuring range of 0-500 kg, a measuring accuracy of 0.01 kg, and a maximum sampling frequency of 30 kHz. The rivet guns are a USATCO rivet gun and an ATLAS rivet gun. The bucking bar consists of a stainless steel bucking bar (stainless steel), a PU damping pad bucking bar (PU), a cylinder bucking bar (cylinder), and a cylinder bucking bar installed with a PU damping pad (PU-cylinder).

During the experiment, a solenoid valve was used to control the input pneumatic pressure and the riveting preload respectively. Before riveting, rivets and a square skin (approximately 200 mm by 30mm) were mounted on the developed riveting experimental platform. The force sensor is placed at the rear end of the bucking bar to measure the squeezing force. Setting the input pneumatic pressure as 0.4 MPa or 0.2 MPa, riveting preload as 30 KG, riveting time as 1 s, four kinds of bucking bar mean peak squeezing force were collected respectively, and each group of experiments was carried out three times.

The mean peak squeezing force on the bucking bars was extracted by the force sensor. The data were analyzed in data processing software. A one-way Analysis of Variance (ANOVA) was performed to statistically compare the Mean peak squeezing force as a function of vibration-damping bucking bars.

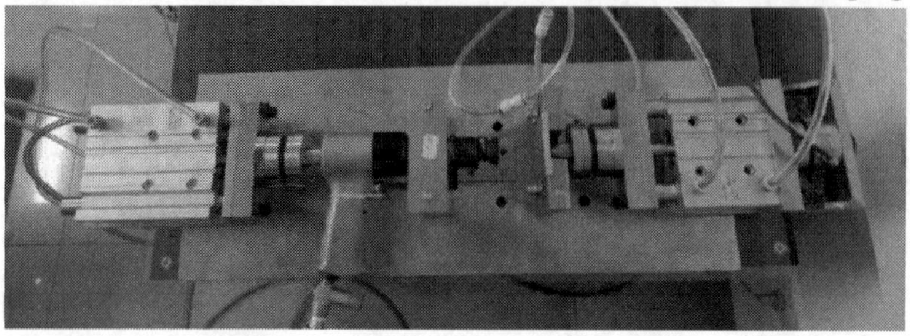

Figure 4. Riveting experiment platform.

4. Result and discussion

The ANOVA indicated that the bucking bar had a significant effect on the mean peak squeezing force when using the USATCO rivet gun (p=0.000) (Figure 5). LSD pairwise comparisons found that the vibration-damping bucking bar resulted in significantly less mean peak squeezing force than the stainless steel bucking bar, but the cylinder bucking bar and the PU-cylinder bucking bar were not significantly different from each other (Figure 5). The same trend was found for the mean peak squeezing force when using the ATLAS rivet gun (p=0.000), where both cylinder bucking bars resulted in prominently less mean peak squeezing force than either the PU bucking bar or the stainless steel bucking bar (Figure 6).

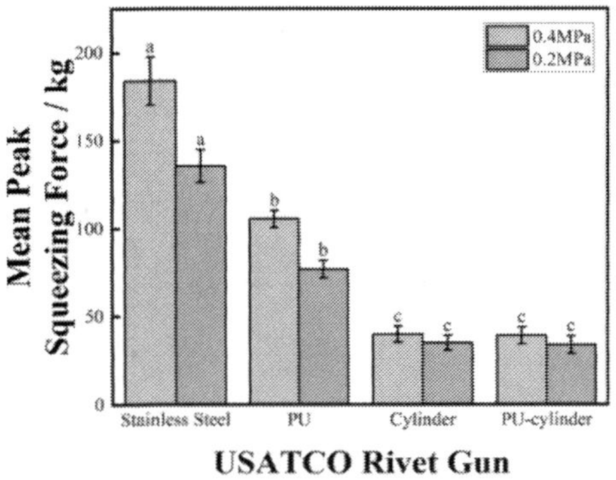

Figure 5. Mean peak squeezing force on the bucking bars when using a UASATCO rivet gun.

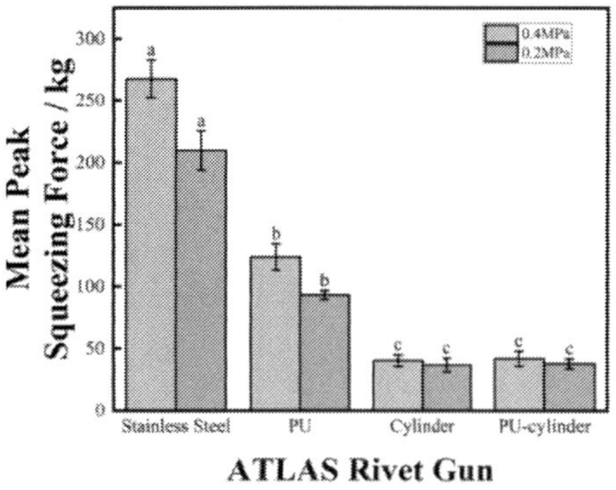

Figure 6. Mean peak squeezing force on the bucking bars when using an ATLAS rivet gun.

During aircraft assembly, the vibrations generated by riveting, which originate primarily from the rivet gun and the bucking bar, are transmitted through the hand to the upper extremity and have been transmitted to the elbow with little or no attenuation. Vibration measured on the rivet gun is lower than that detected on the bucking bar, and operators using the bucking bar are more likely to suffer from HAVS. The risk of operator injury due to vibration exposure can be reduced by controlling the source of vibration, limiting the duration of vibration exposure, or reducing the peak squeezing force. Under the same riveting conditions, the peak squeezing force measured at the rivet gun is higher than the peak squeezing force measured at the bucking bar.

Many types of damping bucking bars are available to reduce the vibration energy from low frequency and high impact tools. Using robots to reduce the vibration exposure of operators during the riveting process is considered a worthwhile method for investigation. Excessive squeezing force can easily cause irreversible structural damage to robots. Reducing the peak squeezing force reduces the chance of robot damage. Therefore, it makes sense to develop tools that can reduce the squeezing force on the robot during the riveting process. The current study investigated the potential to reduce the peak squeezing force on the robot and worker by using different vibration-damping bucking bars.

Four riveting experiments were performed on vibration-damping bucking bars to simulate exposure to representative vibration waves of the riveting tasks. The results of the study showed that the mean peak squeezing force measured for the stainless steel bucking bar was significantly higher than that measured for the other vibration-damping bucking bar. During these four riveting experiments, the mean peak squeezing force measured at the PU bucking bar was reduced by an average of 48.9% compared to the stainless steel bucking bar. The mean peak squeezing force measured at the cylinder bucking bar was reduced by an average of 80.5% compared to the stainless steel bucking bar and 70.4% compared to the PU bucking bar. In addition, the PU-cylinder bucking bar reduced the mean peak squeezing force by 3% compared to the cylinder bucking bar, but this difference was not statistically significant. Thus, it seems that the PU-cylinder bucking bar is very effective in reducing the peak squeezing force. The PU-cylinder bucking bar reduces mean peak squeezing force by 75% compared to stainless steel bucking bars when using a USATCO rivet gun with 0.2 MPa input pneumatic pressure. This value increased from 75% to 84.3% when using the ATLAS rivet gun and 0.4 MPa pneumatic pressure. The PU-cylinder bucking bar shows even greater vibration damping when the squeezing force is increased.

When the rivet gun strikes the rivet, the cylinder reduces the peak squeezing force by absorbing the energy generated by the collision, storing it, and releasing the absorbed energy in the gap between collisions. At the same time, the cylinder can convert part of the kinetic energy during the riveting process into thermal energy, which dissipates into the atmosphere, thereby reducing the squeezing force experienced by the robot and the operator. Typically, the maximum payload capacity of collaborative robots is 20 kg. When using a PU-cylinder bucking bar, the mean squeezing force experienced by the robot and the operator is less than 40 kg, which is very close to the condition that the robot does not suffer irreversible damage during the riveting process. This essentially meets the requirements for human-robot collaborative riveting.

There are some limitations to extrapolating the results of these studies. Firstly, for the different squeezing forces that the robot and the operator may be subjected to under different riveting conditions. If the robot experiences a much greater squeezing force than the operator, the mean peak squeezing force experienced by the robot may exceed its payload capacity. Secondly, in this study, only two types of rivet guns, two input pneumatic pressure, and one riveting preload were investigated. Other rivet guns, different input pneumatic pressure, and riveting preload may cause different levels of vibration.

5. Conclusion

In the aircraft manufacturing process, there is a high level of vibration in the rivet gun and bucking bar, which can lead to hand-arm vibration syndrome in workers as well as damage to the robot over a long time. Compared with stainless steel bucking bars and PU bucking bars, PU-cylinder bucking bars can reduce peak riveting to an appropriate range. Therefore, the use of a PU-cylinder bucking bar for human-robot collaborative riveting may help to minimize vibration and have a positive impact on employee health and robot lifetime.

There are many unfinished aspects of the research in this paper. The long-term effects of different types of riveting operations when using vibration-dampening bucking bars were not studied. Further discussion is also needed on the application of vibration-reducing bucking bars to the actual riveting process to reduce the risk of workers suffering from HAVS disease. This vibration-dampening device can also be extended to other workplaces with vibration exposure, such as reducing the vibration exposure generated by angle grinders.

Reference

[1] Gerger H, Søgaard K, Macri E M, Jackson J A, Elbers R G, van Rijn R M, Koes B, Chiarotto A and Burdorf A 2023 Exposure to hand-Arm vibrations in the workplace and the occurrence of hand-arm vibration syndrome, Dupuytren's contracture, and hypothenar hammer syndrome: a Systematic Review and Meta-Analysis J. Occup. Environ. Hyg. 20 257-67

[2] Shargawi A A, Jorgensen M J, Hakansson N A, Amick R Z and Asmatulu R 2023 Energy absorbency and impact resistance of D3O® materials under dynamic impact loadings J, Aerospace, Eng 10 1-10

[3] Chen Q, Cao Z Q, Sheng X 2012 Study on the miniaturization of electromagnet riveter Mech. Sci. Technol. Aero. Eng. 31 482-6

[4] Jorgensen M J and Khan K S 2008 Use of tungsten to reduce hand-arm vibration exposure in aircraft manufacturing Proc. of the Human Factors and Ergonomics Society Annual Meeting (New York) vol 52 pp 1045–8

[5] Wang L P 2023 Analysis of Force Absorption in Human-Robot-Collaboration Int. Conf. on Intelligent Informatics and Biomedical Sciences (Nanjing) vol 8 pp 314-6

[6] Xu H D, Wen G L, Wu X, Zhang X J 2015 Anti-controlling Hopf-Hopf interaction bifurcations on Poincare Poincaré map of a three-degree-of-freedom vibro-impact system with clearance Vib. Eng. 28 952-9

Research on assisted recovery algorithm of stall state based on signal guidance

Qi Zhang[1*], Bo Gu[2], Tai He[1] and Zili Fang[1]

[1]Chinese Flight Test Establishment, Xi'an 710089, China
[2]Air Force Flight Test Bureau, Xi'an 710089, China

*Email: qizhang0818@126.com

Abstract: Given the lack of quantitative guidance in the current stall recovery process, which leads to the pilot's misoperation or untimely recovery, referring to the energy control theory of complex state recovery, a stall state-assisted recovery algorithm based on pitch angle, engine thrust, and roll angle signal guidance was determined. The structure pilot model and the kinematics simulation model of aircraft at a high angle of attack were established. The effectiveness of the algorithm has been verified by simulation analysis. The man-machine closed-loop simulation analysis results show that the algorithm can guide the pilot in effectively recovering from stall conditions under different stall scenarios and engine failure conditions.

1. Introduction

Thorough research on the stall characteristics of aircraft will be carried out during the design and flight test of aircraft in normal operation. At the same time, the flight manual will also specify the aircraft stall correction method. However, according to a large number of aviation accident statistics, crashes caused by stalls still account for a large proportion of accidents. The main reason is that the current aircraft design is mainly based on stall prevention, such as increasing stall warning, shaking rod, angle of attack limiter, etc. However, stall occurrence is inevitable due to pilot misoperation, complex meteorology, engine failure, and other reasons. The current stall recovery procedures are qualitative descriptions, which require pilots to master the stall characteristics of the aircraft and make timely and accurate judgments in the event of a stall. However, due to the sudden stall phenomenon, the pilot is likely to experience cognitive confusion, emotional tension, and other problems, which in turn lead to manipulation errors or untimely manipulation. In general, the current stall recovery method requires a high level of pilot operation technology and lacks measures to assist the pilot in recovering. Therefore, to ensure the pilot deals with the stall phenomenon accurately and in time, it is necessary to study and determine a quantitative stall correction auxiliary method to guide the pilot in performing the correct manoeuvre in time.

This paper first introduces the general procedure and requirements of stall recovery. Then, regarding the energy control theory of aircraft complex state recovery proposed by NASA, the auxiliary recovery method of stall state based on signal guidance is determined through research, and the definition and constraints of the guidance signal are clarified. It can provide quantitative and clear guidance for pilots to correct stalls. Based on the structural pilot model, the effectiveness of the stall-assisted recovery algorithm under different stall entry modes and engine parking conditions is verified by man-machine closed-loop simulation analysis.

Journal of Physics: Conference Series **2879** (2024) 012056 doi:10.1088/1742-6596/2879/1/012056

2. General procedures and requirements of stall recovery

The determination of the auxiliary recovery algorithm of the stall state should follow the general procedures and requirements of stall recovery.

2.1. General procedure

At present, a large number of theoretical and experimental studies have been carried out by aviation institutions in various countries on the principle and recovery procedure of stall. In general, the general procedure of stall recovery can be summarized into six parts [1-2]:

1) Disconnect autopilot and automatic throttle

For aircraft that do not have the function of stall automatic recovery, it is necessary to ensure that the pilot has sufficient control authority after stall to avoid unintended response caused by automatic driving or automatic throttle because the envelope does not contain the state of high angle of attack.

2) Reduce the angle of attack until the stall alarm disappears

The key to stall recovery is to reduce the angle of attack. Therefore, after the stall occurs, the pilot should control the steering rod to reduce the angle of attack as soon as possible and use the horizontal stabilizer as necessary within the allowable range.

3) Leveling aircraft

In order to avoid the adverse inertial coupling effect, the slope should be flattened as soon as possible after the angle of attack is reduced to the specified value.

4) Adjust thrust

During the stall recovery, the proper use of thrust can effectively reduce the height loss. It is particularly essential for low-altitude stall recovery, such as in the take-off and landing phases. However, the adverse response of aircraft attitude caused by the sharp thrust change should be avoided.

5) Retract the speed reducer and spoiler

It is mainly used to reduce drag and improve the aerodynamic characteristics of aircraft.

6) Softly recover to the desired track

The aircraft is restored to its desired speed and trajectory through gentle manoeuvres.

2.2. Requirements

Stall recovery should follow the following requirements to avoid other derivative risks [3-4]:

1) The overload should be strictly controlled and should not exceed the structural limit; at the same time, it should also avoid being in a negative overload for a long time to prevent engine shutdown.

2) The secondary stall should be avoided.

3) Excessive height loss should be avoided.

3. Stall state auxiliary recovery principle

3.1. Type of guidance signal

In order to determine the quantitative guidance of stall-assisted recovery, the general procedures and requirements of stall recovery are summarized: the stall recovery operation should be under certain constraints to quickly reduce the angle of attack and increase the airspeed. Among them, the change in the angle of attack is mainly realized by the change in the aircraft pitch attitude. The airspeed change directly relates to the aircraft's trajectory angle and engine thrust. The change of aircraft trajectory angle is also mainly realized by attitude change. Therefore, the stall state auxiliary guidance signal can be designed from the aircraft's pitch angle and the engine's thrust to guide the pilot in correcting the stall state. In addition, roll control guidance should also be added to guide the pilot in leveling the aircraft at the right time.

3.2. Determination of guidance signal

3.2.1. Pitch angle guidance signal. According to the aircraft centroid kinematics equation:

$$m\dot{V} = T\cos\alpha - D - mg\sin\gamma \tag{1}$$

$$\frac{T\cos a - D}{m} = \dot{V} + g\sin\gamma \tag{2}$$

In the formula, m is the mass of the aircraft, V is the flight speed, T is the engine thrust, D is the aircraft resistance, α is the angle of attack, and γ is the trajectory angle.

Then, the pitch angle θ can be obtained as:

$$\theta = \alpha + \gamma = \alpha + \arcsin\left(-\frac{1}{g}[\dot{V} - \frac{T\cos a - D}{m}]\right) \tag{3}$$

According to the energy control theory of complex state recovery [5-6], when the efficiency of the aircraft control surface is sufficient, the stall recovery is most obviously affected by the change of height and thrust. Enough energy should be converted into kinetic energy, although the stall recovery should first reduce the angle of attack. However, increasing the airspeed and decreasing the angle of attack does not conflict with manipulation. Accordingly, the design uses the speed change as the guidance control variable to determine the appropriate pitch angle guidance signal. In turn, the aircraft is guided to quickly reduce the angle of attack, increase the airspeed, and finally maintain a stable flight at the required target speed V_r.

Accordingly, the pitch angle guidance signa θ_g is designed as:

$$\theta_g = \alpha + \arcsin\left(-\frac{1}{g}[\dot{V}_r - \frac{T\cos a - D}{m}]\right) = \alpha + \arcsin\left(-\frac{1}{g}[\dot{V}_r - \left(\dot{V} + g\sin\gamma\right)]\right) \tag{4}$$

In the formula, $\dot{V}_r$ is the expected rate of speed change. In order to achieve a rapid increase in speed at the initial stage of stall recovery and to achieve the target speed Vt smoothly after the recovery, the definition is:

$$\dot{V}_r = \frac{1}{T_V}(V_t - V) \tag{5}$$

In the formula, T_V is the speed control time constant, which determines the response speed of the speed guidance. The smaller the value, the faster the response, and the more intense the diving for increasing speed.

For transport aircraft, the pitch angle guidance value should be within the pitch angle limit range to avoid the aircraft entering a complex attitude during the stall recovery process. In addition, to avoid a secondary stall, θ_g should meet the following requirements:

$$\theta_g < \gamma + (\alpha_s - \alpha) \tag{6}$$

In the formula, α_s is the angle of attack of the aircraft stall.

3.2.2. Engine thrust guidance signal. The change in engine thrust can significantly affect the potential and kinetic energy change in stall recovery. Especially for low-height stalls, the engine thrust control guidance signal should be introduced to reduce height loss during stall recovery through a reasonable thrust increase. However, when the thrust direction of the engine is not at the same level as the center of gravity of the aircraft, the thrust change may cause a sharp change in the aircraft's attitude. Therefore, the sharp thrust change should be avoided to avoid unexpected attitude responses. In the design, the thrust guidance signal only requires the engine to provide the thrust required to meet the specified speed change. That is :

$$T_g = \frac{1}{\cos a}\left(m\dot{V}_r + D + mg\sin\gamma\right) \tag{7}$$

Journal of Physics: Conference Series **2879** (2024) 012056 doi:10.1088/1742-6596/2879/1/012056

In addition, the maximum value of thrust guidance should not cause the aircraft to generate head-up torque that cannot be overcome by the aircraft elevator. In actual use, in order to facilitate the pilot's operation, the engine thrust guidance signal should be converted into the throttle rod guidance signal.

3.2.3. Roll angle guidance signal. To control the roll angle, the ultimate goal is to make the aircraft return to the straight flat flight state, and the ultimate goal is to make the roll angle 0 °. However, in the case of stall and near stall, the main goal should be to reduce the angle of attack to prevent the aircraft from adverse yaw and sideslip or other coupling motions due to lateral control. Therefore, the lateral steering guidance must appear after the angle of attack is reduced to a certain safety margin from the stall angle of attack. The specific safety margin should be determined according to the aerodynamic characteristics of the aircraft at a high angle of attack. For the research object of this paper, the safety margin is defined as 5 °. That is, the roll angle guidance signal is:

$$\phi_g=0(\alpha<\alpha_s\text{-}5°) \tag{8}$$

4. Man-machine closed-loop simulation analysis

In order to fully verify the effectiveness of the stall-assisted recovery method proposed in Chapter 3, an accurate kinematics simulation model at a high angle of attack is established [7-8] based on the wind tunnel test data of a certain type of aircraft. According to the characteristics of the task, the pilot simulation model is established. Finally, through the man-machine closed-loop simulation analysis, the effectiveness of the improved algorithm under different stall entry scenarios and engine failure conditions is fully verified.

4.1. Establishment of pilot model

A reasonable pilot model can accurately reflect the control characteristics of the pilot in the mission, which is the key to studying the dynamic characteristics of the man-machine closed-loop system. At present, there has been a more in-depth study on the establishment of the pilot model. The commonly used mathematical models include the McRuer model, optimal control model, neural network model, structure model, and so on. The pilot structure model can well reflect the pilot's adaptive characteristics for different aircraft configurations and command forms. Therefore, the pilot structure model is used in this paper.

The pilot mainly controls the recovery method according to the pitch angle and throttle guidance signal generated by the algorithm to control the aircraft to recover the stall state, similar to tracking control tasks. Therefore, the structural model of the tracking task is selected as the pilot control model. Considering the control process, the pilot mainly controls the input according to the error between the current state of the parameter and the guidance signal, the rate of error change, etc. Accordingly, the structure of the pilot control model is shown in Figure 1. The specific parameters in the model are defined in references [9-10].

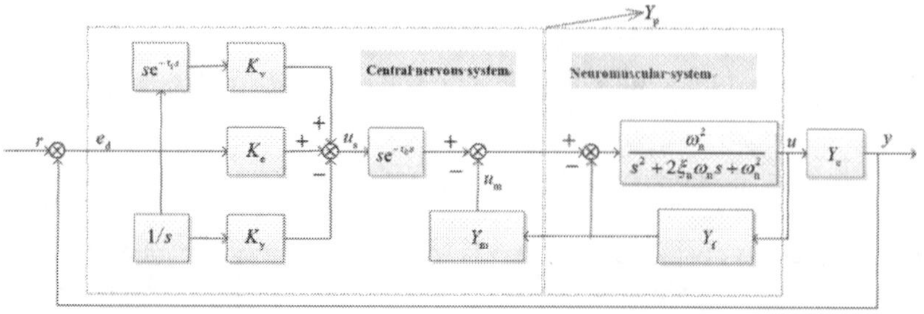

Figire 1. Block diagram of tracking task structure model.

4.2. Simulation settings

In this paper, the initial height of the simulation is defined as the air pressure height of 5000 m, and the aircraft configuration is the cruise configuration. Under this condition, the stall angle of attack of the aircraft is about 22°, and the stall speed is about 200km/h. Taking into account the safety of actual use, the target speed V_t is defined as the indicated airspeed of 340km/h, and the algorithm entry conditions are:

$$\alpha \geq \alpha_s - 2° \tag{9}$$

4.3. Simulation analysis

4.3.1. Slow rod deceleration to stall. The scenario is simulated where the pilot reduces the throttle while slowly pulling the lever to slow down and enter the stall [11]. The simulation results are shown in Figure 2. At the beginning of the simulation, the aircraft maintains a stable flight at a pressure height of Hp = 5000m and a speed of Vi = 350km/h. When t = 5s, the pilot reduces the throttle to slow down, and the tie rod gradually slows down the aircraft with an angle of attack change rate of about 1°/s. When the aircraft's attack angle increases by 20° (red dotted line in the figure), the stall recovery algorithm starts, and the pitch angle and throttle guidance signal begin to appear. Immediately, the pilot performs the recovery operation regarding the guidance signal. During this period, the angle of attack gradually decreases, and the flight speed gradually increases. Finally, the aircraft's angle of attack is stabilized at about 8.5°, and the flight speed is stabilized at about 330km/h. The aircraft successfully recovered from the stall state.

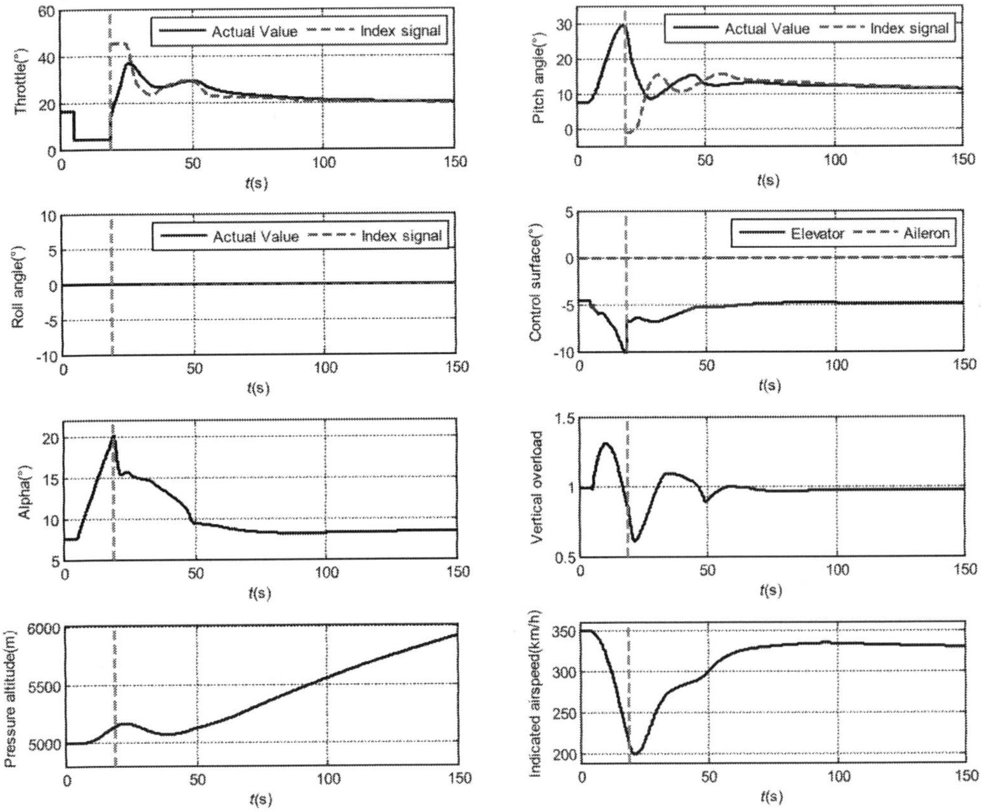

Figure 2. Aircraft state response.

4.3.2. Sharp pull rod to stall. To simulate the scene that the pilot pulls the pilot lever sharply at a high angle of attack to cause the aircraft to stall [12], the simulation results are shown in Figure 3. At the beginning of the simulation, the aircraft maintains stable horizontal flight at a pressure altitude of 5000 m and an indicated airspeed of 230 km/h. T = 5s, the pilot reduces the throttle to slow down and

Journal of Physics: Conference Series **2879** (2024) 012056 doi:10.1088/1742-6596/2879/1/012056

quickly pulls the rod to the end. t ≈ 5.5s, the aircraft's attack angle is increased to 20°, and the algorithm is started. Immediately, the pilot manipulates the driving lever and throttle lever concerning the guidance signal. Initially, the angle of attack continues to increase to a maximum of about 22° under inertia. Then, under the pilot's control, the angle of attack gradually decreases, and the flight speed gradually increases. Finally, the angle of attack is stabilized at about 8.5°, and the flight speed is stabilized at about 327km/h. The aircraft successfully recovered from the stall state.

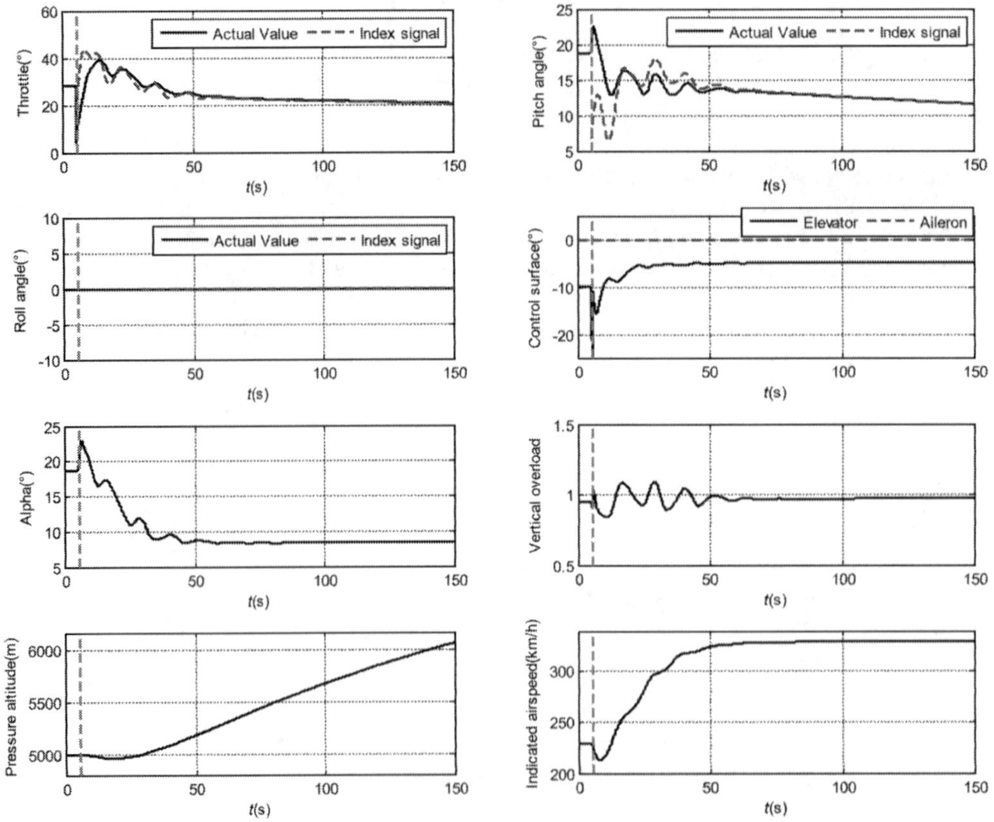

Figure 3. Aircraft state response.

4.3.3. Spiral deceleration to stall. To simulate the scene of the aircraft rod entering the stall during the lateral maneuver, the simulation results are shown in Figure 4. At the beginning of the simulation, the aircraft maintains stable horizontal flight at a pressure altitude of 5000 m and an indicated airspeed of 350 km/h. When t = 5s, the pilot presses the bar to the right to make the aircraft establish a stable slope of about 30°. When t = 20s, the pilot reduces the throttle to slow down, and the tie rod gradually slows down the aircraft with an angle of attack change rate of about 2°/s. When t ≈ 27.3s, the aircraft's angle of attack is increased to 20°, and the algorithm is started. Then, the pilot manipulates the steering lever and the throttle lever with reference to the guidance signal to perform the removal procedure. When the angle of attack is reduced to 17°, the pilot begins to press the bar to correct the slope. Finally, the plane changed the slope. The angle of attack is stable at about 8.3°, and the flight speed is stable at about 332km/h. The aircraft successfully recovered from the stall state.

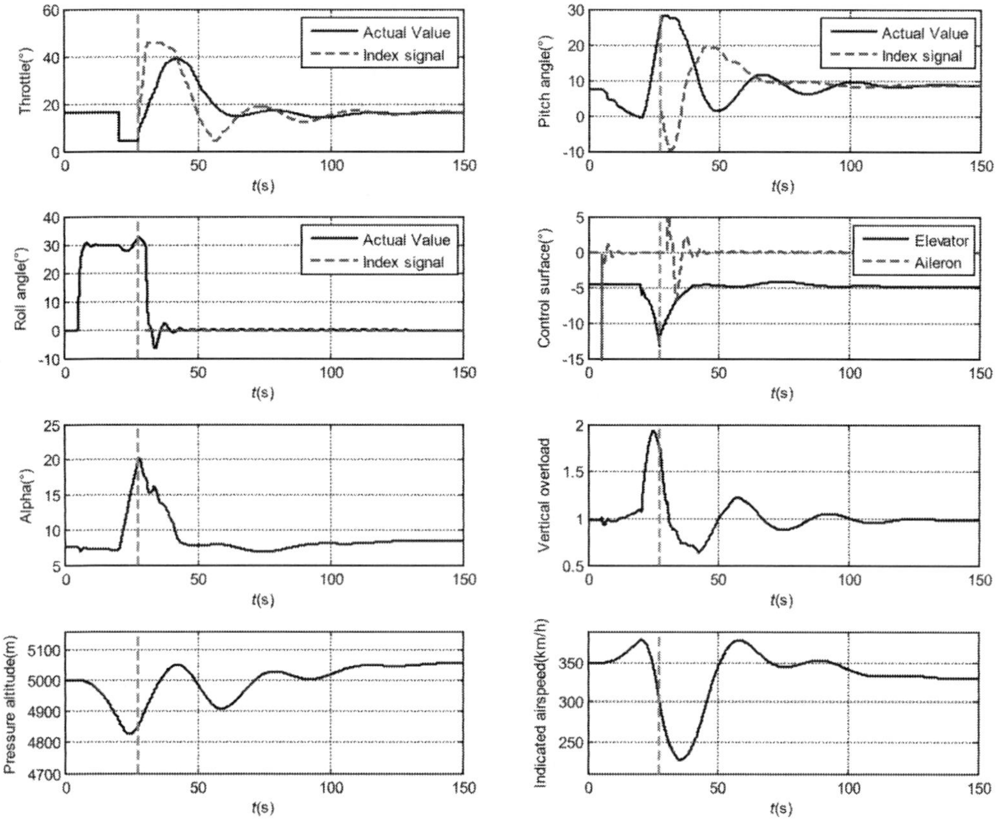

Figure 4. Aircraft state response.

4.3.4. Misoperation after stalling. The scene is simulated where the pilot performs an error correction operation after the aircraft stalls. The simulation results are shown in Figure 5 and Figure 6. Among them, Figure 5 is the phenomenon in which the pilot mistakenly maintains the rod after the rapid rod stalls at a high angle of attack. Figure 6 is the phenomenon in which the pilot mistakenly presses the rod after the rod decelerates to stall. It can be seen from the diagram that the pilot can control the steering rod and the throttle rod regarding the guidance signals in both cases to enable the aircraft to recover from the stall state successfully.

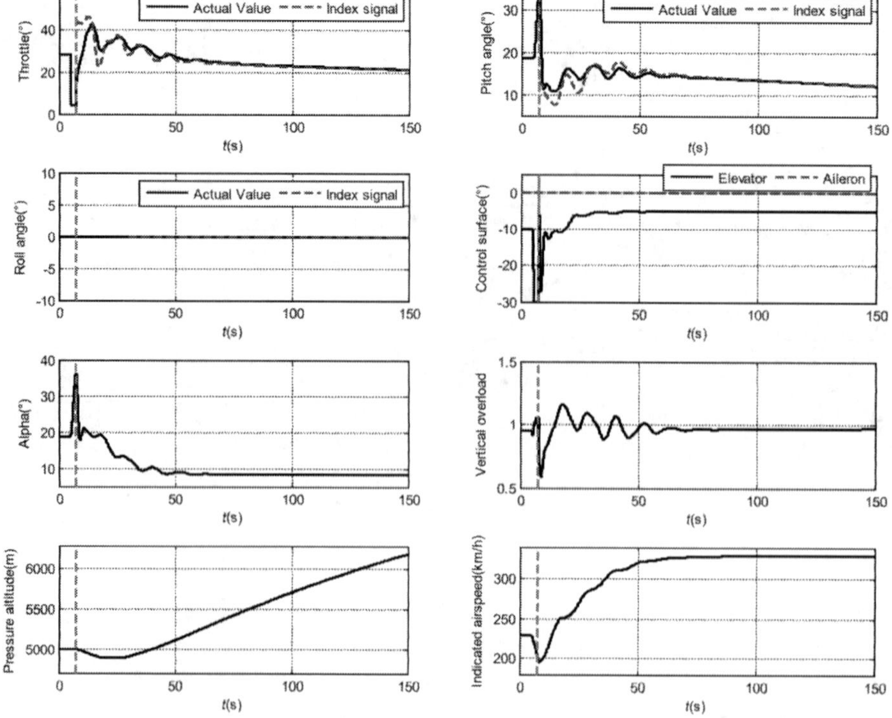

Figure 5. Aircraft state response.

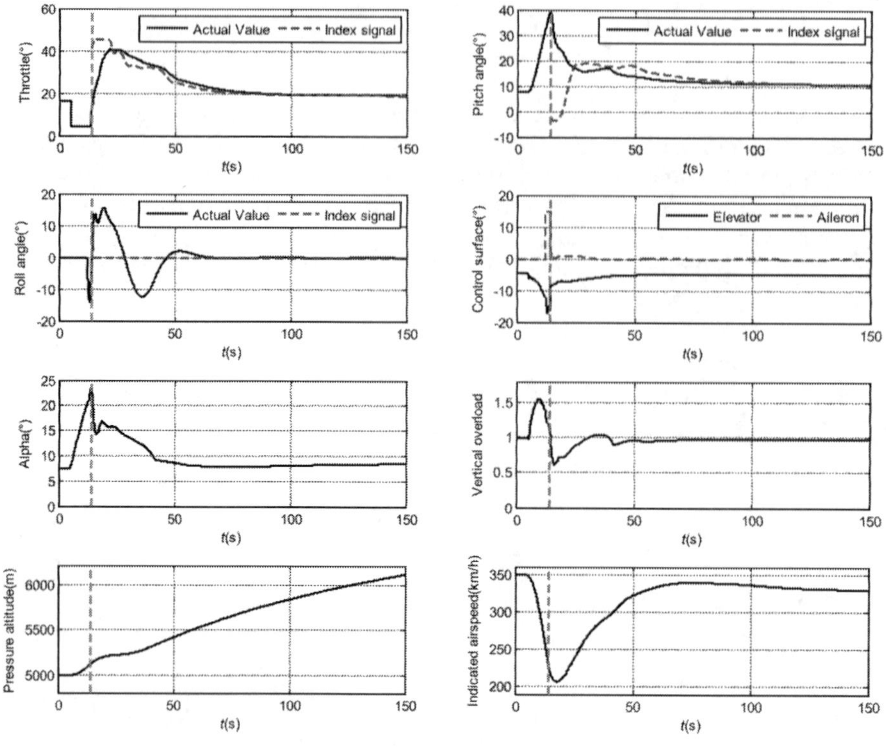

Figure 6. Aircraft state response.

4.3.5. Stall recovery under engine shutdown condition. Considering that at a high angle of attack, engine shutdown may occur due to intake distortion. Therefore, it is necessary to verify the effectiveness of the algorithm proposed in this project under the condition of accidental engine shutdown. The simulation results are shown in Figure 7.

At the beginning of the simulation, the aircraft maintains a stable level flight at a pressure height of 5000 m and an indicated airspeed of 230 km/h. When t = 5s, the pilot quickly pulls the lever to stall the aircraft. When t ≈ 5.5s, the aircraft's attack angle increases to 20°, and the algorithm is triggered. Then, the pilot continues to pull the lever to the end and keep it. When t = 7.5s, the aircraft's angle of attack reaches about 35°, and the engine stops at this time. At the same time, the pilot began to perform the change-out manoeuvre concerning the guidance signal. Then, the aircraft's angle of attack decreases slowly, and the flight speed increases slowly. Finally, the aircraft's angle of attack is stabilized at about 8.5°, and the flight speed is stabilized at about 329km/h. The aircraft successfully recovered from the stall state. However, it can also be seen that the aircraft will have a significant height loss after stalling due to the lack of engine thrust.

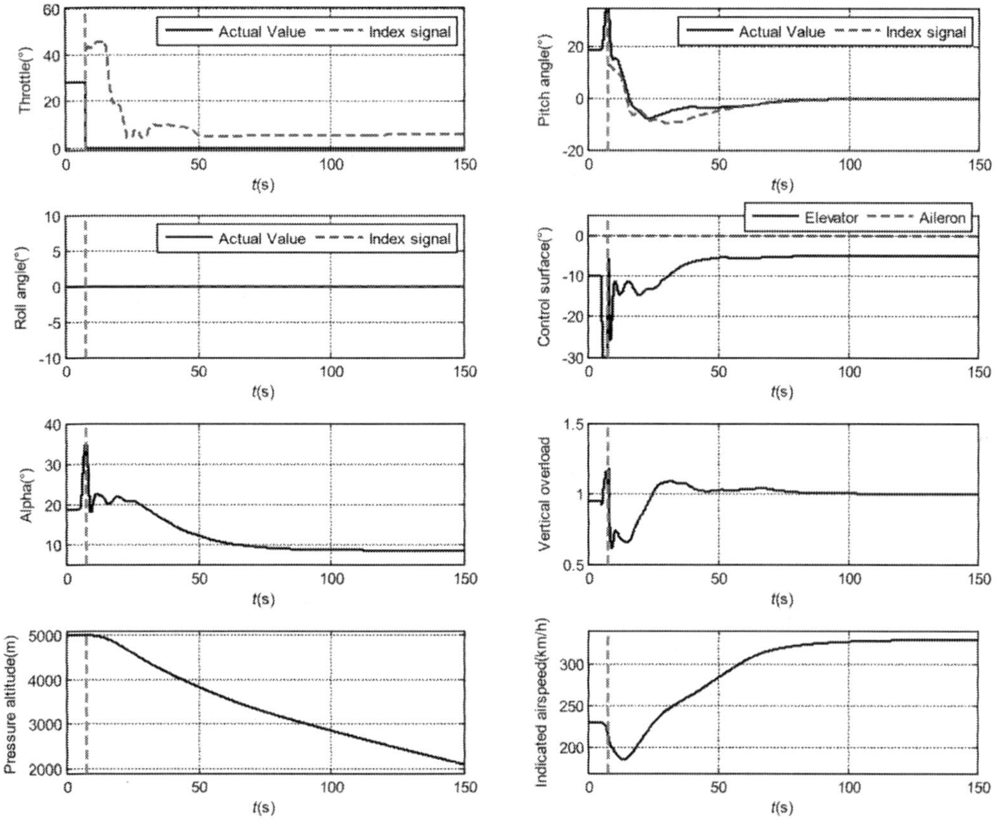

Figure 7. Aircraft state response.

4.4. Effectiveness and robustness analysis

In order to ensure the effectiveness and robustness of the algorithm, factors that may affect the recovery effect should be fully considered. It should be considered that the effect of stall recovery in actual flight is not only related to the entry scene but also closely related to the pilot's handling characteristics. When pilots with different control characteristics perform recovery control, the aircraft's height loss and attitude response will also differ. Reliable algorithms should be able to adapt to different pilot manipulation characteristics. In the simulation analysis, the effectiveness and robustness of the algorithm are fully verified by adjusting the pilot model parameters to simulate the handling characteristics of different pilots. Taking the spiral deceleration to stall scene as an example, the simulation results are shown in Figure 8. From the figure, it can be seen that although the difference in the pilot model will affect the aircraft state response, it can successfully change the stall state under the guidance of the algorithm. During the period, there was no secondary stall and overload overrun phenomenon. After stalling, the aircraft can be restored to a stable state without the problem of entering a complex attitude, which further verifies the effectiveness and robustness of the algorithm.

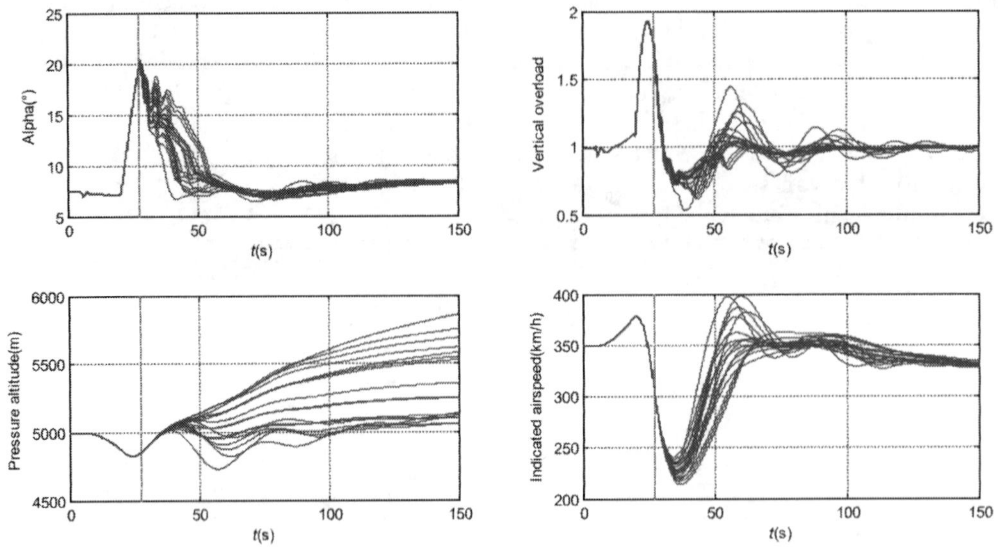

Figure 8. Aircraft state response under different pilot control characteristics.

4.5. Summary

In summary, the results of man-machine closed-loop simulation analysis show that under different stall entry scenarios such as slow rod deceleration to stall, sharp rod deceleration to stall at a high angle of attack, spiral deceleration to stall, and mishandling after stall, the pilot can control the aircraft to recover from stall state according to the guidance signal, and finally basically maintain stable flight at the target speed. During this period, there is no problem with secondary stalls or overload overruns, indicating that the algorithm is effective. In addition, the algorithm can adapt to different pilot manipulation characteristics and has good robustness.

5. Conclusion

The stall state-assisted recovery method based on pitch angle, engine thrust, and roll angle signal guidance determined in this paper can effectively guide the pilot to recover the stall state in different stall scenarios. It can avoid the pilot's wrong recovery operation after the aircraft accidentally enters the stall, which is significant for ensuring flight safety. At the same time, this method does not require a complex calculation process and is easy to implement in engineering. In the future, the algorithm's reliability should be further verified by the actual operation of the pilot in the high-precision ground simulator, and the technical maturity should be improved to lay the foundation for flight test verification.

Reference

[1] Stefan Schuet, Thomas Lombaerts, John Kaneshige, et al., Stall Recovery Guidance Using Fast Model Predictive Control [C], AIAA Guidance, Navigation and Control Conference, 2017.

[2] Richards N. D, Gandhi N, Bateman A. J., et al., Development and Pilot-In-The-Loop Evaluation of Robust Upset-Recovery Guidance [C], AIAA SciTech, No. AIAA-2016-0879, 2016.

[3] YUAN Zhipeng, XUE Yuan, GONG Lei, et al., Research on design and flight test technologies of large aircraft angle of attack protection control law [J]. Flight Dynamics, 2020, 38 (01): 90-94.
[4] Ji Yun, Zhao Hai, Yang Baojun, et al., Protection method of angle of attack and normal acceleration after angle of attack failure for large aircraft [C], China Society of Aeronautics and Astronautics. The 6th China Aeronautical Science and Technology Conference. The First Aircraft Institute of AVIC; 2023:8.
[5] Stepanyan V, Krishnakumar K, Kaneshige J., et al., Stall Recovery Guidance Algorithms Based on Con-strained Control Approaches [C], AIAA SciTech, No. AIAA-2016-0878, 2016.
[6] Richards N. D, Gandhi N, Bateman A. J, et al., Piloted Flight Test Evaluation of Robust Upset-Recovery Guidance [C], Proceedings of the AIAA SciTech 2018 Conference and Exhibit, No. AIAA 2018-1864.
[7] Wang Xiaorui., The research of simulation on stall motion characteristics of a certain type of aircraft [D]. GuangHan: Civil Aviation Flight University of China, 2023.
[8] Xi Guangxu, Liu Yongyi, Wu Chong, et al., Key Technology Research on Stall Spin Simulation Training System of an Aircraft [J]. Journal of System Simulation, 2022, 34(04):878-890.
[9] Zhang Qi, Yuan Guangtian, Gong Pengxiao., Analyse of the man-machine closed-loop system characteristics of aerial tanker [J]. Flight Dynamics, 2019, 167(03):26-31.
[10] Tan Wenqiana, A. V. Efremovb, Qu Xiangjua., A Criterion Based on Closed-loop Pilot-aircraft Systems for Predicting Flying Qualities [J]. Chinese Journal of Aeronautics, 2010, 23(03):511-517.
[11] Su Jianbo, Zhang Jiaqi, Qiu Yuhao, et al., A judgment method for stall in free-flight test of reduced ratio model [J]. Computer Simulation, 2023, 40(02):39-42.
[12] Zhang Lifeng, Su Jianbo, Zhang Jiaqi., Design of a model free-flight test method based on autonomous tie rod, 2019, 39(05):57-60.

Standardized research on multifunctional beacon multi-sensor data acquisition and transmission

Zhuoyi Zhang[1], Kehuai Ji[1], Yu Guo[1], Chuanying Zhu[1,*] and Chen Wang[1]

[1]Tianjin Aids to Navigation Department, Binhai new area, Tianjin, 300456, China

[*]Corresponding author's e-mail: 1751718000@qq.com

Abstract: This paper discusses the research on the standardization of data acquisition and transmission for multifunctional beacon systems. The multifunctional beacon system integrates various types of sensing devices and can provide real-time, comprehensive and accurate marine environment and navigation information for ships and shore-based users. However, the system currently faces challenges such as non-uniform data formats and communication protocols, low hardware standardization and integration, and data and network security risks. In order to solve these problems, this paper puts forward the solution of "standard terminal for data acquisition and transmission", which achieves the compatibility of equipment interface, unified data format, and reliable information security. It lays the foundation for the standardization of multi-purpose beacon system. Through modular design and standardized packaging, the performance, reliability and efficiency of the hardware of multi-purpose beacons are improved, the development and maintenance costs are reduced, and the flexibility of the system is enhanced. The "Standard Terminal for Data Acquisition and Transmission" is applied to the multi-purpose beacon system, and interfaces and integrates with the existing "Operation Management System for Navigation Guarantee" to realize a unified platform for multi-purpose beacons. This research helps to promote the formulation of multi-purpose beacon data format and communication protocol standards at the national or industrial level, provides important reference for the formulation of international standards, and is of great significance in promoting the standardized development and application promotion of multi-purpose beacon technology.

1. Introduction

Multi-functional beacon is a new type of beacon system integrating a variety of sensing equipment. On the basis of retaining the existing navigational guidance function, it provides real-time, comprehensive, accurate, and quantitative navigational guidance information of the sea environment for ships and shore users by carrying various sensing devices such as hydrography, meteorology, and video monitoring. These beacons include light buoys, light piles, light towers, and other forms [1, 2].

The emergence of multifunctional beacon systems brings new technical support for marine traffic safety, marine environment monitoring, and other fields. It can provide marine users with more comprehensive and accurate information about the marine environment and improve navigation safety. At the same time, it can also provide real-time data support for marine environment monitoring, which helps to better understand and protect the marine ecological environment.

However, the current multifunctional beacon system still faces some challenges in practical application. Firstly, the operation platform is relatively decentralized, and the data formats and communication protocols of equipment produced by different manufacturers are not uniform, which makes data integration and sharing difficult. Secondly, the hardware standardization and integration level is low, which makes system maintenance and upgrading inconvenient. In addition, data security and network security risks are also issues that need to be focused on [3, 4].

In order to solve these problems, this paper puts forward a solution of a "standard terminal for data acquisition and transmission," which realizes the compatibility of equipment interfaces, the unification of data formats, information security, and reliability, and lays a foundation for the standardized development of multi-purpose beacon system. Through modular design and standardized packaging, the performance, reliability, and efficiency of the hardware of multi-purpose beacons are improved, the development and maintenance costs are reduced, and the flexibility of the system is enhanced. The "standard terminal for data acquisition and transmission" is applied to the multi-purpose beacon system and interfaces and integrates with the existing "operation management system for navigation security", thus realizing a unified platform for multi-purpose beacons [5, 6, 7].

This research work not only helps to promote the development of multifunctional beacon data format and communication protocol standards at the national or industry level but also provides an important reference for the development of international standards. It is of great significance to promote the standardized development and application promotion of multifunctional beacon technology.

2. Status and problems

2.1 Data format and communication protocols are not uniform

1. Each equipment manufacturer has a custom protocol; each manufacturer is almost different.
2. Some foreign equipment supports the SDI-12 communication standard, but this standard requires external hardware, which is not good to deal with.
3. Positioning equipment and bathymetric equipment support the NMEA-0183 protocol.
4. Some domestic water quality sensors support Modbus protocol, but the data content, high and low bytes, etc., or each manufacturer's respective provisions, are almost not uniform.
5. The manufacturer's protocol is generally rarely modified. If modification will lead to equipment and multi-functional data acquisition terminal does not match, we need to pay attention to the manufacturer's protocol changes. Table 1 shows the different meanings of electronic compass data formats by different manufacturers.

Table 1. Differences in data format of electronic compasses from different manufacturers.

Data format (AA AB BB CC CD DD EE EF FF)	A vender	B vender
AA AB BB	Represents the Heading-angle	Represents the Pitch axis
CC CD DD	Represents the Pitch axis	Represents the Roll axis
EE EF FF	Represents the Roll axis	Represents the Heading-angle

2.2 Low standardization and integration of hardware

At present, the degree of standardization and integration of multifunctional beacon hardware is low, which leads to the following problems:

Poor hardware compatibility: There is a lack of unified standards between the beacon hardware produced by different manufacturers, which leads to the difficulty of interconnection and interoperability between the equipment. It is impossible to realize data sharing and cooperative work.

Difficulty in system maintenance: Due to the lack of uniform hardware standards, it is necessary to maintain different devices independently, which increases the maintenance cost and difficulty, and also reduces the maintenance efficiency.

High development cost: Developers need to adapt to different hardware platforms, which increases the development cost and time and hinders the rapid development and application of multifunctional beacon systems. Figure 1 shows the status quo of non-integrated devices.

Figure 1. Non-integrated multifunctional beacon equipment.

2.3 Data security and cybersecurity risks

In today's digital age, data security, and network security are issues that must be taken seriously by businesses and individuals alike. When using the TCP/IP protocol for public network direct-connection clear-code communication, data is transmitted over the network in clear-text form, which is equivalent to saying your bank card password out loud in a public place, where the information is exposed and highly susceptible to interception, resulting in serious security risks.

2.3.1 Man-in-the-middle attack. Plaintext communication means that all information transmitted over the network, including sensitive data such as usernames and passwords, is not encrypted. This type of communication is highly vulnerable to Man-In-The-Middle (MITM) attacks. An attacker can easily intercept, tamper with, or insert data packets between the two communicating parties without them being able to detect it. For example, if a user logs into a bank account on an unsecured network via explicit communication, an attacker can capture these packets and obtain the user's login credentials.

2.3.2 Sniffing attack. Packets may also be subject to sniffing attacks as they travel through the network. Attackers can use various sniffing tools, such as Wireshark, to capture packets passing through the network. If the data is transmitted in clear text, the attacker can easily read the contents of these packets to obtain sensitive information.

2.3.3 Replay attack. Plaintext communication is also vulnerable to Replay Attacks. An attacker can record the packets during the communication process and then resend these packets at the appropriate time to achieve the purpose of deceiving the system or the user. For example, an attacker can intercept a user's authentication requests and then resend them after the user disconnects in an attempt to log in to the system again.

Therefore, in order to protect data security and network security, it is recommended to use encryption protocols such as SSL/TLS to encrypt data transmission to ensure confidentiality and integrity of data during transmission. Also, the use of technologies such as VPN (Virtual Private Network) can provide additional security for data transmission. For businesses and individual users, understanding and adopting appropriate security measures is an important means of preventing network risks and protecting their interests [8, 9, 10].

3. Solution

3.1 Definition of standard data format

We determine the importance of meteorological, hydrological, water quality, marker status, energy control, and other types of sensory data to nautical users, and then classify the data to determine the data to be collected and content to be displayed by the multifunctional beacon system. For example, the importance of energy control information for beacon users is much lower than that for beacon managers, so it can be determined as a category to be displayed only to beacon managers.

We sort out the data characteristics of meteorological, hydrological, water quality, marker status, energy control, and other types of sensory data, categorize them according to their data characteristics, and define the standard format for each type of data. Table 2 shows the standard data format of wind speed. Table 3 is the unified definition of multifunctional beacon data format.

Table 2. Example of wind speed standard data format.

a	Device address
R1	Wind message query command
Dn	Wind direction minimum (D=degrees)
Dm	Wind direction average (D=degrees)
Dx	Wind direction maximum (D=degrees)
Sn	Wind speed minimum (M=m/s)
Sm	Wind speed average (M=m/s)
Sx	Wind speed maximum (M=m/s)
<cr><lf>	Response terminator

Table 3. Definitions of MFB data formats.

Site name	Device name	Channel type	Meaning	Data unit	Data type	Span
Collection site (Station) An Id 64-bit integer	Acquisition equipment (Device) (devType)1 visibility meter	0x00000001	Average visibility in one minute	m	integer	0~35000
		0x00000002	Ten minutes of average visibility	m	integer	0~35000
		0x00000004	Visibility hardware alarm		integer	==0 R, !=0 F
	Acquisition equipment (Device) (devType)2 Flow velocity profile	0x00000001	Error code		integer	==0 R, !=0 F
		0x00000002	acoustic velocity	m/s	Floating point type	≥0
		0x00000004	direction angle	degree	Floating point type	0.0~360.0
		0x00000008	angle of pitch	degree	Floating point type	-90.0~90.0
		0x00000010	roll angle	degree	Floating point type	-180.0~180.0
		0x00000020	water gage	dbar	Floating point type	≥0
		0x00000040	water temperature	°C	Floating point type	-20.00~50.00
		0x00001000	Probe layer serial number		integer	≥0
		0x00002000	Horizontal flow velocity of the probe layer	mm/s	integer	≥0
		0x00004000	The horizontal direction of the probe layer	degree	Floating point type	0~360.0

Acquisition equipment (Device) (devType)2 Flow velocity profile	0x00000001	direction angle	degree	Floating point type	0.00~360.00
	0x00000002	angle of pitch	degree	Floating point type	-90.00~90.00
	0x00000004	roll angle	degree	Floating point type	-90.00~90.00

3.2 Developing and designing a "standard terminal for data acquisition and transmission"

It has more than 20 data interfaces, compatible with the communication protocols of all existing sensing devices (RS232, RS422, RS485, RJ45, USB, analog signals, etc.). It reserves five spare interfaces. The collected data are automatically recognized, parsed, transformed into standard multifunctional beacon data format, and encrypted and stored with DES, 3DES, AES, etc. It is equipped with two data transmission modes of public network/Beidou 3 and APN/VPDN function to transmit the encrypted data to the shore end server. AIS/VDES data broadcast function is reserved. Figure 2 shows the standard terminal architecture for data acquisition and transmission.

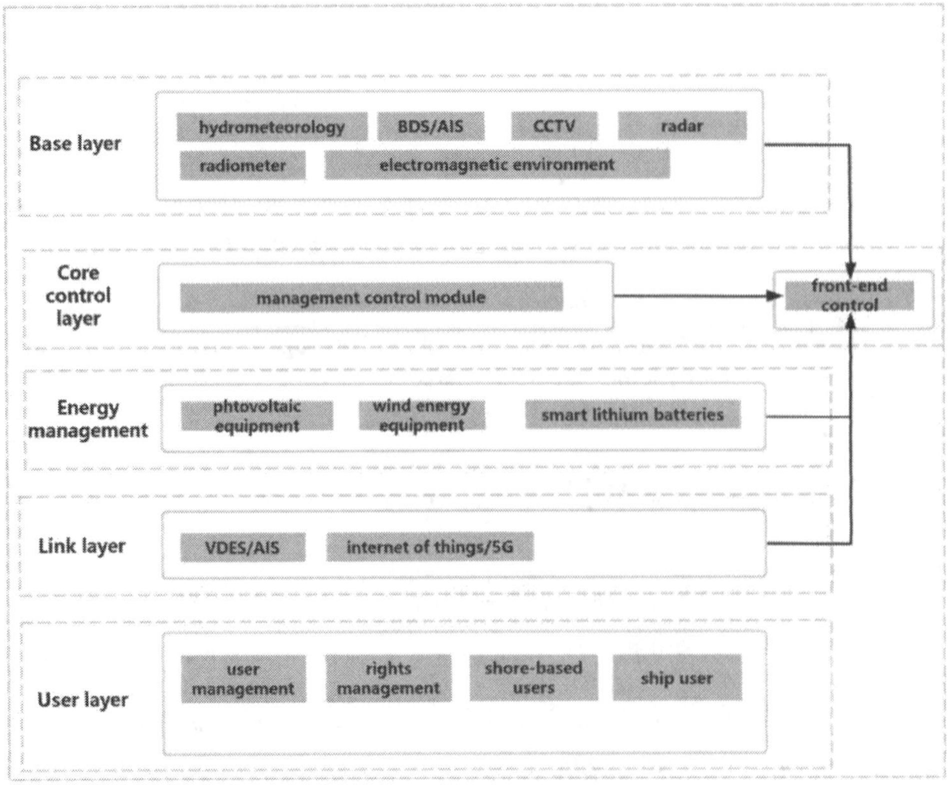

Figure 2. Multi-functional beacon terminal architecture.

Multi-purpose beacon systems typically consist of multiple components, including sensors, data processing units, communication modules, and user interfaces. These components work together to ensure the versatility of the beacon and the transmission of real-time data streams. Figure 3 to Figure 6 illustrates the multifunctional beacon monitoring system.

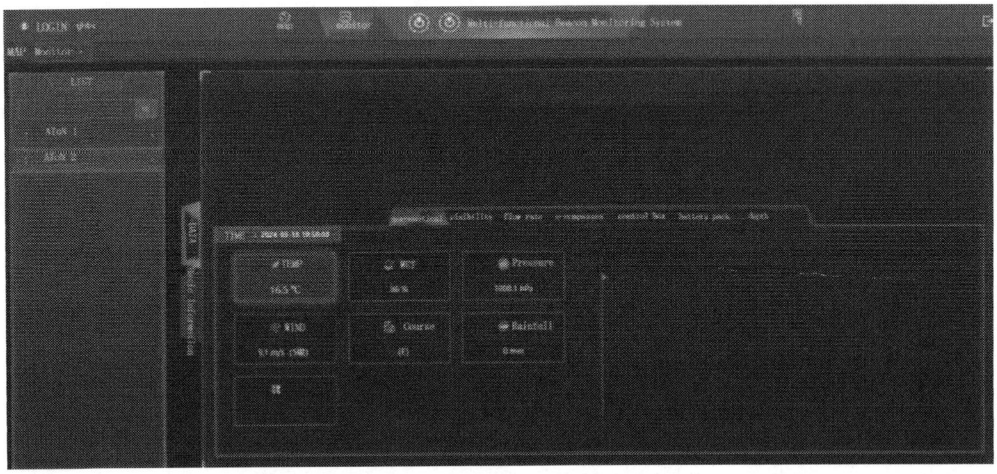

Figure 3. Multi-functional beacon monitoring system.

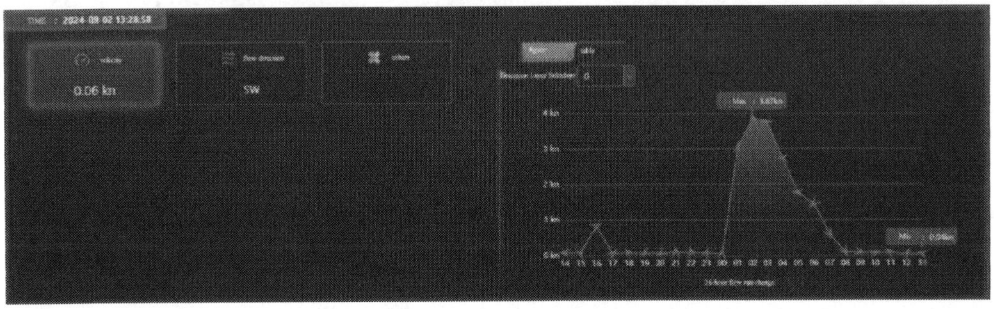

Figure 4. Flow rate display information.

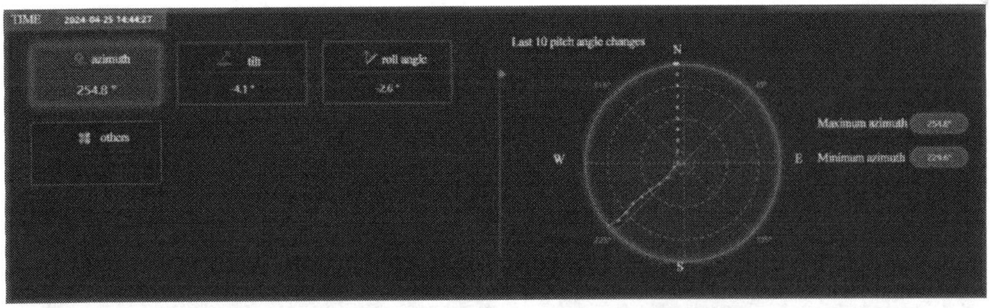

Figure 5. Electronic compass display information.

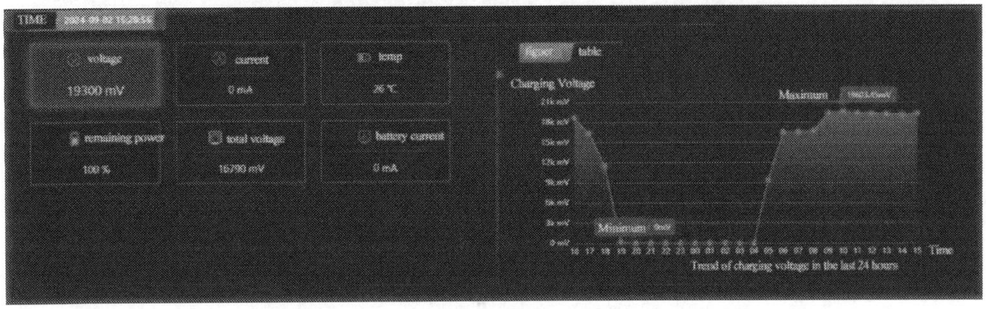

Figure 6. Battery control information.

Journal of Physics: Conference Series **2879** (2024) 012057 doi:10.1088/1742-6596/2879/1/012057

3.3 Promote modularized design

Promoting modular design and standardized packaging of different functional modules is the key to improving the performance, reliability, and efficiency of multifunctional beacon hardware.

Modular design can decompose the beacon system into multiple independent functional modules, such as a positioning module, communication module, data acquisition module, power module, and so on. Each module has a clear function definition and interface specification, which can be independently developed, tested, and maintained and easily combined and upgraded.

Standardized packaging, on the other hand, implies the development of uniform interface standards for each functional module to ensure seamless connection and interoperability between different modules. In this way, modules produced by different manufacturers can be freely combined to build a flexible and diverse beacon system to meet various application requirements.

The advantages of modular design and standardized packaging are:

Improvement of the development efficiency: Developers can focus on the development of a single module, reducing repetitive work and speeding up system development.

Reduction of maintenance costs: Modular design facilitates module replacement and upgrading, reducing maintenance difficulties and costs.

Enhancement of the system reliability: Modular design can isolate faults within a single module, preventing the spread of faults and enhancing the overall reliability of the system.

Enhancement of the flexibility of the system: Modular design can flexibly combine different functional modules according to the actual needs to meet the needs of different application scenarios.

3.4 Unification of the platform of the multi-functional buoy system

We apply the "Standard Terminal for Data Acquisition and Transmission" to multi-functional buoys, lighthouses, and light piles under the jurisdiction, encrypt the data they collect, and transmit them back to the server at the shore end for storage.

The "Navigation Protection Operation Management System" is a technologically mature and comprehensive information protection platform that is compatible with the functions of some multifunctional beacons.

From the perspectives of "one platform, one map" and "living a tight life," we will not build a new "multi-functional navigational system". Still, we will improve some of the functions of the "Operation Management System of Navigational Security" of the Department. Instead of building a new "Multi-functional Beacon System", the "Navigation Guarantee Operation Management System" of our department is improved by improving some of its functions (protocols and key docking), so as to make it a unified platform of the multi-functional beacon system. Figure 7 shows the interface of the Navigation Guarantee Operation Management System. Figure 8 shows the integrated multifunctional beacon.

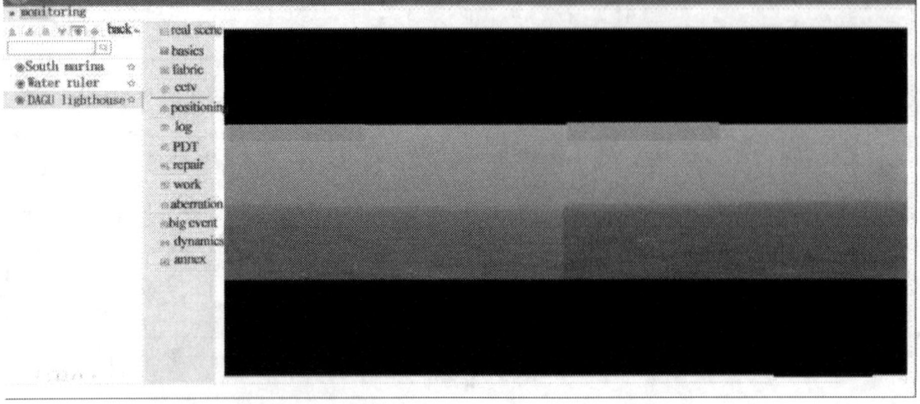

Figure 7. Maritime security operation management system.

Figure 8. Multi-functional beacons.

3.5 Detecting and responding to potential security threats

Firstly, a comprehensive security monitoring system is required, which is capable of collecting and analyzing network traffic, system logs, application logs, and security events in real time. Through the use of Intrusion Detection System (IDS), Intrusion Prevention System (IPS), and Security Information and Event Management (SIEM) tools, abnormal behaviors and potential threats can be detected in a timely manner. Secondly, regular security assessments and penetration tests are conducted to identify vulnerabilities and weaknesses in the system. This includes testing networks, applications, and infrastructure to ensure that vulnerabilities are identified and patched in a timely manner. Additionally, it is vital to establish a response plan. As soon as a security incident is detected, a pre-established emergency response plan should be activated to respond quickly and effectively. This includes isolating affected systems, assessing the extent of damage, notifying relevant parties, restoring services, and conducting post-mortem analyses.

4. Conclusion

In modern navigation systems, the efficient operation of multifunctional beacons is crucial to ensuring maritime traffic safety. In order to achieve this goal, it is necessary to standardize the data formats and communication protocols in the beacon system, ensure the safety and reliability of data transmission, adopt modular design and standardized packaging, and establish a unified multifunctional beacon system platform. These measures can not only improve the efficiency and accuracy of data processing but also enhance the overall safety and reliability of the system.

4.1 Unified standard data format

Standardized data formats and communication protocols are the basis for ensuring the effective flow of data in the multifunctional beacon system. By formulating a unified standard data format, the classification and formatting definition of various types of sensory data can be realized to simplify the data processing process and improve the readability and interoperability of data. At the same time, unifying the communication protocols of multifunctional beacon systems and ensuring the compatibility of interfaces between different devices is the key to achieving seamless communication between devices. This not only helps to reduce compatibility problems between devices but also reduces the complexity and cost of system integration.

4.2 Data encryption
Secure and reliable data transmission is necessary to ensure the safety of navigation. Adopting advanced encryption technology to protect the collected data can effectively prevent the data from being stolen or tampered with in the transmission process and ensure the integrity and confidentiality of the data. In addition, the safe and reliable transmission of data from the front-end collection terminal to the shore-end server is of great significance in reducing the network security risk and guaranteeing the accurate transmission of navigation information.

4.3 Modular design
Modular design and standardized packaging are effective means to improve system development efficiency and maintenance convenience. Promoting the modular design of multifunctional beacon systems and standardized packaging of different functional modules can significantly improve development efficiency and reduce maintenance costs. At the same time, a unified interface standard for each functional module must be formulated to ensure seamless connection and interoperability between different modules, which is crucial for improving the overall reliability and flexibility of the system.

4.4 Multi-functional navigation system platform
Multi-functional Beacon System Platform establishing a unified multi-functional beacon system platform is the key to realizing centralized data management and efficient use. Applying the "standard terminal for data collection and transmission" to multi-functional buoys, light towers, and light piles can realize unified data collection and transmission, ensuring real-time and accurate data. Using the existing "Navigation Guarantee Operation Management System" as the unified platform of multi-functional beacon systems can realize the centralized display and management of data and provide strong technical support for navigation safety.

4.5 Detecting and responding to potential threats
Detecting and responding to potential threats is an important measure to ensure the security and stable operation of the system. By carefully analyzing and evaluating various possible security vulnerabilities and risks, we can detect and take corresponding measures to prevent these potential threats in time. This includes but is not limited to, regular security scanning, vulnerability assessment, and intrusion detection to ensure that vulnerabilities in the system are identified and patched in a timely manner. At the same time, the establishment of a comprehensive set of contingency plans and response mechanisms can enable swift action to be taken to minimize losses when threats occur. In addition, strengthening the security awareness training of staff and enhancing their ability to recognize and respond to potential threats are also important aspects of ensuring overall security. Through these comprehensive measures, we can effectively detect and respond to all kinds of potential threats to ensure the safe and stable operation of the system.

To sum up, the performance and safety of the navigation system can be significantly enhanced by standardizing data formats and communication protocols, ensuring safe and reliable data transmission, adopting modular design and standardized packaging, and establishing a unified multifunctional beacon system platform. These measures will provide a solid technical guarantee for the development of navigation, ensure the safety of maritime traffic, and promote the continuous prosperity of navigation.

References

[1] Chunping Ning, Yulong Wang, Yiming Zheng, et al. Design of online salinity monitoring system for large-scale river and sea canals based on multifunctional beacons [J]. Western Transportation Science and Technology, 2024, (04): 8-10. DOI: 10.13282/j.cnki.wccst.2024.04.004.

[2 Hui Wan, Longle Wu, Yi Yu, et al. Research on the structure and key technology of intelligent navigation protection system [J]. China Navigation, 2024, 47(01): 45-51.

[3] Miner Shan, Youwei Li, Xin Liu, et al. Study on the application of flow restoration based on multifunctional beacon monitoring data [J]. Water Transport Engineering, 2023, (06): 179-184.DOI: 10.16233/j.cnki.issn1002-4972.20230605.033.

[4] The first all-sensor multifunctional beacon put into Fujian sea area [J]. Sensor World, 2023, 29(05): 37.

[5] Yongtao Wang, Haibo Zhang, Zhiyuan Wang, et al. Research on the design and practice of multifunctional beacon terminal in sea area based on AIS [J]. Waterway Port, 2023, 44(02): 231-237.

[6] Miao Gu, XJunjie Wang, Bowen Zhang, et al. Research on ship exhaust monitoring system based on multifunctional beacon platform [J]. China Maritime, 2023, (01): 48-50. DOI: 10.16831/j.cnki.issn1673-2278.2023.01.013.

[7] Feng Zhang. A few thoughts on the construction of multifunctional navigation markers [J]. Navigation, 2023, (04): 45-48.

[8] Yupeng Ma, Xiumin Chu, Mao Zheng, et al. Review and Prospect of Multi-functional Beacon Technology [C]//China Intelligent Transportation Association. Proceedings of the 17th China Intelligent Transportation Annual Conference on Science and Technology. School of Transportation and Logistics Engineering, Wuhan University of Technology; National Engineering Technology Research Centre for Water Transportation Safety, Wuhan University of Technology; Lianyungang Beacon Office, East China Sea Navigation Guarantee Centre, Ministry of Transportation and Communications; 2022: 2. DOI: 10.26914/c.cnkihy.2022.053599.

[9] Xuanqiao Wang. Analysis of multifunctional beacon communication data in Qiongzhou Strait [J]. Pearl River Water Transport, 2022, (20): 72-76. DOI: 10.14125/j.cnki.zjsy.2022.20.034.

[10] Yixin Zhang, Hao Wu, Jashun Huang, et al. Realisation of remote update of a multifunctional beacon collection system based on mobile network [J]. Pearl River Water Transport, 2021, (07): 107-108. DOI: 10.14125/j.cnki.zjsy.2021.07.050.

Journal of Physics: Conference Series **2879** (2024) 012058 doi:10.1088/1742-6596/2879/1/012058

Research on optimization of flatness evaluation method for antenna array of a radar vehicle

Yunping Pan[1,*] and Yirao Yang[1]

[1]School of Mechanical and Electrical Engineering, Wuhan University of Technology, Wuhan, Hubei, 430000, China

[*]Corresponding author's e-mail: panyunping@whut.edu.cn

Abstract: Radar vehicles play an important role in providing radar services in the aviation field. The flatness of the antenna array is one of the key factors in ensuring the detection accuracy of radar vehicles for airborne targets. In this paper, the mechanical structure of a new type of radar antenna truck is designed, the stress analysis under various working conditions is carried out, the mechanical performance of the whole vehicle is checked, and the evaluation method of the flatness of the antenna array is discussed. Using ANSYS to carry out static finite element analysis and using the deformation results of each element node in ANSYS, the evaluation method based on the multiple regression function is applied to the calculation scheme of antenna array flatness. The final results show that the safety factor of the radar vehicle structure can reach more than 1.75 under extreme working conditions and more than 2.0 under other working conditions. It can work safely. Through the optimization of the shape evaluation method, the accuracy of the final shape calculation result is less than 10mm. The precision of shape error processing is improved.

1. Introduction

Vehicle-mounted radar is an important ground radar, and the research of vehicle-mounted radar technology has always been an important research direction in the radar field [1]. The array antenna of radar is composed of several array elements, which is a complex mechanical and electronic system. Its accuracy directly affects key technical and tactical indexes such as radar antenna lobe gain and beam pointing accuracy [3]. Therefore, controlling and ensuring the design accuracy and flatness of the antenna array has become one of the key technologies. As the main load of vehicle-mounted array radar antenna is wind load, it is necessary to verify the maneuverability and safety of the radar antenna under various wind load conditions [2]. Li [2] evaluated the antenna array accuracy of ground-phased array radar through design simulation. Yang et al. [4] introduced the evaluation of flatness error through a mixed Java algorithm. Xin et al. [5] used the graphical method to measure the error of motion flatness.

The above examples all introduce the assessment of the flatness error of different methods, but they are all relatively general. For a large assembly such as a radar antenna truck, the antenna array area is large, the overall strength of the radar truck should be guaranteed. The deformation of the front frame should be within the allowable range. In order to improve the accuracy of error evaluation and ensure the accuracy and reliability of radar, it is necessary to discuss the calculation method of antenna array flatness further.

Journal of Physics: Conference Series **2879** (2024) 012058 doi:10.1088/1742-6596/2879/1/012058

2. Structural mechanics analysis of radar antenna vehicle

2.1. Overall structure

The overall structure of this type of radar antenna truck in the working state when fully lifted is shown in Figure 1 below. It is equipped with a large-volume high-frequency box and front radar antenna. The antenna front is arranged in blocks. This arrangement is convenient for transportation, fast entry into the working state, and can realize rapid contraction under sudden dangerous conditions to ensure the safety of the vehicle [6]. Its structure is shown in Figure 2. A driving motor and a hydraulic station are configured at the subframe as driving plates to expand the antenna array and raise it to the required height. In order to meet its high dead weight and the load brought by the wind load at work, the main load-bearing part is made of steel Strenx700MC. The material parameters are shown in Table 1.

Table 1. Material table of radar antenna vehicle.

Material	Density (kg/m³)	Modulus of Elasticity (GPa)	Poisson's ratio
Strenx700MC	7800	210	0.3

The whole weight of the antenna vehicle is large, which puts forward higher requirements for work safety performance. Now, the work safety performance is analyzed statically. The following radar antenna car model was established. The model was pre-processed. The mesh quality was improved. The stress singularity in the calculation results was reduced. The model was reproduced in ANSYS WORKBENCH.

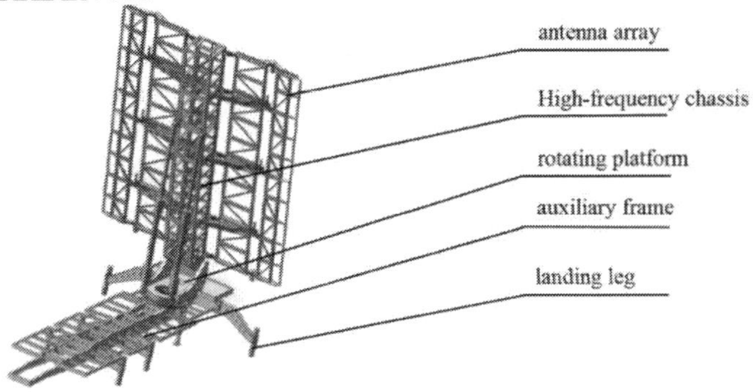

Figure 1. Overall structure model of radar vehicle.

Figure 2. Antenna array.

2.2. Vehicle safety performance check

In order to thoroughly test the safety of the radar vehicle, statics analysis was carried out on sixty operating conditions with different rotation speeds and wind coefficients when the radar vehicle antenna array was lifted 60° and 75°, rotated 0°, 45°, 90°, 135°, and 180°, under front and back wind loads.

The wind load mainly borne by the antenna array is evenly distributed on the windward components of the antenna array. The windward area varies under different lifting angles. The wind load under different working conditions should be calculated according to the following Formula[7]:

$$F_w = C_x q A \tag{1}$$

C_x is the wind power coefficient. The values are shown in Table 2. F_w is wind load (N). A is the windward area of the structure (m^2). q is the dynamic head, and $q = v^2/16$. v is the wind speed (m/s).

Table 2. Wind power coefficient.

Wind speed (m/s)	21.0-22.9	23.0-26.9	27.0-31.9	32.0-38.0
Power coefficient	1.46	1.42	1.40	1.37

The lack of gravity in the model is supplemented by adding equal mass particles at the center of mass.

The six legs at the bottom of the radar car are the fixed support points. The bottom of the large plate is the rotating fulcrum.

2.3. Analysis results

Based on the static analysis results of each working condition, the radar vehicle is in the limit working condition of wind speed of 21.0m/ s-22.9m /s, the maximum stress is 386.35Mpa when lifting 75° and rotating 90°. The safety factor is 1.79. Figure 3 is the setting of its boundary conditions. Figure 4 is the maximum stress point. Figure 5 is the maximum stress results. Figure 6 is the result of the maximum deformation. From the calculation results of each working condition, the safety factor is not less than 2, and the safety factor of the ultimate working condition is not less than 1.75, which meets the requirements of safety work. The maximum deformation of the antenna array is 105.94mm in each working condition, which is within the allowable deformation range. However, in order to ensure the working accuracy of the array radar, it is necessary to evaluate the subsequent flatness error.

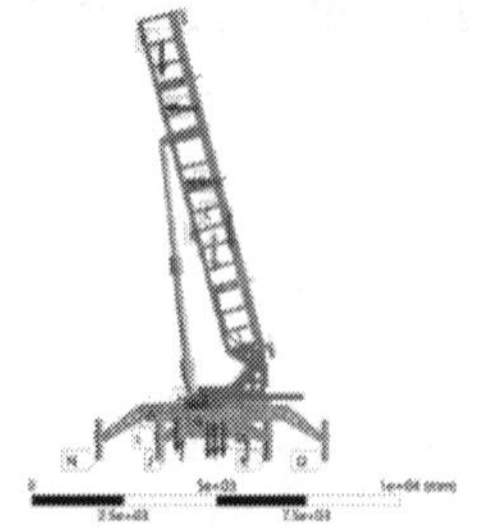

Figure 3. Boundary condition settings.

Figure 4. Maximum stress point.

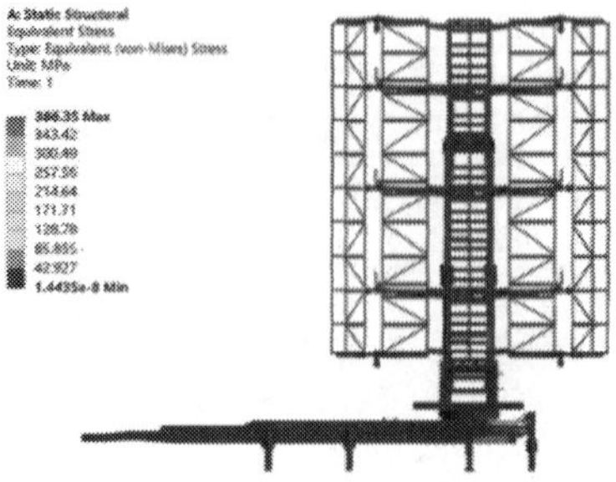

Figure 5. Maximum stress results.

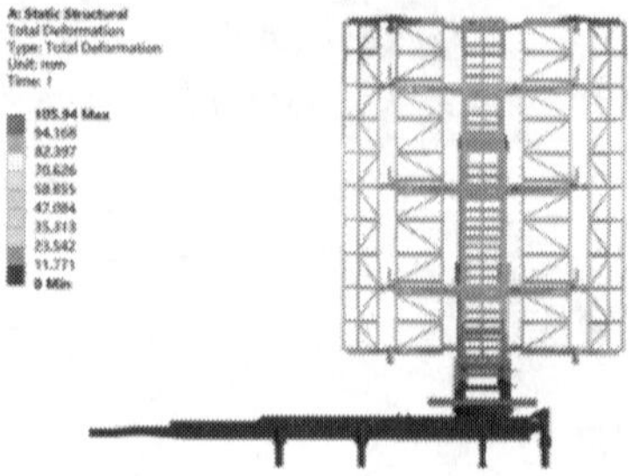

Figure 6. Maximum deformation result.

3. Flatness error calculation

3.1. Flatness evaluation method

Flatness error refers to the change of the actual plane measured to the ideal plane [8]. The antenna array assembly area of this type is 73.5mm². The traditional measurement method is difficult to implement. Considering the use of computer software MATLAB to assist calculation, the optimization algorithm has good convergence and stable process [9].

Based on the calculation results of ANSYS statics analysis, we can obtain the accurate calculation data of the target plane, that is, the coordinate values of each node of the antenna array. Common flatness evaluation methods mainly include the point plane Method, Diagonal Plane Method, diagonal Plane method, and diagonal plane method[10].

Compared with the flatness calculation methods in the table above, the measurement area of the radar antenna array is larger. The measurement points involved in the calculation of the three-far point plane method, the diagonal plane method, and the minimum inclusion area method are three points and four points, respectively. The reference plane uncertainty is larger. It is difficult to observe the distribution of data due to the excessive coordinates of measuring points in the method of minimum inclusion area. It is difficult to set a reasonable arrangement of measuring points because of the large area of the measured plane. The least square method uses the measured data point set on the measured plane to fit the center plane of the least square, which ensures the calculation accuracy of the flatness error. At the same time, the mathematical model is simple and easy to implement through programming. The calculation efficiency is high. Therefore, the least square plane method is selected to evaluate the flatness error in this essay.

3.2. Application of the least square method in the antenna array

When the flatness error is evaluated by the least square method, the base plane is the least square plane. It can be seen from its principle that the final least squares plane fitting result is a relative value. There is no need to strictly specify the three coordinate directions of the reference coordinate system. Therefore, the horizontal opposite direction of the antenna array is taken as the X-axis reference axis of the coordinate system. The derived node data is taken as the point set. The least squares plane of the antenna array is fitted accordingly and taken as the reference plane. Then, we calculate flatness by following these steps:

In the selected coordinate system, the fitting plane can be expressed as $z=ax+by+c$, where a, b, and c are undetermined parameters.

We set the coordinate value of any node as $P_i(x_i, y_i, z_i)$. Then its distance to the fitted plane is as follows:

$$H_i = z - (a_i x + b_i y + c_i) \tag{2}$$

In the formula, the node is positive when it is above the fitting plane and negative when it is below the fitting plane.

According to the least squares principle, when a set of values a, b, c are found, the sum of squares of the distances from each point to the least squares fitting plane $\Delta=\sum_i^n H_i^2$ is minimized. Assuming that the maximum deviation points on both sides of the least square plane in each measured node are (x_u, y_u, z_u) and (x_d, y_d, z_d), respectively, the flatness error can be expressed by the following formula:

$$f = H_u - H_d = \frac{(z_u - z_d) - a(x_u - x_d) - b(y_u - y_d)}{\sqrt{1+a^2+b^2}} \tag{3}$$

3.3. The implementation of the least square method in Matlab

3.3.1. Flatness test calculation under typical working conditions. At first, two typical working conditions in many working conditions were selected for further exploration. The antenna vehicle was in the working state of the array surface, which was fully developed and lifted 75° and 60°. The front of the array was subjected to a 20m/s wind load. The antenna array was rotating at a speed of 12r/min (1.25rad/s). At this time, the number of nodes contained in the antenna array is 19380.

As mentioned in Section 3.2 above, according to the calculation results of 2.3 ANSYS, the three-dimensional coordinates of each node on the surface of the matrix after stress deformation are counted. The data table is constructed in EXCEL. Subsequently, the data in the exported table was sampled. Considering that there were too many measurement points, when the least square method was used to evaluate flatness error, too many sampling points would cause evaluation error, so a quantity interval of 80 was selected for sampling [11]. The regress multiple regression function in Matlab was then called. The least square-fitting plane [12] was finally obtained after several iterations. The visual results are shown in Figure 7 below. The calculated flatness error is =11.98mm, =17.91mm.

Figure 7. Visualization of least square plane fitting results in two typical working conditions.

3.3.2. Flatness calculation under negative wind load and precision reduction. It can be seen from the above calculation, based on the calculated data in Section 2.3, the sampling points in the visualization diagram after sampling are fairly evenly distributed. The least-square plane can be successfully fitted through the calculation of the regress function to calculate the flatness error, thus continuing to evaluate the flatness of the antenna array under other typical conditions. Since the sampled data are all from the matrix of the front, the plane fitting results are only different from the sampling points. Hence, the visualization diagram of the results is no longer listed below. According to the evaluation process in Section 3.2, the preliminary flatness evaluation results are calculated as shown in Table 3:

Table 3. Preliminary flatness evaluation.

Typical condition	Evaluation results
Lift 75°, 60° working state (speed 12r/min), the front face is subjected to 20m/s wind load	f_{75}=11.98mm, f_{60}=17.91mm
Lift 75°, 60° working state (speed 12r/min), the front is subjected to 20m/s wind load	f'_{75}=2.11mm, f'_{60}=4.96mm
Lifting 60° (speed 3r/min) and 75° (speed 6r/min), the front is subjected to a positive wind load of 25m/s	f''_{75}=14.41mm, f''_{60}=20.83mm
Lifting 60° (speed 3r/min) and 75° (speed 6r/min), the front is subjected to 25m/s wind load on the opposite side	f'''_{75}=6.12mm, f'''_{60}=11.09mm

It can be seen that the effect of wind on the flatness of the radar array is very small under the condition of a negative wind load. However, under frontal wind load conditions, the flatness errors calculated by this method differ greatly from the actual conditions, no matter in normal working conditions or in the condition of precision reduction due to high wind speed. Hence, a more accurate flatness evaluation scheme needs to be explored.

3.4. Probe into the improvement of flatness assessment data accuracy

According to the calculation results in Section 2.3, considering that the whole antenna car is a whole in the same coordinate system, the displacement difference of each element grid may be consistent with the constraint conditions in the finite element calculation. There are a large number of mesh cells at the subframe and leg of the antenna car, which will affect the coordinate values after the deformation of the front skeleton and then affect the accuracy of the calculation results of the fitting plane.

Secondly, the values of parameters *a*, *b*, and *c* in the flatness evaluation result of the aforementioned least square method are not exact solutions, but the optimal solution is fitted by the regress multiple regression function in MATLAB software, which is bound to have certain errors, so the value of this optimal solution is greatly affected by the extreme value points of the fitted plane. Considering the subframe and legs as a whole to analyze the antenna array, the result of the flatness calculation is not accurate.

The revised calculation scheme tries to hide the subframe and legs so as to eliminate the influence of frame deformation on plane flatness. With the two typical working conditions selected in Section 3.3, after the improvement scheme, the boundary adjustment boundary conditions and the deformation results are shown in Figure 8.

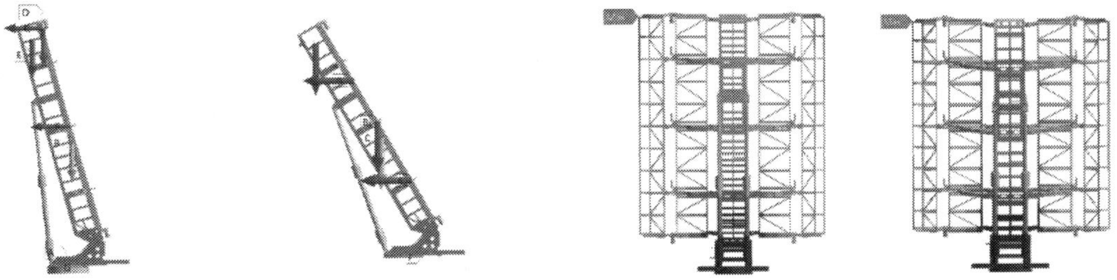

Figure 8. The boundary conditions and maximum deformation results of the improved scheme.

After stress deformation, the coordinate set of antenna plane measurement points is derived. The least square plane is fitted. Finally, the flatness error value of the radar vehicle antenna array are $f_{75}^{*} = 4.4211\text{mm}$, and $f_{60}^{*} = 6.5706\text{mm}$. Before the modification of the scheme, they are $f_{75} = 11.98\text{mm}$ and $f_{60} = 17.91\text{mm}$. The calculation accuracy of the flatness error value is higher, and the redundant calculation part is eliminated, the whole process is simpler and more efficient.

4. Conclusions

Firstly, the static finite element analysis of a new type of radar antenna vehicle is carried out. The results show that the safety factor of all operating conditions is above 1.75 except for some limited operating conditions, the safety factor of other operating conditions is greater than 2, and the overall performance is good. On this basis, the least square flatness evaluation method is applied to the antenna array. The flatness calculation method is discussed. The results show that the calculated flatness errors are all less than 10mm, which ensures the accuracy of error calculation and has better computational efficiency through computer-aided calculation. For large-scale assemblies such as radar antenna arrays and large-area mechanical structures requiring high accuracy of the antenna array, this paper screens the existing flatness evaluation methods and makes some improvements to the measured data extraction scheme, which can provide a certain reference for the subsequent flatness evaluation methods of such large assembly.

Reference

[1] Wang L, Gao D, Zhang L, and Zhou Y, (2023) Modal Analysis of High-Altitude and Heavy-Duty Lifting Mechanism for Vehicle-mounted Radar. Mechanical Engineering & Automation, 5: 81-83.

[2] Li M. (2019) Overview and prospect of antenna array accuracy assurance technology for ground phased array radar. Fire Control Radar Technology, 52: 27-32.

[3] Xia L, Yu J, Liu Z, Yang J and Song R. (2019) Research on wind protection capability of radar antenna array based on Finite Element Analysis. Journal of Ordnance Equipment Engineering, 41: 91-95.

[4] Yang Y, Wang S, Chen Z and Fan X. (2023) Application of Hybrid Jaya Algorithm in Flatness Error Evaluation. Modular Machine Tool & Automatic Manufacturing Technique, 11: 150-153.

[5] Xin L, Zhen B, and Jing W. (2017) Movement flatness error measurement based on an astigmatic method. Applied Optics, 56: 4347-4352.

[6] Chen C, Ding F, Jiang W. (2018) Overall design of a vehicle-mounted radar structure [J]. Machinery and Electronics.36: 56-59.

[7] Jia Z. (2019) Structural Design and Experimental Study of Radar Vehicle-leveling Platform considering Wind Load. Harbin Institute of Technology, Harbin.

[8] Fu C, Fu X. (2009) Form and position tolerance application technology Question and answer. China Machine Press, Beijing.

[9] Ma S, Wu H and Xue S. (2021) Flatness error evaluation program design based on MATLAB. Microcomputer Applications. 37: 77-80.

[10] Yang Y. (2022) Research on flatness error and its measurement uncertainty optimal evaluation methods. Jilin University, Jilin.
[11] Guan P, Yu T, and Wang W. (2011) Research on Flatness Error Measurement of Revolving body End-face. Mechanics and Materials, 44-47: 4002-4006.
[12] Xing M. (2017) Automatic Verification System for Flatness Error of Large Flat. Materials Science and Engineering, 224: 12-20.

Journal of Physics: Conference Series **2879** (2024) 012059 doi:10.1088/1742-6596/2879/1/012059

Research on the meteorological data transmission system of civil aircraft based on BeiDou satellite

Yijia Gao[1,*], Zhan Zhang[1], Jue Wang[1], Tieshuai Li[1] and Jia Chen[1]

[1]COMAC Beijing Aircraft Technology Research Institute, Beijing; 102211, China

*Corresponding author's e-mail: gaoyijia@comac.cc

Abstract: With the official opening of the Beidou-3 global satellite navigation system, Beidou has entered a new era of global services, extending its service area to the whole world. With the advantages of sufficient data volume and high data security, the BeiDou-3 short message satellite communication system can provide global communication services for airlines, realize two-way data transmission between air and land, and effectively reduce the operating costs of airline companies. Aiming at the demand for low-cost civil aircraft weather data transmission proposed by domestic and foreign airline companies, the current development of civil aviation weather information transmission and the current application status of BeiDou in civil aviation were studied, and the BeiDou satellite-based civil aircraft weather data transmission system was designed. The system architecture, detailed functions and data transmission workflow were analyzed, and the message coding structure was designed using bit-based coding technology. Finally, the feasibility of the scheme is verified through the data demand analysis of weather news and the onboard flight test on a domestic aircraft. The results show that t-he BeiDou short message function has the ability to realize the weather data transmission of a civilian aircraft.

1. Introduction

In recent years, under the trend of global warming, the frequency of El Nino and La Nina phenomena has increased, which leads to frequent anomalous and extreme weather in local areas, and weather conditions become unpredictable[1-2]. However, weather conditions are an essential factor affecting the safety of civil aviation flights.

Currently ACARS (Communication Addressing and Reporting System) is the main means for pilots to obtain weather information in the air. It is one of the mandatory avionics equipment installed in modern large and medium-sized civil airliners, playing a huge role in airline operations, air traffic control, and other communication needs. However, research shows that the ACARS system not only has the problem of high cost of use, but also has some technical problems in data transmission between air and ground during flight, such as the system does not support continuous information transmission well during aircraft turning or diving. There are also security problems such as data leakage, data spoofing, entity masquerading, denial of service attacks, etc. For a long time, the core technology of ACARS has been monopolized by a few countries due to the relative lag of air-ground data chain technology in China[3].

As a satellite navigation system with China's independent intellectual property rights, Beidou is one of the four major navigation systems in the world, following GPS of the United States, GALILEO of the European Union and GLONASS of Russia. With the gradual realization of global services of the BeiDou satellite navigation system, the BeiDou short message function has been more and more widely used in various fields such as transportation, land resources, satellite rescue, maritime communication, etc. However, due to the special characteristics of BeiDou civil aviation applications, the related application results are not as significant as those in transportation and other fields[4].

With the delivery of domestic passenger aircraft to domestic and foreign airlines, the use of the Beidou short message function to realize real-time transmission of air weather data has become an increasingly urgent need for domestic and foreign airlines.

2. Development status

2.1. Current status of meteorological data acquisition by civil aviation aircraft

Currently, there are two well-established ways of receiving weather data in the civil aviation fleet:

•Before take-off, pilots can access weather information along the route and at the destination airport through the weather app in the accompanying electronic flight bag (EFB), which can be used to develop a suitable flight plan;

•During the flight, the pilot can obtain the latest weather information through the data chain and voice communication system provided by the aviation weather service provider. The two main ways include the following: automatically receiving weather information broadcasted by the weather center on a regular basis through MCDU to obtain weather information about the airport; the pilot sends a request to the ground to obtain weather data of a certain airport, and the flight division receives the request and sends the latest weather data to the aircraft.

In the second case, the pilot initiates a weather request through the MCDU panel due to aircraft preparation for landing, weather changes, etc. This request is sent to the ACARS ground operator (DSP) via the CMU in the order of VHF/SATCOM/HF by selecting the applicable link and forwarding the request to the flight division via routing. The ground system of the flight division grabs the relevant weather data and packages it after receiving the request, then returns it to the aircraft via the ACARS network via the same data link and displays the weather information via the MCDU or MFD, as shown in Figure 1 in detail.

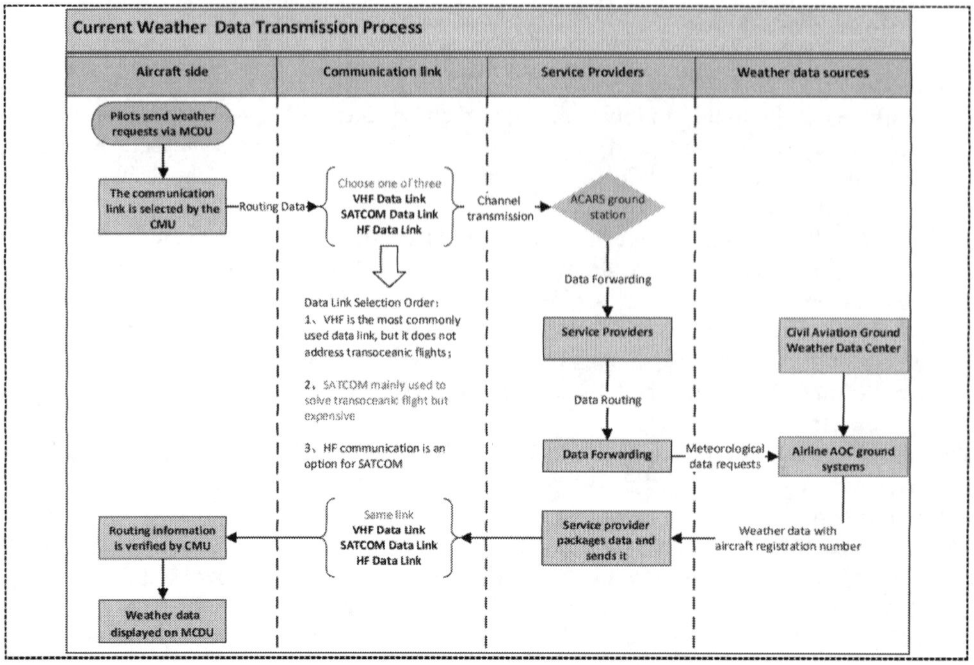

Figure 1. The current process of obtaining weather data for civil aircraft.

The current method of acquiring meteorological data for civilian aircraft has the following drawbacks:

•Information update is not timely: Traditional weather reports mainly rely on data from ground weather stations and flight dispatch centers. But weather conditions change all the time, and these data may be updated slowly, which results in this approach not reflecting the actual situation in a timely manner, bringing certain risks to the flight;

•Imperfect information transmission: The weather transmission method of radio communication is susceptible to weather, equipment and many other factors, which may result in untimely or interrupted information transmission, bringing potential risks to flight safety.

To overcome these shortcomings, the civil aviation industry is gradually adopting more advanced technologies and means, such as Beidou, to improve the update speed, accuracy, and delivery efficiency of weather information to provide pilots with accurate and timely weather information and ensure flight safety[5-7].

2.2. Current status of Beidou applications in civil aviation

In July 2020, the Beidou-3 global satellite navigation system officially opened its services. Beidou stepped into a new era of global services, with the service area extended to the whole world, and the navigation performance and communication capability improved comprehensively. The Beidou-3 short message satellite communication system has global communication capability, sufficient data volume and high data security. It can realize the application of aircraft operation monitoring and control, which can effectively reduce the operating cost of airline companies, enhance the independent controllability of China's airborne equipment, and provide global communication services for airlines.

China has vigorously promoted the application of the Beidou system to replace foreign satellite navigation systems in civil aviation and has mentioned the promotion of the localization of Beidou-related equipment in many policies, among other issues. At present, the use of Beidou's short message communication function to achieve tracking and surveillance of civil aviation aircraft has achieved certain results. 2018, the feasibility of the Beidou system short message communication function in civil aviation applications was verified through a piggyback test carried out on a certain type of domestic aircraft. In December 2019, CAAC issued a Supplemental Type Certificate (STC) for the retrofitting of the Beidou positioning tracking system on Boeing 737-800 aircraft[8-9].

The significance of the above achievements lies in implementing the major national strategy of strengthening the country with science and technology and ensuring the operational safety of the civil aviation fleet. They truly realize the application of the Beidou satellite system in civil aviation. However, from the customer's point of view, the Beidou system is expected to be more effective in terms of revenue and cost for civil aviation operations.

From Table 1, it can be seen that compared to ACARS, Beidou has the advantages of low cost and high data security in the application scenario of air-ground data transmission.

Table 1. Comparison between Beidou system and ACARS.

System	Function	Advantages
ACARS	Mainly through a VHF based bidirectional airborne data communication system, various information exchanges are achieved. Mainly through a VHF based bidirectional airborne data communication system, various information exchanges are achieved. It can also be transmitted through satellite links, but airlines say the price is expensive	1. The system can transmit large amounts of data in real-time without interruption. 2. The system data link is bidirectional communication between ground and air, as well as between air and ground. 3. The ground network is well deployed, so it is widely used.
Beidou	The system combines RNSS and RDSS business models, providing 24/7 positioning and message communication services.	1. The system has good confidentiality and can perform source encryption.

2. The system has a wide coverage area and is not limited by ground base stations.
3. The system has low requirements for equipment, low transmission power, and low equipment prices.
4. The system can synchronously transmit autonomous positioning information.
5. The system can provide two-way emergency communication services.
6. The annual cost of using Beidou terminals is low (in the thousands of yuan range).
7. Convenient networking, with command-type user machines as the core, quickly forming a communication network.

3. System architecture and functions

3.1. System architecture

This system establishes a star-air-ground communication data link through the unique automatic and all-weather transmission of short messages by the BeiDou satellite system. It is able to realize fast communication and airborne flight services, forming an independent meteorological data transmission system. Later, it is also able to carry out the transmission of voice, pictures and other information according to the needs of airlines.

The system includes two parts: the airborne data transmission link and the ground data transmission link.

The airborne link includes the Beidou short message antenna, Beidou airborne transceiver, and Beidou flight display terminal. The Beidou airborne transceiver is connected to the Beidou antenna through RF cables to receive Beidou satellite information and send Beidou short message information. It is connected to the aircraft avionics system through the aircraft's onboard ARINC data bus, and it unidirectionally extracts key data such as aircraft flights. It is connected to the display terminal through the RS422 data bus to transmit and display weather message data, receiving data requests and other information from pilots. The Beidou flight display terminal controls and displays the Beidou airborne transceiver through the RS422 data bus, which is also the channel for human-machine interaction between the system and the pilot.

The ground link consists of ground Beidou antennas, Beidou command aircraft, and ground data service centers. The ground system is mainly used to receive data requests from airborne terminals and send weather data packets to the aircraft. The overall system architecture is shown in Figure 2. This article will mainly focus on the analysis and research of airborne systems.

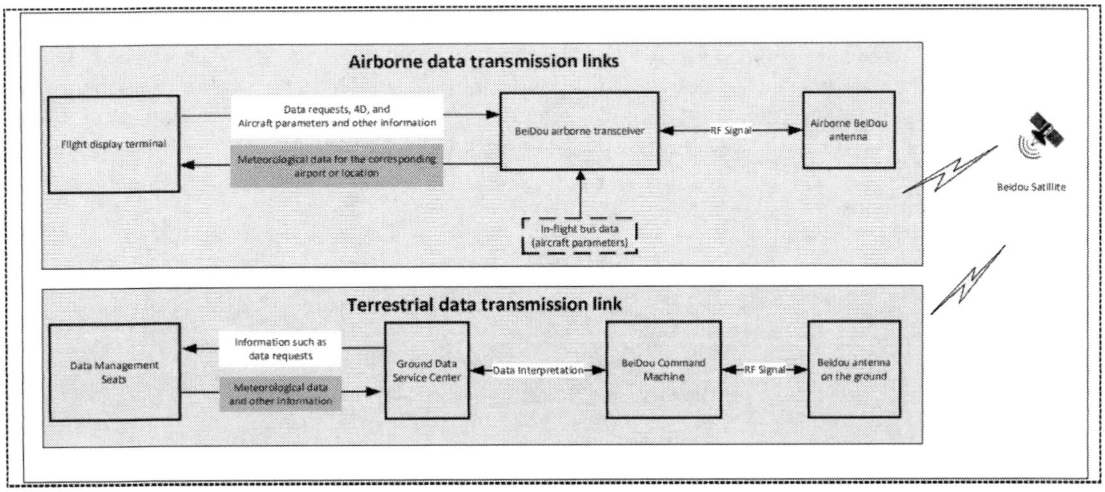

Figure 2. The general architecture of the weather information transmission system based on the BeiDou system.

3.2. System functions

3.2.1. Airborne side. BeiDou Airborne Transceiver. The BeiDou airborne transceiver is capable of sending and receiving BeiDou short message L and S frequency signals based on the BeiDou RDSS short message function, establishing satellite communication between aircraft and airline ground data systems, and realizing the acquisition of real-time or request-based meteorological information. The equipment can also realize the cross-link with the aircraft bus, the acquired aircraft heading, flight number, fuel information, alarm information and other aircraft body information, and automatically transmit the meteorological data request and aircraft body information down to the ground through the short message link. In addition, the device is equipped with various interfaces such as ARINC429 input and output interface, discrete signal interface and Ethernet to support data output to the flight display terminal and export of stored data for extended applications.

Flight Display Terminal. The flight display terminal is an independent display device independent of the aircraft cockpit display system. The main functions include: sending weather data requests by means of pre-set messages, reporting weather conditions independently by means of short message messages, establishing data channels between the display terminal and short message devices, customizing pages for receiving and displaying weather data sent from the ground, etc.

3.2.2. Ground side. The ground end of the system mainly includes antennas, ground commanders, and the ground data service center of the airline. The weather requests, 4D information, aircraft parameters and other related data sent by the aircraft are transmitted to the ground through the Beidou network. Then it is decoded by the BeiDou command plane and transmitted to the ground data service center of the airline. After collecting, processing, analyzing and fusing the data, the ground data service center transmits the corresponding information to each display terminal seat, including data management, system maintenance and data operation seats. Finally, after analysis and decision making, the weather data information is sent to the corresponding aircraft via the Beidou network.

3.3. System workflow

The system is based on Beidou's short message data link for information transfer, including two kinds of data information: short message and weather report data packet. The pilot sends a data request to the ground through a short message, and the ground sends the weather report data packet to the aircraft. Figure 3 shows the workflow of the system to complete one data transfer. Firstly, the pilot sends a weather data request through the flight display terminal during the flight. The message is sent to the airline AOC via the Beidou antenna via the Beidou short message data link and then decoded by the ground command plane and sent to the AOC display terminal seat. The management personnel then packages the weather information containing the aircraft registration number and sends it back to the Beidou transceiver via the Beidou short message data link, which decodes and processes it and sends it to the flight display terminal for the pilot to view. This completes the complete process of weather information transmission.

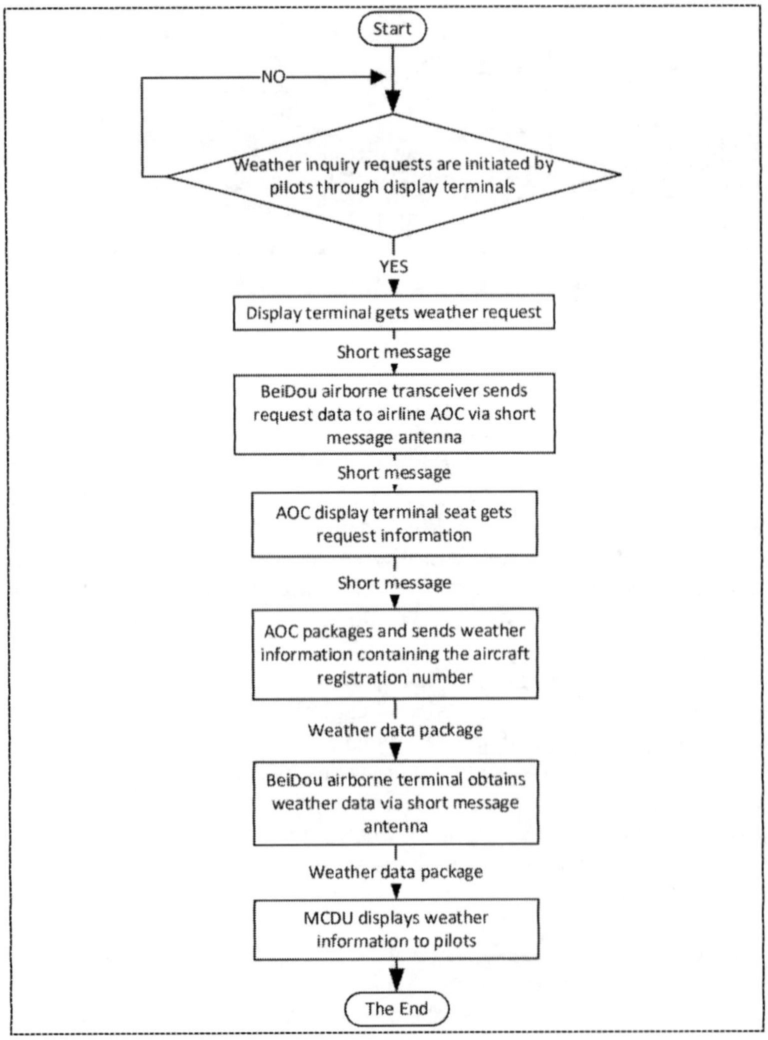

Figure 3. System workflow diagram.

3.4. Data encoding

3.4.1. Bit-based coding techniques. The weather reports currently provided by airlines usually include three formats: routine weather report (METAR), special weather report (SPECI), and weather forecast (TAF). Both METAR and SPECI include the main weather information such as temperature, dew point temperature, wind direction and speed, cloud information, visibility and modified pressure. In addition to this, METAR provides information on precipitation, thunder and other special weather conditions on demand, while SPECI is mandatory for the release of special weather conditions.

According to the current demand for weather information transmission, taking into account the equal applicability of global short message and regional short messages, making full use of the message length space and improving the payload, the system adopts bit-based short message coding technology for the transmission of weather messages. In addition, considering the problem of possible error codes in the transmission process, a data verification mechanism is introduced to guarantee the accuracy of data.

3.4.2. Message contents. Based on the Civil Aviation Meteorological Ground Observation Specification, the data message coding scheme of this system is developed. The message mainly includes three parts: the first part is basic meteorological information, the second part is supplementary meteorological information, and the third part is the verification code.

Basic meteorological information includes place name, time, wind direction and speed, runway visual range, good weather, temperature/dew point, and corrected air pressure.

Supplementary meteorological information includes recent weather, visibility, clouds, precipitation, thunderstorms, hail, smoke, sandstorms, and other special weather conditions.

3.4.3. Basic structure of the message. The coding design was carried out for the basic information and supplementary meteorological information, and the coding structure is shown in Table 2.

Table 2. Meteorological data message transmission structure (Unit: bit).

Type	Content	Length
Basic meteorological information	Message type	8
	Airport	32
	Time	16
	Wind direction and speed	58
	Runway visual range	50
	Current Weather	17
	Cloud situation	28
	Temperature/Dew Point	18
	Correct air pressure	16
	visibility	32
Supplementary meteorological information	Recent Weather	52~260
Data Check	Check Code	16

4. Program feasibility assessment and validation

To fully validate the feasibility of this system solution, technical evaluation and flight verification were conducted.

4.1. Feasibility assessment

As a characteristic application of the BeiDou satellite navigation system, in terms of single data transmission capacity, the regional communication capacity of BeiDou-3 short messages reaches 14,000 bits (1,000 Chinese characters) per time, which can transmit both text, voice and pictures. The global communication service can transmit 560 bits (40 Chinese characters) of data per time. Its regional communication service can provide regional short message service (RSMC) to China and surrounding areas (75 degrees to 135 degrees east and 10 degrees to 55 degrees north) through the L-band and S-band signals of the three GEO satellites in the nominal space constellation of the No. 3 system. The Beidou satellite observation chart is shown in Figure 4, which shows the number of satellites that can be seen by the Beidou satellite navigation system at any location around the world. Different colors in the figure indicate different values (yellow is the largest, and white is the smallest). As can be seen in the figure, the Asia-Pacific and Southeast Asia regions are both yellow, with the highest number of visible satellites.

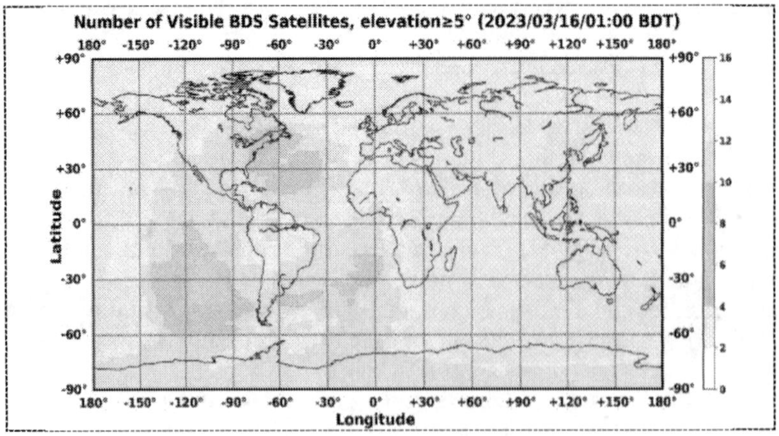

Figure 4. BeiDou satellite observation chart.

As can be seen from the basic structure of the message, the weather message data capacity requirement under normal circumstances is about 59 to 127 bytes, so this solution is completely feasible in areas within the regional communication service range of the Beidou-3 system. It is even possible to develop customized data transmission protocols for airlines, e.g., meteorological cloud maps, etc. However, for areas not in the scope of this service, global short messages need to be used for weather message data transmission. In this way, the data transmission is only 70 bytes at a time, so the data will be sent in packets as needed.

4.2. Flight verification

4.2.1. Test configurations. To preliminarily verify the system's functionality, a short message flight verification test based on the Beidou-2 system was conducted in China, relying on a certain aircraft platform, laying the foundation for the global application of this system. By sending short message data to the ground during the test flight, the basic functions and performance of the system and the stability of the Beidou short message communication link are verified. The system architecture diagram is shown in Figure 5.

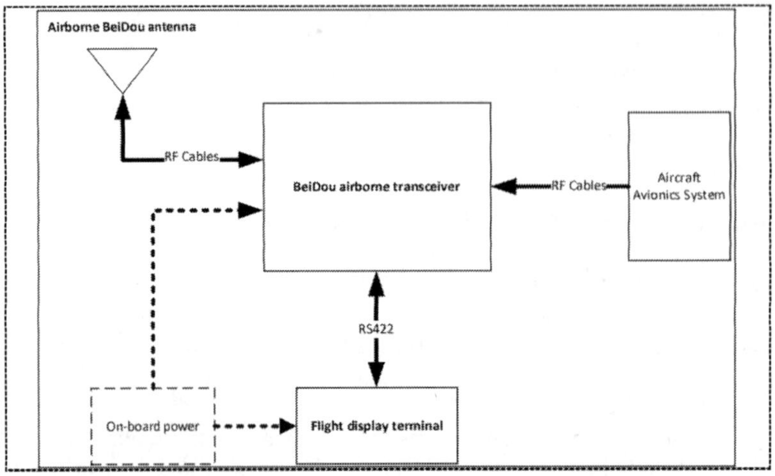

Figure 5. System flight validation architecture diagram.

4.2.2. Test methods. •Straight-line taxiing

1) Round-trip taxiing of the aircraft along the runway centerline;

2) Beidou short message transmission and reception during taxiing; verify that the data transmission is correct;

3) Repeat the straight-line taxiing action at least 5 times.

•Great circle flight

1) The aircraft will fly in a large circle with a radius of 20 nautical miles at an altitude of 10,000ft, with the center of the airport runway as the center of the circle;

2) Sending BeiDou short message messages at 60 seconds intervals during the flight to verify the correctness of data transmission;

3) Repeat the great circle flight at least 3 times.

•Constant altitude flight

1) The aircraft will fly along the centerline of the runway at an altitude of 2000ft, 20nm from the entrance of the runway, and end at 2.5nm on the other side of the runway to complete the constant altitude flight;

2) Sending BeiDou short messages at 60-second intervals during the flight to verify that the data transmission is correct;

3) Repeat the constant altitude flight at least 2 times.

4.2.3. Test conclusion. During the straight-line taxiing, 10,000ft great circle flight and 2,000ft constant altitude test flight, the onboard and ground systems worked normally throughout. The system was able to realize the transmission of necessary message information through Beidou short messages, and the transmission success rate reached more than 95%.

On the one hand, Beidou-3 has higher performance and more stable signal advantages compared to Beidou-2. On the other hand, this system is independent of other onboard avionics systems of the aircraft and will not cause significant changes to the installation of the aircraft. Therefore, based on the above analysis and verification, the feasibility of implementing and applying this system is preliminarily evaluated.

5. Summary and outlook

The Beidou-3 short message has the advantages of integrated communication and navigation, a military-civilian separation system, and secure communication. It has broad application potential in the field of civil aviation's comprehensive applications, where the demand for interconnection is increasingly evident. Typical scenarios with great application prospects include EFB data transmission, air-ground command interaction, emergency situation reporting, air situation mutual transmission, and so on. However, the application of Beidou in the civil aviation field is a long-term strategy, especially in terms of the direct cost-effectiveness of civil aviation operations and Beidou civil aviation international industrial standards. It also requires cooperation and research from host units, industry regulatory units, industry, research institutes, and universities. More practical flight tests need to be carried out to analyze a large amount of actual data, verify key system parameters, and rely on the EUROCAE Beidou dedicated working group to synchronously promote the preparation of relevant international standards, achieving the widespread application of Beidou in the entire civil aviation field.

References

[1] Kauristie, K., Andries, J., Beck, P., Berdermann, J., Berghmans, D., Cesaroni, C., ... & Österberg, K. (2021). Space weather services for civil aviation—Challenges and solutions. Remote Sensing, 13(18), 3685.

[2] Ryley, T., Baumeister, S., & Coulter, L. (2020). Climate change influences on aviation: A literature review. Transport Policy, 92, 55-64.

[3] Nyberg, D. (2021). Development of weather services for aviation personnel.

[4] Li Xin. (2021). Application Prospects and Implementation Strategies of Beidou System in Civil Aviation. Satellite Applications (12), 58-62(in Chinese).
[5] Dai Xuanxuan (2015). Implementation of Message Transmission Function in Civil Aviation Meteorological Database System. Technology Wind (17), 86. doi: 10.19392/j.cnki. 1671-7341. 2015.17.076 (in Chinese).
[6] Wang Fengyun and Zhang Xiaobing (2021). Progress and Innovative Development of Civil Aviation Meteorological Technology. Progress in Meteorological Technology (03), 189-193 (in Chinese).
[7] Chenghang Media (2022). The first domestically produced civilian aircraft equipped with the Beidou system has been officially put into operation. (in Chinese)
[8] He Yuncheng (2019). Application Status and Development Prospects of Beidou System in China's Civil Aviation Industry. Satellite Applications (09), 19-24 (in Chinese).
[9] Wang, J., Zhang, Z., Gao, Y., Guo, J., & Hu, Y. (2022). Application and Promotion Strategy of Beidou Airborne System in Civil Aviation Transport Aircraft. In Proceedings of the 5th China Aeronautical Science and Technology Conference (pp. 1009-1014). Springer Singapore(in Chinese).

Journal of Physics: Conference Series **2879** (2024) 012060 doi:10.1088/1742-6596/2879/1/012060

Three-dimensional configuration tip rotor dynamics analysis

Yan Zhu*, Yuhui Lu, Longtao Xing, Feng Qian, Feng Fan and Zhizhuang Feng

Science and Technology on Rotorcraft Aeromechanics Laboratory, CHRDI, Jingdezhen, China

*e-mail: everyanyan@126.com

Abstract. In order to analyze and study the dynamic characteristics of the 3D blade tip rotor, an aeroelastic dynamic model for a complex 3D blade tip rotor is established based on mechanics principles. The theory of medium deformation beam or large deformation beam is adopted as the blade structural dynamic model, and the quasi-steady and unsteady aerodynamic models (ONERA model) are adopted as the aerodynamic model. The rotor dynamics, aeroelastic stability and rotor loads of three-dimensional blade tip and rectangular blade tip are compared by examples. The analysis shows that both the forward and backward sweep angles have a certain effect on the rotor blade frequency, but the downward dihedral angle has little effect. The 3D tip configuration has the greatest influence on torsion mode frequency, followed by lagging mode frequency, which has little impact on flapping frequency. Reasonable design of forward and backward sweep angles and downward dihedral angles of 3D tip configuration can effectively reduce rotor loads, increase damping margin for aeroelastic stability, and improve stability.

1. Introduction

The blade tip plays an important role in improving the flight performance and maneuverability of the helicopter. Research has shown that swept rotor tips can reduce rotor hub load, vibration, and aerodynamic noise, and improve helicopter flight stability [1~2]. With the wide application of various new materials on helicopter rotors, it is possible to improve the performance of helicopter rotors by changing the shape of the blade tip. Advanced blade tip configuration design has been a hot topic at home and abroad in recent years. Figure 1 shows advanced blade tips such as BERP blade tip and blue-edge blade tip, which all include the characteristics of blade tip sweepback [3].

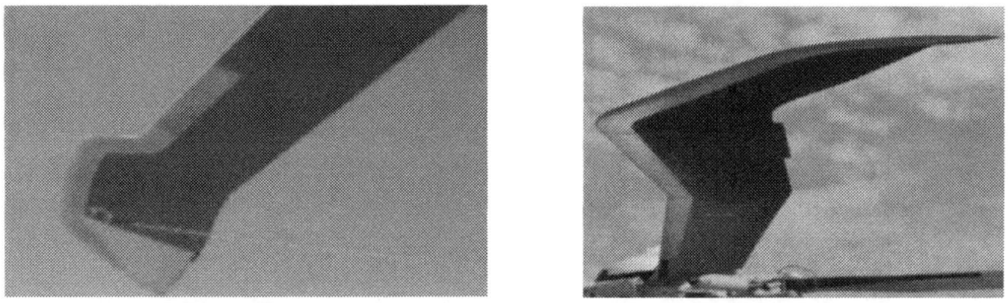

(a) BERP-tip (b) Blue edge tip

Figure 1. Outline drawing of several foreign advanced blade tips.

Changing the tip shape changes the aerodynamic forces in the tip section, thus increasing the coupling between blade bending and torsional motions and having an important effect on rotor aeroelastic performance. In the early research, a simple aeroelastic dynamics model of the blade is mostly used. When modeling the blade tip, it is approximated by deviating its section gravity center from the elastic axis of the section. Wilbur and Sekula [4] analyzed the influence of tip shape on blade torsional aeroelastic response using CAMRAD II, and found that it was mainly the aerodynamic change caused by tip shape change that changed the blade torsional aeroelastic response. The effect of tip sweepback on blade torsional aeroelastic response was far greater than that caused by tip downturning and sharpening. Kumar and Venkatesan [5] established an aerodynamic model of the anti-rotor swept down by blade tip based on Peters -He dynamic inflow theory and modified ONERA dynamic stall theory, obtained the natural frequency and vibration mode of the blade in this configuration by solving the linear undamped structural dynamic equation, and further analyzed that blade tip sweepback will increase the lag bending moment at the blade root. Yang and Deng [6] and Deng [7] made a corresponding study on the aeroelastic stability of swept-back tip rotors, and the results showed that sweep-back increased the first torsion mode frequency of the blade but decreased the first lag damping. Based on the parameters of the BO-105 rotor blade model, Tian [8] carried out the numerical analysis and research on the effect of tip sweepback on rotor dynamic characteristics, dynamic loads and aeroelastic stability. The results showed that tip sweepback had a great impact on the first-order torsion mode frequency of the blade, and the first-order lag damping of the blade decreased with the increase of sweep angle. More sophisticated aeroelastic dynamics models are also needed to fully understand the effect of tip shape on blade dynamics. Chi et al. [9] verified the accuracy of the three-dimensional structural dynamics model of composite blades through experiments, including cross-sectional structural characteristics, non-rotating/rotating inherent characteristics, and rotational strain under vacuum conditions. Huang and Chi [10] studied the aeroelastic characteristics of hovering rotors with a negative twist and swept back tips, and analyzed the effects of negative twist and swept back tips on stress and strain.

In this paper, the rotor dynamics system is modeled based on the mechanics principle for complex three-dimensional blade tip configuration. The finite element model of blade structural dynamics is established by using medium deformation beam or large deformation beam theory, and the aerodynamic model of the rotor is established by using the quasi-steady and unsteady aerodynamic model (ONERA model). The dynamic equations of the rotor system are derived by applying Hamilton's principle and the solution method is given. The rotor dynamics characteristics, aeroelastic stability and rotor loading of the three-dimensional blade tip rotor are compared with those of the rectangular blade tip through numerical example analysis.

2. Three-dimensional blade tip rotor dynamics model

In this paper, combined with the requirements of rotor dynamics analysis for complex 3D rotor configuration, the rotor dynamics system is modeled based on mechanics principles. The motion of the rotor is described in the established coordinate system, and the structural dynamic finite element model of the isolated rotor blade is established. The quasi-steady and unsteady aerodynamic model (ONERA model) is used to model the aerodynamic force. When establishing the dynamic model of the blade tip, the virtual kinetic energy and external force virtual work terms of the blade tip are derived according to the coordinate system relationship of the blade tip to obtain the dynamic equation of the blade tip. Finally, the element matrix of the blade tip is obtained by finite element discretization of the dynamic equation of the blade tip, and the blade tip elements are assembled on the blade using the displacement relationship between the connection points of the blade tip elements and the blade elements. According to the rotor blade dynamics and aerodynamic model, the kinetic energy, potential energy and external load virtual work of the rotor blade are expressed by hub motion and blade motion.

When establishing a dynamic model of a complex 3D blade tip rotor, it is necessary to describe the motion of any point on the 3D blade tip and each part of the blade by establishing multiple coordinate systems, as shown in Figure 2, including inertial coordinate system (X_g, Y_g, Z_g), airframe coordinate system (X_f, Y_f, Z_f), hub coordinate system (X_H, Y_H, Z_H), blade rotation coordinate system (X_k, Y_k, Z_k) and pre-deformation coordinate system (X_s, Y_s, Z_s), blade-deformed coordinate system (X_b, Y_b, Z_b) and tip-deformed local coordinate system (X_i, Y_i, Z_i).

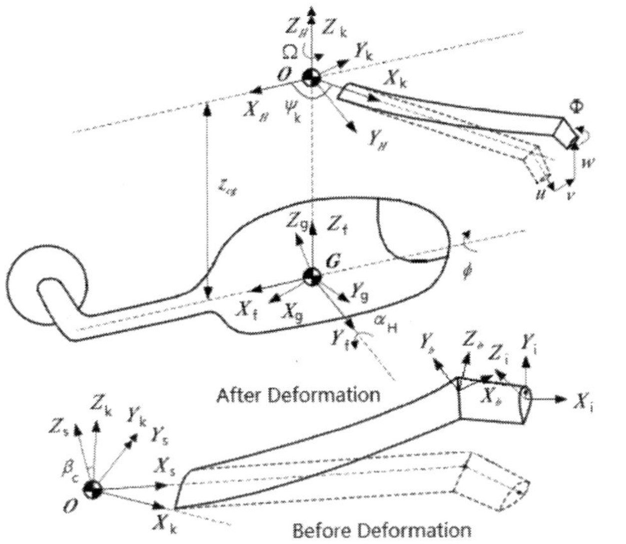

Figure 2. Coordinate system of before and after blade tip deformation.

Different shapes of the blade tip will affect the kinetic energy and aerodynamic virtual work of the blade. The kinetic energy of the blade tip can be obtained by obtaining the position before and after deformation of any profile of the blade tip section, and the velocity at any point of the blade tip section. After converting the coordinates to the inertial coordinate system, the expression of the kinetic energy of the blade tip is obtained. The difference between Hodges nonlinear large-deformed beam and a moderately deformed beam is that the additional increment of tension and torsion caused by bending is added to the tensile and torsional displacement of a large-deformed beam compared with that of the moderately variable beam. The aerodynamic virtual work of the blade tip can be derived by deriving the inflow velocity of each profile of the blade tip and the dynamic aerodynamic characteristic coefficient of the airfoil according to the ONERA unsteady model, so as to derive the aerodynamic load of any airfoil profile of the blade tip. After coordinate conversion to the coordinate system before blade tip deformation, the aerodynamic virtual work of the blade tip part is obtained.

Total rotor deformation potential energy:

$$\delta U = \sum_{r=1}^{N_b} \delta U_b = \sum_{r=1}^{N_b} \frac{1}{2}\int_0^R \iint_A (\sigma_{x_s x_s}\varepsilon_{x_s x_s} + \sigma_{x_s\eta}\varepsilon_{x_s\eta} + \sigma_{x_s\zeta_s}\varepsilon_{x_s\zeta})d\eta d\zeta dr \quad (1)$$

Total rotor kinetic energy:

$$\delta T = \sum_{r=1}^{N_b} \delta T_b = \sum_{r=1}^{N_b} \int_0^R \iint_A \rho\vec{V} \bullet \vec{V} d\eta d\zeta dr \quad (2)$$

Virtual work of total external force of rotor:

$$\delta W_b = \int_0^R (L_u^A\,\delta u + L_v^A\,\delta v + L_w^A\,\delta w + L_{\hat{\phi}}^A\,\delta\hat{\phi})\,dx \quad (3)$$

Rotor dynamics must fully embody the strong coupling and complex nonlinear unsteady aerodynamic characteristics between geometry, aerodynamics, structural elasticity and inertia, which is essentially an aeroelastic problem of the rotor. Based on mechanical modeling, the rotor system dynamics equation is derived from Hamilton's principle. According to the rotor blade dynamics and aerodynamic models, the kinetic energy, potential energy and external load virtual work of the rotor blade are expressed by hub motion [X_H, Y_H, Z_H, ϕ_{XH}, ϕ_{YH}, ϕ_{ZH}] and blade motion [u, v, w, ϕ]. The variational equation of Hamilton's principle for rotor dynamic systems is:

$$\int_{t_1}^{t_2} (\delta U - \delta T - \delta W)dt = 0 \tag{4}$$

where U is the strain energy of an elastic beam established from a strain-displacement field, T is the kinetic energy of the beam established from a motion field, and W is the work of external forces acting on the beam, including aerodynamic and other constrained external forces.

The above rotor aeroelastic dynamics equation is a typical high-dimensional nonlinear periodic motion equation. For this kind of equation, the commonly used numerical solutions include the direct integration method, harmonic balance method and time finite element method.

3. Example analysis

The main parameters of the rotor system are shown in Table 1. The blade tip section adopts a three-dimensional shape (with forward and backward swept down diagonal) and rectangular blade tip respectively, and the profile parameters of the straight section of the blade are the same. The dynamic characteristics, aeroelastic stability and loads of the two types of blade tip rotors are analyzed, and the dynamic performance of the two types of blade tip configurations is compared.

Table 1. Main parameters of rotor.

Contents	Dimensionless
Distance from blade vertical hinge to hub center	0.044
Distance from mass center of one blade to vertical hinge	0.338
Static mass moment of one blade around vertical hinge	0.337
Rotational inertia of a blade around the vertical hinge	0.187
Vertical hinge to damper distance	0.027

3.1. Comparative analysis of rotor dynamic characteristics

Comparative analysis of the frequencies and vibration modes of rotor systems with complex 3D blade tips and rectangular blade tips at different speed states in a vacuum is carried out. The comparison of rotor resonance diagrams is shown in Figure 3. Table 2 shows the frequency values of each order at the rated speed.

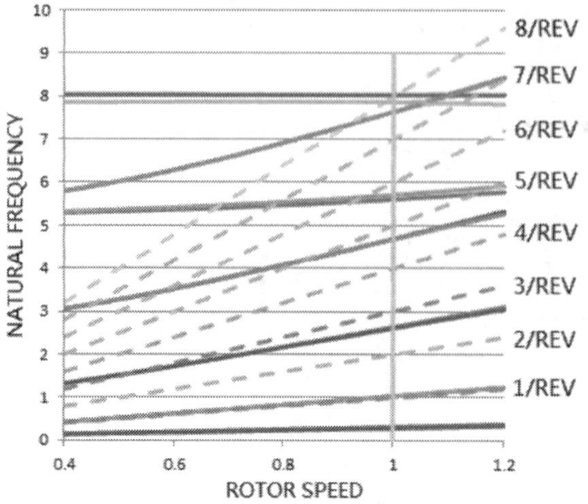

Figure 3. Comparison of rotor resonance diagrams.

Table 2. Frequency comparison at rated speed.

Tip configuration	1 Lag	1 Flap	2 Flap	3 Flap	2 Lag	1 Torsion
Rectangular	0.2843	1.038	2.669	4.773	5.992	6.494
Forward sweep	0.2845	1.037	2.671	4.827	5.963	6.556
Sweepback	0.2847	1.038	2.666	4.777	5.998	6.425
Forward and backward sweep	0.2219	1.018	2.563	4.808	5.959	6.279
Lower reversal	0.2807	1.037	2.667	4.777	5.985	6.492
Sweep back and down	0.2842	1.037	2.667	4.788	5.990	6.416
3D	0.2566	1.019	2.560	4.693	5.980	7.185

By comparing the frequency values of each order at rated speed between rectangular blade tip and different configurations, it can be seen that compared with rectangular blade tip, the first-order lag mode frequency of three-dimensional blade tip configuration decreases slightly, the first-order torsion increases obviously, and other frequencies change little. By studying the influence of forward sweep angle, backward sweep angle and downward dihedral angle on rotor frequency, it is concluded that when the blade tip section sweeps forward or backward, the aerodynamic center of the blade will move forward or backward, resulting in lowering or raising the nose of the blade and changing the angle of attack, thus causing the frequency to change. In the case of a single band lower dihedral angle, the lower dihedral angle has little influence on frequency. Therefore, the blade tip with complex 3D configuration is a combination of forward and backward sweep angles and downward diagonal angles. The forward and backward sweep angles and upward and downward diagonal angles have different degrees of influence on the swing torsional motion frequency of the rotor respectively. The final impact depends on the combination of forward and backward sweep angles and upward and downward diagonal angles. Therefore, it is necessary to comprehensively design all directions of the 3D blade tip to obtain a rotor with reasonable dynamic characteristic performance. The change in blade tip vibration mode can also be seen from the vibration mode comparison in Figure 4.

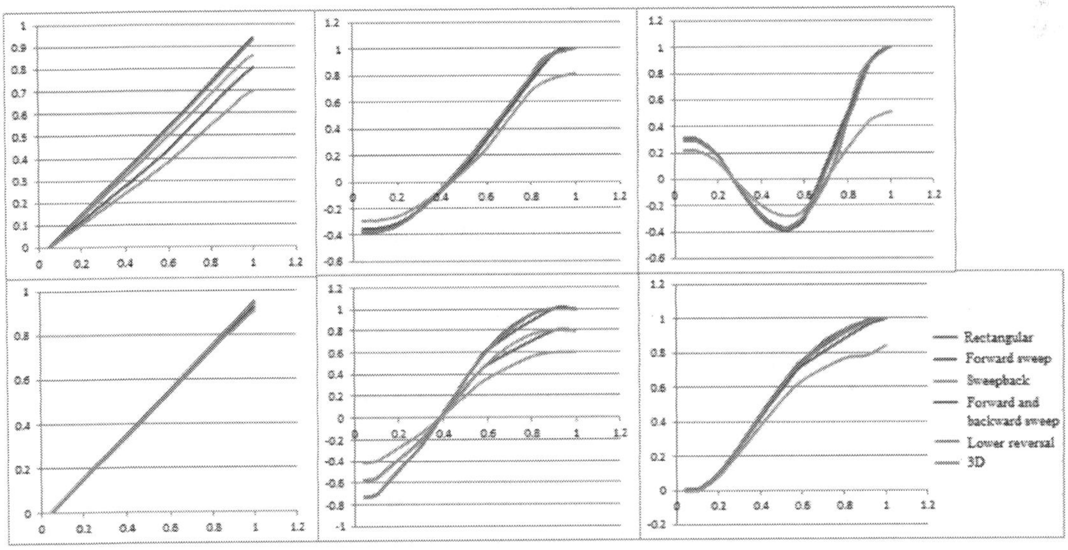

Figure 4. Comparison of vibration modes.

Journal of Physics: Conference Series **2879** (2024) 012060 doi:10.1088/1742-6596/2879/1/012060

3.2. Comparative analysis of rotor aeroelastic stability

By comparing and analyzing the aeroelastic stability of rotors at different speeds in air, Figure 5 and Figure 6 show the frequency curve and damping curve of the rotor system respectively. Figure 7 is a comparison curve of the first-order lag mode damping. It can be seen from the damping curve that the rotor aeroelastic stability of complex 3D blade tip configuration is slightly larger than that of matrix blade tip, so reasonable design of forward and backward sweep and downward dihedral angles of 3D blade tip can obtain better blade configuration dynamic performance compared with rectangular blade.

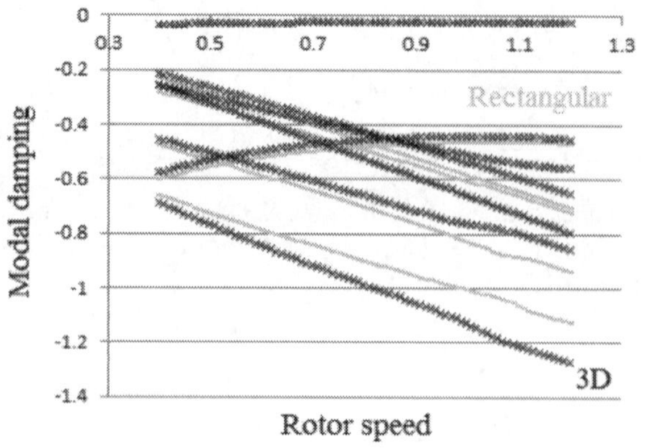

Figure 5. Damping curve.

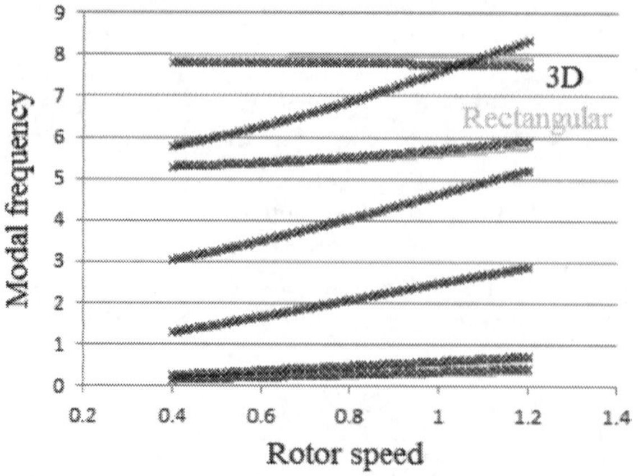

Figure 6. Frequency curve.

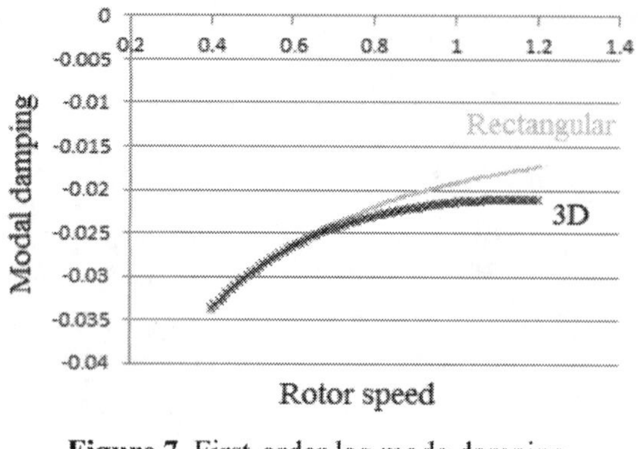

Figure 7. First-order lag mode damping.

Table 3 shows the comparison values of frequencies at rated speed in air. It can be seen from the table that compared with rectangular tips, the torsion frequency of complex three-dimensional blade tip configuration has a greater influence and increases significantly, while the first and second orders of lagging frequencies decrease slightly and the flapping frequency changes little.

Table 3. Frequency comparison at rated speed in air.

Tip configuration	1 Lag	1 Flap	2 Flap	3 Flap	2 Lag	1 Torsion
Rectangular	0.259	0.834	2.522	4.512	5.690	5.839
Three-dimensional	0.2475	0.8164	2.6090	4.6857	5.5127	6.259

3.3. Comparative analysis of rotor load

Comparative analysis of rotor hub load, pull rod load acting on the recliner and rotor load synthesized by blade root load, shimmy damper and pull rod load shows that the loads in hover T=G state are shown in Table 4.

Table 4. Comparison of rotor loads.

	Blade tip	Fx	Fy	Fz	Mx	My	Mz
Hub Loads	Rectangular	0.0086	0.0080	0.1740	0.0344	0.0344	0.0108
(NbΩ)	3D	0.0124	0.0093	0.1175	0.0297	0.0297	0.0090
Hub load (2	Rectangular	0.0026	0.0025	0.1273	0.0127	0.0127	0.0068
NbΩ)	3D	0.0013	0.0019	0.0128	0.0024	0.0024	0.0032
Rod load	Rectangular	--	--	0.0012	0.2355	0.0159	--
(NbΩ)	3D	--	--	0.0230	0.0459	0.0654	--
Rotor	Rectangular	0.0052	0.0552	0.1345	0.1242	0.1375	0.1208
Loads	3D	0.0012	0.0235	0.0105	0.0731	0.0387	0.0549

It can be seen from Table 4 that the hub loads and rotor loads of the 3D tip configuration in this state are much less than those of the rectangular tip, indicating that the rotor with a reasonable design of forward-backward sweep angle and downward dihedral angle is better than that with a rectangular tip.

4. Conclusions

In this paper, the aeroelastic dynamics model of a three-dimensional blade tip rotor is established to study the rotor dynamic characteristics, aeroelastic stability and rotor load. The following conclusions are obtained through analysis by comparing three-dimensional blade tip configuration with rectangular blade tip configuration:

- Both forward and backward sweep angles have a certain effect on the rotor blade frequency, but the downward dihedral angle has little effect. The three-dimensional configuration tip has the greatest effect on torsion frequency, followed by lagging frequency, which has less effect on flapping frequency;
- Reasonable design of forward and backward sweep angles and downward dihedral angles of 3D blade tip configuration can increase the damping margin of rotor aeroelastic stability and improve its stability.
- Reasonable design of a three-dimensional blade tip configuration can effectively reduce rotor loads and make the rotor dynamic performance better than that of a rectangular blade tip configuration.

References

[1] LEISHMAN G J. (2006) Principles of Helicopter Aerodynamics [M]. New York: Cambridge University Press, pp. 292-295.

Journal of Physics: Conference Series **2879** (2024) 012060 doi:10.1088/1742-6596/2879/1/012060

[2] ZHAO Q J, GUO H X (2007) A Study on Aerodynamic and Acoustic Characteristics of Advanced Tip-shape Rotors[J]. Journal of the American Helicopter Society, 52(3):201-213.
[3] JEAN Francois, CEDRIC Lanouette (2021) Lift off of H160 with Blue EdgeTM blade. *The 77th Annual Vertical Flight Society Forum and Technology Display,* Fairfax, USA. pp. 749-754.
[4] Wilbur M L, Sekula M K (2005) The effect of tip geometry on active-twist rotor response. *Proceedings of 61st Annual Forum of the American Helicopter Society.* Fairfax, Virginia, USA: AHS International. pp. 844-855.
[5] M. Rohin Kumar, C. Venkatesan (2016) Effects of Blades Configuration Parameters on Helicopter Rotor Structural Dynamic and Whirl Tower Loads [J]. *The Aeronautical Journal.* pp. 217-290.
[6] Yang Weidong, Deng Jinghui (2003) Research on Aeroelastic Stability of Helicopter Swept-back Tip Rotor [J]. *Journal of Nanjing University of Aeronautics and Astronautics.* (35):248-251.
[7] Deng Jinghui (2023) Aeroelastic stability analysis of rotor blade with complex three-dimensional design. *Journal of Northwestern Polytechnical University.* vol 1. pp. 209-215
[8] Tian Shengyao (2020) Analysis and Research on Dynamic Design of Active Variable Swept Blade Tip Rotor [D]. *Nanjing University of Aeronautics and Astronautics.*
[9] CHI C, DATTA A, CHOPRA I (2021) Three-dimensional Strains on Twisted and Swept Composite Rotor Blades in Vacuum [J]. Journal of Aircraft, 58(1): 1-16.
[10] HUANG Wei, CHI Cheng. (2024) Analysis for Aero-elastic Characteristics of Prop-Rotor in Hover with a Swept Tip. CHINA MECHANICAL ENGINEERING, Vol. 35 No. 2. pp. 191-200.

Journal of Physics: Conference Series **2879** (2024) 012061 doi:10.1088/1742-6596/2879/1/012061

The calculation and evaluation of acceleration errors produced by position movement of acceleration sensors on the seat of a human centrifuge

Yifeng Li[1], Zhao Jin[1], Ke Jiang[1], Lihui Zhang[1], Minghao Yang[1], Baohui Li[1*], Jinghui Yang[1], Zhongzheng Guo[1] and Hongqiu Zhang[1]

[1]Air Force Medical Centre, PLA, Air Force Medical University, Beijing, 100142, China

[*]Corresponding author's e-mail: liyf8886@sina.com

Abstract. With the increasing requirement for pilot anti-overload ability training, the man-machine engineering problems caused by the existing seat on a human centrifuge are becoming increasingly prominent, making it challenging to meet actual requirements and needing to be upgraded. After upgrading, the seat can be quickly adjusted in front, back, up and down. Acceleration sensors fixed on the seat will also generate displacement, which will result in errors. Three-axis acceleration values are a crucial technical index in human centrifuge systems, and their errors need to be within a certain accuracy range in order to make the system operate normally. Otherwise, the system will produce serious consequences such as alarm and shutdown. Analysis shows that the position movement of acceleration sensors on the seat has no impact on the control system, mainly affecting the safety system. This study discusses the situation when acceleration errors in seat adjustment reach the maximum values, and calculates the three-axis acceleration errors in detail. It is observed whether the values are within the set limits by the safety system in order to evaluate whether they have an impact on the safety of the entire system. The result indicates that the acceleration errors caused by the adjustment of the upgraded seat are within a controllable range and will not have any adverse effect on the safety system of the human centrifuge. Therefore, the acceleration errors will not have any adverse effect on the safety of the whole system.

1. Research background

The human centrifuge was put into use in 2005, which makes a significant contribution to the selection and training of anti G endurance for pilots, as well as the physiological and hygienic research and identification of various protective and lifesaving equipment and so on[1][2]. Seat design is a key aspect of centrifuge cabin design, which should be safe and comfortable, with convenient and fast seat adjustment operation, and with a safe and reliable restraint system [3][4][5]. However, with the increasing training demand for pilots' anti-overload ability and the development of new fighter jets and supporting anti G equipment, the human-machine engineering problem caused by seats is becoming increasingly prominent, and the existing equipment is difficult to meet practical needs. In terms of adaptability, convenience, comfort, and safety, seats urgently need to be replaced. The existing seat adopts a mechanical adjustment method, and the position of the seat is basically motionless in actual use. The improved seat belongs to the electric adjustment mode, which can conveniently and quickly adjust the position of the seat and front and back, up and down. The accelerometers are fixedly installed behind the seat headrest and move up and down, forward and backward with the seat. The ideal position of acceleration sensors should be at the centre of the cabin roll and pitch axes. This will result in three-axis acceleration errors, which can cause serious consequences such as system alarms or even shutdown if they exceed a certain range, affecting the normal operation of the human centrifuge. Therefore, it is necessary to calculate and evaluate the three-axis acceleration errors. Usually, when specifying the G accuracy requirements in the z-direction of the centrifuge, the allowable acceleration values in other directions should also be specified at the same time [6][7]. Some scholars have conducted relevant research on centrifuge and its various calculations of acceleration and angular velocities [8-16]. This study first analyzes the situation where the maximum error occurs when the seat is adjusted in any position, calculates the possible maximum error value in this situation, observes whether the value is within the set limits by the safety system, and then evaluates its potential impact on system safety.

2. The impact of the position movement of acceleration sensors on system[6][7]

The impact of acceleration sensors' position movement on the system mainly includes the impact on the control system and the impact on the safety system. The two situations are discussed separately.

2.1. The impact of position movement of acceleration sensors on the control system

The control system here mainly refers to the main drive the control system. The G load measurement of the control system is different from that of the safety system, and this needs to be distinguished. The control system uses acceleration sensors installed at a radius of 8 meters on the main arm of the centrifuge (at the radius of the cabin reference point) to measure the directional angles of cabin roll and pitch angles, and then to calculate the values of three-axis acceleration by using specific formulas. The set G-load value is matched with the G-load value of the cockpit reference point, which is located at the intersection of the cockpit roll axis and pitch axis and is aligned with the cockpit axis. Two levels of limit values, Level 1 and Level 2,are set for each parameter of the control system. For example, +Gx sets Level 1 and Level 2, -Gx sets Level 1 and Level 2. An alarm will be triggered if exceeding Level 1, while an emergency shutdown will be caused if exceeding Level 2. The control system does not use acceleration sensors installed on the seat, so the position movement of the acceleration sensors on the seat has no impact on the control system.

2.2. The impact of position movement of acceleration sensors on the safety system

The safety system controls the operational characteristics and operational limitations of human centrifuge in several aspects during operation. The safety system controls and monitors the power supply system, main drive, roll and pitch drive monitoring, safety interlock, independent safety functions, auxiliary monitoring equipment and so on. The safety system uses three-axis acceleration sensors installed behind the seat backrest. The safety system also sets two levels of limit values for each parameter, Level 1 and Level 2. For example, +Gx sets Level 1 and Level 2, -Gx sets Level 1 and Level 2. An alarm will be triggered if exceeding Level 1, while an emergency shutdown will be caused if exceeding Level 2. Therefore, the position movement of acceleration sensors on the seat has a direct impact on the safety system.

The acceleration sensors installed behind the seat backrest consist of three orthogonal arranged acceleration sensors and play a role in the safe shutdown of the centrifuge. In order to minimize the measurement error, it is necessary to place the sensors at the head position of the subject, which is close to the intersection point of the cockpit roll and pitch axes, making an ideal placement. However, in reality, due to geometric limitations, sensors cannot be accurately installed in that position. Therefore, it is embedded in the seat backrest and close to the subject's head. The angle positions of the sensors are fixed and cannot be manually or automatically adjusted. When the seat angle is 17 degrees, the sensor measuring is perpendicular to the Gz force of the cabin floor. When the seat is located at the longitudinal midpoint, the measured acceleration largely matches the value of the selected training curve. At this time, the error between the set Gz load acceleration and the Gz acceleration measured by the sensor can be ignored. When the seat is far away from the longitudinal midpoint position and has a seat angle of 17 degrees, there will be a systematic error. To simplify operation, safety limits are set separately for each of the two operating modes, "training" mode and "research" mode. In pilot training, safety limits of "training" mode are mainly used. This value is determined by professionals in the field of aviation medicine based on a combination of physiological perspectives, operating modes, and technology. For "studying" operating mode, it is necessary to consider the tolerance of -Gz load produced by physiological induction of the human body for -Gz load. From the general literature on flight physiology, it can be inferred that exposure under -3Gz for more than 5 seconds can lead to fatal consequences. For safety reasons, the -Gz value cannot be set too low. Currently, a Gz curve with a maximum load from-1.5Gz to+9Gz is being used. At present, it is recommended to maintain the original value for the limit value of -Gz.

To study the impact of the position movement of acceleration sensors on the safety system, it is necessary to calculate three-axis acceleration error values in detail, and to compare it with the limit value of the safety system. Then, its impact on the safety system and the entire system is evaluated.

3. The calculation and evaluation of three-axis acceleration errors produced by position movement of sensors

3.1. The calculation and evaluation of z-axis acceleration error

3.1.1. System errors analysis. Random error analysis: When replacing a seat, the safety system remains unchanged, and the previously used accelerometer will continue to be used when a new seat is installed. The directions of the three axes of the cabin are as follows:The downward direction of gravity acceleration is in the positive z-direction, while the opposite direction is in the negative z-direction; The direction of forward from chest to back is x positive, while the opposite is x negative; The direction from right to left is positive y-direction, while the opposite is negative y-direction. When the inclination angle of the seat is 17 degrees, the three axes of the accelerometers and the cabin axis are consistent, and their angular positions accurately correspond to the angular positions of the existing accelerometers on the seat. When the seat inclination angle is not 17 degrees, the sensor coordinate system will rotate accordingly. Assuming the seat inclination angle is θ, the angle between the sensors and the vertical axis is $\theta - 17$ degrees. Figure 1 is a schematic diagram of the installation of three-axis acceleration sensors on the seat. Due to the mechanical adjustment method, the existing seat is relatively cumbersome to adjust. In practical application, the seat is at 17 degrees and has hardly been adjusted. However, the new seat can be quickly adjusted at several different angles, which is its advantage over the existing seat. When the inclination angle of the seat changes, the acceleration sensors also change the inclination angles of the seat. Due to the fact that the sensors installed on the existing seat will be reused and the electromagnetic interference remains unchanged, the difference in geometric dimension will not affect the random error, so the random error of the sensors remains unchanged. The random error generated by the sensors is still within the accuracy range of 5%.

System error analysis: The new seat can be adjusted in the up and down, front and rear, and in inclination angle. Height adjustment does not affect sensor error, as changing the seat height does not affect the sensor's installation position on the seat frame. The error caused by the longitudinal position of the front and rear can be corrected, and the existing seat has also been corrected on this point. However, changing the seat inclination angle will have an impact on a system error, because (1) the position of the sensors relative to the reference point has changed, and (2) the sensors change in three directions relative to the coordinate system of the cabin, i.e. the three axes of the cabin. The centrifuge control system reproduces the required G-acceleration relative to the cockpit coordinate system by controlling the rotation speed of the main arm and the angles of the roll and pitch axes. Through these, the geometric relationship between the integrated sensors on the seat and the cabin reference point can be changed. This can lead to a system error in sensor measurement.

3.1.2. The determination of the generating condition for the maximum value of system error. As analyzed before, height adjustment does not affect sensor error. The change in the longitudinal position of the seat in front and rear, as well as the change in the inclination angle of the seat, has an impact on the system error. When the inclination angle of the seat is 17 degrees, the directions of the coordinate axis of the sensor and the coordinate axis of the cabin are consistent. When the seat inclination angle is not 17 degrees, the sensor coordinate system will rotate accordingly. If the seat tilt angle is θ, the angle between the sensor and the longitudinal axis is $\theta - 17$ degrees. The following Figure 1 is a schematic diagram of the coordinate systems for seats and sensors. In the figure, the blue color represents the coordinate axis of the cabin, and the red color represents the coordinate axis of the sensors. The black solid line in the figure represents the lever arm from the centrifuge axis to the reference point of the cockpit, while the dashed solid line represents the virtual lever arm from the centrifuge axis to the sensors. When the sensors and cabin reference point do not coincide, an angle will be generated between the two lever arms.

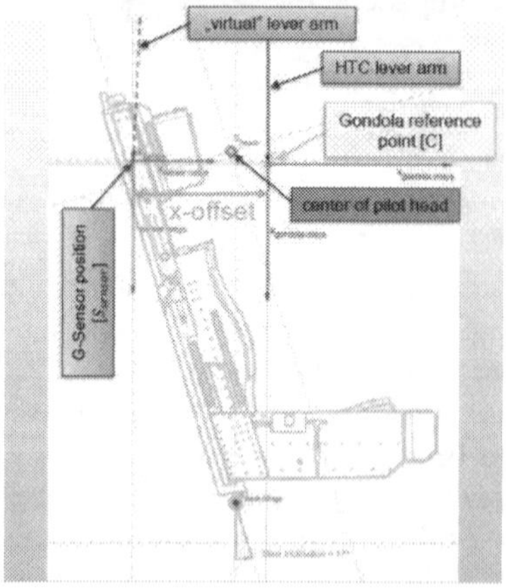

Figure 1. The schematic diagram of the coordinate systems for seat and sensors.

The forward and backward movement of the seat, and the change in the inclination angle of the seat, result in errors in acceleration. Assuming the maximum caster angle is $\theta 1$, and the minimum caster angle is $\theta 2$, it is not difficult to see that, when $\theta 1 - 17 > 17 - \theta 2$. When the seat is moved to the final position and the inclination angle of the seat surface is $\theta 1$, i.e. the seat is at its maximum caster angle, the system error of acceleration generated at this time is the largest.

3.1.3. The calculation and evaluation of system errors. As analyzed before, when the seat is moved to its rear position and the inclination angle of the seat surface is 30 degrees, the acceleration sensors produce the maximum errors relative to the cabin reference point. The virtual lever arm from the centrifuge axis to the sensors which is called r_{sensor} for short, and the centrifuge lever arm from the centrifuge axis to the cockpit reference point which is called r_{htc} for short, will produce the maximum deviation at this time. At this time, there will be a maximum angle deviation between the centrifuge lever arm from centrifuge axis to cockpit reference point which is called r_{htc} for short, and the virtual lever arm from centrifuge axis to sensors which is called r_{sensor} for short.With the G value increasing, the longer the lever arm is, the error between the two lever arms is the greater. The maximum deviation occurs when 9G, so taking the inclination angle of the seat of 30 degrees as an example, the focus is on analyzing the error size of the acceleration when Gz is set to 9G. The symbols are shown in the following Figure 2. In Figure 2, the yellow character C represents the intersection of the roll axis and pitch axis, the red character Csensorrepresents the centre of sensors, and the blue character Chead represents the centre of the head.

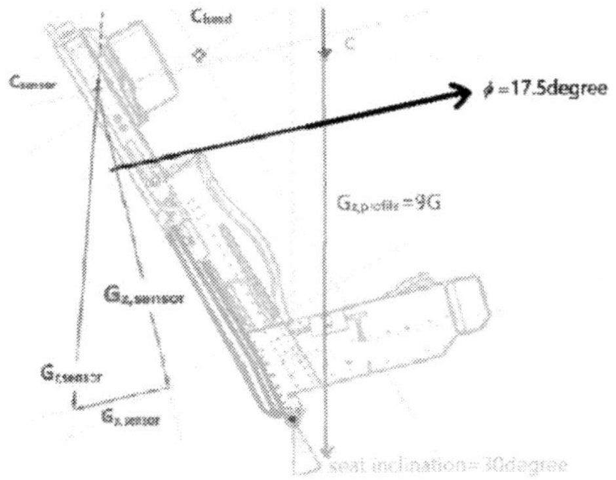

Figure 2. The schematic diagram of z-axis acceleration calculation when inclination angle of the seat is 30 degrees.

Step 1. The angle deviation α between lever arms r_{htc} and r_{sensor}

When the inclination angle of the seat is 30 degrees, it can be known that r_{htc}=8m, $G_{z,profile}$=9G, and the displacement deviations Δx_{30} and Δy_{30} of the sensors in the x and y directions relative to the cabin reference point can be measured. The angle α_{30} between the two lever arms can be calculated by using the below Formula (1). When the inclination angle of the seat is 30 degrees, by calculation it is obtained that α_{30}=4.5 degrees.

$$\alpha = \arctan(\frac{\Delta x}{r_{htc}}) \tag{1}$$

Step 2. The size of the virtual lever arm r_{sensor}

The size of the virtual lever arm $r_{sensor,30}$ can be calculated by using Formula (2). When the inclination angle of the seat is 30 degrees, $r_{sensor,30}$=8.11m.

$$r_{sensor} = \sqrt{(r_{htc} - \Delta y)^2 + \Delta x^2} \tag{2}$$

Step 3. The radial acceleration of centrifuge axis to sensors Gr

The radial acceleration of the centrifuge axis to sensors Gr can be calculated by using Formula (3). When the seat inclination angle is 30 degrees, by calculating it can be obtained that $G_{r,30}$=9.13G. Due to that the virtual lever arm is slightly longer than the centrifuge lever arm, when the set G value is 9G, it leads to the actual Gz load acting on the seat acceleration sensors, that is, G_r is slightly larger than the set G load. Based on Gz and G_r, the percentage of error between the Gz sensors display that is $G_{r,30}$, and the set Gz value can be obtained. The calculated value is between 1% and 2%, which is within the range of 5% of system error and will not have any adverse effects on the centrifuge system.

$$\frac{Gr}{r_{sensor}} = \frac{Gz}{r_{htc}} \tag{3}$$

Step 4. Calculating the angle ϕ between radial acceleration Gr and the z-axis of the sensors

When the inclination angle of the seat θ_{seat} is 30 degrees, due to a 17-degree difference between the sensors and the inclination angle of the seat, at this moment the inclination angle of the sensors θ_{sensor} is 30-17=13 degrees. As known from Step 1, the angle deviation α between the lever arm r_{htc} and r_{sensor}, and α_{30}=4.5 degrees, that is, the angle between Gr and the z-axis of the cabin is

α_{30}=4.5 degrees. According to Formula (4), the angle between Gr and the z-axis sensors ϕ can be calculated. When θ_{seat} =30 degrees, ϕ_{30} =17.5 degrees.

$$\phi = \theta_{sensor} - 17 + \alpha \tag{4}$$

Step 5. Calculating the projection Gr in z-axis direction of sensors $Gz,sensor$

According to Formula (5), the projection of Gr in the z-axis direction of sensors $G_{z,sensor}$ can be calculated, which is the acceleration value of z-axis direction actually perceived by sensors. According to the $G_{r,30}$ value in Step 3 and the angle ϕ in Step 4, it can be determined that when the inclination angle of the seat is 30 degrees, $G_{z,sensor,30}$ =8.76G.

$$G_{z,sensor} = Gr \times \cos\phi \tag{5}$$

Step 6. Calculating the maximum error percentage λ_z in the z-axis direction of the sensors

The maximum percentage of error of the sensors in the z-axis direction λ_z, when the inclination angle of the seat is 30 degrees, can be calculated according to Formula (6). According to the $G_{z,sensor,30}$ value in Step 5, when $G_{z,max,profile}$ takes a maximum value of 9G, it can be calculated that the λ_z value is between 2% and 3%, meeting the range requirement of a 5% acceleration error in the z-axis direction of the centrifuge system, and will not cause any adverse effects on the centrifuge system.

$$\lambda_z = \frac{G_{z,max,profile} - G_{z,sensor}}{G_{z,max,profile}} \times 100\% \tag{6}$$

When the inclination angle of the seat is 13 degrees and 17 degrees, the errors can be calculated sequentially based on the above formulas and geometric relationships.

So the calculated maximum error value ΔGz of the z-axis and the maximum error percentage λ_z of the z-axis are both within the limit range set by the safety system, and they will not have any adverse effect on the safety system.

3.2. The calculation and evaluation of y-axis acceleration error

The sensors are installed at the rear of the seat headrest. Due to the collision with the shaft motor of the height adjustment of the seat, sensors move to the left about 100mm. Here the direction to the left is from the subject's line of sight. Compared to the lever arm r_{htc} of a human centrifuge with an arm of 8m and the virtual lever arm r_{sensor}, the position deviation of sensors in the y-direction is 100mm, which can be ignored. The position deviation and angle deviation of sensors in the y-direction are clearly not a problem.

3.3. The calculation and evaluation of x-axis acceleration error

It is worth noting that there is a significant deviation between Gx and the standard value of 0. This is mainly caused by the misalignment of the sensors and cabin horizontal axis. Similar to the deviation analysis of acceleration of the z-axis, when the seat is moved to the final position and the inclination angle of the seat surface is 30 degrees, acceleration sensors produce the maximum x-axis direction error relative to the cabin reference point. At this time, the maximum angle deviation will occur between the centrifuge lever arm r_{htc} from the centrifuge axis to the cockpit reference point and the virtual lever arm r_{sensor} from the centrifuge axis to the sensors. With the G value increasing, the longer the lever arm is, the greater the angle deviation between the two lever arms is. The maximum deviation occurs when 9G, so the focus is on analyzing the error value of acceleration in the x-axis direction when the inclination angle of the seat is 30 degrees and the set G_z is 9G.

Similar to the analysis of z-axis acceleration error, the calculation Steps 1 to 4 of Gx, are like Steps 1to 4 in the above G_z analysis.

Step 5. The projection $G_{x,sensor}$ of Gr in the x-axis direction of sensors is calculated.

According to the following Formula (7), the $G_{x,sensor}$ of projection Gr in the x-axis direction of sensors can be calculated, which is the acceleration value actually felt by sensors in the x-axis direction. According to the $G_{r,30}$ value in Step 3 and the angle ϕ in Step 4, it can be obtained that $G_{x,sensor,30} = 2.75G$ when the inclination angle of the seat is 30 degrees.

$$G_{x,sensor} = Gr \times \sin\phi \tag{7}$$

Step 6. The maximum error percentage λx of sensors in x-axis direction is calculated.

The maximum error percentage λx of sensors in the x-axis direction when the inclination angle of the seat is 30 degrees can be calculated according to Formula (8). According to the $G_{x,sensor,30}$ value in Step 6, due to $G_{x,\max,profile} = 0$, which leadsto $\lambda x \to \infty$, therefore, the maximum error percentage of sensors in x-axis direction λx is quite large at this time.

From Step **6**, the absolute value of the maximum error of sensors in the x-axis direction is 2.75G,

$$\lambda_x = \frac{G_{x,\max,profile} - G_{x,sensor}}{G_{x,\max,profile}} \times 100\% \tag{8}$$

From this, it can be seen that the maximum error percentage sensors in the x-axis direction are quite large. However, its maximum value of absolute is 2.75G. At present, the error limits values of human centrifuge the safety system for alarm and shutdown in the x and y directions are set based on the absolute values of acceleration errors in the x and y directions. At present, the limit values of absolute values of positive and negative errors for alarm and shutdown in x and y directions are set to be greater than 3G. That is, the maximum deviation of G_x caused by sensor movement is 2.75G, which is within the limit values set by the system, will not cause system alarms or shutdown, and will not have an adverse effect on the system.

4. Discussions

Acceleration errors produced by the position movement of acceleration sensors on the seat of a human centrifuge may have a serious impact on the running of the whole system, so they must make careful calculations and evaluation. The importance of error calculation must be first realized. Second, it is necessary to recognize separately the impacts of the position movement of acceleration sensors on the control system and the safety system. In the study, after analysis, the control system does not use acceleration sensors installed on the seat, so the position movement of the acceleration sensors on the seat has no impact on the control system. The position movement of acceleration sensors has a direct impact on the safety system, so it is very important tocalculate three-axis acceleration error values in detail, and then evaluate their impacts on the safety system and the entire system. Third, in the calculation and evaluation of three-axis acceleration errors, the one important point is to determine the generating condition for the maximum values of system errors. Only when the generating condition is determined correctly can the calculation and evaluation of three-axis acceleration errors be made correctly and the impactsof three-axis acceleration errors on the safety system and the entire system be made correctly. Fourth, the calculation needs to pay attention to the conversion between different coordinate systems, such as the centrifuge three-axis acceleration coordinate system and the three-axis acceleration coordinate system of sensors installed behind the seat backrest. The correct conversion of angles between different coordinate systems plays an important role in the calculation and evaluation of three-axis acceleration errors. Fifth, in the study, the result is that the acceleration errors caused by the adjustment of the upgraded seat will not have any adverse effect on the safety system or the whole system. More centrifuge experiments can be carried out in the future, and it should be observed whether the current acceleration error values will cause system alarms, thus verifying the correctness of the results of this study.

5. Conclusions

This paper first introduces the generation reason for the three-axis acceleration errors of centrifuge, that is, the relationship between three-axis acceleration errors and seat movement in the cabin. Then, a brief analysis is conducted on the impact of the position movement of acceleration sensors on the seat on the control system and the safety system, and the impact of the position movement of acceleration sensors on the safety system is calculated and analyzed in detail. When analyzing the impact of the position movement of acceleration sensorson the safety system, the concepts of random error and systematic error are introduced, and the specific reasons for error generation, the situation of maximum error generated and so on are analyzed. The setting of the safety limits for acceleration errors of the centrifuge system is related to the safe operation of centrifuge-large equipment and cannot be changed arbitrarily. Therefore, it is necessary to make careful calculations and analysis for three-axis acceleration errors. The calculation and analysis results show that, for the errors values of the three-axis acceleration produced by position movement of three-axis acceleration sensors which is caused by the front and back, up and down adjustment of the upgraded seat, both the percentage values and the absolute values are within the allowable control accuracy range, that is, the three-axis acceleration errors values are all within the controllable ranges and will not have adverse effects on the control system and the safety system. Therefore, the three-axis acceleration errors caused by seat movement will not have an adverse effect on the entire system. The calculation and analysis method of sensor acceleration errors in research can provide a reference for similar calculation and analysis of acceleration in centrifuge systems.

References

[1] Geng Xichen, Yan Guiding, Jin Zhao, Research and application of aviation acceleration physiology, FLIGHTSURGEON, 2 0 0 4, 23(5): 189-196.

[2] Xu Yan, Geng Xichen, Wanghong, Jin Zhao, Yan Guiding, +GzPROTECTION PROVIDED BY TILT-BACK SEAT AND GZ-2 ANTI-G SUIT, Chin J Appl Physiol 2003;19(4):368-371.

[3] Cheng Ming, Design of driver seat for civilaircraft, CivilAircraftDesignandResearch, 2009. 02. 011:1674-9804.

[4] Wang Lijing, Yuan Xiugan, Li Yinxia, Wang Yongqing, Analysis on reach envelop of pilot in cockpit, Journal of Beijing University of Aeronautics and Astronautics, 2005, 31(1):42-44.

[5] FENGZhen-an, Study of Special Seat Design of Commercial Aircraft, PACKAGING ENGINEERING, 2009, 30 (9):132-134.

[6] Lu Huiliang, Human Centrifuge and Its Application[M]. National Defence Industrial Press, Beijing, 2004.

[7] M. H. Tsai, M. C. Shih, G-load tracking control of a centrifuge driven by servo hydraulic systems, Proceedings of the Institution of Mechanical Engineers, Part G:Journal of Aerospace Engineering, June 1, 2009 vol. 223 no. 6:669-682.

[8] Maja Lutovac, Vladimir Kvrgi, Goran Ferenc;Zoran Dimi, Jelena Vidakovi. 3D simulator for human centrifuge motion testing and verification. 2013 2nd Mediterranean Conference on Embedded Computing (MECO) IEEE. 2013: 160-163.

[9] Yakup Cengiz, Seref Sagiroglu. System safety of human centrifuge and solving angular velocity of main arm with artificial neural network. 2017 5th International Symposium on Digital Forensic and Security (ISDFS) . 2017:1-6.

[10] A. Winter, N. Mohajer, D. Nahavandi and M. Watson, Kinematic Control of a Human Centrifuge System for Simulation of Aircraft Manoeuvres, 2024 Australian & New Zealand Control Conference (ANZCC), Gold Coast, Australia, 2024, pp. 223-228.

[11] N. Mohajer, A. Winter, T. M. Gregory, D. Nahavandi, M. Watson and S. Nahavandi, Experimental Validation of a High-G Centrifuge System using an Advanced Wireless Human Dummy, 2022 IEEE International Conference on Systems, Man, and Cybernetics (SMC), Prague, Czech Republic, 2022, pp. 2827-2832.

[12] H. R. Morano Okuno, J. Enrique Chong-Quero, G. S. Benitez, R. C. Castillo and A. G. Garcia, Experiences of a "G" Acceleration Generator through Angular Velocities in a Dynamic Design Course, 2023 11th International Conference on Information and Education Technology (ICIET), Fujisawa, Japan, 2023, pp. 114-118,

[13] Y. Liu, Z. Yu and S. Ren, On Pseudo Second-Order Term of Pendulous Integrating Gyro Accelerometer Calibrated in Dual-Axis Precision Centrifuge, in IEEE Transactions on Instrumentation and Measurement, vol. 72, pp. 1-9, 2023, Art no. 1007109.

[14] C. Zhang, T. Xiao, L. Ran, Z. Zhu and L. Song, The Output Nonlinear Mechanism and the Compensation Method of Digital Closed-Loop Quartz Flexible Accelerometer, in IEEE Sensors Journal, vol. 23, no. 18, pp. 21029-21039, 15 Sept. 15, 2023.

[15] J. Shi, H. Gu, C. Shi and T. Ma, Research on AC Servo The control system of Medical Centrifuge Based on Fuzzy Adaptive Fractional PID, 2023 5th International Conference on Intelligent Control, Measurement and Signal Processing (ICMSP), Chengdu, China, 2023, pp. 443-446.

[16] S. Pregl et al. , High-G Acceleration Sensors for the Automotive Industry, 2022 Smart Systems Integration (SSI), Grenoble, France, 2022, pp. 1-6.

Journal of Physics: Conference Series **2879** (2024) 012062 doi:10.1088/1742-6596/2879/1/012062

Experimental research of helicopter blade load in flight based on optical sensing

Lisong Geng[1,*], Zefeng Wang[1], Honglin Zhang[1] and Qianqian Yang[2]

[1]Aircraft Flight Test Technology Research Institute, Chinese Flight Test Research Institute, Xi'an, Shaanxi, 710089, China

[2]Testing Technology Research Institute, Chinese Flight Test Research Institute, Xi'an, Shaanxi, 710089, China

[*]Corresponding author's e-mail: 2254309737@qq.com

Abstract: The actual flight measurement of blade load is an extremely important part of the new helicopter. In order to solve the problems existing in the traditional electrical measurement method, such as weak anti-interference ability, short life, and large added mass, a flight measurement method of blade load based on optical sensing is proposed. Firstly, the strain mode distribution law is obtained by calculating the dynamic response of the real blade. The optimal layout of the optical fiber measurement points is determined. Through theoretical analysis of strain transferring and experimental verification, a highly reliable fusion scheme of blade and fiber is designed to meet the flight requirements. Based on the power line carrier technology, the optical sensing test system for helicopter flight test is developed. The flight test of blade load is completed successfully. The results of optical fiber measurement are consistent with the results of traditional electrical measurement, which indicates that the optical sensing technology has good applicability in the measurement of helicopter blade load. At the same time, it has been proved that the optical measurement system set is reliable and durable. It is of great significance for the leap from electric measurement to optical measurement.

1. Introduction

The rotor system is one of the most important dynamic components of a helicopter; its structure is very complex. The load level of the blade during flight is directly related to the vibration level, flight performance, and structural life, which endangers safety. Determining the blade load is a key step in helicopter development [1-3].

During the flight test of the current helicopter, the blade load is generally measured by electrical measurement. That is, the actual deformation of the structure is reflected by the resistance change of the strain gauge pasted on the surface of the structure. This method has rich engineering experience [4-5]. However, more and more flight test experiences show that the electrical measurement method has some problems, such as short life, weak anti-interference ability, and easy effect on the blade profile. In order to avoid these problems, scholars put forward such as digital image correlation methods [6-7] and others. However, their maturity is not enough to be applied to engineering. Fiber Bragg grating (FBG) measurement came into being.

In terms of applications of FBG sensors, Serafini et al. [8] pasted FBG sensors on the surface of T-22 helicopter blades, achieving load characteristic monitoring of blade flapping, oscillation, and torsion directions. Kim et al. [9] arranged FBG sensors on the surface of helicopter to measure the blade load in hover. But due to the lack of consideration for the roughness and vibration of the blade surface, the results were biased. Wada et al. [10] proposed a dynamic distributed monitoring technology that combines FBG sensors with optical frequency domain reflectometry to measure the time-domain response of blade strain. The results demonstrate the reliability and applicability of FBG-OFDR technology [10]. Weber et al. [11] applied fiber optic sensors to the H135 helicopter in 2021, mainly conducting laboratory and ground driving tests.

The research of FBG sensors in helicopter rotor load measurement has a certain foundation abroad, but it is still in its infancy in the domestic engineering field. Especially in complex multi-axis load, measuring point layout, reliable fusion, flight test measurement, and others, there are many technical problems [12-13]. This paper tries to realize the practical application of optical sensing in blade load measurement.

2. Optical fiber measuring point layout optimization scheme

When the helicopter is flying, the load along the blade chord and radial spatial stress field distribution, in order to accurately obtain the blade load, must first determine the load measuring point layout.

2.1. Analysis of wavelength characteristics of fiber grating

FBG is the most used optical sensor. The principle is that when a light source is fired into the fiber and passes through the grating area, some light of a specific wavelength will be reflected back. The other light will continue to propagate. The change of FBG wavelength offset with strain and temperature is as follows:

$$\frac{\Delta\lambda_B}{\lambda_B} = (1-p_e)\ \varepsilon + (\alpha + \varsigma)\Delta T \tag{1}$$

where ε is the matrix structure strain, P_e is the optical fiber elastic coefficient, α is the thermal expansion coefficient of fiber, and ζ is the thermal optical coefficient of the fiber. In this study, P_e =0.22, α=5x10^{-7}°C^{-1}, and ζ=7x10^{-6}°C^{-1}. Taking the center wavelength of fiber grating as 1550nm, we calculate the wavelength change corresponding to 1000με and 100°C as:

$$\begin{aligned}\Delta\lambda_{BE} &= \lambda_B(1-p_e)\ \varepsilon = 1550\times(1-0.22)\times1000\times10^{-6}nm = 1.21nm\\ \Delta\lambda_{BT} &= \lambda_B(\alpha+\varsigma)\ \Delta T = 1550\times(0.5+7)\times100\times10^{-6}nm = 1.16nm\end{aligned} \tag{2}$$

According to the flight data of previous helicopters, the range of blade flapping and shimmy strain is ±4000με, and the twisting strain is ±3000με. Four optical fibers are arranged on the surfaces. 1# and 3# measure the strain caused by flapping, twisting, and temperature on the upper and lower surfaces, and 2# and 4# measure the strain caused by shimmy, twisting, and temperature. The required wavelengths are shown in Table 1. It can be seen that the total wavelength range of the fiber is 73.76nm. The wavelength bandwidth of 80nm can meet the needs.

Table 1. A wavelength table is required for optical fiber measurement points on the blade.

Type	Strain	Wavelength	Number	Total
Flap/shimmy	±4000με	9.68nm	6	58.08nm
Twist	±3000με	7.26nm	2	14.52nm
Temperature	100 °C	1.16nm	1	1.16nm
Total				73.76nm

2.2. The layout of fiber sensors based on the analysis of blade modal characteristics

While the helicopter is flying, the blade is subjected to cyclic load, and fatigue is very prominent. Therefore, the layout of the optical fiber measuring points must be determined from the aspect of the dynamic characteristics of the blade. By conducting blade structure simulation, the surface strain mode results of the blade under various modes are obtained, and the principles for selecting measurement points and the design scheme for FBG sensors layout are given. The blade profile form is a D-beam double-closed cavity structure. The diagram is shown in Figure 1.

We set the blade to rotate around the Y-axis at a fixed speed of 300r/min. The first seven natural frequencies and modal characteristics of the blade obtained are shown in Table 2. The results show that the errors are less than 10%, which meets the engineering needs. The strain mode diagram is shown in Figure 2, which only shows the first-order result diagram of flapping and shimmy, while the other modes are similar.

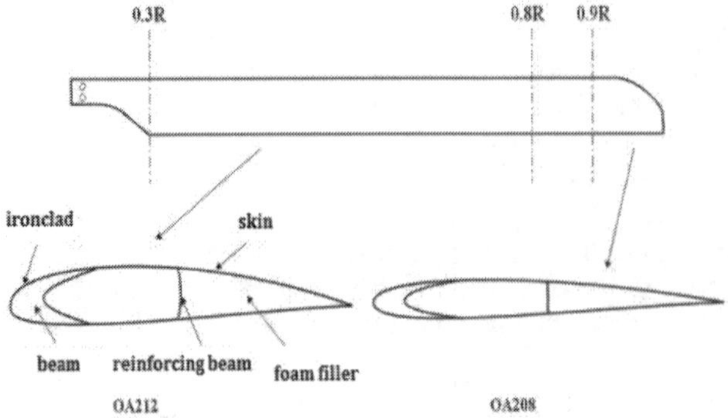

Figure 1. Schematic diagram of helicopter blade geometry.

Table 2. Mode characteristics and natural frequencies of the blade.

Number	Model	Frequency calculated	Frequency designed	Relative error
1	Flap 1st	1.32Hz	1.38Hz	4.35%
2	Shimmy 1st	4.68Hz	4.28Hz	9.35%
3	Flap 2nd	6.75Hz	7.33Hz	7.91%
4	Flap 3rd	19.97Hz	18.89Hz	5.72%
5	Shimmy 2nd	29.09Hz	27.27Hz	6.67%
6	Flap 4th	37.75Hz	35.63Hz	5.95%
7	Twist 1st	47.69Hz	44.09Hz	8.17%

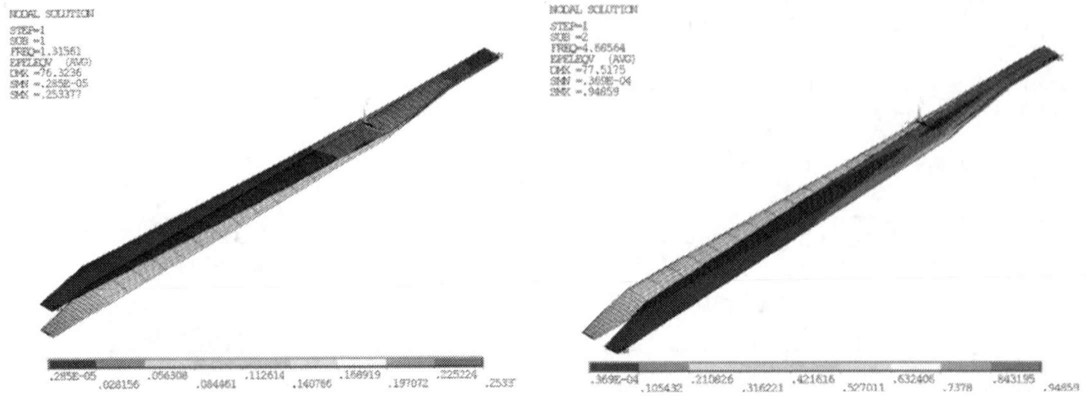

(a) (b)

Figure 2. Characteristics of strain modes of the blade (flapping 1st order and shimmy 1st order).

The strain value of each mode is extracted. The position is obtained. To sum up, the flapping and shimmy points are arranged in sections 0.01R, 0.196R, 0.361R, 0.445R, 0.544R, and 0.666R. The twisting points are in sections 0.051R and 0.155R. The schematic diagram is shown in Figure 3.

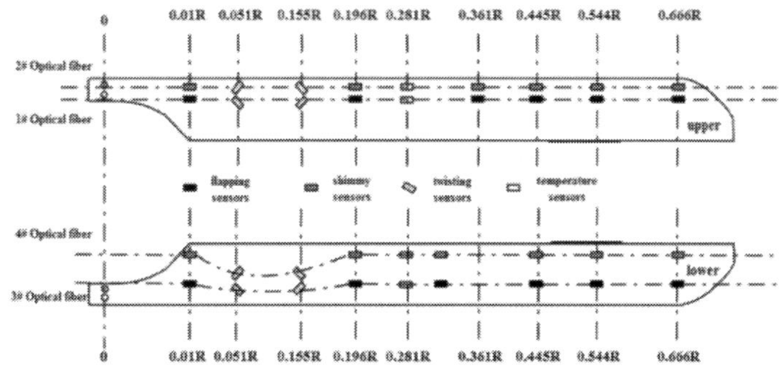

Figure 3. Schematic diagram of the location of measuring points on the blade surface.

3. Blade and optical fiber highly reliable fusion scheme

In order to meet the needs of engineering and ensure flight safety, it is necessary to carry out research on the FBG and blade fusion scheme, that is 1) the strain transfer requirements of the blade structure; 2) It has enough adhesive strength; 3) It minimizes the impact on the aerodynamic characteristics of the original blade.

3.1. Strain transfer calculation of optical fiber on the blade surface

When the blade deforms, the strain is first transferred to the cementing layer in the form of shear and then transferred to the fiber grating by friction [14]. Figure 4 shows the strain transfer model. The length of the bonded layer is $2L$, the width is D, the thickness is H, the thickness of the bonded layer between the fiber and the blade is h, and the length of a single FBG measuring point is l_0.

We select the right part of the symmetry axis of the front view in Figure 4 as the analysis object, establish the coordinate system as shown in Figure 5, and take the dx element segment for force analysis. σ is the normal stress of each part, and τ is the shear stress between each other. For the short writing, g, c, and m are taken as subscripts to represent the fiber grating, cementing layer, and blade matrix, respectively.

The stress analysis of the FBG microelement segment in the x direction can be obtained.

$$\pi r_g^2 \sigma_g = \pi r_g^2 (\sigma_g + d\sigma_g) + 2\pi r_g \tau(x, r_g) dx \tag{3}$$

After a series of assumptions, simplification, and integration, the strain transfer rate α is obtained, as shown in Formula (4). The derivation process has been shown in many literatures [14], so it will not be repeated here.

$$\alpha(x) = \frac{\varepsilon_g}{\varepsilon_m} = 1 - \frac{\cosh(kx)}{\cosh(kL)} \tag{4}$$

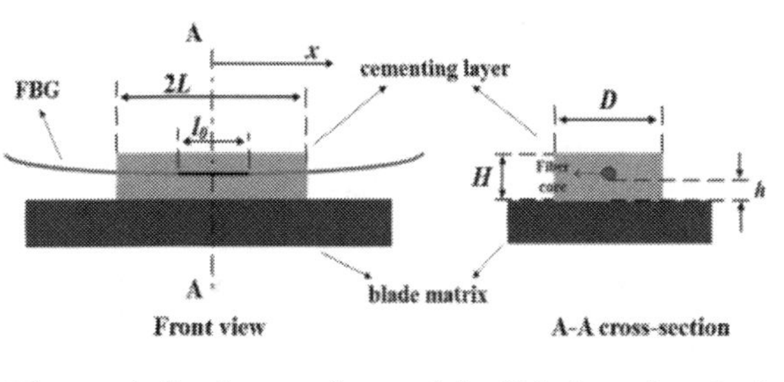

Figure 4. Strain transfer model of blade and optical fiber.

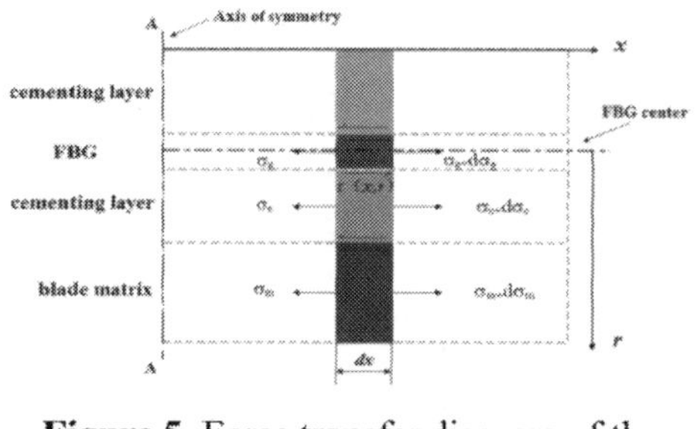

Figure 5. Force transfer diagram of the matrix, adhesive layer, and optical fiber.

In the above formula, the expression for k is as follows.

$$\frac{1}{k^2} = \frac{\pi r_g^2 E_g (r_m - r_g)}{D G_c} \tag{5}$$

It can be seen from Formula (4) that when a surface paste type FBG sensor is installed, the main factors affecting the strain transfer rate are elastic modulus E_g of FBG, radius r_g of FBG, thickness h= R-r_{gm}, shear modulus G_c of cementing layer, width D and length L of layer, etc.

The elastic modulus of FBG sensor are E_g =72GPa and radius r_g =62.5μm. We set the thickness basic value h=1mm, the width basic value D=10mm, the length basic value L=30mm, and the G_c =0MPa ~ 30MPa. When the influence of a certain factor is studied, the other parameters keep the basic value unchanged. Then, according to Formula (4), the strain transmissibility changes, as shown in Figure 6.

As can be seen from Figure 6, the strain transmittance is positively proportional to G_c. However, with the increase of G_c, the rate increases slowly. That is, the contribution of G_c has little effect. The larger D and L are, and the smaller h is, the higher the strain transmittance is.

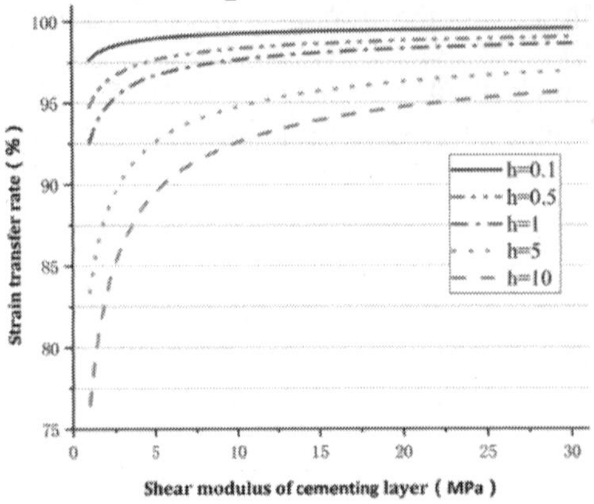

Figure 6. The curve of strain transmittance with cementing thickness and shear modulus.

3.2. Influence of adhesive method on aerodynamic characteristics of the blade

In order to study the effect of the pasting method on the aerodynamic characteristics of the blade, CFD calculations of the normal blade and blade with optical fiber were carried out. The thicknesses of the paste area are 1mm, 2mm, and 3mm. Its string length is 0.5m. The geometric model schematic diagram is shown in Figure 7.

Figure 8 shows the cloud map of the pressure distribution of the airfoil. It can be seen that the macro flow field of the airfoil is not significantly affected after the pasted sensor. There is a very small separation area behind the pasted position, but it does not cause the downstream flow field. Figure 9 shows the aerodynamic calculation. It can be seen that the size of the adhesive layer has little influence on the flow field distribution of the blade. The influence degree is within the allowable range of blade design error.

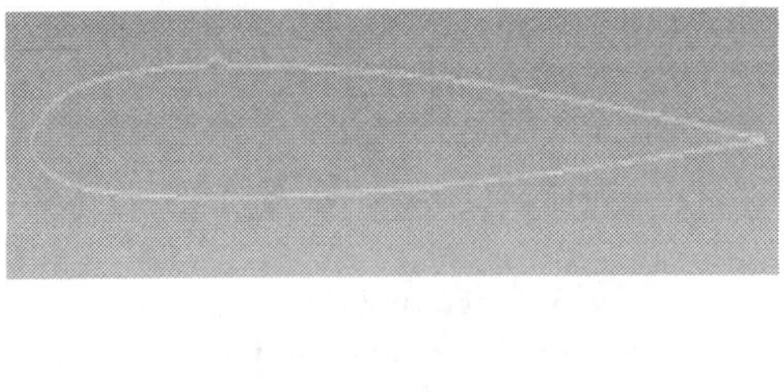

Figure 7. Schematic diagram of the geometry model of the blade section with FBG sensor.

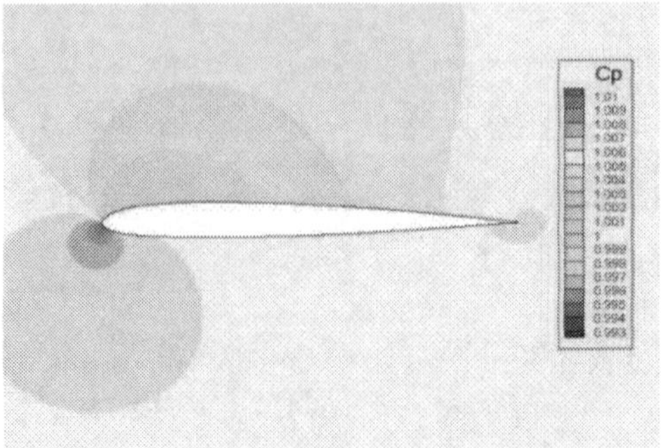

Figure 8. Pressure distribution cloud map of airfoils (H=3mm, L=15mm).

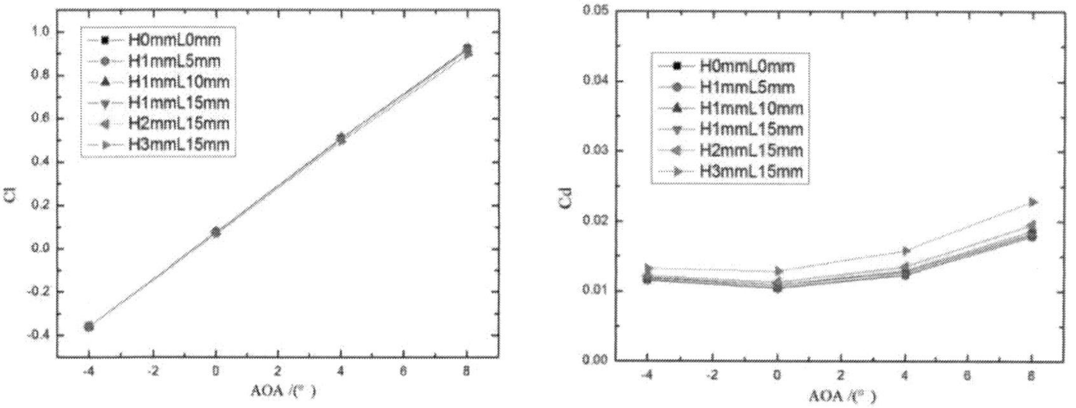

(a) (b)

Figure 9. Aerodynamic calculation curve of the airfoil (lift coefficient C_l and drag coefficient C_d).

4. Flight measurement and analysis of blade load

4.1. Architecture design of the test system

The signal acquisition and transmission of rotating components such as the blade must be completed by a special test system, which can rotate synchronously with the blade and must be able to withstand high-speed rotation, strong vibration, complex electromagnetic environment, etc., which is difficult to develop. The difficulties lie in 1) the power supply of the test system in the rotating state, 2) real-time transmission and recording of large-capacity signals, and 3) the reliability of the system in the complex environment.

The core of the test system is the fiber grating demodulation instrument, which needs to be installed on the hub of the propeller and rotate synchronously with the blade. It is used to collect dynamic optical signals generated by blade motion and convert them into strain signals. Then, the signal is transmitted from the rotating part to the non-rotating part in a specific way and then transmitted to the cabin. Thus, the overall architecture design is formed, as shown in Figure 10.

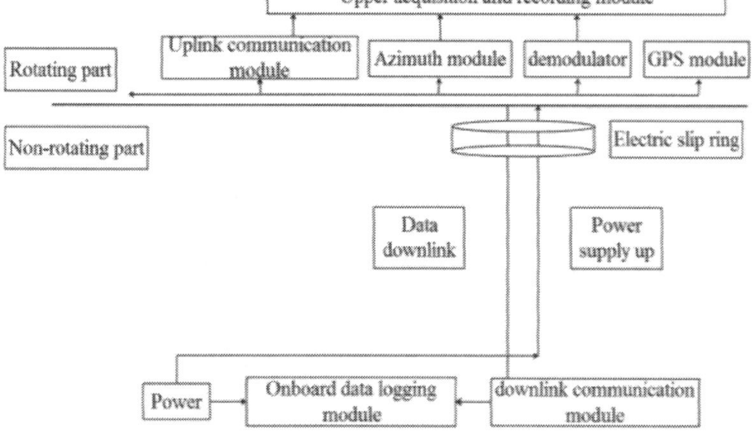

Figure 10. Schematic diagram of the overall structure of the rotor load test system.

The electric slip ring of the anti-deicing system on the helicopter is used to supply power and transmit a signal to the test system. We attach the FBG sensor to the blade. We install the test system on the rotor turntable with an electric slip ring. The test rotating speed of the turntable was set to a certain speed. The blade load measurement under a rotating state was carried out. The time domain curve is shown in Figure 11.

From the time domain, the strain data of the blade are orderly and periodic. From the frequency domain, the signal characteristic frequency is consistent with the rotor speed. There is no other interference frequency, which indicates that the carrier communication technology scheme has good adaptability in the operating environment of a helicopter electric slip ring and meets the engineering requirements of the flight test.

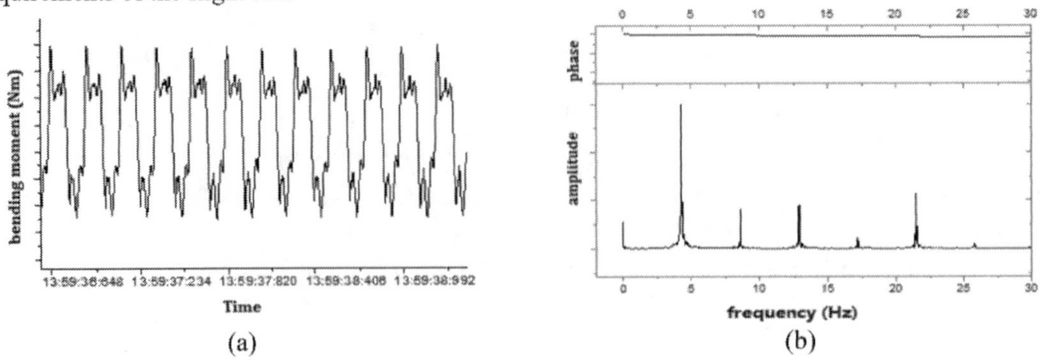

Figure 11. Time domain curve and frequency spectrum of blade rotation test.

4.2. Test flight design

The basic process of flight tests is generally as follows: flight test planning, test modification, load calibration, flight measurement, data analysis, report writing, etc.

The purpose of this paper is to investigate the applicability and reliability of the rotor optical test system in the actual flight. So, the planned flight subjects, flight status, configuration of the testing helicopter, and other conditions are consistent with the traditional electrical measurement method as far as possible in order to better compare the data of the two measurement methods.

The test parameters to be collected are mainly divided into two categories: flight parameters and load parameters. The flight parameters include speed, altitude, attitude angle, blade azimuth angle, overload, etc. The load parameters include six flapping moments, six shimmy moments, two twisting moments, etc.

Before the flight test, the FBG sensors were installed and protected on the blade. The ground load calibration test was completed. The rotor optical sensing test system has been installed on the top of the helicopter rotor hub. Before the formal test flight, the following work needs to be carried out:

1) We conduct a ground power connection test to check the connection of each system.

2) We perform ground rotation to check the working status of each system and carry out rotor dynamic balance adjustment.

3) We conduct a check flight to inspect the working conditions of each part during the flight, adjust the rotor dynamic balance, and carefully inspect the optical fiber connections after the flight to ensure that all test systems are connected reliably and working normally.

Above all, no threat to flight safety is allowed.

4.3. Test flight data analysis

The measured flight subjects of rotor load based on optical sensing include hover, level flight, etc. Figures 12(a) and 12(b) are the time domain diagram and the spectrum diagram of the flapping moment of the 0.361R section of the blade in hover flight. Figures 13(a) and 13(b) are the time domain diagram and the spectrum diagram of the shimmy moment of the 0.361R section of the blade in hover flight.

The time domain diagrams show that the waveform is neat, periodic, and without clutter interference and jump points, which accords with the blade bearing mechanism. The spectrum diagram

doi:10.1088/1742-6596/2879/1/012062

shows that the order of the waveform diagram is clear. The frequency of each order conforms to the theoretical analysis, reflecting the real validity of the data.

It should be noted that the flight states of this test are rich, and the measured points are also plenty, which cannot be listed here. Only the result graphs of individual measurement points are listed. The rules of other data are similar.

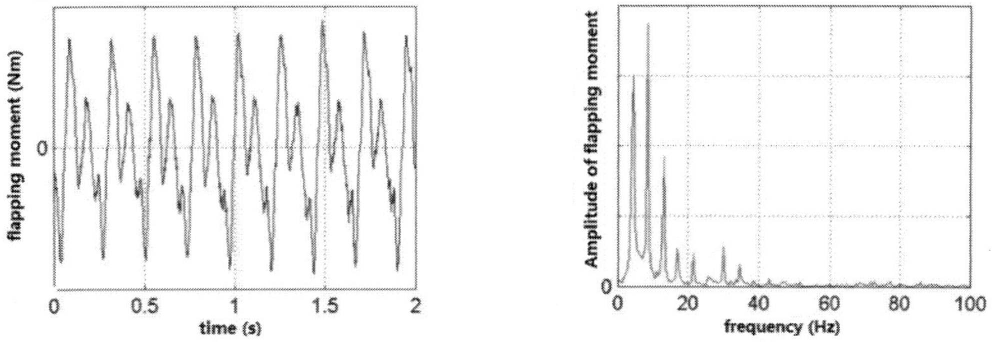

(a) (b)

Figure 12. Time domain curve and frequency spectrum of the flapping moment in hover flight.

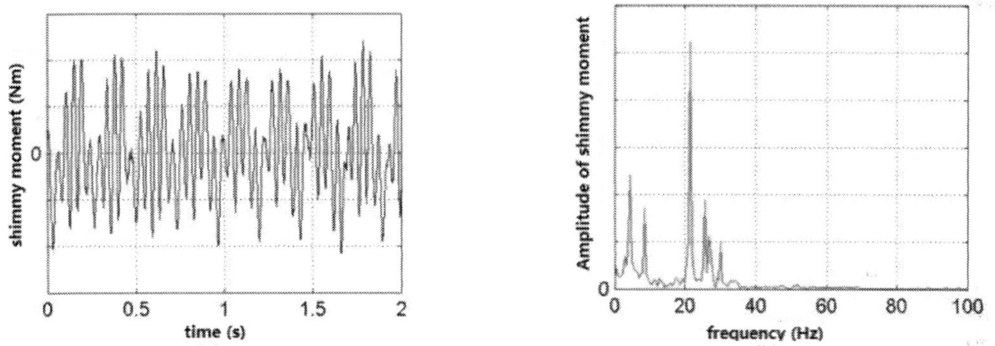

(a) (b)

Figure 13. Time domain curve and frequency spectrum of shimmy moment in hover flight.

Figure 14 and Figure 15 are time domain diagrams of the flapping bending moment of the blade section 0.361R and 0.445R during the steady level flight of the helicopter. As can be seen from the figures, the waveforms of the two sections are basically the same. The magnitude is similar, which accords with theoretical analysis, indicating that the data are consistent.

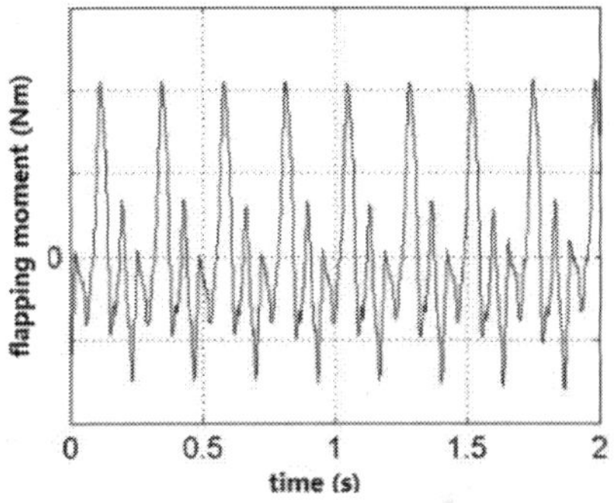

Figure 14. Time domain diagram of the flapping moment of blade 0.361R section.

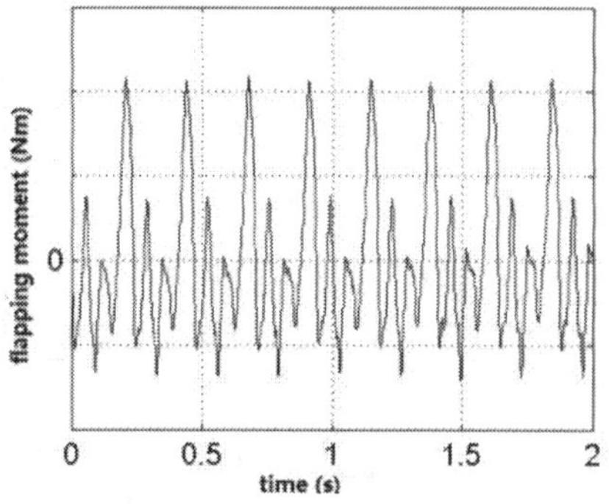

Figure 15. Time domain diagram of the flapping moment of blade 0.445R section.

Journal of Physics: Conference Series **2879** (2024) 012062 doi:10.1088/1742-6596/2879/1/012062

4.4. Comparison between traditional electrical method and optical sensing method

In order to further verify the correctness of FBG sensing blade load measurement technology, comparison and analysis are made with the measured flight data of blade load based on traditional electrical measurement methods. Figure 16 and Figure 17 show the time domain curve comparison of the flapping moment and shimmy moment of the 0.361R section of the blade when the forward ratio μ is 0.18 in steady-level flight.

It can be seen that the time domain curve variation trend of the results obtained by the optical measurement method and the traditional electrical measurement method is basically the same, which indicates that the FBG sensor has a good frequency response following characteristics in the helicopter blade load measurement. The optical measurement system developed in this paper has good adaptability in engineering applications.

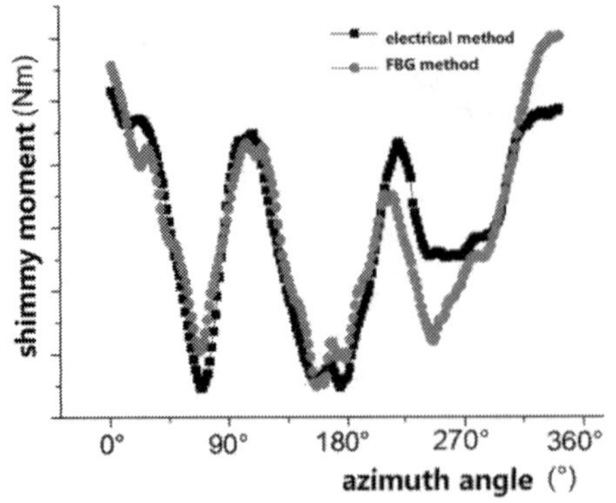

Figure 16. Comparison of the shimmy moment of blade 0.361R section measured by two methods (μ=0.18).

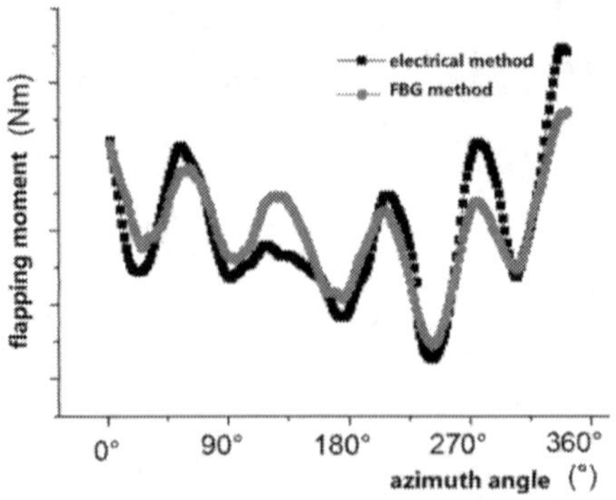

Figure 17. Comparison of the flapping moment of blade 0.361R section measured by two methods (μ=0.18).

From the test results, it can be seen that compared to traditional electrical measurement methods, the data measured using FBG sensors is smoother and less cluttered, which proves that optical sensors have stronger anti-electromagnetic interference capabilities.

In terms of cost-effectiveness, although FBG sensors are more expensive than resistance strain gauges, they have good fatigue performance. They can achieve long-term structural health monitoring of helicopter blades, effectively ensuring flight safety and avoiding greater losses to national property and personnel lives. From this perspective, FBG sensors are very economical.

5. Summary and outlook

This paper breaks through the layout design and fusion of fiber grating sensors on helicopter blades, optical sensing test system design, flight test measurement, and other technologies. It verifies the feasibility and effectiveness of this technical route and is highly significant for the leap of helicopter blade load measurement from electrical to optical measurement.

This article realizes the application of FBG sensors in the flight measurement of helicopter main blade load, but there are still some limitations, as follows:

(1) For the sake of ensuring flight safety, this article only conducted steady-state flight tests such as hovering and level flight and did not conduct flight tests under large maneuverability. Further research is needed to investigate the applicability under different flight conditions.

(2) Compared to resistance strain gauges, theoretical studies have shown that FBG sensors have good fatigue performance. However, due to the fact that this article only conducted about 10 hours of flight testing and did not conduct long-term fatigue tests, it is impossible to provide the fatigue life of FBG sensors from an experimental perspective. Further related research is needed in the future.

(3) This article only conducted research on the application of FBG sensors on the main rotor and did not investigate the use of FBG on the tail rotor blades. The smaller size of the tail rotor structure poses a challenge for the development of testing systems.

(4) Due to the unsuitability of the helicopter and measured data in this project for public publication, no actual test photos were provided in the article, nor were the specific values of the measured data displayed. Only the trend of data changes was shown. We apologize for any inconvenience caused.

Acknowledgments
This paper is one of the phased achievements of the fund project of the State Key Laboratory of Helicopter Rotor Dynamics—"Helicopter Rotor Fault Diagnosis Method Based on Information Fusion Technology"(project number: 61422202104).

References

[1] Junyan Yang, Yaming Zhang and Bin Chen. "Helicopter load manual," Beijing: Aviation Industry Press, 1991: pp. 1-10.

[2] Zhang Honglin, Cheng Weizhen, Xia Pinqi. Engineering modeling and flight testing of the dynamic loads on rotor structure based on distributed fiber Bragg grating sensing measurement. Chinese Journal of Applied Mechanics. pp. 1-14[2023-07-20]. http://kns.cnki.net/kcms/detail/61.1112.O3. 20230720. 1022.004.html.

[3] Yan Zhiqiang. Application of Fiber Optic Sensor Technology in Helicopter Flight Test. China Science and Technology Information, 2023, (19): 72-74.

[4] Ekaterinaris J A, et al. Numerical investigation of airfoil dynamic stall in simultaneous harmonic oscillatory and translator motion. AIAA-97-0615, 1997.

[5] Jie Wu, Weidong Yang, Zhihao YU. A comparative study of load calculation methods for rotor blade structures. Journal of Vibration and Shock, 2014, 33(7): 210-214.

[6] J. Sirohi, M.S. Lawson. Measurement of helicopter rotor blade deformation using digital image correlation. Optical Engineering, 2012, 51(4): 1-8.

[7] S. Rizo-Patron, J. Sirohi. Operational modal analysis of a rotating cantilever beam using high-speed digital image correlation. 57th AIAA/ASCE/AHS/ASC Structures, Structural Dynamics, andMaterial Conference, 4-8 January, California, USA, 2016, pp. 1–19.

[8] J. Serafini, G. Bernardini, I. Mattioni, et al. Non-invasive dynamic measurement of helicopter blades. Journal of Physics: Conference Series 882, 2017: 1-12.

[9] K Kim, W C Kwon, J Y Yoon, et al. Determination of local thrust coefficients of an aerospace component with the fibre Bragg grating array. Aerospace Engineering, 224: 331-338.

[10] Daichi Wada, Hirotaka Igawa, Tokio Kasai. Vibration monitoring of a helicopter blade model using the optical fiber distributed strain sensing technique. Applied Optics, 2016, 55(25): 6953-6959.

[11] Simone Weber, Thomas Kissinger, Edmond Chehura, et al. Application of fibre optic sensing systems to measure rotor blade structural dynamics. Mechanical Systems and Signal Processing, 2021, 158: 1-17.

[12] Xiaoming Sun, Su Shiwei, Yin Huaguang. Location of Abnormal Vibration Source in Helicopter Repair Based on Optical Fiber Sensing Detection. Equipment Environmental Engineering, 2023, 20(05): 128-137.

[13] Yaxing Wu, Jie Zeng. Research on 3D Shape Reconstruction Method of Helicopter Blade. Structure & Environment Engineering, 2023, 50(02): 54-59.

[14] Li W Y, CHENG C C, LO Y L. Investigation of strain transmission of surface-bonded FBGs used as strain sensors. Sensors and Actuators A: Physical, 2009, 149(2): 201-207.

Journal of Physics: Conference Series **2879** (2024) 012063 doi:10.1088/1742-6596/2879/1/012063

Uniaxial tensile mechanical properties of NEPE propellant under wide temperature and wide strain rate

Fan Liu[1], Hongfu Qiang[1], Xueren Wang[1], Zhejun Wang[1*] and Chuanjun Lu[1]

[1]Zhijian Laboratory, Rocket Force University of Engineering, Xi'an, 710025, China

[*]Corresponding author's e-mail: qinglongzaitian@126.com

Abstract: In order to gain insight into the mechanical properties of the NEPE propellant under a range of temperature and strain rate conditions, the uniaxial tensile mechanical property test of NEPE propellant with a wide temperature range of -40°C~70°C and a wide strain rate range of $0.0012s^{-1}$~$1s^{-1}$ has been carried out based on the in-situ testing device of solid propellant microfabrication under complex working conditions and the supporting high and low temperature integrated machine. Additionally, the morphology of the solid particles of the propellant was observed and analyzed during stretching using the microscope equipment provided by the stretching machine. The results show that it has been demonstrated that both temperature and strain rate exert a more pronounced influence on the mechanical properties of propellants; at low temperatures, it has been demonstrated that the stress-strain curve of the propellant exhibits a more pronounced inflection point as the strain rate increases; the maximum tensile strength of the propellant demonstrates a gradual increase with a reduction in temperature and an increase in strain rate. Furthermore, a linear double logarithmic relationship is observed with the strain rate. In contrast, the maximum elongation exhibits no discernible pattern of change with temperature and strain rate; when observed during stretching, it was found that the main forms of propellant damage destruction at low temperatures were dehumidification between the particles and the matrix and particle rupture.

1. Introduction

Solid rocket motors are widely used as the main energy power plant for rocket weapons and are used in rockets, tactical missiles, and launch vehicle boosters. With the increasing demand for the performance of solid rocket motors in severe environments, the mechanical properties of solid rocket motor pillars under various types of loads in the course of service have also put forward higher requirements [1].

At the same time, the mechanical properties of solid propellants may undergo a significant alteration, potentially by a factor of several or even tens, in response to changes in temperature and strain rate [2]. Therefore, the impact of temperature and strain rate on the mechanical properties of propellants has been the subject of extensive investigation by researchers from a range of international research institutions. In their investigation, Shekhar [3] and Zalewski and Wolszakiewicz [4] identified that variations in the mechanical properties of the propellant are significantly influenced by both temperature and the composition of the propellant formulation. Liu et al. [5] discovered that the strain rate exerts an influence on the mechanical properties of propellant. Furthermore, they analyzed the impact of tensile rate on the section morphology of propellant based on the results obtained from scanning electron microscopy. Zhao et al. [6] and Zhu et al. [7] found that the "coupling" effect of high temperature and slow pulling, low temperature, and fast pulling exacerbated the occurrence of internal damage in the propellant. As the solid rocket motor in airborne missiles and other equipment is applied more widely, equipment mounted on the ambient temperature has become diversified, which leads to the solid propellant on the ambient temperature of the adaptability of the requirements of the gradual increase in the range of its use of the temperature from -40~50°C gradually increased to -55~70°C [8]. There is a paucity of research concerning the mechanical behavior of NEPE propellant subjected to uniaxial tensile testing under conditions of wide temperature (-40°C~70°C) and wide strain rate ($0.0012s^{-1}$~$1s^{-1}$). It is, therefore, of great significance to conduct an experimental study into the mechanical properties of NEPE propellant in these conditions.

In this study, based on the in-situ testing device of solid propellant microscopic view under complex working conditions and the supporting high and low temperature integrated machine, a uniaxial tensile mechanical property test under wide temperature and wide strain rate conditions was carried out for NEPE propellant to study the influence of temperature and strain rate on the characteristics of propellant stress-strain curves and the rule of change in the mechanical properties and to establish the maximum tensile strength master curve of the propellant based on the method of GJB 770B-2005[9]. At the same time, the observation and analysis of the morphological changes of solid particles on the surface of the propellant during tensile was carried out according to the microscope equipment attached to the equipment.

2. Experimental section

2.1. Test specimens and test equipment

The solid propellant test piece used in the test was provided by a research institute. The total mass fraction of solid filler particles Ammonium Perchlorate (AP), Aluminum Powder (Al), and Octojin (HMX) was 80%. The solid propellant slurry was made by mixing, casting, and vulcanizing. The dimensions of the test piece were fabricated according to the standard dumbbell-type B specimen, as shown in method 413.1 in GJB 770B-2005. Figure 1 shows the NEPE propellant specimens and their dimensions.

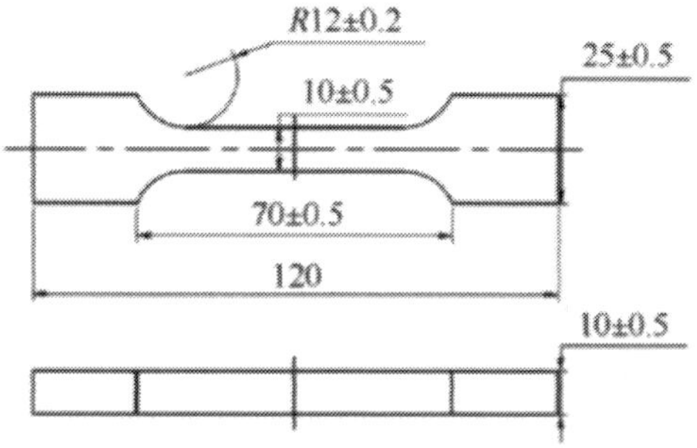

Figure 1. Dimensional drawing of NEPE propellant test piece.

In this study, uniaxial tensile tests of NEPE propellant under wide temperature and wide strain rate conditions are carried out based on the in-situ testing device for solid propellant microfabrication under complex working conditions and the accompanying high- and low-temperature all-in-one machine and a standard dumbbell-type specimen of solid propellant.

The test machine used for this test is the in-situ testing device of solid propellant microfine view under complex working conditions and the supporting high and low-temperature integrated machine, as shown in Figure 2, as shown in machine main technical indicators: maximum test force: ≥ 1000N; temperature range: -60°C~80°C; maximum uniaxial tensile rate: ≥ 4200mm/min; pressurized pressure: 0 MPa ~10 MPa, can achieve the requirements of this paper in the temperature of -40°C~70°C and the strain rate of $0.0012s^{-1}$~$1s^{-1}$ under the loading.

Figure 2. In-situ testing device for solid propellant microfabrication under complex working conditions and high and low-temperature integrated machine.

2.2. Experimental program

Through the set six temperatures (-40°C, -20°C, 0°C, 20°C, 50°C, 70°C) and six strain rates (0.0012s^{-1}, 0.0120s^{-1}, 0.24s^{-1}, 0.12s^{-1}, 0.24s^{-1}, 1s^{-1}) conditions, furthermore, the in-situ testing device of solid propellant microfabrication is employed under complex working conditions, in conjunction with the high- and low-temperature integrated machine, to conduct uniaxial tensile mechanical property testing of NEPE propellant across a range of temperatures and strain rates, three groups of repetitive tests are carried out for each working condition. The subsequent analysis of the test data is based on the mean value of the results obtained from the three experimental groups, in order to guarantee the reliability of the test results.

3. Test results and analysis

3.1. Analysis of propellant stress-strain curves

The data obtained from the aforementioned tests has been utilized to derive the stress-strain curve of the propellant, the results of which are presented in the form of Figure 3. It should be noted that the scaling factor employed in this analysis is represented by C_σ. The following characteristics can be derived from an analysis of Figure 3:

(1) The graph illustrates that the temperature and strain rate exerts a markedly pronounced influence on the alteration of the stress-strain curve of the propellant. (2) Under the condition of low strain rate at room temperature, it is found from Figure. 3 d) that the stress-strain curve of the propellant has no obvious dehumidification point. The analysis suggests that this may be related to the material composition and structure of NEPE propellants, as NEPE propellants usually use high molecular weight polymer binders, and these polymers have good elasticity and toughness, and are able to deform over a wide range; in addition, the homogeneous distribution of the solid particles in NEPE propellants avoids the concentration of stresses and may also result in the absence of obvious dehumidification points on the stress-strain curves. High and low temperatures as well as high strain rates, on the other hand, can directly or briefly affect the properties of the polymer binder to change, which can lead to significant dehumidification points in the stress-strain curve. (3) At a high temperature of 70°C, the trend of the "stress plateau region" of the propellant can be found to start gradually decreasing under high strain rate conditions. The analysis suggests that the high temperature environment leads to a reduction in the strength and modulus of the propellant, which leads to an increase in strain at a rate greater than the increase in stress, and consequently to a stress-softening phase.

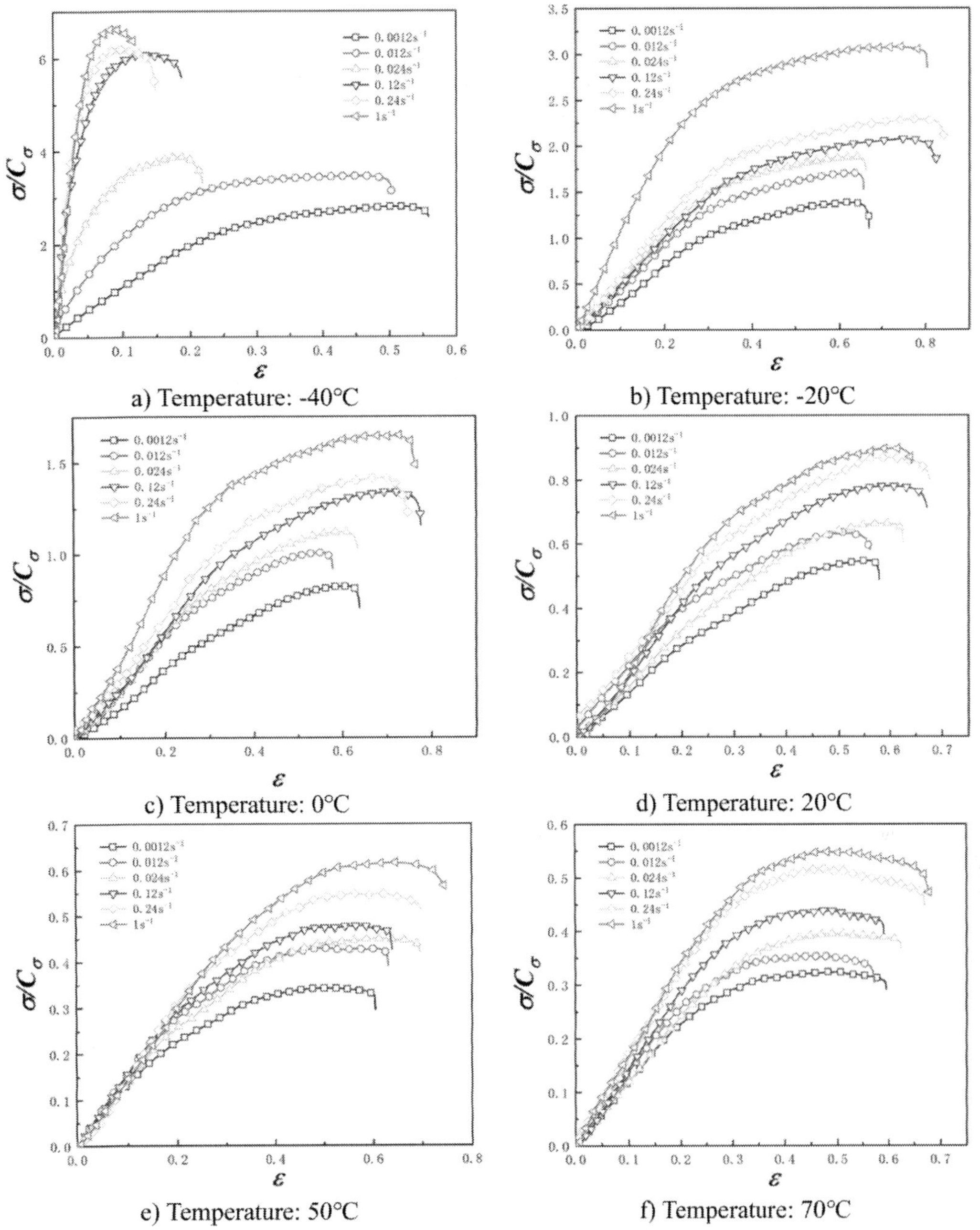

a) Temperature: -40°C b) Temperature: -20°C

c) Temperature: 0°C d) Temperature: 20°C

e) Temperature: 50°C f) Temperature: 70°C

Figure 3. Stress-strain curves of propellants.

3.2. Analysis of mechanical properties of propellants

According to the methodology outlined in Method 413.1 of GJB 770B-2005, 'Test Methods for Gunpowder', the mechanical property parameters of the propellant were calculated. Figure 4 illustrates the variation of these parameters with respect to temperature and strain rate.

From a) in Figure 4, it can be observed that the maximum tensile strength of the propellant exhibits a more pronounced variation in accordance with the established temperature and strain rate parameters, with the decrease of temperature and the increase of strain rate and gradually increasing, and with the strain rate of the linear double logarithmic relationship. Taking 20°C as the reference temperature, the maximum tensile strength of the propellant gradually decreases with the decrease in temperature, which is mainly due to the fact that with the decrease of temperature, the molecular chain segments are gradually frozen, which restricts the mobility of molecular chains. It is difficult for the matrix to expand, which exhibits the increase of the maximum tensile strength; at the same time, the maximum tensile strength of the propellant gradually decreases as the temperature increases, which is mainly due to the softening phenomenon of the propellant material due to the increase in temperature, the adhesive matrix is in a visco-elastic state, and the molecular chain movement is active, which leads to a gradual decrease in the yield stress.

From b) in Figure 4, the variation law of maximum elongation with respect to temperature and strain rate is more complicated; however, it can be observed that the maximum elongation at -40°C exhibits a tendency to diminish gradually as the strain rate is increased.

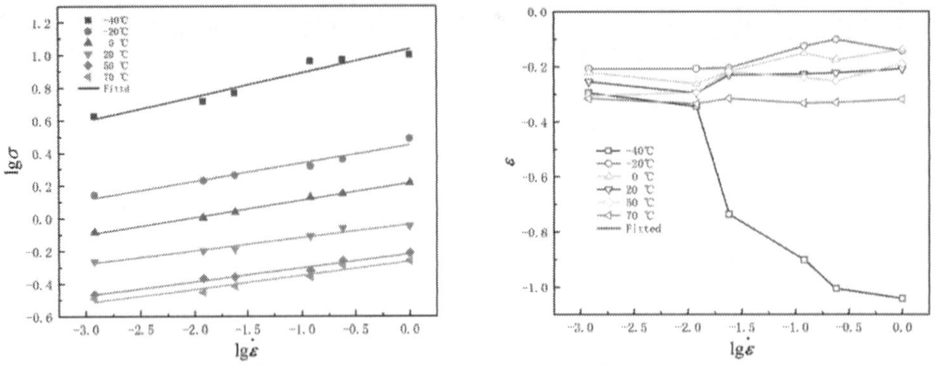

a) Maximum tensile strength b) Maximum elongation

Figure 4. Variation of mechanical property parameters with temperature and strain rate.

3.3. Maximum tensile strength master curve analysis

The maximum tensile strength master curve of the propellant was established by utilizing Method 413.3 of GJB 770B-2005. Initially, the logarithmic value of the temperature-to-strength conversion factor for the master curve representing the maximum tensile strength is determined through the graphing method. Subsequently, the horizontal coordinate is expressed in accordance with the formulation presented in Equation (1):

$$\lg\frac{1}{R\times\alpha_T}=\lg L_0-\lg V-\lg\alpha_T \qquad (1)$$

where L_0 is the initial engineering scale distance of the specimen, i.e., 70mm; V is the tensile rate in *mm/min*.

Then the longitudinal coordinate is defined as $\lg\left[\sigma\cdot(T_S/T)\right]$. Finally the experimental data are shifted to get the scatter plot of the main curve, which is fitted with a linear function to get the fitted relation of the main curve of the maximum tensile strength. The main curve plot and the relation are shown in Figure 5 and Table 1 below.

In Table 1 and Figure 5, it can be observed that there is a linear decrease in the main curve of maximum tensile strength, with a gradual decrease in value at high temperatures and low strain rates; the value in question displays a gradual increase under low-temperature and high-strain-rate conditions; however, analyzing with the stress-strain curve graph of -40°C in Figure 3, the maximum allowable elongation of the propellant is lower when the strain rate is high and the temperature is low. This increases the likelihood of a solid rocket motor failure of the pillars. In parallel, the plotted master curve of maximum tensile strength for the propellant can be used as an effective predictive tool in estimating the maximum tensile strength of the propellant under different conditions.

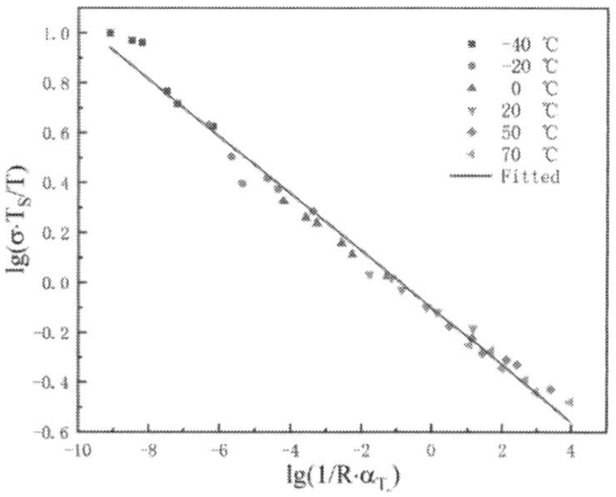

Figure 5. Maximum tensile strength master curve.

Table 1. Relationship equation for maximum tensile strength master curve.

Mechanical parameter	Fitting the relational equation	Correlation coefficient/R^2
Maximum tensile strength	$\lg(\sigma \cdot T_S/T) = -0.10144 - 0.11448\lg(1/R \cdot \alpha_{TS})$	0.9876

3.4. Evolution of surface solid particle morphology during propellant stretching

The in-situ testing device for solid propellant microfabrication under complex working conditions is equipped with a high-precision microscope, which can effectively observe the surface of solid propellant. Among them, Figure 6 shows the morphology of the solid particles on the propellant surface under the conditions of low temperature -20°C and strain rate $0.024s^{-1}$ with the increase of strain.

The dehumidification of the solid particles on the propellant surface can be seen more clearly in Figure 6. The dehumidification of the particles becomes more and more prominent as time passes. At 25s, the particles at the propellant interface shown basically appear to be dehumidified, which can be seen by the red and yellow labeling in Figure 6; and the particle rupture of the solid particles under the influence of stretching is indicated by the green wireframe in Figure 6. Therefore, the main forms of damage destruction of the propellant during low-temperature stretching are the phenomena of dehumidification between the particles and the matrix, as well as particle rupture.

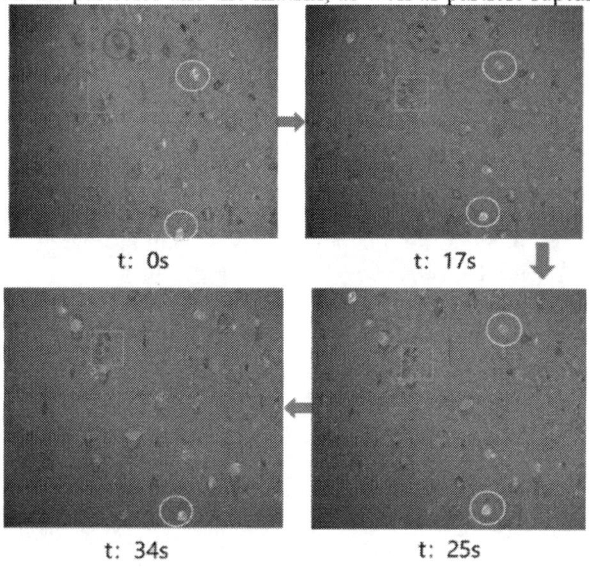

Figure 6. Evolution of solid particle dehumidification on the propellant surface.

4. Conclusion

(1) The propellant has obvious temperature and strain rate correlation; with 20°C as the reference temperature, as the temperature decreases and the strain rate increases, the uniaxial tensile stress-strain curve of the propellant begins to show a more obvious inflection point; the maximum tensile strength of the propellant increases with the decrease of temperature and the increase of strain rate. There is a linear bi-logarithmic relationship with the strain rate; in the condition of low temperature and high strain rate, the under the condition of low temperature and high strain rate are, the higher the tensile strength of propellant is, while the maximum elongation has no obvious change rule.

(2) A master curve of the maximum tensile strength of propellant has been established, which can be used to make predictions about the strength of propellant under different conditions.

(3) It can be found through observation that the propellant is basically free of damage in the initial state, while the damage phenomenon gradually appears with the beginning of stretching; under the conditions of low temperature -20°C and strain rate $0.024s^{-1}$, there are two main cases of damage destruction form of propellant in the stretching process: the phenomena of dehumidification between particles and substrate and the rupture of particles.

References

[1] Hou Xiao, Zhang Xu, Liu Xiangyang, et al. Solid rocket motor pillar structural integrity progress [J]. Journal of Astronautics and Astronautics, 2023, 44(04): 566-579.
[2] Gligorijević N, Živković S, Subotić S, Rodić V, Gligorijević I. Effect of cumulative damage on rocket motor service life [J]. Journal of Energetic Materials, 2015, 33(4), 229-259.
[3] Shekhar H. Effect of Temperature on Mechanical Properties of Solid Rocket Propellants [J]. Defence Science Journal, 2011, 61(6): 529-533.
[4] Zalewski R, Wolszakiewicz T. Analysis of Uniaxial Tensile Tests for Homogeneous Solid Propellants under Various Loading Conditions [J]. Central European Journal of Energetic Materials, 2011, 8(4): 223-231.
[5] LIU Yuanxiang, HU Shaoqing, ZHANG Yajun, et al. Mechanical properties of modified bilayered propellants at different tensile rates [J]. Propulsion Technology, 2022, 43(01): 364-369.
[6] ZHAO Min, WANG Xiaoying, YANG Jie, et al. Study of temperature and strain rate effects on mechanical properties of NEPE propellants [J]. Chemical Propellants and Polymer Materials, 2023, 21(04): 64-68.
[7] ZHU Guocui, YUAN Shen, LIU Changyi et al. Quasi-static tensile mechanical properties of HTPE/AP/Al/RDX propellants [J]. Journal of Fire Explosives, 2021, 44(05): 686-692.
[8] Li Fei, Ru Feng, Zhao Jungui et al. Research on all-weather natural environment laboratory for missile weapons [J]. Missile and Space Delivery Technology, 2016, (03): 40-43.
[9] GJB 770B-2005. Test methods for gunpowder [S]. Commission of Science, Technology and Industry for National Defense, 2005.

Torque distribution of distributed drive electric vehicle based on efficiency map of drive system

Weichen Ren[1,*] and Yanfeng Li[1]

[1]Qingdao Engineering Vocational College, Qingdao, Shandong, 266112, China

[*]Corresponding author's e-mail: weichen519@qq.com

Abstract. In order to optimize the system efficiency of distributed drive electric vehicles, based on the MAP of the efficiency of the drive motor system, the objective function of the optimal efficiency is established, and the problem of the no-load loss of the drive motor is considered. A torque distribution control strategy method considering the no-load loss of the motor is proposed. Then, AVL CRUISE and Simulink are used to establish a joint simulation platform, and a comparative simulation test analysis is carried out under ECE conditions. The simulation results show that the torque distribution strategy considering the loss makes the distributed drive motor work in a relatively high efficiency range. Compared with the torque distribution strategy without considering the loss, the power consumption is reduced by 3.4%, which effectively improves the vehicle economy. The research results can provide a reference for system efficiency optimization of distributed drive electric vehicles.

1. Introduction

In the face of energy crisis and environmental pollution, the development of various types of new energy vehicles has risen to the national strategic level. Distributed drive electric vehicles have become a new research hotspot in the field of new energy vehicles due to their high transmission efficiency and independent control of driving wheel torque [1]. At present, in view of the research on the optimal efficiency of the driving system of the distributed drive electric vehicle, the research hotspots at home and abroad mostly focus on the rational distribution of the torque of the distributed drive system, so that the front and rear motors work in the higher efficiency area respectively [2]. The methods are mainly based on the MAP of the driving system efficiency and the system loss model [3].

Theoretically, this method based on the system loss model is globally optimal. However, the above method has its own limitations. The loss of the motor changes in real time during the operation of the vehicle, it is necessary to accurately identify the parameters of the motor online, which becomes very complicated. The method based on the system efficiency MAP does not have the above problems, and the motor efficiency MAP is relatively easy to obtain. Therefore, many scholars at home and abroad have conducted related research based on the efficiency MAP of the drive system, but such research rarely considers the problem of no-load loss of the drive motor [4].

Therefore, based on the MAP of driving motor system efficiency, a torque distribution control strategy method considering motor no-load losses is proposed by establishing an objective function to maximize the efficiency of the driving motor system. Then, AVL CRUISE and Simulink are used to establish a joint simulation platform, and a comparative simulation test analysis is carried out under ECE conditions.

2. Optimal torque distribution based on efficiency map of drive system

2.1 Torque distribution objective function without considering motor loss

When the distributed drive vehicle works in a straight line without yaw moment control, the concept of drive system efficiency η_i is introduced. η_i refers to the ratio of output power P_{out} to input power P_{in} at the speed n_i and torque T_i of the drive system. The objective function J_e of maximizing the efficiency of the drive system is:

$$J_e = \max\left\{ Tn \Big/ \left[\frac{\lambda Tn}{\eta_f} + \frac{(1-\lambda)Tn}{\eta_r} \right] \right\} \tag{1}$$

where n is the wheel speed; λ is the distribution torque coefficient of the front axle single side motor; η_f and η_r are the motor efficiency of the front axle unilateral motor and the rear axle unilateral motor at the current speed and torque, respectively.

The front axle torque distribution coefficient λ represents the relationship between the front axle distribution torque T_1 and the total demand torque T. When $T_1 = T/4$, the total torque is evenly distributed among the four motors, and the torque distribution coefficient λ is defined as 0.5; when $T_1 = 0$, the distribution torque of the motor on the left and right sides of the front axle is 0, the required torque is all distributed to the left and right motors of the rear axle, and the torque distribution coefficient λ is 0. When $T_1 = T/2$, it means that the distribution torque of the motors on the left and right sides of the rear axle is 0, the required torque is all distributed to the two motors on the left and right sides of the front axle, and the torque distribution coefficient λ is 1. The distributed electric vehicle studied is equipped with four drive motors with the same parameters. The distributed drive electric vehicle studied in this paper selects four surface-mounted permanent magnet synchronous motors (PMSM) with a maximum power of 8 kw as the driving force source. Figure 1 is the efficiency MAP of the drive motor. It can be seen from the figure that the maximum speed of the modified motor is 800 r/min and the maximum torque is 165 N·m.

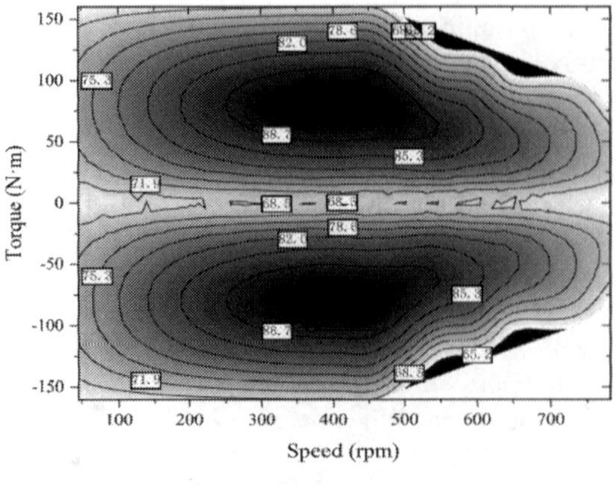

Figure 1. Map of drive motor efficiency.

In the research based on the system loss model, it can be found that when the drive system is in a steady state, the loss model of the inverter has little effect on the final analysis conclusion. Therefore, only the influence of the drive motor on the economy of the vehicle is considered.

When the target torque T of the drive system is given and the output power P_{out} of the drive system is constant, the maximum efficiency η_i of the drive system is equivalent to the minimum input power P_{in} of the system. Therefore, the minimum input power of the drive system is taken as the research goal, and the output power Tn of the drive system is kept unchanged. Equation (1) can be simplified as:

$$J_e = \min\left(\frac{\lambda Tn}{\eta_f} + \frac{(1-\lambda)Tn}{\eta_r}\right) \tag{2}$$

When the vehicle is running in a straight line without yaw control, the vertical load may shift backward when accelerating. According to the friction elliptic curve of the tire, in order to avoid vehicle skidding, the distribution torque of the motor on the left and right sides of the rear axle should be greater than that of the motor on the left and right sides of the front axle. Therefore, the objective function without considering the loss can be expressed as:

$$\begin{cases} J_e = \min\left(\dfrac{\lambda Tn}{\eta_f} + \dfrac{(1-\lambda)Tn}{\eta_r}\right) \\ s.t.\ \ 0 \le \lambda \le 0.5 \\ \quad 0 \le T \le T_{\max} \\ \quad 0 \le 0.5T\lambda \le T_{\max 1} \\ \quad 0 \le 0.5T(1-\lambda) \le T_{\max 2} \\ \quad 0 \le n \le n_{\max} \end{cases} \tag{3}$$

where $T_{\max 1}$ and $T_{\max 2}$ are the maximum torque of the front shaft drive motor and the rear shaft drive motor at the current speed; $T_{\max}$ and $n_{\max}$ are the maximum external characteristics of the drive motor.

2.2 Torque distribution objective function considering motor loss

When the torque distribution coefficient λ of the current axle motor is 0, the total demand torque of the vehicle is all distributed to the rear axle motor, and the power of the front axle motor in the objective function J_e becomes 0. However, in practice, there is no clutch device between the low-speed external rotor motor and the hub shaft. When the torque of the vehicle is fully distributed to the rear axle motor, the front axle motor will be towed by the vehicle, and there is no-load power loss. Therefore, when the torque distribution coefficient λ is 0, the objective function should be:

$$J_e = P_{\text{loss\_0}} + \frac{Tn}{\eta_r} \tag{4}$$

where $P_{\text{loss\_0}}$ refers to the no-load loss of a single motor of the front axle at the current speed n. The no-load loss change of the drive motor is shown in Figure 2.

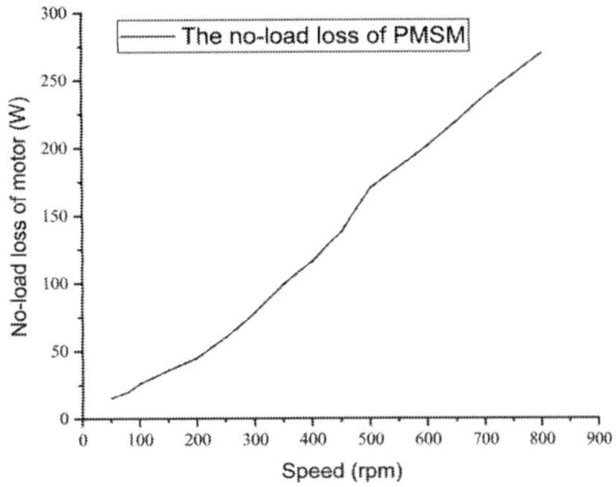

Figure 2. No-load loss variation of drive motor.

In summary, the objective function considering loss can be expressed as:

$$\begin{cases} J_e = \min\left(\dfrac{\lambda Tn}{\eta_f} + \dfrac{(1-\lambda)Tn}{\eta_r}\right) & 0 < \lambda \leq 0.5 \\ J_e = P_{\text{loss\_0}} + \dfrac{Tn}{\eta_r} & \lambda = 0 \\ s.t. \quad 0 \leq T \leq T_{\max} & \\ \quad 0 \leq 0.5T\lambda \leq T_{\max 1} & \\ \quad 0 \leq 0.5T(1-\lambda) \leq T_{\max 2} & \\ \quad 0 \leq n \leq n_{\max} & \end{cases} \tag{5}$$

2.3 Solution of optimal torque distribution strategy

Using MATLAB/Simulink software to program the m function, the three-dimensional data in the motor efficiency MAP is globally optimized [5]. The final torque distribution coefficient of the motor drive system without considering the loss and considering the loss is shown in Figure 3.

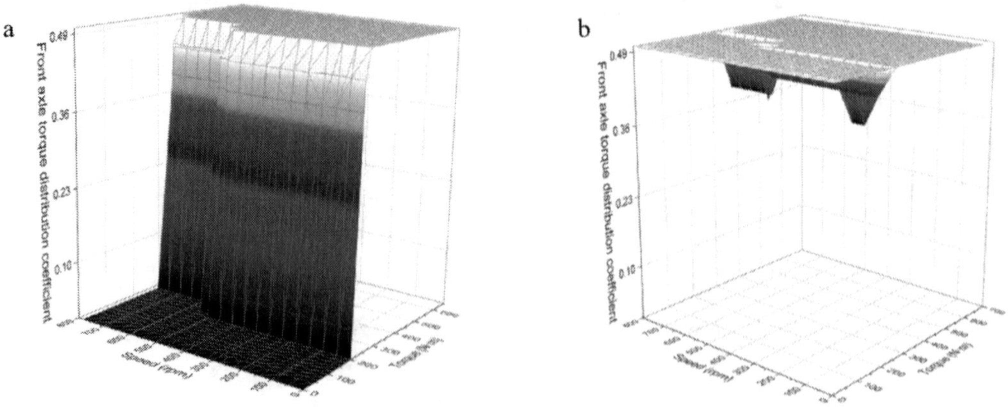

Figure 3. (a) Without considering the loss of the motor; (b) Considering the loss of the motor.

In Figure 3(b), the front axle torque distribution coefficient is basically maintained at around 0.5, which means that the strategy considering the loss is more inclined to distribute the total demand torque equally. In Figure 3(a), the strategy without considering the loss is more inclined to allocate to the rear wheel motor when the torque requirement is low.

3. Simulation analysis

As shown in Figure 4, based on AVL CRUISE, the model of distributed drive electric vehicle is built.

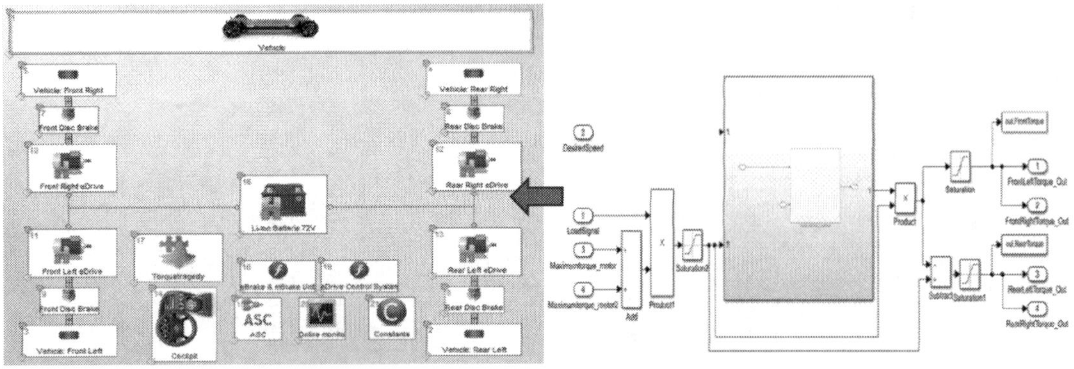

Figure 4. Vehicle model based on AVL CRUISE.

The simulation results of the power consumed by the torque distribution strategy considering loss and the torque distribution strategy without considering loss are shown in Figure 5. It can be observed intuitively that the power consumption of the torque distribution strategy considering the loss is 0.4548 kWh when driving four ECE cycles, and the power consumption of the torque distribution strategy without considering the loss is 0.4707 kWh when driving four ECE cycles. Compared with the scheme without considering the loss, the power consumption of the power battery can be reduced by 3.4%. Through simulation, it is verified that the torque distribution strategy considering loss can reduce the energy consumption of the vehicle to a certain extent compared with the torque distribution strategy without considering loss, which proves that the torque distribution strategy considering loss is better.

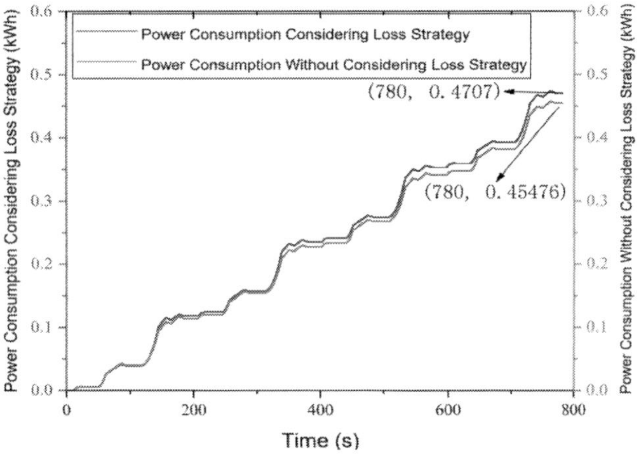

Figure 5. Vehicle power consumption under two strategies.

Figure 6 shows the comparison of the power losses of the left drive motor of the front axle and the left motor of the rear axle under the two strategies. By comparison, it can be found that the losses of the front axle motor and the rear axle motor are different under two different control strategies.

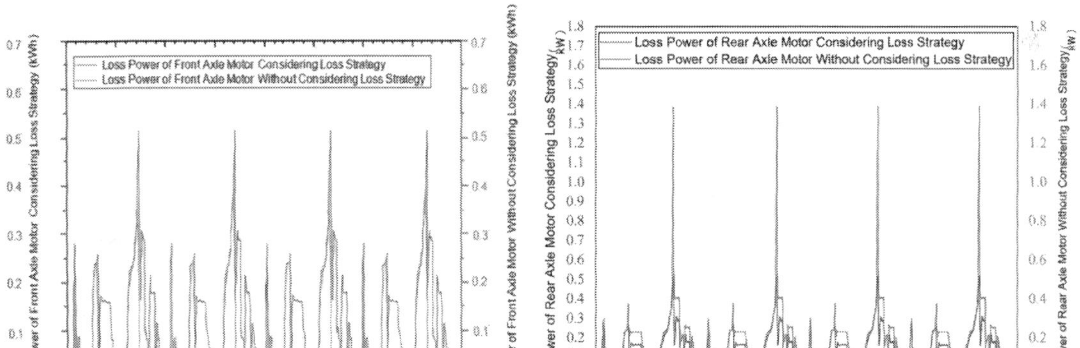

Figure 6. (a) Power loss diagram of front axle motor; (b) Power loss diagram of rear axle motor.

Through the further analysis of the loss of the front axle motor under different strategies in Figure 5(a), it can be found that in the torque distribution scheme without considering the loss, the front axle motor hardly participates in the torque distribution task when the total demand torque is small. At this time, the no-load loss of the motor is much smaller than the power loss of the motor during load operation, and the total loss of the front axle motor is relatively small. Compared with the strategy without considering the loss, the torque distribution strategy considering the loss also participates in the torque distribution task, so the total loss is relatively large, but the larger area is also limited to the case where the total demand torque of the vehicle is small.

The loss of the rear axle motor under different strategies can be analyzed in Figure 5(b). The loss of the rear axle motor without considering the loss strategy is large when the total demand torque of the vehicle is small, especially when the motor speed is high and the torque is small. The difference between the loss of the rear axle motor and the loss of the rear axle motor considering the loss strategy reaches a peak. Considering that the reason for this result may be that without considering the loss strategy, the total demand torque of the vehicle is allocated to the rear axle motor at this time, resulting in the rear axle motor being in a state of extremely low efficiency.

Through the above analysis, it can be seen that the torque distribution strategy considering the loss makes the front axle motor and the rear axle motor jointly undertake the task of torque distribution, especially when the total demand torque of the vehicle is relatively small, because this will make the loss distribution of the two axles more average. When the loss is not considered, and when the total demand torque of the vehicle is small, the torque is all distributed to the rear axle motor. The reason for considering this result is to misunderstand that the average torque distribution method will cause the torque that each motor can allocate. Too small, resulting in low efficiency of the drive motor, only allocated to the rear axle bilateral electric opportunity to improve the efficiency of the rear axle motor and avoid the loss of the front axle motor. The actual results show that this will make the rear axle motor torque larger, and the motor efficiency is too low, resulting in more power loss. In addition, the front axle motor is also lost in the no-load state, and the overall loss is more.

4. Conclusion

Based on the limitation of the system loss model method, a method of establishing the efficiency maximization objective function by driving system efficiency MAP when considering the no-load loss of the motor is proposed.

The results of both methods show that when the distributed electric vehicle equipped with four drive motors with the same parameters is driving in the steady-state straight line condition, the average distribution of the total torque to the four drive motors can reduce the vehicle power consumption by 3.4% and effectively improve the efficiency of the drive system. This provides some reference value for the economy research of distributed drive electric vehicles.

References

[1] Cheng Yin. Research on torque distribution strategy based on energy efficiency of distributed drive system [D]. Zhejiang University, 2023.

[2] Song Bingtan, Sun Binbin, Ge Wenqing, Li Bo, Zhang Tiezhu, Gao Tian. Research on anti-skid control strategy for dual-motor electric vehicle drive [J]. Modern Manufacturing Engineering, 2021 (10): 11-16.

[3] Dong Jiachen, Gao Qinhe, Liu Zhihao. Dynamic modeling and driving/steering cooperative control of distributed drive vehicles considering dynamic hysteresis [J/OL]. Journal of Beijing University of Aeronautics and Astronautics, 1-15 [2024-08-28].

[4] Wen Chuanxin, Wang Peixin, Hua Wei. Research status and development trend of electric vehicle drive system [J]. Micromotor, 2019, 52 (10): 103-109.

[5] Mikle D, Bat'a M. Torque vectoring for an electric all-wheel drive vehicle[J]. IFAC-Papers OnLine, 2019, 52(27): 163-169.

Research of signal processing and detection of satellite telemetry data based on EMD and MBSE

Dong Tian[1,2], Dongxue Yang[2], Weina Xia[2], Shuai Wang[1] and Chunsheng Xu[2,*]

[1]Beijing Institute of Technology, Beijing, China
[2]ITNS, CAST, Beijing, China

[*]Corresponding author's e-mail: xcstju@126.com

Abstract. Based on the characteristics of abnormal data and high Signal-to-Noise Ratio in satellite in-orbit telemetry data, the empirical mode decomposition algorithm is used to remove the abnormal data, and the method is effective. By using the empirical mode decomposition algorithm, trend prediction of telemetry data can also be achieved, and its practicality can be demonstrated through signal construction and development. The combination of the empirical mode decomposition algorithm for telemetry data processing with the satellite's digital model constructed by MBSE software has expanded the application scope of the digital satellite model.

1. Introduction

While the satellite is in orbit, operations staff obtains performance parameters or environmental information (such as temperature) of onboard equipment through telemetry data, which is used to determine whether the equipment is working properly or to obtain data for calculating certain performance parameters. The accuracy and effectiveness of telemetry data, as the sole basis for interpreting equipment status, are crucial. However, in engineering practice, there are various abnormal situations in satellite telemetry data [1], and there are also various methods for processing abnormal data [2].

A time-frequency analysis algorithm - Empirical Mode Decomposition (EMD) [3] was proposed by Huang et al. from NASA in 1998. The EMD algorithm has been widely used in various fields, such as fault diagnosis [4], time-frequency analysis [5], and trend prediction [6], and has also been applied in the aerospace field [7]. This article attempts to use the EMD algorithm to process telemetry data in order to obtain correct data for the status judgment of onboard equipment.

Model based systems engineering (MBSE) can solve some problems in the aerospace field using traditional documents to design satellites, such as dispersion of design information, poor correlation, and difficulty in traceability. With the application and promotion of MBSE in the aerospace field, it has been applied in multiple scenarios, such as satellite scheme iteration and design simulation [8, 9]. In this paper, MBSE is combined with the EMD algorithm to construct a digital model for anomaly detection and processing of telemetry data of satellites.

2. The EMD algorithm and its application

2.1. Introduction to EMD algorithm

The EMD algorithm is mainly used to extract the Intrinsic Mode Function (IMF). In order to obtain the IMF of the original signal $x(t)$, the general algorithm of EMD is shown in Figure 1.

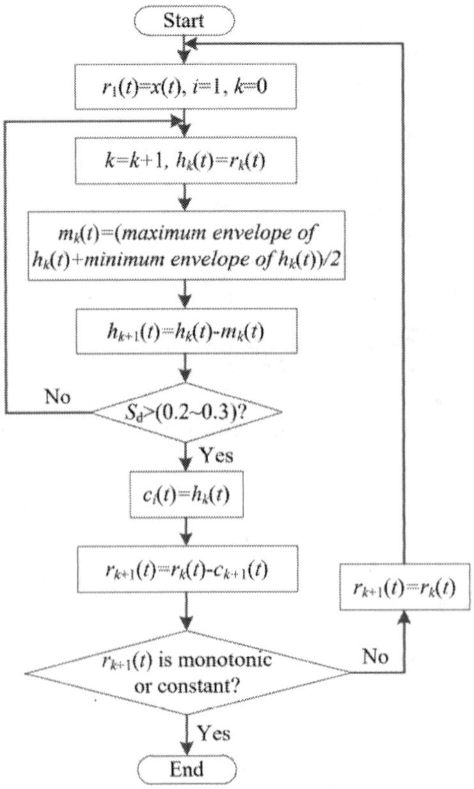

Figure 1. Flow chart of the EMD algorithm.

In Figure 1, $c_i(t)$ is the ith IMF, and S_d is calculated according to the following equation:

$$S_d = \sum_{t=0}^{T} \frac{|h_{1(k-1)}(t) - h_{1k}(t)|^2}{h_{1k}^2(t)} \tag{1}$$

where T represents the duration of the signal, and h_1 (k-1) (t) and h_{1k} (t) are the time series of two consecutive processing results in the screening IMF process. S_d is usually taken as 0.2~0.3.

2.2. Application of the EMD algorithm

2.2.1 Removing abnormal data. Based on the EMD algorithm, the noise component is determined by comparing the correlation coefficients between each IMF and its original signal:

$$\rho\ (i) = R_{i,\ sig}(\tau) \quad i=1, 2, ..., \text{number of IMFs}. \tag{2}$$

where $R_{i,sig(\tau)}$ indicates the cross-correlation between the i_{th} IMF component and its original signal. The size of $\rho(i)$ can be used as a marker for data anomaly detection: when the Signal-to-Noise Ratio(abbreviated as SNR) is high, the smaller the value of $\rho(i)$ is, the greater the likelihood that the i_{th} IMF is abnormal data is; conversely, the larger the value of $\rho(i)$ is, the greater the likelihood that the i_{th} IMF is abnormal data is. In the scenario of satellite telemetry data processing, the SNR of telemetry data is high, so the cross-correlation coefficient of abnormal data has a small $\rho(i)$. A threshold ρ_{thd} is determined, and when the IMF's $\rho(i)$ is less than the threshold, the signal is removed and the remaining IMF signals are superimposed and restored to obtain a normal signal.

A telemetry data *sig*1 is constructed as shown in Figure 2, which is a time series signal that is greatly affected by the surrounding environment and produces many abnormal jumps.

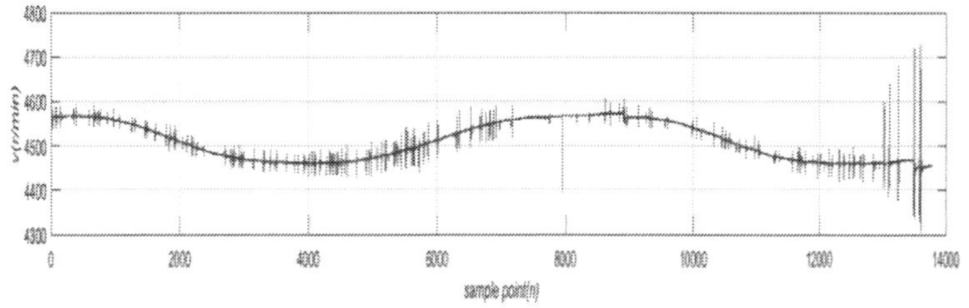

Figure 2. Time domain waveform of cumulative signal $sig1$.

The EMD decomposition on the $sig1$ is performed as shown in Figure 3.

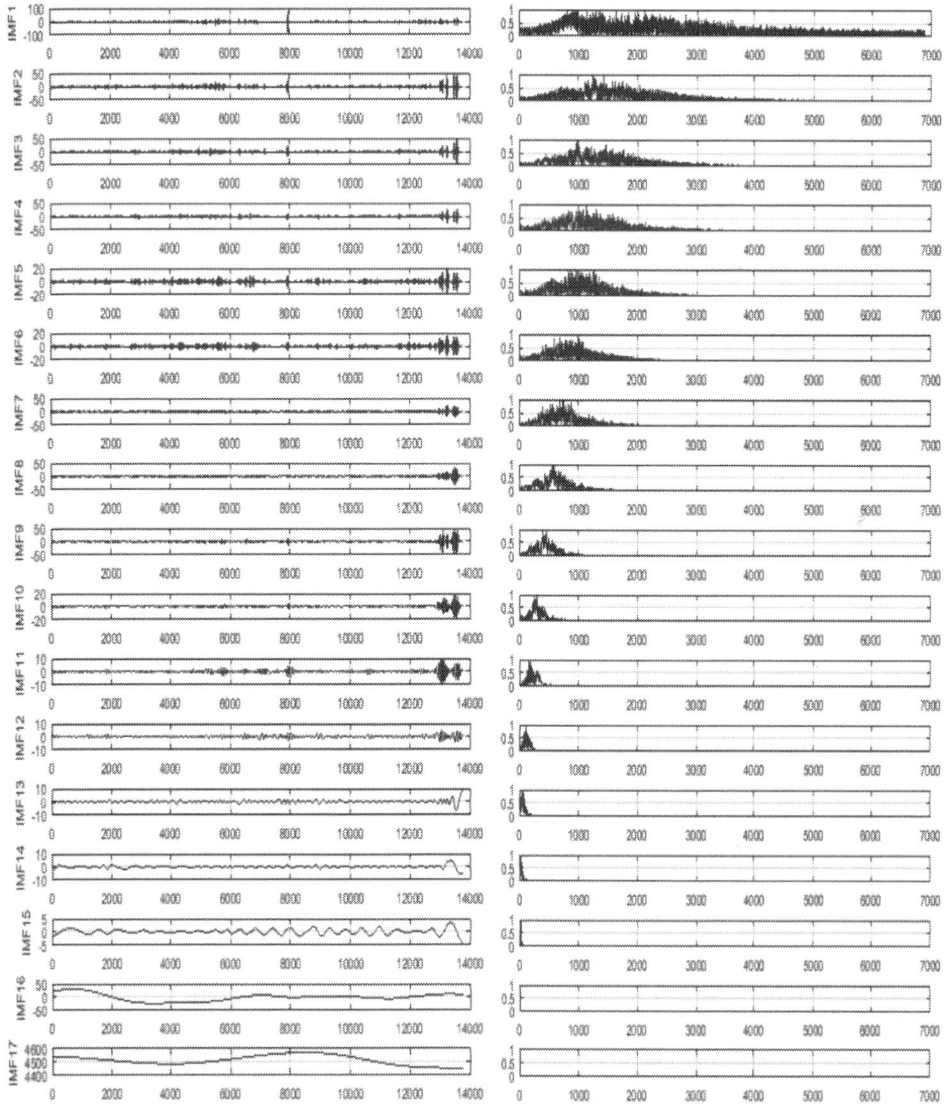

Figure 3. Time domain waveforms and amplitude spectra of IMF components using EMD decomposition.

The EMD decomposition results obtained a total of 17 IMF components. The right side of Figure 3 is the normalized amplitude spectrum of IMFs, from which it can also be seen that the first few IMF components obtained by EMD decomposition are noise-like time series with a high rate of change. The $\rho(i)$ between IMFs and telemetry data is shown in Figure 4.

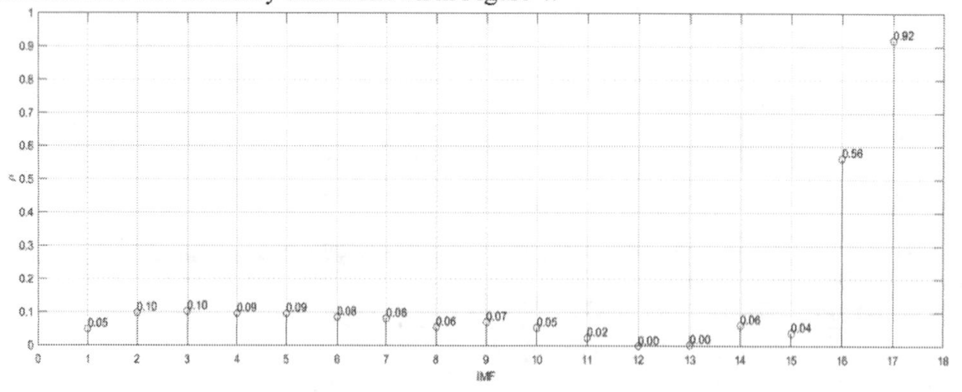

Figure 4. Cross correlations between $sig1$ and IMF components.

If ρ_{thd} is set to 0.50, the IMF components whose correlation coefficient is smaller than ρ_{thd} will be excluded. The data sequence which is reconstructed from the remaining IMF components is shown by the red line in the figure below.

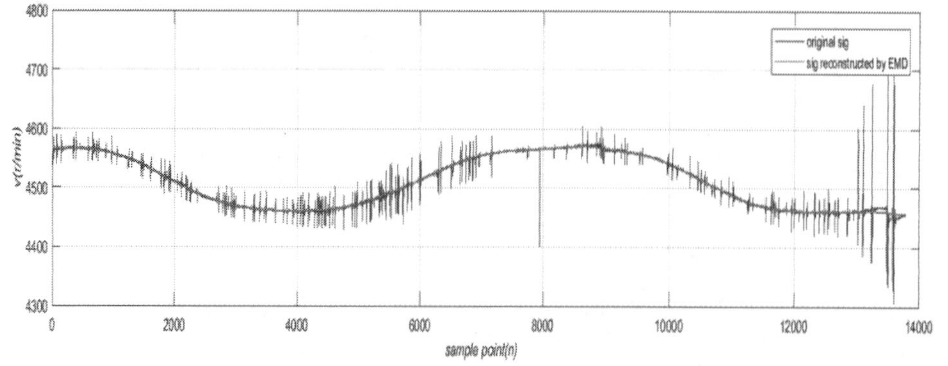

Figure 5. Time domain waveforms reconstructed by EMD.

From Figure 5, it can be seen that the abnormal data in telemetry data sig1 has been removed. There is no need to worry about removing the data that truly indicates telemetry anomalies here, as the normal range of telemetry data has already been set. If the data exceeds the normal range, it will automatically trigger an anomaly alarm. In addition, physically speaking, abnormal phenomena do not occur instantaneously in telemetry data and return to normal quickly. Therefore, if the correlation coefficient of satellite telemetry data exceeds a certain threshold, the corresponding IMF component will not miss effective information.

2.2.2 Removing trend items. When analyzing telemetry data, in addition to the need to eliminate abnormal data, there is also a requirement with practical application value, which is the trend prediction of telemetry data. The function can predict a certain parameter and prepare precautionary measures. Taking the calibration subsystem of the satellite antenna as an example, the pointing accuracy of the antenna is related to the azimuth and elevation differences, while the pointing deviation is closely related to factors such as the accuracy of the satellite structure, antenna installation accuracy, sensor installation accuracy, and the ambient temperature of the satellite. Although the pointing accuracy can be calculated and analyzed theoretically, the trend of the pointing accuracy cannot be quantified due to certain unpredictability. According to the characteristics of the EMD algorithm, the last IMF component generally contains the temporal trend information of the processed data. The trend information of the satellite's telemetry data can be extracted to assist the satellite engineer in preparing calibration strategies in advance when the satellite is in orbit.

A telemetry data *sig*2 (representing the azimuth difference of the antenna pointing mechanism) is constructed and the EMD decomposition is performed to obtain 11 IMF components. After removing abnormal data, the reconstructed signal obtained is shown in Figure 6. In Figure 6, the black dashed line represents the slowly changing trend term obtained from the IMF component. From Figure 6, it can be seen that the azimuth difference of the antenna pointing mechanism tends to increase and gradually converge over time.

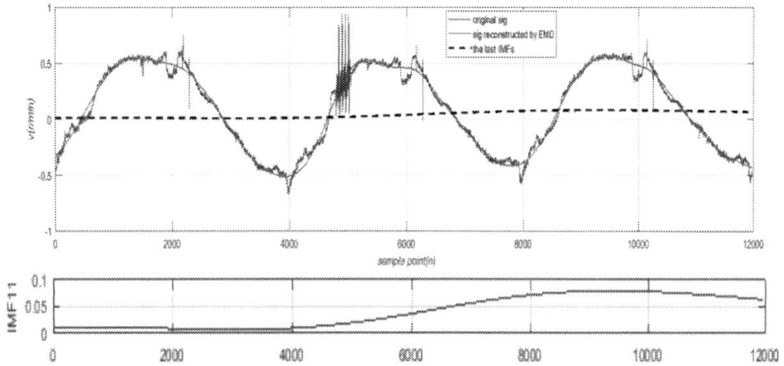

Figure 6. Signal *sig*2 and reconstructed by EMD.

3. MBSE modeling

A telemetry data processing model is built using MBSE modeling software, and the model is combined with the EMD calculation program developed using MATLAB to process the specified telemetry data in the MBSE modeling software. The MBSE model is shown in Figure 7.

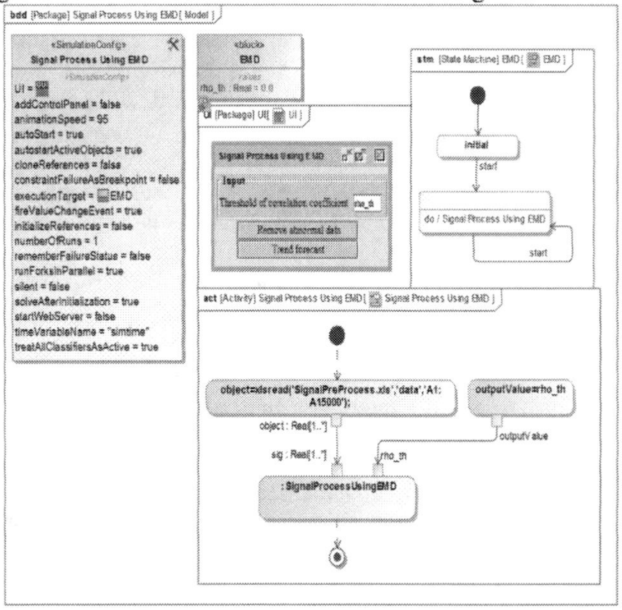

Figure 7. A telemetry data processing model using MBSE software.

The model shown in the above figure adopts several diagrams included in the SysML language: Module Definition Diagram (BDD), State Machine Diagram (STM), and Activity Diagram (ACT), which are respectively used to implement the top-level architecture of the design model, provide startup signals for calling the EMD algorithm, and provide telemetry data and correlation coefficient thresholds required for the EMD algorithm. "SignalProcessUsingEMD" refers to the EMD algorithm processing program developed by MATLAB, which can effectively combine the algorithm processing program with the MBSE model. Additionally, using the UI interface provides ρ_{thd} and configures simulation parameters. "Remove abnormal data" and "Trend forecast" respectively represent calling different telemetry data processing modes: the former represents removing abnormal data, while the latter extracts slowly changing trend items.

The above method of integrating MATLAB requires submitting the MATLAB code to other engineers. In view of this situation, the MATLAB code can be generated as a ". p" file, which can hide the program code and achieve intellectual property protection.

4. Conclusion

This article combines the EMD algorithm with the processing requirements of satellite telemetry data, providing a practical method for removing abnormal data. In addition, the EMD algorithm can also perform trend prediction analysis. By using this algorithm, we can extract slowly changing trend terms, thereby achieving prediction of telemetry data change trends and taking targeted measures in advance. The above ideas are applied to the MBSE modeling software to construct a data processing mode, expanding the digital satellite models to the field of monitoring scenarios for satellites in orbit.

Based on the implementation of the preliminary engineering application, the follow-up plan will introduce advanced signal processing technologies, such as Neural Networks, to improve the accuracy of the EMD algorithm.

References

[1] Peng Xiyuan, Pang Jingyue, Peng Yu and Liu Datong 2016 "Review on any detection of spacecraft telemetry data," Chinese Journal of Scientific Instrument, 37(9), pp. 1929-1945.

[2] Song Ge, Liang Jun, Liu Datong and Peng Yu 2015 "Anomaly detection of condition monitoring with predicted uncertainty for aerospace applications," 12th International Conference on Electronic Measurement & Instruments (ICEMI 2015), Qingdao, China, July, pp. 16-19.

[3] Huang N E, Shen Zheng, Long S R, et al 1998 "The empirical mode decomposition and the Hilbert spectrum for non-linear and non-stationary time series analysis," Proc R. Soc. London, A, pp. 903-995.

[4] Chen Junbai, Liu Yongzhi, Chen Yong and Nie Kai 2021 "Fault diagnosis of airborne fuel pump based on EMD and SVM," Journal of Beijing University of Aeronautics and Astronautics, 47(8), pp. 1687-1696.

[5] Liu Xikang, Ding Zhifeng, Li Yuan and Liu Zhiguang 2023 "Application of EMD to GNSS time series periodic term processing," Geomatics and Information Science of Wuhan University, 48(1), pp. 135-145.

[6] Jin Ji, Wang Bin, Yu Min and Wang Wenbo 2021 "Short-Term wind speed prediction based on EMD optimized by Lorenz equation," Acta Energiae Solaris Sinica, 42(6), pp. 342-348.

[7] Qi Lei, Liang Zhenxin, Ding Hongbing, Zheng Yue, Rui Xiaobo and Zhang Yu 2022 "A recognition method of spacecraft leakage based on EMD-WPD feature fusion of acoustic emission signal," Journal of Vibration and Shock, 41(4), pp. 110-116.

[8] Jiao Hongchen, Lei Yong, Zhang Hongyu, Zhang Guobin and Wang Yaodong 2021 "Research on modeling and design method of spacecraft system based on MBSE," System Engineering and Electronics, 43(9), pp. 2516-2525.

[9] Li Zhonghua, Gao Xiaoliang, Qiang Lidong and Liu Zhengnan 2022 "Model based system engineering method for design of satellite telecommunications & telecommunications simulator," Computer Applications and Software, 39(11), pp. 344-349.

Journal of Physics: Conference Series **2879** (2024) 012066 doi:10.1088/1742-6596/2879/1/012066

A star identification algorithm based on 1D Conv Neural Network

Yang Zhong[1] and Hao Wang[1,2,3,*]

[1]Micro-Satellite Research Center, Zhejiang University, Hangzhou 310027, China
[2]Huanjiang Labotoray, Zhuji 311899, China
[3]Key Laboratory of Micro-Nano Satellite Research Zhejiang Province, Hangzhou 310027, China

zhoytou@gmail.com(Y.Z.)
*Corresponding author's e-mail: roger@zju.edu.cn

Abstract. A novel star identification algorithm based on 1D-CNN for celestial navigation is introduced. This approach constructs a distinctive star pattern using radial and angular separations and cyclic angles. Several tests have been conducted on simulated star images, demonstrating the proposed algorithm's superior performance over conventional algorithms and high resilience to three error types, including spurious stars, positional deviations, and magnitude uncertainties.

1. Introduction

The star identification algorithm is a core process for star trackers to enable spacecraft's autonomous navigation. It aims to accurately calculate the position and orientation of a spacecraft by determining the stars observed in the field of view (FOV). Typically, star identification approaches can be categorized into three main types, i.e., subgraph isomorphism, pattern identification, and neural network approach [1]. The idea of subgraph isomorphism approaches is to treat the star image as a weighted undirected graph, with stars as vertices and angular distance as edge weights. Then, the task of star identification transforms into finding the most similar subgraph in the database. Well-known examples include the triangle algorithm and its enhanced versions. The pattern identification methods assign each star a distinctive pattern according to the relative positions of its neighbouring stars and determine the star by seeking the closest match in the pattern library [2]. Examples of these algorithms involve the grid algorithm [3], the log-polar transform algorithm [4], and the radial and cyclic algorithm [5]. Algorithms of the last category introduce neural networks into the problem, using preprocessing to extract image features as model training input, e.g., modified polestar method with NN [6] or directly using images as input, e.g., Spider-CNN [7]. Employing raw image data as input to neural networks circumvents the necessity for manual feature extraction, enabling the model to learn and capture salient information autonomously through an end-to-end paradigm. However, these approaches typically require larger models with increased parameter counts, consequently imposing heightened computational resource demands. This constraint poses challenges for the practical deployment of such models on star trackers.

This paper introduces a novel star identification algorithm based on a one-dimensional convolutional neural network (1D-CNN) using radial and cyclic features as input. Unlike the NN star identification method, which directly uses images as input, the proposed algorithm outperforms them in terms of memory and time consumption. With the novel feature extraction, the proposed method enables the extraction of more information to generate more reasonable results. For example, compared to the modified polestar method with NN [6], the proposed method utilizes the radial distribution of neighbouring stars over each ring, namely the distance information after the log-polar transform. Still, the proposed method extracts the angular information besides that, leading to its better performance in later experiments. Moreover, in contrast to the approach proposed by Zhang et al. [5], the presented method employs multiple neighbouring stars as the initial reference for constructing cyclic patterns, thereby mitigating the risk of ambiguous matches. Experimental evaluations demonstrate the algorithm's robust performance in three distinct error modalities: spurious stars, positional deviations, and magnitude uncertainties.

2. Proposed algorithm

2.1. Star pattern construction

The neighbour stars $N_i(i = 1, 2, 3...)$ within a circular area defined by radius R around the reference star S are selected to construct the star patterns. As shown in Figure 1(b), the circular region is divided into equally spaced concentric rings, and the number of stars in each ring is utilized as a radial feature.

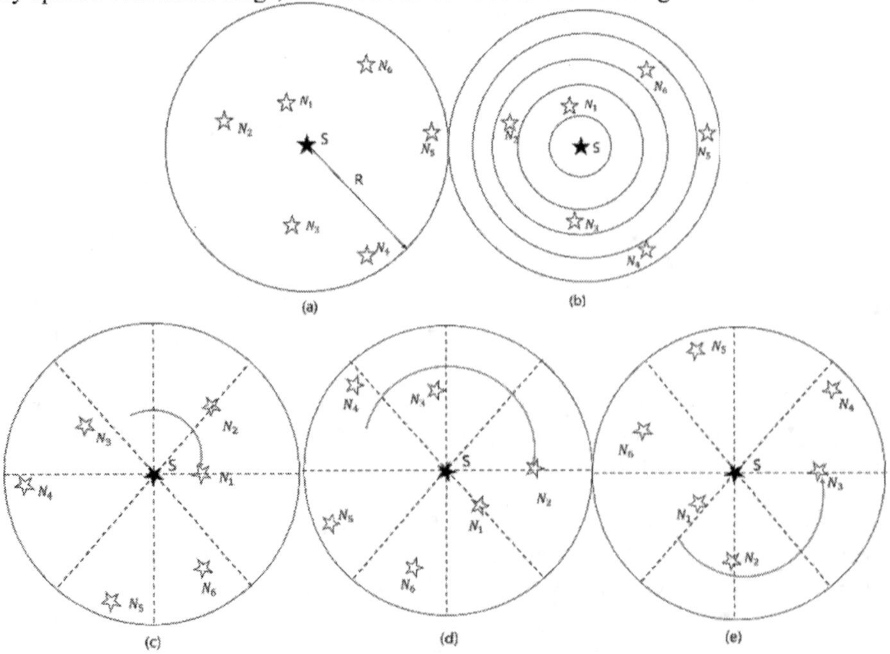

Figure 1. Construction of the radial and cyclic features.

Analogous to the radial feature, the cyclic feature represents the number of stars within each evenly divided sector. To ensure rotational and translational invariance of the cyclic features, the three nearest neighbour stars to the main star S are regarded as the orientation stars. The image is then rotated until the orientation stars are aligned with the horizontal axis. The curve with an arrow in Figure 1 denotes the directions of rotation. Furthermore, the multiple choices for the starting side when generating cyclic features also mitigate the influence of spurious stars on this feature extraction process, thereby enhancing the underlying robustness.

2.2. Neural network architecture

As discussed in Section 2.1, the cyclic features consist of multiple related sequences, making them well-suited for processing with 1D-CNNs. This is because 1D convolutional kernels are adept at exploiting strong correlations in sequential data to extract high-level features [8]. Consequently, a 1D-CNN architecture has been designed to leverage both the cyclic and radial features for predicting the identification of the primary star.

Unlike the four-staged match method proposed by Zhang et al. [5], the approach introduced here directly integrates both radial and cyclic features, rather than utilizing radial features solely for the initial match and cyclic features solely for reducing redundant candidates in subsequent steps due to uncertainty. As depicted in Figure 2, the two feature inputs are processed in parallel streams and subsequently fused to obtain the final result.

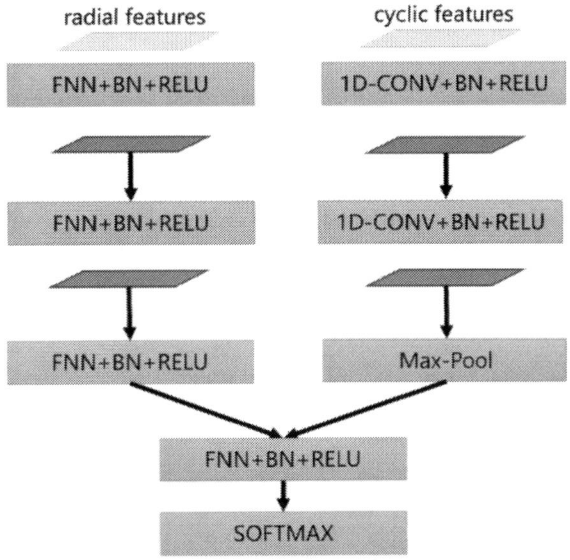

Figure 2. Neural network architecture.

Specifically, the radial feature input is passed through a fully connected layer with 500 units and a ReLU activation function. Concurrently, the cyclic feature input undergoes a series of one-dimensional convolutions, batch normalizations, and a max-pooling operation. The kernel size convolution and max-pool layers are set to 3 and 2, respectively. The outputs of the two branches are then concatenated and fed into a subsequent fully connected layer with 1000 units, batch normalization, and a ReLU activation. Finally, the processed features are mapped to the output dimension through a linear layer, yielding the predicted central star identification.

This parallel processing of radial and cyclic features within a unified CNN architecture allows for effective feature fusion. It leverages the strengths of both convolutional and fully connected layers in extracting meaningful representations from the respective input modalities.

3. Simulation and analysis

3.1. Dataset generation

The basic star catalogue used in this study is the Smithsonian Astrophysical Observatory (SAO) [9] star catalogue, with a maximum visual magnitude of 6.0 Mv set as the threshold for star tracker detection. Redundant guide stars are discarded in densely populated areas to ensure a uniform distribution over the celestial sphere. This is because an excessive number of stars within FOV typically does not significantly impact recognition results or contribute to attitude determination accuracy. Conversely, too many stars can increase recognition time and introduce redundancy in matching. Following the removal of double stars and redundant guide stars, a guide star catalogue comprising 2418 stars is derived, as depicted in Figure 3. Subsequently, training samples are generated.

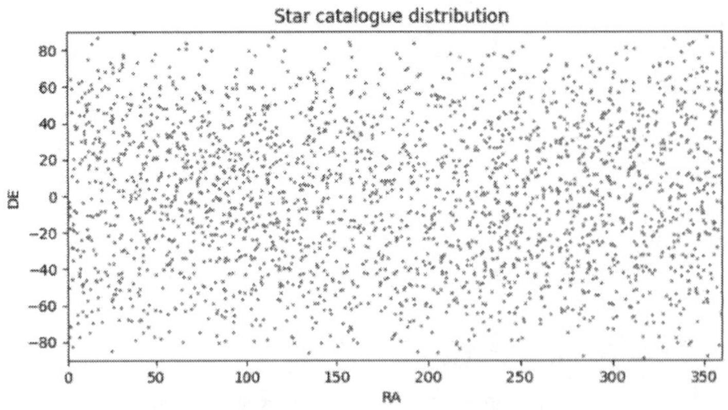

Figure 3. Star catalogue.

The parameters for dataset generation are listed in Table 1. The number of samples for each class in the training dataset is consistent. That is because the asymmetrical distribution of samples can lead to unsatisfactory classification model performance [10].

Table 1. Parameters of dataset generation.

Parameter	Value
Resolution of image	1024×1024
Field of view	15°×15°
Focus	5.8mm
Instrumental threshold	6.5Mv
The radius of the neighbouring region	6°
Number of class	2418
Number of samples for each class	100
Number of rings	50
Number of sectors	16

3.2. Robustness performance test

Different noise levels are added to the star images for testing to evaluate underlying robustness compared to conventional pattern match approaches under the same environment. The baseline methods for contrast involve the grid algorithm [3], the log-polar transform algorithm [4], and the polestar-based NN algorithm [6]. The pattern radius for these algorithms is 6°. Other parameters, including the number of grids for both the grid and log-polar algorithm and the number of bins for the polestar-based NN algorithm, are set to 60 and 25, respectively. The recognition rates of each method are the statistical results of identifying 3000-star images randomly generated from the entire celestial sphere.

Figure 4 shows the change curve of the recognition rate when the position noise variance changes from 0 to 2.5 pixels. From the statistical results, the recognition rate of this method is 99.24% when the position noise variance is 2.5 pixels, which is approximately 30% higher than the grid algorithm under the same experimental conditions.

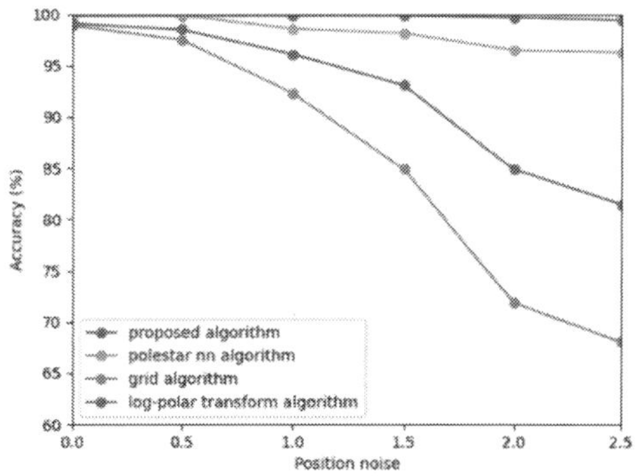

Figure 4. Performance comparison under position noise.

As depicted in Figure 5, the proposed method has a significantly slower decline in performance compared to the other three algorithms, indicating its superior robustness to magnitude noise. Notably, the recognition rate of the new approach achieves 98.11% under a magnitude noise variance of 0.5.

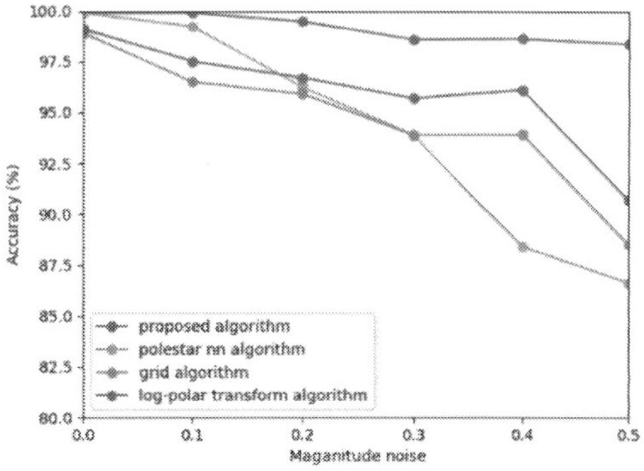

Figure 5. Performance comparison under magnitude noise.

The recognition rate curves in Figure 6 show that both methods are affected by the false star noise. However, the proposed algorithm is more reliable than the grid algorithm. As the number of false stars increases from 0 to 5, the recognition rate of the proposed algorithm only decreases from 99.9% to 99.56%.

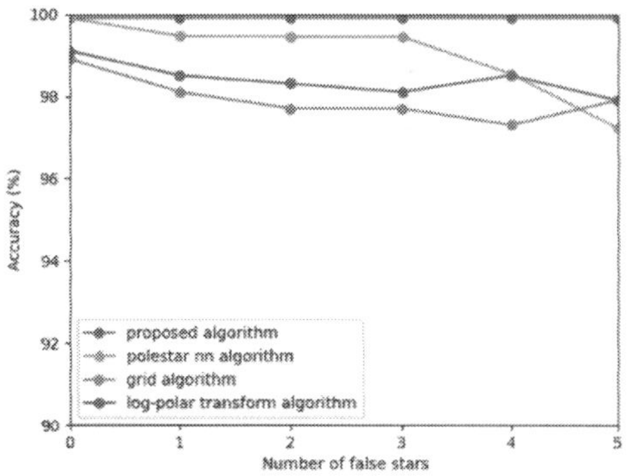

Figure 6. Performance comparison under false star noise.

4. Summary

This study presents a star identification algorithm based on the 1D-CNN. The approach constructs star patterns using radial and cyclic attributes extracted from neighbouring stars' positions and then inputs

Journal of Physics: Conference Series **2879** (2024) 012066 doi:10.1088/1742-6596/2879/1/012066

them to the 1D-CNN model to identify the tested star. Specifically, radial features are derived from the number of neighbouring stars in concentric rings, while cyclic features represent star counts in equally divided sectors. Compared to conventional pattern construction methods like the grid algorithm [3] or the radial and cyclic algorithm [5], using the number of stars instead of binary indicators to mark the existence of stars in a region provides more information for later steps to harness. Moreover, the three nearest neighbouring stars are selected as the orientation stars to ensure rotational and translational invariance of the cyclic features, thereby undermining the effect of false stars and enhancing the underlying robustness. The test results indicate that the proposed algorithm is highly reliable and outperforms the conventional grid algorithm under three error types.

References

[1] Yang, Shaofei, et al. "Robust and efficient star identification algorithm based on 1-D convolutional neural network." IEEE Transactions on Aerospace and Electronic Systems 58.5 (2022): 4156-4167.

[2] Rijlaarsdam, David, et al. "A survey of lost-in-space star identification algorithms since 2009." Sensors 20.9 (2020): 2579.

[3] Padgett, Curtis, and Kenneth Kreutz-Delgado. "A grid algorithm for autonomous star identification." IEEE Transactions on Aerospace and Electronic Systems 33.1 (1997): 202-213.

[4] Wei X., Zhang G., Jiang J. Star identification algorithm based on log-polar transform. J. Aerosp. Comput. Inf. Commun. 2009, 6, 483–490.

[5] Zhang Guangjun, Xinguo Wei, and Jie Jiang. "Full-sky autonomous star identification based on radial and cyclic features of star pattern." Image and Vision Computing 26.7 (2008): 891-897.

[6] Rijlaarsdam, David, et al. "Efficient star identification using a neural network." Sensors 20.13 (2020): 3684.

[7] Jiang Jie, Lei Liu, and Guangjun Zhang. "Star identification based on spider-web image and hierarchical CNN." IEEE Transactions on Aerospace and Electronic Systems 56.4 (2019): 3055-3062.

[8] G. Xu, T. Ren, Y. Chen, and W. Che. A one-dimensional CNN-LSTM model for epileptic seizure recognition using EEG signal analysis Front. Neurosci., vol. 14, 2020, Art. no. 1253. [Online]. Available: https://www.frontiersin.org/article/10.3389/ fnins.2020.578126

[9] M. W. Myers and C. B. Sande Sky2000-master star catalog-star catalog database Bull. Amer. Astronomical Soc., vol. 191, no. 128, p. 95, 1997.

[10] Xue J, Ma J. Extreme Sample Imbalance Classification Model Based on Sample Skewness Self-Adaptation. Symmetry. 2023; 15(5):1082. https://doi.org/10.3390/sym15051082.

AUTHOR INDEX

Institute of Physics
Dirac House, Temple Back
Bristol BS1 6BE UK

ISSN: 1742-6588
ISBN 979-8-3313-0808-7

CURRAN ASSOCIATES INC.
proceedings.com

12th International Conference on Compressors and their Systems (Compressors 2021)

IOP Conference Series: Materials Science and Engineering
Volume 1180

London, United Kingdom
6 - 8 September 2021

IOP
Institute of Physics